J. Cuny   H. Ehrig   G. Engels
G. Rozenberg   (Eds.)

Graph Grammars and Their Application to Computer Science

5th International Workshop
Williamsburg, VA, USA
November 13-18, 1994
Selected Papers

Springer

Series Editors

Gerhard Goos, Karlsruhe University, Germany

Juris Hartmanis, Cornell University, NY, USA

Jan van Leeuwen, Utrecht University, The Netherlands

Volume Editors

Janice Cuny
University of Oregon, Dept. of Computer and Information Science
Eugene, OR 97403, USA

Hartmut Ehrig
Technische Universität Berlin, Fachbereich Informatik
Franklinstr. 28-29, D-10587 Berlin, Germany

Gregor Engels
Grzegorz Rozenberg
Rijksuniversiteit Leiden, Vakgroep Informatica
Niels Bohrweg 1, 2333 CA Leiden, The Netherlands

Cataloging-in-Publication data applied for

Die Deutsche Bibliothek - CIP-Einheitsaufnahme

Graph-grammars and their application to computer science : ... international workshop ; proceedings. - Berlin ; Heidelberg ; New York ; London ; Paris ; Tokyo ; Hong Kong ; Barcelona ; Budapest : Springer.
 Bd. 1 (1979) u.d.T.: Graph-grammars and their application to computer science and biology
5. Williamsburg, VA, USA, November 13 - 18, 1995.
 Selected papers. - 1996
  (Lecture notes in computer science ; Vol. 1073)
  ISBN 3-540-61228-9 (Berlin ...)
NE: GT

CR Subject Classification (1991): F.4.2-3, I.1.1, I.2.4, I.5.1, J.3, J.5.1

ISBN 3-540-61228-9 Springer-Verlag Berlin Heidelberg New York

This work is subject to copyright. All rights are reserved, whether the whole or part of the material is concerned, specifically the rights of translation, reprinting, re-use of illustrations, recitation, broadcasting, reproduction on microfilms or in any other way, and storage in data banks. Duplication of this publication or parts thereof is permitted only under the provisions of the German Copyright Law of September 9, 1965, in its current version, and permission for use must always be obtained from Springer-Verlag. Violations are liable for prosecution under the German Copyright Law.

© Springer-Verlag Berlin Heidelberg 1996
Printed in Germany

Typesetting: Camera-ready by author
 SPIN 10512902   06/3142 - 5 4 3 2 1 0   Printed on acid-free paper

Lecture Notes in Computer Science   1073
Edited by G. Goos, J. Hartmanis and J. van Leeuwen

Advisory Board: W. Brauer   D. Gries   J. Stoer

Springer
*Berlin
Heidelberg
New York
Barcelona
Budapest
Hong Kong
London
Milan
Paris
Santa Clara
Singapore
Tokyo*

Preface

This volume consists of papers selected from the contributions presented at the Fifth International Workshop on Graph Grammars and Their Application to Computer Science that took place in Williamsburg, Virginia, November 13-18, 1994. The papers underwent an additional refereeing process to result in the 37 papers presented here (out of 78 papers presented at the workshop). The workshop was supported by EC-US Exploratory Collaborative Activity ECUS035; in particular this support has allowed many European researchers to travel to Williamsburg. The organizers of the workshop and the editors of this volume gratefully acknowledge this support. They are also indebted to NSF for providing financial support for the American organizers of the workshop.

The generic term "graph grammars" covers the whole spectrum of methods and techniques for investigating the structure of graphs and graph transformations. The motivations for and the applications of graph grammars come from very diverse areas ranging from very theoretical to very practical. These areas include pattern recognition and generation, software specification and development, data bases, massively parallel computer architectures, computer animation, developmental biology, analysis of concurrent systems, and many others. Due to such diverse applications and motivations the research in graph grammars is conducted by people with very different backgrounds.

The International Workshop on Graph Grammars and Their Application in Computer Science provides a communication forum for the graph grammar community. The workshop in Williamsburg celebrated the 25th anniversary of the publication of the two papers that initiated the whole area: one by J.L. Pfaltz and A. Rosenfeld, and one by H.J. Schneider. All three authors were present in Williamsburg, and A. Rosenfeld and H.J. Schneider gave invited talks. Also Prof. R. Siromoney gave an invited talk concerning the development of graph-grammars in India during 25 years.

The papers in this volume are grouped into 9 sections representing 9 research areas that are quite active today. The papers surely reflect current trends in graph-grammars.

We are grateful to H.-J. Kreowski who, together with the editors, served on the Program Committee for the Williamsburg Workshop. We are also indebted to all the referees of papers selected for this volume. The referees were: S. Arnborg, R. Banach, M. Bauderon, D. Blostein, F. Brandenburg, A. Corradini, B. Courcelle, J. Cuny, J. Engelfriet, G. Engels, D. Fracchia, F. Gadducci, R. Heckel, H.J. Hoogeboom, D. Janssens, W. Kahl, J.-W. Klop, H.-J. Kreowski, N. Lagergren, C. Lautemann, A. Maggiolo-Schettini, M. Nagl, F. Parisi-Presicce, G. Paun, A. Proskurowski, P. Prusinkiewicz, J.-C. Raoult, J. Rekers, L. Ribeiro, G. Rozenberg, A. Salomaa, H.J. Schneider, A. Schürr, G. Taentzer, A. Wagner, E. Wanke.

February 1996

Janice Cuny (Eugene, Oregon)
Hartmut Ehrig (Berlin)
Gregor Engels (Leiden)
Grzegorz Rozenberg (Leiden)

Contents

1. Rewriting Techniques

R. Alberich, P. Burmeister, F. Roselló, G. Valiente, B. Wojdyto
A Partial Algebras Approach to Graph Transformation   1

R. Banach
The Contractum in Algebraic Graph Rewriting   16

M. Bauderon
A Category-Theoretical Approach to Vertex Replacement: The Generation of Infinite Graphs   27

D. Blostein, H. Fahmy, A. Grbavec
Issues in the Practical Use of Graph Rewriting   38

A. Corradini, H. Ehrig, M. Löwe, U. Montanari, J. Padberg
The Category of Typed Graph Grammars and its Adjunctions with Categories of Derivations   56

D. Plump, A. Habel
Graph Unification and Matching   75

2. Specification and Semantics

H.-J. Kreowski, S. Kuske
On the Interleaving Semantics of Transformation Units - A Step into GRACE   89

A. Maggiolo-Schettini, A. Peron
A Graph Rewriting Framework for Statecharts Semantics   107

A. Schürr
Programmed Graph Transformations and Graph Transformation Units in GRACE   122

3. Software Engineering

H. Ehrig, G. Engels
Pragmatic and Semantic Aspects of a Module Concept for Graph Transformation Systems   137

M. Nagl, A. Schürr
Software Integration Problems and Coupling of Graph Grammar Specifications   155

L.M. Wills
Using Attributed Flow Graph Parsing to Recognize Clichés in Programs   170

4. Algorithms and Architectures

M.D. Derk, L.S. DeBrunner
Reconfiguration Graph Grammar for Massively Parallel, Fault Tolerant Computers   185

F. Drewes
The Use of Tree Transducers to Compute Translations Between Graph Algebras   196

K. Skodinis, E. Wanke
The Bounded Degree Problem for Non-Obstructing eNCE Graph Grammars   211

5. Concurrency

K. Barthelmann
Process Specification and Verification   225

A. Corradini, H. Ehrig, M. Löwe, U. Montanari, F. Rossi
An Event Structure Semantics for Graph Grammars with Parallel Productions   240

A. Corradini, F. Rossi
Synchronized Composition of Graph Grammar Productions   257

D. Janssens
The Decomposition of ESM Computations   271

M. Korff, L. Ribeiro
Formal Relationship between Graph Grammars and Petri Nets   288

G. Taentzer
Hierarchically Distributed Graph Transformation   304

6. Graph Languages

U. Aßmann
On Edge Addition Rewrite Systems and their Relevance to Program
Analysis   321

F.J. Brandenburg, K. Skodinis
Graph Automata for Linear Graph Languages   336

B. Courcelle, G. Sénizergues
The Obstructions of a Minor-Closed Set of Graphs Defined by Hyperedge
Replacement can be Constructed   351

J. Engelfriet, J.J. Vereijken
Concatenation of Graphs   368

C. Kim, T.E. Jeong
HRNCE Grammars - A Hypergraph Generating System with an eNCE Way
of Rewriting   383

R. Klempien-Hinrichs
Node Replacement in Hypergraphs: Simulation of Hyperedge Replacement,
and Decidability of Confluence   397

7. Patterns and Graphics

J. Dassow, A. Habel, S. Taubenberger
Chain-Code Pictures and Collages Generated by Hyperedge Replacement   412

F. Parisi-Presicce
Transformations of Graph Grammars   428

G. Zinßmeister, C.L. McCreary
Drawing Graphs with Attribute Graph Grammars   443

A. Zündorf
Graph Pattern Matching in PROGRES   454

8. Structure and Logic of Graphs

S. Arnborg, A. Proskurowski
A Technique for Recognizing Graphs of Bounded Treewidth with
Application to Subclasses of Partial 2-Paths 469

B. Courcelle
The Definition in Monadic Second-Order Logic of Modular Decompositions
of Ordered Graphs 487

A. Ehrenfeucht, T. Harju, G. Rozenberg
Group Based Graph Transformations and Hierarchical Representations of
Graphs 502

9. Biology

F.D. Fracchia
Integrating Lineage and Interaction for the Visualization of Cellular
Structures 521

J. Lück, H.B. Lück
Cellworks with Cell Rewriting and Cell Packing for Plant Morphogenesis 536

P. Prusinkiewicz, L. Kari
Subapical Bracketed L-Systems 550

Author Index 565

A Partial Algebras Approach to Graph Transformation

R. Alberich[1], P. Burmeister[2], F. Roselló[1], G. Valiente[1], and B. Wojdyło[3]

[1] Mathematics and Computer Science Department, University of the Balearic Islands, Spain
[2] Fachbereich Mathematik, Arbeitsgruppe Allgemeine Algebra, Technische Hochschule Darmstadt, Germany
[3] Department of Mathematics, Kopernikus University, Toruń, Poland

During the last years, a good deal of work on the algebraic approach to graph transformation, both in the single and double pushout cases, has been based on the understanding of graphs as many-sorted unary total algebras, and it has even been generalized to such algebras. A good exponent of this approach is [8].

This understanding of graphs as total algebras (with source and target unary operations) works quite smoothly for graphs and hypergraphs. Unfortunately, and to our taste, it is not so suitable for dealing with higher-order hypergraphs [13] (*graphs* henceforth; see Definition 2.1 below). Higher-order hypergraphs allow the abstraction of arcs to arcs, which is accomplished by abstraction mappings, and their development has been motivated by the need of locally encoding global structure, such as paths and trees, in a hypergraph itself, for instance for the structural verification of knowledge bases [11, 12, 13]. The presence of the abstraction mappings, however, causes the consideration of a too large and involved set of sorts and a too messy overloading when graphs are understood as total algebras.

On the other hand, it seems natural to interpret graphs as partial two-sorted algebras with countably many possible source, target and abstraction operations. This was the first motivation for us to study graph transformations from the point of view of partial algebras. But a second motivation soon appeared to us on the way. The theory of partial algebras is richer in concepts than the one of total algebras: for instance, it furnishes several different notions of subalgebras and (total and partial) homomorphisms, which in applications can be used to modelize mappings that forget or add information. And some of these concepts imported from the theory of partial algebras could be of some use for somebody in the graph transformation world.

So, the purpose of this paper is to introduce the basic language of partial algebras in the graphs context, as well as some results on partial algebras relevant to the double and single pushout algebraic approaches to graph transformation. We do so at two levels: at the level of unary partial algebras, and at the one of graphs (as special partial algebras).

We devote to each level a section. In Section 1, we study double and single pushout algebraic transformation of unary partial algebras. The richness of the theory of partial algebras entails here the existence of two different approaches to double pushout rewriting, one using plain homomorphisms and another using closed homomorphisms (Definition 1.3), and (at least) three different approaches to single pushout rewriting, using conformisms, q-conformisms and cdq-conformisms (Definition 1.5). Then, in Section 2, we translate all notions and results from unary partial algebras to (higher-order hyper)graphs, understanding them as special cases of partial algebras. Unfortunately, for graphs we lose one of the approaches to single pushout transformation, the one based on q-conformisms, but we still have the remaining two single pushout approaches and both double pushout approaches to transformation. It is worth mentioning here that the double pushout graph transformation using closed homomorphisms, and the single pushout graph transformation using cdq-conformisms, correspond to the usual double and single pushout approaches to graph transformation using graph morphisms and partial graph morphisms, respectively, while the other two approaches seem to be novel.

This paper should be viewed as a preamble to a series of papers [5, 4] on the topic of algebraic transformation of unary partial algebras. While some of the main results in this paper are based on results already existing in the literature, and due to some of the authors of this note, in the aforementioned forthcoming papers a new approach is developed, specially in the single pushout case [4], closer to the usual constructions in the graph transformation community. On the other hand, those papers contain detailed proofs of most of the results announced here.

The lack of space preventing us from giving all main definitions about partial algebras in their most general form, we only introduce (in Section 1) the notions used in this paper, and simply for *graph structures* (signatures of unary algebras [8]), and we refer the reader to [2], or the same author's recent survey [3], for the general definitions, or for any other concept on partial algebras not defined here. An introduction to the basic language of the theory of partial algebras is also included, as an Appendix, in [5].

To end this introduction, we recall the definitions of rule and derivation in the double and single pushout approaches to algebraic transformation.

A *double-pushout transformation rule* in a category $\mathcal{C}$ is a pair of morphisms
$$P = (\mathbf{L} \xleftarrow{l} \mathbf{K} \xrightarrow{r} \mathbf{R})$$
in $\mathcal{C}$, and it can be applied to an object $\mathbf{G}$ when there exists a morphism $m : \mathbf{L} \to \mathbf{G}$ and a diagram

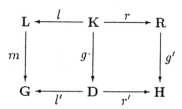

such that both squares are pushout squares. In other words, when there exists a *pushout complement* D for l and m (that is, an object D and two morphisms $g : K \to D$ and $l' : D \to G$ such that G, together with m and l', is the pushout of l and g), and then a pushout of r and g. In this case, the object H is said to be *derived* from G by the application of rule P through morphism m.

So, in order to develop a (general) double pushout approach to algebraic transformation in some category $\mathcal{C}$, one must find necessary and sufficient conditions for the existence of the pushout of two morphisms with the same source (of course, it is better if one proves that the category has all binary pushouts), as well as necessary and sufficient conditions (usually called *gluing conditions*) for the existence of pushout complements of two composable morphisms. And, since in general a derived object from an object G by the application of a rule P need not be unique, even up to isomorphism, one has to find sufficient conditions to guarantee such uniqueness up to isomorphism, or to characterize by means of some universal property one of these derived objects. Due to the uniqueness up to isomorphism of pushouts (through a universal property), it is usually enough to study the uniqueness of the pushout complement, or to characterize one of them by means of some universal property.

A *single-pushout transformation rule* in a category $\mathcal{C}$ is simply a morphism $f : L \to R$ (usually, a partial morphism of some kind), and it can be applied to an object G when there exists a morphism $m : L \to G$ and a pushout square in $\mathcal{C}$

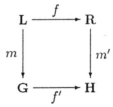

In this case, the object H is said to be *derived* from G by application of rule $f : L \to R$ through morphism m. So, in order to develop a (general) single pushout approach to algebraic transformation in some category $\mathcal{C}$, it is clearly enough to prove that $\mathcal{C}$ has all binary pushouts. Notice moreover that such derived object from an object G by the application of a rule f is always unique up to isomorphism.

1. Unary Partial Algebras

As we have mentioned, we devote this first section to generalize to partial unary algebras the results given in [7, 8] for total unary algebras. For the convenience of the reader, we introduce the notions of the theory of partial algebras relevant to this goal.

Let $\Gamma = (S, \Omega, \eta)$ be in the sequel a *graph structure*, that is, a type of many-sorted unary algebras, with set of sorts S, set of operation symbols Ω and arity function $\eta : \Omega \to S \times S$.

A *partial Γ-algebra* is a structure $\mathbf{A} = (A, (\varphi^{\mathbf{A}})_{\varphi \in \Omega})$ where $A = (A_s)_{s \in S}$ is an S-set and, for every $\varphi \in \Omega$, say with $\eta(\varphi) = (s, s')$, the *operation* $\varphi^{\mathbf{A}}$ is a partial function $\varphi^{\mathbf{A}} : A_s \to A_{s'}$, with domain $\operatorname{dom} \varphi^{\mathbf{A}}$. The algebra $\mathbf{A}$ is usually said to be *supported* on the S-set A. We shall generically refer to partial Γ-algebras, for Γ any graph structure, as *unary partial algebras*.

Definition 1.1. *Let $\mathbf{A} = (A, (\varphi^{\mathbf{A}})_{\varphi \in \Omega})$ and $\mathbf{B} = (B, (\varphi^{\mathbf{B}})_{\varphi \in \Omega})$ be two partial Γ-algebras, with $B \subseteq A$ (i.e., $B_s \subseteq A_s$ for every $s \in S$).*

$\mathbf{B}$ is a weak subalgebra *of $\mathbf{A}$ when, for every $\varphi \in \Omega$, if $b \in \operatorname{dom} \varphi^{\mathbf{B}}$ then $b \in \operatorname{dom} \varphi^{\mathbf{A}}$ and $\varphi^{\mathbf{B}}(b) = \varphi^{\mathbf{A}}(b)$.*

$\mathbf{B}$ is a relative subalgebra *of $\mathbf{A}$ when it is a weak subalgebra and, for every $\varphi \in \Omega$, say with $\eta(\varphi) = (s, s')$, and for every $b \in B_s$, if $b \in \operatorname{dom} \varphi^{\mathbf{A}}$ and $\varphi^{\mathbf{A}}(b) \in B_{s'}$ then $b \in \operatorname{dom} \varphi^{\mathbf{B}}$ (and then, of course, $\varphi^{\mathbf{B}}(b) = \varphi^{\mathbf{A}}(b)$).*

$\mathbf{B}$ is a closed subalgebra *of $\mathbf{A}$ when it is a weak subalgebra and, for every $\varphi \in \Omega$, say with $\eta(\varphi) = (s, s')$, and for every $b \in B_s$, if $b \in \operatorname{dom} \varphi^{\mathbf{A}}$ then $b \in \operatorname{dom} \varphi^{\mathbf{B}}$ (and then, of course again, $\varphi^{\mathbf{B}}(b) = \varphi^{\mathbf{A}}(b)$).*

Notice that, if $\mathbf{A} = (A, (\varphi^{\mathbf{A}})_{\varphi \in \Omega})$ is a partial Γ-algebra and $B \subseteq A$, then in principle there may be many weak subalgebras $\mathbf{B}$ of $\mathbf{A}$ supported on B, while there is only one such weak subalgebra that is a relative subalgebra: the one with operations the restrictions to B of the operations in $\mathbf{A}$. Moreover, there is one (and only one) such weak subalgebra that is closed iff[1] B is *closed* (i.e., when for every $\varphi \in \Omega$, say with $\eta(\varphi) = (s, s')$, if $b \in \operatorname{dom} \varphi^{\mathbf{A}} \cap B_s$ then $\varphi^{\mathbf{A}}(b) \in B_{s'}$), and in this case the only closed subalgebra supported on B is the relative subalgebra supported on B.

Example 1.1. If Γ is a graph structure with a single sort (which we shall henceforth omit in practice) and a single unary operation symbol φ, and $\mathbf{A}$ is a partial Γ-algebra supported on $A = \{a_1, a_2, a_3\}$ and with $\varphi^{\mathbf{A}}$ given by $\varphi^{\mathbf{A}}(a_1) = a_2$, $\varphi^{\mathbf{A}}(a_2) = a_3$, then

- The partial Γ-algebra $\mathbf{B}$ supported on $B = \{a_1, a_2\}$ and with $\varphi^{\mathbf{B}}$ undefined everywhere, is a weak (but not relative) subalgebra of $\mathbf{A}$.
- The partial Γ-algebra $\mathbf{B}'$ supported on $B' = \{a_1, a_3\}$ and with $\varphi^{\mathbf{B}}$ undefined everywhere, is a relative (but not closed) subalgebra of $\mathbf{A}$.
- The partial Γ-algebra $\mathbf{B}''$ supported on $B'' = \{a_2, a_3\}$ and with $\varphi^{\mathbf{B}''}$ given by $\varphi^{\mathbf{B}''}(b_2) = b_3$ is a closed subalgebra of $\mathbf{A}$.

The closed subsets of a partial algebra $\mathbf{A}$, with the usual inclusion, form a complete lattice. In particular, given a subset $X \subseteq A$, there exists the least closed subset of $\mathbf{A}$ containing X, which we shall denote by $C_{\mathbf{A}}(X)$. Let us recall in the following lemma a description for such $C_{\mathbf{A}}(X)$ in the case of unary algebras (see [2, §3.4] for the general case).

---

[1] if and only if

Lemma 1.1. *Let $\Gamma = (S, \Omega, \eta)$ be a graph structure, let $\mathbf{A} = (A, (\varphi^{\mathbf{A}})_{\varphi \in \Omega})$ a partial Γ-algebra, and let $X = (X_s)_{s \in S}$ be a subset of A.*

For every $s \in S$, set $X_s^{(0)} = X_s$ and for every $n \geq 0$ set

$$X_s^{(n+1)} = X_n \cup \{\varphi^{\mathbf{A}}(a) \mid \varphi \in \Omega,\ \eta(\varphi) = (s_1, s),\ a \in \text{dom}\ \varphi^{\mathbf{A}} \cap X_{s_1}^{(n)}\}$$

Then $C_{\mathbf{A}}(X) = (C_{\mathbf{A}}(X)_s)_{s \in S}$ with $C_{\mathbf{A}}(X)_s = \bigcup_{n \geq 0} X_s^{(n)}$, $s \in S$.

Definition 1.2. *Let $\mathbf{A} = (A, (\varphi^{\mathbf{A}})_{\varphi \in \Omega})$ be a partial Γ-algebra, and let θ be an equivalence relation on A, i.e., an S-indexed family of equivalence relations $\theta_s \subseteq A_s \times A_s$.*

θ is a congruence *on $\mathbf{A}$ when for every $\varphi \in \Omega$, say with $\eta(\varphi) = (s, s')$, if $(a, a') \in \theta_s$ and $a, a' \in \text{dom}\ \varphi^{\mathbf{A}}$ then $(\varphi^{\mathbf{A}}(a), \varphi^{\mathbf{A}}(a')) \in \theta_{s'}$.*

A congruence θ is closed *when for every $\varphi \in \Omega$, say with $\eta(\varphi) = (s, s')$, if $(a, a') \in \theta_s$ and $a \in \text{dom}\ \varphi^{\mathbf{A}}$ then $a' \in \text{dom}\ \varphi^{\mathbf{A}}$, and of course $(\varphi^{\mathbf{A}}(a), \varphi^{\mathbf{A}}(a')) \in \theta_{s'}$.*

Given a congruence θ on a partial Γ-algebra $\mathbf{A} = (A, (\varphi^{\mathbf{A}})_{\varphi \in \Omega})$, we define the quotient *algebra*

$$\mathbf{A}/\theta = \left(A/\theta, (\varphi^{\mathbf{A}/\theta})_{\varphi \in \Omega}\right)$$

(where $(A/\theta)_{s \in S} = A_s/\theta_s$ for every $s \in S$) in the following way: for every $\varphi \in \Omega$, say with $\eta(\varphi) = (s, s')$, $[a]_{\theta_s} \in \text{dom}\ \varphi^{\mathbf{A}/\theta}$ iff there exists some $a' \in \text{dom}\ \varphi^{\mathbf{A}}$ such that $(a, a') \in \theta_s$, and in this case $\varphi^{\mathbf{A}/\theta}([a]_{\theta_s}) = [\varphi^{\mathbf{A}}(a')]_{\theta_{s'}}$.

In the case of a closed congruence, $[a]_{\theta_s} \in \text{dom}\ \varphi^{\mathbf{A}/\theta}$ iff $a \in \text{dom}\ \varphi^{\mathbf{A}}$.

The congruences on a partial algebra $\mathbf{A}$, with the usual inclusion, form a complete lattice, of which the closed congruences form an initial segment. In particular, given a set X of pairs of elements of A, there exists the least congruence on $\mathbf{A}$ containing X, and if there exists some closed congruence on $\mathbf{A}$ containing X then there also exists the least such closed congruence containing it. In the following lemma, which can be considered partial algebras folklore, we give a description of such least congruences (again, only in the unary case).

Lemma 1.2. *Let $\Gamma = (S, \Omega, \eta)$ be a graph structure. Let $\mathbf{A} = (A, (\varphi^{\mathbf{A}})_{\varphi \in \Omega})$ be a partial Γ-algebra. For every $s \in S$, let $X_s \subseteq A_s \times A_s$ be a set of ordered pairs of elements of A of sort s.*

For every $s \in S$, set $\theta_{s,0} = \Delta_{A_s} \cup X_s \cup \{(a, a') \mid (a', a) \in X_s\}$ (where Δ_{A_s} denotes the diagonal relation on A_s), and for every $n \geq 0$ set

$$\theta_{s,n+1} = \theta_{s,n} \cup \{(a, a') \in A_s \times A_s \mid \exists a'' \in A_s : (a, a''), (a'', a') \in \theta_{s,n}\}$$
$$\cup \{(a, a') \in A_s \times A_s \mid \exists \varphi \in \Omega,\ \eta(\varphi) = (s_1, s),\ \exists (a_1, a_1') \in \theta_{s_1, n}$$
$$\text{such that}\ a = \varphi^{\mathbf{A}}(a_1), a' = \varphi^{\mathbf{A}}(a_1')\}$$

and let finally $\theta_s = \bigcup_{n \geq 0} \theta_{s,n}$.

Then $\theta = (\theta_s)_{s \in S}$ is the least congruence on $\mathbf{A}$ containing $X = (X_s)_{s \in S}$.

Moreover, if there exists some closed congruence containing X then θ is closed, and it is the least closed congruence containing X.

Definition 1.3. *Let $\mathbf{A} = (A, (\varphi^{\mathbf{A}})_{\varphi \in \Omega})$ and $\mathbf{B} = (B, (\varphi^{\mathbf{B}})_{\varphi \in \Omega})$ be two partial Γ-algebras, and let $f : A \to B$ be a mapping of S-sets, i.e., an S-indexed family of mappings $f_s : A_s \to B_s$.*

f is a (plain) homomorphism of partial Γ-algebras when for every $\varphi \in \Omega$, say with arity $\eta(\varphi) = (s, s')$, if $a \in \mathrm{dom}\,\varphi^{\mathbf{A}}$ then $f_s(a) \in \mathrm{dom}\,\varphi^{\mathbf{B}}$ and $f_{s'}(\varphi^{\mathbf{A}}(a)) = \varphi^{\mathbf{B}}(f_s(a))$.

A homomorphism $f : \mathbf{A} \to \mathbf{B}$ is closed when (with the previous notations) if $f_s(a) \in \mathrm{dom}\,\varphi^{\mathbf{B}}$ then $a \in \mathrm{dom}\,\varphi^{\mathbf{A}}$.

So, for instance, the inclusion of a weak subalgebra is a homomorphism of partial algebras, and the inclusion of a closed subalgebra is a closed homomorphism. Another typical example of homomorphism is given by any mapping between partial algebras, when the source algebra is *discrete* (i.e., when all its operations are undefined everywhere), but in this case, such mapping yields not a closed homomorphism, unless no operation is defined in the target algebra on any element of the image. A last (and very important) example of homomorphism: if θ is a congruence on $\mathbf{A}$ then the natural mapping $\mathrm{nat}_\theta : A \to A/\theta$ is a homomorphism $\mathbf{A} \to \mathbf{A}/\theta$, and it is closed when θ is a closed congruence.

Let Alg_Γ and $\mathrm{C\text{-}Alg}_\Gamma$ be the categories of partial Γ-algebras having as morphisms the usual homomorphisms and the closed homomorphisms, respectively. The following theorem can be deduced from some general results given in [6].

Theorem 1.1. *Let $\mathbf{A} = (A, (\varphi^{\mathbf{A}})_{\varphi \in \Omega})$ and $\mathbf{B} = (B, (\varphi^{\mathbf{B}})_{\varphi \in \Omega})$ be two partial Γ-algebras.*

Let $\mathbf{A} + \mathbf{B} = (A \sqcup B, (\varphi^{\mathbf{A}+\mathbf{B}})_{\varphi \in \Omega})$ be the partial Γ-algebra defined in the following way:

- *$A \sqcup B = (A_s \sqcup B_s)_{s \in S}$ where, for every $s \in S$, $A_s \sqcup B_s$ stands for the disjoint union of A_s and B_s.[2]*
- *For every $\varphi \in \Omega$, $\varphi^{\mathbf{A}+\mathbf{B}} = \varphi^{\mathbf{A}} \sqcup \varphi^{\mathbf{B}}$, in the sense that $\mathrm{dom}\,\varphi^{\mathbf{A}+\mathbf{B}} = \mathrm{dom}\,\varphi^{\mathbf{A}} \sqcup \mathrm{dom}\,\varphi^{\mathbf{B}}$ and if $a \in \mathrm{dom}\,\varphi^{\mathbf{A}}$ (resp., $b \in \mathrm{dom}\,\varphi^{\mathbf{B}}$) then $\varphi^{\mathbf{A}+\mathbf{B}}(a) = \varphi^{\mathbf{A}}(a)$ (resp., $\varphi^{\mathbf{A}+\mathbf{B}}(b) = \varphi^{\mathbf{B}}(b)$)*

i) The algebra $\mathbf{A} + \mathbf{B}$, together with the obvious monomorphisms $i_{\mathbf{A}} : \mathbf{A} \hookrightarrow \mathbf{A} + \mathbf{B}$ and $i_{\mathbf{B}} : \mathbf{B} \hookrightarrow \mathbf{A} + \mathbf{B}$, is the coproduct of $\mathbf{A}$ and $\mathbf{B}$ both in Alg_Γ and $\mathrm{C\text{-}Alg}_\Gamma$. Such partial Γ-algebra $\mathbf{A} + \mathbf{B}$ is usually called the disjoint sum of $\mathbf{A}$ and $\mathbf{B}$.

---

[2] For instance, we can take $A_s \sqcup B_s = A_s \times \{0\} \cup B_s \times \{1\}$. Nevertheless, in the sequel we shall identify the elements of A_s and B_s with their images in $A_s \sqcup B_s$, so that the actual definition of such a disjoint union is unrelevant.

Let now $f : \mathbf{K} \to \mathbf{A}$ and $g : \mathbf{K} \to \mathbf{B}$ be two homomorphisms (resp., closed homomorphisms) of partial Γ-algebras. Let $\theta(\mathbf{K})$ be the least congruence on $\mathbf{A} + \mathbf{B}$ containing

$$\left(\{ (f_s(x), g_s(x)) \in A_s \times B_s \subseteq (A_s \sqcup B_s)^2 \mid x \in K_s \} \right)_{s \in S}$$

ii) The quotient $(\mathbf{A} + \mathbf{B})\big/_{\theta(\mathbf{K})}$, together with the homomorphisms

$$f' = \mathrm{nat}_{\theta(\mathbf{K})} \circ i_{\mathbf{B}} : \mathbf{B} \to (\mathbf{A} + \mathbf{B})\big/_{\theta(\mathbf{K})}, \; g' = \mathrm{nat}_{\theta(\mathbf{K})} \circ i_{\mathbf{A}} : \mathbf{A} \to (\mathbf{A} + \mathbf{B})\big/_{\theta(\mathbf{K})}$$

is the pushout of f and g in Alg_Γ (resp., in C-Alg_Γ).

Example 1.2. Let Γ be a graph structure with a single sort and a single operation symbol φ. Let $\mathbf{K}$ be a partial Γ-algebra supported on $K = \{k_1, k_3\}$ with operation $\varphi^{\mathbf{K}}$ undefined everywhere, let $\mathbf{A}$ be a partial Γ-algebra supported on $A = \{a_1, a_2, a_3\}$ with operation $\varphi^{\mathbf{A}}$ given by $\varphi^{\mathbf{A}}(a_1) = a_2$, $\varphi^{\mathbf{A}}(a_2) = a_3$, and let $\mathbf{B}$ be a partial Γ-algebra supported on $B = \{b_1, b_2, b_3\}$ with operation $\varphi^{\mathbf{B}}$ given by $\varphi^{\mathbf{B}}(b_1) = b_3$, $\varphi^{\mathbf{B}}(b_3) = b_2$. Let $f : \mathbf{K} \to \mathbf{A}$ and $g : \mathbf{K} \to \mathbf{B}$ be the homomorphisms defined by $f(k_1) = a_1$, $f(k_3) = a_3$, and $g(k_1) = b_1$, $g(k_3) = b_3$, respectively. Then the pushout in Alg_Γ of f and g is given by a partial (actually, total) Γ-algebra $\mathbf{H}$ supported on $H = \{x, y\}$ and with with operation $\varphi^{\mathbf{H}}$ given by $\varphi^{\mathbf{H}}(x) = \varphi^{\mathbf{H}}(y) = y$, together with homomorphisms $g' : \mathbf{A} \to \mathbf{H}$, defined by $g'(a_1) = x$ and $g'(a_2) = g'(a_3) = y$, and $f' : \mathbf{B} \to \mathbf{H}$, defined by $f'(b_1) = x$ and $f'(b_2) = g'(b_3) = y$.

Now that we know that every pair of homomorphisms, or closed homomorphisms, $f : \mathbf{K} \to \mathbf{A}$ and $g : \mathbf{K} \to \mathbf{B}$ have a pushout, we must study when, given two homomorphisms (resp., closed homomorphisms) $f : \mathbf{K} \to \mathbf{A}$ and $m : \mathbf{A} \to \mathbf{D}$, does a *pushout complement* for them (in the sense of the Introduction) exist. In particular, we look for the right "gluing condition", the necessary and sufficient condition on f and m for the existence of such pushout complement.

Definition 1.4. *Let $f : \mathbf{K} \to \mathbf{A}$ and $m : \mathbf{A} \to \mathbf{D}$ be two homomorphisms of partial Γ-algebras.*

We say that m satisfies the identification condition *w.r.t.[3] f when for every $s \in S$, if $a \in A_s$, $a' \in A_s - C_{\mathbf{A}}(f(K))_s$, and $m_s(a) = m_s(a')$ then $a = a'$.*

We say that m satisfies the dangling condition *w.r.t. f when for every $\varphi \in \Omega$, say with $\eta(\varphi) = (s, s')$, and for every $x \in \mathrm{dom}\, \varphi^{\mathbf{D}}$, if $\varphi^{\mathbf{D}}(x) \in m_{s'}(A_{s'} - C_{\mathbf{A}}(f(K))_{s'})$ then $x \in m_s(A_s - C_{\mathbf{A}}(f(K))_s)$.*

We say that m satisfies the relative closedness condition *w.r.t. f when for every $\varphi \in \Omega$, say with $\eta(\varphi) = (s, s')$, and for every $x \in \mathrm{dom}\, \varphi^{\mathbf{D}}$, if $x = m_s(a)$ with $a \in A_s - C_{\mathbf{A}}(f(K))_s$ then $a \in \mathrm{dom}\, \varphi^{\mathbf{A}}$.*

---

[3] with respect to

We say that m satisfies the gluing condition *w.r.t. f when it satisfies simultaneously the dangling, the identification and the relative closedness conditions.*

If f is closed then $f(K)$ is a closed subset of $\mathbf{A}$ [2, Prop. 3.1.9] and thus $C_\mathbf{A}(f(K)) = f(K)$. Therefore, the formulation of the gluing condition can be simplified in the case of closed homomorphisms by replacing everywhere $C_\mathbf{A}(f(K))$ by $f(K)$ and removing moreover the relative closedness condition, which becomes obviously superfluous when m is closed.

The following result can be proved by means of some diagram chasing.

Theorem 1.2. *i) Let $\mathbf{D}$, together with $m : \mathbf{A} \to \mathbf{D}$ and $d : \mathbf{B} \to \mathbf{D}$, be the pushout in Alg_Γ or in C-Alg_Γ of $f : \mathbf{K} \to \mathbf{A}$ and $g : \mathbf{K} \to \mathbf{B}$. Then m satisfies the gluing condition w.r.t. f.*

ii) Conversely, let $f : \mathbf{K} \to \mathbf{A}$ and $m : \mathbf{A} \to \mathbf{D}$ be two homomorphisms (resp., closed homomorphisms) of partial Γ-algebras such that m satisfies the gluing condition w.r.t. f.

Let $\mathbf{B}$ be the relative subalgebra of $\mathbf{D}$ supported on the (always closed) set $B = (D - m(A)) \cup m(C_\mathbf{A}(f(K)))$. Then $\mathbf{B}$, together with the embedding $d : \mathbf{B} \to \mathbf{D}$ and the homomorphism $g : \mathbf{K} \to \mathbf{B}$ consisting of $m \circ f$ considered as a homomorphism from $\mathbf{K}$ to $\mathbf{B}$, is a pushout complement for f and m in Alg_Γ (resp., in C-Alg_Γ).

iii) The pushout complement in Alg_Γ (resp., in C-Alg_Γ) of two homomorphisms (resp., closed homomorphisms) $f : \mathbf{K} \to \mathbf{A}$ and $m : \mathbf{A} \to \mathbf{D}$, with m satisfying the gluing condition w.r.t. f, is unique up to isomorphism iff f is closed and injective.

For instance, if we find a pushout complement of the homomorphisms $f : \mathbf{K} \to \mathbf{A}$ and $g' : \mathbf{A} \to \mathbf{H}$ given in Example 1.2, following the recipe given in this theorem, we obtain the same $\mathbf{H}$, together with the identity $\mathbf{H} \to \mathbf{H}$ as the "horizontal arrow" and the composition $g' \circ f : \mathbf{K} \to \mathbf{H}$ as the "vertical arrow." This is very different from the pushout complement $\mathbf{B}$ of f and g' in Example 1.2. So, in general, the pushout complement of two homomorphisms $f : \mathbf{K} \to \mathbf{A}$ and $m : \mathbf{A} \to \mathbf{D}$ need not be unique if f is not closed and injective. But it turns out that the pushout complement of f and m described in (ii) can be characterized by means of a universal condition among all pushout complements of f and m (see [5] for details).

Starting with these results, one can develop two different double pushout approaches to partial Γ-algebra transformation, using plain homomorphisms and closed homomorphisms. It is worth mentioning (see again [5] for details) that double pushout algebraic transformation in Alg_Γ and C-Alg_Γ using production rules with both left- and right-hand side homomorphisms closed and injective satisfies HLR-conditions HLR2* and HLRI, and therefore it satisfies the same good properties that for instance double-pushout graph transformation using production rules with both graph morphisms injective [9].

On the other hand, in this paper we also consider three different types of partial homomorphisms between partial Γ-algebras: the conformisms, the quomorphic conformisms, q-conformisms for short, and the closed-domain quomorphic conformisms, cdq-conformisms for short. The first two types being quite popular in the literature of partial algebras (see for instance [2, 6], where q-conformisms are called closed quomorphisms), the latter are introduced (to our knowledge) in [13], and they generalize to partial algebras the partial homomorphisms of total algebras of [8]. Later than the first version of this note, cdq-conformisms (under the more appealing name of partial closed homomorphisms) have been thoroughly studied in [1]. It is worth mentioning that we do not consider here another popular type of partial homomorphism of partial algebras, the quomorphisms, because they give a category that does not have all binary pushouts (see [6]). They have been studied in [14], and also some results on them are given in [4].

Definition 1.5. *Let* $\mathbf{A} = (A, (\varphi^{\mathbf{A}})_{\varphi \in \Omega})$ *and* $\mathbf{B} = (B, (\varphi^{\mathbf{B}})_{\varphi \in \Omega})$ *be two partial Γ-algebras, and let* $f : \mathbf{A} \to \mathbf{B}$ *be a partial mapping of S-sets, with domain* Dom f *(i.e., an S-indexed family* $f_s : A_s \to B_s$ *of partial mappings, each* f_s *with domain* (Dom $f)_s =$ Dom f_s).

f *is a* conformism *when it is a closed homomorphism from a weak subalgebra of* $\mathbf{A}$ *supported on* Dom f. *Equivalently, when for every* $\varphi \in \Omega$, *say with* $\eta(\varphi) = (s, s')$, *and* $a \in$ (Dom $f)_s$, *if* $f_s(a) \in$ dom $\varphi^{\mathbf{B}}$ *then* $a \in$ dom $\varphi^{\mathbf{A}}$ *and* $\varphi^{\mathbf{A}}(a) \in$ (Dom $f)_{s'}$, *and then* $\varphi^{\mathbf{B}}(f_s(a)) = f_{s'}(\varphi^{\mathbf{A}}(a))$.

f *is a* quomorphic conformism, q-conformism *for short, when it is a closed homomorphism from the relative subalgebra of* $\mathbf{A}$ *supported on* Dom f. *Equivalently, when for every* $\varphi \in \Omega$, *say with* $\eta(\varphi) = (s, s')$, *and for every* $a \in$ (Dom $f)_s$, $a \in$ dom $\varphi^{\mathbf{A}}$ *and* $\varphi^{\mathbf{A}}(a) \in$ (Dom $f)_{s'}$ *iff* $f_s(a) \in$ dom $\varphi^{\mathbf{B}}$, *and in this case* $\varphi^{\mathbf{B}}(f_s(a)) = f_{s'}(\varphi^{\mathbf{A}}(a))$.

f *is a* closed-domain quomorphic conformism, cdq-conformism *for short, when it is a closed quomorphism with* Dom f *a closed subset of* $\mathbf{A}$. *Equivalently, when for every* $\varphi \in \Omega$, *say with* $\eta(\varphi) = (s, s')$, *and for every* $a \in$ (Dom $f)_s$, $a \in$ dom $\varphi^{\mathbf{A}}$ *iff* $f_s(a) \in$ dom $\varphi^{\mathbf{B}}$, *and in this case* $\varphi^{\mathbf{A}}(a) \in$ (Dom $f)_{s'}$ *and* $\varphi^{\mathbf{B}}(f_s(a)) = f_{s'}(\varphi^{\mathbf{A}}(a))$.

Example 1.3. Let Γ be a graph structure with a single sort and a single operation symbol φ. Let $\mathbf{A}$ be a partial Γ-algebra supported on $A = \{a_1, a_2, a_3\}$ and with operation $\varphi^{\mathbf{A}}$ given by $\varphi^{\mathbf{A}}(a_1) = a_2$, $\varphi^{\mathbf{A}}(a_2) = a_3$. Then

- The partial mapping $f : A \to A$ given by $f(a_1) = a_1$, $f(a_2) = a_3$ is a conformism, but clearly not a q-conformism (it is a closed homomorphism from the discrete weak subalgebra of $\mathbf{A}$ supported on $\{a_1, a_2\}$, which is not a relative subalgebra of $\mathbf{A}$).
- The partial mapping $f : A \to A$ given by $f(a_1) = a_2$, $f(a_2) = a_3$ is a q-conformism, but clearly not a cdq-conformism (it is a closed homomorphism from the relative subalgebra of $\mathbf{A}$ supported on $\{a_1, a_2\}$, which is not a closed subset).

– The partial mapping $f : \mathbf{A} \to \mathbf{A}$ given by $f(a_2) = a_2$, $f(a_3) = a_3$ is a cdq-conformism (it is a closed homomorphism from the closed subalgebra of $\mathbf{A}$ supported on $\{a_2, a_3\}$).

Let CF-Alg$_\Gamma$, QC-Alg$_\Gamma$ and CDQC-Alg$_\Gamma$ be the categories of partial Γ-algebras having as morphisms conformisms, q-conformisms and cdq-conformisms, respectively. It is clear that the (strict) inclusions C-Alg$_\Gamma \subset$ CDQC-Alg$_\Gamma \subset$ QC-Alg$_\Gamma \subset$ CF-Alg$_\Gamma$ hold. Notice moreover that totally defined q-conformisms (and cdq-conformisms) from $\mathbf{A}$ to $\mathbf{B}$ are always closed homomorphisms $\mathbf{A} \to \mathbf{B}$, while a totally defined conformism from $\mathbf{A}$ to $\mathbf{B}$ need not be a closed homomorphism $\mathbf{A} \to \mathbf{B}$.

We have now the following theorem, concerning the pushouts of these different types of conformisms of partial algebras. In it, the cases of CF-Alg$_\Gamma$ and QC-Alg$_\Gamma$ can be deduced from general results forthcoming in part III of [6] (in preparation), and the case of CDQC-Alg$_\Gamma$ can be deduced from general results given in [1].

Theorem 1.3. *Let* $\mathbf{A} = (A, (\varphi^{\mathbf{A}})_{\varphi \in \Omega})$ *and* $\mathbf{B} = (B, (\varphi^{\mathbf{B}})_{\varphi \in \Omega})$ *be two partial Γ-algebras.*

i) The disjoint sum $\mathbf{A} + \mathbf{B}$ of $\mathbf{A}$ and $\mathbf{B}$ is also their coproduct in CF-Alg$_\Gamma$, QC-Alg$_\Gamma$ *and* CDQC-Alg$_\Gamma$.

Let now $f : \mathbf{K} \to \mathbf{A}$ and $g : \mathbf{K} \to \mathbf{B}$ be two conformisms, with domains Dom f *and* Dom g *respectively. Let $\theta(\mathbf{K})'$ be the least equivalence relation on $\mathbf{A} + \mathbf{B}$ containing*

$$\Big(\{(f_s(x), g_s(x)) \in A_s \times B_s \subseteq (A_s \sqcup B_s)^2 \mid x \in \mathrm{Dom}\, f_s \cap \mathrm{Dom}\, g_s\}\Big)_{s \in S}$$

Let $D = (D_s)_{s \in S}$ be given by

$$D_s = \{x \in A_s \sqcup B_s \mid f_s^{-1}([x] \cap A_s) \cup g_s^{-1}([x] \cap B_s) \subseteq \mathrm{Dom}\, f_s \cap \mathrm{Dom}\, g_s\},\ s \in S$$

(where the equivalence classes are taken w.r.t. $\theta(\mathbf{K})'$), and let $\theta(\mathbf{K})$ be the restriction of $\theta(\mathbf{K})'$ to D. Let $\mathbf{D}$ be the relative subalgebra of $\mathbf{A} + \mathbf{B}$ supported on D.

ii) Let $\mathbf{D}_{\theta(\mathbf{K})}$ be the partial Γ-algebra supported on $D/_{\theta(\mathbf{K})}$ and with operations defined as follows: for every $\varphi \in \Omega$

$$\mathrm{dom}\, \varphi^{\mathbf{D}_{\theta(\mathbf{K})}} = \{[x] \mid [x] \subseteq \mathrm{dom}\, \varphi^{\mathbf{D}}\}$$

(where the equivalence classes $[x]$ are taken with respect to $\theta(\mathbf{K})$) and if $[x] \in$ dom $\varphi^{\mathbf{D}_{\theta(\mathbf{K})}}$ then $\varphi^{\mathbf{D}_{\theta(\mathbf{K})}}([x]) = [\varphi^{\mathbf{D}}(x)]$.

Then $\mathbf{D}_{\theta(\mathbf{K})}$, together with the conformisms $f' = nat_{\theta(\mathbf{K})} \circ i_{\mathbf{B}} : \mathbf{B} \to \mathbf{D}_{\theta(\mathbf{K})}$ and $g' = nat_{\theta(\mathbf{K})} \circ i_{\mathbf{A}} : \mathbf{A} \to \mathbf{D}_{\theta(\mathbf{K})}$ (both with domain D), is the pushout of f and g in CF-Alg$_\Gamma$.

iii) Let now f and g be q-conformisms. Then $\theta(\mathbf{K})$ is a closed congruence on $\mathbf{D}$ and the quotient $\mathbf{D}/_{\theta(\mathbf{K})}$, together with f' and g' defined as in (ii), which now are q-conformisms, is their pushout in QC-Alg$_\Gamma$.

iv) Let now f and g be cdq-conformisms. Let D_c be the greatest closed subset of $\mathbf{A} + \mathbf{B}$ contained in D, let $\mathbf{D}_c$ be the closed subalgebra of $\mathbf{A} + \mathbf{B}$ supported on D_c, and let $\theta(\mathbf{K})_c$ be the restriction of $\theta(\mathbf{K})$ to $\mathbf{D}_c$. Then $\theta(\mathbf{K})_c$ is a closed congruence on $\mathbf{D}_c$ and the quotient $\mathbf{D}_c/\theta(\mathbf{K})_c$, together with the cdq-conformisms $f'_c = nat_{\theta(\mathbf{K})_c} \circ i_\mathbf{B} : \mathbf{B} \to \mathbf{D}_c/\theta(\mathbf{K})_c$ and $g'_c = nat_{\theta(\mathbf{K})_c} \circ i_\mathbf{A} : \mathbf{A} \to \mathbf{D}_c/\theta(\mathbf{K})_c$, is the pushout of f and g in CDQC-Alg$_\Gamma$.

Example 1.4. Let Γ be a graph structure with a single sort and a single operation symbol φ. Let $\mathbf{K}$ be a partial Γ-algebra supported on the set $K = \{a, b\}$ and with $\varphi^\mathbf{K}$ total and given by $\varphi^\mathbf{K}(a) = \varphi^\mathbf{L}(b) = b$. Let $\mathbf{A}$ and $\mathbf{B}$ be both equal to $\mathbf{L}$.

Let $f : \mathbf{K} \to \mathbf{A}$ be the cdq-conformism with domain Dom $f = \{b\}$ and $f(b) = b$. Let $g : \mathbf{K} \to \mathbf{B}$ be the (total) closed homomorphism with $g(a) = g(b) = b$.

A simple computation, using the previous theorem, shows that the pushout (object) of f and g in QC-Alg$_\Gamma$ and CF-Alg$_\Gamma$ is a discrete Γ-algebra with two elements, while their pushout in CDQC-Alg$_\Gamma$ is the empty Γ-algebra.

From this theorem, one can develop three different single pushout algebraic approaches to partial Γ-algebra transformation, using these three types of partial homomorphisms. The one using cdq-conformisms can be viewed as the straightforward generalization to partial Γ-algebras of the single pushout approach to *total* Γ-algebras transformation using partial homomorphisms developed in [8].

Moreover, it is worth mentioning (see [4]) that

- while the approach using partial closed homomorphisms is related to the double pushout approach using closed homomorphisms in the usual way (for instance, as the single pushout approach using partial homomorphisms is related to the usual double pushout approach in [8]), the other two ones seem to have no equivalent in the double pushout approach
- the pushout of two q-conformisms in QC-Alg$_\Gamma$ and in CF-Alg$_\Gamma$ is the same (we have just shown in Example 1.4 that the pushout of two cdq-conformisms can be different in CDQC-Alg$_\Gamma$ than in QC-Alg$_\Gamma$ or CF-Alg$_\Gamma$).

2. Graphs

Let us recall the definition of graph we are considering in this paper. (In this definition, $X^{*\infty}$ and X^* stand respectively for the set of all finite and countably infinite words, and the set of all finite words over a set X.)

Definition 2.1. *A* higher-order hypergraph *(or simply* graph*) is a structure*

$$\mathbf{G} = (G_V, G_E, s^\mathbf{G}, t^\mathbf{G}, a^\mathbf{G})$$

where G_V and G_E are two arbitrary sets (of nodes and arcs, respectively) and $s^\mathbf{G}, t^\mathbf{G}, a^\mathbf{G}$ are mappings

$$s^\mathbf{G} : G_E \to G_V^{*\infty}$$
$$t^\mathbf{G} : G_E \to G_V^{*\infty}$$
$$a^\mathbf{G} : G_E \to G_E^{*\infty}$$

called the source, target and abstraction mappings respectively.

When G_V and G_E are finite sets, and $s^\mathbf{G}(e), t^\mathbf{G}(e) \in G_V^*$ and $a^\mathbf{G}(e) \in G_E^*$ for every $e \in G_E$, then we say that the graph $\mathbf{G}$ is finite.

Such a graph can be understood as a partial algebra in the following way. Let us consider the graph structure $\Sigma_G = (S_G, \Omega_G, \eta_G)$ with $S_G = \{V, E\}$,

$$\Omega_G = \{s_i \mid i \in \mathbb{N}\} \cup \{t_i \mid i \in \mathbb{N}\} \cup \{a_i \mid i \in \mathbb{N}\}$$

and $\eta_G : \Omega_G \to S_G \times S_G$ defined by

$$\eta_G(s_i) = \eta_G(t_i) = (E, V), \quad i \in \mathbb{N}$$
$$\eta_G(a_i) = (E, E), \quad i \in \mathbb{N}$$

A partial Σ_G-algebra is then a structure

$$\mathbf{G} = \left(\{G_V, G_E\}, (s_i^\mathbf{G}, t_i^\mathbf{G}, a_i^\mathbf{G})_{i \in \mathbb{N}}\right)$$

with $s_i^\mathbf{G}, t_i^\mathbf{G} : G_E \to G_V$, $a_i^\mathbf{G} : G_E \to G_E$, $i \in \mathbb{N}$, partial mappings.

Such a partial Σ_G-algebra is then a *graph* when it satisfies the following condition:

dom $s_{i+1}^\mathbf{G} \subseteq$ dom $s_i^\mathbf{G}$, dom $t_{i+1}^\mathbf{G} \subseteq$ dom $t_i^\mathbf{G}$ and dom $a_{i+1}^\mathbf{G} \subseteq$ dom $a_i^\mathbf{G}$, for all $i \in \mathbb{N}$

and such a graph is *finite* when, furthermore, G_V and G_E are finite, and only a finite number of operations $s_i^\mathbf{G}$, $t_i^\mathbf{G}$ and $a_i^\mathbf{G}$ have non-empty domain (in the usual terminology of partial algebras, when it is *totally finite*).

It turns out that every graph (resp., finite graph) in the sense of Definition 2.1 can be understood as a graph (resp., finite graph) in the sense of partial algebras, and conversely. It makes sense, then, to translate the main concepts of partial algebras to the language of graphs. In the case of subalgebras and total and partial homomorphisms, this yields the following dictionary.

Proposition 2.1. *Let* $\mathbf{G} = (G_V, G_E, s^\mathbf{G}, t^\mathbf{G}, a^\mathbf{G})$, $\mathbf{G}' = (G_V', G_E', s^{\mathbf{G}'}, t^{\mathbf{G}'}, a^{\mathbf{G}'})$ *be two graphs.*

a) Assume that $G_V' \subseteq G_V$ *and* $G_E' \subseteq G_E$. *Then* a.1) **G'** is a weak subalgebra[4] of **G** iff, for every $e \in G'_E$, there exists some $w \in G_V^{*\infty}$ such that $s^{\mathbf{G}}(e) = s^{\mathbf{G'}}(e) \cdot w$, and similar conditions for target and abstraction mappings[5].

a.2) **G'** is a relative subalgebra of **G** iff, for every $e \in G'_E$, there exists some $w \in (G_V - G'_V)^{*\infty}$ such that $s^{\mathbf{G}}(e) = s^{\mathbf{G'}}(e) \cdot w$, and similar conditions for target and abstraction mappings.

a.3) **G'** is a closed subalgebra of **G** iff, for every $e \in G'_E$, we have that $s^{\mathbf{G'}}(e) = s^{\mathbf{G}}(e)$, and similar conditions for target and abstraction mappings.

b) Let $f = (f_V : G'_V \to G_V, f_E : G'_E \to G_E)$ be a pair of total mappings. Then

b.1) f is homomorphism from **G'** to **G** iff, for every $e \in G'_E$, there exists some $w \in G_V^{*\infty}$ such that $s^{\mathbf{G}}(f_E(e)) = f_V^{*\infty}(s^{\mathbf{G'}}(e)) \cdot w$, and similar conditions for target and abstraction mappings[6].

b.2) f is a closed homomorphism from **G'** to **G** iff, for every $e \in G'_E$, $s^{\mathbf{G}}(f_E(e)) = f_V^{*\infty}(s^{\mathbf{G'}}(e)) \in G_V^{*\infty}$, and similar conditions for target and abstraction mappings.

c) Let $f = (f_V : G'_V \to G_V, f_E : G'_E \to G_E)$ be a pair of partial mappings. Let D_V and D_E be the domains of f_V and f_E respectively.
If f is a conformism then by definition it is a closed homomorphism from some weak subalgebra **D** of **G'** supported on $D = \{D_V, D_E\}$, which is the relative subalgebra supported on D if f is a q-conformism and it is a closed subalgebra if f is a cdq-conformism. Then

c.1) With the previous notations, **D** is always a graph.

c.2) f is a conformism from **G'** to **G** iff, for every $e \in D_E$, there exists some $w' \in (G'_V)^{*\infty}$ such that $s^{\mathbf{G'}}(e) = s^{\mathbf{D}}(e) \cdot w'$ and $s^{\mathbf{G}}(f_E(e)) = f_V^{*\infty}(s^{\mathbf{D}}(e))$, and similar conditions for target and abstraction mappings.

c.3) f is a q-conformism from **G'** to **G** iff, for every $e \in D_E$, there exists some $w' \in (G'_V - D_V)^{*\infty}$ such that $s^{\mathbf{G'}}(e) = s^{\mathbf{D}}(e) \cdot w'$ and $s^{\mathbf{G}}(f_E(e)) = f_V^{*\infty}(s^{\mathbf{D}}(e))$, and similar conditions for target and abstraction mappings.

c.4) f is a cdq-conformism from **G'** to **G** iff, for every $e \in D_E$, $s^{\mathbf{G'}}(e) \in D_V^{*\infty}$ and $s^{\mathbf{G}}(f_E(e)) = f_V^{*\infty}(s^{\mathbf{D}}(e))$, and similar conditions for target and abstraction mappings.

---

[4] Of course, this should read "the partial Σ_G-algebra corresponding to **G'** is a weak subalgebra ..." As we have mentioned, we shall systematically make the abuse of language of using the partial algebras jargon to talk about graphs in the sense of Definition 2.1, and viceversa, henceforth without any further mention

[5] The concatenation $w_1 \cdot w_2$ of two words $w_1, w_2 \in X^{*\infty}$ is defined iff $w_1 \in X^*$ or $w_2 = \lambda$, the empty word.

[6] Given a mapping $f : X \to Y$, we denote by $f^{*\infty}$ its extension to $X^{*\infty} \to Y^{*\infty}$.

In the case of finite graphs, to obtain the corresponding translations it is enough to delete all superscripts ∞ in this proposition. Details for this finite case can be found in [10].

In particular, closed subalgebras of graphs, closed homomorphisms of graphs and cdq-conformisms of graphs correspond to the usual notions of subgraphs, morphisms of graphs and partial morphisms of graphs (the latter in the sense of [8]), while the other concepts seem to be new in the literature on graphs. Moreover, we want to mention that every closed Σ_G-subalgebra of a graph is again a graph, while a relative or weak Σ_G-subalgebra of a graph need not be in general a graph.

Let now **Gra**$_h^\infty$, **Gra**$_c^\infty$, **Gra**$_{cf}^\infty$, **Gra**$_{qc}^\infty$ and **Gra**$_{cdqc}^\infty$ (resp., **Gra**$_h$, **Gra**$_c$, **Gra**$_{cf}$, **Gra**$_{qc}$ and **Gra**$_{cdqc}$) be the full subcategories of Alg$_{\Sigma_G}$, C-Alg$_{\Sigma_G}$, CF-Alg$_{\Sigma_G}$, QC-Alg$_{\Sigma_G}$ and CDQC-Alg$_{\Sigma_G}$ with objects the graphs (resp., finite graphs) considered as partial Σ_G-algebras. Then we have

Theorem 2.1. *i)* **Gra**$_h$ *and,* **Gra**$_h^\infty$ *are closed under the construction of pushouts and pushout complements in* Alg$_{\Sigma_G}$. *Both categories are co-complete.*

ii) **Gra**$_c$ *and* **Gra**$_c^\infty$ *are closed under the construction of pushouts and pushout complements in* C-Alg$_{\Sigma_G}$. *Both categories are co-complete.*

iii) **Gra**$_{cdqc}$ *and* **Gra**$_{cdqc}^\infty$ *are closed under the construction of pushouts in* CDQC-Alg$_{\Sigma_G}$. *Both categories are co-complete.*

iv) **Gra**$_{cf}$ *and* **Gra**$_{cf}^\infty$ *have all binary pushouts, but they need not be equal, in general, to the corresponding pushouts in* CF-Alg$_{\Sigma_G}$ *(see below for details). Both categories are co-complete.*

v) *Neither* **Gra**$_{qc}$ *nor* **Gra**$_{qc}^\infty$ *have all binary pushouts.*

In particular, points (i)–(iv) allow to develop, beside the usual double and single pushout approaches to graph transformation ([7, 8]), another double pushout and another single pushout approach, independent of each other.

It remains to describe the pushout of two conformisms of graphs. Let $f : \mathbf{K} \to \mathbf{G}$ and $g : \mathbf{K} \to \mathbf{G}'$ be two conformisms of graphs, with domains Dom f and Dom g respectively. We shall understand these graphs as partial Σ_G-algebras.

Let **H**, together with conformisms $g' : \mathbf{G} \to \mathbf{H}$ and $f' : \mathbf{G}' \to \mathbf{H}$ be the pushout of f and g as conformisms of partial Σ_G-algebras. So, **H** is a partial Σ_G-algebra, but it need not be a graph. Let $\mathbf{H}_0$ be the greatest weak subalgebra of **H** that is a graph. It is given by the weak subalgebra of **H**, with the same sets of nodes and arcs that **H**, and with source, target and abstraction operations defined as follows: for every arc $e \in H_E$, $e \in$ dom $s_k^{\mathbf{H}_0}$ iff $e \in$ dom $s_i^{\mathbf{H}}$ for every $i \leq k$, and similar definitions for target and abstraction operations.

It turns out that the mappings g' and f' are conformisms $g' : \mathbf{G} \to \mathbf{H}_0$ and $f' : \mathbf{G}' \to \mathbf{H}_0$, and a simple diagram chasing (using Proposition 2.1.(c.1)) shows that $\mathbf{H}_0$, together with the conformisms g' and f', is the pushout of f and g in **Gra**$_{cf}^\infty$ (or **Gra**$_{cf}$, if **G**, **H** and **H**' are finite graphs).

References

1. Ricardo Alberich, Manuel Moyà, Francesc Rosselló, and Llorenç Sastre. A note on a paper by Peter Burmeister and Boleslaw Wojdylo. Preprint, 1995.
2. Peter Burmeister. A Model Theoretic Oriented Approach to Partial Algebras. Introduction to Theory and Application of Partial Algebras—Part I. In *Mathematical Research*, volume 32. Akademie-Verlag, Berlin, 1986.
3. Peter Burmeister. Partial Algebras. An Introductory Survey. In *Algebras and Orders: Proceedings of the NATO Advanced Study Institute and Séminaire de Mathématiques supérieures on Algebras and Orders*, volume 389 of *NATO ASI Series C*, pages 1–70. Kluwer Academic Publishers, 1993.
4. Peter Burmeister, Miquel Monserrat, Francesc Rosselló, and Gabriel Valiente. Algebraic Transformation of Unary Partial Algebras II: Single-Pushout Approach. Preprint, 1995.
5. Peter Burmeister, Francesc Rosselló, Joan Torrens, and Gabriel Valiente. Algebraic Transformation of Unary Partial Algebras I: Double-Pushout Approach. Submitted to Theoretical Computer Science, 1995.
6. Peter Burmeister and Boleslaw Wojdylo. The meaning of basic category theoretical notions in some categories of partial algebras. Technical Report TR 1350, Fachbereich Mathematik, Technische Hochschule Darmstadt, January 1991. Appeared as Parts I and II in *Demonstratio Mathematica* XXV (1992), pages 583–602 and 973–994 respectively.
7. Hartmut Ehrig, Martin Korff, and Michael Löwe. Tutorial introduction to the algebraic approach of graph grammars based on double and single pushouts. In Hartmut Ehrig, Hans-Jörg Kreowski, and Grzegorg Rozenberg, editors, *Proceedings 4th International Workshop on Graph-Grammars and their Application to Computer Science*, volume 532 of *Lecture Notes in Computer Science*, pages 24–37, Berlin, 1991. Springer-Verlag.
8. Michael Löwe. Algebraic approach to single-pushout graph transformation. *Theoretical Computer Science*, 109:181–224, 1993.
9. Julia Padberg. Survey of high-level replacement systems. Technical Report 93-8, Fachbereich Informatik, Technische Universität Berlin, March 1993.
10. Francesc Rosselló and Gabriel Valiente. Partial algebras for the graph grammarian. In Gabriel Valiente and Francesc Rosselló, editors, *Proceedings Colloquium on Graph Transformation and its Application in Computer Science*, pages 107–115, Mallorca, 1994. Technical Report UIB-DMI-B-19 (1995).
11. Gabriel Valiente. On knowledge base redundancy under uncertain reasoning. In Bernadette Bouchon-Meunier, Llorenç Valverde, and Ronald Yager, editors, *Proceedings International Conference on Information Processing and Management of Uncertainty in Knowledge-Based Systems IPMU-92*, volume 682 of *Lecture Notes in Computer Science*, pages 321–329, Berlin, 1993. Springer-Verlag.
12. Gabriel Valiente. Verification of knowledge base redundancy and subsumption using graph transformations. *The International Journal of Expert Systems: Research & Applications*, 6(3):341–355, September 1993.
13. Gabriel Valiente. *Knowledge Base Verification using Algebraic Graph Transformations*. PhD thesis, University of the Balearic Islands, December 1994.
14. Annika Wagner and Martin Gogolla. Defining Operational Behavior of Object Specifications by Attributed Graph Transformations. To appear in *Fundamenta Informatica*, 1994.

The Contractum in Algebraic Graph Rewriting

R. Banach[1]

Computer Science Department, Manchester University,
Manchester, M13 9PL, U.K.

Abstract

Algebraic graph rewriting, which works by first removing the part of the graph to be regarded as garbage, and then gluing in the new part of the graph, is contrasted with term graph rewriting, which works by first gluing in the new part of the graph (the contractum) and performing redirections, and then removing garbage. It is shown that in the algebraic framework these two strategies can be reconciled. This is done by finding a natural analogue of the contractum in the algebraic framework, which requires the reformulation of the customary "double pushout" construction. The new and old algebraic constructions coexist within a pushout cube. In this, the usual "outward" form of the double pushout appears as the two rear squares, and the alternative "inward" formulation as the two front squares. The two formulations are entirely equivalent in the world of algebraic graph rewriting. An application illustrating the efficacy of the new approach to the preservation of acyclicity in graph rewriting is given.

1   Introduction

Algebraic graph rewriting has a relatively long history and forms a mature body of knowledge with applications in many areas of computer science. Both the applications and the theory continue to expand in many directions. From the large literature on the subject we might mention Ehrig (1979, 1986), and Ehrig et al. (1990). See also T.C.S. (1993).

Term graph rewriting arose rather more recently, (Barendregt et al. (1987)), and its application is typically rather more focussed, principally at intermediate and lower level descriptions of implementations of functional languages and similar systems; though the amount of work in related areas is expanding. See eg. Banach (1994), and Sleep et al. (1993).

Algebraic graph rewriting works by the well known "double pushout" construction. In this construction, the first step of a rewrite, once a redex has been located, is to remove the part of the graph that is to be garbaged by the rewrite, leaving a suitable hole. Then the new part of the graph is glued into the hole, yielding the result. In term graph rewriting by contrast, the first step of a rewrite, once the redex has been located, is to glue into the graph some new structure, called the contractum; then to change the shape of the graph by redirecting arcs. Finally, the garbage is removed.

Thus one goes about things in the opposite order in the two models of rewriting; and so one question of interest, is whether there is any relationship between the two approaches. Now the algebraic approach has been used to address some of the problems of direct interest to the term graph rewriting community, (Habel et al. (1988), Hoffman and Plump (1988), Plump (1993)), so one might speculate that the two approaches are not so far apart.

---

1. Email: rbanach@cs.man.ac.uk

The aim of this paper is to show that the strategy of the term graph rewriting approach can be used to reformulate the algebraic approach into a construction entirely equivalent to the original double pushout construction, but having much of the superficial appearance of the term graph rewriting construction. In particular, the new construction allows a precise notion of contractum and of contractum building to be formulated within the algebraic graph rewriting world.

The rest of this paper is as follows. Section 2 reviews the details of the conventional double pushout construction for a suitable class of graphs. Section 3 describes term graph rewriting and highlights the contrast between it and the algebraic approach. Section 4 gives the new construction in the algebraic world, shows that it is entirely equivalent to the original construction, and argues that it displays the features required for it to be regarded as incorporating a convincing analogue of the term graph contractum concept. Section 5 presents a simple example of the new approach, while section 6 presents an application of the approach by proving a theorem on the preservation of acyclicity in the rewriting of directed graphs, which would have been somewhat more inconvenient to establish in the conventional approach. Section 7 concludes.

2   Algebraic Graph Rewriting

Algebraic graph rewriting originated as a way of manipulating the objects in a specific category of graphs; one whose objects have coloured nodes and coloured edges, with source and target functions mapping each edge to its source and target. However the underlying algebraic construction is very general and can be adapted to many other categories of graph-like systems (see Ehrig et al. (1991a,b, 1993)). Since the main point that this paper makes is algebraic in nature, it too can be adapted to many such categories. However, rather than seek the greatest possible generality in the presentation, by heavy use of universal algebra, we will pick a fairly simple category of graphs to work with, and the reader will be quickly able to construct the appropriate generalisations as required.

Let $\mathcal{DG}$ be the category of directed graphs and graph morphisms. An object G of $\mathcal{DG}$ is a pair $\langle N_G, A_G \rangle$ where N_G is a set of nodes and $A_G \subseteq N_G \times N_G$ is a set of arcs built from N_G, i.e. a set of ordered pairs of N_G. An arrow $g : G \to H$ of $\mathcal{DG}$ is a map $g : N_G \to N_H$ such that

(x, y) is an arc of $G \Rightarrow (g(x), g(y))$ is an arc of H.

Like many categories of graph-like systems, $\mathcal{DG}$ has all pushouts. Thus if $f : K \to X$ and $g : K \to Y$ are two arrows, their pushout is the graph $P = \langle N_P, A_P \rangle$ given by:

$N_P = N_X \uplus N_Y / \approx$ where $\uplus$ is disjoint union, and $\approx$ is the smallest equivalence relation such that $x \approx y$ if there is a $k \in N_K$ such that $x = f(k)$ and $y = g(k)$.

$A_P = \{([x]_P, [y]_P) \mid \exists u \in [x]_P, v \in [y]_P$ such that $[(u, v) \in A_X$ or $(u, v) \in A_Y\}$ where we have not distinguished between $u \in N_X$ and the tagged version of $u \in N_P$

And the arrows $f^* : Y \to P$ and $g^* : X \to P$ are obvious.

Algebraic graph rewriting is given by the double pushout construction. Rules are given by a pair of arrows in $\mathcal{DG}$

$L \xleftarrow{l} K \xrightarrow{r} R$

with $l : K \to L$ injective. (For categories of graph-like systems, there is normally a natural notion of injectivity that is used; in our case it is ordinary set-theoretic injectivity). A redex for a rule

$L \leftarrow K \rightarrow R$ is an arrow $g : L \rightarrow G$, and the rewrite proceeds by constructing the diagram below where both squares are pushouts.

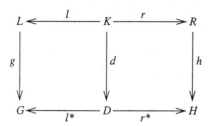

The construction is a two stage process.

Intuitively, the first stage of the construction removes the g image of L from G, except for the $g \circ l$ image of K, which provdes the interface for the second stage. In the second stage, a copy of R is glued into the "hole" left behind by the first stage; the edge of the hole being the aforementioned $g \circ l(K)$.

The first stage attempts to construct the object D and the arrows $d : K \rightarrow D$, $l^* : D \rightarrow G$, such that the left square is a pushout. D is known as the pushout complement and is not guaranteed to exist even if (as is the case here) $\mathcal{DG}$ has all pushouts. It is standard lore in algebraic graph rewriting theory that a unique smallest pushout complement exists if

(INJ-O) $l : K \rightarrow L$ is injective.

(IDENT-O) $\{x, y\} \subseteq N_L$ and $g(x) = g(y) \Rightarrow [\, x = y,$ or $\{x, y\} \subseteq l(N_K)\,]$.

(DANGL-O) $(x, y) \in A_G - g(A_L)$, and $\{x, y\} \cap g(N_L) \neq \emptyset \Rightarrow \{x, y\} \cap g(N_L) \subseteq g(l(N_K))$.

(INJ-O), which we have assumed already, ensures that a pushout complement with unique $d(K)$ exists if one exists at all. (IDENT-O) ensures that the pushout of l and d is in fact G, by ensuring that the pushout is never forced to try to map distinct nodes of L into the same node of G, other than as instructed by d — something the pushout definition above can never accomplish. (DANGL-O) ensures that D is actually an object of $\mathcal{DG}$, so that when the g image of $(L - l(K))$ is removed from G, no arc is left dangling without a source or target node. These remarks make a little more sense when we see the explicit construction of D.

$N_D = N_G - g(N_L - l(N_K))$,

$A_D = A_G - g(A_L - l(A_K))$,

The arrow $d : K \rightarrow D$ is given by

$d : K \rightarrow D : x \mapsto g(l(x))$

with the obvious extension to arcs. Arrow $l^* : D \rightarrow G$ is just the inclusion on $N_G - g(N_L - l(N_K))$, again with the obvious extension to arcs.

3   Term Graph Rewriting

Just as most applications of algebraic graph rewriting use categories of objects with a richer structure than $\mathcal{DG}$, so too with term graph rewriting, where normally, the category is that of term graphs, i.e. graphs consisting of nodes and arcs, where the nodes are labelled by the symbols from some alphabet, and the out-arcs of each node are labelled by consecutive positive integers $[1 \ldots n]$, each node having an arity as in term rewriting. Other markings may adorn the nodes and arcs depending on the application.

We will however continue to work with $\mathcal{DG}$, which contains (almost) enough structure to enable us to achieve our algebraic objectives for term graph rewriting, albeit in a more austere setting.

In fact we will work with the category $\mathcal{DG}^{(*)}$, whose objects and arrows are those of $\mathcal{DG}$, except that each non-empty object G, optionally has a distinguished node, the root of G, root_G. In fact $\mathcal{DG}$ occurs as a full subcategory of $\mathcal{DG}^{(*)}$. Each object G of $\mathcal{DG}$ occurs both "as is" in $\mathcal{DG}^{(*)}$, and also in a collection of objects with roots, once for each choice of root from N_G. We can write such objects as (G, root_G) when we want to highlight the root, writing (G, ε) if we want to emphasise that G does not have a root.

An arrow $g : G \to H$ in $\mathcal{DG}^{(*)}$ is like an arrow in $\mathcal{DG}$ except that if G has a root root_G, then H must have one, root_H, and we must have

$$g(\text{root}_G) = \text{root}_H.$$

Under these circumstances, readers can check that $\mathcal{DG}^{(*)}$ has all pushouts of $f : K \to X$ and $g : K \to Y$ unless if X and Y both have roots, root_X and root_Y, and root_X, root_Y do not both occur in the same equivalence class in the usual formula for the set-theoretic pushout of $f : N_K \to N_X$ and $g : N_K \to N_Y$. This is more than adequate for our needs in the rewriting construction.

A rule Q is now given by a pair $\langle incl : L \to P, Red \rangle$. The first component is the inclusion of an object L of $\mathcal{DG}^{(*)}$ into another object P. Neither L nor P may have a root. Red is a set of pairs $\langle x, y \rangle$ of nodes, such that $x \in N_L$ and $y \in N_P$.

A redex for a rule $Q = \langle incl : L \to P, Red \rangle$ is an arrow $g : L \to (G, \text{root}_G)$ where G must have a root, root_G, except that we must have

(LIVE) Each node $g(x)$, (and arc $(g(x), g(y))$) occurring in the image of a redex $g : L \to (G, \text{root}_G)$ is accessible from root_G.

When we say that x is accessible from r, we mean of course that there is a directed path from r to x in the graph.

Rewriting is a three stage process. Intuitively, the first stage of a rewrite glues a copy of P into G along L. This is just an honest pushout of g and $incl$ which always exists by our remarks above. The second phase, redirection, takes all in-arcs of nodes $g(x)$ where $\langle x, y \rangle \in Red$, and redirects them so that they become in-arcs of $g'(y)$ (where g' is the extension of g provided by the pushout of the first stage). Having done this, the third phase removes everything not accessible from the root, completing the rewrite.

In more detail, stage one constructs the following pushout, whose existence is unproblematic in $\mathcal{DG}^{(*)}$, since of the three graphs involved in $incl$ and g, only G has a root. Obviously G' has a root, such that $incl'(\text{root}_G) = \text{root}_{G'}$.

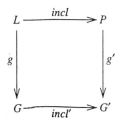

$P - L$, which generally contains dangling arcs, is called the contractum of the rule, and the pushout construction just mentioned, is called contractum building, as up to isomorphism, the pushout is just the process of gluing a copy of the contractum into G. This paper is mainly concerned with finding an analogue of this process in the algebraic world.

The second stage requires a further condition to hold. Let

$$Red' = \{\langle g'(x), g'(y)\rangle \mid \langle x, y\rangle \in Red\}$$

The condition is that Red' is the (set-theoretic) graph of a function.

(FUNC) $\langle x', y'\rangle \in Red'$ and $\langle x', z'\rangle \in Red' \Rightarrow y' = z'$

Assuming (FUNC) holds, it makes unambiguous sense to redirect all in-arcs of LHS members of Red', and make them point to the corresponding RHS nodes. This gives a graph G''.

$N_{G''} = N_{G'}$,

$A_{G''} = (A_{G'} - A^L{}_{Red'}) \cup A^R{}_{Red'}$,

$\text{root}_{G''} =$ If $\langle \text{root}_{G'}, y\rangle \in Red'$ for some $y \in N_{G'}$ then y else $\text{root}_{G'}$

where

$A^L{}_{Red'} = \{(t, g'(x)) \in A_{G'} \mid$ for some $g'(y)$, there is a $\langle g'(x), g'(y)\rangle \in Red'\}$,

$A^R{}_{Red'} = \{(t, g'(y)) \mid$ there is a $(t, g'(x)) \in A^L{}_{Red'}$ and $\langle g'(x), g'(y)\rangle \in Red'\}$.

Note that where an arc $(t, g'(x))$ in G' is redirected to $(t, g'(y))$ and there was already a $(t, g'(y))$ arc in G', the two become one arc in G''. (This is at variance with the usual situation in term graph rewriting.) Note also that, unlike in algebraic graph rewriting, where the only nodes and arcs of G manipulated by the rewrite are in the redex, there is no (DANGL)-like condition to prevent the node t in an arc $(t, g'(x))$ which is to be redirected, from being outside $g'(L)$. This is because the removal of arcs and introduction of new ones implicit in redirection, do not involve any removal of nodes, the only origin of any threat of dangling arcs.

Thus far, rewriting can only increase the size of a graph. To enable graphs to shrink, i.e. for rewriting to be able to garbage collect, the third stage defines the graph H by

$N_H = \{x \in N_{G''} \mid x$ is accessible from $\text{root}_{G''}\}$,

$A_H = \{(x, y) \in A_{G''} \mid \{x, y\} \subseteq N_H\}$,

$\text{root}_H = \text{root}_{G''}$.

Thus the third, or garbage collection stage, discards anything not accessible from the root of G''. H is the result of the rewrite. Note that H is such that any redex $h : M \rightarrow H$ for the first stage of the next rewrite automatically satisfies (LIVE).

It is worth noting at this point that whereas garbage collection is a purely local phenomenon in algebraic graph rewriting — the garbage is collected during the construction of the pushout complement, in term graph rewriting garbage collection is a global phenomenon — being defined by a condition over the whole of G''.

4   The Algebraic Contractum and the Pushout Cube

The basic differences between algebraic graph rewriting and term graph rewriting should now be clear. The former collects garbage first, and then replaces it with the new stuff, while the latter glues in the new stuff first, and only after redirection does the garbage get collected.

To bring the two styles of rewriting closer together, we recast algebraic graph rewriting into a form where the basic sequence of steps conforms more closely to that in term graph rewriting. Essentially we point out how contractum building can be done in the algebraic style.

To do so we employ a simple trick. Let $L \xleftarrow{l} K \xrightarrow{r} R$ be an algebraic rule. It consists of two arrows of $\mathcal{DG}$ with common domain K. Therefore we can form the pushout

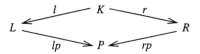

In brief, we show that we can reformulate conventional algebraic graph rewriting using rules of the form $L \xleftarrow{l} K \xrightarrow{r} R$, into a new construction, using rules of the form $L \xrightarrow{lp} P \xleftarrow{rp} R$, and that this new form embodies a credible version of contractum building as the first stage of the rewriting process, allowing a closer comparison with term graph rewriting. We will call the original form of algebraic rules and the rewriting construction that goes with them, the outward form, and the new form and construction, the inward form. Both are named after the direction of the horizontal arrows. The whole thing turns on the construction of the following pushout cube.

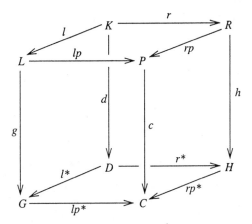

In this cube, the colimit of $l : K \to L$, $r : K \to R$ and $d : K \to D$, in which all squares are pushouts, we see the conventional construction in the two rear faces, while the new construction will emerge as the two front faces. In each case we start with G, construct an intermediate graph (either D or C) and then finally construct H.

Since we work from left to right through the cube in both cases, the first stage of the inward form will be an honest pushout of the redex $g : L \to G$, and of the LHS branch of the inward rule $lp : L \to P$. This is the algebraic equivalent of contractum building, comparable to the first stage in term graph rewriting. As in section 3, we can call $P - lp(P)$ which in general will contain dangling arcs, the contractum of the rule; and the graph C constructed by the pushout, is the analogue of the graph G' in term graph rewriting.

After this "contractum building" the inward form of the algebraic rule forms a pushout complement of $c : P \to C$ and $rp : R \to P$, to give the result of the rewrite H. The conditions for this to work, are similar to those needed in constructing D in the outward form of the rule.

Now we turn to the technical details of the new construction. Because $\mathcal{DG}$ has small colimits, up to isomorphism, the pushout cube given above really does commute as required. A particular consequence of this is that the choice of unique smallest pushout complement in the outward form of rewriting corresponds to a similar choice of unique smallest pushout complement in the inward form. The main facts about inward and outward rewriting are the following.

Theorem 4.1 Inward and outward rewriting are dual in the following sense. Let $g : L \to G$ be an arrow of $\mathcal{DG}$, serving as redex. Then statement (I) below which ensures the existence of an outward rewrite, and statement (II) below which ensures the existence of an inward rewrite, are equivalent.

(I) There is an outward rule $l : K \to L, r : K \to R$ satisfying

(INJ-O) $l : K \to L$ is injective.

(IDENT-O) $\{x, y\} \subseteq N_L$ and $g(x) = g(y) \Rightarrow [\, x = y$, or $\{x, y\} \subseteq l(N_K)\,]$.

(DANGL-O) $(x, y) \in A_G - g(A_L)$, and $\{x, y\} \cap g(N_L) \neq \emptyset \Rightarrow \{x, y\} \cap g(N_L) \subseteq g(l(N_K))$.

(II) There is an inward rule $lp : L \to P, rp : R \to P$ satisfying

(SURJ-I) $P = \langle N_P, A_P\rangle = \langle lp(N_L) \cup rp(N_R), lp(A_L) \cup rp(A_R)\rangle$.

(INJ-I) $rp : R \to P$ is injective; $lp : L \to P$ is injective on $L - lp^{-1}(rp(R))$.

(IDENT-I) $\{x, y\} \subseteq N_L$ and $g(x) = g(y) \Rightarrow [\, x = y$, or $\{lp(x), lp(y)\} \subseteq rp(N_R)\,]$.

(DANGL-I) $(x, y) \in A_G - g(A_L)$, and $\{x, y\} \cap g(N_L) \neq \emptyset \Rightarrow$
                $\{x, y\} \cap g(N_L) \subseteq g(lp^{-1}(rp(N_R)))$.

Proof sketch. The theorem claims that statement (II) is sufficient to guarantee that an inward rewrite of $g : L \to G$ exists. Since $\mathcal{DG}$ has all pushouts, we merely need to check that the analogous conditions for the pushout complement of $rp : R \to P$ and $c : P \to C$ hold. These are:

(INJ-I)^C $rp : R \to P$ is injective.

(IDENT-I)^C $\{x, y\} \subseteq N_P$ and $c(x) = c(y) \Rightarrow [\, x = y$, or $\{x, y\} \subseteq rp(N_R)\,]$.

(DANGL-I)^C $(x, y) \in A_C - c(A_P)$, and $\{x, y\} \cap c(N_P) \neq \emptyset \Rightarrow \{x, y\} \cap c(A_P) \subseteq c(rp(N_R))$.

Clearly (INJ-I) $\Rightarrow$ (INJ-I)^C. For (IDENT-I) $\Rightarrow$ (IDENT-I)^C we pull a collection of elements in G which witness $c(x) = c(y)$ up along g and use (IDENT-I) to get the result. That (DANGL-I) $\Rightarrow$ (DANGL-I)^C, is a relatively straightforward diagram chase.

Thus the pushout complement of $rp : R \to P$ and $c : P \to C$ exists under the conditions stated. These conditions turn out to be necessary as well as sufficient, so we have a set of conditions for the existence of inward rewrites, expressed soley in terms of the redex and the arrows in the rule. The remainder of the argument is as follows.

(I) $\Rightarrow$ (II). Suppose we have the hypotheses of (I). Form the pushout of $l : K \to L, r : K \to R$ giving $lp : L \to P, rp : R \to P$. Then (SURJ-I) is immediate, and (INJ-I) follows from (INJ-O). Also (IDENT-O) $\Rightarrow$ (IDENT-I) and (DANGL-O) $\Rightarrow$ (DANGL-I) by easy diagram chases.

(II) $\Rightarrow$ (I). Suppose we have the hypotheses of (II). We need to construct the top pushout square "in reverse". It follows from work of Ehrig and Kreowski (1979) that (SURJ-I) and (INJ-I) are sufficient for this to be done in an essentially unique way so we get $l : K \to L, \, r : K \to R$ from $lp : N_L \to N_P, rp : N_R \to N_P$ and (INJ-O) follows from (INJ-I). Now to get (II) $\Rightarrow$ (I) it is sufficient to show (IDENT-I)^C $\Rightarrow$ (IDENT-O) and (DANGL-I)^C $\Rightarrow$ (DANGL-O). These are again easy diagram chases. ☺

For full details of the proof see Banach (1996). We immediately find:

Theorem 4.2 Let $g : L \to G$ be a redex. Let r^O be an outward rule satisfying 4.1.(I) and r^I be an inward rule satisfying 4.1.(II), and such that r^O and r^I form a pushout in $\mathcal{DG}$. Then H can be derived from G using r^O iff H can be derived from G using r^I.

5 An Example

We present a short example of the preceding considerations. Below is an outward rule $L \xleftarrow{l} K \xrightarrow{r} R$ of $\mathcal{DG}$, with numbered nodes carrying the morphism information.

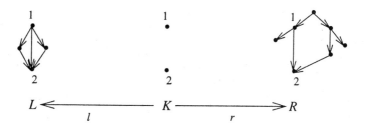

Forming the pushout of these two arrows, we arrive at the corresponding inward form of the rule

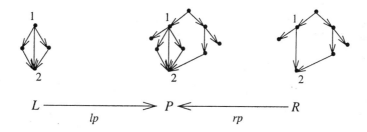

When we apply this to the following graph G we get the sequence

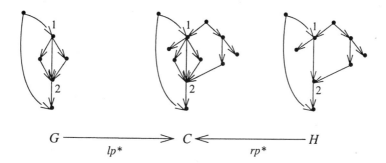

In this two step sequence (the completion of the pushout $G \leftarrow D \rightarrow H$ in conventional outward rewriting, as the reader can check), we first bolt in the contractum, and then remove what needs to be removed in terms of the image of the LHS graph L. Note that the inward form embodies a small optimisation compared with the outward form, namely that the outward form first removes the arc $(1, 2)$ in the construction of D, and then replaces it when the pushout with R is performed; this does not happen in the inward form. Of course one can prevent this inefficiency in the outward form by including the the arc $(1, 2)$ in K, but this essentially says that outward rules ought to have the property of being a pullback image, as well as what we already demand of them.

6 An Application: Acyclic Rewriting

In this section we give a brief presentation of a topic where we claim that the inward rewriting approach has some advantages over the conventional outward form. Since the two approaches are entirely equivalent by theorems 4.1 and 4.2, there is of course nothing here that cannot ultimately be done in the outward style.

Acyclicity is an important safety property of computing systems. Deadlock avoidance in resource allocation systems is the classic example, but many other safety properties can be represented in terms of the acyclicity of some directed structure that models the state of the system as it evolves. We give here a simple theorem that guarantees that rewriting of an acyclic graph via a suitable rule preserves acyclicity. We stick to the category $\mathcal{DG}$ whose objects possess an obvious notion of acyclicity, and to the notation of the pushout cube.

Theorem 6.1 Let $\Gamma = lp : L \to P \leftarrow R : rp$ be an inward rule with lp (as well as rp) injective, and let $g : L \to G$ be a redex with g injective, in an acyclic graph G. Let $K^* \subseteq P$ be the subgraph given by $K^* = lp(L) \cap rp(R)$, and let Γ have the property:

(Π) P is acyclic; and for every directed path π in P between nodes a and b of K^*, there is a path θ in P between the same nodes a and b, but lying entirely within $lp(L)$.

Let H be the result of rewriting G via the rule Γ. Then H is acyclic.

Proof. Let H arise via the arrows $lp^* : G \to C \leftarrow H : rp^*$ of the pushout cube. We claim that C is acyclic; which is sufficient since H arises via the inverse homomorphism $C \leftarrow H : rp^*$ and inverse homomorphisms preserve acyclicity.

To substantiate the claim, suppose there was a cycle $\Omega = [x,...,x]$ in C. The cycle cannot lie entirely in $lp^*(G)$ since lp is injective, and neither can it lie entirely in $c(P)$ since g is injective, by properties of pushouts. Therefore it must lie in $lp^*(G) \cup c(P)$, and split into $[\alpha_1,..., \beta_1], [\beta_1,..., \alpha_2], ...,$ with $[\alpha_i,..., \beta_i] \subseteq (c(P) - (lp^*(G) - c(K^*)))$, and $[\beta_i,..., \alpha_{i+1}] \subseteq (lp^*(G) - (c(P) - c(K^*)))$, (and cyclically), and with the $\{\alpha_i, \beta_i\} \subseteq c(K^*) = lp^*(G) \cap c(P)$. Since c is injective, each $[\alpha_i,..., \beta_i]$ is the c image of a unique path $\pi_i = [a_i,..., b_i] \subseteq P$ with $\{a_i, b_i\} \subseteq K^*$, whence property (Π) supplies us with a corresponding path $\theta_i \subseteq lp(L)$ also from a_i to b_i. Replacing each $[\alpha_i,..., \beta_i]$ by $c(\theta_i) \subseteq lp^*(G)$ in the cycle Ω, gives us a cycle entirely in $lp^*(G)$, a contradiction. ☺

One can see that to achieve the same thing in the outward approach would be somewhat more cumbersome. The inverse homomorphism of the pushout complement of the outward approach is less useful than that of the inward approach since one adds material to the graph subsequently, and one has to check that the new material does not inadvertently close a cycle. All the pieces required for the argument are present in the outward approach to be sure, but they lie scattered about in a number of different graphs so that building the contradiction is little less easy. Furthermore, in the inward approach, it is a lot more convenient to check whether a particular rule satisfies the condition (Π) by inspecting a diagram of the intermediate graph P; one can check whether any path π between nodes (a, b) of K^* and straying outside of $lp(L)$, has an alternative route between a and b entirely within $lp(L)$, at a glance.

An easy induction now gives:

Theorem 6.2 Let G_0 be an acyclic initial graph, and let $\mathcal{R}$ be an inward rule system in which each rule consists of a pair of injective arrows satisfying condition (Π) above. Then every graph generated from G_0 by rewriting injective redexes using rules from $\mathcal{R}$ is acyclic.

Theorem 6.1 applies to the example described in the previous section as is easily seen. The graph G is acyclic, the rule employed and the redex satisfy the relevant conditions, and as a consequence, the graph H is acyclic too.

As with all safety properties, by working harder and inventing more subtle invariants of the objects of interest, one can generalise and strengthen the above results in a number of different ways. In addition, one can adapt the arguments to suit graph rewriting in other categories of graphs. However to do so would take us far outside the scope of this paper.

7 Conclusions

In the previous sections, we have reviewed double pushout algebraic graph rewriting and term graph rewriting, both from a conveniently uncluttered perspective, that of the category $\mathcal{DG}$. By an algebraic trick, we were able to reformulate the former construction from its original outward form, into a new inward form, that bore comparison with term graph rewriting. Nevertheless, one should not try to push the analogy too far. Algebraic graph rewriting is "equational" in a way that term graph rewriting is not. Specifically, if in an algebraic rewrite, node x is to be merged with node y, and node y is to be merged with node z, then a pushout will ensure that in-arcs to all three nodes end up at the same node of the result. However, in term graph rewriting, if $\langle x, y \rangle$ and $\langle y, z \rangle$ are two redirections, then the in-arcs of x end up at y, and the in-arcs of y and z end up at z. To emulate the algebraic behaviour we would need $\langle x, z \rangle$ and $\langle y, z \rangle$. Thus there are phenomena in term graph rewriting that do not correspond to algebraic graph rewriting.

Finally, we showed off the potential advantages that the new approach has in certain areas, by giving a simple theorem on the preservation of acyclicity. The category of graphs used $\mathcal{DG}$, is too austere for the given result to be of immediate and great value in real world applications, but the argument used in the proof is one that stands generalisation to more complex categories whose objects and arrows are much more suitable for representing real applications. Further elaboration of these ideas will appear in other papers.

References

Banach R. (1996); Locating the Contractum in the Double Pushout Approach, Theoretical Computer Science **156**, to appear.

Banach, R. (1994); Term Graph Rewriting and Garbage Collection Using Opfibrations, Theoretical Computer Science **131**, 29-94.

Barendregt H.P., van Eekelen M.C.J.D., Glauert J.R.W., Kennaway J.R., Plasmeijer M.J., Sleep M.R. (1987); Term Graph Rewriting, *in*: Proc. PARLE-87, de Bakker J.W., Nijman A.J., eds., Lecture Notes in Computer Science **259** 141-158, Springer, Berlin.

Ehrig H. (1979); Introduction to the Algebraic Theory of Graph Grammars (A survey), *in*: Lecture notes in Computer Science **73**, 1-69, Springer, Berlin.

Ehrig H. (1986); A Tutorial Introduction to the Algebraic Approach of Graph Grammars, *in*: Third International Workshop on Graph Grammars, Lecture Notes in Computer Science **291**, 3-14, Springer, Berlin.

Ehrig H., Habel A., Kreowski H-J., Parisi-Presice F. (1991a); From Graph Grammars to High Level Replacement Systems, *in*: Fourth Int. Workshop on Graph Grammars and their Applications to Computer Science, Ehrig, Kreowski, Rozenberg (eds.), Lecture Notes in Compter Science **532**, 269-291, Springer, Berlin.

Ehrig H., Habel A., Kreowski H-J., Parisi-Presice F. (1991b); Parallelism and Concurrency in High Level Replacement Systems, Mathematical Structures in Computer Science **1**, 361-404.

Ehrig H., Kreowski H-J. (1979); Pushout Properties: An Analysis of Gluing Constructions for Graphs, Mathematische Nachrichten **91**, 135-149.

Ehrig H., Kreowski H-J., Taentzer G. (1993); Canonical Derivations for High-Level Replacement Systems, in: Graph Transformations in Computer Science, Schneider, Ehrig (eds.), Lecture Notes in Computer Science **776**, 152-169, Springer, Berlin.

Habel A., Kreowski H., Plump D. (1988); Jungle Evaluation, *in*: Proc. Fifth Workshop on Specification of Abstract Data Types, Sannella D., Tarlecki A., eds., Lecture Notes in Computer Science **332**, Springer, Berlin.

Hoffman B., Plump D. (1988); Jungle Evaluation for Efficient Term Rewriting, *in*: Proc. International Workshop on Algebraic and Logic Programming, Mathematical Research **49**, Akademie-Verlag, Berlin.

Plump D. (1993); Hypergraph Rewriting: Critical Pairs and Undecidability of Confluence, *in*: Term Graph Rewriting: Theory and Practice, Sleep et al. (eds.), John Wiley.

Sleep M.R., Plasmeijer M.J., van Eekelen M.C.J.D. (eds.) (1993); Term Graph Rewriting: Theory and Practice, John Wiley.

T.C.S. (1993); Special Issue of Selected Papers of the International Workshop on Computing by Graph Transformation, Bordeaux, France, 1991. Theoretical Computer Science, **109**, Nos. 1-2.

A Category-Theoretical Approach to Vertex Replacement: The Generation of Infinite Graphs*

Michel Bauderon

Laboratoire Bordelais de Recherche en Informatique
Université Bordeaux I
33405 Talence Cedex, France
bauderon@labri.u-bordeaux.fr

Abstract : To define NLC grammars for vertex replacement, we provide a categorical framework based on pullbacks in the category of graphs, whose major feature is probably that the *connection relation is embedded within the rewriting rule* and not distinct of it. We then indicate how it can be used to describe the generation of infinite graphs by recursive equations.

1 Introduction

Graphs can be considered from two points of view : as set of vertices linked by edges or as sets of edges glued by vertices. Each has led to a different kind of graph rewriting systems, edge rewriting (or hyperedge replacement, shortly HR [6,11]) and node rewriting (NLC or vertex replacement, shortly VR [12,8]).

In both cases, context-free rewriting systems can be defined and turned into systems of equations which can be used to describe either sets of finite graphs or infinite graphs. In two earlier papers ([1], [2]), we have considered infinite graphs generated by systems of recursive equations within the context of edge replacement (more precisely infinite hypergraphs generated by hyperedge replacement). We are now interested in vertex rewriting systems as defined in [12] or [7] : at each step, one single node u is rewritten into a graph U whose vertices can be linked only to immediate neighbours of u, in a way defined by the labelling of the vertices in both U and the original graph.

Although the class of graphs and the rewriting mechanisms are different, the technical issues remain the same. Actually, to properly define infinite objects and their generation, one basically needs two notions :

- a proper definition of substitution allowing to solve equations by iteration,
- a reasonable topology to guarantee that infinite sequences generated by substitution will converge.

---

* This work has been supported by the Esprit BRA "Computing with graph transformations"

This second point is normally handled (for words or trees) using complete metric spaces or complete partial order structures. Unfortunately, as we already pointed out in [1], no such structure is available on infinite graphs unless too restrictive conditions are imposed (such as finite degree for each vertex which would prevent from handling complete graphs in a uniform way : the first example we shall give in this paper involves the countable complete graph). But we showed that the graph morphisms which give the intuitive notion of order in the case of finite graphs are sufficient to provide us with a suitable notion of convergence, namely that of a cocomplete category. Systems of recursive equations are then solved in a fairly straightforward way, associating with a system an ω-continuous functor and using a standard fixpoint theorem for ω-continous functors.

Of course we tried to follow a similar approach to study infinite graphs generated by vertex replacement. But the first problem we met was that unlike the case of HR rewriting, no categorical framework was available to describe VR rewriting and therefore, we had no *a priori* solution to associate a functor with a VR rewriting. Moreover, trying to extend directly the existing formalism was clearly impossible : indeed, the treatment of systems in the HR case relies on the construction of an "increasing sequence" of approximants where the n-th approximant is in a *canonical way* a *subgraph* of the $n+1$-th. A very simple example in the VR case shows that this is meaningless in this framework : take the rule which replaces an a-labelled node by K_2 with a-labelled nodes (see Figure 1), and connects all its nodes to all those of the context. It is clear that iterative application starting with the one node graph will successively yield the complete graphs $K_2, K_3, ..., K_n, ...$ and should converge towards the countable complete graph K_ω (see Figure 2). But obviously, there is *no canonical way* to consider any of the K_i as a subgraph of the graph K_{i+1}.

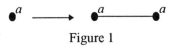

Figure 1

The only reasonable property we could perceive was that the graph K_{i+1} *projects* onto K_i (in a sense to made more precise later in this paper), by simply reducing the replacing graph to the node it had replaced (dotted ellipse and arrows in Figure 2).

If we now look at rewriting, it turns out that in the HR case, the basic mechanism is pushout, whose older name *amalgamated sum* is more suitable to understand the basics of the framework : pushout rewriting mainly adds together existing items to create a new object, possibly identifying (*amalgamating*) some of those items. It

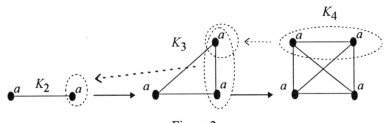

Figure 2

cannot *create* the new edges we need in the VR case : only a *product* can do it, or some part of it as described by a pullback mechanism.

Both remarks suggest that in order to solve our problem, we have to look at it exactly the other way round - in a categorical sense. Indeed, it must by now be clear what the main properties of the framework we need must be :

- VR-rewriting can be described through standard categorical operations (actually *pullbacks*) with good functorial properties,
- the category is complete and has a terminal object, in order to provide us with a "fixed point theorem",
- systems of equations can be interpreted as continuous functors and therefore solved by fixpoint iteration.

In section 2 and 3 of this paper we provide a categorical treatment for vertex replacement based on pullbacks in the category of graphs, whose major feature is probably that the *connection relation is embedded within the rewriting rule* and not distinct of it. To our knowledge, this is a totally new approach, and it sheds a new light on graph rewriting.

In section 4, we show that this framework actually allows us to solve recursive equations defined by VR-rewriting.

To keep within size limitations, we describe in this paper only the minimum of the formalism that we need to convince the reader of its intrinsic interest and to achieve our basic objective : describe infinite graphs generated by vertex replacement. Nevertheless, since products, pullbacks and limits are much less familiar than their co-counterparts, we include enough of their definitions to help the reader to avoid refering to specialized texts in category theory such as [13].

In a companion paper [3], we show that pullback rewriting can describe both NLC and NCE rewriting and that it encompasses edge rewriting (both HR rewriting and double pushout rewriting), thus providing a unified framework for graph rewriting.

2 Basic properties of Graphs

In this paper, we shall consider simple undirected graphs, possibly with loops.

Definition 1: A *graph* is a pair $G = <V, E>$, where V is the set of vertices and $E \subseteq V \times V$ is the set of edges. An edge between vertices u and v will be denoted by $[u, v]$. A node is *reflexive* if $[u, u] \in E$. A graph if *totally reflexive* if all its nodes are reflexive. A *graph morphism* $h : G \to G'$ is a pair $h = <h_V, h_E>$ where the components $h_V : V \to V'$ and $h_E : E \to E'$ are such that $h_E([u,v]) = [h_V(u), h_V(v)]$.

It is well known that the good properties of graph morphisms turn the set of graphs into a category that we shall denote by $\mathfrak{G}$. We make *no assumption* about finiteness of V and E, not even on countability.

Proposition 1: The category $\mathfrak{G}$ has arbitrary products and equalizers. The graph with one vertex and one edge is a terminal object simply denoted by $\odot$. It is a neutral element for the product. □

We shall not give the proof of this result which is classic, but simply enumerate the descriptions of the objects that the proof would build. If G_1 and G_2 are two graphs, their categorical product G (also known as the *Kronecker product* c.f. [10]) is classically defined by its sets of vertices V and edges E, in the following way:

- $V = V_1 \times V_2$,
- $E = \{[u_1 u_2, v_1 v_2] / [u_1, v_1] \in E_1 \wedge [u_2, v_2] \in E_2\}$.

The definition of the corresponding projections $\pi_i : G \to G_i$ is quite obvious. The product is defined by the following universal property: for any pair $h_i : H \to G_i$ of arrows, there exists a unique arrow $h : H \to G$ such that $h_i = \pi_i \circ h$. This extends in an obvious way to arbitrary products.

The equalizer of two morphisms f and g is simply given by the inclusion of the subgraph where $f = g$.

The unique arrow from any graph G to $\odot$ is the arrows which sends all nodes and all edges of G respectively into the unique node and edge of $\odot$. Hence $\odot$ is a terminal object. It is easily checked from the definition of the product that $\odot$ is a unit. Note that the unique vertex of $\odot$ creates all the vertices of the product, while the unique edge creates all the edges.

Remark: An infinite product may be uncountable. The simplest example is that of the product of a countable number of copies of the discrete graph with two vertices.

Corollary: The category $\mathfrak{G}$ has arbitrary limits (is *complete*). In particular, $\mathfrak{G}$ has pullbacks. □

It is a standard result that a category has limits if and only if it has both products and equalizers. If $G_0 \xleftarrow{g_1} G_1 \xleftarrow{g_2} \ldots \xleftarrow{g_n} G_n$ is a chain diagram (i.e. a contravariant functor on the category ω whose objects are the natural numbers and whose arrows are given by the usual order relation) of graphs $G_n = \langle V_n, E_n \rangle$, its limit is the graph G_ω defined by its sets of vertices V and edges E:

$$V = \{(u_0, u_1, \ldots, u_n) / \forall n > 0, u_n \in V_n \wedge g_n(u_n) = u_{n-1}\},$$

$$E = \{[(u_0, \ldots, u_n, \ldots), (v_0, \ldots, v_n, \ldots)] \in V \times V / \forall n \geq 0, g_{n+1}([u_{n+1}, v_{n+1}]) = [u_n, v_n]\}$$

In the same way, the pullback of two graph morphisms $f_i : G_i \to K, i = 1,2$ is a pair of arrows $h_i : H \to G_i, i = 1,2$ where H is the subgraph of the product consisting of exactly those items (nodes and vertices) on which the f_i coincide. The pullback has the following universal property that we shall use in the next proposition: for any pair of arrows $h'_i : H' \to G_i, i = 1,2$, such that the $f_i \circ h'_i$ coincide, there exists a unique arrow $h': H' \to H$ such that $h'_i = h_i \circ h$ (a similar universal property is valid for the limit defined above).

Note also that if K is the terminal graph ⊙, the pullback is equal to the product.

To ease the intuition of the reader who might be unfamiliar with pullbacks, the diagram in Figure 3 shows the pullback of two arrows from K_3 to a totally reflexive copy of K_2 given by the obvious projections : the vertical edges project onto the corresponding loop, while the horizontal edges remain unchanged. The full graph with nine nodes and a lot of dotted edges in the upper left corner is the product of the two copies of K_3, while the pullback consists only of the bold part (four nodes and four edges).

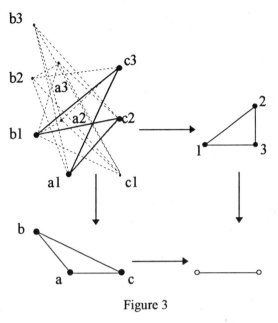

Figure 3

Proposition 2 : Let $r: R \to P$ be a given arrow in 𝔊. Then, pullback with r defines a chain-continuous functor F i.e. a functor with transforms a chain diagram with limit G into a new chain diagram with limit FG.

Proof : More precisely, let $g: G \to P$ be any arrow in 𝔊 and let $G \xleftarrow{h} H \xrightarrow{l} R$ be the pullback of r and g. We set $H = FG$ and $h = Fg$. Now, if φ is any arrow $G' \to G$, it defines by composition, an arrow $g' : G' \to P$. Let then H' be the corresponding pullback graph. We let $FG' = H'$ and $F\varphi$ be the unique arrow from H' to H given by the pullback property of H. The assertion that F is functorial is an easy consequence of the fact that in the diagram of Figure 4, all squares are pullbacks and both triangles are commutative. The only thing to prove is that the left parallelogram is a pullback. Let

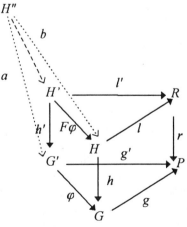

Figure 4

$G' \xleftarrow{a} H'' \xrightarrow{b} R$ be a pair of arrow such that $h \circ b = \varphi \circ a$. Since all diagrams commute, one has $g \circ \varphi \circ a = g \circ h \circ b = r \circ l \circ b$, and since H is a pullback, there exists a unique arrow u such that $\varphi \circ a = h' \circ u$ and $l \circ b = l \circ u$. The last equation implies that $u = b$. Looking at the second pullback square, one has : $r \circ b' = r \circ l \circ b = g \circ h \circ b = g \circ \varphi \circ a = g' \circ a$ and there exists a unique arrow $h'' : H'' \to H'$ such that : $l \circ b = b' = l' \circ h''$

and $a = h' \circ h''$.

Since $l' = l \circ F\varphi$, one has $l \circ b = l \circ F\varphi \circ h''$ and the unicity of b proves that $b = F\varphi \circ h''$, which concludes the first part of the proof.

Let now $G_0 \xleftarrow{g_1} G_1 \xleftarrow{g_2} \ldots \xleftarrow{g_n} G_n$ be a chain diagram with limit G_ω. From the previous result, it is clear that F transforms this diagram into another diagram $FG_0 \xleftarrow{Fg_1} FG_1 \xleftarrow{Fg_2} \ldots \xleftarrow{Fg_n} FG_n$. To complete the proof, it is enough to check that the limit of this diagram is precisely FG_ω. Once again, this is a matter of simple diagram chasing. □

3 Vertex Replacement

A vertex replacement rewriting rule is usually defined by *separately* giving the graph to be substituted to a node with a certain label and a connection relation which specifies the way its nodes will be linked to the neighbours of the rewritten node. We now translate this rewriting mechanism into our new setting where all the items of the traditional NLC mechanism will be integrated within the rewriting rule itself. In this section, we shall only describe that part of the formalism which is strictly necessary to obtain our results. A more comprehensive treatment will be found in [3].

Let $\mathbb{N}$ be the set of non negative integers and $\mathbb{N}*$ be the set of positive integers.

Definition 2 : Let A be the infinite totally reflexive graph with vertices $\{-1\} \cup \mathbb{N}$ and edges $\{[-1, n] / n \in \mathbb{N}\} \cup \{[0, n] / n \in \mathbb{N}*\} \cup \{[m, n] / m,n \in \mathbb{N}^*\}$. A will be called the *alphabet graph*.

In other words, A is obtained by taking the countable reflexive complete graph K_ω and two extra nodes called 0 and -1 and linking both of them to all nodes in K_ω. For convenience, we shall call 0 the *context*, -1 the *unknown* and all other nodes the *letters*. Considering the graph A will allow us to take into account an arbitrary number of distinct letters. If we only need a finite number m of such letters, we can restrict to its subgraph A_m, where $\mathbb{N}$ is replaced by the finite interval $[1..m]$ (see Figure 5 for the graph A_3).

As a matter of fact, our approach is such that we

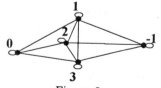

Figure 5

shall not really need letters or labels, since the labelling of nodes will be provided by morphims into A, but we shall sometimes color the drawings with labels to make them more intuitive.

Definition 3 : Let G be a graph and u be a vertex of G. A *label a on u* is a morphism $a : G \to A$ such that $a^{-1}(-1) = \{u\}$ and for each $i \in \mathbb{N}$, either $a^{-1}(i)$ is empty or it consists of immediate neighbours of u. An *unknown* is a label on a reflexive node.

Intuitively, a label on u distinguishes between u, its immediate neighbours which are mapped to the letters in A and the rest of the graph, which is indistinctly mapped onto the context in A.

Reflexivity is a technical condition to ensure that the unknown will actually be replaced by the right hand side during the pullback rewriting (due to the fact that $\odot$ is a unit).

Definition 4 : A *VR-rule* is a morphism $r : R \to A$ where $\#r^{-1}(0) = 1$ and for $i \in \mathbb{N}$ all the $r^{-1}(i)$ have at most a single element. A *production* is a pair (a, r) where a is an unknown and r is a rewriting rule.

Intuitively, $r^{-1}(-1)$ is the graph to be substituted to the rewritten node, and the edges $r^{-1}([i,-1])$ describe the connection relation of the node rewriting rule, $a^{-1}(-1)$ is the node x to be rewritten, $a^{-1}(n)$ for $n \in A$ are those neighbours which will be connected to the rewritten graph according to the connection described by $r^{-1}(n)$, and $a^{-1}(0)$ are the nodes of G which will not be affected by the rewriting (this is why we talk about the context of x).

Definition 5 : The *application of r to G at a* is the pullback of r and a. Let $\underline{G}$ denote the graph built as a pullback.

We now show that our rewriting mechanism actually encompasses NLC rewriting. We first describe how any NLC rewriting rule in the sense of [12] can be described by a VR-rule in the sense of Definition 4.

We first let $\underline{A}$ be an *"ordinary countable alphabet"*, whose letters we consider to be enumerated : $a_1,...,a_n,...$ Let $\rho = (a, X, C)$ be an NLC rule, where $a \in \underline{A}$ is the label of the nodes where ρ can be applied, X is the right hand side of the rule and C the connection relation over the alphabet A. We let R be the following graph :

- vertices : those of X, one i for each letter a_i in $\underline{A}$, an extra one u,
- edges : those of X, one from i to a vertex v labelled by b in X iff $(a,b) \in C$, one from each i to u.

The morphism $r : R \to A$ is defined :

- on the nodes by : $r(u) = 0$, $r(i) = i$ and $r(v) = -1$ for each node v coming from X,
- on the edges simply by being a morphism.

Definition 6 : The rule $r : R \to A$ is the VR-rule associated with the NLC rule ρ.

A very simple example is provided by the NLC rule whose rewriting part is represented on Figure 1, and where the connecting relation states that all nodes of the right hand side graph (K_2) will be linked will all nodes in the original graph.

The corresponding VR-rule is represented on Figure 6 (where we have simply drawn the alphabet with one letter which is all what we need for this very simple example).

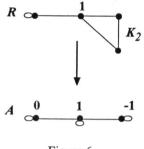

Figure 6

Proposition 3 : Let $\rho = (X, C)$ be an NLC rule and r be the associated VR-rule. Then the application of ρ and r to a graph G both define the same graph $\underline{G}$. □

Note that conversely, any VR-rule r can be decomposed into the basic items of an NLC rule : $\{r^{-1}(-1)\}$ is the right hand side of the rule, and the edges between the elements of $\{r^{-1}(-1)\}$ and those of $\{r^{-1}(i)\}$, $i \in \mathbb{N}$ define the connection relation. For further reference, let us formalise that into the following definition.

Definition 7 : For any VR-rule r, $r^{-1}(-1)$ will be called the *right hand side* of the rule r, $r^{-1}(0)$ its *left hand side* and the $r^{-1}(i)$, $i \in \mathbb{N}$ will be called the *link vertices*.

Proposition 4 : The correspondance $G \to \underline{G}$ defined by the rewriting rule r or ρ is a chain continuous functor.

Proof : This is simply a restatement of Proposition 2 to this specific context. □

4 Recursive Equations

The framework we have described in the previous sections provides us with a convenient notion of substitution. We now need to identify an appropriate notion of convergence. It is not a surprise that resolution of systems of VR-recursive equations will rely on the following standard technical result :

Proposition 5 : Let **C** be a complete category with a terminal object. A chain-continuous endofunctor F of **C** has fixpoints (i.e., objects c such that c is isomorphic to Fc). Its terminal fixpoint can be computed by iterative application to the terminal object. □

To turn a rewrite rule into a recursive equation, one simply needs to designate the substitution node(s) in the right hand side and to specify which connection rule will be applied to its neighbours during the next application.

More precisely, let $r : R \to A$ be a VR rule, and $a : R \to A$ be an unknown on R.

According to Proposition 4, r defines a chain-continuous functor F. Application of F to $\odot$ defines a unique arrow $F\odot \to \odot$. By composition with the other projection, the unknown a creates an unknown on $F\odot$ which we shall denote by a as well. This first application simply produces the right hand side of the rule in the traditional sense, together with the vertex where the rule will have to be applied.

Figure 7

The second step of the computation will now be the application of F to the unknown $a: F\odot \to A$, yielding a new unknown $a: FF\odot \to A$, (again denoted with the same letter) to which the rewriting can be further applied (Figure 7).

But in the meanwhile, the computation creates the projection arrows $F\odot \to \odot$ and $FF\odot \to F\odot$. Further iteration of the process will clearly generate a chain diagram with general term the unique arrow $F^{n+1}\odot \to F^n\odot$ defined by the pullback. Since the category of graphs is complete this diagram has a limit which is a solution of the given recursive equation.

Let us look at an example and consider the rewrite rule described in Figure 1 which generates the complete graphs. As noted in the introduction, our theory must justify the fact that recursive application of the rule should generate the countable complete graph.

The corresponding VR-rule is the rule described in Figure 6. It is quite clear that applying this rule to the terminal graph yields the graph K_2. Selecting one of the a-labelled nodes in K_2 to apply the rule again will yield K_3, etc... together with the unique projection given by the pullback mechanism. It remains simply to show that the countable complete graph is the limit of this diagram. This is routine checking from the definition of the limit.

A complete treatment of the resolution of systems of recursive equations in the framework will rely on the following facts :

- definition of a new alphabet with an arbitrary number of distinct unknowns (see Figure 8 for an alphabet with three letters and two unknown), mainly by replacing $\mathbb{N}$ by $\mathbb{Z}$, set of all integers,

- definition of graphs with multiple unknowns,

- definition of simultaneous applications of several rules to the same graph,

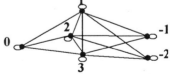

Figure 8

- transformation of a system of NLC rules into *a single VR rule*.

This involves some more definitions and work and cannot be developed here any further (see [5] for more details). An important point to note is that this improvment of the formalism will yield a framework of a genuine definition of parallel graph rewriting.

5 Conclusion

This paper must be merely seen as a starting point to further research. It introduces a new approach to vertex replacement which has been developed while trying to give a rigorous definition of the generation of infinite graphs generated by vertex replacement. This new framework is quite interesting *per se* since it closes a gap between the formalisms describing HR and VR system : so far both could be described either by set theoretical definitions describing the substitution mechanism or by an algebraic formalism giving a clean description of the generated languages, but only HR systems could be described within a categorical framework. This new setting raises a number of issues which will be topics for further investigation :

- completely describe the formalism we have just hinted at in this short note, relate it precisely to the NLC approaches of [7] and [12] and see how both kinds of rewriting can live together in the same category and how they do interact,
- study in more details the generation of infinite graphs along the lines set in [2], i.e. considering systems of recursive equations and compare the sets of graphs generated by both approaches.

These points are dealt with in more details respectively in [3-5].

Acknowledgement : It is a great pleasure to thank Michael Smyth for a very inspiring discussion about the respective merits of direct and inverse limits.

6 References

1   M. Bauderon, Infinite Hypergraphs I. Basic properties, *Theor. Comput. Sci.* 82 (1991) 177-214

2   M. Bauderon, Infinite Hypergraphs II. Systems of recursive equations, *Theor. Comput. Sci.* 103 (1993) 165-190

3   M. Bauderon, A Uniform approach to graph rewriting : the pullback approach Research Report 1058-95, LaBRI, Bordeaux I University 1995, *WG'95, to appear, Lect. Notes in Comp. Sci.*

4   M. Bauderon, Parallel rewriting of graphs through the pullback approach, *SEGRAGRA'95, to appear, Elect. Notes in Theor. Comp. Sci.*

5   M. Bauderon, Infinite graphs generated by vertex replacement, *in preparation*, LaBRI, Bordeaux I University, 1995

6   M. Bauderon, B. Courcelle, Graph expressions and graph rewriting, *Math. Systems Theory* 20 (1987), 83-127

7   B. Courcelle, An axiomatic definition of context-free rewriting and its applications to NLC graph grammars, *Theor. Comput. Sci.* 55 (1987), 141-181.

8   J. Engelfriet, G. Rozenberg, Graph grammars based on node rewriting : an introduction to NLC grammars, *Lect. Notes in Comp. Sci.* N° 532, 1991, 12-23.

9   H.Ehrig, Introduction of the algebraic theory of graph grammars, in Graph Grammars and their applications to Computer Science, *Lect. Notes in Comp. Sci.* N° 73, 1979, 1-69

10 M. Farzan, D.A. Waller, Kronecker products and local joins of graphs, *Can. J. Math.*, Vol. XXIX, No 2 1977, 255-269

11 A. Habel, Hyperedge Replacement : Grammars and Languages, *Lect. Notes in Comp. Sci.* N° 643, 1992.

12 D. Janssens, G. Rozenberg, Graph grammars with node label controlled rewriting and embedding, *Lect. Notes in Comp. Sci.*, 153 (1982), 186-205.

13 S. McLane, *Categories for the working mathematician*, Springer, Berlin, 1971

Issues in the Practical Use of Graph Rewriting

Dorothea Blostein, Hoda Fahmy, Ann Grbavec

Department of Computing and Information Science
Queen's University, Kingston, Ontario, Canada K7L 3N6
blostein@qucis.queensu.ca

Abstract. Graphs are a popular data structure, and graph-manipulation programs are common. Graph manipulations can be cleanly, compactly, and explicitly described using graph-rewriting notation. However, when a software developer is persuaded to try graph rewriting, several problems commonly arise. Primarily, it is difficult for a newcomer to develop a feel for how computations are expressed via graph rewriting. Also, graph-rewriting is not convenient for solving all aspects of a problem: better mechanisms are needed for interfacing graph rewriting with other styles of computation. Efficiency considerations and the limited availability of development tools further limit practical use of graph rewriting. The inaccessible appearance of the graph-rewriting literature is an additional hindrance. These problems can be addressed through a combination of "public relations" work, and further research and development, thereby promoting the widespread use of graph rewriting.

1. Introduction

Graph rewriting has the potential to be useful in a large variety of applications. Graphs provide an expressive and versatile data representation. Typically, nodes represent objects or concepts, and edges represent relationships among them. In addition, hierarchical relationships can be depicted by node-nesting [Hare88] [SiGJ93]. Auxiliary information is expressed by adding attributes to nodes or edges. Given the widespread use of graphs as a data representation, it is natural that graph manipulations form the basis of many useful computations. Graph manipulations can be represented implicitly, embedded in a program that, among other things, constructs or modifies a graph. Alternatively, graph manipulations can be represented explicitly, using clearly-delineated graph rewriting rules that modify a host graph. The explicit use of graph-rewriting rules offers several advantages. Graph rewriting provides an abstract, high-level representation of a solution to a computational problem. Also, the theoretical foundations of graph rewriting assist in proving correctness and convergence properties.

Despite this potential, graph rewriting has not attained widespread practical use. To discover the reasons for this, it is helpful to consider an outside viewpoint:

> Mr. and Mrs. Maggraphen manage a small software house in Bavaria. Most of their important data structures are graphs. Currently, all of their programs are written in C, with much of the code devoted to graph manipulations.
>
> The Maggraphens are planning for the future, and want to switch from C to a graph-rewriting language.

The Maggraphens are enthusiastic about graph rewriting, but have many questions. To begin with, important practical considerations arise. Will the graph-rewriting language be fast enough? Are there tools for developing, displaying, and debugging graph-rewrite rules? Suppose, optimistically, that the answer to both questions is "Yes".

This research is supported by Canada's Natural Sciences and Engineering Research Council.

Even so, there is another major hurdle: the Maggraphens can't imagine how to recast their C programs in terms of graph rewriting. They desperately need small-scale advice (how to formulate individual rewrite rules) and large-scale advice (how to organize a collection of rules). Let us consider a sampling of their questions. (Figure 1 shows our terminology.)

Graph g   A directed or undirected graph. Nodes and/or edges may be labeled and may have associated attributes.

Graph Rewrite Rule   A rule specified by:
- $g_l \rightarrow g_r$ g_l and g_r are unattributed graphs. During rule application, an attributed subgraph g_l^{host} (isomorphic to g_l) is replaced by g_r^{host} (a subgraph created to be isomorphic to g_r).
- **Embedding Information**
  Calculates post-embedding edges from pre-embedding edges (defined below). Embedding information can be textual or graphical. Gluing models specify embedding with a gluing isomorphism.
- **Application Condition**   (Optional)
  Defines conditions on attribute values or host-graph structure. These conditions must hold for rule application to proceed.
- **Attribute Transfer Function** (Optional)
  Assigns attribute values to g_r^{host}, using attribute values in g_l^{host}.

Host Graph g   The graph to which a rule is being applied.

g_l^{host} A subgraph of the host graph g, isomorphic to g_l. In some models, g_l^{host} must be an *induced* subgraph: if an edge of g connects two nodes of g_l^{host}, then that edge must be part of g_l^{host}.

RestGraph   The graph $g - g_l^{host}$. (The "-" operator denotes removal of all nodes and edges of g_l^{host} and all edges with one or both endpoints in g_l^{host}.)

g_r^{host} A subgraph isomorphic to g_r; used to replace g_l^{host}.

Pre-embedding Edges   the set of edges joining g_l^{host} to RestGraph

Post-embedding Edges   the set of edges joining g_r^{host} to RestGraph

Figure 1. Our terminology for graph rewriting. These definitions assume the use of subgraph isomorphism, where some models actually allow for a general graph morphism.

---

2. Mrs. Maggraphen: *We are new to graph rewriting. Where do we start?*

The Maggraphens are looking to us, the graph-rewriting community, as a source of information about how to express computations in graph rewriting. Consider an analogous change from C to Lisp programming: avid C programmers who cannot use Lisp effectively (due to a C mindset that dominates their approach to programming), can absorb "Lisp culture" by immersing themselves in an environment of experienced Lisp programmers. These same C programmers, in attempting to learn graph rewriting, may have trouble locating sources of "graph-rewrite culture". The graph-rewriting community should make an effort to promote such a culture, to allow newcomers to quickly develop a proper mindset for performing practical, effective computations using graph rewriting. Relevant materials include the following:

- Accessible written expositions about the practical use of graph rewriting: systems organizations, styles of computation, etc.
- Easily-available tools for creating, editing, executing, debugging graph rewriting systems (Section 4).
- Examples of non-trivial, practical uses of graph rewriting. Complete, executable systems are most helpful. These illustrate various computational styles in which graph rewriting may be used. (Relevant references, discussed in [BlFG95], include: software engineering [EnLS87] [ELNSS92] [LoKa92] [Pfei90], syntactic pattern recognition [Fu82], document image analysis [Bunk82a] [FaBl93] [GrBl95] [CoTV93], 3D object recognition [LiFu89], visual programming environments [EgPM92], diagram editors [Gött92] [DoTo88], databases [EhKr80], and semantic networks [EhHK92]. Further discussion is given by [Panel91].)

The fostering of a graph-rewriting culture will go far toward the popularization of graph rewriting.

3. Mr. Maggraphen: *In C, we use standard algorithms (searching, sorting, hashing) and algorithm-design methods (divide-and-conquer, dynamic programming, greedy algorithms). What is the equivalent to this in graph rewriting?*

Currently, we have little to offer the Maggraphens, in terms of graph-rewrite-oriented techniques for algorithm design or analysis. We have few libraries of standard graph-rewriting code. (An inspiring example is given by the parameterized graph-rewrite rules for abstract-syntax-tree manipulation reported in [ELNSS92]).

We need to develop specialized algorithm design techniques, geared toward graph rewriting as the primitive operation. Precedents for such specialized algorithm design techniques include VLSI design (with area*time used as a cost function) and optical computing (where primitive operations include Fourier transform, convolution, union and intersection of figures, coordinate transforms).

4. Mrs. Maggraphen: *What development tools are available?*

As everyone is well aware, practical use of graph rewriting depends heavily on the availability of development and debugging tools. Unfortunately, construction of these tools is a time-consuming, complex task, due to the need to combine textual and diagrammatic elements, the need to provide readable displays of large graphs, and the need to visualize the interactions among graph rewriting rules. Development of graph-rewrite debugging techniques is an interesting and challenging research topic. Currently it is difficult even to define what kind of tools are needed to support widespread practical use of graph rewriting. This will become clearer over time, as the improving set of available tools allow us to gather more extensive experience with executable graph-rewriting systems.

For the reader interested in experimenting with graph rewriting, here is a brief list of graph-rewriting environments. The first two environments are mature enough to be in widespread use, and are under active further development. The remaining environments may become available for general use. Our apologies if this list is incomplete.
- PROGRES provides extensive facilities for ordered graph rewriting [NaSc91] [ELNSS92]. Contact andy@i3.informatik.rwth-aachen.de to obtain this software.

- GraphEd [Hims91] provides extensive graph-display capabilities, and supports a limited form of graph-rewriting (direct-derivation steps of context-free rewrite rules). Contact himsolt@fmi.uni-passau.de to obtain this software.
- Pfeiffer describes development plans for a graphical editing environment for algebraic graph rewriting [Pfei90]. In the meantime, a textual representation of a graph grammar is compiled into C.
- A prototype implementation of algebraic graph transformation is described in [LöBe93]. At that time, the tool performed direct derivation steps in the single-pushout approach.
- Göttler [Gött92] mentions a succession of implementations for executing ordered graph rewriting (Y and X notation); a new C implementation is under development, including a graphical editor for X notation rules.

5. Mr. Maggraphen: *Can graph rewriting be efficient? Isn't subgraph-isomorphism testing intractable?*

This question readily comes to mind, but we can give some reassurance. It is true that subgraph-isomorphism testing is an NP-complete problem in general, but various factors make it tractable in a graph-rewriting system. Firstly, it is often possible to express a computation using small subgraphs on the left-hand-side of rewrite rules. Secondly, node labels, edge labels, and directed edges drastically reduce the search space for isomorphic subgraphs. Finally, some graph-rewriting systems have certain phrases that frequently appear in application conditions; these can be exploited to greatly reduce the search space for isomorphic subgraphs that meet the application condition. The optimization of subgraph-isomorphism testing is discussed in [BuGT91] [Zünd94].

Of course, graph rewriting should not be marketed as a fast style of computation: the von Neumann architecture (geared toward instruction fetch and execution, with a bottleneck between processor and memory), is not well-suited to the interpretation of graph rewriting. Strong demand could motivate the development of a new computer architecture with graph-rewriting as a fundamental operation. First we would need to develop suitable graph-rewriting architectures in software, and thus popularize graph rewriting as a style of computation. Special-purpose graph-rewriting hardware may sound far-fetched, but consider neural-network computations as an analogy: years of research with software-implemented neural-net architectures have now resulted in commercially-available neural-net architectures implemented as VLSI circuits.

6. Mrs. Maggraphen: *How can we organize rewrite rules?*

The graph-rewriting literature reports on various methods of organizing a collection of graph-rewrite rules: unordered, ordered and event-driven graph-rewriting systems, as well as graph grammars (Table 1). This taxonomy arose from our efforts to organize our reading of the graph rewriting literature. (This literature is confusing because many systems are called "grammars", whether they define a graph-language or not.) An understanding of these systems-organizations provide a helpful starting point in the process of deciding how a computation could be expressed as graph rewrite rules.

The choice of system organization greatly affects the number of rewrite-rule applications that must be tried during execution. Parsing with a grammar normally requires backtracking, and frequent testing of inapplicable rules. In contrast, an ordered graph rewriting system can directly transform an input graph into an output graph, with a limited number of production rules under consideration at any given time [Bunk82a]. Event-driven graph-rewriting systems can be highly time-efficient, applying rules only

System Components	System Execution
Unordered Graph-rewriting System	
A set of graph-rewrite rules.	Rewrite the given host graph (choosing nondeterministically among applicable rules) until no further rules apply.
Graph Grammar	
A set of graph-rewrite rules (*productions*). A start graph. A designation of labels as terminal or nonterminal.	In *generative* use, rewrite the start graph to obtain a terminal graph (no non-terminal labels.) The set of generatable terminal graphs is the *language* of the grammar. For *recognition*, parse the given graph: find a sequence of rewrite-rules that derive the given graph from the start graph.
Ordered Graph-rewriting System	
A set of graph-rewrite rules. A control specification (provides complete or partial ordering of rule-application).	Rewrite the given host graph (choosing nondeterministically among applicable rules consistent with the control specification) until a final state in the control specification is reached.
Event-driven Graph-rewriting System	
A set of graph-rewrite rules. An externally-arising sequence of events.	Rewrite the given initial host graph: rewrite rules are executed in response to events.

Table 1. Four organizations for graph-rewriting systems.

in direct response to external actions. Thus, if an application is such that it can be implemented using event-driven graph-rewriting, then likely it can run with acceptable time-efficiency. If the application calls for ordered (or partially ordered) graph rewriting without backtracking, then it may well run with acceptable efficiency. If the application calls for graph grammar use, then careful grammar and parser construction (context free, if possible) are necessary if there is to be hope of parsing speeds allowing large-scale practical use. In any case, graph rewriting can be useful even if it does not provide an acceptably efficient implementation: a practical software development cycle can include the use of graph rewriting to form an executable specification (e.g. [ZüSc92]).

We now briefly review the practical use of these four system organizations.

<u>Unordered graph rewriting</u>

An excellent example of unordered graph rewriting is provided by Δ-rewriting [KaLG91] [LoKa92]. The rewriting system is given an initial host-graph (e.g. the quicksort example of [LoKa92, p. 177] uses a list of numbers to be sorted, the specification of the Actor language of [KaLG91, p. 484] uses a graph compiled from an Actor program). This initial host-graph is transformed via graph-rewrite rules, either infinitely (as in the dining philosophers example of [LoKa92, p. 112]), or with termination (as in the quicksort example). The *platform* concept used to modularize Δ-rewriting is discussed in Section 8. Unfortunately, no Δ-rewriting environment is available; current experience is limited to paper-based descriptions of Δ-rewriting systems.

Graph grammars

In a pure graph grammar, productions can be listed in any order, but order-dependence often arises in practice. Once a developer has chosen a particular parser, the developer is usually aware of the order in which the parser tries alternatives. The developer may make use of this to design a smaller or faster graph grammar. For example, Anderson [Ande77] uses a set-based "coordinate grammar" to recognize mathematical notation. He describes his reliance on production-rule ordering to distinguish an input "cos" as a word denoting a trigonometric function, rather than as an implied multiplication denoting "c*o*s". It would be possible to rewrite the grammar to avoid this order dependence, but the grammar would increase in size and complexity. The drawback of such order dependence is that the language is no longer defined by the grammar alone, but arises through the interaction of the grammar with a particular parser.

In addition to order-dependence, there is the issue of reversibility. Can a given grammar be used both for recognition and generation? While a pure grammar is reversible, in practice non-reversible constructs like application conditions and attribute computations are common. Reversibility is desired in various domains, but difficult to achieve. For example, there is on-going research into reversible string-grammars for natural language processing [Strz90]. On a related note, a graph grammar with non-reversible rules is limited to either bottom-up or top-down parsers.

Practical use of graph grammars is seriously hampered by the high complexity of parsing. Sub-exponential parsers have been developed for certain restricted classes of graph grammars. A selection of parsing references are as follows. Kaul presents a linear-time precedence parser for a special class of context free graph-grammars [Kaul83]. Bunke and Haller describe an extension of Early's parser for context-free plex languages [BuHa92]; this parser permits left-recursion and is capable of recognizing partial structures. Recently, a parsing algorithm applicable to context-sensitive graph grammars has been developed [ReSc94]. Egar et al. use a graph-grammar parser in the design of a visual programming environment for clinical protocols [EgPM92]. Lin and Fu recognize three-dimensional objects (in two-dimensional images) using a semantic-directed top-down backtrack parser for plex grammars [LiFu89]. Collin et al. interpret dimensions in engineering drawings using a plex-grammar parser that mixes top-down and bottom-up processing [CoTV93]. A chart-based parser for hierarchical graphs is discussed in [MaKl92]. More recently, Klauck reports on a heuristically-driven chart parser and it's application to CAD/CAM [Klau94]. On a related note, Henderson and Samal discuss efficient parsing of stratified shape grammars, building on the table-driven methods used for LR(k) string grammars [HeSa86]; these techniques might be relevant to graph-grammar parsing.

Ordered graph rewriting

For many computations it is convenient to order, or partially order, a collection of rewrite rules. For example, Bunke recognizes circuit diagrams by first applying a collection of noise-reduction rules [Bunk82a]. It is critical that these noise-reduction rules be applied first, and exhaustively, before application of rules for recognition of transistors, capacitors, and so on. Similarly, a recognition approach for music notation [FaBl93] uses ordered recognition stages, each of which consists of three ordered phases (*Build* creates edges, *Weed* removes inconsistent edges, and *Incorporate* prunes the graph while adding semantic information to attributes). Graph applications in software engineering have made extensive use of ordered graph rewriting (e.g. [ELNSS92]).

Various forms of ordered graph rewriting are possible, depending on the use of non-determinism and backtracking:

- A completely deterministic system results from pairing a deterministic control specification with the use of cursor-nodes (also called demon nodes) to indicate the desired host graph location for rule application. Determinism is desirable in editing applications, where end-users expect a deterministic response to an editing command (e.g. [Gött92]).

- Partially ordered rewrite systems, without backtracking, have been used for software engineering (e.g. [ELNSS92]) and diagram recognition (circuit-diagrams [Bunk82a], music-notation [FaBl93] [Fahm95], math-notation [GrBl95]). In the diagram recognition work, the control specification orders the phases that make up the recognition process; rules within a phase are unordered or partially ordered, and all non-deterministic alternatives lead to a desired result.

- Partially ordered rewrite systems, with backtracking, can be expressed in the PROGRES language [ZüSc92]. The PROGRES interpreter automatically backtracks in the search for a successful path through the control specification: alternate matches for g_l^{host}, and alternate control paths, are tried as needed. This allows straightforward coding of classical AI search problems as a partially-ordered collection of rewrite rules.

Control specifications can be expressed in a variety of forms, including lists, diagrams, or text. The simplest control specification associates two sets with each production rule. The Success set lists the possible production(s) to try after successful application of the current production. The failure set lists productions to try after unsuccessful application of the production. This can be specified in tabular form [Fu82], which quickly becomes difficult to read. Diagrammatic control specifications (*control diagrams*) are used by [Bunk82a], with extensions by [DoTo88] [FaBl93] and others. For example, a *block condition* allows the control diagram to test attribute values of any nodes involved in the most recent production [DoTo88]. To permit more flexible control constructs, the control specification can take a textual form, similar to an imperative programming language. For example, PROGRES provides deterministic and non-deterministic versions of And, Or, Loop [ZüSc92][ELNSS92], in addition to encapsulation tools such as transactions and subdiagrams.

<u>Event-driven graph rewriting</u>

Whereas ordered graph rewriting systems provide an internally-imposed ordering of the rewrite rules, event-driven systems have an externally-imposed ordering, arising from the ordering of external events. This is illustrated by the library system of Ehrig and Kreowski [EhKr80]. An external event, such as loaning, returning, or ordering a library book, results in the invocation of a corresponding rewrite rule. Parameters provide the rewrite rule with information describing the details of the event. The authors mention an anticipated need for control structures within a single transaction.

Ordered graph rewriting can be used to regulate event-driven graph rewriting. In the Forrester-diagram editor of [DoTo88], the control specification defines which editing events are legal at any given point. Events not foreseen by the control specification are disallowed, resulting in an error message to the user. A similar structure is used by the diagram editors described in [Gött92].

7. Mr. Maggraphen: 𝓗ow do we choose a graph-rewriting mechanism?

A large variety of graph-rewriting mechanisms have been investigated. No one rewriting mechanism is universally suitable. Practical choice of a rewriting mechanism depends on the application, on the availability of tools, and on personal taste. Relevant factors include the power of the embedding, formal properties of rewrite rules, readability and intellectual manageability, and efficiency of rule application.

Power of the Embedding

Complex embedding mechanisms permit significant graph inspection and graph manipulation during the embedding step. Conversely, highly-restricted embedding mechanisms, such as the invariant embedding of the gluing models, are inconvenient for expressing certain common graph operations such as node deletion (Figure 2).

The choice of an embedding mechanism involves a tradeoff between using fewer, but complex, rewrite rules versus using a larger number of simpler rules. Up to now, we have few practical examples of graph-rewriting systems that make heavy use of complex embeddings. It appears that many software designers find it is easier or more natural to express a computation using more rules of a restricted embedding type.

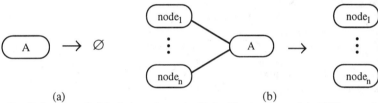

(a)   (b)
Figure 2 Delete an A-labeled node and all incident edges. (a) With an elementary embedding mechanism. (b) With a gluing model. The invariant embedding necessitates that g_1 be expanded to include all edges incident on the A-labeled node. A set of rewrite rules is used to enumerate each possible configuration of incident edges. (The "..." notation, denoting variable repetition of nodes and edges, is adapted from [EhHK92]. Similarly, Δ-notation uses *-groups, which denote zero or more occurrences of starred graph elements, to implement node-deletion [KaLG91, p. 478]. A Δ-rule that deletes a node is syntactic shorthand for an infinite collection of Δ-rules that meet the gluing condition.)

Formal Properties of Rewrite Rules

Formal properties of graph rewriting are practically important. The strong theoretical foundations of the gluing models can offer significant advantages. For example, algebraic graph rewriting simplifies construction of proofs about the integrity of a database system, as illustrated by the library-transaction system of [EhKr80].

Using rewrite rules with formally-characterized properties, graph rewriting can provide a formal definition of graph classes; examples include the class of well-formed Forrester diagrams [DoTo88] and the class of well-formed semantic networks [EhHK92].

Readability and Intellectual Manageability

Readability of rewrite rules affects intellectual manageability, system development time, ease of maintenance, and ease of debugging. It can be particularly difficult to present complex embeddings in a readable way. Since textual embedding specifications can be difficult to read, various diagrammatic notations have been proposed (Figure 3). Visual presentation can be simplified by avoiding the duplication of graph-parts common to g_l and g_r (Figure 4). In our opinion, these diagrammatic depictions are advantageous for embeddings of intermediate complexity:

- Elementary embeddings can be specified textually, and are easily perceived from visually-corresponding nodes in g_l and g_r (Figure 5). Similarly, gluing isomorphisms are effectively conveyed by the visual correspondence of g_l and g_r nodes, as in [EhHK92].
- Embeddings that are more complex than the elementary type (e.g., they involve testing of node-labels in RestGraph, or following of edges in RestGraph) are easier to perceive if a diagrammatic notation is used instead of a textual one.
- Selected embedding paths that are very long and highly complex benefit from textual rather than diagrammatic depiction. An example is the use of the PROGRES "path" construct, which permits extensive searching and testing of the host-graph, as part of the embedding process [ELNSS92].

Some applications require complex embeddings, others don't. In our experience, major difficulties arise not in the formulation of individual rewrite rules, but in the structuring of a large collection of rules that interact in a desired way.

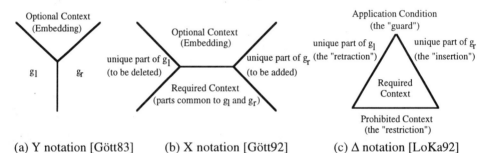

(a) Y notation [Gött83]   (b) X notation [Gött92]   (c) Δ notation [LoKa92]

Figure 3 Three diagrammatic notations for graph-rewrite rules. In Y and X notations, the embedding is shown as optional context: these diagrammatic depictions of embedding are used *if* they match in the host graph. The required context must match in order for the rewrite rule to be applied. In Δ notation, the center of the Δ is used both for required and optional context, with a * placed next to the optional parts. (Elements of a * group may occur zero, one or more times.) The prohibited context depicts host-graph structure that must not be present; restrictions on labels and attributes are expressed textually in the guard.

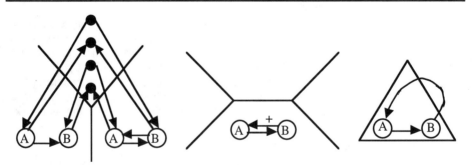

Figure 4 Graph-rewrite rules to add a second edge between an A-labeled node and a B-labeled node. Avoiding duplication of graph-parts common to g_l and g_r shrinks the drawing of g_l and g_r, and greatly reduces the graphical depiction of the embedding. (The Y-notation rule appears in [Gött92, Fig. 14].)

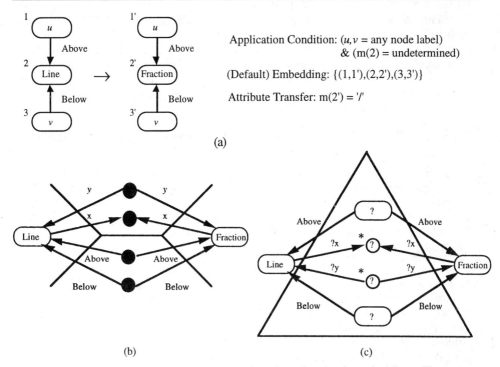

Figure 5 Textual (a) versus graphical (b, c) depiction of a simple embedding. These are three notations for a graph-rewrite rule to replace a *Line*-labeled node by a *Fraction*-labeled node, in the context of incoming *Above* and *Below* edges (as used in [GrBl95]). (a) The analogous embedding is conveyed by similarly-denoted nodes in visually-corresponding places; this is reinforced by the textual description "{(1,1'), (2,2'), (3,3')}". (b) In X-notation, the embedding is conveyed as optional context. One filled-in node (indicating arbitrary node label) and two edges depict a node-correspondence. Since directed edges are used, this must be repeated for incoming and outgoing edges. (c) In Δ-notation, the embedding is conveyed similarly, using *-groups to indicate 0 or more occurrences of the starred structures.

Isomorphisms versus General Graph Morphisms

Selection of a rewrite mechanism involves choosing isomorphisms or general morphisms for finding a subgraph g_l^{host} matching g_l. The utility of general graph morphisms is illustrated by small examples in the literature ([EhHK92, p. 560], [KrRo90, p. 200]). However, general morphisms could easily result in unexpected matches. We would be interested to hear of the use of general graph morphisms in large-scale system; debugging of such rewrite systems could be difficult.

A useful compromise is to allow the rule-author to selectively and explicitly indicate where general morphisms may be used. For example, Δ-rewriting uses subgraph isomorphism, but with a label-subscript notation (called a *fold*) to explicitly indicate groups of nodes which can optionally be matched to a single host-graph node [KaLG91] [LoKa92]. The utility of this construct is demonstrated by a rule to insert an element into a circular list: one rule works for circular lists of any length ≥ 1.

Extensions to the Rewrite Mechanism

Many extensions to rewrite mechanisms are useful in practice [BlFG95]. These include hierarchical label organization; calculation of attribute values; application conditions; parameters to graph-rewrite rules; variable node and edge labels within rewrite rules; variable graph structure within rewrite rules (e.g. optional or repeated nodes and/or edges). While all of these extensions are useful in certain applications, care must be used to select only the features necessary to cleanly express the graph transformations needed in a given application.

8. Mrs. Maggraphen: How do we modularize a graph-rewriting system?

A graph-rewriting system that is constructed in a modular way is easier to design, implement, debug, and maintain. Various aspects of a graph-rewriting system can be modularized -- the host-graph structure, the rewrite rules, the control specification. This is an active research area. Selected approaches to modularization are listed below. Several of these approaches can be used in combination.

Modular specification of host-graph structure

A description of allowable host-graph structure provides a foundation for the design of a graph-rewriting system. For example, the graph scheme in PROGRES defines statically-declarable graph properties [ELNSS92]. The graph scheme defines a class hierarchy for node labels and edge labels (multiple inheritance is allowed). Based on this, edge typing information is declared: for each edge-label, define what node-types are admissible at the endpoints of the edge. This static type information allows useful compile-time and run-time checks on graph-rewrite rules and on host-graph structure.

Host-graph triggers

This method of modularization is proposed for an unordered graph-rewriting system (wherein a host-graph is nondeterministically transformed by a set of graph-rewrite rules, with no control specification). To allow the designer to divide a large problem into more manageable subproblems, Δ-rewrite systems use *platforms* of related rules [LoKa92] [ToKa94]. These platforms are defined via specially labeled nodes called *trigger* nodes. To define a platform, choose a new trigger label. Every rewrite rule in the platform contains this trigger node in g_l (i.e., in the required context or retraction). If some rewrite rule wishes to invoke rules in a particular platform P, the rewrite rule adds the P trigger to the host graph. This satisfies one of the preconditions of rule-application from platform P, and thus may result in execution of a P-platform rule. The label of a trigger node is a tuple of arbitrary structure, and can include parameters to influence the resultant application of a G-platform rule. This style of computation has been used to solve (on paper) a variety of specification and concurrency problems.

Modular control specification

In an ordered graph-rewriting system, the control specification can be structured in a modular way. For example, PROGRES provides transactions and subdiagrams as encapsulation tools [ZüSc92]. Ordering can be used to structure the computation into phases; for example, Build-Constrain-(Rank)-Incorporate recognition stages are used in [FaBl93] [GrBl95].

Two-level rewrite rules

Generic graph-rewrite rules (expressed as graphs) can be transformed via meta-rules, to produce executable rewrite rules. This has been used in a system to describe legal database transactions [GöHi94]: complex transactions are conveniently expressed as a hyperproduction, which is transformed by a metaproduction to produce the final production. This construct allows general operations to be expressed generically, as a hyperproduction, and then used in a variety of ways. For example, a hyperproduction for the manipulation of geometric objects can be specialized (via metaproductions) to treat polylines or rectangles.

Modules of rewrite rules arising from host-graph locality

In many applications, a host graph can be represented hierarchically, with an abstract level, as well as a refined level (consisting of local graphs and interfaces). In this case, graph productions can be modularized, with some modules transforming local graphs, others changing interfaces or the global graph, and yet others changing the graph hierarchy (split or join local graphs) [EhEn94] [Taen94].

Inheritance

Inheritance is a powerful tool for layering in object-oriented system design. Several forms of inheritance can be used within a graph-rewriting system; some examples are mentioned earlier in this list, as well as in [EhEn94].

Import-Export-Interface

As described in [EhEn94], graph transformations can be organized into modules, where each module has an import interface, local operations, and an export interface. This is challenging to implement, because imported graph-rewrite rules are known by name only.

9. Mr. Maggraphen: *How can we design a graph-rewriting system to accommodate evolving host-graph structure?*

The Maggraphens are producing software for clients with changing needs. Thus they need to plan for evolution of their graph-rewriting system. Adding a new feature may require extensions to the host-graph representation; for example, new node labels and edge labels may be introduced. When this happens, the Maggraphens expect most of their old rewrite rules to continue to work properly, and they want it to be clear which of the old rules must be updated in response to the expanded host-graph representation. Many aspects of a rewrite system bear on this problem, such as the use of graph schemes to statically declare permissible host-graph structure [ELNSS92]. Here we consider only the effect of choosing induced versus non-induced subgraph matching. (If g_l^{host} is an induced subgraph of g, then g_l^{host} must include all local edges of g, i.e. all edges of g that connect two g_l^{host} nodes. A non-induced subgraph may omit some or all of these edges. This is illustrated in Figure 6.)

Compared to non-induced subgraphs, induced subgraphs meet more stringent matching criteria, and provide more information about local host-graph structure. The following consequences result.

- Using induced subgraphs increases the number of rewrite rules: g_l cannot match unless the rule-author has anticipated all the edges present in that part of the host graph. Various edge-configurations must be enumerated in separate graph-rewrite rules (where a single non-induced rewrite rule could suffice).
- Non-induced subgraphs require extra application conditions, necessary to ensure the absence of certain host-graph edges.
- Implicit edge-deletion is a major pitfall of non-induced subgraphs. Edges present in host-graph but not mentioned in g_l are deleted by rule application (Figure 6).

These points become particularly significant in case of host-graph evolution. Consider the addition of a new type of edge, with the new edge-label "Grow". Ideally, the old graph-rewrite rules should continue functioning as before, so that we merely need to create a few new rules that directly process the Grow edges. Both induced and non-induced subgraphs disappoint us.

- Using induced subgraphs, the presence of a Grow edge prevents application of any of the old rules. The old rules must be replicated, to enumerate all possible permutations of Grow edges that might occur in the g_l^{host} area.
- Using non-induced subgraphs, the old graph-rewrite rules continue to apply, but they perform implicit Grow-edge deletion. Rewrite rules apply whether or not a Grow-edge is present, but if a Grow-edge was present before rule application, it is no longer present after rule application.

These problems are independent of the embedding mechanism, arising similarly in all gluing and embedding models that use removal of g_l^{host} during the rewriting step. Improved semantics can be defined by using non-induced subgraph matching and avoiding node deletion where possible. (If g_l^{host} and g_r^{host} contain corresponding nodes, then these nodes are identified, rather than removing the g_l^{host} node and replacing it with the g_r^{host} node.) Such incomplete removal of non-induced subgraphs is provided in the definition of structured graph rewriting [KrRo90], and in the current PROGRES language [Schü91, p. 652]. (These semantics evolved over time: an earlier PROGRES reference describes the removal from host-graph of the complete subgraph corresponding to the non-induced g_l^{host} [EnLS87, p. 192]). Many graph-rewriting papers give scant mention of their choice to use induced or non-induced subgraph matching. This issue is important both theoretically and practically.

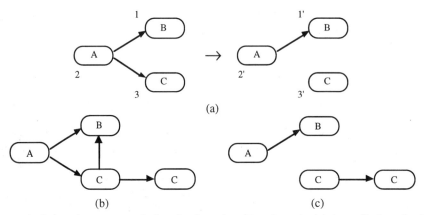

Figure 6 Induced versus non-induced subgraphs. Rewrite rule (a) is applied to the host graph (b). If an induced g_l^{host} is required, the isomorphism test fails and the rewrite rule cannot be applied. If non-induced subgraph matching is used, a suitable g_l^{host} is found and replaced, resulting in the new host graph (c). (We use the standard removal and replacement of g_l^{host}, as in the LEARRE steps: Locate, establish Embedding Area, Remove, Replace, Embed [Roze87].) Note the implicit edge-deletion in (c): the edge from the C-labeled node to the B-labeled node is removed in host-graph, an effect that may or may not have been anticipated by the author of rewrite-rule (a).

10. Mrs. Maggraphen: *Can hierarchical graphs be rewritten?*

Hierarchical host-graph structures arise naturally in many applications. In a strict definition of hierarchical graphs, all edges must connect siblings, or connect a parent and a child node. However, many practical problems cannot be modeled without additional edges that cross the hierarchy, for example to connect "cousin" nodes. The presence of such hierarchy-crossing edges greatly complicates the construction of tools for hierarchical graph rewriting. Various notations for hierarchical graph structures are described in [Hare88] [SiGJ93]. Hierarchical structure assists in the display of a large graph. Zoom-in and zoom-out operations reduce the graph to manageable proportions for viewing, or delimit selected portions of the graph for processing.

It is possible to consider hierarchical graphs as merely a notational device pertaining to graph display: a hierarchically-structured graph can easily be translated into a flat graph, with the addition of special edges to indicate parent/child relationships in the hierarchy. However, a full implementation of hierarchical-graph rewriting must give many special considerations to these edges. There is significant interest in the topic of hierarchical graph rewriting. Relevant references include a chart-based parser for hierarchical-graphs [MaKl92]; abstract graphs in a prototype algebraic-rewrite environment [LöBe93]; graphs where node labels can be graphs themselves [Schn93]; flat host-graph structure with hierarchy-expressing rewriting rules used to zoom in and out [EhHK92] and to manage and display a derivation [Hims94]; use of hierarchical graphs in a formal approach to plan generation [ArJa94]; use of hierarchically distributed graph transformations [Taen94].

11. Mr. Maggraphen: *A lot of our C code performs graph inspections. How can we translate this into graph-rewrite rules?*

The Maggraphens' current software freely mixes graph-inspection operations with graph-manipulation operations. Their graph inspection operations test local or global host-graph properties; examples include searching for a short path between nodes, or testing whether a graph is bipartite. The Maggraphens are concerned about the feasibility of translating to a pure graph-rewriting language. It is true that some host-graph inspection is performed during a graph-rewriting step (find g_l^{host}, find embedding edges, test the application condition). But these host-graph inspections accompany or follow subgraph-isomorphism testing, making it clumsy and expensive to express graph inspections that should be undertaken before the subgraph-isomorphism test.

More direct methods for expressing host-graph inspections are desirable. The designers of PROGRES recognize this, providing a variety of graph-inspection language constructs [ELNSS92]. Statically-declarable graph properties are defined in the graph scheme; these include the class hierarchy for node labels and edge labels, as well as restrictions on the source- and target-node-labels for edges with a particular edge label. In addition to this static construct, a variety of dynamic graph-inspection constructs are provided. General control structures direct the application of graph tests and graph productions [ZüSc92]. A rule's g_l can be augmented with *path* constructs, permitting complex, far-reaching examination of graph structure as part of the localization of g_l^{host}. Independent of rewrite-rule application, path descriptions can be used to compute values for derived attributes. The applicability of a rewrite rule (or a subprogram of rules) can be tested without executing it. Global on-going graph inspection is proposed in [NaSc91]: global runtime conditions are used to state host-graph conditions that should always (or never) hold.

In summary, practically-usable graph rewriting languages must provide general facilities for graph inspection. Different language constructs may be suitable for unordered, grammar-based, or ordered graph-rewriting environments.

12. Mrs. Maggraphen: *What about our user-interface and image-processing code? We want to leave that coded in C.*

Graph rewriting is a suitable formalism for expressing only part of the Maggraphens' computation. To encourage widespread use of graph rewriting, we need convenient methods to combine graph rewriting with other styles of computation. This is an interesting research topic. A few possible approaches include combining graph rewriting with a blackboard architecture (with the host graph stored as part of the blackboard); combining graph rewriting with methods for performing major computations on attributes (where attributes can be complex entities such as tables or lists or even other graphs); using graph rewriting with or on top of a standard programming language (as is already being done with some ordered graph-rewriting systems such as PROGRES [ZüSc92]).

13. The Maggraphens: *Thanks for the information. We'll probably continue to use C...*

Currently we cannot advise the Maggraphens to stake their financial future on graph rewriting as their tool for product development. We hope that this situation will change, so that in perhaps ten years time we could give different advice. Here's what we have to do to achieve this.

- Make it less difficult for an outsider to learn how to use graph rewriting in a practical application. The Maggraphens' experience mirrors our own: as we set out to apply graph rewriting to diagram recognition [FaBl93] [GrBl95] [Fahm95], we found it hard to figure out how to organize our computation. A careful reading of the literature was only of limited help: we found extensive discussion of graph-rewriting mechanisms, but little discussion of systems issues, and few examples of significantly-large graph-rewriting systems. Currently, the graph-rewriting literature appears confusing and uninviting to an outsider.

- Disseminate the graph-rewriting research/experience that is currently available. Graph rewriting is an intuitive, widely appealing concept, and outsiders are continually reinventing it. (Several attendees at Williamsburg invented graph rewriting during the course of their research, only later to discover that there already existed research on this subject, and thus found their way to the workshop. Other reinventors of graph-rewriting never find us. This should not be happening for a research community that has a decades-long history.) The profile of graph-rewriting must be raised. One important goal is to have graph-rewriting included in the standard undergraduate computing science curriculum. A few lectures' worth of material can be included in a data-structures or algorithms course, where graph-representation techniques and graph-inspection algorithms are already taught. Alternatively, graph grammars can be introduced in a formal languages class.

- Develop a better sense for which applications (or parts of applications) are suitable for implementation via graph rewriting. (We found an enthusiastic atmosphere at the Williamsburg conference: all sorts of computer-science applications were eagerly characterized as "yes, yes, graph grammars would be a great way to solve that problem".) We need to develop guidelines for identifying when graph rewriting use is advisable, and we need to develop methods for integrating graph rewriting into systems that use other styles of computation as well.

- Continue to develop and refine environments for graph rewriting. We are delighted that the PROGRES environment (and other environments to follow) are sufficiently mature to be generally usable. (When we began our diagram-recognition work, we found that the [Bunk82a] software was not in a state to be reused. Thus we had to create our own modest graph-rewriting environment; this took time, and the poor quality of the executing environment hampered our debugging and testing. We are happy that now, if we interest other colleagues in graph rewriting, we can direct them to existing graph-rewriting environments!)

In summary, our current situation is this. We are very enthusiastic about graph rewriting as a style of computation, and we are eager to convince other researchers to use graph rewriting. However, when we do succeed in convincing someone to try graph rewriting, we are left in the awkward position of being flooded with Maggraphen-type questions, few of which we can answer satisfactorily. Let us continue to work toward giving graph rewriting the widespread use it deserves.

References

[Ande77] R. Anderson, "Two Dimensional Mathematical Notation," in *Syntactic Pattern Recognition, Applications*, K. S. Fu editor, Springer 1977, pp. 147-177.

[ArJa94] O. Arnold and K. Jantke, "Therapy Plans as Hierarchically Structured Graphs," in [IWGG94], pp. 338-343.

[BlFG95] D. Blostein, H. Fahmy, A. Grbavec, "Practical Use of Graph Rewriting," Technical Report No. 95-373, Computing and Information Science, Queen's University, Jan 1995.

[Bunk82a] H. Bunke, "Attributed Programmed Graph Grammars and Their Application to Schematic Diagram Interpretation," *IEEE Pattern Analysis and Machine Intelligence*, Vol. 4, No. 6, Nov. 1982, pp. 574-582.

[Bunk82b] H. Bunke, "On the Generative Power of Sequential and Parallel Programmed Graph Grammars," *Computing*, Vol. 29, 1982, pp. 89-112.

[BuGT91] H. Bunke, T. Glauser, T. Tran, "An Efficient Implementation of Graph Grammars Based on the RETE Matching Algorithm," in [IWGG91], pp. 174-189.

[BuHa92] H. Bunke and B. Haller, "Syntactic Analysis of Context-Free Plex Languages for Pattern Recognition," in *Structured Document Image Analysis*, Eds. Baird, Bunke, Yamamoto, Springer 1992, pp. 500-519.

[CoTV93] S. Collin, K. Tombre, P. Vaxiviere, "Don't Tell Mom I'm Doing Document Analysis; She Believes I'm in the Computer Vision Field," *Proc. Second Intl. Conf. on Document Analysis and Recognition*, Tsukuba, Japan, Oct. 1993, pp. 619-622.

[DoTo88] J. Dolado, F. Torrealdea, "Formal Manipulation of Forrester Diagrams by Graph Grammars," *IEEE Trans. Systems, Man and Cybernetics* 18(6), pp. 981-996, Nov 1988.

[EgPM92] J. Egar, A. Puerta, M. Musen, "Automated Interpretation of Diagrams for Specification of Medical Protocols," *AAAI Symposium: Reasoning with Diagrammatic Representations,* Stanford University, March 1992, p 189-192.

[EhKr80] H. Ehrig and H. Kreowski, "Applications of Graph Grammar Theory to Consistency, Synchronization, and Scheduling in Data Base Systems," *Information Systems*, Vol. 5, pp. 225-238, 1980.

[EhHK92] H. Ehrig, A. Habel, H. Kreowski, "Introduction to Graph Grammars with Applications to Semantic Networks," *International Journal of Computers and Mathematical Applications*, Vol. 23, No 6-9, pp. 557-572, 1992.

[EhEn94] H. Ehrig and G. Engels, "Pragmatic and Semantic Aspects of a Module Concept for Graph Transformation Systems," in [IWGG94], pp. 157-168.

[EnLS87] G. Engels, C. Lewerentz, W. Schafer, "Graph Grammar Engineering: A Software Specification Method," in [IWGG87], pp. 186-201.

[ELNSS92] G. Engels, C. Lewerentz, M. Nagl, W. Schafer, A. Schürr, "Building Integrated Software Development Environments Part 1: Tool Specification," *ACM Trans. Software Engineering and Methodology*, Vol. 1, No. 2, Apr. 1992, pp. 135-167.

[FaBl93] H. Fahmy and D. Blostein, "A Graph Grammar Programming Style for Recognition of Music Notation," *Machine Vision and Applications*, Vol 6, No 2, pp. 83-99, 1993.

[Fahm95] H. Fahmy, "Reasoning in the Presence of Uncertainty via Graph Rewriting," Ph.D. Thesis, Computing and Information Science, Queen's University, March1995.
[Fu82] K. S. Fu, *Syntactic Pattern Recognition and Applications*, Prentice Hall 1982.
[Gött83] H. Göttler, "Attribute Graph Grammars for Graphics," in [IWGG83], pp. 130-142.
[Gött87] H. Göttler, "Graph Grammars and Diagram Editing," in [IWGG87], pp. 216-231.
[GöGN91] H. Göttler, J. Gunther, G. Nieskens, "Use Graph Grammars to Design CAD-Systems!" in [IWGG91], pp. 396-410.
[Gött92] H. Göttler, "Diagram Editors = Graphs + Attributes + Graph Grammars," *International Journal of Man-Machine Studies*, Vol 37, No 4, Oct. 1992, pp. 481-502.
[GöHi94] H. Göttler and B. Himmelreich, "Modeling of Transactions in Object-Oriented Databases by Two-level Graph Productions," in [IWGG94], pp. 151-156.
[GrBl95] A. Grbavec and D. Blostein, "Mathematics Recognition Using Graph Rewriting," *Third International Conference on Document Analysis and Recognition*, Montreal, Canada, August 1995, pp. 417-421.
[Hare88] D. Harel, "On Visual Formalisms," *Communications of the ACM*, Vol 31, No 5, pp. 514-530, May 1988.
[HeSa86] T. Henderson and A. Samal, "Shape Grammar Compilers," *Pattern Recognition*, Vol 19, No 4, pp. 279-288, 1986.
[Hims91] M. Himsolt, "GraphEd: An Interactive Tool for Developing Graph Grammars," in [IWGG91], pp. 61-65.
[Hims94] M. Himsolt, "Hierarchical Graphs for Graph Grammars," in [IWGG94], pp. 67-70.
[IWGG79] *Intl. Workshop on Graph Grammars and Their Application to Computer Science and Biology*, LNCS Vol . 73, V. Claus, H. Ehrig, G. Rozenberg Eds, Springer, 1979.
[IWGG83] *Second Intl. Workshop on Graph Grammars and Their Application to Computer Science*, LNCS Vol. 153, H. Ehrig, M. Nagl, G. Rozenberg Eds, Springer, 1983.
[IWGG87] *Third Intl. Workshop on Graph Grammars and Their Application to Computer Science*, LNCS Vol. 291, Ehrig, Nagl, Rozenberg, Rosenfeld Eds, Springer, 1987.
[IWGG91] *Fourth Intl. Workshop on Graph Grammars and Their Application to Computer Science*, LNCS Vol. 532, H. Ehrig, H. Kreowski, G. Rozenberg Eds, Springer, 1991.
[IWGG94] Pre-proceedings of the *Fifth Intl. Workshop on Graph Grammars and Their Application to Computer Science*, Williamsburg, VA, Nov. 1994. Full versions of selected papers appear in this volume.
[KaLG91] S. Kaplan, J. Loyall, S. Goering, "Specifying Concurrent Languages and Systems with Δ-grammars," in [IWGG91], pp. 475-489.
[Kaul83] M. Kaul, "Parsing of Graphs in Linear Time," in [IWGG83], pp. 206-218.
[Klau94] C. Klauck, "Heuristic Driven Chart-Parsing," in [IWGG94], pp. 107-113.
[KrRo90] H.-J. Kreowski, G. Rozenberg, "On Structured Graph Grammars, I, II" *Information Sciences*, Vol. 52, 1990, pp. 185-210, 210-246.
[LiFu89] W. Lin and K.S. Fu, "A Syntactic Approach to Three-Dimensional Object Recognition," *IEEE Trans. Systems Man and Cybernetics*, **16**(3), May 1986, pp. 405-422.
[LöBe93] M. Löwe, M. Beyer, "AGG -- An Implementation of Algebraic Graph Rewriting," *Fifth Intl. Conf. on Rewriting Techniques and Applications*, Montreal, Canada, June 1993, in LNCS 690, Springer, pp. 451-456.
[LoKa92] J. Loyall and S. Kaplan, "Visual Concurrent Programming with Delta-Grammars," *Journal of Visual Languages and Computing*, Vol 3, 1992, pp. 107-133.
[MaKl92] J. Mauss and C. Klauck, "A Heuristic Driven Parser Based on Graph Grammars for Feature Recognition in CIM," *Advances in Structural and Syntactic Pattern Recognition*, Ed. H. Bunke, World Scientific, 1992, pp. 611-620.
[NaSc91] M. Nagl, A. Schürr, "A Specification Environment for Graph Grammars," in [IWGG91], pp. 599-609.
[Panel91] "Panel Discussion: The Use of Graph Grammars in Applications," in [IWGG91], pp. 41-60.
[Pfei90] J. Pfeiffer, "Using Graph Grammars for Data Structure Manipulation," *Proc. 1990 IEEE Workshop on Visual Languages*, pp. 42-47.
[ReSc94] J. Rekers and A. Schürr, "Parsing for Context-Sensitive Graph Grammars," in [IWGG94], pp. 89-94.
[Roze87] G. Rozenberg, "An Introduction to the NLC Way of Rewriting Graphs," in [IWGG87], pp. 55-70.
[Schü91] A. Schürr, "PROGRESS: A VHL-Language Based on Graph Grammars," in [IWGG91], pp. 641-659.

[Schn93] H. Schneider, "On categorical graph grammars integrating structural transformations and operations on labels," *Theoretical Computer Science*, Vol. 109, 1993, pp. 257-275.
[SiGJ93] G. Sindre, B. Gulla, H. Jokstad, "Onion Graphs: Aesthetics and Layout," *Proc. 1993 IEEE Symposium on Visual Languages*, Bergen, Norway, Aug. 1993, pp. 287-291.
[Strz90] T. Strzalkowski, "Reversible Logic Grammars for Natural Language Parsing and Generation," *Canadian Computational Intelligence Journal*, **6**(3), pp. 145-171, 1990.
[Taen94] G. Taentzer, "Hierarchically Distributed Graph Transformations," in [IWGG94], pp. 430-435.
[ToKa94] W. Tolone and S. Kaplan, "A Semantic Definition for Introspect using Δ-Grammars," in [IWGG94], pp. 418-423.
[ZüSc92] A. Zündorf and A. Schürr, "Nondeterministic Control Structures for Graph Rewriting Systems," *Proc 17th Intl. Workshop on Graph-Theoretic Concepts in Computer Science WG91*, LNCS Vol 570, Springer Verlag, 1992.
[Zünd94] A. Zündorf, "Graph Pattern Matching in PROGRES," in [IWGG94], pp. 174-178.

The Category of Typed Graph Grammars and its Adjunctions with Categories of Derivations

A. Corradini[1], H. Ehrig[2], M. Löwe[2], U. Montanari[1], J. Padberg[2]

[1]Università di Pisa
Dipartimento di Informatica
Corso Italia 40, I-56125 Pisa (Italia)

[2]Technical University of Berlin
Institut für Software und Theoretische Informatik
Franklinstraße 28 / 29, 10587 Berlin (Germany)

Abstract. Motivated by the work which has been done for Petri-nets, the paper presents a categorical approach to graph grammars "in the large". In the large means, that we define categories of graph grammars, graph transition systems, and graph derivation systems which embody the notion "grammar", "direct derivation", and "derivation", respectively, as they are defined in the classical algebraic theory. For this purpose we introduce a suitable notion of graph grammar morphism on "typed graph grammars" in analogy to Petri-nets. A typed graph grammar is a grammar for typed graphs which is a slight generalization of the standard case. The main result shows that the three categories are related by left-adjoint functors. We discuss the relationship of our results to similar results obtained in the Petri-net field, and applications to entity/relationship models.

1 Introduction

As shown in [CELMR 94b], algebraic graph grammars of the double-pushout type [Ehr 87] are a proper generalization of Petri-nets [Rei 85]. By contrast to Petri-nets, they allow to specify structured states, i.e. graphs instead of place vectors, and context dependent transitions, where part of the precondition is read but not consumed.

However, by contrast to other formalisms for the specification of concurrent and distributed systems (like Petri-nets, Event Structures, Asynchronous Transition Systems and others [NPW 81, Bed 88, MM 90, SNW 93] graph grammars lack a formal categorical treatment in the large. This is mainly due to the absence of a natural notion of graph grammar morphism, which allows to compare and translate different grammars, their induced direct derivations and derivation sequences. For the time being all categorical analysis of graph grammars has been in the small, i.e. local to a given grammar, like for example analysis for sequential and parallel independence of certain derivation steps within one grammar. In this paper we are not interested in the analysis of the internals of the categories of derivations induced by some grammar but in the analysis of the relationships among all these categories "in the small".

For this analysis we also apply a categorical approach for which we borrow the general outline from [MM 90]. Three categories will be defined: The first is **GraGra**, having grammars as objects and grammar morphisms as arrows. For a suitable notion of grammar morphism we need to pass from standard graph grammars to typed graph grammars [CMR 95], which are a slight generalization of the classical approach (compare section 2). The usefulness of these grammars is demonstrated by two examples, namely the modelling of concrete Petri-nets (P/T-nets with distinguished tokens) and Entity/Relationship specifications as typed graph grammars. The second and third category, namely the categories of direct derivations **Gra TS** and general derivations **GraCat**, will be defined as "specialization" of **GraGra** by requiring additional algebraic properties. The objects of **GraTS** are transition systems, i.e. grammars whose rule set is closed under direct derivations, the objects of **GraCat** are derivation systems which are "categories of derivations", i.e. grammars whose rule set is additionally closed w.r.t. sequential composition of (direct) derivations. Due to this set-up we obtain forgetful functors U:**GraTS** → **GraGra** and V:**GraCat** → **GraGra**. The main result of the paper shows that both functors have left-adjoints (section 4), which allow to construct free graph transition systems and derivation categories over a given graph grammar. This is a complete analogy to the results in [MM 90] obtained for Petri-nets.

The conclusion section (section 5) is dedicated to the comparison of the treatment of graph grammars presented here and the theory about Petri-nets in [MM 90]. Especially we point out work that remains to be done in order to show that the Petri-net case is a special case of the presented general theory for graph grammars and to consider more general types of graph grammar morphisms.

2 Typed Graph Grammars

This section reviews the basic notions and definitions of the algebraic approach to graph transformation [Ehr 87], introduces the concept of typed graph grammars which is a proper generalization of the classical approach [CMR 95, Löw 94] and demonstrates the usefulness of the typing mechanism by means of two examples, namely the modelling of (concrete) P/T-nets and entity/relationship diagrams by graph grammars (compare [CELMR 94b] and [CL 95] respectively).

The basis of the algebraic approach to graph transformation is the category of graphs **Graph**.

Definition 1 (category of graphs) A *directed graph* $G=(V, E, s, t)$ consists of a set of vertices V, a set of edges E, and two mappings $s,t:E \rightarrow V$ which provide a source and a target vertex for every edge. A *graph morphism* $f:G \rightarrow H$ is a pair of functions $(f_V:G_V \rightarrow H_V, f_E:G_E \rightarrow H_E)$ which is compatible with the graph structure, i.e. $f_V \circ s^G = s^H \circ f_E$ and $f_V \circ t^G = t^H \circ f_E$. Graphs and graph morphisms constitute a category which is called **Graph** and has all limits and colimits.

Graph transformation rules are specified by pairs of co-initial injective arrows $L \leftarrow K \rightarrow R$ in [Ehr 87], describing the precondition for rule application L, its postcondition R, and the context graph K which specifies the "gluing items" of the rule, i.e. the objects which are read during rule application but not consumed. Therefore, graph transformation rules are members of the category of spans over **Graph** which is defined below.

In the following, we assume a fixed choice of pullbacks for the category **Graph** which induces the category of spans over graphs.

Definition 2 (choice of pullbacks) In a category C which has all pullbacks, a *choice of pullbacks* is a fixed pullback (PB(f, g), i(f, g), j(f, g)) for each pair of arrows f:A → C and g:B → C in C, satisfying the following properties for identities and composition:

(1) PB(f:A → C, id$_C$) = (C, id$_C$, f)   preservation of identities

(2) PB(f, g ∘ h) = PB(j(f, g), h)
    i(f, g ∘ h) = i(f, g) ∘ i(j(f, g), h)
    j(f, g ∘ h) = j(j(f, g), h)   compositionality,

and similar properties for id$_C$ and g ∘ h in the first component.

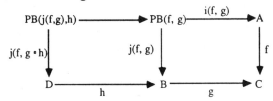

Definition 3 (category of concrete spans) A (concrete span s in a category C is a a pair of co-initial arrows (l:K → L, r:K → R), The five components of a span s are denoted by L(s), the source of s, R(s), the target of s, K(s), the interface of s, l(s), the left-hand side of s, and r(s), the right-hand side of s.

$$s: L(s) \xleftarrow{l(s)} K(s) \xrightarrow{r(s)} R(s)$$

If C has a fixed choice of pullback, we denote by **Span(C)** the category of concrete spans over C having the same objects as C and as morphisms from A to B all spans s such that L(s) = A and R(s) = B. The identity span for each object A is given by (id$_A$:A → A, id$_A$:A → A) and the composition of two spans s1, s2 with L(s2) = R(s1) is given by the pullback construction s1; s2 = (l(s1) ∘ i(r(s1), l(s2)):PB(r(s1), l(s2)) → L(s1), r(s2) ∘ j(r(s1), l(s2)):PB(r(s1), l(s2)) → R(s2)) using our choice of pullbacks. We choose the composition symbol ";" in order to distinguish span composition in C from the composition of the underlying morphisms in C.

It is easy to check that the category axioms for the identities and the composition of spans follows from the corresponding properties for the given choice of pullbacks. Note that our definition of spans over a given category C is different from the standard categorical notion of partial morphism over C, since we distinguish two spans which only differ up to isomorphism of their interfaces. We stick in this paper to more concrete categories (also in the following) since naive abstraction of morphism up to isomorphism can cause severe problems in defining a suitable composition of arrows representing computations, compare [MM 90, DMM 89, CELMR 94a] for these problems and possible solutions. Using these techniques it should also be possible to work out the theory without a fixed choice of pullbacks, but this is left to future research.

On this background, graph transformation rules in the sense of [Ehr 87] are special spans over the category **Graph**, namely those with monomorphic morphisms for the left and right hand side. A span is monomorphic if both components are. The subcategory of all monomorphic spans over a category C will be denoted by **MSpan(C)**. **MSpan(C)** ⊆ **Span(C)** since pullbacks preserve monomorphisms. Direct derivations are generated by double-pushout constructions in the algebraic approach to graph transformation. We capture this situation by the notion of span morphism and span DPO-morphisms.

Definition 4 (span morphism) Given two spans s1, s2 ∈ **Span(C)**, a *span morphism* t:s1 → s2 is a triple of arrows (t_L:L(s1) → L(s2)), t_K:K(s1) → K(s2), t_R:R(s1) → R(s2)) between sources, targets and interfaces such that the resulting squares commute, i.e. $t_L \circ l(s1) = l(s2) \circ t_K$ and $t_R \circ r(s1) = r(s2) \circ t_K$. A *span DPO-morphism* is a span morphism such that the resulting squares (1) and (2) are pushouts in C. Spans with span morphisms and with span DPO-morphisms form two (vertical) categories, for which identities and composition are defined component wise.

$$\begin{array}{ccccc} L(s1) & \xleftarrow{l(s1)} & K(s1) & \xrightarrow{r(s1)} & R(s1) \\ {}^{t_L}\downarrow & (1) & {}^{t_K}\downarrow & (2) & \downarrow{}^{t_R} \\ L(s2) & \xleftarrow{l(s2)} & K(s2) & \xrightarrow{r(s2)} & R(s2) \end{array}$$

With these notions the double-pushout approach to graph transformation of [Ehr 87] can be summarized as follows:

Definition 5 (double-pushout graph transformation) A *double-pushout graph grammar* GG = (G_0, R) consists of a start graph G_0 ∈ **Graph** and a set of rewrite rules R which are elements of **MSpan(Graph)**. A *direct derivation* in G is a span DPO-morphism t:r → s with r ∈ R, written L(s) ⇒ R(s) via (r, t). A *derivation* d is a sequence of direct derivation G1 ⇒ G2 ⇒ .. ⇒ Gn via (r1,...,rn-1, t1,...,tn-1), short G ⇒$_*$ Gn,. The *generated language* L(GG) is given by the set of graphs which are derivable from the startgraph, i.e. L(GG) ={G|G_0 ⇒$_*$ Gn}.

Compared to [Ehr 87], this is not the whole story. In general the algebraic approach to graph transformation is applied to *labeled graphs*, i.e. graphs which come equipped with two label functions l_1:V → M and l_2:E → M for vertices and edges into a predefined set of labels. The labels can be considered type information for every object in a graph since several different objects in a graph can have the same label or type and transformation rules are required to preserve labels and labellings of objects. Thus labels play the same role for graph objects as places for tokens in a Petri-net framework (see below), entity and relationship types for entities and relationships in semantical data modelling (see below) or types for variables in programming languages. This means that given a type m ∈ M the population of this type in a labeled graph is the preimage w.r.t. l_1 and l_2, i.e. $l_1^{-1}(m) \cup l_2^{-2}(m)$.

One major disadvantage of this form of typing is that it is not structural in the sense that types do not distinguish between vertices and edges and edge types do not prescribe source and target types for the source and target vertices of the instances. While the first can be repaired by two separate label sets for vertices and edges the latter requires a different set-up for types, where the type information itself carries graphical structure. Introducing such a fine grain typing mechanism into graph grammars leads to the notion of *typed graph grammars*. Typed graph grammars use a fixed graph TG ∈ **Graph**, called type graph, to represent the type information and the typing of another graph G in TG is given by a morphism g:G → TG. This passing from untyped graphs to typed graphs means passing from the category **Graph** to the comma category (**Graph** ↓ **TG**).

Since **Graph** is complete and co-complete, (**Graph** ↓ **TG**) has all limits and colimits as well and the construction of pushouts and pullbacks coincide in **Graph** and (**Graph** ↓ **TG**) up to the additional typing information which is uniquely induced by mediating morphism in the case of colimits and morphism composition in the case of limits. Summarizing this discussion, we define typed graph grammars[1] formally:

Definition 6 (typed graphs and graph grammars) For any graph G ∈ **Graph**, we denote the comma category (**Graph** ↓ G) short by **G-Graph**. A *typed graph grammar* TGG = (TG, G_0, P, Π) consists of a *type graph* TG ∈ **Graph**, a *start graph* G_0 ∈ **TG-Graph**, a set of *production names* P and a mapping Π:P → MSpan(TG-Graph) associating with each production name its *semantics* as a rule in the double-pushout sense. We also write p:s if Π(p) = s. Note that the introduction of a name concept into the notion of graph grammars allows for two different productions, which are in fact production names, to have the same semantics.

The concept of typed graph grammars adds to the expressive power of graph transformation systems as the following two examples show.

Example 7 (semantics for entity / relationship models) The original idea for the graph grammar model for entity / relationship specifications [Chen 76] which is presented below, stems from [CLWW 94, CL 95]: each entity / relationship diagram is translated to a type graph, called E/R, and the data base states conforming to a diagram are specified by E/R-typed graphs. Transformations of data base states can then be specified by E/R-typed graph productions which yield a natural model for transactions on the entity / relationship model. These ideas are explained below using the following sample entity / relationship diagram. It contains three

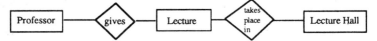

entity types, namely "Professor", "Lecture", and "Lecture Hall" and two relationship types "gives" and "takes place in" relating entities of type "Professor" with entities of type "Lecture" respectively entities of type "Lecture" with entities of type "Lecture Hall".

This diagram translates into a type graph E/R as follows:
$V_{E/R}$ = {Professor, Lecture, Lecture Hall}
$E_{E/R}$ = {gives, takes place in}
$s^{E/R}$::= gives ⟼ Professor, takes place in ⟼ Lecture
$t^{E/R}$::= gives ⟼ Lecture, takes place in ⟼ Lecture Hall

In entity / relationship modelling a state conforming to a diagram is given by a set for each entity type and a relation for each relationship type whose "type" conforms to the connection of the relationship type to entity types in the diagram. This corresponds to a graph which is typed in E/R, i.e. t:G → E/R. In such a typed graph the set of entities of type e is given by the preimage of t, i.e. by $t^{-1}(e)$. The relations are also encoded via preimages of the relationship type r, i.e by $\{(s^G(x), t^G(x)) | x \in t^{-1}(r)\}$. Having data base states represented as typed graphs, transaction schemes can be specified by typed graph productions. A suitable set for our example could be:
hire professor (p:Professor)
fire professor (p:Professor)

---

[1] Our notion of types and typed grammars shall not be confused with the notion of type within type theory. Although we are aware of this "name clash" we stick to the notion "typed graph grammars" which have been originally introduced under this name in [CMR 95].

new course (p:Professor, lh:Lecture Hall)

As an example, the semantics of the last transaction scheme could look like the following span (l:K → L, r:K → R) where l and r are inclusions.

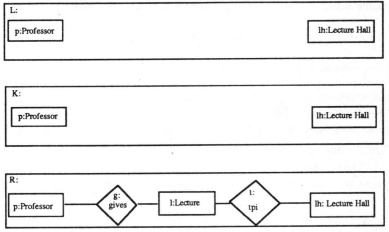

The typing of L, K, and R in the typed graph E/R is given by the specification right of the colon for each entity and relationship type.

Transactions result from applying transaction schemes to actual data base states, i.e. to E/R-typed graphs. This is achieved by the double-pushout mechanism explained above. Transaction histories are simply sequences of transactions in this set-up. □

A special case of these graph grammar models for entity/relationship models can be interpreted as Petri-nets, namely those whose type graph is discrete and whose productions are consumptive, i.e. have an empty interface.

Example 8 (interpretation of typed graph grammars as P/T-nets) Let TGG = (TG, G_0, P, Π) be a typed graph grammar over a discrete type graph TG (i.e. a graph consisting of vertices only). Furthermore assume that TGG is TG-finite, i.e. each preimage under a typing morphism in G_0 or in p ∈ P of a vertex in TG is finite, and that all productions p ∈ P have empty interfaces, i.e. $K(\Pi(p)) = \emptyset$. Then TGG can be interpreted as a P/T-net. A P/T-net N = (S, T, pre, post, $\underline{m}$) consists of a set of places S, a set of transitions T, two mappings pre, post:T → $S^\oplus$ into the free commutative monoid over S, and an initial marking $\underline{m} \in S^\oplus$, compare for example [MM 90]. The interpretation of a discrete consumptive graph grammar TGG = (TG, G_0, P, Π) provides a P/T-net N_{TGG}, whose components are defined as follows:

$S_{TGG} = V_{TG}$,
$T_{TGG} = P$,

and let #:|TG-Graph| → $S_{TGG}^\oplus$ be the following mapping

for each t:G → TG ∈ **TG-Graph**: $\#(t) = \oplus_{v \in V_{TG}} |t^{-1}(v)| \cdot v$ in:

$pre^{TGG}(p) = \#L(\Pi(p))$
$post^{TGG}(p) = \#R(\Pi(p))$
$\underline{m}_{TGG} = \#G_0$

It is obvious that we have a direct derivation G $\Rightarrow_p$ H in TGG if and only if #G[p>#H is a transition switching in N_{TGG}, and that different graph grammars are mapped to the same net, namely those grammars that differ only in the names of the vertices in the production and the start graph. This equivalence on discrete consumption graph grammars which is induced by the interpretation in P/T-nets will be further discussed in section 5. □

3 The Category of Typed Graph Grammars

The typing mechanism that has been presented in the previous section is not only useful if complex examples shall be specified by graph grammars but also enables the definition of graph grammar morphisms turning typed graph grammars into a category. The basic idea for these morphisms stems from the Petri-net world as well. For this reason, we briefly review the notion of Petri-net morphism as given [MM 90]: A net morphism from h:N1 → N2 from a net N1 = (S1, T1, pre1, post1, m1) into a net N2 = (S2, T2, pre2, post2, m2) is a pair ($h_S:S1^\oplus \to S2^\oplus$, $h_T:T1 \to T2$) where h_S is a monoid homomorphism between the two free commutative monoids and h_T is a function between the sets of transitions such that (1) the initial marking is preserved, i.e. m2 = h_S(m1) and (2) sources and targets of transitions are preserved, i.e. for each t ∈ T1 :pre2(h_T(t)) = h_S(pre1(t)) and post2(h_T(t)) = h_S(post(t)). Note that this sort of morphisms allows to map places of N1 to linear combinations of places in N2 via h_S (including deletion, i.e. mapping to the unit in $S2^\oplus$). And this mapping of places almost completely specifies the whole morphism up to a choice among transitions having exactly the same pre- and post-conditions.

As we have seen in example 8, the role of the place set has been taken by the type graph in the graph grammar case. Hence if we think of generalizing Petri-net morphisms to the field of graph transformations, we must find a way of mapping a type graph TG1 to a type graph TG2 that mimics a mapping of the elements in TG1 to "linear combinations" of elements in TG2. Candidates for such mappings are spans between type graphs which are the combination of morphisms for which we have the following results:

Proposition 9 (functors among comma categories) Let C be a category and A, B be objects of C.
1. Each arrow f:A → B induces a functor f$^>$:(C ↓ A) → (C ↓ B) defined as f$^>$(h:X → A) = f • h on objects and f$^>$(m) = m on arrows.
2. If we have a fixed choice of pullbacks for C, each arrow f:A → B induces a functor f$^<$:(C ↓ B) → (C ↓ A) defined as f$^<$(h:X → B) = (i(f, h):PB(f, h) → A) by a pullback construction on objects and on arrows f$^<$(k:(h:X → B) → (h':X' → B)) is given by the uniquely determined arrow from PB(f, h) to PB(f, h') satisfying i(f, h') • f$^<$(k) = i(f, h) and j(f, h') • f$^<$(k) = k • j(f, h). Moreover the resulting square with morphisms k, f$^<$(k), j(f, h), and j(f, h') becomes a pullback.

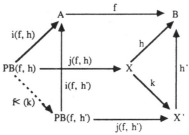

3. As a consequence, each pair of coinitial arrows or span $s = (l:K \to A, r:K \to B)$ in **C** induces a functor $<s>:(C \downarrow A) \to (C \downarrow B)$ defined by $<s> = r^> \circ l^<$, if **C** has a fixed choice of pullbacks.

Proof: The functor property of $f^<$ follows directly from the uniqueness of mediating pullback arrows. The resulting square becomes a pullback due to well-known pullback decomposition properties. □

The functor $f^<$ is attractive for our purpose since it realizes "multiplication" in **TG-Graph** in the following sense: if a type $b \in B$ has two different preimages under $f:A \to B$, the pullback construction of $f^<$ duplicates every object typed in b. Even multiplication with zero is included, if a type $b \in B$ has no preimage under f. In this case the pullback of $f^<$ eliminates each instance of b. But with $f^<$ alone there is no way of mapping one type to a sum of different types. This can be achieved if $f^<$ is followed by another construction of the "$g^>$-type": if we want to map a type $a \in A$ onto $2c + 3d$ in B, we first generated 5 copies of a in an intermediate type graph C by a suitable morphism $f:C \to A$ and the pullback functor $f^<$. Then we retype the 5 copies by another morphism $g:C \to B$ which maps 2 copies of a in C to the type c in B and the three remaining copies to the type d. Thus spans between type graphs seem to be the right choice to generalize monoid homomorphism in the case of Petri-nets to the case of graph transformation. A more detailed discussion of this correlation is given at the end of the section by a continuation of example 8.

Definition 10 (category of typed graph grammars) A graph grammar morphism $f:G1 \to G2$ from a typed grammar $G1 = (TG1, G1_0, P1, \Pi1)$ to a grammar $G2 = (TG2, G2_0, P2, \Pi2)$ is a pair (f_{TG}, f_p) where f_{TG} is a span between TG1 and TG2, i.e. $f_{TG} = (f_2:K \to TG1, f_R:K \to TG2)$, and $f_p:P1 \to P2$ in a mapping of production names such that
(1) the initial graph is preserved, i.e. $<f_{TG}>(G1_0) = G2_0$ and
(2) the semantics of productions is preserved, i.e. $\Pi2(f_p(p)) = <f_{TG}>(\Pi1(p))$ for all $p \in P1$.
    (Here the functor $<f_{TG}>$ is extended to arbitrary diagrams.)

Given two graph grammar morphisms $f1:G_1 \to G_2$ and $f2:G_2 \to G_3$, their composition $f2 \circ f1:G_1 \to G_3$ is defined by $f2 \circ f1 = (f1_{TG}; f2_{TG'} f2_p \circ f1_p)$. The identity on a grammar $G = (TG, G_0, P, \Pi)$ is given by $id_G = ((id_{TG}, id_{TG}), idp)$. Due to the presupposed properties of the underlying choice of pullbacks in **Graph**, these definitions satisfy the category axioms. Hence we obtain a category of typed graph grammars **GraGra**.

The central property of the graph grammar morphisms defined above is that they preserve direct derivations which is shown in the following proposition.

Proposition 11 (preservation of direct derivations) If $f = (f_{TG}, fp)$ is a morphism from $G1 = (TG1, G1_0, P1, \Pi1)$ to $G2 = (TG2, G2_0, P2, \Pi2)$ and d is a direct derivation in G1 using production p, then $<f_{TG}>(d)$ is a direct derivation in G2 with production $f_P(p)$.

Proof: A direct derivation d in G1 with production p is a span DPO-morphism $t:\Pi1(p) \to s$, compare definition 5. If $\Pi1(p) = (l:K \to L, r:K \to R)$, the situation is depicted in the background plane of the following diagram where t is represented by (t1, t2, t3) and s by $(g:D \to G, h:D \to H)$ such that the two resulting

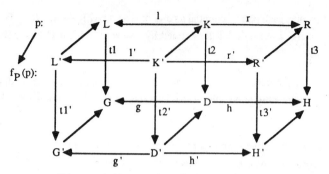

Diagram 1: Preservation of Direct Derivations

squares are pushouts. If we apply the functor $<t_{TG}>$ to this diagram we obtain diagram 1, where $(l':K' \to L', r':K' \to R') = <f_{TG}>(\Pi 1(p)) = \Pi 2(f_P(p))$ by definition of morphisms such that the front plane of diagram 1 becomes a direct derivation in G2 with production $f_P(p)$ if the two front squares are pushouts. This follows from the 3-cube lemma (compare lemma 12 appendix A) since the back squares are pushouts and all squares except the back and the front are pullbacks in **Graph** due to the definition of $<f_{TG}>$ and proposition 9.2. ◻

In the rest of the section, we show that our choice of morphisms specializes to the Petri notion if we restrict ourselves to TG-finite discrete and consumptive grammars as in example 8. And we further discuss how the defined morphisms have to be interpreted in the entity / relationship model of example 7.

Example 12 (Petri-nets; example 8 continued) In example 8, a special class of typed graph grammars have been interpreted as P/T-nets, namely those grammars which are TG-finite, discrete and have empty interfaces. We call this class "net grammars" in the following and denote this subcategory of **GraGra** by **GraNet**. Here we sketch how the object assignment # of example 8 can be extended to a functor #:**GraNet** → P/T-net. For this we must translate morphisms in **GraNet** to Petri-net morphisms:

If f:G1 → G2 is a GraNet morphism from G1 = (TG1, G1$_0$, P1, Π1) to G2 = (TG2, G2$_0$, P2, Π2) consisting of the span $f_{TG} = (f_L:K \to TG1, f_R:K \to TG2$ on types) and the mapping $f_P:P1 \to P2$ on productions, we denote by #f = (f_S, f_T) the following Petri-net morphism from #G1 = (S1 = V_{TG1}, T1 = P1, pre1, post1, #G1$_0$) to #G2 = (S2 = V_{TG2}, T2 = P2, pre2, post2, #G2$_0$):

$f_S:S1^\oplus \to S2^\oplus$ is the monoid homomorphism uniquely induced by

the following mapping of generating elements $s \in S1:f_S(s) = \oplus_{s' \in S2} k_{s'} \cdot s'$

where $k_{s'} = |\{k \in K \mid f_L(k) = s$ and $f_R(k) = s'\}|$

$f_T:T1 \to T2$ coincides with f_P. By properties of the explicit pullback construction in the category of sets it can be shown that #f satisfies the two conditions for net morphisms, and that identities are mapped to identities and compositions to compositions. Therefore # becomes a functor from **GraNet** to P/T-net.

Additional properties of this functor will be discussed in the conclusion section. ◻

Example 13 (scheme transformation; example 7 continued) In the framework of graph models for entity / relationship specification as it is described in example 7, morphisms between typed graph grammars provide a tool for the comparison of database states which conform to different models. The transition from one diagram to another diagram via some morphism can be interpreted as scheme evolution in time. With the induced translation of graphs and spans, such a scheme evolution takes care that the actual database state and the actual set of translation schemes is consistent with the actual scheme.

Consider for example an entity/relationship diagram which is designed to control the salary information for professors as is depicted below. The corresponding graph can be considered as a type graph for a second typed graph grammar.

The set of transaction schemes should consist of the productions
hire professor (p: Professor)
fire professor (p: Professor)
increase salary (p: Professor)
decrease salary (p: Professor)
new course (p: Professor)

There is an obvious morphism from the typed grammars of example 7 to this grammar, which is given by the span of inclusions of the graph corresponding to the "gives"-relation into the graphs corresponding to the two entity relationship diagrams used as type graphs, and the identity mapping of production names. From the semantical condition of such morphisms, we can conclude that the production "new course" in the target grammar has the following form

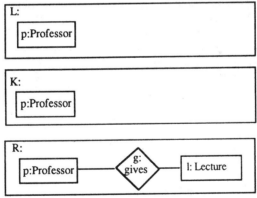

Note that spans between entity / relationship diagrams exactly look like transactions on the state level. Thus considering type graphs as states on a meta-level, we could design graph grammars which transform scheme information according to a rule base. This could be the starting point for an integrated framework for database evolution based on graph rewriting theory, compare [CL 95]. □

4 Categories of Graph Transition Systems and Categories of Derivations

Following the outline of [MM 90], we now consider special graph grammars, i.e. graph grammars whose set of productions has additional algebraic structure. With these special systems, it is possible to consider all (direct) derivations within a grammar as a grammar itself. This gives us the chance to relate grammars to their direct derivations and derivation sequences on one semantical level, namely on that within the category **GraGra**. The first specialization of graph grammars are *graph transition systems* which are grammars whose rule set is closed under derived productions.

Definition 14 (category of graph transition system) A graph transition system TS = (TG, G_0, P, Π, /) consists of a typed graph grammar (TG, G_0, P, Π) and a partial operation /:P × DPO-Morph → P on the production names which satisfies the following axioms:

(1) p/t is defined for each p ∈ P and each span DPO-morphism t ∈ DPO-Morph
 if $\Pi(p)$ = s and t:s → s', such that $\Pi(p/t)$ = s'
(2) p/id = p   <production axiom>
(3) (p/t1) / t2 = p/(t2 ∘ t1)   <derived production axiom>

Morphisms between transition systems f:TS1 → TS2 are typed graph grammar morphisms between the underlying graph grammars which additionally satisfy
(4) f_P(p/t) = f_P(p)/<f_{TG}>(t)   <preservation of production structure>

The category of graph transition systems and their morphisms is denoted by **GraTS**. □

Note that (3) and (4) are well-defined since pushouts are composable which guarantees that t2 ∘ t1 in (3) is a span DPO-morphism and <f_{TG}> preserves direct derivation (compare proposition 11) providing a span DPO-morphism <f_{TG}>(t) in (4).

The axioms (2) and (3) and the condition (4) for morphisms are purely syntactical, i.e. restrict the mapping of production names only, as the following lemma shows.

Lemma 15 For every transition system TS resp. transition system morphism f:TS1 → TS2 we have without assuming axioms (2) and (4) in definition 14:
(a) Π(p/id) = Π(p)
(b) Π((p/t1)/t2) = Π(p/(t2 ∘ t1))
(c) Π_2(f_P(p/t)) = Π_2(f_P(p)/<f_{TG}>(t))

Proof (a) is obvious. In (b) we let Π(p) = s and t1:s → s1 and t2:s1 → s2. The we conclude Π((p/t1)/t2) = s2 by def. 14(1) and Π(p/t2 ∘ t1) = s2 again by def.14(1). Let t:s → s1 in (c), then we obtain Π_2(f_P(p/t)) = <f_{TG}> (Π_1(p/t)) = <f_{TG}> (s1) = Π_2(f_P(p)/<f_{TG}>(t)) since <f_{TG}>(t):<f_{TG}>(s) → <f_{TG}>(s1) (see lemma 17(a) below). □

As we have said above graph transition systems are special graph grammars because there is the obvious forgetful functor U:**GraTS** → **GraGra** which assigns to each transition system the underlying grammar and maps the morphisms identically. U simply forgets about the algebraic structure on the production names and turns all names, either atomic or composed, into atomic names. The first main result of this paper shows that U has a left-adjoint TS:**GraGra** → **GraTS** which generates the free transition system over a grammar.

Theorem 16 (free transition system) The forgetful functor U:**GraTS** → **GraGra** has a left-adjoint TS:**GraGra** → **GraTS**.

Proof Given a graph grammar $G = (TG, G_0, P, \Pi)$, we construct $TS(G) = (TG, G_0, P', \Pi', /)$ as follows:
(i) Construct P^* and Π^* inductively by:
   (a) $p \in P \Rightarrow p \in P^*$ and $\Pi^*(p) = \Pi(p)$
   (b) $p \in P^*$, $\Pi^*(p) = s$, and $t: s \to s'$ span DPO-morphism $\Rightarrow <p, t> \in P^*$ and $\Pi^*(p, t) = s'$
   (c) there is nothing else in P^*.
(ii) Let P' be $P^*/_\equiv$, the quotient of P^* w.r.t. the congruence $\equiv$ generated by the production and the derived production axiom (def. 14(2) + (3)), and $\Pi'[p] = \Pi^*(p)$. This is well-defined due to lemma 15.
(iii) Define $[p]/t = [<p, t>]$ for each [p] with $\Pi'([p]) = s$, and $t: s \to s'$.
This is well-defined because $\equiv$ is congruence. Hence TS(G) is a graph transition system.

The universal morphism $u: G \to U(TS(G))$ is given by
(iv) $u_{TG} = (id: TG \to TG, id: TG \to TG)$ and
(v) $u_P: P \to P'$ by $u_P(p) = [p]$.

In order to prove freeness of TS we have to show that for each graph grammar morphism $f: G \to U(T)$ into the forgetful image of some transition system T, there is a unique transition system morphism $g: TS(G) \to T$ such that the diagram below commutes

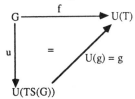

We define $g = (g_{TG}, g_P)$ by
(vi) $g_{TG} = f_{TG}$ and

(vii) $g_P([p])$ $\begin{cases} f_P(p) \text{ if } p \in P \\ g_P([p'])/<f_{TG}>(t) \text{ if } p = <p', t> \end{cases}$

Well-definedness of g can be shown using the axioms (2) and (3) of T in definition 14 and the properties $<f_{TG}>(id) = id$ and $<f_{TG}>(t2 \bullet t1) = <f_{TG}>(t2) \bullet <f_{TG}>(t1)$ (see lemma 17(b), (c)). g is a morphism of graph transition systems because it satisfies (4) of definition 14 by (vii) and (1), (2) of definition 10 because f preserves the initial graph and we are able to show
$$\Pi_T(g_P([p])) = <f_{TG}>(\Pi'([p])) \text{ for all } [p] \in P'.$$
In fact we have for $p \in P$ using the corresponding property for f
$$\Pi_T(g_P([p])) = \Pi_T(f_P(p)) = <f_{TG}>(\Pi(p)) = <f_{TG}>(\Pi'([p]))$$
and for $p = <p', f>$ with $t: s \to s'$ and lemma 17(a):
$$\Pi_T(g_P([p])) = \Pi_T(g_P[p']/<f_{TG}>(t)) = <f_{TG}>(s') = <f_{TG}>(\Pi'[p])$$
The diagram commutes since

(viii) $g_P \cdot u_P(p) = g_P([p]) = f_P(p)$ by (vii) and
(ix) $u_{TG}; g_{TG} = g_{TG} = f_{TG}$ since u_{TG} is an identity span and the underlying choice of pullbacks preserves identities, compare def. 2(2).

The transition morphism g is uniquely determined by (vii) because the commutativity of the diagram above enforces the first case in (vii) and the property (4) of definition 14 for morphisms on transition systems requires the second case since $[<p', t>] = [p']/t$. □

Lemma 17 (Span-Transition-Lemma) Given a span f_{TG} between TG1 and TG2 then we have
(a) $<f_{TG}>(t):<f_{TG}>(s) \to <f_{TG}>(s')$ for each span DPO-morphism $t:s \to s'$
(b) $<f_{TG}>(id) = id'$ for the identities $id:s \to s$ and $id':<f_{TG}>(s) \to <f_{TG}>(s)$
(c) $<f_{TG}>(t2 \cdot t1) = <f_{TG}>(t2) \cdot <f_{TG}>(t1)$ for $t1:s1 \to s2$, $t2:s2 \to s3$
(d) $<f_{TG}>(s; s') = <f_{TG}>(s); <f_{TG}>(s')$ if s; s' is defined
(e) $<f_{TG}>(t; t') = <f_{TG}>(t); <f_{TG}>(t')$ if t; t' is defined

Remark Let $t:s \to s1$ and $t':s' \to s1'$ with $R(s) = L(s')$ $R(s1) = L(s')$ and $t3 = t1'$ for $t = (t1, t2, t3)$ and $t' = (t1', t2', t3')$ then t; t' is defined by t; t' = $(t1, t^*, t3')$ where t^* is the induced morphism between the interfaces K^* and H^* of the composed spans s; s' and s1; s1' (see also proof of lemma 19).

Proof (a) and (b) are direct consequences of proposition 11, while (c) - (e) can be obtained similar to the proof of proposition 11 using the fixed choice of pullbacks and the composition property and the 3-cube lemmata A2 in the appendix.

In the next step we pass from transition systems to categories of derivations by requiring an additional composition operator on transition systems. In these categories whole derivation sequences are represented by productions and the semantics is given by the "derived production" for a derivation sequence which is defined by the composition of the spans which constitute the sequence.

Definition 18 (category of derivation systems) A derivation system for typed graphs DS = (TG, G_0, P, Π, /, id, ;) consists of a transition system (TG, G_0, P, Π, /), a family of identity constants in P id = $(id_G)_{G \in \mathbf{Graph}}$ with $\Pi(id_G) = (id_G:G \to G, id_G:G \to G)$, and a partial operator for sequential composition ;:$P \times P \to P$ such that the following axioms are satisfied:
(1) The operator ; is defined for each pair (p, p') if $R(\Pi(p)) = L(\Pi(p'))$ with semantics
    $\Pi(p; p') = \Pi(p); \Pi(p')$
(2) p; $id_{R(\Pi(p))} = p = id_{L(\Pi(p))}$; p   <identity axiom>
(3) (p1; p2); p3 = p1; (p2; p3)   <associativity of ; >
(4) (p; p') / (t; t') = (p / t); (p' / t')   <compatibility of / and ; >

A derivation system morphism f:DS1 $\to$ DS2 is a transition system morphism which additionally satisfies
(5) $f_P(id_G) = id_{<f_{TG}>(G)}$ <preservation of identities>
(6) $f_P(p; p') = f_P(p); f_P(p')$ <preservation of sequential composition>

The category of derivation systems and derivation system morphisms is denoted by **GraCat**.

□

Again these axioms and conditions do not affect the semantics.

Lemma 19 For every derivation system DS resp. derivation system morphism f:DS1 → DS2 we have without assuming the axioms (2) - (6) of definition 18
(a) $\Pi(p; id_{R(\Pi(p))}) = \Pi(p) = \Pi(id_{L(\Pi(p))}; p)$
(b) $\Pi((p1; p2); p3) = \Pi(p1; (p2; p3))$
(c) $\Pi((p; p') / (t; t')) = \Pi(p / t; p' / t')$
(d) $\Pi2(f(id_G)) = \Pi2(id_{<f_{TG}>(g)})$
(e) $\Pi2(f_P(p; p')) = \Pi2(f_P(p)); \Pi2(f_P(p'))$

Proof (a) and (b) follow from "preservation of identities" and "associativity" of the choice of pullbacks. For (c) consider the following diagram

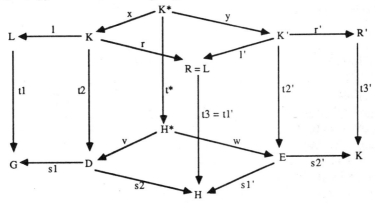

Here $p = (l:K \to L, r:K \to R)$, $p' = (l':K' \to L', r':K' \to R')$, $t = (t1, t2, t3)$, $t' = (t1', t2', t3')$. We have $t3 = t1'$ since $t; t'$ is defined. The construction of $t; t'$ is as follows: take the pullback of r and l' which provides K^*, take the pullback of $s2$ and $s1'$ which provides H^* and let t^* be the unique (pullback) arrow satisfying $v \cdot t^* = t2 \cdot x$ and $w \cdot t^* = t2' \cdot y$. Since pullbacks preserve monomorphisms x, y, v, and w are injective. Now the cube with the middle of the diagram has pullbacks of injective morphism as top and bottom and two pushouts in the front. With the 3-cube lemma (compare appendix A lemma A2), the two back squares become pushouts as well. Hence we can define $t; t' = (t1, t^*, t3')$. With this definition (c) is obvious.

The property (d) follows from the general semantical property of morphisms $\Pi(f_P(p)) = <f_{TG}>(\Pi(p))$ and preservation of identities by the choice of pullbacks. For
(e) we have $\Pi2(f_P(p; p')) =$ \<general property for morphisms\>
$<f_{TG}>(\Pi1(p; p')) =$ \<definition 17(1)\>
$<f_{TG}>(\Pi1(p); \Pi1(p')) =$ \<lemma 17(d)\>
$<f_{TG}>(\Pi1(p)); <f_{TG}>(\Pi1(p')) =$ \<general property of morphisms\>
$\Pi2(f_P(p)); \Pi2(f_P(p'))$

With these preparations, we are able to show the second central result of the paper, namely that the forgetful functor which maps derivation systems to their underlying grammar has a left-adjoint.

Theorem 20 (free derivation system) The forgetful functor V:**GraCat** → **GraGra**, which maps each system (TG, G_0, P, Π, /, id, ;) to the underlying grammar (TG, G_0, P, Π) and each morphism identical to its underlying grammar morphism, has a left adjoint DS:**GraGra** → **GraCat**.

Proof The free derivation system DS(G) for a grammar G = (TG, G_0, P, Π) is constructed in analogy to TS(G) in the proof of theorem 16:

(i) Construct P^* and Π^* inductively as follows:
   (a) $p \in P \Rightarrow p \in P^*$ and $\Pi^*(p) = \Pi(p)$
   (b) $p \in P^*$, $\Pi^*(p) = s$, and $t{:}s \to s'$ span DPO-morphism $\Rightarrow <p, t> \in P^*$ and $\Pi^*(<p, t>) = s'$
   (c) $G \in$ **Graph** $\Rightarrow \underline{id}_G \in P^*$ and $\Pi^*(\underline{id}_G) = (id_G{:}G \to G, id_G{:}G \to G)$
   (d) $p, p' \in P^*$ with $R(\Pi^*(p)) = L(\Pi^*(p')) \Rightarrow <p; p'> \in P^*$ and $\Pi^*<p; p'> = \Pi^*(p); \Pi^*(p')$

(ii) Let P' be $P^*/_\equiv$ where $\equiv$ is generated by the axioms (2) and (3) of definition 14 and (2)-(4) of definition 17 and
   (e) $p \equiv p' \Rightarrow <p, t> \equiv <p', t>$
   (f) $p \equiv p', q \equiv q' \Rightarrow <p; q> \equiv <p'; q'>$
   provided that all terms on the right-hand side are defined.
   Define Π' for P' by $\Pi'([p]) = \Pi^*(p)$. This is well-defined due to lemma 18 and the fact that $\Pi^*(<p, t>) = \Pi^*(p, t')$ if $t = t'$ and $\Pi^*<p ; q> = \Pi^*<p' ; q>$ if $\Pi^*(p) = \Pi^*(p')$ and $\Pi^*(q) = \Pi^*(q')$ because we have a fixed choice of pullbacks.

(iii) Let DS(G) = (TG, G_0, P', Π', /, id, ;) where the operators / , id, ; are well-defined using (e) and (f) by
   (g) $[p] / t = [<p, t>]$ if $<p, t>$ is defined
   (h) $id_G = [\underline{id}_G]$
   (k) $[p] ; [p'] = [<p ; p'>]$ if $<p ; p'>$ is defined

(iv) The universal morphism u:G → V(DS(G)) is again the pair consisting of $u_{TG} = (id_{TG}, id_{TG})$ and u_P mapping $p \in P$ to $[p] \in P'$.

(v) The extension of a **GraGra**-morphism f:G → V(D), i.e. g:DS(G) → D, can then be defined by
   (m) $g_{TG} = f_{TG}$

   (n) $g_P([p]) = \begin{cases} f_P(p) & \text{if } p \in P \\ id_{<f_{TG}>(G)} & \text{if } p = id_G \\ g_P(p') / <f_{TG}>(t) & \text{if } p = <p', t> \\ g_P(p') ; g_P(p'') & \text{if } p = <p' ; p''> \end{cases}$

Well-definedness, morphism properties and uniqueness of g are following similar to the proof of theorem 16 using lemma 17. □

The free transition system and the free derivation system provides in our example of entity / relationship modelling the semantical model for all transaction and transaction histories resp. In the Petri-net case it is open how far our construction coincides with the standard constructions for transition systems and derivation categories as they are proposed in [MM 90]. The problem is an issue for further research and briefly discussed in the conclusion.

5 Conclusion

We have introduced a categorical presentation "in the large" for the algebraic approach to graph transformation. The central issue is the right notion of graph grammar morphism which has been given in section 3. With these morphisms, we obtain the category **GraGra** of typed graph grammars and with some additional structural requirements for objects and morphisms the categories of direct derivations and of derivation sequences. The central results in section 4 show that there is a free system of direct derivations for each grammar.

For the presentation we have chosen, it was necessary to change slightly the basic set-up for graph grammars and to choose a notion for morphisms. For the first we choose typed graph grammars as they have been introduced in [CMR 95]. The typing mechanism of these grammars requires that every graph and every rule is typed in a fixed type graph TG by a homomorphism resp. by a triple of morphism for the rule's left-hand, right-hand, and interface graphs. It is easy to check that the classical approach to double-pushout graph transformation [Ehr 87] can be embedded into this framework by choosing the final graph as type graph in the case of unlabeled graphs or a type graph TG built up by the label alphabets Σ_E, Σ_V, i.e. $TG_E = \Sigma_V \times \Sigma_V \times \Sigma_E$, $TG_V = \Sigma_V$, $s^{TG} = p1$, $t^{TG} = p2$ in the case of labeled graphs. Thus, typed graph grammars can be considered as generalization of untyped graph grammars. On the first look, our morphism notion seems a little arbitrary. It has been motivated by the morphisms on Petri-nets in [MM 90]. Since graph grammars are "generalizations" of P/T-nets [CEMLR 94b], our notion of morphism should specialize to that of [MM 90] if we restrict the class of graph grammars to those which can be interpreted as P/T-nets. That this is true is demonstrated in example 12. That the set-up is reasonable in general has been motivated by the entity/relationship examples 7 and 13, where morphism correlate to scheme transformations.

The first interesting issue for future research is the comparison of the results in section 4 to those in the Petri-net field: Can the functor #:**GraNet** → **P/T-Net** be specialized for transition and derivation systems such that it maps to transition systems and categories of derivations in Petri-net? Furthermore it remains to be investigated if these functors $\#_{TS}$ and $\#_{DS}$ are compatible with the free constructions TS and DS on graph grammars restricted to graph nets resp. $TS_{P/T}$ and $DS_{P/T}$ on Petri-net, i.e. if $\#_{TS} \circ TS = TS_{P/T} \circ \#$ and $\#_{DS} \circ DS = DS_{P/T} \circ \#$. If these results can be obtained we get a strong justification for our framework on the graph grammar side as a proper generalization of Petri-nets.

Another issue for future research is to investigate the functor #:**GraNet** → **P/T-Net** in detail. As we have mentioned above, it maps different graph grammars to the same net. Hence it induces an equivalence on **GraNet**. Unfortunately this equivalence is not an equivalence of isomorphic objects due to our choice of category of spans: since we have concrete spans, i.e. spans that differ if the interface object is different, we get as isomorphisms pairs of identities only. And it is obvious that # identifies grammars which cannot be related by identity typed graph morphisms. For a result of the type "#(G1) = #(G2) if and only if G1 $\cong$ G2", we have to redo the theory presented above in the category of abstract spans, which corresponds to the standard categorical notion of partial morphism if the left-hand side is monomorphic. In this case, a span between type graphs does not induce a strict functorial translation of the typed objects and rules, but is determined up to isomorphism only. Thus, it has to be checked that all constructions and results of this paper are invariant under isomorphisms for this framework. (Compare [CELMR 94a], that this is not always an easy task and cannot be taken for granted.)

Since P/T-nets are abstract graph grammars due to the equivalence induced by #, it is interesting to ask which variant of P/T-nets directly corresponds to grammars in **GraNet**. A candidate for this type are nets whose tokens can be pairwise distinguished and in which transition switching makes explicit reference to the names of the token which take part in the switching process. These nets, between P/T-nets and high-level-nets, shall be easily defined by just rephrasing the notions in the graph grammar world of GraNet in terms of transitions, places, and tokens.

Finally it is an interesting topic for future research, (1) to investigate the relation of the free derivation system over G to the classical notion of sequential derivation in [Ehr 87] for G, compare definition 5, and (2) to extend the results to more general notions of morphisms especially morphisms $f:G1 \rightarrow G2$ which allow productions in G1 to be mapped to derived productions in G2, i.e. $fp:P1 \rightarrow P_{TS(G2)}$.

Appendix A: 3-Cube Lemmata

For the 3-cube lemmata in **Graph** we use the following characterization of pullbacks and pushouts in **Set**. They carry over to **Graph** since limits and colimits are componentwise constructed in **Graph**. Unfortunately the 3-cube lemmata in the literature [ER 80] do not imply parts (i) and (ii) of our lemma A2.

Lemma A1 (characterization of pullbacks and pushouts in Set The following commuting diagram

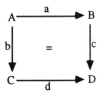

(i) is a pullback in **Set** if and only if
    (a) $d(x) = c(y) \Rightarrow \exists z \in A \; x = b(z), y = a(z)$
    (b) $a(x) = a(y), b(x) = b(y) \Rightarrow x = y$
(ii) is a pushout in **Set** if and only if
    (c) c and d are jointly surjective
    (d) every preimage set under c, i.e. $c^{-1}(x)$, consist either of a single element or there is a finite or infinite enumeration, $(x_i)_{i \in I}$ of $c^{-1}(x)$, i.e. $c^{-1}(x) = \{x_i / i \in I\}$, such that for each i there are y_i and y'_i in A with $x_i = a(y_i)$, $x_{i+1} = a(y'_i)$, and $b(y_i) = b(y'_i)$
    (e) symmetrical (d)-condition for function d
    (f) $d(x) = z = c(y) \Rightarrow \exists x' \in d^{-1}(z), y' \in c^{-1}(z) : x' = b(w), y' = a(w)$

Proof The properties follow directly from explicit constructions of pushouts and pullbacks in Set.

For the other direction let in (i) $a':X \rightarrow B$ and $b':X \rightarrow C$ be given such that $c \circ a' = d \circ b'$. Construct $u:X \rightarrow A$ by $u(x) = y$ such that $a(y) = a'(x)$ and $b(y) = b'(x)$. By property (a) u is total,

Lemma A2 (3-cube lemmata) Consider the following commuting 3-cube diagram in

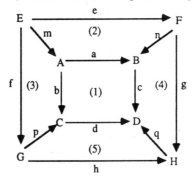

Set. We have the following 3-cube lemmata:
(i)   If (1) is pushout and (2)-(5) pullbacks, then the outer square, i.e. (1)+(2)+(3)+(4)+(5), is pushout.
(ii) If (1) and the outer square are pullbacks of injective morphisms and (4) and (5) are pushouts, then (2) and (3) are pushouts.
(iii) If the outer square is pullback of injective morphisms and (2)-(5) are pushouts, then (1) is pullback.

Proof
(i) We show the characterizing properties (c)-(f) of A1 (ii):
  (c) h and g are together surjective since $q(x) = c(y)$ or $q(x) = d(z)$, because c and d are together surjective. Then property (a) for pullbacks for (4) and (5) implies preimage of x for g or h resp.
  (d) Let $X = g^{-1}(X)$ such that $|x| > 1$, by lemma A1(b) for (4) n is injective on X and provides $n(X) \subseteq c^{-1}(q(x))$. By A1(a) for (4) it follows $n(X) = c^{-1}(q(x))$. By A1(d) for (1) we get enumeration $(b_i)_{i \in I}$ for n(X) such that $b_i = a(z_i)$, $b_{i+1} = a(z'_i)$, $b(z_i) = b(z'_i)$ for $i \in I$. Fix any $i \in I$: Since $d(b(z_i)) = d(b(z'_i)) = q(x)$, A1(a) for (5) provides $b(z_i) = p(w_i)$, same property for (3) provides $w_i = f(v_i)$ such that $m(v_i) = z_i$ and $w_i = f(v'_i)$ such that $m(v'_i) = z'_i$. Since we have $n \circ e = a \circ m$ let $e(v_i) = t_i$ with $n(t_i) = b_i$ and $(t_i)_{i \in I}$ the desired enumeration for X. In fact we have $e(v'_i) = t_{i+1}$ using $n(e(v'_i)) = (a\, m(v'_i)) = a(z'_i) = b_{i+1} = n(t_i + 1)$ and injectivity of n on X and $f(v_i) = w_i = f(v'_i)$ which implies property A1(d).
  (e) The proof is symmetrical to (d).
  (f) Let $g(x) = z = h(y)$, then $c(n(x)) = d(p(y)) = q(z)$. By A1(f) for (1) there is $w \in c^{-1}(q(z))$ and $v \in d^{-1}(q(z))$ with $w = a(u)$ and $v = b(u)$. With A1(a) for (4) and (5), $w = n(r)$ such that $g(r) = g(x)$, i.e. $r \in g^{-1}(x)$, and $v = p(s)$ such that $h(s) = h(y)$, i.e. $s \in h^{-1}(y)$. Same property for (2) and (3) provides $e1 \in E$ such that $e(e1) = r$ and $m(e1) = u$ respectively $e2 \in E$ such that $f(e2) = s$ and $m(e2) = u$. It remains to show that $e1 = e2$, which follows from A1 (b) for (4) + (2) since $m(e1) = u = m(e2)$ and $g \circ e(e1) = g(r) = g(x) = h(y) = h(s) = h(f(e2)) = g \circ e(e2)$.

(ii) We show that (2) is pushout. The proof for (3) is symmetrical. For (c) we use that c(x) = q(y) or c(x) = d(z) by property (c) for (5). In the first case x = n(u) because g and c are injective. In the second case x = a(v) since (1) is pullback.

(d) reduces to n(x) = n(y) ⇒ x = y or x = e(p), y = e(q) with m(p) = m(q). Let n(x) = n(y) for x ≠ y. Since g is injective we have g(x) ≠ g(y) but q(g(x)) = q(g(y)). Hence property (1) applied to (5) implies u, v ∈ G with h(u) = g(x), h(v) = g(y). Now property (a) of the outer diagram implies p, q ∈ E with f(p)) u, e(p) = x and f(q) = v, e(q) = y. But this implies m(p) = m(q) using injectivity of a.

(e) Symmetric to (d). (f)
Let a(x) = n(y). Then d(b(x)) = q(g(y)). Since (5) is pushout, we obtain z ∈ G with p(z) = b(x) and h(z) = g/y. Outersquare is pullback. Thus z = f(u) and y = e(u). We need now m(u) = x, which we get by injectivity of b and b(m(u)) = p(f(u)) = p(z) = b(x).

(iii) Proof is achieved by checking the characterizing condition (a) for (1). Since (b) is trivial for injective morphisms. Let c(x) = d(y) and (I) c(x) = q(z), then the required property follows from the same property for the outer square. If (II) c(x) ∉ q(H), we obtain that x = a(z) because c ∘ a and g are jointly surjective because (4) + (2) is pushout. □

References

[Bed 88]   M. A. Bednarcsyk: *Categories of asynchronous systems*, Ph.D. Thesis, University of Sussex, Report no. 1/88, 1988.

[CELMR 94a]   A. Corradini, H. Ehrig, M. Löwe, U. Montanari and F. Rossi: *Abstract Graph Derivations in the Double-Pushout Approach*, LNCS 776, Springer-Verlag, 1994, 86-103.

[CELMR 94b]   A. Corradini, H. Ehrig, M. Löwe, U. Montanari and F. Rossi: *An Event Structure Semantics for Safe Graph Grammars*, in Proc. IFIP Conf. PROCOMET'94 (Conference Edition), San Miniato, Italy, 1994.

[Chen 76]   P. P. Chen: *The Entity-Relationship Model: Toward a Unified View of Data*, ACM Transactions on Database Systems, 1(1):9-36, 1976.

[CL 95]   I. Claßen, M. Löwe: *Scheme Evolution in Object-Oriented Models: A Graph Transformation Approach*, to appear Proc. Workshop on Formal Methods at the ICSE'95, Seattle (U.S.A.), 1995.

[CLWW 94]   I. Claßen, M. Löwe, S. Wasserroth, J. Wortmann: *Static and Dynamic Semantics of E/R Models Based on Algebraic Methods*, Integration von semiformalen und formalen Methoden für die Spezifikation von Software-Systemen (B. Wolfinger, ed.), Springer-Verlag, Informatik aktuell, 1994, 2-9.

[CMR 95]   A. Corradini, U. Montanari and F. Rossi: *Graph Processes*, accepted for publication in Fundamenta Informaticae.

[DMM 89]   P. Degano, J. Meseguer, U. Montanari: *Axiomatizing Net Computations and Processes*, in Proc. 4th Annual Symp. on Logic in Comp. Sci., Asilomar, CA, USA, 1989, 175-185.

[Ehr 87]   H. Ehrig: *Tutorial Introduction to the Algebraic Approach of Graph-Grammars*, LNCS 291, Springer-Verlag, 1987, 3-14.

[ER 80]   H. Ehrig, B.K. Rosen: *Parallelism and Concurrency of Graph Manipulation*, TCS 11 (1980), 247-275.

[Löw 94]   M. Löwe: *Von Graphgrammatiken zu Petri-Netzen und zurück*, Tagungsband Alternative Konzepte für Sprachen und Rechner 1994 (F. Simon, eds.), Univ. Kiel, FB Informatik, Nr. 9412, 79-82.

[MM 90]   J. Meseguer, and U. Montanari: *Petri-Nets are Monoids*, in Info. and Co., 88(1990) 105-155.

[NPW 81]   M. Nielsen, G. Plotkin and G. Winskel: *Petri-Nets, Event Structures and Domains, Part 1*, in Theoret. Comp. Sci. 13 (1981), 85-108.

[Rei 85]   W. Reisig: *Petri Nets: An Introduction*, EATCS Monographs on Theoretical Computer Science, Springer-Verlag, 1985.

[SNW 93]   V. Sassone, M. Nielsen and G. Winskel: *Relationship between models of concurrency*, in Proc. REX'93, 1993.

Graph Unification and Matching*

Detlef Plump
Universität Bremen[†]

Annegret Habel
Universität Hildesheim[‡]

Abstract

A concept of graph unification and matching is introduced by using hyperedges as graph variables and hyperedge replacement as substitution mechanism. It is shown that a restricted form of graph unification corresponds to solving linear Diophantine equations, and hence is decidable. For graph matching, transformation rules are given which compute all (pure) solutions to a matching problem. The matching concept suggests a new graph rewriting approach which is very simple to describe and which generalizes the well-known double-pushout approach.

1 Introduction

The unification of expressions plays an important role in many areas of computer science, for example in theorem proving, logic programming, type inference, and natural language processing (see the surveys on unification of Knight [15], Jouannaud and Kirchner [14], and Baader and Siekmann [1]). While expressions are usually considered as strings or trees, efficient unification algorithms often use graph representations of expressions in order to exploit the sharing of common subexpressions. This applies, for example, to the algorithms of Huet [11], Paterson and Wegman [18], Corbin and Bidoit [6], and Jaffar [12]. Thus it seems natural to abstract from implementation details and to study a concept of *graph unification* in its own right.

With a somewhat different motivation, Parisi-Presicce, Ehrig, and Montanari [17] consider graph rewriting with unification. In their approach, variables are instantiated by graph morphisms and graph rewriting is defined by double-pushouts of graph morphisms. In the introduction of [17], the authors identify the following problem: "One of the basic concepts missing is a suitable notion of variable and of graph unification." We also address this problem and propose

---

*Work partially supported by ESPRIT Basic Research Working Group COMPUGRAPH II, # 7183, and by Deutsche Forschungsgemeinschaft, # Kr 964/2-2.

[†]Address: Fachbereich Mathematik und Informatik, Universität Bremen, Postfach 33 04 40, 28334 Bremen, Germany. E-mail: det@informatik.uni-bremen.de.

[‡]Address: Institut für Informatik, Universität Hildesheim, Marienburger Platz 22, 31141 Hildesheim, Germany. E-mail: habel@informatik.uni-hildesheim.de.

a solution which we consider as natural and simple: *hyperedges* serve as graph variables and *hyperedge replacement* is used to substitute graphs for variables.

In this paper, we study this idea in its pure form, without tailoring graph unification to the implementation of term unification. The latter requires both restriction to a smaller graph class (term graphs) and consideration of a graph equivalence more liberal than isomorphism. A corresponding adaptation of the present approach can be found in [10].

Although the general graph unification problem is easy to state (determine, for any hypergraphs G and H, whether there is a substitution making G and H isomorphic), we do not know whether the problem is decidable or not. Below we solve a restricted form of the graph unification problem, solve the general graph matching problem (unification in the case where one graph is variable-free), and define a substitution-based graph rewriting approach:

- In the restricted case where variables are not attached to nodes, the graph unification problem is equivalent to the problem of finding non-negative integer solutions to a set of linear Diophantine equations. This in turn is known to be equivalent to the restricted AC1 unification problem for first-order terms, where terms are built up from an associative-commutative function symbol, a unit element, and free constants.

- There is a graph matching algorithm based on transformation rules which computes all solutions (without useless variables) to a given matching problem.

- Substitution-based graph rewriting is introduced. In this approach, rules are just pairs $\langle L, R \rangle$ of hypergraphs which are applied by matching L to a given hypergraph and by applying the corresponding substitution to R. This approach allows to copy and delete non-local subgraphs, and generalizes the well-known double-pushout approach.

2 Hypergraphs and substitutions

Let Σ be a set the elements of which are called *labels*. Each label l comes with a natural number $type(l) \geq 0$. A *hypergraph* over Σ is a system $G = \langle V_G, E_G, lab_G, att_G \rangle$ consisting of two finite sets V_G and E_G of *nodes* and *hyperedges*, a labelling function $lab_G \colon E_G \to \Sigma$, and an attachment function $att_G \colon E_G \to V_G^*$ which assigns a string of nodes to a hyperedge e such that $type(lab_G(e))$ is the length of $att_G(e)$.

A *hypergraph morphism* $f \colon G \to H$ consists of two functions $f_V \colon V_G \to V_H$ and $f_E \colon E_G \to E_H$ that preserve labels and attachment to nodes, that is, $lab_H \circ f_V = lab_G$ and $att_H \circ f_E = f_V^* \circ att_G$. (The extension $f^* \colon A^* \to B^*$ of a function $f \colon A \to B$ maps the empty string to itself and $a_1 \ldots a_n$ to $f(a_1) \ldots f(a_n)$.) The morphism f is *injective* (*surjective*) if f_V and f_E are injective (surjective), and is an *isomorphism* if it is injective and surjective. In the latter case, G and H are *isomorphic*, which is denoted by $G \cong H$.

A *pointed hypergraph* $\langle G, p_G \rangle$ is a hypergraph G together with a sequence $p_G = v_1 \ldots v_n$ of pairwise distinct nodes from G. We write $type(G)$ for the number n of points and P_G for the set $\{v_1, \ldots, v_n\}$. If convenient, we denote a pointed hypergraph only by its hypergraph component.

We assume that there is a distinguished subset Var of Σ the elements of which are used as *variables* and which are denoted by $x, y, z, x_1, x_2, \ldots$ The set of variables occurring in a hypergraph G is denoted by $Var(G)$, and we write $G^\ominus$ for the hypergraph obtained from G by removing all hyperedges labelled with variables.

Definition 1 A *substitution pair* x/G consists of a variable x and a pointed hypergraph G such that $type(x) = type(G)$. A *substitution* is a finite set $\sigma = \{x_1/G_1, \ldots, x_n/G_n\}$ of substitution pairs such that $x_1, \ldots, x_n$ are pairwise distinct. The *domain* of σ is the set $Dom(\sigma) = \{x_1, \ldots, x_n\}$.

Given a hypergraph H and a hyperedge e in H labelled with x, the application of a substitution pair x/G to e proceeds in two steps: (1) Remove e from H, yielding the hypergraph $H - \{e\}$, and (2) construct the disjoint union $(H - \{e\}) + G$ and fuse the ith node in $att_H(e)$ with the ith point of G, for $i = 1, \ldots, type(x)$. (Note that if $att_H(e)$ contains repeated nodes, then the corresponding points in G are identified.)

The application of a substitution σ to a hypergraph H yields the hypergraph $H\sigma$ which is obtained by applying all substitution pairs in σ simultaneously to all hyperedges with label in $Dom(\sigma)$.

Given two substitutions σ and τ, the *composition* $\sigma\tau$ is the substitution $\{x/G\tau \mid x/G \in \sigma\} \cup \{y/H \in \tau \mid y \notin Dom(\sigma)\}$. By associativity of hyperedge replacement (see [9]), the composition has the following important property.

Lemma 2 *For all hypergraphs G and substitutions σ and τ, $G(\sigma\tau) \cong (G\sigma)\tau$.*

3 Unification

After a notion of substitution is introduced, it is obvious how to define unification and matching. In this section we show that graph unification is decidable for a restricted class of problems, and we give sufficient conditions for the existence of solutions in the general case.

Definition 3 A *unification problem* $G =^? H$ consists of two hypergraphs G and H. A *solution* to this problem is a substitution σ such that $G\sigma$ and $H\sigma$ are isomorphic.

Example 4 The unification problem in Figure 1 consists of two flow graphs with variables S and B (standing for statements and boolean expressions). The rectangles represent hyperedges of type 2 which are connected with their first and second attachment node by a line and an arrow, respectively. Rhombi represent hyperedges of type 3, where arrows labelled with t and f point to the second and

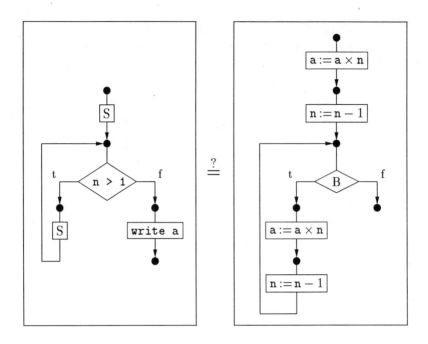

Figure 1: A unification problem

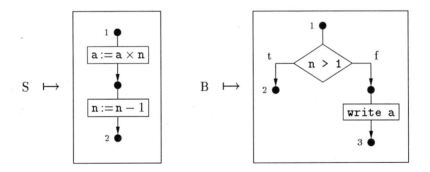

Figure 2: A solution to the problem in Figure 1

third attachment node, respectively. A solution to this unification problem is the substitution shown in Figure 2 (where the numbered nodes are the points of the hypergraphs). Another solution is obtained if the node above the hyperedge labelled with "`write a`" becomes the third point in the right hypergraph.

The *general unification problem* is the problem to decide, for any unification problem $G =^? H$, whether it has a solution or not. Note that for hypergraphs without variables, the general unification problem is the hypergraph isomorphism problem.

We do not know whether the general unification problem is decidable or not. Below we establish decidability for the case that all variables have type 0, that is, for the case that variable hyperedges are not attached to any nodes. Even this strongly restricted kind of unification turns out to be non-trivial. Essentially, here unification is equivalent to solving a system of linear Diophantine equations.

Definition 5 The *type* of a unification problem $G =^? H$ is the maximal type of the variables in G and H. If there are no variables, then the type is $\perp$.

Consider a unification problem $G =^? H$ of type 0. Let $C_1, \ldots, C_m$ be the non-isomorphic, connected, variable-free components and $x_1, \ldots, x_k$ be the variables occurring in G and H. We represent G and H as disjoint unions of their connected components: $G = \sum_{i=1}^{k} a_i X_i + \sum_{j=1}^{m} b_j C_j$ and $H = \sum_{i=1}^{k} a'_i X_i + \sum_{j=1}^{m} b'_j C_j$, where a_i, a'_i, b_j, b'_j are non-negative integers and X_i stands for a hyperedge labelled with x_i. Now assign to G and H the following set $Eq(G =^? H)$ of linear Diophantine equations:

$$Eq(G =^? H) = \left\{ \sum_{i=1}^{k} (a_i - a'_i) x_{i,j} = (b'_j - b_j) \ \middle| \ 1 \leq j \leq m \right\}.$$

Theorem 6 *A unification problem $G =^? H$ of type 0 has a solution if and only if the linear Diophantine equations in $Eq(G =^? H)$ have a solution in non-negative integers.*

Thus, the general unification problem of type 0 reduces to the problem of solving homogeneous and inhomogeneous linear Diophantine equations in non-negative integers. Efficient algorithms for the latter have recently been given by Clausen and Fortenbacher [4] and by Boudet, Contejean, and Devie [3, 5]. This problem is equivalent to the unification problem for expressions built up from an associative-commutative function symbol, a unit element, and free constants (see [14]).

Proof of Theorem 6. "Only if": Let $\sigma = \{x_1/F_1, \ldots, x_k/F_k\}$ be a solution to $G =^? H$. Without loss of generality, we may assume that $F_1, \ldots, F_k$ are variable-free (otherwise $\sigma\tau$ is as required, where τ removes all variable hyperedges). Let $C_1, \ldots, C_m, C_{m+1} \ldots, C_n$ be the non-isomorphic, connected, variable-free components occurring in $F_1, \ldots, F_k, G$, and H. Then, for $i = 1, \ldots, k$, F_i can be represented as the disjoint union of its connected components: $F_i = \sum_{j=1}^{n} c_{i,j} C_j$, where the $c_{i,j}$ are non-negative integers. By setting $b_j = 0$ for $j = m+1, \ldots, n$ we get the following:

$$G\sigma = \sum_{i=1}^{k} a_i \left(\sum_{j=1}^{n} c_{i,j} C_j \right) + \sum_{j=1}^{n} b_j C_j = \sum_{j=1}^{n} \sum_{i=1}^{k} (a_i c_{i,j} + b_j) C_j$$

$$H\sigma = \sum_{i=1}^{k} a'_i \left(\sum_{j=1}^{n} c_{i,j} C_j \right) + \sum_{j=1}^{n} b'_j C_j = \sum_{j=1}^{n} \sum_{i=1}^{k} (a'_i c_{i,j} + b'_j) C_j$$

Since $C_1, \ldots, C_n$ are non-isomorphic, connected components, $G\sigma \cong H\sigma$ implies $\sum_{i=1}^{k}(a_i c_{i,j} + b_j) = \sum_{i=1}^{k}(a'_i c_{i,j} + b'_j)$ for $j = 1, \ldots, n$. Thus, for $j = 1, \ldots, m$, $(c_{1,j}, \ldots, c_{k,j})$ is a solution of $\sum_{i=1}^{k}(a_i - a'_i) x_{i,j} = (b'_j - b_j)$ in non-negative integers.

"If": Let, for $j = 1, \ldots, m$, $(c_{1,j}, \ldots, c_{k,j})$ be a solution of $\sum_{i=1}^{k}(a_i - a'_i) x_{i,j} = (b'_j - b_j)$ in non-negative integers. Consider the substitution $\sigma = \{x_i / \sum_{j=1}^{m} c_{i,j} C_j \mid i = 1, \ldots, k\}$. Then:

$$G\sigma = \sum_{i=1}^{k} a_i \left(\sum_{j=1}^{m} c_{i,j} C_j \right) + \sum_{j=1}^{m} b_j C_j = \sum_{j=1}^{m} \left(\sum_{i=1}^{k} a_i c_{i,j} + b_j \right) C_j$$

$$H\sigma = \sum_{i=1}^{k} a'_i \left(\sum_{j=1}^{m} c_{i,j} C_j \right) + \sum_{j=1}^{m} b'_j C_j = \sum_{j=1}^{m} \left(\sum_{i=1}^{k} a'_i c_{i,j} + b'_j \right) C_j$$

Since, for $j = 1, \ldots, m$, $(c_{1,j}, \ldots, c_{k,j})$ is a solution of $\sum_{i=1}^{k}(a_i - a'_i) x_{i,j} = (b'_j - b_j)$, we get $\sum_{i=1}^{k} a_i c_{i,j} + b_j = \sum_{i=1}^{k} a'_i c_{i,j} + b'_j$. It follows $G\sigma \cong H\sigma$, that is, σ is a solution to $G =^? H$. □

We now come back to unification problems of arbitrary type. In order to give sufficient conditions for the existence of solutions, we consider substitutions that reduce the type of unification problems.

Call a non-empty substitution $\sigma = \{x_1/G_1, \ldots, x_n/G_n\}$ *simplifying* if, for $i = 1, \ldots, n$, G_i is a hypergraph with $type(x_i)$ nodes and either no hyperedge or one variable hyperedge of type less than $type(x_i)$. Given a unification problem $G =^? H$, we denote by $Simplify(G =^? H)$ the set of all unification problems $G\rho =^? H\rho$ such that ρ is simplifying.

Proposition 7 *A unification problem $G =^? H$ has a solution if some problem in $Simplify(G =^? H)$ has a solution.*

Proof. Let σ be a solution to a problem $G\rho =^? H\rho$ in $Simplify(G =^? H)$, where ρ is a simplifying substitution. Then, by Lemma 2, $G(\rho\sigma) \cong (G\rho)\sigma \cong (H\rho)\sigma \cong H(\rho\sigma)$ and hence $\rho\sigma$ is a solution to $G =^? H$. □

There are two particularly simple cases in which Proposition 7 guarantees that a problem $G =^? H$ has a solution: (1) $G^\ominus =^? H^\ominus$ has a solution (i.e. $G^\ominus \cong H^\ominus$) and (2) $G^0 =^? H^0$ has a solution, where G^0 and H^0 are obtained from G and H by substituting type-0 variables for all variables. The following proposition, which gives a syntactic condition for the existence of a solution, can also be seen as a consequence of Proposition 7.

Proposition 8 *A unification problem $G =^? H$ has a solution if there are variables $x \in Var(G) - Var(H)$ and $y \in Var(H) - Var(G)$ such that x and y occur only once in G and H, respectively.*

Proof. Define a substitution $\sigma = \{z/\sigma(z) \mid z \in Var(G) \cup Var(H)\}$ by

$$\sigma(z) = \begin{cases} type(x)^\bullet + H^\ominus & \text{if } z = x, \\ type(y)^\bullet + G^\ominus & \text{if } z = y, \\ type(z)^\bullet & \text{otherwise,} \end{cases}$$

where for each $n \geq 0$, $n^\bullet$ is a hypergraph with n nodes and no hyperedges. Then σ is a solution to $G =^? H$ since $G\sigma \cong G^\ominus + H^\ominus \cong H\sigma$. □

The converse of Proposition 7 does not hold. In fact, there are unification problems of arbitrary type that cannot be solved via simplification.

Proposition 9 *For each $n \geq 1$, there is a unification problem $G =^? H$ of type n that has a solution while no problem in $Simplify(G =^? H)$ has a solution.*

Proof. The unification problem $G =^? H$ given in Figure 3 obviously has a solution. But each problem in $Simplify(G =^? H)$ is of the form $G' =^? H$, where G' contains at least one isolated node (a node to which no hyperedge is attached). Since H does not contain isolated nodes and is variable-free, there is no solution to $G' =^? H$. □

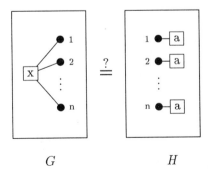

Figure 3: Unification problem for the proof of Proposition 9

4 Matching

We now turn from unification to matching (with variables of arbitrary type).

Definition 10 A *matching problem* $G \leq^? H$ consists of two hypergraphs G and H such that H is variable-free. A *solution* to this problem is a substitution σ such that $G\sigma \cong H$. The solution is *pure* if $Dom(\sigma) = Var(G)$.

The assumption that H is variable-free can be made without loss of generality, for if it is violated one considers a problem $G \leq^? H'$ in which H' is obtained from H by renaming each variable into a new non-variable label. Then the solutions to $G \leq^? H$ can be obtained by applying the inverse transformation to the hypergraphs in the solutions to $G \leq^? H'$.

In the following we do not distinguish between substitutions that are equal up to isomorphism. More precisely, two substitution pairs x/G and x/H are considered to be equal if there is an isomorphism $f: G \to H$ with $f^*(p_G) = p_H$. (Recall that p_G and P_G denote the sequence and the set of points, respectively, of a pointed hypergraph G.)

Definition 11 Let $G \leq^? H$ be a matching problem and e be a hyperedge in G with $lab_G(e) = x \in Var(G)$. Then a pointed hypergraph C with $type(C) = type(x)$ is a *candidate* for e if there is an injective hypergraph morphism $in: G^\ominus \to H$ and a hypergraph morphism $f: C \to H$ such that the following conditions are satisfied:

(1) $in^*(att_G(e)) = f^*(p_C)$,

(2) the subhypergraphs $in(G^\ominus)$ and $f(C)$ of H overlap only in $f(P_C)$, that is, $in(G^\ominus) \cap f(C) = f(P_C)$,

(3) for all items c, c' in C with $c \neq c'$, $f(c) = f(c')$ implies $c, c' \in P_C$,

(4) no hyperedge in $E_H - f(E_C)$ is incident to a node in $f(V_C - P_C)$.

Let MATCH be the set of transformation rules shown in Figure 4. These rules

TRY	$\langle G \leq^? H, \sigma \rangle$	$\Rightarrow$	$\langle G\{x/C\} \leq^? H, \sigma\{x/C\}\rangle$ if $x \in Var(G)$ and C is a candidate for some x-labelled hyperedge in G
FAIL 1	$\langle G \leq^? H, \sigma \rangle$	$\Rightarrow$	F if there is no injective morphism $G^\ominus \to H$
FAIL 2	$\langle G \leq^? H, \sigma \rangle$	$\Rightarrow$	F if $Var(G) = \emptyset$ and $G \not\cong H$

Figure 4: The rule set MATCH

operate on pairs of matching problems and substitutions. The transformation relation defined by repeated application of MATCH rules is denoted by $\Rightarrow^*_{\text{MATCH}}$. By the following result, all pure solutions of a problem $G \leq^? H$ can be obtained by computing all MATCH derivations starting from the pair $\langle G \leq^? H, \emptyset \rangle$.

Theorem 12 *A substitution σ is a pure solution to a matching problem $G \leq^? H$ if, and only if, $\langle G \leq^? H, \emptyset \rangle \Rightarrow^*_{\text{MATCH}} \langle M \leq^? H, \sigma \rangle$ such that no MATCH rule is applicable to $\langle M \leq^? H, \sigma \rangle$.*

The proof of this result is given in two parts which establish soundness and completeness of the MATCH rules.

Lemma 13 (soundness) *If $\langle G \leq^? H, \emptyset \rangle \Rightarrow^*_{\text{MATCH}} \langle M \leq^? H, \sigma \rangle$ such that no MATCH rule is applicable to $\langle M \leq^? H, \sigma \rangle$, then σ is a pure solution to $G \leq^? H$.*

Proof. Since FAIL 1 is not applicable, there is an injective morphism $M^\ominus \to H$. Then $Var(M) = \emptyset$ as otherwise the attachment nodes of a variable hyperedge would form a candidate for this hyperedge, implying that TRY were applicable. It follows that M and H are isomorphic, because FAIL 2 is not applicable.

Now consider an arbitrary derivation $\langle A \leq^? B, \emptyset \rangle \Rightarrow^*_{\text{TRY}} \langle A' \leq^? B, \alpha \rangle$. By using the definition of TRY, an easy induction on the length of this derivation shows that $A\alpha \cong A'$ and $Dom(\alpha) \subseteq Var(A)$ hold. Thus $G\sigma \cong M \cong H$ and $Dom(\sigma) = Var(G)$, that is, σ is a pure solution to $G \leq^? H$. □

Lemma 14 (completeness) *If σ is a pure solution to a matching problem $G \leq^? H$, then $\langle G \leq^? H, \emptyset \rangle \Rightarrow^*_{\text{TRY}} \langle G\sigma \leq^? H, \sigma \rangle$.*

Proof. We proceed by induction on the size of $Var(G)$. If $Var(G) = \emptyset$, then $G \cong H$ and the TRY derivation of length 0 is as required. Now let $Var(G) \neq \emptyset$ and suppose, as induction hypothesis, that the proposition holds for every matching problem $G' \leq^? H$ with $|Var(G')| < |Var(G)|$. Consider some substitution pair x/C in σ and let $\tau = \sigma - \{x/C\}$. By Lemma 2 and the fact that C is variable-free,

$$\begin{aligned}(G\{x/C\})\tau &= G(\{x/C\}\tau) \\ &= G(\{x/C\} \cup \tau) \\ &= G\sigma \\ &\cong H.\end{aligned}$$

So τ is a pure solution to $G\{x/C\}) \leq^? H$. Hence, by induction hypothesis,

$$\langle G\{x/C\} \leq^? H, \emptyset \rangle \Rightarrow^*_{\text{TRY}} \langle (G\{x/C\})\tau \leq^? H, \tau \rangle.$$

Moreover, a straightforward induction on the length of derivations shows that every derivation $\langle A \leq^? B, \emptyset \rangle \Rightarrow^*_{\text{TRY}} \langle A' \leq^? B, \alpha \rangle$ can be transformed into a derivation $\langle A \leq^? B, \beta \rangle \Rightarrow^*_{\text{TRY}} \langle A' \leq^? B, \beta\alpha \rangle$, where β is an arbitrary substitution. Thus

$$\langle G\{x/C\} \leq^? H, \{x/C\} \rangle \Rightarrow^*_{\text{TRY}} \langle (G\{x/C\})\tau \leq^? H, \{x/C\}\tau \rangle.$$

It is easy to check that C is a candidate for each x-labelled hyperedge in G, implying $\langle G \leq^? H, \emptyset \rangle \Rightarrow_{\text{TRY}} \langle G\{x/C\} \leq^? H, \{x/C\} \rangle$. Therefore, by transitivity of $\Rightarrow^*_{\text{TRY}}$,

$$\langle G \leq^? H, \emptyset \rangle \Rightarrow^*_{\text{TRY}} \langle (G\{x/C\})\tau \leq^? H, \{x/C\}\tau \rangle.$$

This completes the proof since $(G\{x/C\})\tau \cong G\sigma$ and $\{x/C\}\tau = \sigma$. □

Corollary 15 *There is an algorithm that computes for every matching problem all pure solutions.*

Proof. Every sequence of MATCH applications terminates since for every step $\langle G \leq^? H, \sigma \rangle \Rightarrow_{\text{TRY}} \langle G' \leq^? H, \sigma' \rangle$, G' contains fewer variables than G. Moreover, from every pair $\langle G \leq^? H, \sigma \rangle$ only finitely many pairs can be generated by a TRY step. For there are, up to isomorphism, only finitely many candidates for each variable hyperedge in G. It follows that the set of MATCH derivations starting from a pair $\langle G \leq^? H, \emptyset \rangle$ is finite, and hence it can be computed. Call a derivation in this set successful if it ends with a pair to which no MATCH rule can be applied. By Theorem 12, the substitutions in the final pairs of the successful derivations are the pure solutions of $G \leq^? H$. □

5 Substitution-based graph rewriting

The above considerations suggest a hypergraph rewriting mechanism based on matching. The approach is very simple to describe and needs neither pushouts nor embedding instructions: A *rule* is just a pair $\langle L, R \rangle$ of hypergraphs which is applied to a hypergraph G by (1) choosing a solution σ to the matching problem $L \leq^? G$ and (2) applying σ to R.

Example 16 Figure 5 shows a rule for program transformation on flow graphs,

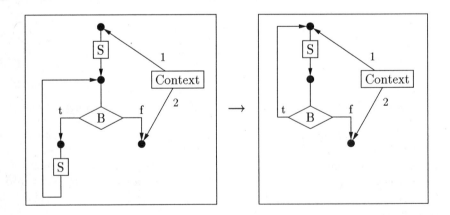

Figure 5: A rule

where B, S, and Context are variables. The rule can transform a pattern of the form "S; while B do S" into "repeat S until ¬B".

In order to state that substitution-based rewriting generalizes the well-known double-pushout approach [7], recall that a *DPO-rule* $r = \langle L \leftarrow K \rightarrow R \rangle$ consists of two hypergraph morphisms with a common domain. The application of r to a hypergraph G consists of the construction of the two hypergraph pushouts shown in Figure 6. Without loss of generality we may assume that the vertical

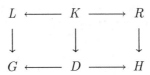

Figure 6: A double-pushout

morphisms in these pushouts are injective, since r can be translated into a finite set $S(r)$ of DPO-rules such that each application of r corresponds to an application of a rule in $S(r)$ obeying the injectivity condition and vice versa.

Given a DPO-rule $r = \langle L \leftarrow K \rightarrow R \rangle$, we construct a rule $r^\oplus = \langle L^\oplus, R^\oplus \rangle$ as follows: Let $K^\oplus$ be the extension of K by a variable-hyperedge of type $|V_K|$ which is attached to all nodes of K. Then $L^\oplus$ is the pushout object of L and $K^\oplus$ along K, and $R^\oplus$ is the pushout object of R and $K^\oplus$ along K. When $r^\oplus$ is applied to a hypergraph G, the variable in $L^\oplus$ is replaced by the context of L in G.

Theorem 17 *Let G and H be hypergraphs and r be a DPO-rule. Then H results from an application of r to G in the double-pushout approach if, and only if, H results from an application of $r^\oplus$ to G in the substitution-based approach.*

So the double-pushout approach is included in the substitution-based approach. But the latter is a proper generalization of the former as there are rules which remove or copy non-local parts of the hypergraphs they are applied to. This is simply achieved by omitting or duplicating variables from the left-hand side of a rule in the right-hand side. For instance, the rule in Example 16 cannot be simulated by a double-pushout rule. This is because there are applications of the rule with arbitrarily large subhypergraphs matched by the S-variables. So the rule can remove a subhypergraph of unbounded size whereas the number of items removed by a double-pushout rule is bounded by the size of its left-hand side.

Proof of Theorem 17. Let H result from an application of r to G in the double-pushout approach as shown in Figure 6. Let K_0 be the discrete hypergraph obtained from K by removing all hyperedges, and D_0 be the hypergraph obtained from D by removing the images of all hyperedges of K. Then there are inclusion morphisms $K_0 \to K$ and $D_0 \to D$, and $K \to D$ can be restricted to $K_0 \to D_0$. By assumption, the vertical morphisms in Figure 6 are injective. So D is the pushout object of K and D_0 along K_0. Correspondingly, G and H are the pushout objects of L and D_0 along K_0, and of R and D_0 along K_0, respectively.

Now let $\sigma = \{x/\langle D_0, p_0 \rangle\}$, where x is the label of the variable hyperedge in $K_0^\oplus$, and p_0 is the image of the attachment sequence of this hyperedge. Then, obviously, $K_0^\oplus \sigma \cong D_0$. The following relation between pushouts and substitutions is easy to show: if D is the pushout object of B and C along A, and $A^\oplus \sigma \cong B$, then $C^\oplus \sigma \cong D$. Hence we obtain $L^\oplus \sigma \cong G$ and $R^\oplus \sigma \cong H$, that is, H results from an application of $r^\oplus$ to G in the substitution-based approach.

Conversely, let H result from an application of $r^\oplus$ to G in the substitution-based approach. Then $L^\oplus \sigma \cong G$ and $R^\oplus \sigma \cong H$ for some substitution σ. Let $D = K^\oplus \sigma$, G' be the pushout object of L and D along K, and H' be the pushout object of R and D along K. By the above mentioned relation between pushouts and substitutions, we have $G' \cong L^\oplus \sigma \cong G$ and $H' \cong R^\oplus \sigma \cong H$. Thus, H results from an application of r to G in the double-pushout approach. □

6 Conclusion

The decidability of the general graph unification problem is an intriguing theoretical problem. It is easy to state but turns out to be non-trivial already for the very restricted case of type 0 variables. The general case may be hard to solve in view of Makanin's extremely complex proof of the decidability of word unification [16] (see also [13, 19]).

On the other hand, it is not clear whether word unification can be simulated by graph unification. The problem is that for the obvious graph representation of words as chains of edges, unifying graph substitutions need not preserve this structure and hence may not correspond to word substitutions. For example, the word unification problem **ax** $=^?$ **by** would have a graph solution (see Proposition 8) although the two words are not unifiable. For similar reasons it is unclear whether graph unification can simulate second-order term unification. Since the latter is undecidable [8], this simulation would show that graph unification is undecidable.

Instead of representing words or terms as graphs, one can also go the reverse direction. Bauderon and Courcelle [2] define three graph operations by which hypergraphs can be represented as first-order terms over a given set of typed labels. They give eleven equations which equate each two different terms that represent the same hypergraph. By providing such terms with variables for hyperedges, graph unification can be rephrased as term unification modulo the equational theory induced by the eleven equations.

References

[1] Franz Baader and Jörg H. Siekmann. Unification theory. In D. M. Gabbay, C. J. Hogger, and J. A. Robinson, editors, *Handbook of Logic in Artificial Intelligence and Logic Programming*, pages 41–125. Oxford University Press, 1994.

[2] Michel Bauderon and Bruno Courcelle. Graph expressions and graph rewritings. *Mathematical Systems Theory*, 20:83–127, 1987.

[3] Alexandre Boudet, Evelyne Contejean, and Hervé Devie. A new AC unification algorithm with an algorithm for solving systems of Diophantine equations. In *Proc. Annual IEEE Symposium on Logic in Computer Science*, pages 289–299, 1990.

[4] Michael Clausen and Albrecht Fortenbacher. Efficient solution of linear Diophantine equations. *Journal of Symbolic Computation*, 8:201–216, 1989.

[5] Evelyne Contejean. An efficient incremental algorithm for solving systems of linear Diophantine equations. *Information and Computation*, 113:143–172, 1994.

[6] Jacques Corbin and Michel Bidoit. A rehabilitation of Robinson's unification algorithm. In R.E.A. Mason, editor, *Proc. Information Processing 83*, pages 909–914, 1983.

[7] Hartmut Ehrig. Introduction to the algebraic theory of graph grammars. In *Proc. Graph-Grammars and Their Application to Computer Science and Biology*, pages 1–69. Springer Lecture Notes in Computer Science 73, 1979.

[8] Warren D. Goldfarb. The undecidability of the second-order unification problem. *Theoretical Computer Science*, 13:225–230, 1981.

[9] Annegret Habel. *Hyperedge Replacement: Grammars and Languages*, volume 643 of *Lecture Notes in Computer Science*. Springer-Verlag, 1992.

[10] Annegret Habel and Detlef Plump. Unification, rewriting, and narrowing on term graphs. In *Proc. SEGRAGRA '95*, Electronic Notes in Theoretical Computer Science. Elsevier, 1995. To appear.

[11] Gérard P. Huet. Résolution d'equations dans les langages d'ordre $1,2,\ldots,\omega$. Thèse d'état, Université de Paris VII, 1976.

[12] Joxan Jaffar. Efficient unification over infinite terms. *New Generation Computing*, 2:207–219, 1984.

[13] Joxan Jaffar. Minimal and complete word unification. *Journal of the ACM*, 37(1):47–85, 1990.

[14] Jean-Pierre Jouannaud and Claude Kirchner. Solving equations in abstract algebras: A rule-based survey of unification. In Jean-Louis Lassez and Gordon Plotkin, editors, *Computational Logic: Essays in Honor of Alan Robinson*. MIT Press, 1991.

[15] Kevin Knight. Unification: A multidisciplinary survey. *ACM Computing Surveys*, 21(1):93–124, 1989.

[16] G.S. Makanin. The problem of solvability of equations in a free semigroup. *Matematiceskiĭ Sbornik*, 103:147–236, 1977. In Russian. English translation in *Math. USSR Sbornik*, 32:129–198, 1977.

[17] Francesco Parisi-Presicce, Hartmut Ehrig, and Ugo Montanari. Graph rewriting with unification and composition. In *Proc. Graph-Grammars and Their Application to Computer Science*, pages 496–514. Springer Lecture Notes in Computer Science 291, 1987.

[18] M.S. Paterson and M.N. Wegman. Linear unification. *Journal of Computer and System Sciences*, 16:158–167, 1978.

[19] Klaus U. Schulz. Word unification and transformation of generalized equations. *Journal of Automated Reasoning*, 11:149–184, 1993.

On the Interleaving Semantics of Transformation Units – A Step into GRACE

Hans-Jörg Kreowski, Sabine Kuske*

Universität Bremen, Fachbereich 3
Postfach 33 04 40, D-28334 Bremen
email: {kreo,kuske}@informatik.uni-bremen.de

Abstract. The aim of the paper is to introduce the notion of a transformation unit together with its interleaving semantics and to study it as a means of constructing large graph transformation systems from small ones in a structured and systematic way. A transformation unit comprises a set of rules, descriptions of initial and terminal graphs, and a control condition. Moreover, it may import other transformation units for structuring purposes. Its semantics is a binary relation between initial and terminal graphs which is given by interleaving sequences. As a generalization of ordinary derivations, an interleaving sequence consists of direct derivation steps interleaved with calls of imported transformation units. It must obey the control condition and may be seen as a kind of structured derivation. The introduced framework is independent of a particular graph transformation approach and, therefore, it may enhance the usefulness of graph transformations in many contexts.

1 Introduction

The significance of graphs and rules in many areas of computer science is evident: On the one hand, graphs constitute appropriate means for the description of complex relationships between objects. Trees, forests, Petri nets, circuit diagrams, finite automata, flow charts, data flow graphs, and entity-relationship diagrams are some typical examples. On the other hand, rules are used to describe "permitted" manipulations on objects like, for example, in the areas of functional and logic programming, formal languages, algebraic specification, theorem proving, and rule-based systems.

The intention of bringing graphs and rules together – motivated by several application areas – has led to the theory of *graph grammars* and *graph transformation* (see [CER79, ENR83, ENRR87, EKR91, SE93] for a survey). A wide spectrum of approaches exists within this theory and some of them are implemented (see, for example, PROGRES [Sch91a, Sch91b], Graph^{Ed} [Him91], Dactl [GKS91], and AGG [LB93, TB93]).

With the aim of enhancing the usefulness of graph transformation, we propose

---

*This work has been supported by COMPUGRAPH II, ESPRIT Basic Research Working Group 7183.

a new approach-independent structuring method for building up large systems of graph transformation rules from small pieces. The method is based on the notion of a transformation unit and its interleaving semantics. A transformation unit is allowed to use other units such that a system of graph transformation rules can be structured hierarchically and existing transformation units can be re-used. The transformation unit is a basic concept of the new graph and rule centered language GRACE that is being developed by researchers from Aachen, Berlin, Bremen, and Leiden (see also [Kre95, Kus95, Sch95]). Nevertheless, the notion is meaningful in its own right because – independently of GRACE – it can be employed as a structuring principle in most graph transformation approaches one encounters in the literature where graph transformation is also called graph rewriting.

The paper is organized as follows. Section 2 introduces the notion of a transformation unit together with its interleaving semantics. In Section 3, the concepts of a transformation unit are illustrated with an example. Section 4 presents how some operations on binary relations can be modelled by suitable operations on transformation units. In Section 5, some normal forms of transformation units are considered. The paper ends with some concluding remarks. To avoid wrong expectations, we would like to point out that the goal of the paper is to shed some light on the usefulness of the introduced structuring method rather than to come up with deep theory.

2 Transformation Units with Interleaving Semantics

The key operation in graph rewrite approaches is the transformation of a graph into a graph by applying a rule such that sets of rules specify graph transformations by iterated rule applications. This derivation process is highly non-deterministic in general and runs on arbitrary graphs which is both not always desirable. For example, if one wants to generate graph languages, one may start in a particular axiom and end with certain terminal objects only. Or if a more functional behaviour is required, one may prefer to control the derivation process and to cut down its non-determinism. The latter can be achieved by control mechanisms for the derivation process like application conditions or programmed graph transformation (see, e.g., [Bun79, Nag79, EH86, KR90, MW91, Sch91a, SZ91, Kre93, LM93, HHT95], cf. also [DP89] for regulation concepts in string grammars) and the former by the use of graph class expressions that specify subclasses of graphs. Moreover, in practical cases, one may have to handle hundreds or thousands of rules which cannot be done in a transparent and reasonable way without a structuring principle.

To cover all these aspects, we introduce the notion of a transformation unit that allows to specify new rules, initial and terminal graphs, as well as a control condition, and to import other transformation units. Semantically, a transformation unit describes a graph transformation, i.e. a binary relation on graphs given by the interleaving of the graph transformations with each other and with rule applications. Moreover, interleaving sequences must start in initial graphs, end in terminal graphs and satisfy the control condition. If nothing is imported, the interleaving semantics coincides with the derivation relation.

To make the concept independent of a particular graph rewriting framework, we assume an abstract notion of a graph transformation approach comprising a class of graphs, a class of rules, a rule application operator, a class of graph class expressions, and a class of control conditions. The semantic effect of control conditions is described depending on so-called environments that associate binary relations on graphs to identifiers. In this way, it can be defined without forward reference to transformation units. Intuitively, one may think of an environment as rules with their corresponding direct derivation relations (cf. 2.5).

2.1 Graph Transformation Approach

A *graph transformation approach* is a system $\mathcal{A} = (\mathcal{G}, \mathcal{R}, \Longrightarrow, \mathcal{E}, \mathcal{C})$ where

- $\mathcal{G}$ is a class of *graphs*,
- $\mathcal{R}$ is a class of *rules*,
- $\Longrightarrow$ is a *rule application operator* yielding a binary relation $\Longrightarrow_r \subseteq \mathcal{G} \times \mathcal{G}$ for every $r \in \mathcal{R}$,
- $\mathcal{E}$ is a class of *graph class expressions* such that each $e \in \mathcal{E}$ specifies a subclass $SEM(e) \subseteq \mathcal{G}$, and
- $\mathcal{C}$ is a class of *elementary control conditions* over some set ID of identifiers such that each $c \in \mathcal{C}$ specifies a binary relation $SEM_E(c) \subseteq \mathcal{G} \times \mathcal{G}$ for each mapping $E: ID \longrightarrow \mathcal{P}(\mathcal{G} \times \mathcal{G})$.

A pair $(G, G') \in \Longrightarrow_r$, usually written as $G \Longrightarrow_r G'$, is called a *direct derivation* from G to G' through r. For a set $P \subseteq \mathcal{R}$ the union of all relations $\Longrightarrow_r$ ($r \in P$) is denoted by $\Longrightarrow_P$ and its reflexive and transitive closure by $\stackrel{*}{\Longrightarrow}_P$. A pair $(G, G') \in \stackrel{*}{\Longrightarrow}_P$, usually written as $G \stackrel{*}{\Longrightarrow}_P G'$, is called a *derivation* from G to G' over P. A mapping $E: ID \longrightarrow \mathcal{P}(\mathcal{G} \times \mathcal{G})$ is called an *environment*. In the following, we will use boolean expressions over $\mathcal{C}$ as *control conditions* with elementary control conditions as basic elements and disjunction, conjunction, and negation as boolean operators. Moreover, we assume the constant *true*. The semantic relations of elementary control conditions are easily extended to boolean expressions by

$SEM_E(true) = \mathcal{G} \times \mathcal{G}$,
$SEM_E(e_1 \vee e_2) = SEM_E(e_1) \cup SEM_E(e_2)$,
$SEM_E(e_1 \wedge e_2) = SEM_E(e_1) \cap SEM_E(e_2)$,
$SEM_E(\overline{e}) = \mathcal{G} \times \mathcal{G} - SEM_E(e)$.

The set of control conditions over $\mathcal{C}$ is denoted by $\mathcal{B}(\mathcal{C})$.

Note that we refer to the meaning of graph class expressions and control conditions by the overloaded operator SEM. This should do no harm because it is always clear from the context which is which.

All the graph grammar and graph rewrite approaches one encounters in the literature provide notions of graphs and rules and a way of directly deriving a graph from a graph by applying a rule (cf. e.g. [Ehr79, Nag79, JR80, Cou90, KR90, Sch91b, Him91, Hab92, Löw93]). Therefore, all of them can be considered as graph transformation approaches in the above sense, if one chooses the components $\mathcal{E}$ and $\mathcal{C}$ in some standard way. The singleton set $\{all\}$ with $SEM(all) = \mathcal{G}$ may provide the only graph class expression, and the class of elementary control conditions may be empty. Non-trivial choices for $\mathcal{E}$ and $\mathcal{C}$ are discussed in the next paragraphs.

2.2 Graph Class Expressions

There are various standard ways to choose graph class expressions that can be combined with many classes of graphs and hence used in many graph transformation approaches.

1. In most cases, one deals with some kind of finite graphs with some explicit representations. Then single graphs (or finite enumerations of graphs) may serve as graph class expressions. Semantically, each graph G represents itself, i.e. $SEM(G) = \{G\}$. The axiom of a graph grammar is a typical example of this type.

2. A graph G is *reduced* with respect to a set of rules $P \subseteq \mathcal{R}$ if there is no $G' \in \mathcal{G}$ with $G \underset{r}{\Longrightarrow} G'$ and $r \in P$. In this way, P can be considered as a graph class expression with $SEM(P) = RED(P)$ being the set of all reduced graphs with respect to P. Reducedness is often used in term rewriting and term graph rewriting as a halting condition.

3. If $\mathcal{G}$ is a class of labelled graphs with label alphabet Σ, then a set $T \subseteq \Sigma$ is a suitable graph class expression specifying the graph class $SEM(T) = \mathcal{G}_T$ consisting of all graphs labelled in T only. This way of distinguishing terminal objects is quite popular in formal language theory.

4. Graph theoretic properties can be used as graph class expressions. In particular, monadic second order formulas for directed graphs or hypergraphs or undirected graphs are suitable candidates (see e.g. [Cou90]).

2.3 Control Conditions

Every description of a binary relation on graphs may be used as a control condition. Here, we introduce some typical examples.

1. Let $E \colon ID \to \mathcal{P}(\mathcal{G} \times \mathcal{G})$ be an environment. Then E can be extended to the powerset of the set of strings over ID in a natural, straight-forward way. $\widehat{E} \colon \mathcal{P}(ID^*) \to \mathcal{P}(\mathcal{G} \times \mathcal{G})$ is defined by $\widehat{E}(L) = \bigcup_{w \in L} \overline{E}(w)$ for $L \subseteq ID^*$ where $\overline{E} \colon ID^* \to \mathcal{P}(\mathcal{G} \times \mathcal{G})$ is recursively given by $\overline{E}(\lambda) = \Delta \mathcal{G}$ [1], and $\overline{E}(xv) = E(x) \circ$

---

[1] $\Delta \mathcal{G}$ denotes the identity relation on $\mathcal{G}$.

$\overline{E}(v)$ [2] for $x \in ID$ and $v \in ID^*$. Hence, every grammar, every automaton and every expression specifying a language over ID can serve as a control condition. If it is not necessary to distinguish between syntax and semantics, we may represent a control condition of this language type by the language itself with $SEM_E(L) = \widehat{E}(L)$ for all $L \subseteq ID^*$ and $E : ID \to \mathcal{P}(\mathcal{G} \times \mathcal{G})$. In this case, the class of elementary control conditions is $\mathcal{P}(ID^*)$.

2. In particular, the class of regular expressions over ID can be used for this purpose. For explicit use below, $REG(ID)$ is recursively given by $\emptyset, \epsilon \in REG(ID)$ [3], $ID \subseteq REG(ID)$, and $(e_1 ; e_2), (e_1 | e_2), (e^*) \in REG(ID)$ if $e, e_1, e_2 \in REG(ID)$. Outermost parentheses of regular expressions will be omitted in the following.

3. Let $\mathcal{E}$ be a class of graph class expressions. Then each pair $(e_1, e_2) \in \mathcal{E} \times \mathcal{E}$ defines a binary relation on graphs by $SEM((e_1, e_2)) = SEM(e_1) \times SEM(e_2)$ and, therefore, it can be used as a control condition which is independent of the choice of an environment, i.e. $SEM_E((e_1, e_2)) = SEM((e_1, e_2))$ for all environments E.

2.4 Transformation Units

A transformation unit encapsulates a specification of initial graphs, a set of transformation units to be used, a set of rules, a control condition, and a specification of terminal graphs.

Let $\mathcal{A} = (\mathcal{G}, \mathcal{R}, \Longrightarrow, \mathcal{E}, \mathcal{C})$ be a graph transformation approach. A *transformation unit* over $\mathcal{A}$ is a system $trut = (I, U, R, C, T)$ where $I, T \in \mathcal{E}$, U is a set of transformation units over $\mathcal{A}$, $R \subseteq \mathcal{R}$, and $C \in \mathcal{B}(\mathcal{C})$. The components of $trut$ may be denoted by $U_{trut}, I_{trut}, R_{trut}, C_{trut}$, and T_{trut}, respectively.

This is meant as a recursive definition of the set $\mathcal{T}_\mathcal{A}$ of transformation units over $\mathcal{A}$. Hence, initially, U must be chosen as the empty set yielding unstructured transformation units without import that may be used in the next iteration, and so on.

If I specifies a single graph (cf. 2.2.1), U is empty, and C is the constant *true*, one gets the usual notion of a graph grammar (in which approach ever) as a special case of transformation units.

2.5 Interleaving Semantics

The operational semantics of a transformation unit is a graph transformation, i.e. a binary relation on graphs containing a pair (G, G') of graphs if, first, G is an initial graph and G' is a terminal graph, second, G' can be obtained from G by interleaving direct derivations with the graph transformations specified by the used transformation units, and third, the pair is allowed by the control condition.

Let $trut = (I, U, R, C, T)$ be a transformation unit over the graph transformation approach $\mathcal{A} = (\mathcal{G}, \mathcal{R}, \Longrightarrow, \mathcal{E}, \mathcal{C})$. Assume that the set ID of identifiers associated

---

[2] Given $\varrho, \varrho' \subseteq \mathcal{G} \times \mathcal{G}$, the sequential composition of ϱ and ϱ' is defined as usual by $\varrho \circ \varrho' = \{(G, G'') | (G, G') \in \varrho \text{ and } (G', G'') \in \varrho' \text{ for some } G' \in \mathcal{G}\}$.

[3] While $\emptyset$ denotes the empty set $\{\}$, the expression ϵ denotes the regular set $\{\lambda\}$. We prefer a direct reference to $\{\lambda\}$ rather than to use $\emptyset^*$.

to $\mathcal{C}$ contains the disjoint union of U and R. Let the interleaving semantics $SEM(t) \subseteq \mathcal{G} \times \mathcal{G}$ for $t \in U$ be already defined. Let $E(trut): ID \to \mathcal{P}(\mathcal{G} \times \mathcal{G})$ be defined by $E(trut)(r) = \underset{r}{\Longrightarrow}$ for $r \in R$, $E(trut)(t) = SEM(t)$ for $t \in U$, and $E(trut)(id) = \{\}$, otherwise. Then the *interleaving semantics SEM(trut)* of *trut* consists of all pairs $(G, G') \in \mathcal{G} \times \mathcal{G}$ such that

1. $G \in SEM(I)$ and $G' \in SEM(T)$,
2. there are graphs $G_0, \ldots, G_n \in \mathcal{G}$ with $G_0 = G$, $G_n = G'$ and for $i = 1, \ldots, n$, $G_{i-1} \underset{R}{\Longrightarrow} G_i$ or $(G_{i-1}, G_i) \in SEM(t)$ for some $t \in U$,
3. $(G, G') \in SEM_{E(trut)}(C)$.

The sequence of graphs in point 2 is called an *interleaving sequence in trut from G to G'*. Let $RIS(trut)$ denote the binary relation given by interleaving sequences, i.e. the set of all pairs of graphs $(G, G') \in \mathcal{G} \times \mathcal{G}$ such that there is an interleaving sequence in *trut* from G to G' as in point 2. Then the interleaving semantics of *trut* is defined as the intersection of $RIS(trut)$ with $SEM(I) \times SEM(T)$ and $SEM_{E(trut)}(C)$. Note that all three relations may be incomparable with each other. For example, $(G, G') \in SEM_{E(trut)}(C)$ does not imply in general that there is an interleaving sequence in *trut* from G to G', and vice versa.

A control condition C specifies a binary predicate depending on other binary graph relations through the notion of environments, but independent of a particular transformation unit. As a component of *trut*, only the *environment of trut* given by $E(trut)$ is effective, meaning that C can restrict the semantics by specifying certain properties of the direct derivation relations of rules in *trut*, the interleaving semantics of imported transformation units, and the interrelation of all of them. If transformation units are used in a specification language, it will be more realistic to assume that ID is a countable set of predefined identifiers out of which the elements of U and R are named, rather than to assume that U and R are subsets of ID. But we prefer here to avoid an explicit naming mechanism because it is not essential for the introduced concepts.

The definition of the interleaving semantics follows the recursive definition of transformation units. Hence, its well-definedness follows easily by an induction on the structure of transformation units provided that the import structure is acyclic. Initially, if U is empty, an interleaving sequence is just a derivation such that one gets in this case

$$SEM(trut) = \underset{R}{\overset{*}{\Longrightarrow}} \cap (SEM(I) \times SEM(T)) \cap SEM_{E(trut)}(C).$$

In other words, interleaving semantics generalizes the ordinary operational semantics of sets of rules given by derivations. If, furthermore, C is *true* and I is a single graph, then *trut* is a graph grammar (cf. 2.4), and the following holds for its interleaving semantics:

$$SEM(trut) = \{(I, G) \mid G \in L(trut)\}$$

with $L(trut) = \{G \in SEM(T) \mid I \underset{R}{\overset{*}{\Longrightarrow}} G\}$. In this sense, the interleaving semantics covers the usual notion of graph languages generated by graph grammars.

2.6 Application Sequences

In interleaving sequences, rules are applied and imported transformation units are called in some order. Such sequences of applied rules and called transformation units help to clarify the role of control conditions of the language type as defined in 2.3.1.

Let $trut = (I, U, R, C, T)$ be a transformation unit over the graph transformation approach $\mathcal{A} = (\mathcal{G}, \mathcal{R}, \Longrightarrow, \mathcal{E}, \mathcal{C})$. Assume that U and R are disjoint subsets of the set ID associated to $\mathcal{C}$. Then $x_1 \ldots x_n \in (U \cup R)^*$ ($x_i \in U \cup R$) is called an *application sequence* of $(G, G') \in \mathcal{G} \times \mathcal{G}$ if there is an interleaving sequence $G_0, \ldots, G_n$ with $G_0 = G$, $G_n = G'$ and, for $i = 1, \ldots, n$, $G_{i-1} \underset{x_i}{\Longrightarrow} G_i$ if $x_i \in R$ and $(G_{i-1}, G_i) \in SEM(x_i)$ if $x_i \in U$. In the case $n = 0$, the application sequence is the empty string λ. Using these notions and notations, the following observation states that a language over $U \cup R$, used as a control condition due to 2.3.1, controls the order in which rules are applied and imported transformation units are actually used.

Observation

Let $\mathcal{C} = \mathcal{P}(ID^*)$ be the class of control conditions of language type, and let $trut = (I, U, R, L, T)$ with $L \subseteq (U \cup R)^* \subseteq ID^*$. Then for all $G, G' \in \mathcal{G}$, the following statements are equivalent.

1. $(G, G') \in SEM(trut)$.
2. $(G, G') \in SEM_{E(trut)}(L) \cap SEM(I) \times SEM(T)$.
3. There is an application sequence w of (G, G') with $w \in L$ and $(G, G') \in SEM(I) \times SEM(T)$.

Proof. Let $(G, G') \in SEM(trut)$. Then by definition, there is an interleaving sequence in $trut$ from G to G', $(G, G) \in SEM_{E(trut)}(L)$, and $(G, G) \in SEM(I) \times SEM(T)$. Hence, point 1 implies point 2.

To show that point 2 implies point 3 and that point 3 implies point 1, we prove first the following claim: $(G, G') \in SEM_{E(trut)}(L)$ iff there is an application sequence $w \in L$ of (G, G').

By definition, we have $(G, G') \in SEM_{E(trut)}(L) = \widehat{E(trut)}(L)$ iff $(G, G') \in \overline{E(trut)}(w)$ for some $w \in L$. We show now by induction on the structure of w that $(G, G') \in \overline{E(trut)}(w)$ iff w is an application sequence of (G, G').

If $w = \lambda$, we get $(G, G') \in \overline{E(trut)}(\lambda) = \Delta\mathcal{G}$, i.e. $G = G'$ which defines the interleaving sequence with $n = 0$ and $G_0 = G = G'$ and the corresponding application sequence λ (and vice versa).

Assume now that the statement holds for $v \in (U \cup R)^*$.

And consider $w = xv$ with $x \in U \cup R$. Then $(G, G') \in \overline{E(trut)}(xv) = E(trut)(x) \circ \overline{E(trut)}(v)$, and there is some $\overline{G} \in \mathcal{G}$ with $(G, \overline{G}) \in E(trut)(x)$ and $(\overline{G}, G') \in \overline{E(trut)}(v)$. The latter implies by induction that v is an application sequence of $(\overline{G}, G')$ such that there is an interleaving sequence $G_0, \ldots, G_n$ with $\overline{G} = G_0$ and $G' = G_n$. The former means $G \underset{x}{\Longrightarrow} \overline{G}$ if $x \in R$ and $(G, \overline{G}) \in SEM(x)$

if $x \in U$. Altogether, $G, G_0, \ldots, G_n$ defines an interleaving sequence with xv as corresponding application sequence. Conversely, an application sequence xv of (G, G') is related to an interleaving sequence $G_0, \ldots, G_n$ with $G = G_0$, $G' = G_n$ and, in particular, $G_0 \underset{x}{\Longrightarrow} G_1$ if $x \in R$ or $(G_0, G_1) \in SEM(x)$ if $x \in U$ such that $(G, G_1) \in E(trut)(x)$ in any case. Moreover, v is an application sequence of (G_1, G_n) because $G_1, \ldots, G_n$ is an interleaving sequence. By induction hypothesis, we get $(G_1, G') \in E(trut)(v)$. The composition yields $(G, G') \in E(trut)(x) \circ E(trut)(v) = E(trut)(xv)$. This completes the proof of the claim.

From the just proved claim follows directly that point 2 implies point 3.

Furthermore, let $w \in L$ be an application sequence of (G, G') with $(G, G') \in SEM(I) \times SEM(T)$. Then by definition, we have that there is an interleaving sequence in $trut$ from G to G' with $(G, G') \in SEM(I) \times SEM(T)$, and by the claim, $w \in SEM_{E(trut)}(L)$. Hence, $(G, G') \in SEM(trut)$. This completes the proof. □

2.7 Initial and Terminal Graphs

As pointed out in 2.3.3, two graph class expressions form a control condition. Hence, the specifications of initial and terminal graphs may be handled as a control condition. In the interleaving semantics of a transformation unit, the product of initial and terminal graphs as well as the relation specified by the actual control condition must be intersected with the relation established by the interleaving sequences. This proves the following observation.

Observation
Let $trut = (I, U, R, C, T)$ be a transformation unit over $\mathcal{A} = (\mathcal{G}, \mathcal{R}, \Longrightarrow, \mathcal{E}, \mathcal{C})$. Then $trut' = (all, U, R, C \wedge (I, T), all)$ is a transformation unit over $\mathcal{A}' = (\mathcal{G}, \mathcal{R}, \Longrightarrow, \{all\}, \mathcal{C} \cup \mathcal{E} \times \mathcal{E})$ with $SEM(trut) = SEM(trut')$.

This means that the components I and T of a transformation unit are not necessary. Nevertheless, we keep them because we would like to distinguish between input and output conditions and other control conditions explicitly and to emphasize the different intuitions behind.

3 Butterfly Networks – an Example

In this section, we illustrate the concepts of transformation units by specifying the set of butterfly networks. Such high-bandwidth processor organizations – well-known from the area of VLSI theory – are well suited for performing highly parallel computations (see e.g. Ullman [Ull84] and Lengauer [Len90]).

A *butterfly network* of size k for some $k \in \mathbb{N}$ consists of $(k+1)$ *ranks* of 2^k nodes each. Let v_{ir} be the ith node on rank r and let $b_{i_1} \ldots b_{i_k}$ be the binary representation of i ($0 \leq i < 2^k$, $0 \leq r \leq k$). Then for $r > 0$, v_{ir} is directly connected to v_{jr-1} if either $i = j$ or $b_{i_1} \ldots b_{i_k}$ is equal to the binary representation of j up to b_{i_r}. In the following, we call the nodes on rank 0 *bottom nodes* and label them with b. All other nodes of a butterfly network are unlabelled.

To make the paper self-contained, we tailor a graph transformation approach for this example. This approach can be easily expressed in many of the general graph transformation approaches in the literature, like, e.g., PROGRES (see [Sch91a]) or graph grammars with negative application conditions (see [HHT95]).

- A *graph* is a system $G = (V_G, E_G, l_G, m_G)$ where V_G is the set of *nodes*, E_G is the set of *edges* being 2-element subsets of V_G, $l_G: V_G \to C$, and $m_G: E_G \to C$ are labelling functions for nodes and edges respectively with $C = \{*, b, c\}$. The symbol $*$ stands for *unlabelled*, b for *bottom*, and c for *copied*. Subgraph relation and isomorphy are defined and denoted in the usual way.

- A *rule* is a triple $r = (N, L, R)$ of graphs with $L \subseteq N$ and $V_L \subseteq V_R$, i.e. L is a subgraph of N, and each node of L is a node of R, but its label in L may be different from that in R.

- A *rule* $r = (N, L, R)$ is applied to a graph G according to the following steps.
  1. Choose $L' \subseteq G$ with $L' \cong L$.
  2. Check the negative context condition: It fails if there is any $N' \subseteq G$ with $N' \cong N$ and $L' \subseteq N'$.
  3. Remove $E_{L'}$ from G.
  4. Glue the remaining graph with R by merging each node of $v \in V_L (\subseteq V_R)$ with the corresponding node in L' and labelling it with $l_R(v)$.

- We use the constant *all*, the graph consisting of a single b-labelled node and sets of rules (to specify reduced graphs) as graph class expressions.

- We use regular expressions over the alphabet $\{copy, next\_rank, copy\_items, delete\}$ as elementary control conditions.

In the following, transformation units are presented by indicating the components with respective keywords. Trivial components (i.e. no import, no rules, the graph class expression *all*, and the control condition *true*) are omitted. A rule $r = (N, L, R)$ is represented in the form $N \to R$ where the items of N that do not belong to L are dotted. Two nodes in r are drawn with the same shape and fill style if and only if they are equal. The label $*$ is not depicted.

The transformation unit *butterfly* uses the transformation units *copy* and *next_rank* and applies them in that order arbitrarily often. The initial graph is the butterfly network of size 0. After k calls of *copy* and *next_rank* a butterfly network of size k is generated.

butterfly
    initial: ●b
    uses: *copy, next_rank*
    conds: (*copy* ; *next_rank*)*

If the input graph of the transformation unit *copy* is a butterfly network B of size k, it is copied and, additionally, a c-labelled edge is inserted between each bottom node and its copy. The output graph of *copy* without the c-labelled edges and the b-labels corresponds, roughly speaking, to a butterfly network of size $k + 1$ where the bottom nodes together with all connections to them are missing. The transformation unit *next_rank* transforms this graph into a butterfly network of size $k + 1$ by adding these missing bottom nodes and connecting edges.

The transformation unit *copy* uses the transformation units *copy_items* and *delete* which are applied exactly once in this order. If the butterfly network B is the input of the transformation unit *copy_items*, then B is transformed into a graph B' by copying each node and edge of B together with its label and generating a c-labelled edge between each node of B and its copy. The transformation unit *delete* removes each c-labelled edge provided that it connects $*$-labelled nodes. Note that the node labels and the c-labelled edges, used in *copy_items*, *delete*, and *next_rank*, serve to control rule application: They prevent undesired multiple rule applications to the same subgraph and mark subgraphs rules may be applied to. The term *reduced* in the terminal components of *copy_items*, *delete*, and *next_rank* indicates that the terminal graphs are reduced with respect to the actual set of rules. This means all the rules of the respective transformation unit are applied as long as possible.

copy
    uses: *copy_items, delete*
    conds: *copy_items ; delete*

copy_items
    rules:

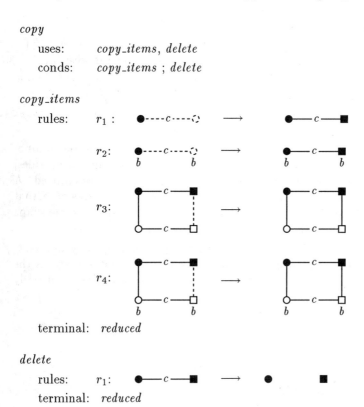

    terminal: *reduced*

delete
    rules: r_1:
    terminal: *reduced*

next_rank
   rules:

terminal: *reduced*

Figure 1 shows an interleaving sequence of *butterfly* generating the butterfly networks of size 0, 1, and 2, where $G \longrightarrow_{trut} G'$ means that $(G, G') \in SEM(trut)$ for some transformation unit *trut* and some graphs G, G'.

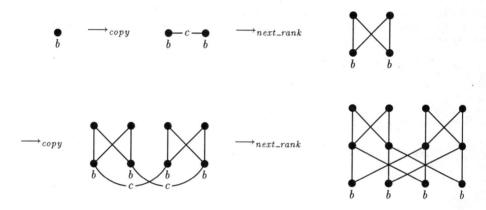

Figure 1: An interleaving sequence of *butterfly*

By induction on the length of the interleaving sequences in *butterfly* and on the size of butterfly networks (using some lemmata concerning the imported transformation units) one can prove the following correctness result.

$(\bullet b, G) \in SEM(butterfly)$ iff G is a butterfly network of some size.

4 Operations on Transformation Units

The concept of transformation units may be seen as an operation on transformation units that describes the interleaving of the semantic relations of the imported transformation units with each other and with a derivation relation. This somewhat complicated operation on binary relations is motivated by the idea that transformation units encapsulate hierarchically sets of rules and the interleaving semantics generalizes the derivation process accordingly. But there are many other operations on binary relations like union, conversion, complement, transitive closure, etc. that may be of interest in various situations. Hence, one may wonder whether and how certain operations on binary relations can be achieved by suitable operations on transformation units. We show in this section that various standard operations can be specified in terms of transformation units.

4.1 Operations Without Conversion

The definition of the interleaving semantics of transformation units is based on the union, the sequential composition and the reflexive and transitive closure of relations on graphs. If regular expressions are employed as control conditions, these operations can be modelled as constructions on transformation units in an obvious way, because the effect of regular expressions is directly related to them.

Observation
Let $\mathcal{A} = (\mathcal{G}, \mathcal{R}, \Longrightarrow, \mathcal{E}, \mathcal{C})$ be a graph transformation approach where $REG(ID) \subseteq \mathcal{C}$. Then for all $t, t' \in \mathcal{T}_\mathcal{A}$, one can construct transformation units $trans(t)$, $refl(t)$, $union(t, t')$, $sc(t, t') \in \mathcal{T}_\mathcal{A}$ such that $SEM(trans(t)) = SEM(t)^+$ [4], $SEM(refl(t)) = SEM(t) \cup \Delta\mathcal{G}$, $SEM(union(t, t')) = SEM(t) \cup SEM(t')$, and $SEM(sc(t, t')) = SEM(t) \circ SEM(t')$,

Proof. Define $trans(t)$, $refl(t)$, $union(t, t')$, $sc(t, t')$ as follows.

$trans(t)$			$refl(t)$	
uses:	t		uses:	t
conds:	$t\,;(t^*)$		conds:	$t\,|\,\epsilon$

$union(t, t')$			$sc(t, t')$	
uses:	t, t'		uses:	t, t'
conds:	$t\,|\,t'$		conds:	$t\,;\,t'$

For a regular expression $c \in \mathcal{C}$, let $L(c)$ be the language described by c. Then by definition, we get $\widehat{SEM_{E(trans(t))}}(t\,;(t^*)) = \widehat{E(trans(t))}(L(t\,;(t^*))) = \widehat{E(trans(t))}(t) \circ \widehat{E(trans(t))}(L(t^*)) = SEM(t) \circ \widehat{E(trans(t))}(L(t^*))$. By induction on the structure of the words in $L(t^*)$, we get $\widehat{E(trans(t))}(L(t^*)) = (\widehat{E(trans(t))}(t))^*$ [4]. Hence, $\widehat{SEM_{E(trans(t))}}(t\,;(t^*)) = SEM(t) \circ SEM(t)^* = SEM(t)^+$. Moreover, from the observation in 2.6 it follows that for all $(G, G') \in \widehat{SEM_{E(trans(t))}}(t\,;(t^*))$, there is an interleaving sequence in $trans(t)$ from G to G'. Hence, $SEM(trans(t)) = SEM(t)^+$. The correctness of $refl(t)$, $union(t, t')$, $sc(t, t')$ can be shown analogously. □

Remark
There is a more general version of this observation for arbitrary regular expressions.

Let $e \in REG(X)$ for $X \subseteq \mathcal{T}_\mathcal{A} \cap ID$. Let $\varrho_x \subseteq \mathcal{G} \times \mathcal{G}$ for $x \in X$, and $E_\varrho : ID \to \mathcal{P}(\mathcal{G} \times \mathcal{G})$ given by $E_\varrho(x) = \varrho_x$ for $x \in X$ and $E_\varrho(x) = \emptyset$, otherwise.

---

[4] For a binary relation R, R^+ denotes the transitive closure of R and R^* denotes its reflexive and transitive closure.

And consider the transformation unit $rho(X)$ that uses X and gets e as control condition. Then we have $SEM(rho(X)) = \widehat{E_\varrho}(L(e))$ where $L(e)$ is the language described by e.

4.2 Operations with Conversion

Under the assumption that all rules of a transformation unit are invertible and that there is a suitable control condition, the conversion and the symmetric closure of the binary relation on graphs induced by a transformation unit can also be modelled. Moreover, if one puts together the reflexive, symmetric and transitive closures, one gets a transformation unit specifying the equivalence induced by the semantic relation of a given transformation unit.

Observation

Let $\mathcal{A} = (\mathcal{G}, \mathcal{R}, \Longrightarrow, \mathcal{E}, \mathcal{C})$ be a graph grammar approach in which, for each $r \in \mathcal{R}$, there is an $r^{-1} \in \mathcal{R}$ such that $(\underset{r}{\Longrightarrow})^{-1} = \underset{r^{-1}}{\Longrightarrow}$ [5], and let $t \in \mathcal{T_A}$. Then one can construct transformation units $conv(t)$, $sym(t)$ and $equiv(t) \in \mathcal{T_A}$ such that $SEM(conv(t)) = SEM(t)^{-1}$, $SEM(sym(t)) = SEM(t) \cup SEM(t)^{-1}$, and $SEM(equiv(t)) = equiv(SEM(t))$ [5], provided that there is a control condition $C_{conv(t)} \in \mathcal{B}(\mathcal{C})$ with $SEM_{E(conv(t))}(C_{conv(t)}) = (SEM_{E(t)}(C_t))^{-1}$.

Proof (by induction on the recursion depth of t). Let $I_{conv(t)} = T_t$, $U_{conv(t)} = \{conv(t') \mid t' \in U_t\}$, $R_{conv(t)} = \{r^{-1} \mid r \in R_t\}$, $T_{conv(t)} = I_t$, and assume that $SEM_{E(conv(t))}(C_{conv(t)}) = (SEM_{E(t)}(C_t))^{-1}$ for some $C_{conv(t)} \in \mathcal{B}(\mathcal{C})$. Then by induction on the length of derivations, we get $\underset{R_{conv(t)}}{\overset{*}{\Longrightarrow}} = (\underset{R_t}{\overset{*}{\Longrightarrow}})^{-1}$. Hence, by definition of $conv(t)$, it follows that $SEM(conv(t)) = SEM(t)^{-1}$ if $U_t = \{\}$. Assume that the statement holds for all $t' \in U_t$. Then, for $G_0, \ldots, G_n \in \mathcal{G}$ we have $G_0, \ldots, G_n$ is an interleaving sequence in $conv(t)$ iff, for $i = 1, \ldots, n$, $G_{i-1} \underset{R_{conv(t)}}{\overset{*}{\Longrightarrow}} G_i$ or $(G_{i-1}, G_i) \in SEM(conv(t'))$ for some $t' \in U_t$ iff, for $i = 1, \ldots, n$, $G_i \underset{R_t}{\overset{*}{\Longrightarrow}} G_{i-1}$ or $(G_i, G_{i-1}) \in SEM(t')$, for some $t' \in U_t$ iff $G_n, \ldots, G_0$ is an interleaving sequence in t. By definition of $conv(t)$, it follows that $SEM(conv(t)) = SEM(t)^{-1}$.

Moreover, define $sym(t)$ and $equiv(t)$ as follows.

$sym(t)$		$equiv(t)$	
uses:	$union(conv(t), t)$	uses:	$trans(refl(sym(t)))$
conds:	$union(conv(t), t)$	conds:	$trans(refl(sym(t)))$

Because of the correctness of $conv(t)$ and the observation in 4.1, we get $SEM(sym(t)) = SEM(t) \cup SEM(t)^{-1}$ and $SEM(equiv(t)) = equiv(SEM(t))$. □

Remark

If C_t contains no negations and only context-free grammars as elementary

---

[5] For a binary relation R, R^{-1} denotes the conversion of R, i.e. $R^{-1} = \{(G, G') \mid (G', G) \in R\}$, and $equiv(R)$ denotes the equivalence closure of R.

control conditions (in the sense of 2.3.1), it can be shown that $C_{conv(t)}$ can be constructed from C_t such that $SEM_{E(conv(t))}(C_{conv(t)}) = (SEM_{E(t)}(C_t))^{-1}$ holds.

5 Normal Forms of Transformation Units

In this section, we present two unary operations on transformation units each of which constructs a normal form without changing the interleaving semantics. The presented normal forms give some information of how the different components of a transformation unit are related to each other. It should be noticed that the result in 2.7 is also of this type, because it presents a way to get rid of the initial and terminal graph specifications as extra components.

5.1 Encapsulating Rules

Let $\mathcal{A} = (\mathcal{G}, \mathcal{R}, \Longrightarrow, \mathcal{E}, \mathcal{C})$ be a graph transformation approach. Then, for each rule $r \in \mathcal{R}$, consider the transformation unit $encapsulate(r)$ with r as rule and regular control condition. It is easy to see that $SEM(encapsulate(r)) = \underset{r}{\Longrightarrow}$.

Based on this fact, the application of a rule $r \in \mathcal{R}$ corresponds to the call of $encapsulate(r)$. Hence, for each $t \in \mathcal{T}_\mathcal{A}$, one can construct a new transformation unit $t' \in \mathcal{T}_\mathcal{A}$ such that each rule of R_t is encapsulated in a used transformation unit of t', and t' has the same interleaving semantics as t provided that there is a suitable control condition.

To formulate this observation we need the set $FLAT(t)$ containing all transformation units occurring in the import structure of t, i.e.

$$FLAT(t) = U_t \cup (\bigcup_{t' \in U_t} FLAT(t'))$$

Observation
Let $\mathcal{A} = (\mathcal{G}, \mathcal{R}, \Longrightarrow, \mathcal{E}, \mathcal{C})$ be a graph transformation approach and let $t \in \mathcal{T}_\mathcal{A}$. Then one can construct a transformation unit $disperse(t) \in \mathcal{T}_\mathcal{A}$ with $R_{disperse(t)} = \{\}$, and, for $t' \in FLAT(disperse(t))$, either $R_{t'} = \{\}$ or $t' = encapsulate(r)$ for some $r \in R_t$ such that $SEM(disperse(t)) = SEM(t)$, provided that there is a control condition $C_{disperse(t)} \in \mathcal{B}(\mathcal{C})$ with $SEM_{E(disperse(t))}(C_{disperse(t)}) = SEM_{E(t)}(C_t)$.

Proof. Define $I_{disperse(t)} = I_t$, $U_{disperse(t)} = \{encapsulate(r) \mid r \in R_t\} \cup \{disperse(t') \mid t' \in U_t\}$, $R_{disperse(t)} = \{\}$, and $T_{disperse(t)} = T_t$.
Then it can be shown by induction on the recursion depth of t that for all $G, G' \in \mathcal{G}$, there is an interleaving sequence in t from G to G' iff there is an interleaving sequence in $disperse(t)$ from G to G'. Hence, $SEM(t) = SEM(disperse(t))$ if $SEM_{E(disperse(t))}(C_{disperse(t)}) = SEM_{E(t)}(C_t)$ for some $C_{disperse(t)} \in \mathcal{B}(\mathcal{C})$. □

Remark
If C_t contains only grammars of type 0 as elementary control conditions, it can

be shown that $C_{disperse(t)}$ can be constructed from C_t such that the condition $SEM_{E(disperse(t))}(C_{disperse(t)}) = SEM_{E(t)}(C_t)$ holds.

5.2 Flattening Transformation Units

Transformation units can be flattened meaning that if the control condition behaves properly one can get rid of the import structure by putting together all the occurring rules.

Observation
Let $\mathcal{A} = (\mathcal{G}, \mathcal{R}, \Longrightarrow, \mathcal{E}, \mathcal{C})$ be a graph transformation approach and let $t \in \mathcal{T_A}$ such that, for each $(G, G') \in SEM_{E(t)}(C_t)$, there is an interleaving sequence in t from G to G'. Then one can construct a transformation unit $flatten(t) \in \mathcal{T_A}$ with $U_{flatten(t)} = \{\}$, such that $SEM(flatten(t)) = SEM(t)$, provided that there is a control condition $C_{flatten(t)} \in \mathcal{B}(\mathcal{C})$ with $SEM_{E(flatten(t))}(C_{flatten(t)}) = SEM_{E(t)}(C_t)$.

Proof. Define $I_{flatten(t)} = I_t$, $U_{flatten(t)} = \{\}$, $R_{flatten(t)} = R_t \cup (\bigcup_{t' \in FLAT(t)} R_{t'})$, $T_{flatten(t)} = T_t$, and assume that $SEM_{E(flatten(t))}(C_{flatten(t)}) = SEM_{E(t)}(C_t)$ for some $C_{flatten(t)} \in \mathcal{B}(\mathcal{C})$.

By assumption, for all $(G, G') \in SEM_{E(t)}(C_t)$, there is an interleaving sequence in t from G to G'. Moreover, $SEM(I_{flatten(t)}) \times SEM(T_{flatten(t)}) = SEM(I_t) \times SEM(T_t)$. By induction on the recursion depth of t, it can be shown that for each interleaving sequence in t from G to G', $G \stackrel{*}{\underset{R_{flatten(t)}}{\Longrightarrow}} G'$. Altogether, we get $SEM(flatten(t)) = SEM(t)$. □

Remark
If C_t contains only context-free grammars as elementary control conditions and if for all $t' \in FLAT(t)$, $SEM(I_{t'}) = SEM(T_{t'}) = \mathcal{G}$, it can be shown that $C_{flatten(t)}$ can be constructed from C_t such that $SEM_{E(flatten(t))}(C_{flatten(t)}) = SEM_{E(t)}(C_t)$. Note that in this case, $(G, G') \in SEM_{E(t)}(C_t)$ implies that there is an interleaving sequence in t from G to G' such that all assumptions of the observation above are fulfilled.

6 Conclusion

In this paper, the notion of a transformation unit together with its interleaving semantics has been introduced, illustrated and studied with respect to operations on transformation units and some normal forms. Transformation units provide an approach-independent structuring method for building up large graph transfomation systems from small ones. The very first results indicate that transformation units may also be helpful tools for proving correctness of graph transformation systems with respect to given binary relations on graphs.

As mentioned in the introduction, transformation units are intended to be one of the basic concepts of the new graph and rule centered language GRACE which is under development by the BrAaBeLei-initiative (consisting of researchers from Bremen, Aachen, Berlin, and Leiden). The adequate use of transformation units in GRACE (and outside) requires further investigations and considerations including the following points and questions.

- Case-studies may help to gather experience with handling large sets of rules.
- Each construction in section 4 or 5 builds transformation units from transformation units where the shape of all resulting ones depends only on the imported transformation units. Hence, the construction can be described syntactically by a single transformation unit if one allows identifiers as formal parameters instead of imported transformation units. A particular result of the construction is then obtained by replacing formal parameters by actual transformation units. It may be convenient to introduce such a concept of parameterized transformation units explicitly.
- The recursion depth of transformation units and the lengths of interleaving sequences provide induction principles. Under which assumptions and how can these be turned into proof rules and a proof theory that may lead to a proof system for GRACE?
- A transformation unit describes a single binary relation on graphs by using other binary relations possibly. This excludes n-ary relations for $n \neq 2$ and sets or families of relations as results. Hence, one may wonder which further concepts of modularization should be investigated and how they may coexist with transformation units.
- And, finally, one should study how the notion of transformation units is related to the few other structuring principles for graph rewriting systems encountered in the literature like, for example, the module concepts proposed by Ehrig and Engels (see [EE93]), by Taentzer and Schürr (see [TS95]), or the notion of a transaction introduced by Schürr and Zündorf in the framework of PROGRES (see [SZ91]).

Acknowledgement. We are grateful to the two anonymous referees for their hints on possible improvements of the paper, to Frank Drewes for his suggestions concerning the environment concept in the semantics of control conditions, and to our friends in the BrAaBeLei group – in particular, Ralf Betschko, Hartmut Ehrig, Gregor Engels, Annegret Habel, Berthold Hoffmann, Jürgen Müller, Detlef Plump, Andy Schürr, and Gabriele Taentzer – for fruitful discussions on the subject of this paper.

References

[Bun79] H. Bunke. Programmed graph grammars. In Claus et al. [CER79], 155–166.

[CER79] V. Claus, H. Ehrig, G. Rozenberg, eds. Graph Grammars and Their Application to Computer Science and Biology, Lecture Notes in Computer Science 73, 1979.

[Cou90]   B. Courcelle. Graph rewriting: An algebraic and logical approach. In J. van Leeuwen, ed., Handbook of Theoretical Computer Science, volume Vol. B., 193–242. Elsevier, Amsterdam, 1990.

[DP89]   J. Dassow, G. Păun. Regulated Rewriting in Formal Language Theory, volume 18 of EATCS Monographs on Theoretical Computer Science. Springer-Verlag, 1989.

[Ehr79]   H. Ehrig. Introduction to the algebraic theory of graph grammars. In Claus et al. [CER79], 1–69.

[EE93]   H. Ehrig, G. Engels. Towards a module concept for graph transformation systems. Technical Report 93-34, Leiden, 1993.

[EH86]   H. Ehrig, A. Habel. Graph grammars with application conditions. In G. Rozenberg, A. Salomaa, eds., The Book of L, 87–100. Springer-Verlag, Berlin, 1986.

[EKR91]   H. Ehrig, H.-J. Kreowski, G. Rozenberg, eds. Graph Grammars and Their Application to Computer Science, Lecture Notes in Computer Science 532, 1991.

[ENR83]   H. Ehrig, M. Nagl, G. Rozenberg, eds. Graph-Grammars and Their Application to Computer Science, Lecture Notes in Computer Science 153, 1983.

[ENRR87]   H. Ehrig, M. Nagl, G. Rozenberg, A. Rosenfeld, eds. Graph-Grammars and Their Application to Computer Science, Lecture Notes in Computer Science 291, 1987.

[GKS91]   J.R.W. Glauert, J.R. Kennaway, M.R. Sleep. Dactl: An experimental graph rewriting language. In Ehrig et al. [EKR91], 378–395.

[Hab92]   A. Habel. Hyperedge replacement: Grammars and languages. Lecture Notes in Computer Science 643, 1992.

[HHT95]   A. Habel, R. Heckel, G. Taentzer. Graph grammars with negative application conditions. Fundamenta Informaticae, 1995. To appear.

[Him91]   M. Himsolt. Graph-Ed: An interactive tool for developing graph grammars. In Ehrig et al. [EKR91], 61–65.

[JR80]   D. Janssens, G. Rozenberg. On the structure of node-label-controlled graph languages. Information Sciences 20, 191–216, 1980.

[Kre93]   H.-J. Kreowski. Five facets of hyperedge replacement beyond context-freeness. In Z. Ésik, ed., Fundamentals of Computation Theory, Lecture Notes in Computer Science 710, 69–86, 1993.

[Kre95]   H.-J. Kreowski. Graph grammars for software specification and programming: An eulogy in praise of GRACE. In Rosselló and Valiente [RV95], 55–61.

[KR90]   H.-J. Kreowski, G. Rozenberg. On structured graph grammars, I and II. Information Sciences 52, 185–210 and 221–246, 1990.

[Kus95]   S. Kuske. Semantic aspects of the graph and rule centered language GRACE. In Rosselló and Valiente [RV95], 63–69.

[Len90]   T. Lengauer. VLSI theory. In J. van Leeuwen, ed., Handbook of Theoretical Computer Science, volume A. Elsevier Science Publishers B.V., 1990.

[LM93]   I. Litovsky, Y. Métivier. Computing with graph rewriting systems with priorities. Theoretical Computer Science 115, 191–224, 1993.

[Löw93]   M. Löwe. Algebraic approach to single-pushout graph transformation. Theoretical Computer Science 109, 181–224, 1993.

[LB93]   M. Löwe, M. Beyer. AGG — an implementation of algebraic graph rewriting. In C. Kirchner, ed., Rewriting Techniques and Applications, Lecture Notes in Computer Science 690, 451–456, 1993.

[MW91]   A. Maggiolo-Schettini, J. Winkowski. Programmed derivations of relational structures. In Ehrig et al. [EKR91], 582–598.

[Nag79]   M. Nagl. Graph-Grammatiken: Theorie, Anwendungen, Implementierungen. Vieweg, Braunschweig, 1979.

[RV95]   F. Rosselló, G. Valiente, eds. Proceedings Colloquium on Graph Transformation and its Application in Computer Science, Technical Report UIB-DMI-B-19. University of the Balearic Islands, 1995.

[SE93]   H.J. Schneider, H. Ehrig, eds. Graph Transformations in Computer Science, Lecture Notes in Computer Science 776, 1993.

[Sch91a] A. Schürr. Operationales Spezifizieren mit programmierten Graphersetzungssystemen. Deutscher Universitäts-Verlag, Wiesbaden, 1991.

[Sch91b] A. Schürr. PROGRES: A VHL-language based on graph grammars. In Ehrig et al. [EKR91], 641–659.

[Sch95]   A. Schürr. Programmed graph transformations and graph transformation units in GRACE, 1995. This volume.

[SZ91]   A. Schürr, A. Zündorf. Nondeterministic control structures for graph rewriting systems. In G. Schmidt, R. Berghammer, eds., Graph-Theoretic Concepts in Computer Science, Lecture Notes in Computer Science 570, 48–62, 1991.

[TB93]   G. Taentzer, M. Beyer. Amalgamated graph transformation systems and their use for specifying AGG – an algebraic graph grammar system. In Schneider and Ehrig [SE93], 380–394.

[TS95]   G. Taentzer, A. Schürr. DIEGO, another step towards a module concept for graph transformation systems. Electronic Notes in Theoretical Computer Science, 1995. To appear.

[Ull84]   J.D. Ullman. Computational Aspects of VLSI. Computer Science Press, Rockville, MD, 1984.

A Graph Rewriting Framework for Statecharts Semantics*

Andrea Maggiolo-Schettini[1] and Adriano Peron[2]

[1] Dipartimento di Informatica, Università di Pisa, Corso Italia 40, 56125 Pisa, Italy.
E mail: maggiolo@di.unipi.it
[2] Dipartimento di Matematica e Informatica, Università di Udine, Viale delle Scienze 206, 33100 Udine, Italy.
E mail: peron@dimi.uniud.it

Abstract. The purpose of the paper is to show that graph rewriting is a suitable environment to formalize semantics of specification languages with dynamic features. This is exemplified considering Statecharts in a variant allowing dynamic creation of processes. Graph rewriting rules give the semantics of Statecharts in such interpretation. Standard Statecharts semantics can be recovered by suitably restricting derivation sequences that describe behaviours in the dynamic semantics. More abstract semantics, better suited to study equivalences and general properties, can be obtained from graph rewriting semantics.

1 Introduction

The purpose of this paper is to show that graph rewriting is a suitable environment to formalize semantics of specification languages with dynamic features, like run-time creation of processes.

The formalism in which we are interested is Statecharts, a specification language for reactive systems introduced originally in [3]. (We shall call "Statecharts" the language and "statecharts" the specifications in the language Statecharts.) Statecharts are finite state machines having the appeal of visual formalisms like state-transitions diagrams and Petri nets, but, differently with respect to both of them, they offer facilities of hierarchical structuring of states and modularity, which allow high level description and stepwise development. A wide spectrum of semantic variants have been considered (for a survey see [1]). Here we consider a new interpretation of the formalism allowing the description of dynamic creation of processes that is not supported by previous semantics.

In [6] we have shown that statecharts with standard interpretations can be expressed by translation into graph rewriting systems and the evolution of a statechart described by means of derivation sequences. Now, in this paper, we use graph rewriting rules to give the operational semantics of Statecharts in the new dynamic interpretation. (That graph rewriting is a natural technique

---

* Research partially supported by the COMPUGRAPH Base Research Esprit Working Group n. 7183 and by Esprit BRA 8130 LOMAPS.

to express semantics of languages with dynamic creation of processes has been shown firstly in [5].) Two significant Statecharts standard semantics can be recovered by suitably restricting derivation sequences that describe behaviours in the dynamic semantics. In this sense our formulation of semantics offers a general framework for expressing, studying and comparing different semantics of Statecharts. Now, sometimes, in order to analyze the behaviour of a distributed system, more abstract models are needed which emphasize causality, conflicts and concurrency phenomena. Event structures (see [9]) can provide models with these properties. As shown in [8], there is a link between graph rewriting and event structures. So, following [8], we show how one can derive event structure semantics of Statecharts starting from the given rewriting semantics. This suggests a general methodology for system description, having graph rewriting as its basis.

In section 2 we define the syntax as well as the standard and dynamic semantics of Statecharts. In section 3 we give a translation of statecharts into graph rewriting systems and characterize dynamic and non-dynamic interpretations. In section 4 we give a partial order semantics of Statecharts induced by the graph rewriting semantics.

2 Statecharts

Statecharts is a visual specification formalism which enriches state-transition diagrams by a tree-like structuring of states, explicit representation of parallelism and communication among parallel components. States are either of type AND, called *and- states*, or of type OR, called *or-states*. An *or-state* has a privileged immediate substate (*default* state). Each transition is labelled by a set of signals enabling the performance of the transition (the *constraint*) and a set of signals which are communicated when the transition is performed (the *action*). In the original version signals, both communicated by an "external world" and due to transition performances, are broadcast and are global. Here we introduce the possibility of expressing a kind of locality of communication, namely we allow to restrict the set of states where a communicated signal can be received. Syntactically, we associate with a state the subset of signals which it can receive, intending that a signal received by a state can be sensed by all of its substates as well.

Definition 1. A *statechart* Z is a tuple

$$(B_Z, \rho_Z, \phi_Z, \delta_Z, T_Z, in_Z, out_Z, P_Z, \chi_Z, \sigma_Z), \text{ where:}$$

1. B_Z is the non-empty, finite set of *states*.
2. $\rho_Z : B_Z \to \mathcal{P}(B_Z)$ is the *hierarchy function*[3]; for $s \in B_Z$, $\rho_Z^*(s)$ denotes the least $S \subseteq B_Z$ such that $s \in S$ and $\rho_Z(s') \subseteq S$ for all $s' \in S$, and $\rho_Z^+(s)$ denotes $\rho_Z^*(s) - \{s\}$; ρ_Z describes a tree-like structure, namely:

---

[3] $\mathcal{P}(A)$ denotes the powerset of A.

(a) There exists a unique $b \in B_Z$, denoted $root_Z$, s.t. $\rho_Z^*(b) = B_Z$.
(b) $b \notin \rho_Z^+(b)$, for $b \in B_Z$.
(c) If $\rho_Z^*(b) \cap \rho_Z^*(b') \neq \emptyset$, then either $b' \in \rho_Z^*(b)$ or $b \in \rho_Z^*(b')$, for $b, b' \in B_Z$.
A state b is *basic* iff $\rho_Z(b) = \emptyset$.
3. $\phi_Z : B_Z \to \{AND, OR\}$ is the (partial) *state type function* defined only for all non-basic states.
4. $\delta_Z : B_Z \to B_Z$ is the (partial) *default function* defined only for all or-states, so that $b' = \delta_Z(b)$ implies $b' \in \rho_Z(b)$.
5. T_Z is the finite set of *transitions*.
6. $in_Z, out_Z : T_Z \to B_Z - \{root_Z\}$ are the *target* and *source functions*.
7. P_Z is the finite set of *signals*.
8. $\chi_Z : T_Z \to \mathcal{P}(P_Z) \times \mathcal{P}(P_Z)$ is the *labelling function*; the first component of $\chi_Z(t)$ is denoted by $Sig(t)$ and called the *constraint* of t, the second component of $\chi_Z(t)$ is denoted by $Act(t)$ and called the *action* of t.
9. $\sigma_Z : B_Z \to \mathcal{P}(P_Z)$ is the *locality function*.

Example 1. In Fig.1 we show an example of statechart. The graphical convention is that states are represented by boxes, and the box of a substate of a state is drawn inside the area of the box of that state. The area of an and-state is partitioned by dashed lines. Each element of the partition represents a parallel component. The default substate of an or-state is pointed to by an arrow without source. For each transition t, $Sig(t)$ and $Act(t)$ are listed below the statechart together with the locality function. The statechart in Fig.1 describes a system executing an interpretation cycle (component E of B). The system shows a *prompt* and waits for an instruction *instr* (transition t_3) and executes the instruction when required (transitions t_4 and t_5). A signal *close* causes ceasing the activity of the system interpretation cycle (transition t_2). The example may help us to give an idea of the two kinds of semantics of Statecharts we are going to formalize. One, called *non-dynamic semantics*, is based on the idea that being in an and-state means being in all its components simultaneously, and that being in an or-state means being in one and only one of its component. With reference to the example, if the system is in state W (and therefore also in states A, B, D, F, E and in one of the substates of E), when a signal *open* is communicated, t_1 is performed, state B is exited together with all its components. Now, by performing t_1 states A, B, D, E, F and G are entered anew, and thus performing t_1 acts as a restart of the system. The other semantics, called *dynamic semantics*, provides that when transition t_1 is performed, state E is not exited together with F and reentering state B creates a new active copy of state E which is supposed to evolve in parallel with the existing one. In this case, performing transition t_1 acts as a dynamic creation of a process executing another interpretation cycle (something like a forking of processes). Being in a certain state, the system may react to broadcast signals that are in the locality (defined by the locality function). In the case of the dynamic semantics, copies of the same state may be stimulated with different set of signals.

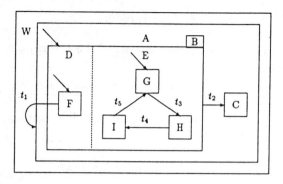

$t_1, \{open\}, \emptyset \ t_2, \{close\}, \emptyset \ t_3, \emptyset, \{prompt\} \ t_4, \{instr\}, \emptyset \ t_5, \emptyset, \{exec\}$
$\sigma(E) = \{instr, prompt, exec\} \ \sigma(A) = \{close, open\} \ \sigma(X) = \emptyset \text{ for } X \notin \{E, A\}$

Fig. 1. A statechart.

We introduce now the non-dynamic semantics. When a state b is entered a subset of its substates are entered consistently with the requirement that if an and-state (resp.: an or-state) is entered, then all (resp.: exactly one) of its immediate substates are entered. In particular, one of these sets of states is that induced by the default function and called *default closure of b*. For $K \subseteq B_Z$, the *default closure of K*, denoted by $\Downarrow_Z K$, is the least superset of K such that

1. $\rho_Z(d) \subseteq \Downarrow_Z K$, for any and-state $d \in \Downarrow_Z K$.
2. $\delta_Z(d) \in \Downarrow_Z K$, for all or-states $d \in \Downarrow_Z K$, such that $\rho_Z(d) \cap K = \emptyset$.

A *configuration* is a pair $C = (D, \mathcal{E})$, where $D \subseteq B_Z$ and $\mathcal{E} : D \to \mathcal{P}(P_Z)$, such that $\mathcal{E}(b) \subseteq \sigma_Z(b)$, for $b \in D$, is the *environment function*, allowing to establish the signals that are available in each state of D. Actually, for $e \in P_Z$ and $b \in B_Z$, assume $loc(e, b) = b$ if $e \in \sigma_Z(b)$ and $loc(e, b) = b'$ where b' is the lowest ancestor of b such that $e \in \sigma_Z(b')$, otherwise. The state $loc(e, b)$ is the state closest to b, upward in the hierarchy, to which signal e is associated by the locality function. Now, signal e is *available at* state b iff $e \in \mathcal{E}(loc(e, b))$.

Given a configuration $C = (D, \mathcal{E})$, the set of signals given by $\mathcal{E}$ enable performing from C a set of transitions whose source states are in D and whose constraining signals are available at source states (a *microstep*). For a transition t, $lca_Z(t)$ is the lowest common ancestor of $in_Z(t)$ and $out_Z(t)$ such that $in_Z(t), out_Z(t) \in \rho_Z^+(lca_Z(t))$. A signal e in the constraint of a transition t is available at $out_Z(t)$ if e is associated by $\mathcal{E}$ with the lowest ancestor of $lca_Z(t)$ having signal e in its local environment. Performing a transition t causes exiting all the entered states up to $lca_Z(t)$; in particular, exiting a component of an and-state means exiting all of its parallel components simultaneously. Now, for $b' \in \rho_Z^*(b)$, let us denote with $B_{b'}^b$ the set containing b, b' and the substates of b having b' as a substate. The set of states entered due to a transition is $\Downarrow_Z B_{in_Z(t)}^{lca_Z(t)}$. Transitions in a microstep must be pairwise *consistent* in the sense that they must act on different parallel components of the statechart, namely, two transitions t and t' are *consistent* if $\rho_Z^*(lca_Z(t)) \cap \rho_Z^*(lca_Z(t')) = \emptyset$ and the lowest common ancestor of $lca_Z(t)$ and $lca_Z(t')$ is an and-state. Performing a microstep Ψ may change the set of entered states and give a new environment function. A signal e, broadcast when a transition t is performed, is available at any state b such that $loc(e, b) = loc(e, lca(t))$. So, after performing Ψ, a new set of transitions might be enabled and a chain reaction might occur. The sequence of microsteps from $C = (D, \mathcal{E})$ represents an instantaneous reaction triggered by $\mathcal{E}$. To ensure boundeness of such reaction one may require either that a transition may occur at most once in the sequence (see [7]) or that every two transitions in the sequence are consistent (*standard semantics* in [4]). In the former case sequences of transitions can be performed in the same step, whereas in the latter case each pair of performed transitions affect components that are in parallel.

Definition 2. A transition t is *enabled* in $C = (D, \mathcal{E})$ iff $out_Z(t) \in D$ and e is available at $out_Z(t)$, for all $e \in Sig(t)$. A *microstep* from C is a set Ψ of pairwise consistent transitions enabled in C.
The configuration *reached from C by* a microstep Ψ is $C' = (D', \mathcal{E}')$, with:

1. $D' = (D - \bigcup_{t \in \Psi} \rho_Z^{\pm}(lca_Z(t))) \cup \bigcup_{t \in \Psi} \Downarrow_Z B_{in_Z(t)}^{lca_Z(t)}$.
2. $\mathcal{E}' : D' \to \mathcal{P}(P_Z)$ is as follows:
   $\mathcal{E}'(b) = \emptyset$ if $b \in \bigcup_{t \in \Psi} \Downarrow_Z B_{in_Z(t)}^{lca_Z(t)}$
   $\mathcal{E}'(b) = \mathcal{E}(b) \cup \bigcup_{t \in \Psi} \{e \in Act(t) : b = loc(e, lca_Z(t))\}$, otherwise.

A (possibly null) sequence $\mathcal{S} = \Psi_0 \Psi_1 \ldots \Psi_n$ of pairwise disjoint microsteps is a *step* from a configuration C to a configuration C' iff there exists a sequence $C_0 \ldots C_{n+1}$ of configurations such that $C_0 = C$, $C_{n+1} = C'$, and C_{i+1} is reached from C_i by microstep Ψ_i, for $1 \leq i \leq n$.
A step $\Psi_0 \Psi_1 \ldots \Psi_n$ is *standard* iff transitions in $\bigcup_{i=1}^n \Psi_i$ are pairwise consistent.
A step (resp.: standard step) $\mathcal{S}$ from C to C' is *maximal* iff, for each microstep Ψ from C', $\bigcup_{i=0}^n \Psi_i \cap \Psi \neq \emptyset$ (resp.: either $\bigcup_{i=0}^n \Psi_i \cap \Psi \neq \emptyset$ or transitions in $\bigcup_{i=0}^n \Psi_i \cup \Psi$ are not pairwise consistent).

Now, we can assume that a statechart represents a system that is stimulated by an "external world" at each instant of a discrete time domain. The *behaviour* of a statechart starts (at time 0) from an initial configuration $C_0 = (D_0, \mathcal{E}_0)$. The environment function $\mathcal{E}_0$ describes the signal given by the external world. The statechart reacts instantaneously (a maximal step) getting to a configuration $C'_0 = (D'_0, \mathcal{E}'_0)$. At time 1 the external world supplies the set of states D'_0 with new signals (environment function $\mathcal{E}_1$) and a new step may be performed from the configuration $C_1 = (D_1, \mathcal{E}_1)$, with $D_1 = D'_0$, and so on.

Definition 3. A sequence of steps $\mathcal{S}_0 \ldots \mathcal{S}_n$, is a *behaviour* (resp.: *standard behaviour*) from an initial configuration C iff there exists a sequence of configurations $C_0 = (D_0, \mathcal{E}_0) \ldots C_n = (D_n, \mathcal{E}_n)$, where $C_0 = C$ and $\mathcal{S}_i$ is a maximal step (resp.: maximal standard step) from $C_i = (D_i, \mathcal{E}_i)$ to the configuration $C'_i = (D'_i, \mathcal{E}'_i)$ with $D'_i = D_{i+1}$, for all $0 \leq i \leq n$, and $\mathcal{S}_n$ is a step (resp.: standard step) from C_n.

Example 2. The pair $C_0 = (D_0, \mathcal{E}_0)$, with $D_0 = \{W, A, B, D, E, F, G\}$, $\mathcal{E}_0(A) = \{open\}$ and $\mathcal{E}_0(b) = \emptyset$ for $b \neq A$, is an initial configuration for the statechart of Fig.1. The sequence of microsteps $\{t_1\}, \{t_3\}$ is a maximal step from C_0 to $C'_0 = (D'_0, \mathcal{E}'_0)$, with $D'_0 = \{W, A, B, D, E, F, H\}$, $\mathcal{E}'_0(E) = \{prompt\}$ and $\mathcal{E}'_0(b) = \emptyset$ for $b \neq E$. Note that $\mathcal{E}'_0(A) = \emptyset$ (whereas $\mathcal{E}_0(A) = \{open\}$) since state A is exited and then entered anew by performing t_1.

Consider now the environment $\mathcal{E}_1 : D_1 \to \mathcal{P}(P_Z)$ with $D_1 = D'_0$ and such that $\mathcal{E}_1(E) = \{instr\}$, $\mathcal{E}_1(b) = \emptyset$ for $b \neq E$. Then the sequence of microsteps $\{t_4\}, \{t_5\}, \{t_3\}$ is a maximal step from the configuration $(D_1, \mathcal{E}_1)$ to the configuration $(D'_1, \mathcal{E}'_1)$ with $D'_1 = D_1$, and $\mathcal{E}'_1(E) = \{instr, prompt, exec\}$ and $\mathcal{E}'_1(b) = \emptyset$ for $b \neq E$.

The idea of the dynamic semantics is that statechart states are task descriptions and entering a state means activating a task. As, in general, one may want to activate more copies of the same task, one may need to have more instances of the same state to be simultaneously entered. In order to obtain this, we modify the effect of performing a transition t, namely we assume that it causes exiting only the set of states up to the source of the transition (and not the set of states up to $lca_Z(t)$). Since now more copies of the same state can be entered simultaneously, it is natural to describe the structure of entered states by a graph[4] having nodes labelled by states and whose edges model the hierarchy of states.

Definition 4. A *(dynamic) configuration* is a pair $C = (G, \mathcal{ED})$, where:

1. G is a connected graph over B_Z having exactly one node labelled by $root_Z$ and such that, for each edge $a \in E_G$, $l_G(tg_G(a)) \in \rho_Z(l_G(sc_G(a)))$.
2. $\mathcal{ED} : V_G \to \mathcal{P}(P_Z)$ is a function such that $\mathcal{ED}(v) \subseteq \sigma_Z(l_G(v))$, for $v \in V_G$.

Note that the structure imposed on B_Z forces a tree structure also on G (we shall call G a *tree-graph*). A tree-graph H is *induced* by a set of symbols $R \subseteq B_Z$, with ρ_Z restricted to R describing a tree, if V_H is in a one-to-one correspondence with R. A configuration $C = (G, \mathcal{ED})$ is *initial* if G is induced by $\Downarrow_Z \{root_Z\}$.

Different nodes with the same label represent different instances of the same state in the configuration. If v and v' are nodes with the same label, one may have $\mathcal{ED}(v) \neq \mathcal{ED}(v')$. In particular, the external world may prompt different instances of the same state selectively. Differently from the standard case, the same transition can be performed from different instances of the same state.

Definition 5. For a configuration $C = (G, \mathcal{ED})$, a *transition occurrence* is a pair $(v, t) \in V_G \times T_Z$ with $l_G(v) = out_Z(t)$. A transition occurrence (v, t) is *enabled* in C if $e \in \mathcal{ED}(v')$ with v' the node in the path from the root of G to v such that $l_G(v') = loc(e, out_Z(t))$, for all $e \in Sig(t)$. Two transition occurrences (v, t) and (v', t') are *consistent* if v and v' are not connected by a path in G.

---
[4] For us a (node labelled) graph over A is a tuple (V, E, sc, tg, l) with V the set of nodes, E the set of edges, $sc, tg : E \to V$ giving source and target of an edge, respectively, $l : V \to A$ labelling vertices over A.

In order to describe how a configuration is changed by performing a transition, we introduce some basic operations on tree-graphs.

For a tree-graph G and a subtree-graph H of G, $K = G - H$ is the sub-tree-graph of G with $V_K = V_G - V_H$ and $E_K = \{e \in E_G : sc_G(e), tg_G(e) \in V_K\}$.

For graphs $G, H_1, \ldots, H_k$ and partial injective maps $f_i : V_G \to V_{H_i}$ such that $l_G(v) = l_{H_i}(f_i(v))$, for $v \in V_G$ and f_i defined for v, $+_{f_1,\ldots,f_k}(G, H_1, \ldots, H_k)$ is the graph (V, E, sc, tg, l) with:

1. $V = \{0\} \times V_G \cup \bigcup_{1 \leq i \leq k}(\{i\} \times V_{H_i} - f_i(V_G))$
2. $E = \{0\} \times E_G \cup \bigcup_{1 \leq i \leq k}(\{i\} \times E_{H_i} - \{a \in E_{H_i} :$ there is $a' \in E_G$ s.t. $f_i(sc_G(a')) = sc_{H_i}(a)$ and $f_i(tg_G(a')) = tg_{H_i}(a)\})$
3. $sc((i, a)) = \begin{cases} (0, sc_G(a)) & \text{if } i = 0 \\ (0, sc_G(f_i^{-1}(a))) & \text{if } i \neq 0 \text{ and } f_i^{-1}(a) \text{ is defined} \\ (i, sc_{H_i}(a)) & \text{otherwise.} \end{cases}$
4. Functions tg and l are defined analogously.

A node $v \in V_{+_{f_1,\ldots,f_k}(G,H_1,\ldots,H_k)}$ comes from a node $v' \in V_G$ (resp.: $v' \in V_{H_i}$) if $v = (0, v')$ (resp.: $v = (i, v')$). The same notion can be extended in the obvious way to a node "coming from a node of a graph" to which a sequence of graph compositions $+$ are applied.

Given a transition occurrence (v, t) enabled in a configuration $C = (G, \mathcal{ED})$, the set of state instances (i.e. nodes of G) exited by performing the transition t is the subtree of G rooted in v. The performance of t causes entering instances of the states $\Downarrow_Z B_{in_Z(t)}^{lca_Z(t)}$. Note that if two transition occurrences are consistent, performing one of them does not causes exiting the source state instance of the other, and so they may be performed simultaneously.

Definition 6. A *(dynamic) microstep* Ψ from a configuration C is a set of pairwise consistent transition occurrences enabled in C. The configuration *reached* by $\Psi = \{(v_1, t_1), \ldots, (v_k, t_k)\}$ is $C' = (G', \mathcal{ED}')$, where:

1. $G' = +_{f_1,\ldots,f_k}(G - G_{v_1} - \ldots - G_{v_k}, H_1, \ldots, H_k)$, where G_{v_i} is the subtree-graph of G rooted in v_i, $f_i = \{(v_i, \bar{v}_i)\}$, $\bar{v}_i$ is the vertex of G in the path from $root_G$ to v_i labelled by $lca_Z(t_i)$ and H_i is the tree-graph induced by $\Downarrow_Z B_{in_Z(t_i)}^{lca_Z(t_i)}$, for $1 \leq i \leq k$.
2. $\mathcal{ED}' : V_{G'} \to \mathcal{P}(P_Z)$ is as follows:
   $\mathcal{ED}'((0, v)) = \mathcal{ED}(v) \cup \bigcup_{(v',t) \in \Psi}\{e \in Act(t) : l_G(v) = loc(e, out_Z(t))$ and there is a path in G from v to $v'\}$,
   $\mathcal{ED}'((i, v)) = \emptyset$, for $i \neq 0$.

A node $v' \in V_{+_{f_1,\ldots,f_k}(G-G_{v_1}-\ldots-G_{v_k}, H_1, \ldots, H_k)}$ is added by the transition occurrence $(v, t) \in \Psi$ if $f_j = \{(v, \bar{v})\}$, for some $1 \leq j \leq k$ and v' comes from a node of H_j. A sequence $\Psi_0 \ldots \Psi_k$ of microsteps, where Ψ_i is a microstep from a configuration $(G_i, \mathcal{ED}_i)$ to $(G_{i+1}, \mathcal{ED}_{i+1})$ ($0 \leq i \leq k$), is a *step* if there is no causal cycle among transition occurrences, namely, there is not a sequence $(v_1, t_1) \in \Psi_{m_1}, \ldots, (v_s, t_s) \in \Psi_{m_s}$ with $t_1 = t_s$, $1 \leq m_1 < \ldots < m_s \leq k$ and such that either v_{i+1} is added by (v_i, t_i) or comes from a node added by (v_i, t_i). The sequence above is a *standard step* if for each $(v,t) \in \bigcup_{0 \leq i \leq k} \Psi_i$, v comes from a node of G. The notions of *behaviour* and *standard behaviour* are analogous to those of the non-dynamic case.

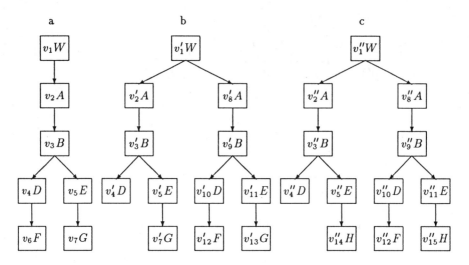

Fig. 2. Configurations.

Since there cannot be any causal cycle among transition occurrences in a step, two occurrences of the same transition occur in a step only if they affect different insances of the same state (it is the natural extension of boundeness of step imposed in Def.2 in the non-dynamic case).

Example 3. The pair $C_0 = (G_0, \mathcal{ED}_0)$, where G_0 is the graph of Fig.2.a and $\mathcal{ED}_0(v_2) = \{open\}$ and $\mathcal{ED}_0(v) = \emptyset$ for $v \neq v_2$ is an initial configuration graph for the statechart of Fig.1. The set $\{(v_6, t_1)\}$ is a microstep from C_0 to $C'_0 = (G'_0, \mathcal{ED}'_0)$, where G'_0 is the graph in Fig.2.b and $\mathcal{ED}'_0(v'_2) = \{open\}$, $\mathcal{ED}'_0(v) = \emptyset$ for $v \neq v'_2$. The set $\{(v'_7, t_3), (v'_{13}, t_3)\}$ is a microstep from C'_0 to $C''_0 = (G''_0, \mathcal{ED}''_0)$ where G''_0 is the graph in Fig.2.c and $\mathcal{ED}''_0(v''_2) = \{open\}$, $\mathcal{ED}''_0(v''_5) = \mathcal{ED}''_0(v''_{11}) = \{prompt\}$ and $\mathcal{ED}''_0(v) = \emptyset$ otherwise. The sequence of microsteps $\{(v_6, t_1)\}, \{(v'_7, t_3), (v'_{13}, t_3)\}$ is a maximal step from C_0 to C''_0. Now, we have two interpretation processes simultaneously active and ready to execute an instruction. If the next step is performed starting from the configuration $C_1 = (G''_0, \mathcal{ED}_1)$ with $\mathcal{ED}_1(v''_5) = \{instr\}$ and $\mathcal{ED}_1(v) = \emptyset$ otherwise (in particular $\mathcal{ED}_1(v_{11}) = \emptyset$), then only one of the two interpretation cycles evolves. The external world can prompt different instances of the same state selectively, whereas internal communication cannot distinguish between different state instances.

3 Rewriting Semantics of Statecharts

We want now to translate statecharts into graph rewriting systems with the idea of characterizing different semantics by means of suitable restrictions on derivation sequences. In [8] a notion of concrete graph rewriting is considered. This approach is closely related to the algebraic double pushout approach, but avoids the difficulties of using graphs "up to isomorphism" when one wants to relate derivations of graph rewriting to event structures (see [2]). So, we use concrete graph rewriting, referring to [8] for definitions. To a configuration $C = (G, \mathcal{ED})$ a *configuration graph* corresponds. Since G is a graph, we simply link to a node v of G a node for each signal in $\mathcal{ED}(v)$.

We need a preliminary definition. Let $C = (G, \mathcal{ED})$ be a configuration. For $v \in V_G$ and $Q \subseteq P_Z$, let $Env(v, Q)$ be the graph defined as follows:

1. $V_{Env(v,Q)} = Q \cup \{l_G(v)\}$ (assume that $l_G(v) \notin Q$).
2. $E_{Env(v,Q)} = \{(v, e) : e \in Q\}$;
3. $sc_{Env(v,Q)}(v, e) = v$; $tg_{Env(v,Q)}(v, e) = e$.
4. $l_{Env(v,Q)}$ is the identity function.

A *graph for a configuration* $C = (G, \mathcal{ED})$ (with $V_G = \{v_1 \ldots v_n\}$) is

$$Conf(C) = +_{f_1 \ldots f_n}(G, Env(v_1, \mathcal{ED}(v_1)), \ldots, Env(v_n, \mathcal{ED}(v_n)))$$

with $f_i = \{(v_i, l_G(v_i))\}$, for $1 \leq i \leq n$.

The graph for the configuration $(G, \mathcal{ED})$ of Ex.3 is shown in Fig.3.a. Note that $Env(v_2, \mathcal{ED}(v_2))$ is a graph consisting of two nodes labelled by A and *open*, respectively, and an edge from the former to the latter. Instead, $Env(v_i, \mathcal{ED}(v_i))$ for $i \neq 2$, is a graph consisting of a single node labelled by $l_G(v_i)$.

To a statechart a set of productions corresponds. Of these productions a set is in a one-to-one correspondence with the set of the statechart transitions. For a transition t, we denote with p_t the *transition production* for t. Production p_t is suitably obtained from two subproductions: the *structural production* $S(p_t)$ and the *communication production* $C(p_t)$. The structural production simulates entering and exiting states as a consequence of performing t. The communication production verifies enabling conditions required for performing t and simulates signal broadcast.

The structural production $S(p_t)$ is (L, K, R) with $L \supset K \subset R$, where:

1. L is a graph induced by $B_{out_Z(t)}^{lca_Z(t)}$.
2. K is the graph induced by $B_{b'}^{lca_Z(t)}$ with $out_Z(t) \in \rho_Z(b')$ and $K \subset L$;
3. R is a graph isomorphic to $+_f(K, H)$, where H is the graph induced by $\Downarrow_Z B_{in_Z(t)}^{lca_Z(t)}$, $f = \{(v, v') : l_K(v) = l_H(v') = lca_Z(t)\}$, and $K \subset R$.

The communication production $C(p_t)$ is (L, K, R) with $L \supset K \subseteq R$, where:

1. $L = +_{f_1, \ldots, f_n}(G, Env(v_1, Q_1), \ldots, Env(v_n, Q_n))$, where:
   G is the graph induced by $\bigcup_{e \in Sig(t)} B_{out_Z(t)}^{loc(e, out_Z(t))} \cup \bigcup_{e \in Act(t)} B_b^{loc(e, lca_Z(t))}$,
   $V_G = \{v_1, \ldots, v_n\}$, $Q_i = \{e \in Sig(t) : l_G(v_i) = loc(e, out_Z(t))\}$, b is the state such that $out_Z(t) \in \rho_Z(b)$ and $f_i = \{(v_i, l_G(v_i))\}$, for $1 \leq i \leq n$.

2. K is the graph $+_{f_1,\ldots,f_m}(G, Env(v_1, Q_1), \ldots, Env(v_m, Q_m))$ where:
   G is the graph induced by $\bigcup_{e \in Sig(t)} B_b^{loc(e, out_Z(t))} \cup \bigcup_{e \in Act(t)} B_b^{loc(e, lca_Z(t))}$,
   $V_G = \{v_1, \ldots, v_n\}$, $Q_i = \{e \in Sig(t) : l_G(v_i) = loc(e, b)\}$, b is the state such that $out_Z(t) \in \rho_Z(b)$ and $f_i = \{(v_i, l_G(v_i))\}$, for $1 \leq i \leq n$.
3. $R \supseteq K$ is a graph isomorphic to $+_{f_1,\ldots,f_n}(K, Env(v_1, Q_1), \ldots, Env(v_n, Q_n))$, where: $V_G = \{v_1, \ldots, v_n\}$,
   $Q_i = \{e \in Act(t) : l_G(v_i) = lca(e, lca_Z(t))\}$, $f_i = \{(v_i, l_G(v_i))\}$, for $1 \leq i \leq n$.

The *transition production* $p_t = (L, K, R)$ for a transition t is obtained from $S(p_t) = (L_S, K_S, R_S)$ and $C(p_t) = (L_C, K_C, R_C)$ in the following way: $L = +_{f_L}(L_S, L_C)$, $K = +_{f_K}(K_S, K_C)$ and $R = +_{f_R}(R_S, R_C)$, where f_L (resp.: f_K, f_R) is the correspondence between nodes in L_S and L_C (resp.: in K_S and K_C, in R_S and R_C) having the same label. In Fig.3 the production $p_{t_1} = (L, K, R)$ for the transition t_1 of Fig.1 is shown: Fig.3.b (resp.: Fig.3.c, Fig.3.d) shows the graph L (resp.: K, R).

We have a set of auxiliary productions called *delete-state productions*. A delete-state production for a state b, denoted by pds_b, when applied, removes a (leaf) node labelled by b from the configuration graph. The *delete-state production* for a state b is $pds_b = (L, K, R)$, with $L \supset K = R$, where L is the graph induced by $\{b, b'\}$, with $b \in \rho_Z(b')$ and $K = R$ is the graph induced by $\{b\}$.

We have also a set of auxiliary productions called *delete-signal productions*. A delete-signal production for a signal $e \in \sigma_Z(b)$, denoted by $pdsi_{e,b}$, when applied, removes from the configuration graph a node labelled by e and connected to a node labelled by b. The *delete-signal production* for a state b and a signal $e \in \sigma_Z(b)$ is $pdsi_{e,b} = (L, K, R)$, with $L \supset K = R$, where $L = Env(v, \{e\})$ with $\{v\} = V_K$ and $K = R$ is the graph induced by $\{b\}$.

Finally, we have a set of *external-world productions* which simulate the prompts from the external world. The external-world production for a state b and a signal $e \in \sigma_Z(b)$ is $pew_{e,b} = (L, K, R)$, with $L = K \subset R$, where $K = L$ is the graph induced by $\{b\}$ and $R = Env(v, \{e\})$ with $\{v\} = V_K$.

Now, we characterize sequences of derivations corresponding to steps and standard steps, and we call them *rewriting steps* and *standard rewriting steps*, respectively. The change of environment (prompts from the external world) which takes place in between two consecutive steps is simulated by a sequence of derivations by delete-signal and external-world productions. The two boundedness criteria introduced for steps and standard steps are naturally expressed in terms of sequential dependence of derivations. So, *rewriting behaviours* are sequences of maximal rewriting steps, alternated with sequences of derivations by delete-signal and external-world productions.

As usual, Church-Rosser properties are exploited for defining a notion of equivalence among derivation sequences, namely two derivation sequences are *equivalent* if they can be reduced one to the other by a suitable sequence of swaps of pairs of consecutive sequential independent derivations. For a derivation sequence s, a derivation d *enables* a derivation d', if there exists a derivation sequence $s' = d_1 \ldots d_n$ equivalent to s such that d corresponds to d_k, d' corresponds to d_m (with $k < m$) and d_i, d_{i+1} are sequential dependent deriva-

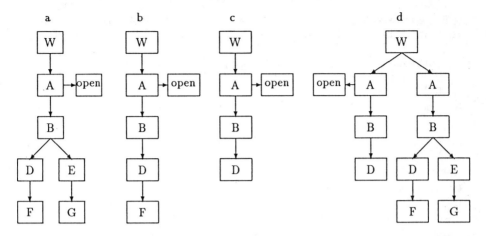

Fig. 3. A configuration (a) and the production for transition t_1 (b,c,d).

tions, for all $k \leq i \leq m - 1$. Derivations by a set U of transition productions are *parallel* in a derivation sequence s iff there exists a derivation sequence $s' = G_0 \stackrel{p_1}{\Longrightarrow} \ldots \stackrel{p_n}{\Longrightarrow} G_n$ equivalent to s such that there are $k \leq m \leq n$ and $p_i \in U$ iff $k \leq i \leq m$ and for each permutation π of indices $k \ldots m$, there is a derivation sequence $G_{k-1} \stackrel{S(p_{\pi(k)})}{\Longrightarrow} \ldots \stackrel{S(p_{\pi(m)})}{\Longrightarrow} G_m$ equivalent to $G_{k-1} \stackrel{S(p_k)}{\Longrightarrow} \ldots \stackrel{S(p_m)}{\Longrightarrow} G_m$.

Now, derivations by transition productions which are parallel in a derivation sequence correspond to performances of transition occurrences which are consistent. Note that we have to consider only derivation dependences originated from the structural part of productions disregarding dependences originated by the communication part.

Definition 7. A sequence s of transition and delete derivations[5] is a *rewriting step* (resp.: *standard rewriting step*) iff

1. Each delete derivation enables a transition derivation.
2. Transition derivations by the same transition production (resp.: all of transition derivations) are parallel.

A rewriting step (resp.: standard rewriting step) s from G is *maximal* iff each derivation sequence $s.s'$ (with s' a nonempty derivation sequence) is not a rewriting step (resp.: standard rewriting step).

A derivation sequence s from a configuration graph G is a *rewriting behaviour* (resp.: *standard rewriting behaviour*) iff $s = s_1.w_1.s_2.w_2 \ldots w_{n-1}.s_n$, where s_i is a maximal rewriting step (resp.: maximal standard rewriting step), w_i is a sequence of delete-signal and external-world derivations, for $1 \leq i < k$, and s_n

---

[5] For a transition (resp.: a communication, an external-world, a delete- state, a delete-signal, a delete) derivation we shall mean a derivation by a transition (resp.: a communication, an external-world, a delete- state, a delete-signal, a delete-state or delete-signal) production.

is a (possibly empty) rewriting step (resp.: standard rewriting step). The set of rewriting behaviours from an initial configuration graph G is written $\mathcal{B}(G)$.

Due to condition 1 of Def.7, deleting production can be applied only to enable the application of a transition production, namely to remove, from the configuration graph, the instances of substates of the source of the corresponding transition (and related signals). On (standard) rewriting steps, condition 2 imposes boundeness constraints corresponding to those of Def.6 on (standard) steps (i.e. the absence of dependence cycles among transition occurrences).

We establish now the formal correspondence between steps from a configuration C and rewriting steps from a graph $Conf(C)$. In the considered translation, transition derivations may attach more copies of the same signal to the environment local to a vertex. So, we introduce a notion of correspondence between graphs and configurations which allows to forget multiple copies of a signal. (This could be avoided by complicating the structure of the configuration graph.)

Given a configuration $C = (G, \mathcal{ED})$, a graph H *corresponds to* C if:

1. The subgraph H' of H induced by nodes labelled by elements of B_Z is isomorphic to G (let $f : V_G \to V_{H'}$ be the one-to-one correspondence).
2. $\mathcal{ED}(v) = \{l_H(v') \in P_Z : \text{there is } e \in E_H \text{ s.t. } sc_H(e) = f(v), tg_H(e) = v'\}$.

In order to simulate a microstep (i.e. the simultaneous performance of a set of transitions) we exploit parallelism theorems of the adopted algebraic approach to graph replacement (see [8]). In this case, a derivation by means of a sum of transition productions (when independence conditions allow it) corresponds to performing simultaneously a set of transitions.

A sequence of microsteps $\Psi_0 \ldots \Psi_k$ from C to C' *corresponds* to a sequence s of derivations from G to G' if:

1. C corresponds to G and C' corresponds to G'.
2. $s = s_0 \ldots s_k$ and s_i is a sequence of derivations where one derivation is by $p_{t_{1_i}} + \ldots + p_{t_{m_i}}$ with $\Psi_i = \{t_{1_i} \ldots t_{m_i}\}$ all the others are delete derivations.

Theorem 8. *Each maximal (standard) step from a configuration C has a corresponding maximal (standard) rewriting step from any configuration graph G that corresponds to C.*
For each maximal (standard) rewriting step s from a configuration graph G there is a maximal (standard) rewriting step s' equivalent to s and a maximal (standard) step from C which corresponds to s', for any configuration C such that G corresponds to C.

Non-dynamic semantics of Def.2 can be obtained by considering sequences of derivations where a transition derivation removing the vertex labelled by the source of the transition t is followed by deletions, by means of delete-state derivations, of all the vertices labelled by the ancestors of the source of the transition up to $lca_Z(t)$.

A delete-state derivation d by the production $p = (L, K, R)$ in a derivation sequence s *synchronizes* with the transition derivation d' in s by the transition production p' with $S(p') = (L', K', R')$ in s iff there exists a derivation sequence $s' = d_1 \ldots d_n$ equivalent to s such that d' corresponds to d_k, d corresponds to d_m ($1 \leq k < m$), there exists a morphism $g : L \to K'$, $d_{k+1} \ldots d_m$ is a sequence of delete derivations and derivation d_i enables d_m for all $1 \leq k < m$. Moreover, the transition derivation d' is *maximally synchronized* in s if, for each delete state production $p'' = (L'', K'', R'')$ such that there is a morphism $h : L'' \to K'$, there is a delete derivation d'' by p'' in s which synchronizes with d'.

Definition 9. A sequence of derivations s is a *synchronized* (resp.: *standard synchronized*) *rewriting step* iff:

1. There are not derivations by the same production (resp.: all transition derivations are parallel).
2. Each delete-state derivation either synchronizes with a transition derivation or enables a derivation which synchronizes with a transition derivation or enables a transition derivation.
3. Each delete-signal derivation enables a delete-state derivation.
4. Each transition derivation in s is maximally synchronized.

A notion of *(standard) synchronized rewriting behaviour* can be defined as a sequence of (standard) synchronized rewriting steps analogously to the notion of (standard) rewriting behaviour of Def.7.

Proposition 10. *If s is a standard synchronized rewriting step, each pair of (different) transitions corresponding to transition derivations in s are consistent.*

The next theorem establishes a correspondence between the semantics and the standard semantics of Def.2 and the given characterizations in terms of synchronized rewriting behaviours. In the following, the notions of configuration, step and standard step are the ones provided for non-dynamic semantics. We need to redefine the notion of correspondence between configurations (in the non-dynamic case) and configuration graphs. The notion of correspondence for sequences of derivations and sequences of microsteps from a (non-dynamic) configuration C is similar to that of the dynamic case.

A configuration $C = (D, \mathcal{E})$ *corresponds* to a configuration graph G if:

1. The subgraph H of G induced by nodes labelled by elements of B_Z is isomorphic to the tree-graph induced by set of states D.
2. $\mathcal{E}(b) = \{l_G(v') \in P_Z : \text{there is } e \in E_G \text{ s.t. } sc_G(e) = v, tg_G(e) = v', l_G(v) = b\}$.

Theorem 11. *Each maximal (standard) step from a configuration C has a corresponding maximal synchronized (standard) rewriting step from any configuration graph G such that G corresponds to C.*
For each maximal synchronized (standard) rewriting step s from a configuration graph G there is a maximal synchronized rewriting (standard) step s' equivalent to s and a maximal (standard) step from C which corresponds to s', for any configuration C which corresponds to G.

4 Partial Order Semantics of Statecharts

Following [8], we associate an event structure (see [9]) with the set of rewriting behaviours starting from a given initial configuration graph. An event structure consists of a set of events, a partial order and a conflict relation (compatible with the partial order) on the set of events. When the behaviour of a system is specified by an event structure, events are the events that may possibly occur, the partial order describes causality between events and the conflict relation describes incompatibility of events. Events that neither causally depend nor are in conflict may occur in parallel or in an arbitrary order as well.

Let us consider the set of events in our case. In [8], on derivation sequences an equivalence relation is defined where all derivation sequences that end with the same derivation (up to permutation of independent derivations) become equivalent. In this way, an event represents a derivation. We do not want to have events corresponding to all direct derivations but only events corresponding to transition derivations. So, we consider equivalent two behaviours that end with the same *single rewriting step*, namely a sequence $d_1 \ldots d_k$ where d_i is a delete production for $1 \leq i \leq k-1$ and d_k is a transition derivation.

Definition 12. For an initial configuration graph G, $\equiv_G \subseteq \mathcal{B}(G) \times \mathcal{B}(G)$ is the least equivalence such that, for $s, t \in \mathcal{B}(G)$ and a and b single rewriting steps:

1. $s.a.b \equiv_G s.b'$, if there exist two single rewriting steps a' and b' corresponding to a and b, respectively, such that $a.b$ is equivalent to $b'.a'$.
2. $s.a \equiv_G t.a$, if s and t are equivalent;
3. $s \equiv_G t$, if there is a sequence of swaps leading from s to t such that no pair of transition productions are commuted.

$\mathcal{F}(G)$ is the quotient set of $\mathcal{B}(G)$ with respect to $\equiv_G$.

On events we define two relations, a partial order describing *structural dependence* and a relation describing *communication dependence*. The wanted causality relation is the union of the two relation. The performance of a transition t' structurally depends on the performance of t if the source of t' is entered by performing t. The performance of a transition t' depends by communication on the performance of t if they are structurally independent and the performance of t broadcasts a signal which enables performing t' (in the same step). The former partial order is obtained investigating sequential dependences of transition derivations in rewriting behaviours limiting ourselves to structural subproductions. The latter order is obtained investigating dependences of transition derivations in rewriting steps, limiting ourselves to communication subproductions.

For a single rewriting step $s = d_1 \ldots d_k$, where $p_k = (L, K, R)$ is the transition production in d_k applied by morphism g, $struct(s)$ (resp.: $comm(s)$) is the derivation sequence $d_1 \ldots d_{k-1}.d'_k$, where where $S(p_k) = (L', K', R')$ (resp.: $C(p_k) = (L', K', R')$) is the production applied in d'_k by the morphism g' such that $g'(v') = g(v)$ for any $v' \in V_{L'}$, $v \in V_L$ where v comes from v'.

Definition 13. The relation $\leq_G^S \subseteq \mathcal{F}(G) \times \mathcal{F}(G)$ is such that $(w, w') \in \leq_G^S$ iff either $w = w'$ or there exists a derivation sequence $s'.a'.s''.a'' \in w'$, with a', a'' single rewriting steps and s', s'' derivation sequences, such that $s'.a' \in w$ and there exists no derivation sequence equivalent to $s'.a'.s''.struct(a'')$ where a' and $struct(a'')$ are commuted.

The relation $\leq_G^C \subseteq \mathcal{F}(G) \times \mathcal{F}(G)$ is such that $(w, w') \in \leq_G^C$ iff either $w = w'$ or $(w, w') \notin \leq_G^S$ and there exists a derivation sequence $s'.a'.s''.a'' \in w'$, with a', a'' single rewriting steps and s', s'' derivation sequences, such that $a'.s''.a''$ is a rewriting step, $s'.a' \in w$ and there exists no derivation sequence equivalent to $s'.a'.s''.comm(a'')$ where a' and $comm(a'')$ are commuted.

The relation $\#_G \subseteq \mathcal{F}(G) \times \mathcal{F}(G)$ is such that $(w, w') \in \#_G$ iff $(w, w'), (w', w) \notin \leq_G^S \cup \leq_G^C$, $w \neq w'$, and there is not a derivation sequence $s.a.b \in w$ and a derivation sequence $s.a \in w'$, with a and b single rewriting steps, such that there is a derivation sequence equivalent to $a.b$ where a and b are commuted.

Theorem 14. *The triple* $(\mathcal{F}(G), \leq_G^S \cup \leq_G^C, \#_G)$ *is an event structure.*

References

1. von der Beek, M.: A Comparison of Statecharts Variants, LNCS 863, Springer, Berlin, 1994, pp. 128–148.
2. Corradini, A., Ehrig, H., Löewe, M., Montanari, U., Rossi, F.: Note on Standard Representation of Graphs and Graph Derivations, Springer LNCS 776, 1994, pp. 104-118.
3. Harel, D.: Statecharts: A Visual Formalism for Complex Systems, Science of Computer Programming 8 (1987), pp. 231–274.
4. Harel, D., Pnueli, A., Schmidt, J., P., Sherman, R.: On the Formal Semantics of Statecharts, Proc. 2nd IEEE Symposium on Logic in Computer Science, IEEE CS Press, New York, 1987, pp. 54–64.
5. Janssens, D., Rozenberg, G.: Actor Grammars, Mathematical Systems Theory 22 (1989), pp. 75–107.
6. Maggiolo-Schettini, A., Peron, A.: Semantics of Full Statecharts Based on Graph Rewriting, Springer LNCS 776, 1994, pp. 265–279.
7. Peron, A., Maggiolo-Schettini, A.: Transitions as Interrupts: A New Semantics for Timed Statecharts, Springer LNCS 789, 1994, pp. 806–821.
8. Schied, G.: On Relating Rewriting Systems and Graph Grammars to Event Structures, Springer LNCS 776, 1994, pp. 326–340.
9. Winskel, G.: An Introduction to Event Structures, Springer LNCS 354, 1989, pp. 364–397.

Programmed Graph Transformations and Graph Transformation Units in GRACE

Andy Schürr

Lehrstuhl für Informatik III, RWTH Aachen,
Ahornstr. 55, D-52074 Aachen, Germany

Abstract. Rule-based languages attract more and more attention as a high-level mechanism for the description of complex transformation processes on graph-like data structures. Unfortunately, pure rule-based approaches are not well prepared for expressing any kind of procedural knowledge. Therefore, various extensions were proposed which regulate the application of rewrite rules. This paper compares already existing regulation mechanisms and proposes a new approach based on control flow graphs. This approach is the first one, where complex control structures inherit all the properties of single rewrite rules, thereby allowing a smooth transition from a rule-oriented to an imperative programming paradigm. Finally, we will show how all reviewed regulation mechanisms may be defined using a very small set of basic concepts and a recently developed new fixpoint theorem. Having such a common formal background offers the opportunity to combine different regulation mechanisms within the future multi-paradigm graph grammar programming environment GRACE.

1 Introduction

Graphs play an important role within many application areas of computer science. Furthermore, rule-based systems are often used to describe complex transformation or inference processes on a very high level. Although graphs and rule-based systems are quite popular, their combination in the form of *graph rewriting systems* or *graph grammars* were more or less unknown to the majority of computer scientists for a long time.

Nowadays the situation is gradually improving with the appearance of system implementations like AGG [11], GraphED [8], PAGG [7], and our system PROGRES [16, 17, 22]. There exist even plans to combine the efforts of different groups in Bremen, Aachen, Berlin, and Leiden to develop a *GRaph Centered Environment*, called *GRACE* [9, 10], which encompasses at least the functionality of the systems AGG and PROGRES and supports various graph grammar approaches within a common framework. The dream behind GRACE is to simplify the selection of a "suitable" graph grammar approach for a given application domain and to allow even the combination of different approaches within a single specification if necessary. It is our hope that such an environment makes the potentials of graph grammars and graph rewriting systems more visible and applications of them more popular.

Within this future environment and within currently existing systems like PAGG and PROGRES, additional means are or should be offered which allow us to regulate the application of a large set of graph rewrite rules and to divide such a set of rewrite rules into manageable subsets with well-defined interfaces and hidden implementation details.

Many quite different proposals exist how to *control application of rewrite rules*, with each of them having its specific (dis-)advantages [1, 14, 4, 13, 19], as for instance:
- Apply rules as long as appropriate or as long as possible in any order; this is the *standard semantics* of rewriting systems.
- Introduce *rule priorities* and prefer applicable rules with higher priorities at least in the case of overlapping matches (redices).
- Use *regular expressions* or even complex graph rewriting programs to define permissible derivation sequences.
- Draw *control flow graphs* and interpret them as graphical representations of rule controlling programs.

Taking the goals of GRACE into account it seems to be useful to develop a *common framework* for the definition and comparison of *rule regulation mechanisms*. This is a necessary prerequisite for offering a number of rule regulation mechanisms in GRACE, which complement each other, and for defining the semantics of specifications which use a combination of offered mechanisms. In our opinion, recursively defined control structures of programmed graph grammars are the most powerful approach to regulate rewriting processes [5, 7, 19]. Their semantics are definable by means of so-called "*basic control flow (BCF) operators*" and a recently developed new fixpoint theorem [15, 18]. Unfortunately, these programs emphasize the imperative programming paradigm and have a textual representation. Both properties are inconsistent with the overall goal of visual and rule-oriented programming with graph rewriting systems.

Within this paper, we will first recall the definition of BCF operator expressions and their accompanying fixpoint semantics definition introduced in [18]. Furthermore, we will compare various rule controlling approaches and discuss their relationships to BCF expressions. Finally, we will show how BCF expressions may even be used to define the semantics of a very powerful new kind of graph rewriting control flow graphs. These *control flow graphs* have about the same expressiveness as graph rewriting programs in [19], and preserve the rule-oriented as well as visual character of pure graph rewriting systems as far as possible.

2 Graph Transactions and Transformation Units

So-called *programmed graph grammars* were already suggested many years ago in [1, 14]. Nowadays, they are a fundamental concept of the graph grammar specification languages PAGG [7] and PROGRES [17, 19]. Experiences with using these languages and their predecessors for the specification of software engineering tools [6, 20] showed that their control structures should possess the following properties:

(1) *Boolean nature:* the application of a programmed graph transformation either succeeds or fails — like the application of a single graph rewrite rule — and, depending on success or failure, execution may proceed along different control flow paths.
(2) *Atomic character:* a programmed sequence of graph rewrite steps modifies a given host graph if and only if all its intermediate rewrite steps do not fail.
(3) *Consistency preserving:* programmed graph rewriting has to preserve a graph's consistency with respect to a given set of separately defined integrity constraints.

(4) *Nondeterministic behavior:* the nondeterminism of a single rewrite rule — it replaces any match of its left-hand side — should be preserved as far as possible on the level of programmed graph transformations.

(5) *Recursive definition:* for reasons of convenience as well as expressiveness, programmed graph transformations should be allowed to call each other without any restrictions including any kind of recursion.

Without conditions (1) through (4) we would have difficulties to replace a complex graph rewrite rule by an equivalent sequence of simpler rewrite rules, and without condition (5), the manipulation of recursively defined data structures would be quite cumbersome.

In the sequel, programmed graph transformations with the above mentioned properties will be termed *graph transactions*. The usage of the term "transaction" underlines the most important properties of programmed graph transformations: *atomicity* and *consistency*. The usually mentioned additional properties of transactions, *isolation* and *duration*, are irrelevant as long as parallel programming is not supported and backtracking cancels the effects of already successfully completed transactions (cf. section 3).

As the reader may imagine, it is very difficult to develop a sound formal definition of graph transactions. Until now, proposals for programmed graph rewrite systems compromised either their boolean nature [14, 13] or their atomic as well as nondeterministic character [7], or they were not able to provide a meaningful semantics for all constructible programs [17]. It is even more difficult to develop efficiently working implementations of graph transactions. Our system PROGRES is the first and as far as we know the only representant of a programmed graph rewrite system implementation with the above mentioned properties [19, 21, 22].

Still missing in any (implemented) graph rewrite system are additional means for the afore mentioned decomposition of rule sets and rule controlling programs into separate components. Therefore, proposals have been made to introduce so-called *graph transformation unit*s as the fundamental concept of GRACE [10]. These units realize binary relations between graphs in a similar way as graph rewrite rules and graph transactions. Furthermore, they have a hidden body, which consists of a set of own graph rewrite rules, a set of imported (graph rewrite rules from other) transformation units, and so-called control conditions which regulate the application of rewrite rules.

A precise definition of syntax and semantics of GRACE control conditions is still subject to ongoing research activities [9]. Currently, rule priorities, graph transactions, and control flow graphs are considered as possible candidates. To summarize, graph transactions of PROGRES and application conditions of transformation units in GRACE share many properties, and the same approach should be used to define their intended semantics.

3 Basic Control Flow Operators

The definition of a fixed set of control structures for graph transactions is complicated by contradicting requirements. From a *theoretical point of view* the set of offered control structures should be as small as possible, and we should be allowed to combine them without any restrictions. But from a *practical point of view* control structures are

```
<Transaction>   ::=   <TransactionId> "=" <BCFExpr> ;
<BCFExpr>       ::=   <BasicAction> | <ActionCall> | <BCFTerm> ;
<BasicAction>   ::=   "skip" | "loop" ;
<ActionCall>    ::=   <RuleId> | <TransactionId> ;
<BCFTerm>       ::=   "def" "(" <BCFExpr> ")" | "undef" "(" <BCFExpr> ")" |
                      "(" <BCFExpr> ";" <BCFExpr> ")" |
                      "(" <BCFExpr> "[]" <BCFExpr> ")"
                      "(" <BCFExpr> "&" <BCFExpr> ")";
```

Fig. 1: Syntax of graph transactions and BCF expressions

needed which are easy to use and to understand, which cover all frequently occurring control flow patterns, and the application of which may be directed by a number of context-sensitive rules.

Therefore, it was quite natural to distinguish between basic control flow operators of an underlying theory of recursively defined graph transactions and more or less complex control structures of a higher level programming language. Starting with a formal definition of basic control flow operators, in the sequel termed *BCF operators*, we should then be able to define the meaning of arbitrary control structures by translating them into equivalent BCF expressions.

Figure 1 contains the definition of BCF expressions and of graph transactions as functional abstractions of BCF expressions. It distinguishes between *basic actions* like

- skip, which represents the always successful identity operator and relates a given graph G to itself, and
- loop, which neither succeeds nor terminates for any given graph and represents therefore "crashing" or forever looping computations,

calls of basic rewrite rules or other graph transactions, and finally between two unary and three binary *BCF operators* with

- def(A) as an action which succeeds applied to a given graph G, whenever A applied to G produces a defined result, and returns G itself,
- undef(A) as an action which succeeds applied to a graph G, whenever A applied to G terminates with failure, and returns G itself,
- (A ; B) as an action which is the sequential composition of A and B, i.e. applies first A to a given graph G and then B to any "suitable" result of B,
- (A [] B) as an action which represents the nondeterministic choice between the application of A or B, and
- (A & B) as an action which returns the intersection of the results of A and B, which requires an equality or isomorphy testing operator on graphs.

Note that the operators suggested above are intentionally similar to those proposed by Dijkstra [3] and especially to those presented by Nelson [15] with one essential difference: due to the boolean nature of basic graph rewrite rules and complex graph transactions, we are not forced to distinguish between side effect free boolean expressions and state modifying actions. This has the consequence that complex guarded commands of the form

$$(\text{Cond}_1 \rightarrow \text{Body}_1 \, [] \, \ldots \, [] \, \text{Cond}_n \rightarrow \text{Body}_n)$$

are no longer necessary (such a command executes a nondeterministically selected body with a valid boolean condition; it fails if all conditions are invalid). They may replaced by expressions like

(<u>def</u>(Cond$_1$) ; Body$_1$) [] ... [] (<u>def</u>(Cond$_n$) ; Body$_n$)

where "<u>def</u>(Cond$_i$)" tests either the applicability of a single graph rewrite rule or of a graph transaction without modifying the given input graph. It is even possible to write

(Cond$_1$; Body$_1$) [] ... [] (Cond$_n$; Body$_n$)

if Cond$_1$ through Cond$_n$ are side effect free expressions or if graph modifying side effects of their execution are even required.

Furthermore, BCF expressions offer almost all possibilities for combining binary relations, the semantic domain of graph transactions. There are operators for intersection, union, and concatenation. The *missing difference operator* is not supported due to the following reasons: A \ B1 is a subrelation of A \ B2, if B2 is a subrelation of B1. As a consequence, the nonmonotonic difference operator had to be excluded in order to be able to come up with a sound definition for recursively defined graph transactions (cf. section 4). It would be possible to define a difference operator where refinements of its second argument are simply prohibited. This corresponds to the requirement that the termination of the evaluation of its second argument must be guaranteed in advance. But such an extensions seems to be unnecessary, taking into account that we are able to define the semantics of all aforementioned rule regulation mechanisms without any needs for building differences.

Having motivated our reasons for the selection of two unary and three binary BCF operators, we are now prepared to discuss the intricacies of their intended semantics. A first problem comes with the definition of the meaning of "(A ; B)" as "apply B to *any suitable* result of A". Let us assume that the application of a graph transformation A to a graph G has three possible results named G1, G2, and G3, respectively. Furthermore, let us assume that the graph transformation B applied to G1 fails but applied to G2 and G3 succeeds. In this case we may either select G2 or G3 but not G1 as a suitable result of the application of A. This means that we need knowledge about future states of an ongoing graph transformation process in order to be able to discard those possible results of a single transformation step, which cause failure of the overall transformation process.

It should be quite obvious that, in general, this kind of clairvoyant nondeterminism requires either a breadth-first search implementation or a depth-first search implementation with backtracking out of "dead-ends". Note that the realization of a *breadth-first search* strategy requires the maintenance of eventually very large sets of graphs, where each graph may itself be a large storage consuming data structure. Therefore, we were forced to adhere to a *depth-first search* semantics while implementing the control structures of the language PROGRES, although breadth-first search strategies are better prepared to deal with certain termination problems.

Furthermore, we were even forced to vote for a *depth-first search semantics* for the BCF operators of this section. Choosing a breadth-first search semantics in theory and a depth-first search semantics in practice results in formal definitions which cannot be used at all for reasoning about or explaining the termination behavior of implemented graph transactions.

Another problem comes with the definition of expressions like "(A [] B)" where A loops forever applied to a certain graph G but B has a well-defined set of possible results. Having a depth-first search semantics in mind, we are forced to define the outcome of the expression "(A [] B)" as being either a nonterminating computation or any defined result produced by B.

This means that the kind of *nondeterminism* we are going to define is not "angelic" but more or less "*erratic*". Using backtracking we are able to discard (nondeterminstic) selections which lead to defined failures of basic actions or graph rewrite rules but not selections which cause nonterminating computations.

4 Fixpoint Semantics for Graph Transactions

After an informal introduction of graph transactions as named BCF expressions we will now define their intended semantics: this is a *semantic function* from the syntactic domain of BCF expressions onto the range of (extended) binary relations over graphs. In order to be able to deal with recursion and nondeterminism in the presence of an atomic sequence operator ";"[1] we had to follow the lines of [15]. There, a new form of the fixpoint theorem is used to give an axiomatic definition of nondeterministic commands.

For the sequel, a certain class of (abstract) graphs $\mathcal{G}$ and a set of rewrite rules $\mathcal{R}$ is expected to be defined elsewhere together with an equality/isomorphy testing operator "≅". Furthermore, we demand the existence of a function $\mathcal{S}: \mathcal{R} \rightarrow \mathcal{G} \times \mathcal{G}$ which maps rewrite rules onto binary relations over graphs and which has the following meaning:

$(G1, G2) \in \mathcal{S}[\![r]\!] :\Leftrightarrow$ *r applied to G1 may produce a graph G2.*

$\neg \exists G: (G1, G2) \in \mathcal{S}[\![r]\!] :\Leftrightarrow$ *r is not applicable to G1*

Note that we assume for simplicity reasons only that the attempt to apply a rewrite rule r to a given graph G always terminates with either returning an arbitrary graph out of the set of all possible results or aborting without any results. This is no longer true for graph transactions which may either return or loop forever applied to a given graph. Therefore, we need an appropriate *extension of the semantic domain* of binary relations over graphs:

$$\mathcal{D} := 2^{\mathcal{G} \times (\mathcal{G} \cup \{ \infty \})}.$$

The semantics of a graph transaction is a binary relation between graphs, where the symbol "∞" in a second component is intended to represent *potentially nonterminating computations*. The word "potential" includes computations with unknown effects, i.e. computations which either return still unknown results or abort or loop forever. A relation S_∞ for instance, which maps any given graph G onto "∞", represents a computation with completely unknown outcomes.

Now, we are ready to define a semantic function $\mathcal{S}$ from the syntactic domain of BCF expressions of section 3 onto the semantic domain $\mathcal{D}$ inductively:

Definition 1 (semantics of BCF expressions)
With A, B ∈ <BCFExpr> our function $\mathcal{S}$: <BCFExpr> → $\mathcal{D}$ is inductively defined as follows:

---

1. The sequence operator is not (chain) continuous in its 2nd argument. Therefore, the original version of the fixpoint theorem, proved in [12], does not work in our case.

(1) $(G, G') \in S[\![\underline{skip}]\!] :\Leftrightarrow\ G = G'$
(2) $(G, G') \in S[\![\underline{loop}]\!] :\Leftrightarrow\ G' = \infty$.
(3) $(G, G') \in S[\![\underline{def}(A)]\!] :\Leftrightarrow$
   $\exists G'' \neq \infty : ((G,G'') \in S[\![A]\!] \wedge\ G = G') \vee ((G,\infty) \in S[\![A]\!] \wedge\ G' = \infty)$.
(4) $(G, G') \in S[\![\underline{undef}(A)]\!] :\Leftrightarrow$
   $((\neg \exists G'' : (G,G'') \in S[\![A]\!]) \wedge\ G = G') \vee ((G,\infty) \in S[\![A]\!] \wedge\ G' = \infty)$.
(5) $(G, G') \in S[\![(A\ ;\ B)]\!] :\Leftrightarrow$
   $(\exists G'' \neq \infty : (G,G'') \in S[\![A]\!] \wedge\ (G'',G') \in S[\![B]\!]) \vee ((G,\infty) \in S[\![A]\!] \wedge G' = \infty)$.
(6) $(G, G') \in S[\![(A\ [\!]\ B)]\!] :\Leftrightarrow$
   $(G, G') \in S[\![A]\!] \vee\ (G, G') \in S[\![B]\!]$.
(7) $(G, G') \in S[\![(A\ \&B)]\!] :\Leftrightarrow$
   $(G' = \infty \wedge ((G, \infty) \in S[\![A]\!] \vee\ (G, \infty) \in S[\![B]\!]))$
   $\vee ((G, G') \in S[\![A]\!] \wedge \exists\ G'' : (G, G'') \in S[\![B]\!] \wedge\ G' \cong G'')$. ☐

The definitions above are rather straightforward with the exception of the treatment of the operators <u>def</u> and <u>undef</u>. The expressions <u>def</u>(A) and <u>undef</u>(A) loop forever if A returns not a single defined result but loops forever. Furthermore, <u>def</u>(A) may return its input if A returns at least one defined result, even if A may loop forever. It terminates with failure if A terminates with failure. On the other hand, <u>undef</u>(A) returns its input if and only if A fails, and it fails if and only if A has at least one defined result and may not loop forever.

Therefore, <u>undef</u> is *stricter* than <u>def</u> with respect to the treatment of looping computations. From a practical point of view, this distinction may be justified as follows:
- Often, we would like to know whether A computes at least one result without evaluating all possible execution paths of A after a successful path has been found.
- On the other hand, answering the question whether A fails is not possible without taking all execution paths of A into account, thereby running into any nonterminating execution branch of A.

Nevertheless, a *strict version* of <u>def</u> might be useful, too, which checks whether a subexpression returns *always* a defined result. Its definition is no longer independent from the definition of <u>undef</u> but may be given as follows:

Definition 2 (strict version of def operator)
A strict version of the <u>def</u> operator may be defined as follows with $A \in$ *<BCFExpr>:*
   $(G, G') \in S[\![\underline{def}^s(A)]\!] :\Leftrightarrow\ (G, G') \in S[\![\underline{undef}(\underline{undef}(A))]\!]$
$\Leftrightarrow\ (\neg \exists G'' : (G, G'') \in S[\![\underline{undef}(A)]\!]) \wedge\ G = G'$
   $\vee\ (G, \infty) \in S[\![\underline{undef}(A)]\!] \wedge\ G' = \infty$
$\Leftrightarrow\ \exists G'' \neq \infty : (G, G'') \in S[\![A]\!]) \wedge\ (G, \infty) \notin S[\![A]\!] \wedge\ G = G'$
   $\vee\ (G, \infty) \in S[\![A]\!] \wedge\ G' = \infty$. ☐

For the proof that all defined BCF operators are monotonic with respect to a suitable partial order on $\mathcal{D}$ — a necessary prerequisite for any fixpoint theorem — the reader is referred to [18]. Based on this proof we are able to define a fixpoint semantics for a given set of graph transactions, which may call each other in an arbitrary manner.

Please note that the treatment of *n transactions* instead of one requires the introduction of a more complex semantic domain $\mathcal{D}^n$ and a corresponding semantic function. This function takes n binary relations as input, which are already computed approximations of n transactions; it produces better approximations in the form of n binary relations as output. Applying fixpoint theory to this extended setting the semantics of the i-th transaction is defined as the i-th component of their common least fixpoint, an n-dimensional vector over $\mathcal{D}$. For the proof of the *existence of unique least fixpoints* the reader is again referred to [18].

Definition 3 (fixpoint semantics of graph transactions)
Let E_1 to $E_n \in$ <BCFExpr> be BCF expressions which contain at most calls to graph rewrite rules in $\mathcal{R}$ and calls with transaction identifiers t_1 through t_n. These expressions may be used to define n (mutually recursive) graph transactions $t_1 = E_1, \ldots, t_n = E_n$. The common unique least fixpoint ($S[\![t_1]\!], \ldots, S[\![t_n]\!]) \in \mathcal{D}^n$ of these graph transactions defines their semantics and may be approximated as follows:

$$S^0 := (S_\infty, \ldots, S_\infty) \in \mathcal{D}^n \quad \text{with } S_\infty \text{ being the relation } \mathcal{G} \times \{\infty\},$$
$$S^{k+1} := (S[\![E_1]\!][S_1^k, \ldots, S_n^k], \ldots, S[\![E_n]\!][S_1^k, \ldots, S_n^k]),$$

where $S[\![E_i]\!]: \mathcal{D}^n \to \mathcal{D}$ takes already computed approximations for $t_1, \ldots, t_n$ as input and yields a new approximation for t_i by applying definition 1 to expression E_i. ❑

The definition above completes the presentation of BCF expressions and their abstraction to graph transactions, which have the five *required properties* of section 2:

(1) The operators "<u>def</u>" and "<u>undef</u>" allow us to test success and failure of any subcomputation and to continue along different execution paths depending on their results.
(2) The definition of the operator ";" guarantees that the execution of any graph transaction succeeds or fails as a whole.
(3) Furthermore, ";" guarantees that failing graph transactions do not return an inconsistent intermediate computation result but keep their input graph unmodified.
(4) The operator "[]" supports the construction of nondeterminstic computations on the level of control structures.
(5) And graph transaction may call each other without any restrictions.

5 Fixpoint Semantics for Various Programming Approaches

Having presented a hopefully minimal as well as sufficiently large set of BCF operators, we still have to show their appropriateness for the definition of *complex control structures*. For this purpose, figure 2 contains a selection of typical examples and their semantic definitions. The first five lines define slightly simplified control structures of the language PROGRES [19]:

<u>abort</u>	$\hat{=}$	<u>undef</u>(<u>skip</u>) .
<u>do</u> A <u>or</u> B <u>end</u>	$\hat{=}$	(A [] B) .
<u>do</u> A <u>and</u> B <u>end</u>	$\hat{=}$	((A ; B) [] (B ; A)) .
<u>if</u> A <u>then</u> B <u>else</u> C <u>end</u>	$\hat{=}$	((<u>def</u>(A) ; B) [] (<u>undef</u>(A) ; C)) .
<u>while</u> A <u>do</u> B <u>end</u>	$\hat{=}$	T = ((<u>def</u>(A) ; (B ; T)) [] <u>undef</u>(A)) .
<u>begin</u> A; B <u>end</u>	$\hat{=}$	((<u>def</u>(A) [] <u>def</u>(B)) ; (A [] <u>undef</u>(A)) ; (B [] <u>undef</u>(B))) .
<u>try</u> A <u>else</u> B <u>end</u>	$\hat{=}$	<u>if</u> A <u>then</u> A <u>else</u> B <u>end</u> .
<u>repeat</u> A <u>end</u>	$\hat{=}$	<u>while</u> A <u>do</u> A <u>end</u> .
(A > B)	$\hat{=}$	<u>repeat</u> <u>try</u> A <u>else</u> B <u>end</u> <u>end</u> .

Fig. 2: Proposals for complex control structures

- The first line introduces a new basic action "<u>abort</u>", which never succeeds but always terminates, as a complement of "<u>skip</u>".
- The first proposed complex control structure "<u>do</u> A <u>or</u> B <u>end</u>" executes either A or B and fails if neither A nor B are applicable.
- The following complex control structure "<u>do</u> A <u>and</u> B <u>end</u>" tries to execute A and B in any possible order and fails if neither A before B nor B before A succeeds.
- The next control structure "<u>if</u> A <u>then</u> B <u>else</u> C <u>end</u>" tests whether A would be applicable; it executes B if A is applicable and executes C in case of defined failure of A.
- The fifth line defines conditional iteration: B is executed as long as A is applicable. The iteration process terminates successfully as soon as A is no longer applicable, it fails whenever A is applicable but B not, and it loops forever if A never fails.

The next three lines of figure 2 are an attempt to define the semantics of control structures in PAGG [7]. Its sequential composition operator *compromises* the required *atomic character* of graph transactions in favor of an efficiently working implementation. They have to be read as follows[2]:

- The construct "<u>begin</u> A; B <u>end</u>" requires sequential application of A followed by B. Its definition is rather complicated since its execution fails only if both A and B fail. Otherwise, the failing subexpression is skipped and the execution process continues.
- The construct "<u>try</u> A <u>else</u> B <u>end</u>" is a convenient shorthand for "<u>if</u> A <u>then</u> A <u>else</u> B <u>end</u>" which may even be used with more than two arguments in the general case.
- The construct "<u>repeat</u> A <u>end</u>" is a convenient shorthand for "<u>while</u> A <u>do</u> A <u>end</u>". It may also be used with more than one argument between "<u>repeat</u>" and "<u>end</u>", but this is just another shorthand for nesting a "<u>begin/end</u>" statement within "<u>repeat/end</u>".

Using the nonatomic sequence statement and offering no means which are equivalent to the previously introduced BCF operators <u>def</u> and <u>undef</u> reduces the expressiveness of PAGG significantly, but has the advantage that an implementation has no needs for backtracking and undoing already performed graph modifications.

Another recently developed graph grammar programming approach [5] has about the same properties and deficiencies as PAGG and is, therefore, not explicitly mentioned in figure 2. It offers merely a combination of the already presented <u>or</u> of PROGRES, the <u>if</u>-<u>then</u>-<u>else</u> with a restriction to boolean conditions, and finally the nonatomic <u>begin</u>/<u>end</u> of PAGG.

Finally, the last line of figure 2 defines the semantics of rule priorities with "(A > B)" as apply "A and B as often as possible and prefer A if both A and B may be applied". This *definition of rule priorities* is just an *approximation* of their usual semantics. According to their definition over here, the execution of B is blocked as long as A is executable, even in the case were A and B modify disjoint parts of a given graph. Unfortunately, a more liberal definition of rule priorities is beyond the capabilities of BCF expressions. This is a principle problem of our approach, since enlarging an already nonempty set of redices (matches) within a graph for a rule A may reduce the set of allowed redices of another graph rewrite rule B with lower priority. This is a kind of nonmonotonic behavior, we are not able to deal with.

---

2. The original definition uses "<u>sapp</u>" instead of "<u>begin</u>", "<u>capp</u>" instead of "<u>try</u>" and "<u>wapp</u>" instead of "<u>repeat</u>".

Beside these more or less graph grammar oriented programming approaches many other approaches for regulating the application of rewrite rules do exist. The most popular ones are presented in [4] together with an in-depth discussion of their expressiveness and related decidability problems. So-called *matrix grammars* offer means for defining rewriting processes with the following flow of control:

$$T = (((r_{1,1} ; \ldots ; r_{1,k(1)}) [] \ldots [] (r_{n,1} ; \ldots ; r_{n,k(n)})) ; (T [] \underline{skip})),$$

where all $r_{i,j}$ are simple rewrite rules. They combine the usual "apply as long as needed" semantics of unregulated rewriting systems with the idea that a sequence of rewrite rules instead of a single rewrite rule should be applied. The reader should notice that the semantics of sequential application of matrix grammars fulfills our atomicity requirement, but all other required properties for control structures in section 2 are still missing.

So-called *programmed grammars* in [4] are another step towards really useful control structures. They are equivalent to sets of graph transactions as they were defined over here, where each transaction T has the following form:

$$T = (r ; (S_1 [] \ldots [] S_n) [] (\underline{undef} (r) ; (F_1 [] \ldots [] F_n))$$

with r being a simple rewrite rule, and $S_1, \ldots S_m$, $F_1, \ldots, F_n$ being other transactions. In this way programmed grammars allow the definition of sequences of rule applications and they even support branching in the flow of control depending on success or failure of single rewrite rules. But without any support for functional abstraction they do not allow branching in the flow of control depending on success or failure of complex rewriting processes.

6 Fixpoint Semantics for Control Flow Graphs

In the previous section typical exponents of various programmed rewriting approaches were presented on a more or less informal level. In all cases denotational semantics definitions were sketched by translating their control structures into sets of graph transactions with a well-defined fixpoint semantics definition. But all discussed approaches — except graph transactions themselves and their more readable form in the language PROGRES — had severe deficiencies. And even graph transactions with their low level text representation compromise the *visual nature* of programming with graph rewrite rules.

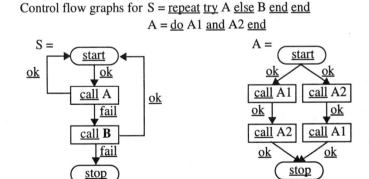

Fig. 3: Examples of control flow graphs

An obvious solution for this problem is to replace text-oriented control structures by some kind of *control flow graphs*. But until now, all suggested variants of control flow graphs were either less expressive than our graph transactions or came without a sound denotational semantics definition:
- Bunke proposed in [1] control flow graphs, which are not allowed to call each other in an arbitrary manner, i.e. do not properly support functional abstraction.
- Nagl introduced control flow graphs in [14], where a node may be a call to another control flow graph, but where branching depending on success or failure of subgraph calls is not supported.
- And the author of this paper suggested in [17] control flow graphs without the above mentioned deficiencies, but also without a nonoperational semantics definition.

In the sequel, we will show how BCF expressions may be used to define a fixpoint semantics for the control flow graphs of [17]. These control flow graphs are equivalent to graph transactions with strict $\underline{def}^{\,S}$ operators and without the intersection operator &. Figure 4 contains an example of a control flow graph with two subgraphs and their textual counterparts. Their main properties are:
- Any node in a subgraph should be reachable from its unique <u>start</u> node and should have a path to its unique <u>stop</u> node.
- The execution of a (sub-)graph begins at its <u>start</u> node and ends at its <u>stop</u> node.
- All remaining nodes have an inscription which is either <u>nop</u> (for no operation has to be performed) or the <u>call</u> of a simple rewrite rule or another control flow subgraph.
- The execution of a node either succeeds or fails; executions of <u>start</u>, <u>stop</u>, and <u>nop</u> nodes always succeed without any graph modifying effects.
- After a successful execution of a node we have to follow one of its outgoing <u>ok</u> edges; multiple <u>ok</u> edges reflect nondeterministic computations and absence of <u>ok</u> edges partially defined computations.
- After a failing execution of a node we have to follow one of its outgoing <u>fail</u> edges; multiple <u>fail</u> edges reflect again nondeterministic computations and absence of <u>fail</u> edges again partially defined computations.

For a more detailed explanation of this kind of control flow graphs, including an operational semantics definition, the reader is referred to [17]. A *denotational semantics definition* may be provided by translating a control flow graph into an equivalent set of graph transactions. Any node of the control flow graph will be translated into a transaction with calls to other transactions reflecting its outgoing <u>ok</u> and <u>fail</u> edges:

Definition 4 (semantics of control flow graphs)
Any control flow graph node n with label "<u>call</u> A" or "<u>start</u>" or "<u>nop</u>", outgoing <u>ok</u> edges to nodes $n_1, \ldots, n_j$, and outgoing <u>fail</u> edges to nodes $m_1, \ldots, m_k$ is translated into:

$n = (\,S\,;\,(n_1\,[]\,\ldots\,[]\,n_j\,)\,)\,[]\,(\,\underline{undef}(S)\,;\,(m_1\,[]\,\ldots\,[]\,m_k\,)\,)$, *if $j > 0, k > 0$*
and into $n = (\,S\,;\,(n_1\,[]\,\ldots\,[]\,n_j\,))$, *if $j > 0, k = 0$*
$n = (\,\underline{undef}(S)\,;\,(m_1\,[]\,\ldots\,[]\,m_l\,))$, *if $j = 0, k > 0$*
and into $n = abort := \underline{undef}(skip)$, *if $j = 0, k = 0$*
where $S = A$, *if inscription of node is "<u>call</u> A"*
and $S = \underline{skip}$, *otherwise.*

Finally, any node n with inscription "<u>stop</u>" is translated into the transaction
$n = \underline{skip}$. ❑

Theorem 5 (**equivalence of graph transactions and control flow graphs**)
Any set of graph transactions without the intersection operator & and with a strict def ^{s} operator (cf. def. 1 and def. 2) may be translated into an equivalent set of control flow graphs according to definition 4 and vice versa.

Proof. The translation of any control flow graph into an equivalent BCF expression is an essential part of definition 4. The reverse direction, the translation of BCF expressions without & and def occurrences, is defined by the graph rewrite rules in fig. 4. These rewrite rules transform a node which has the definition of a graph transaction as its label into an equivalent control flow graph. Rewrite rules are specified in a PROGRES like notation with double dashed nodes representing embedding rules. The rewrite rule TranslateUndef redirects for instance any adjacent ok and fail edge of the rewritten node "undef(A)" to the new node "call T". ❏

Please note that the given proof is just a sketch. Still missing is (the more or less trivial) part which checks (by induction) that the output of the translation of any BCF expression E into a control flow graph followed by the translation back into another BCF expression E' preserves indeed its semantics, i.e. that $\mathcal{S}[\![E]\!] = \mathcal{S}[\![E']\!]$.

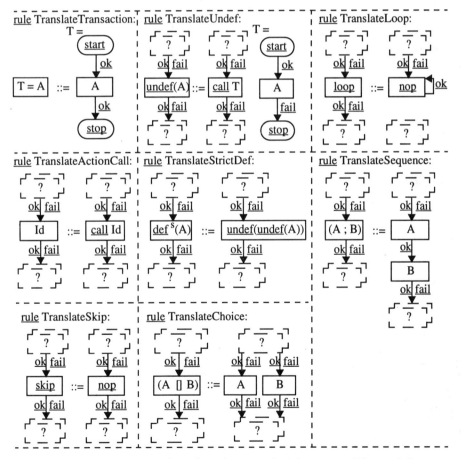

Fig. 4: Translation of graph transactions into control flow graphs

7 Conclusion

This papers constitutes a *new framework* for defining and comparing various approaches which support programming with (graph) rewrite rules. Based on this framework, advantages and disadvantages of previously suggested programming formalisms were discussed. The results of this discussion are summarized in table 1 with column headlines referring to required rule programming properties in section 2.

Motivated by apparently existing deficiencies of reviewed approaches, a new type of control flow graphs was introduced, which allows a smooth transition from rule-oriented programming to the imperative programming paradigm and which supports *visual programming with graph rewrite rules* on a very high level.

Based on the definition of BCF expressions (graph transactions) and their fixpoint semantics we were even able to relate all listed rule regulation approaches — with the exception of those presented in [12] — to the formal notation of BCF expressions:

- The last line of fig. 2 sketches the translation of simple rule priorities into BCF expressions.
- The translations of matrix grammars as well as programmed grammars in [4] are part of the running text in section 5.
- The lines 6, 7, and 8 of fig. 2 deal with control structures of programmed attributed graph grammars in [7].
- The translation of PROGRES control structures into control flow graphs was already presented in [17, 19].

	Sequential Control Flow	Nondeterm. Control Flow	Success & Failure Test	Recursive Calls	Paradigm & Notation
Simple Rule Priorities (here)	no	no	no	no	purely decl. & text-oriented
Matrix Grammars [4]	atomic sequence	no	no	no	decl./imperative & text-oriented
Programmed Grammars [4]	nonatomic sequence	no	rules only	tail recursion	decl./imperative & text-oriented
Prog. Attributed Graph Grammars [7]	nonatomic sequence	no	rules only	yes	imperative & text-oriented
Prog. Graph Rewriting Systems [5]	nonatomic sequence	yes	no	yes	imperative & text-oriented
PROGRES [19] & BCF Expressions [18]	yes	yes	yes	yes	imperative & text-oriented
Prog. Deriv. of Rel. Structures [13]	yes	yes	no	yes	parallel imper. & text-oriented
Control Flow Diagrams in [2]	yes	yes	no	no	imperative & graphical
Control Flow Diagrams in [14]	yes	yes	rules only	yes	imperative & graphical
New Control Flow Diagrams (here & [17])	yes	yes	yes	yes	imperative & graphical

Table 1: A comparison of various regulated rewriting approaches

- The lines 2 through 5 of fig. 2 sketch how programmed graph transformation systems in [5] must be handled.
- The translation of our new type of control flow graphs into BCF expressions was presented in def. 4.
- And all remaining forms of control flow graphs in [1, 2], [14], and [17, 19] are just subsets of the new type of control flow graphs.

Please note that all presented work within the previous sections had its focus on *sequential rewriting processes*. This is no longer true for so-called "programmed derivations of relational structures" in [13] which are also mentioned in table 1. This approach supports *parallel programming with rewrite rules*, but does not offer any constructs for testing whether a complex graph transformation returns a defined result or not. Parallel composition of subtransformations as well as testing their applicability are both very valuable means and require completely different formal treatments. Parallel composition, on one hand, with its "interleaving semantics" excludes the definition of a program's semantics as a function of the semantics of its subprograms. Testing success or failure of subprograms, on the other hand, enforces the introduction of a special symbol "∞" for nonterminating computations and the usage of a rather complicated partial order for the resulting semantic domain. Further investigations are necessary to check whether a combination of both approaches is possible or not.

Finally, we should mention that all presented results in this paper are valid for any kind of rewriting approach, where single rewrite rules define binary relations over a given domain of objects. Nevertheless, we have used the terms "graph" and "graph rewrite rules" throughout this paper — instead of more generic terms — in order to emphasize the graph oriented background of our current research activities. Having developed such a common framework for rather different rule regulation mechanisms, which is independent from a specific (graph) data model or a selected (graph) rewriting approach, offers the opportunity to combine different regulation mechanisms as needed within a future *multi-paradigm graph grammar environment* GRACE [9, 10].

References:

1. Bunke H.: *Sequentielle und parallele programmierte Graphgrammatiken*, Dissertation, Technical Report IMMD-7-7, Universität Erlangen, Germany (1974)
2. Bunke H.: *Programmed Graph Grammars*, in: Proc. Int. Workshop on Graph-Grammars and Their Application to Computer Science and Biology, LNCS 73, Springer Verlag, Germany (1979), 155-166
3. Dijkstra E.W.: *Guarded Commands, Nondeterminacy, and Formal Derivation of Programs*, in CACM, no. 18, acm Press, USA (1975), 453-457
4. Dassow J., Paun G.: *Regulated Rewriting in Formal Language Theory*, EATCS 18, Springer Verlag, Germany (1989)
5. Dörr H.: *Efficient Graph Rewriting and Its Implementatiion*, Dissertation, LNCS 922, Springer Verlag, Germany (1995)
6. Engels G., Lewerentz C., Nagl M., Schäfer W., Schürr A.: *Building Integrated Software Development Environments*, in: ACM TOSEM, vol. 1, no. 2, acm Press, USA (1992),135-167

7. Göttler H.: *Graphgrammatiken in der Softwaretechnik*, IFB 178, Springer Verlag, Germany (1988)
8. Himsolt M.: *GraphED: An Interactive Graph Editor*, in: Proc. STACS 89, LNCS 349, Springer Verlag, Germany (1988), 532-533
9. Kreowski H.J., Kuske S.: *On the Interleaving Semantics of Transformation Units - A Step into GRACE*, in: Proc. 5th Int. Workshop Workshop on Graph-Grammars and Their Application to Computer Science, same volume
10. Kreowski H.J.: *Graph Grammars for Software Specification and Programming: An Eulology in the Praise of GRACE*, in: Proc. Colloquium on Graph Transformations and its Applications in Computer Science, Technical Report B-19, Universitat de les Illes Balears, Departament de Ciencies Matematiques i Informatica (1994), 55-62
11. Löwe M., Beyer M.: *AGG - An Implementation of Algebraic Graph Rewriting*, LNCS 690, Proc. 5th Int. Conf. on Rewriting Techniques and Applications, Springer Verlag, Germany (1993)
12. Manna Z.: *Mathematical Theory of Computation*, New York: McGraw-Hill, USA (1974)
13. Maggiolo-Schettini A., Winkowski J.: *Programmed Derivations of Relational Structures*, in: Proc. 4th Int. Workshop on Graph-Grammars and Their Application to Computer Science, LNCS 532, Springer Verlag, Germany (1991), 582-598
14. Nagl M.: *Graphgrammatiken*, Vieweg Verlag, Germany (1979)
15. Nelson G.: *A Generalization of Dijkstra's Calculus*, in ACM Transactions on Programming Languages and Systems, vol. 11, no. 4, acm Press, USA (1989), 517-561
16. Schürr A.: *PROGRES: A VHL-Language Based on Graph Grammars*, in: Proc. 4th Int. Workshop on Graph-Grammars and Their Application to Computer Science, LNCS 532, Springer Verlag, Germany (1991), 641-659
17. Schürr A.: *Operationales Spezifizieren mit programmierten Graphgrammatiken*, Dissertation, RWTH Aachen, Deutscher Universitätsverlag, Germany (1991)
18. Schürr A.: *Logic Based Programmed Structure Rewriting Systems*, appears in: Special Issue on Graph Transformation Systems, Fundamenta Informaticae, North-Holland
19. Schürr A., Zündorf A.: *Nondeterministic Control Structures for Graph Rewriting Systems*, in Proc. WG'91 Workshop in Graphtheoretic Concepts in Computer Science, LNCS 570, Springer Verlag, Germany (1992), 48-62
20. Westfechtel B.: *Revisionskontrolle in einer integrierten Softwareentwicklungs-Umgebung*, Dissertation, RWTH Aachen, Informatik-Fachberichte 280, Springer Verlag, Germany (1991)
21. Zündorf A.: *Implementation of the Imperative/Rule Based Language PROGRES*, Technical Report AIB 92-38, RWTH Aachen, Germany (1992)
22. Zündorf A.: *Eine Entwicklungsumgebung für PROgrammierte GRaphErsetzungs-Systeme*, Dissertation, RWTH Aachen (1995)

Pragmatic and Semantic Aspects of a Module Concept for Graph Transformation Systems

Hartmut Ehrig

Technical University of Berlin
Dept. of Computer Science
Franklinstr. 28/29, D-10587 Berlin
Germany
email: ehrig@cs.tu-berlin.de

Gregor Engels

Leiden University
Dept. of Computer Science
P.O. Box 9512, NL-2300 RA Leiden
The Netherlands
email: engels@wi.leidenuniv.nl

Abstract. The paper presents a conceptual framework for a module concept for graph transformation systems from a software engineering as well as from a theoretical point of view. The basic idea is to reuse concepts, which are known within or without the graph grammar field, to structure large specifications. These are the concept of distributed graph transformation systems, the concept of inheritance of specifications, and the import-export-interface concept. All these concepts are presented in a uniform framework based on the syntactical notion of a graph class specification and its semantics given by a graph transformation system. This is the basis for an explicit integration of these concepts and a corresponding specification language, to be discussed in a subsequent paper.

1 Introduction

Graph grammars and graph transformation systems have been developed for about 25 years with various applications in computer science. More recently some approaches to graph transformations have been used for the specification of software development environments ([ELS 87], [ELNSS 92]) and different kinds of software systems. A great drawback while using graph transformation systems for specification purposes is that a suitable module concept for graph transformation systems is missing. In fact, a module concept is a must for any specification technique from a software engineering point of view, at least if large software systems have to be specified. It is the intention of our work to discuss a conceptual framework for a module concept for graph transformation systems. We introduce the syntactical notion of a graph class specification, the semantics of which is given by a graph transformation system. Using those graph class specifications as basic specification units, we investigate different types of interrelations between them.

We illustrate this by discussing a model for a simple problem situation. The model describes a world of bikes, which are produced by employees of a manufacturer and bought by persons in a bike shop. Figure 1 shows an instantiation of this model.

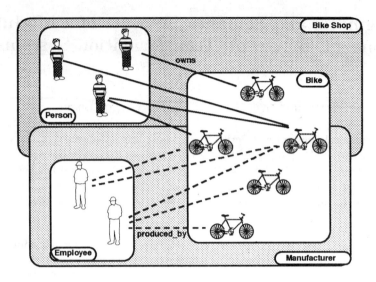

Fig. 1. An instantiation of the sample model

This problem situation can be regarded as a large graph, where the structure of each object, i.e. persons, employees, and bikes, is represented itself as a graph. Furthermore, the overall situation can be represented by a graph, where additional edges connect these object graphs and express, for instance, a "produced-by" or "owns" relationship. We will use this example in the following sections to discuss in more detail how graphs and interrelations between them can be specified in a modular way.

It is clear that a satisfying module concept for graph transformation systems will not be developed within one step. The approach presented in this paper, is to try to reuse existing concepts, which have been successfully applied within or outside the graph grammar field, in order to define a conceptual framework for a module concept. It is not the idea of this paper to present already a concrete specification language which offers the notion of a module as one of its language constructs. We will discuss separately different aspects of an intended integrated conceptual framework within this paper. The full integration is still to be done.

There are a few other approaches for defining a module concept for graph transformation systems. An interesting one is developed in the context of the GRACE initiative [Kr 95], where in a bottom-up manner existing concepts in the graph grammer field are extended to define so-called transformation units.

The paper is organized as follows:
In section 2, the above introduced example will be discussed in more detail in order to illustrate three different concepts for structuring large specifications. These are the distributed state graph concept, the inheritance concept and the import-export-interface concept. Syntax and semantics of each of these concepts will be defined in sections 4 to 6. As a common basis, section 3 gives the definition of syntax and semantics of graph class specifications. Section 7 concludes

the paper with a discussion on future work, especially the development of a specification language based on graph transformations including the modularization concepts introduced in this paper.

2 Pragmatic Aspects for Structuring Concepts

In this section, we review three structuring concepts known from different specification techniques, and discuss how they can be reused to structure graph class specifications.

2.1 Distributed State Graph Concept

The main idea of specifications with graph transformations is to use graphs for the representation of states of objects, and graph transformations to model state changes. Since most systems in practice tend to have a large number of objects which will lead to large state graphs it seems reasonable to distinguish a set of small local state graphs as decomposition of each global state graph. This means the global state is distributed into local states which vice versa can be glued together via a suitable graph, called interface to reestablish the global state if necessary. This idea makes sense for graph transformation systems if the graph rules can be decomposed into local sets of rules which operate on the local state graphs only. In most cases some graph rules may remain which have an effect on different local state graphs at the same time. For such rules we have to reestablish the global state graph before we are able to apply them. In this way we obtain a distributed graph class specification consisting of $n \geq 2$ local and one global graph class specification. These local and global graph class specifications can be considered as different modules within the distributed system. This idea has been used in the algebraic approach to graph grammars in [Ehr et al 87] and [EL 91].

In our running example, we represent objects like persons or bikes by local state graphs. Each state graph resembles the structural information of such an object, as, e.g., the name, birthday and address information of a person. The global state graph has additional edges to connect these local state graphs. As an example see Figure 2, where an edge labelled "owns" connects a person state graph with a bike state graph.

Typical local operations are "to change the address of a person", or "to modify the size of a bike". While these two local operations may be executed in parallel, an operation like "to buy a new bike" is a global operation, which works on the global state graph.

Thus, the specification of this situation can be structured into two local graph class specifications for PERSON and BIKE and one global graph class specification for BIKE SHOP.

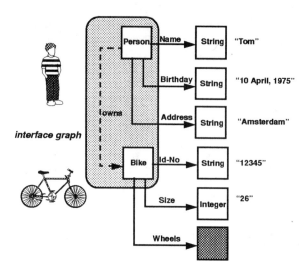

Fig. 2. A global state graph connecting two local state graphs

2.2 Inheritance Concept

The basic idea of the inheritance concept used in object-oriented system design (e.g. [RBP 91]) is to support a layered approach to system design in the following sense: It is intended to build on already specified components - drawing on the services they provide - by "augmenting" them - by extending their state space and/or the operations on it - while preserving their properties or redefining some of them. If the properties are preserved in the augmented component we have inheritance (without overriding) between the corresponding specification components. If properties from different components are inherited we speak of multiple inheritance.

In our running example, the use of inheritance would offer the opportunity to reuse the specification of a PERSON for the specification of an EMPLOYEE. This makes a specification much shorter, as all parts defining the structure and behaviour of a person have not to be repeated in the definition of an employee. This means that an employee is a person, who has as additional behaviour for instance an operation like "to produce a bike".

2.3 Import-Export-Interface Concept

The idea to use modules with import and export interfaces is due to Parnas ([Par72]) and is used in the algebraic module specification concept ([EM90]). The main idea is to define an export part of a module description, e.g. the name of a type and the signature of operations, and to hide the remaining part, e.g. the realization of a type and of operations, in the encapsulated body of a module. Modules are interconnected by a uses relationship in order to import exported resources of another module. Imported resources are used to realize exported resources of a module. As an example, we have already seen in figure 2 the graph model of a bike object. The structure of wheels is drawn as a black box indicating that it is not known within the model of a bike. This means that the type for wheels is defined and realized elsewhere and only imported in BIKE to be able to declare a link to such a structure. Within the realization of an operation of BIKE like "to replace a wheel", a modification of this part of a bike has to be done by using an operation imported from WHEEL. On the specification level, the module (class) specifications are linked by a uses relationship.

This idea of importing operations to realize new ones, requires a powerful mechanism to define new graph transformation rules, especially in the case that the imported rules are given by names only, i.e. without tight semantics. It requires graph rules and transformations with loose semantics, which is not available in any of the known approaches to graph grammars and transformations up to now. In section 3, we make a proposal for a notion of graph class specifications which allow to have loose or tight semantics. Based on this kind of graph class specifications and corresponding graph transformation systems we are able to discuss the possibilities of an import-export-interface-approach for graph class specifications in section 6.

2.4 Modular Specification Using Different Structuring Concepts

In figure 3, we show the architecture of a modular specification for our running example using all the structuring concepts discussed above. Each module is represented by a rectangle. Modules are linked together by a circled "+" symbol expressing a distributed graph class specification (e.g. PERSON, BIKE, BIKE SHOP), by a triangle expressing an inheritance relationship (EMPLOYEE is_a PERSON), or an edge expressing a uses relationship (BIKE uses WHEEL).

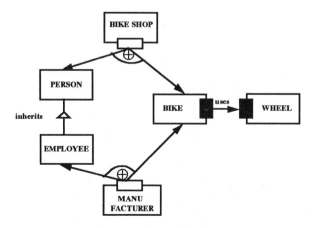

Fig. 3. A sample specification architecture

These notions may be considered as a first sketch for a graphical specification architecture definition language. A concrete specification language, however, is not the intended goal of this paper.

3 From Graph Grammars to Graph Class Specifications and Transformation Systems

In this section, we want to present a general notion of graph class specifications and graph transformation systems which on one hand covers most of the graph grammar approaches in the literature and on the other hand is a suitable basis to formulate the three alternatives for a module concept of graph transformation systems. In fact, there are already some attempts to unify different graph grammar approaches, like $LEARRE$ ([Ro 87], [KR 90]). Although this kind of $LEARRE$-grammars covers really most of the approaches in the literature it is not general enough to cover programmed graph replacement systems in the sense of $PROGRES$ [Sch 91] or graph rules with loose semantics which are necessary for our import-export-interface concept (see 2.3). Moreover we would like to include the idea of typing for nodes and edges in [Sch 91] leading to graph types and also the idea of distributed graph derivations [Ehr et al 87] including composition ($JOIN$) and decomposition ($SPLIT$) of graphs as graph procedures. All these considerations have influenced us to present the following notion of graph signatures and graph class specifications, which includes a very general notion of constraints, similar to constraints for algebraic specifications in [EM 90].

First we define graph class signatures and graph class specifications as syntactical level of graph transformation systems.

3.1 Definition (Graph Class Signature, Constraints, and Graph Class Specification)

1. A *graph class signature* $GSIG = (T, P)$ consists of
   T: set of types for graphs and/or graph morphisms
   P: family of graph class procedure symbols $p : w1 \longrightarrow w2$ for $w1, w2 \in T^*$
2. For each graph class signature $GSIG$ we assume to have a class CC, called *class of all constraints over $GSIG$*.
3. A *graph class specification* $GSPEC = (T, P, C)$ consists of a graph class signature $GSIG = (T, P)$ and a set $C \subseteq CC$, called *set of constraints*.

Remarks and Interpretation

1. Similar to algebraic signatures $SIG = (S, OP)$ in [EM 85], T and P in a graph class signature $GSIG = (T, P)$ are sets of formal symbols, i.e. pure syntax, but graph class procedure symbols may have arity $n = \mid w1 \mid \geq 0$ and coarity $m = \mid w2 \mid \geq 0$.
2. The set C of constraints (and similar the class CC) is a set of formal symbols which is intended to define constraints for the semantics of types $t \in T$

and procedure symbols $p \in P$. The constraints C can be used to fix the semantics for all types and procedure symbols of $GSPEC$ (tight semantics) or to formulate only properties for the semantics (loose semantics).

Now we are able to define graph transformation systems as semantics of graph class signatures and graph class specifications:

3.2 Definition ($GSIG-$ and $GSPEC-$Graph Transformation Systems, Satisfaction Relation)

1. Given a graph class signature $GSIG = (T, P)$ a $GSIG-$graph transformation system $GTS = ((D_t)_{t \in T}, (p_{GTS})_{p \in P})$ consists of
   D_t: domain of graphs and/or graph morphisms for each $t \in T$
   $p_{GTS}: D_{w_1} \rightsquigarrow D_{w_2}$ graph class procedures for each $p : w1 \longrightarrow w2$ in P
   where $D_w = D_{t_1} \times ... \times D_{t_n}$ for $w = t_1...t_n \in T^*$, $D_\lambda = \{1\}$, a distinguished one-element set, and p_{GTS} is a relation between D_{w_1} and D_{w_2}.
   For each n-tuple $(G_1, ..., G_n) \in D_{w_1}$ and m-tuple $(H_1, ..., H_m) \in D_{w_2}$ with $(H_1, ..., H_m) \in p_{GTS}(G_1, ..., G_n)$ we have a *graph transformation* $(G_1, ..., G_n) \Longrightarrow (H_1, ..., H_m)$ via p_{GTS}.
2. Given the class CC of all constraints over a graph class signature $GSIG$ (see 3.1.2), we assume to have a relation $\models$, called *satisfaction relation*, between $GSIG-$graph transformation systems GTS and constraints $c \in CC$, where $GTS \models c$ means that GTS satisfies the constraint c.
3. Given a graph class specification $GSPEC = (T, P, C)$ with $C \subseteq CC$, a $GSPEC-$graph transformation system GTS is a $GSIG$-graph transformation system which satisfies the constraints C, i.e., for all $c \in C$ $GTS \models c$.
4. The set of constraints C of $GSPEC = (T, P, C)$ is called *tight*, if there is exactly one $GSPEC-$graph transformation system GTS satisfying C. Otherwise it is called *loose*.

Remarks and Interpretation

Our notion of graph transformation systems is based on the idea that graph transformations are defined by procedures on suitable domains of graphs. The idea is that these graph procedures are defined via graph rules or other graph manipulating mechanisms (see example 3.3). In fact, our notion is flexible enough to cover also all kinds of attributed graphs [Sch 91a], graph structures [Löw 93] or relational structures which have been considered in the framework of graph grammars and high-level replacement systems [EHKP 91].

3.3 Examples (Graph Transformation Systems)

1. Given an algebraic graph grammar $GG = (S, RULES)$ where S is a start graph and $RULES$ a set of graph rules $r : L \longleftarrow K \longrightarrow R$, we obtain a graph class specification $GSPEC(GG) = (T, P, C)$ with a single type t, graph class procedure symbols $p_r : t \longrightarrow t$ for all $r \in RULES$ and $p_{init} : \lambda \longrightarrow t$ and

constraints "Σ-graph", $c_r (r \in RULES)$ and c_{init} defined as follows:
$T = \{t\}$
$P = \{p_{init}\} \cup \{p_r \mid r \in RULES\}$
$C = \{\Sigma\text{-graph}, c_{init}\} \cup \{c_r \mid r \in RULES\}$
The essential point is now that the satisfaction relation $\models$ makes sure that for each $GSPEC(GG)$-graph transformation system
$GTS = ((D_t)_{t \in T}, \{p_{init_{GTS}}\} \cup \{p_{r_{GTS}} \mid r \in RULES\})$ the domain D_t is the class of all Σ-graphs, i.e. $D_t = \Sigma\text{-GRAPHS}$, and the interpretation of the other constraints c_{init} and c_r makes sure that our start graph is S and the graph class procedures $p_{r_{GTS}}: \Sigma\text{-GRAPHS} \leadsto \Sigma\text{-GRAPHS}$ correspond to applications of the rules $r \in RULES$ in the sense of algebraic graph grammars. For this purpose we define $\models$ as follows:
$GTS \models \Sigma\text{-graph} \iff D_t = \Sigma\text{-GRAPHS}$
$GTS \models c_{init} \iff p_{init_{GTS}} = \{(1, S)\}$
$GTS \models c_r \iff p_{r_{GTS}} = \{(G, H) \mid G \Longrightarrow_{DPO} H \text{ via } r\}$
where $G \Longrightarrow_{DPO} H$ via r means that the graph rule r is applicable to G in the DPO-approach (see [Ehr 79]) and leads to the graph H, i.e., there are pushouts

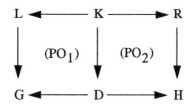

in the category of Σ-graphs where the top row is equal to r, and H is determined by r and the pushouts (PO_1) and (PO_2). In this case a direct graph derivation $G \Longrightarrow_{DPO} H$ via r in the DPO-approach leads to a graph transformation $G \Longrightarrow H$ via $p_{r_{GTS}}$ in GTS and vice versa.

The graph transformation system GTS is completely determined by the graph specification $GSPEC(GG)$ and hence by the given graph grammar GG. Note, however, that there may be different direct graph derivations $G \Longrightarrow_{DPO} H$ via r which are represented by one pair $(G, H) \in \{p_{r_{GTS}} \mid r \in RULES\}$ and hence by one graph transformation $G \Longrightarrow H$ via $p_{r_{GTS}}$ in GTS only.

2. In fact, example 1 using the algebraic DPO-approach depends only on a well-defined notion of direct graph derivation $G \Longrightarrow_{DPO} H$ via a rule r. Hence it can be extended to any other graph grammar approach like the single pushout approach [Löw 93], the NLC-approach [Ro 87], or any other LEARRE-graph grammar GG (see [Ro 87] and the introduction to section 3).

3. A programmed graph replacement system in the sense of Schürr [Sch 91] can also be translated into a graph class specification and graph transformation system in our sense. The idea of different types for nodes and edges

can be reflected by different types for the corresponding graph domains. A programmed graph rule of type $p : t_1 \longrightarrow t_2$ leads to a graph class procedure $p_{GTS} : D_{t_1} \rightsquigarrow D_{t_2}$ and a corresponding graph transformation $G_1 \Longrightarrow G_2$ via p_{GTS} if $G_1 \in D_{t_1}, G_2 \in D_{t_2}$ and the program of graph rules is applicable to G_1 and all intermediate graphs G are obtained by partial execution of the program. The graph class procedure may lead to $k > 0$ different graph transformations $G_1 \Longrightarrow G_{2_k}$ via p_{GTS} according to the nonexistence or multiple existence of intermediate occurrences.

4. Composition and decomposition of graphs in the sense of JOIN and SPLIT for distributed graph grammars [Ehr et al 87] or distributed graph transformation systems are graph class procedures of the following types
$JOIN\colon t_1...t_n t_{int} \longrightarrow t_0$, and
$SPLIT\colon t_0 \longrightarrow t_1...t_n t_{int}$
where t_k corresponds to the types of the global graph grammar or transformation system for $k = 0$, to that of a local one for $k = 1,...,n$ and t_{int} is the type of the interface graph and graph morphisms (see section 4 for more details).

5. If we have a graph class specification $GSPEC = (T, P, C)$ where C consists of the constraint "Σ-graph" only we have a typical example of a graph class specification with loose semantics. In fact, for a $GSPEC$-graph transformation system GTS in this case a graph procedure for $p : w_1 \longrightarrow w_2$ in P may be any relation $p_{GTS} : \Sigma-\text{GRAPHS}^n \rightsquigarrow \Sigma-\text{GRAPHS}^m$ for $n =\mid w_1 \mid$ and $m =\mid w_2 \mid$.

4 Distributed Graph Class Specifications

According to the discussion of the distributed state graph concept in 2.1, we extend this idea to graph class specifications and graph transformation systems in the sense of section 3. For simplicity we assume that each local specification has one type only and that there is one interface type t_{int} only in order to express the interaction of all the local components. We start with the definition of syntactical aspects.

4.1 Definition (Distributed Graph Class Specifications)

A *distributed graph class specification*
$\quad DGSPEC = (GSPEC_1, ..., GSPEC_n, GSPEC_0)$
consists of $n \geq 2$ *local graph class specifications*
$\quad GSPEC_i = (\{t_i\}, P_i, C_i)\ i = 1,...,n$
and a *global graph class specification*
$\quad GSPEC_0 = (T_0, P_0, C_0)$
with
$\quad T_0 = \{t_0, t_1, ..., t_n, t_{int}\}$
$\quad P_0 = \bigcup_{i=1}^{n} P_i \cup \{JOIN, SPLIT\} \cup PO'$
$\quad JOIN : t_1...t_n t_{int} \longrightarrow t_0$

$SPLIT : t_0 \longrightarrow t_1...t_n t_{int}$
$PO' =$ set of procedure symbols $p_0 : t_0 \longrightarrow t_0$ and $p_{int} : t_{int} \longrightarrow t_{int}$
$C_0 = \bigcup_{i=1}^{n} C_i \cup \{c_{dist}\} \cup CO'$
$c_{dist} =$ distribution constraint for $JOIN$ and $SPLIT$ with fixed semantics
$CO' =$ set of constraints for t_0, t_{int} and PO'.

Remarks and Interpretation

The idea is to allow in the local components $GSPEC_i$ graph procedure sequences p_i of type i which are executed in parallel with an interface graph procedure p_{int} until the local state graphs are joined by application of the JOIN procedure. The application of the JOIN procedure requires synchronization. Then we may have a global graph procedure sequence p_0 of type t_0 until we may be able to apply the SPLIT procedure to obtain an n-tuple of local state graphs with interface. Now we are able to apply again local procedures $(p'_1, ..., p'_n)$ in parallel with an interface procedure p'_{int} which requires synchronization in order to apply again the JOIN procedure.

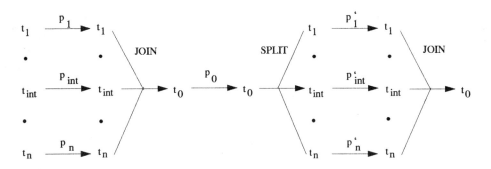

The semantics of a distributed graph class specification is defined as follows. We assume to have a fixed class of graphs and graph morphisms, which can be considered as objects and morphisms of a category **GRAPHS**. This allows to express the glueing of graphs by colimits in the category **GRAPHS**.

4.2 Definition (Distributed Graph Transformation Systems)

1. Given a distributed graph class specification $DGSPEC$ as in 4.1, a *distributed DGSPEC-graph transformation system*
   $DGTS = (GTS_1, ..., GTS_n, GTS_0)$
   consists of
   $GTS_i =$ local $GSPEC_i$-graph transformation system $(i = 1, ..., n)$
   $GTS_0 =$ global $GSPEC_0$-graph transformation system which, restricted to $GSPEC_i$, coincides with GTS_i (i = 1,..., n)
   All domains $D_t(t \in T_0)$ of graphs or graph morphisms are included in the given category **GRAPHS**.

2. The distribution constraint c_{dist} for $DGTS$ has the following meaning fixing the interpretation of JOIN and SPLIT, i.e. $DGTS \models c_{dist}$ iff the following condition $DIST$ is satisfied for the graph procedures
$JOIN_{GTS_0} : D_{t_1} \times ... \times D_{t_n} \times D_{t_{int}} \rightsquigarrow D_{t_0}$
$SPLIT_{GTS_0} : D_{t_0} \rightsquigarrow D_{t_1} \times ... \times D_{t_n} \times D_{t_{int}}$
$DIST$: For all $G_i \in D_{t_i} (i = 0, ..., n)$ and $(G_{int}, f_1, ..., f_n) \in D_{t_{int}}$ we have $G_0 \in JOIN_{GTS_0}(G_1, ..., G_n, (G_{int}, f_1, ..., f_n))$ and $(G_1, ..., G_n, (G_{int}, f_1, ..., f_n)) \in SPLIT_{GTS_0}(G_0)$
iff f_i are morphisms in **GRAPHS** with $fi : G_{int} \longrightarrow G_i (i = 1, ..., n)$ and G_0 is colimitobject in **GRAPHS** in the following diagram:

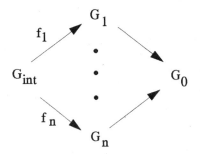

4.3 Example

In the running example discussed in section 2.1, we have sketched a distributed graph class specification with two local graph class specifications for PERSON and BIKE and one global graph class specification for BIKE SHOP. The interface, local state, and global state graphs shown in figure 2 correspond to the graphs G_{int}, G_1, G_2, and G_0 in the colimit diagram above where all morphisms are inclusions. Accordingly, the local operations modify only the local state graphs G_1 and G_2, while the global operation ("to buy a new bike") is an operation on G_0.

5 Graph Class Specifications with Inheritance

In this section we give a first approach how to formalize the inheritance concept outlined in 2.2 in the framework of graph class specifications presented in section 3. The main idea is to introduce morphisms on graph class signatures and graph class specifications, and graph inheritance morphisms. The idea of an inheritance morphism is to state explicitly which graph types and graph procedures should be inherited from the old to the new specification. This can be formalized in graph class specifications by the fact that the constraints for the corresponding types and procedures are equivalent in the old and the new specification. A specification morphism from an old to a new specification allows to redefine types and procedures in such a way that the new constraints imply the old ones, but the new ones may require additional properties. This makes sense especially if the old specification has loose semantics, i.e., the semantics of types and procedures is not yet fixed by the constraint. If the old specification has tight semantics, then the corresponding types and procedures in the new specification have the same semantics, but the new specification may include new types and operations not available in the old one. This notion of inheritance morphisms is motivated by "superposition morphisms" in [FR 92] which are based on modal logics in order to specify reactive systems. In our case we don't want to fix a specific logic by using very general constraints as introduced in section 3. But in this section we have to be slightly more specific. We assume that we have separate constraints for each type and each procedure symbol in a specification and that constraints can be translated via signature morphisms.

5.1 General Assumptions for Constraints

For each graph class specification $GSPEC = (T, P, C)$ we assume that the constraints C are given by $C = \{constr(t) \mid t \in T\} \cup \{constr(p) \mid p \in P\}$.

5.2 Definition (Graph Class Signature Morphisms, Graph Class Specification and Inheritance Morphisms)

1. Given graph class signatures $GSIG_i = (T_i, P_i), (i = 1, 2)$ a *graph class signature morphism* $f : GSIG_1 \longrightarrow GSIG_2$ is a pair $f = (f_T : T_1 \longrightarrow T_2, f_P : P_1 \longrightarrow P_2)$ of functions satisfying the following condition:
   For all $p_1 : u_1 \longrightarrow v_1$ in $GSIG_1$ we have $f_P(p_1) : f_T^*(u_1) \longrightarrow f_T^*(v_1)$ in $GSIG_2$ with $f_T^* : T_1^* \longrightarrow T_2^*$, where f_T^* is the free extension of f_T.
2. Given graph class specifications $GSPEC_i = (GSIG_i, C_i)$ $(i = 1, 2)$, a *graph class specification morphism* $f : GSPEC_1 \longrightarrow GSPEC_2$ is a graph class signature morphism $f : GSIG_1 \longrightarrow GSIG_2$ satisfying the following condition:
   For all $t_1 \in T_1$ and $p_1 \in P_1$ we have
   - translated constraints $f(constr_1(t_1))$ and $f(constr_1(p_1))$ which satisfy the satisfaction condition in the sense of [EM 90]
   - $constr_2(f_T(t_1)) \Longrightarrow f(constr_1(t_1))$ and
     $constr_2(f_P(p_1)) \Longrightarrow f(constr_1(p_1))$
     where the implication $c \Longrightarrow c'$ means that for all $GSIG$-graph transformation systems GTS we have: $(GTS \models c) \Longrightarrow (GTS \models c')$.
3. A graph class specification morphism $f : GSPEC_1 \longrightarrow GSPEC_2$ is called *inheritance morphism* if the implications above are equivalences, i.e. for all $t \in T_1$ and $p \in P_1$ we have:
   $constr_2(f_T(t)) \Longleftrightarrow f(constr_1(t))$ and $constr_2(f_P(p)) \Longleftrightarrow f(constr_1(p))$.

Remarks and Interpretation

1. A graph class signature morphism allows to establish a formal relationship between two graph class signatures $GSIG_1$ and $GSIG_2$, where types and procedure symbols of $GSIG_1$ are related to those of $GSIG_2$ by the functions f_T and f_P. This is an important first step in order to consider inheritance. The condition for graph class signature morphisms ensures that graph class procedure symbols are mapped in a type compatible way.
2. The graph class specification morphism condition means that the constraints for $f(t_1)$ and $f(p_1)$ in $GSPEC_2$ are at least as strong as the corresponding constraints for t_1 and p_1 in $GSPEC_1$. In fact, we may require additional properties for $f(t_1)$ and $f(p_1)$, which corresponds to *redefinition* of t_1 and p_1. In the case of inheritance, the constraints in $GSPEC_1$ and $GSPEC_2$ have to be equivalent. This means that in this case no redefinition is possible.

5.3 Example

In the running example given in section 2.2, we have discussed graph class specifications PERSON and EMPLOYEE where the inheritance corresponds to an inheritance morphism from PERSON to EMPLOYEE. Assuming tight semantics this means that the graph transformation system GTS_2 for EMPLOYEE inherits all properties from the graph transformation system GTS_1 for PERSON, and the restriction of GTS_2 is equal to GTS_1. This restriction can be formally defined as reduct according to the following definition.

5.4 Definition (Reduct)

Given a graph class signature morphism $f : GSIG_1 \longrightarrow GSIG_2$ and a $GSIG_2$-graph transformation system GTS_2, there is a $GSIG_1$-reduct $GTS_1 = V_f(GTS_2)$ of GTS_2 defined by
$GTS_1 = ((D_t)_{t \in T_1}, (p_{GTS_1})_{p \in P_1})$ with
$D_t = D_{f(t)}$ in GTS_2 and $p_{GTS_1} = f(p)_{GTS_2}$ for $t \in T_1, p \in P_1$.

The following reduct property shows that the reduct of a $GSPEC_2$-graph transformation system GTS_2 becomes a $GSPEC_1$-graph transformation system GTS_1, i.e. satisfies the constraints of $GSPEC_1$.

5.5 Lemma (Reduct Property for Graph Transformation Systems)

Given a graph specification morphism $f : GSPEC_1 \longrightarrow GSPEC_2$ and a $GSPEC_2$-graph transformation system GTS_2, the $GSIG_1$-reduct $GTS_1 = V_f(GTS_2)$ of GTS_2 is a $GSPEC_1$-graph-transformation system, called $GSPEC_1$-*reduct* of GTS_2.

Proof

1. Type compatibility of f implies that GTS_1 is well defined.

2. GTS_1 satisfies the constraints C_1 according to the graph class specification morphism condition and the general satisfaction condition in 5.2.2
$GTS_2 \models f(constr_1(x)) \iff V_f(GTS_2) \models constr_1(x)$ for all $x \in T_1 \cup P_1$.

In order to define graph class specifications with import and export interfaces in the next section, we have to extend our constructions above to categories and functors of graph transformation systems in the following sense.

5.6 Remarks (Categories and Functors of Graph Transformation Systems)

1. In analogy to algebraic signature and specification morphisms in [EM 85 and 90] we have defined graph class signature and specification morphisms leading to categories **GSIG** and **GSPEC**. Similar to SIG-homomorphisms for SIG-algebras we are able to define GSIG-morphisms between GSIG-graph transformation systems. This leads to a category **Cat(GSIG)** and, by restriction to GSPEC-graph transformation systems, to a category **Cat(GSPEC)**. The reduct property in 5.4 defines a contravariant functor Cat from **GSPEC** to the category **CatCat** of all categories leading to a specification frame (**GSPEC**, Cat) in the sense of [EG 94]. This functor Cat is a restriction of a corresponding functor Mod from **GSIG** to **CatCat**. Due to the general assumptions 5.1, we have an institution (**GSIG**, Constr, Mod) (see [GB 92] where Constr(GSIG) is the class CC of all constraints over GSIG (see 3.1). These classes of constraints for GSIG in **GSIG** together with translation of constraints (see 5.1.2) are defining a functor Constr from **GSIG** to the category of classes. More precisely this should be called "quasi-institution" because institutions allow only "sets" of constraints instead of classes. In fact, the specification frame (**GSPEC**, Cat) is induced by the institution (**GSIG**, Constr, Mod) in the sense of [EG 94]. The general framework of institutions and specification frames allow to apply several general constructions to graph class signatures, specifications and graph transformation systems which will partly be discussed in the next section and discussed in more detail in a subsequent paper.
2. The reduct constructions can be extended to functors $V_f : \mathbf{Cat(GSIG_2)} \longrightarrow \mathbf{Cat(GSIG_1)}$, resp. $V_f : \mathbf{Cat(GSPEC_2)} \longrightarrow \mathbf{Cat(GSPEC_1)}$, which are usually called forgetful functors because they "forget" part of the structure.

6 Graph Class Specifications with Import and Export Interfaces

In this section we give a first approach how the import-export-interface concept proposed in 2.3 can be applied to graph class specifications and graph transformation systems. Although this concept is the most difficult one to apply to convential graph grammars, it seems to be straight forward once we have the general definitions of graph class specifications and graph transformation systems presented in section 3. The main new aspect is the notion of a "constructive graph class specification morphism" $f : GSPEC_1 \longrightarrow GSPEC_2$ which allows to construct for each $GSPEC_1$–graph transformation system GTS_1 a new system GTS_2 satisfying the specification $GSPEC_2$. In the case of algebraic specification morphisms $f : SPEC_1 \longrightarrow SPEC_2$ there is always a free construction $F_f : Cat(SPEC_1) \longrightarrow Cat(SPEC_2)$ (see [EM 85/90]). In the case of graph class specification morphisms we assume to have a construction only for specific morphisms, depending on the kind of constraints in $GSPEC_2$. The construction is some functor which allows to construct domains and graph class procedures for GTS_2 from those of GTS_1 using graph transformation mechanisms like parallelism, concurrency, amalgamation, distributed derivations, programmed graph transformations and transformation units (see [Kr 95]).

In this section, we need the notions of categories and functors of graph transformation systems as introduced in remark 5.6 of the last section. Using these notions, we can define constructive graph class specification morphisms and graph class specifications with interfaces as follows:

6.1 Definition (Constructive Graph Class Specification Morphism)

A *constructive graph class specification morphism* (f, F_f) from $GSPEC_1$ to $GSPEC_2$ is a graph class specification morphism $f : GSPEC_1 \longrightarrow GSPEC_2$ together with a functor $F_f : \mathbf{Cat(GSPEC_1)} \longrightarrow \mathbf{Cat(GSPEC_2)}$.

6.2 Definition (Graph Class Specifications with Interfaces)

1. A *graph class specification with interfaces*
   $IGSPEC = (PAR, EXP, IMP, BOD, e, i, s, v)$
   consists of graph class specifications PAR *(parameter interface)*, EXP *(export interface)*, IMP *(import interface)*, and BOD *(body specification)*, and graph class specification morphisms such that the following diagram commutes

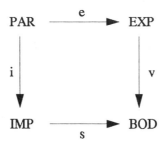

and a construction F_s such that (s, F_s) is a constructive graph class specification.

2. The *semantics* of $IGSPEC$ is given by the categories $\mathbf{Cat}(INT)$ of the interface specifications $INT = PAR, EXP, IMP$, called *loose semantics*, and the functor $SEM = V_v \circ F_s : \mathbf{Cat}(IMP) \longrightarrow \mathbf{Cat}(EXP)$, called *constructive semantics*.

Remarks and Interpretation

Since (s, F_s) is a constructive graph class specification morphism we have a functor $F_s : \mathbf{Cat}(IMP) \longrightarrow \mathbf{Cat}(BOD)$ which can be combined with the forgetful functor $V_v : \mathbf{Cat}(BOD) \longrightarrow \mathbf{Cat}(EXP)$ (see remark of 5.6.2) to obtain the functor $SEM : \mathbf{Cat}(IMP) \longrightarrow \mathbf{Cat}(EXP)$. This means that we obtain for each IMP-graph transformation system GTS an EXP-graph-transformation system $SEM(GTS)$ where the export procedures are constructed from the import procedures in GTS using perhaps additional body procedures and the construction mechanisms defined by the constructive body constraints.

6.3 Remark (Interconnection Mechanisms)

Similar to algebraic module specifications in [EM 90] we are able to define several interconnection mechanisms between graph class specifications with interfaces, especially composition, union and actualization, leading from given components to new graph class specifications with interfaces.

6.4 Example

In the running example sketched in section 2.3, BIKE can be considered as a graph class specification with import interface represented by the black box in figure 3. WHEEL would be another graph class specification with export interface also represented by a black box. The interconnection mechanism composition (see 6.3) allows to match the import interface of BIKE with the export interface of WHEEL.

7 Discussion and Conclusions

Following the pragmatic aspects and the running example in section 2, we have presented in sections 3 to 6 a conceptual, formal framework for a module concept for graph transformation systems. What is still missing, is a concrete application of the introduced notions to an existing graph grammar based specification approach. Candidates for that are, for instance, the algebraic DPO (double pushout) or SPO (single pushout) approach ([Ehr et al. 87], [EL 91]), or the PROGRES language [Sch 91]. Parts of the introduced notions have already been applied to these approaches or are already a constituent of them. For instance, the notion of constraints is already incorporated in PROGRES, as it is possible to define type constraints using Σ-formulas over Σ-graph signatures (see chapter 3 of [Sch91a]) or constructive constraints by using programmed graph replacement systems.

It is due to future work to define a concrete specification language based on graph transformation systems, which combines all concepts presented in this paper in a uniform module concept. In fact, there seems to be a close relationship between the distributed state graph concept and the union concept for graph specifications with interfaces, because different components of the union might be considered as local graph specifications with local state graphs. The union concept however integrates the local components via a shared subspecification, the distributed concept via a global specification.

There is also a close relationship between the inheritance concept and the import-export-interface concept, because both are based on the same kind of graph specifications with a mixed form of loose and tight semantics and suitable notions of graph specification morphisms. In fact, an intermediate step could be graph specifications with formal parameters, similar to parameterized algebraic specifications in [EM 85].

If this specification language is available, the example introduced in section 2, can be developed in more detail in order to get a complete specification. At the moment, it served only as an illustration for the requirements for a sophisticated module concept.

Acknowledgements

A first version of this paper was developed during the visit of H. Ehrig in Leiden in October 1993 in connection with his Kloosterman professorship. Some of the ideas were born during previous meetings of members of the ESPRIT Basic Research Working Group COMPUGRAPH, especially the BrAaBeLei-meetings. We are most grateful to the University of Leiden and the ESPRIT BRA's COMPUGRAPH, COMPASS and ISCORE for direct and indirect support of this paper. We are also grateful to the anonymous referees for their valuable comments and suggestions.

References

[EE 93]   Ehrig, H., Engels, G.: Towards a Module Concept for Graph Transformation Systems. Technical Report 93-34, Vakgroep Informatica, Rijksuniversiteit Leiden, October 1993

[EG 94]   Ehrig, H., Große-Rhode, M.: Functional Theory of Parameterized Specifications in a General Specification Framework. TCS 135 (1994), 221-266.

[EHKP 91] Ehrig, H., Habel, A., Kreowski, H.J., Parisi-Presicce, F.: From Graph Grammars to High Level Replacement Systems, LNCS 532, Springer, Berlin, 269-291

[Ehr 79]   Ehrig, H.: Introduction to the Algebraic Theory of Graph Grammars, Springer LNCS 73 (1979), 1-69

[Ehr et al 87] Ehrig et al: Towards Distributed Graph Grammars, in [ENRR 87], 86-98

[EL 91]   Ehrig, H., Löwe, M.: Parallel and Distributed Derivations in the Single Pushout Approach, TCS, 123-143 (1993)

[ELNSS 92] G. Engels, C. Lewerentz, M. Nagl, W. Schäfer, A. Schürr: Building Integrated Software Development Environments, Part I: Tool Specification. ACM Transactions on Software Engineering and Methodology, Vol.1, No.2, April 1992, 135-167

[ELS 87] Engels, G., Lewerentz, L., Schäfer, W.: Graph Grammar Engineering: A Software Specification Method, in [ENRR 87], 186-201

[EM 85/90] Ehrig, H., Mahr, B.: Fundamentals of Algebraic Specifications 1/2. EATCS-Monographs in TCS, Springer 1985/1990

[ENRR 87] Ehrig., H, Nagl, M., Rozenberg, G., Rosenfeld, A.: Graph Grammars and their Application to Computer Science, Springer LNCS 291 (1987)

[FR 92] Fiadeiro, J.L., Reichwein, G.: A Categorical Theory of Superposition, Technical Report, Department of Mathematics, University of Lisbon 17/92 (1992)

[GB 92] Goguen, J.A., Burstall, R.M.: Institutions: Abstract Model Theory for Specification and Programming, JACM 39,1, 1992, 95-146

[Kr 95] Kreowski, H.-J.: Graph Grammars for Software Specification and Programming: An Eulogy in Praise of GRACE. In G. Valiente Feruglio, F. Rossello Llompart (Eds.): Colloquium on Graph Transformation and Its Application in Computer Science. Technical Report B-19, June 1995, Universitat de les Illes Balears, Mallorca (1995), 55-62

[KR 90] Kreowski, H.-J., Rozenberg, G.: On Structured Graph Grammars. Information Sciences 52, 185-210 (1990)

[Löw 93] Löwe, M.: Algebraic Approach to Single Pushout Graph Transformations, TCS 109, 181-224 (1993)

[Par 72] Parnas, D.C.: A Technique for Software Module Specification with Examples, CACM 15, 5 (1972), 330-336

[RBP 91] Rumbaugh, J., Blaha, M., Premerlani, W., Eddy, F., Lorensen, W.: Object-Oriented Modeling and Design, Prentice-Hall 1991

[Ro 87] Rozenberg, G.: An Introduction to the NLC way of rewriting graphs. In [ENRR 87], 55-66

[Sch 91a] Schürr, A.: Operationales Spezifizieren mit programmierten Graphersetzungssystemen, DUV 1991

[Sch 91b] Schürr, A.: PROGRESS: A VHL-language based on graph grammars. In H. Ehrig, H.-J. Kreowski, G. Rozenberg (eds.): Proceedings of the 4th International Workshop on Graph Grammars and Their Application to Computer Science, Bremen, March 1990, Springer, LNCS 532, Berlin 1991, 641-655

Software Integration Problems and Coupling of Graph Grammar Specifications

Manfred Nagl / Andy Schürr

Lehrstuhl Informatik III, Aachen University of Technology,
Ahornstr. 55, D−52056 Aachen, Germany

Abstract. This paper discusses graph grammar (gragra) specification and (de)composition problems which are related to the realization of integrated development environments. Gragras are used as an operational specification method for formally describing the effect of tools on internal configurations of such an environment. We propose a new approach for decomposing large unstructured gragra specifications into coupled modules (subspecifications), sketch how the interfaces of interconnected modules look like, and how coupling can be described by means of correspondence relationships. Furthermore, we discuss a number of technical difficulties coming up with the usage of correspondence relationships for coupling gragra modules.

1 Introduction

Various *integrated development environments* for different fields of *constructive development* have been investigated and corresponding prototypes have been built in our group [8]. In the IPSEN 1.0 prototype [4, 25] a tightly integrated programming environment has been developed. The 2.0 prototype was an integrated software development environment, consisting of a design, a programming, and a documentation environment [15]. Further prototypes contain a requirements specification environment [9, 11, 18] and an administration environment for management aspects of an integrated and distributed project [13]. The administration environment manages cooperative development processes and was developed for quite another context, namely for a posteriori integration of CIM components in the SUKITS project [6]. Nevertheless, it may also be applied to software development. Finally, the graph grammar specification environment PROGRES offers means for studying and implementing these integrated environments for *software development, CIM,* and *formal specification* [27, 31].

All environments handle the development, maintenance, and evaluation of *complex overall configurations* [22, 23] consisting of extended technical configurations (in software engineering: requirements spec, design specs, program modules etc. belonging to working areas requirements engineering, design, implementation etc.). These configurations consist of subconfigurations (e.g. for subsystems), *documents* (a subsystem design spec), *increments* within documents (a type declaration in a module interface), and many fine-grained consistency *relations* within but especially between different documents (e.g. consistency of imports and exports within a design spec or between different design spec documents). *Fine-grained integration* within and between different documents was studied in the above prototypes ranging from intradocument consistency relations to interdocument relations within the subconfiguration of one working area (as design) to interdocument relations between different subconfigurations of different working areas (as between requirements engineering and design).

In order to build such environments a product *reuse machinery* [8] has been introduced consisting of an architectural framework for all these environments and basic components (as the underlying object storage [14] etc.). The major part of realizing such systems still consists of building up various data structures for internal documents for logical, representation, integration, and administration purposes together with their mutual relations. Especially, it has to be determined which complex internal transactions are induced by command activations of the user and which internal consequences result. So, overall configurations have their pendant in *internal configurations* [23] of integrated environments which support development and maintenance of these configurations by groups of developers.

In order to study these internal configurations carefully before realizing them, *graph grammar specifications* [7, 19, 27] were written. They operationally specify internal documents but also their mutual consistency relations. A methodology for writing specs was developed, called *graph grammar engineering* [2, 3, 20]. Starting with specifications of single internal logical document classes [4], it was later on extended to handle subconfigurations and fine-grained relations between subconfigurations [30, 18], the latter being based on triple graph grammars [28].

Up to IPSEN 2.0 such specifications were *hand-written*, and were mechanically translated by hand to get equivalent efficient code (process reuse). For a few years, we *build up, maintain* and, recently, also *execute* such specifications using the above mentioned graph grammar specification environment, and there is a machinery to *generate* corresponding pieces of code automatically [27, 31].

So, the long term goal of our projects [8] is to *mechanize* the *production* of *integrated* and *distributed environments* for cooperative development and different application areas by (a) extending the architectural framework, (b) its basic components in direction of distributedness [10, 14], (c) developing a parameterizable administration component for handling management aspects [23], (d) specifying the components of internal configurations which are specific to an application, and (e) generating efficient code for such specifications to be plugged into the code of the framework.

At the moment our prototypes are more or less ahead of this mechanical and formalized proceeding inasmuch as they show integrated behavior, however, with a big bunch of hand-written code. So, what we need in the future, are *appropriate extensions* of the underlying specification language, the specification method, the specification environment, and the generator machinery. This paper focuses on the problems we have found in writing large specifications which give input to the future development of all the other topics. For reasons of simplicity we restrict ourselves to the application area of software engineering environments in the sequel.

2 Specification Problems

In the following we sketch the *specification problems* we have detected when building intelligent and integrated development environments. We do this to fix the realm of problems and to introduce what is regarded in this paper. The range is from local to global problems within the internal overall configuration, from application specific to generic procedures and their parameterization.

(1) Even one *logical* internal *document* is a rather complex graph class. It can be built up from a tree part (abstract syntax tree), consistency relations (e.g. for context sensitive syntax relations) and in the case of an executable document from a part supporting and realizing runtime behavior. In any case, general and reusable specification portions can be factorized out (e.g. for general abstract syntax tree handling) and used for writing a specific specification. This was outlined in [22] and successfully demonstrated for the internals of the PROGRES environment in [31].

(2) For any kind of integration specifications (e.g. for an incremental integration tool from requirements engineering to design) it is true that the integration tool need not and must not see all the internals of involved documents. So, what is needed is a *view concept* hiding unimportant details and offering the information needed for the specific integration problem. We are now giving examples for integration problems.

(3) For any logical document there may exist different representations (e.g. in graphical, tabular, or text form). Conversely, also for different logical documents there may exist one combined representation. Representations in the above projects are also internally handled as graphs. Therefore, *logical* and *representation* documents have to be *coupled*, in order to describe so-called incremental unparsers and parsers. Applying the view concept, a specific unparser or parser needs only read or write access to views of the involved logical and representation documents.

(4) Within an overall configuration and, correspondingly, an internal configuration many integration problems occur. The first is sometimes called *horizontal integration* [19]. This means, that different documents of one working area (e.g. requirements engineering) are seen as parts of one subconfiguration. Two forms of horizontal integration can be identified: Integrating different perspectives (as e.g. process and data modelling within requirements engineering), the corresponding cooperative development sometimes being called concurrent engineering, or the integration between a master document (e.g. a design defining a subsystem) and its dependent document (e.g. the detailed design of that subsystem).

(5) *Vertical integration* is integration of documents belonging to different working areas and, therefore, to different subconfigurations (e.g. integration between requirements engineering and design) [9, 17, 18, 30]. New forms of cooperation avoiding unnecessary sequentiality are sometimes called simultaneous engineering. Although being quantitatively more complex the situation is often conceptually easier to handle: In vertical integration usually master and dependent documents may be distinguished, such that document updates must be propagated in one direction only (from master to the dependent documents).

(6) Development environments mainly give product support by offering intelligent tools for the construction/change of a (sub)configuration, thereby guaranteeing structural integrity. The process itself is left open and the developer or the group of developers may carry out quite different processes to deliver the same product. In so-called process-centered environments [29] the process itself is explicitly stored for analysis, evaluation, retracing, etc. purposes. Following such an approach means that internally *process* and *product* descriptions have to be *integrated*.

(7) The administration configuration of an overall configuration contains coarse-grained descriptions of information contained in the extended technical configuration, but also of the processes, resources etc. in order to control and to support the cooperation of technical developers. This is only possible, if both *layers* (technical and administration layer) are *integrated*. This is necessary, for example, to describe the behavior of a management tool, which, after the design, automatically inserts all the necessary implementation, documentation, etc. tasks into a process net. The corresponding information has to be delivered by a technical design analysis tool.

(8) As outlined in [23] the overall configuration of a software system is changing at project runtime, which is called *dynamics* (e.g. the administration configuration is changed according to evolving technical details). Even more, *parametrization and instantiation mechanisms* are needed for object, type, and rule patterns to adapt the administration configuration to needs of a specific project or a whole family of projects. In principal, dynamic problems do also occur on technical layer and do influence each other. As a consequence, a layered model for an overall configuration was introduced. In graph grammar terms this means that the schema of such a configuration as well as the rules and pregiven patterns change at project runtime. The problem is even more complicated as in cooperation between companies different models for overall configurations within different subprojects can be used and the *different overall configurations* have to be *integrated*.

Problem (1) was introduced above as being solved. This is, however, only true for different *specification portions* which nevertheless work on one big graph. What is missing at the moment is to have a suitable graph grammar language which has syntactical expressiveness for specification portions in the sense of *specification modules* and their interconnections, reflecting graph classes, abstractions on graph classes, and their integration to configuration classes. This is even more important for problems (2) to (8). First steps in this direction are given in [1, 12].

The spec problems (1) to (8) have to do with integration and cannot be handled adequately by means of traditional module interconnection concepts: (1) Abstraction/encapsulation cannot be total as in usual object-based design approaches, where e.g. the realization of a stack is completely hidden (whether with an array, linked list on the heap, form of file, etc.). (2) This is due to the fact that integration and coupling needs internals: Not only documents A and B have to be integrated but it has to be said how integration looks like. (3) Especially, in order to realize fine-grained relations between different documents, increments of one document have to be related to increments of the other one. That means that the existence of increments, their compositions etc. must be available for the integration specification. Thereby, the specificator has to be very careful, what to reveal and what to hide.

From the above list of *problems* we address (1) to (5) in this *paper*. We do this by first taking a simplified example, which is studied in the next section to work out the specification problems and their formulation in more detail. This discussion delivers the necessary understanding needed for the following discourse. This discourse is on two levels: We first sketch which specification module interconnection patterns we need in order to handle the regarded problems. In the following section we discuss which spec coupling problems arise when defining a module concept for large specifications, which is able to handle the above problems, and match the specification interconnection patterns.

3 The View Problem and a Possible Solution

We are now discussing the above mentioned *specification problem (2)* taking a *small* (and toy) *example*. This is done in order to give detailed treatment in a limited space. Nevertheless, we can clearly state the problem, describe technical details behind it, suggest a solution, and propose a language for describing this solution.

Remember that we distinguish between logical documents being abstract data types or graph classes containing the knowledge of user documents, their representations which are kept separately, and the corresponding coupling. We only regard logical documents in this section. Logical documents offer data services for tools on the corresponding graph class (editing, analysis, instrumenting, executing, monitoring in addition for executable documents). There are good reasons, the discussion of which is outside the scope of this paper, that in order to have a natural behavior of tools and their corresponding representations to the user, there nearly has to be a 1-1-mapping between representations and logical document behavior. To follow this argument and to accomplish tool and unparser unspecific logical document construction we define *different views* on a logical document which are *customized* to corresponding tools and representations/unparsers.

The example we take in the following is a editing/analysis step of a tool for a simplified programming-in-the-large language [21]. The corresponding logical document of the tool is a graph describing the structure of a design consisting of modules and interconnections. We define two *views* for the logical document, namely a *textual* and a *graphical* one.

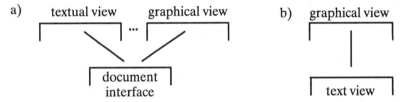

Fig. 1: Two views on a logical document necessary for defining corresponding tools (and parsers as well as unparsers): a) situation as it should be, b) simplified

The *situation* as it *should be* is depicted in fig. 1.a. The logical programming-in-the-large document has a general interface; specific view interfaces are derived for text and graphics following the above arguments. For the discussion in this section we use a *simplified* situation as given in fig. 1.b. This is due to the fact (1) that textual representation of software engineering documents are often syntactically and semantically richer than the corresponding graphical representations, and (2) that we started in IPSEN with the development of textual tools and later developed graphical tools. Therefore, fig. 1.b reflects, what we did.

```
system S is
    module A is
        imports:  from B imp F;
        exports:  type T,
                  proc Cr (...);
        ...
    end A;
    module B is
        exports:  proc F (...);
        ...
    end B;
    ...
end S;
```

      a) Text representation   b) Graphical representation

Fig. 2: A small design example: a) text representation, b) graphical representation

Situations of figs. 1.a but also of 1.b are not easy to handle. A change of the textual view may also change the graphical view, a change of the graphical view does change the textual view. So, both views are *tightly* coupled. This means that the arcs of fig. 1 represent *bidirectional coupling* of views.

Fig. 2 describes the information of a tiny *design spec* in *textual* and *graphical* form which we have to encode in a logical graph class and the corresponding views. In the text representation of (fig. 2.a) a system is a list of modules (here A, B) and interrelations (A and B have export interfaces, A is importing one resource of B). In the graphical representation (fig. 2.b) a system is a box with its constituents being inner boxes. Type and other exports are drawn separately. Import relations are unspecific, only by saying that some resources of B are imported by A.

Fig. 3 contains two *graph views* which meet the needs of the *textual* and the *graphical representations* of fig. 2. Prefix T_ in node labels stands for textual, G_ for graphical, suffixes or infixes M for module, Sy for System, T for type, P for procedure, R for resource, Id for identifier, L for list, D for declaring, A for applied occurrence, E for export, and I for import. Node attributes are given in quotations. Composition edge labels (consists-of relation) are prefixed by To. Dashed edges represent bindings of applied identifier occurrences to their declarations.

The *differences* between *both views* are as follows: (a) Identifiers in fig. 3.a are represented as structure (ToId edge, DSyId node, together with a node attribute), in fig. 3.b only as attribute value. Formal parameter lists are given in the text view (however not shown), whereas in the graphical view only the resource identifier appears. Import lists in the text view import specific resources which, by corresponding links, are bound to corresponding export interfaces. In the graphical view an edge, node, edge triple only says that some imports are needed. Exports in the graphical view contain different lists for type and procedure exports, whereas the textual view has a single list for all type and procedure exports.

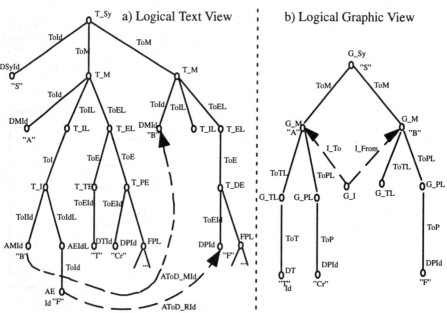

Fig. 3: Logical text and graphical view for the example of fig. 2

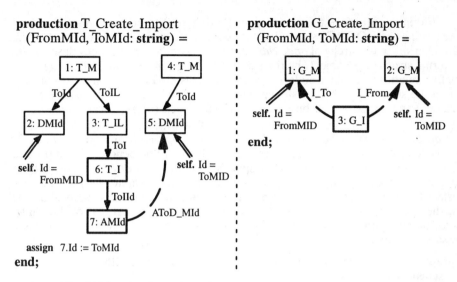

Fig. 4: Editor/analyzer operations inserting import for textual and graphical view

Fig. 4 gives *productions* describing the *effect* of a *textual* or *graphical* editor/analyzer *operation*, each *inserting* a *new import*. In the textual case a new import list is inserted, the insertion of specific imports within the import list is due to another rule which is not shown. In the graphical case an unspecific import is inserted. Both rules of fig. 4.a and b are denoted in PROGRES [26, 27] using the following abbreviation: Nodes and edges, drawn as usual, form the production's left hand side, those in bold additional nodes (edges) its right hand side. So, thin-line nodes and edges are identically replaced. The construct self.Id = ... accesses the identifier attribute of a specific node (the target of the corresponding double arrow) and compares its value with a given parameter value.

The next fig. 5 outlines what we think has to be the interface of both graph views or, more general, the *interface of a gragra spec module*. Such an interface contains all static and dynamic structural information about a class of graphs which is accessible for any client module. The language in which fig. 5 is denoted is the first proposal for a suitable extension of PROGRES.

graph schema T_View;
  (* structural part *)
  node class T_comp;
  node type T_M, T_Sy **is a** T_comp;
  node class DeclCompId
    attribute Id: **string**;
  node type DMId, DSyId
    is a DeclCompId;
  edge type ToId: T_comp → DeclCompId;
  edge type ToM: T_Sy → T_M;
  ...
  (* operational part
    productions as in fig 4.a *)

graph schema G_View;
  (* structural part *)
  node class G_comp
    attribute Id: **string**;
  node class Relship;
  node type G_Sy, G_M
    is a G_comp;
  node type G_I **is a** Relship;
  edge type ToM: G_Sy → G_M;
  edge type ToTL : G_M → G_TL;
  edge type ToPL: G_M → G_PL;
  ...
  (* + productions as in fig 4.b *)

Fig. 5: The interface of a spec portion

Within the interface the object types (node types) to be seen should be fixed. According to PROGRES similarities of *node types* are determined by constructing an inheritance hierarchy of *node classes*. Node classes and node types describe the internal structure of objects by their derived or explicitly set *attributes*. *Edge types* define relations between node types or node classes. Further parts (not shown) could be statical graph consistency relations [30] or generalized object types or classes to comprise objects built up from other objects (hierarchical objects).

Besides the structural part, the operational part has to be given. A first set of *productions* describes how graph classes are *built up*, additional productions for tools describe *changes* due to *tool activations*. It should be denoted that a production is nothing else than a procedure specification in the interface of a usual module, however describing not only the syntax (formal parameter list) but also the effect (dynamic semantics). It is up to the specificator that the structural as well as the operational part of a gragra modul interface is only that part of the knowledge of the graph structure needed by clients and, therefore, abstracts from unnecessary implementation details (see discussion in section 2).

graph interface correspondences between T_Spec, G_Spec with
    schema correspondences
        node class   (* on type level: *)
           T_Comp, DeclCompId $\sim$ G_Comp,   (* node classes,   *)
        ...   (* node/edge types *)
        node type
           T_Sy, DSId $\sim$ G_Sy,
           T_M, DMId $\sim$ G_M,
           T_I $\sim$ G_I,
        ...
        edge type
           ToM $\sim$ ToM,
        ...
    operation correspondences   (* on object/operation level *)
        production
           T_CreateImport $\sim$ G_CreateImport **with**
           1, 2 $\sim$1,
           4, 5 $\sim$2,
           7$\sim$3;

Fig. 6: Relating spec portions to each other by means of correspondences

In fig. 6 we outline now how both views can be related to each other using again an extension of PROGRES. This is done by so-called *correspondences*. Correspondences relate first similar schema declarations within two different interfaces to each other and establish thereby a relation between the static structures of two different classes of graphs. Such a relation is not always a 1-1 mapping, but relates in the general case structural element groups of one graph class to single elements within another graph class or vice-versa. This is necessary for handling our view coupling problems. In fig. 6, for example, node classes T_Comp and DeclCompId are mapped onto one class, thereby mapping node-edge-node structures of fig. 3.a onto single nodes in 3.b (triples of T_Sy-node, ToId-edge, and DSyId-node in fig. 3.a correspond to a single node of type G_Sy in the graph view of 3.b). In the same way as node classes, node types, and edge types can be related to each other, productions and their left and right hand side elements have to be coupled, thereby defining a kind of *pair graph grammar* [28]. The application of coupled productions to coupled graph instances creates or deletes related node (and edge) instances and manipulates thereby fine-grained intergraph relationships.

Graph interface *correspondences are declarative*. They replace the manual implementation of the interface of G_Spec in the body of G_Spec using the interface of T_Spec and, in order to make a bidirectional coupling, also conversely. By the way, implementing both G_Spec by means of T_Spec and T_Spec by means of G_Spec in order to be able to propagate changes *in both directions* is not a feasible solution. More details about gragra module interfaces and their interconnections are given in section 5.

4. Specification Interconnection Patterns

In this section we give an abstract formulation how the problems (1) to (5) of section 2 can be solved by coupling gragra modules, thereby only giving the statical structure of spec module interconnections. So, in terms of [21], we give architectural patterns for composing gragra specifications. By the way, it is our belief that these patterns also apply to other specification approaches (different from gragra specs) and to software architectures. The reader should remember that we are discussing integration problems which have to be solved by specific architectural patterns (see section 2). It is our goal to use interface correspondences to define all kinds of coupling gragra specs which occur in the following patterns (of fig. 7).

Fig. 7.a explains the *composition* of a *logical document*. There is an underlying abstract syntax tree (AST) structure describing the composition of objects from others. This is enriched by further consistency relations to describe the context sensitive syntax, set-use data relations, control flow, etc., forming the static semantics in a wider sense. Runtime support (for executable documents) needs preparation for execution and the execution machinery. From bottom to middle layer we have various extensions of the underlying AST structure, the top layer sums up the interfaces of layer 1 and 2.

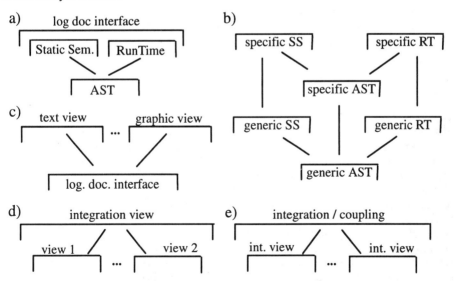

Fig. 7: Spec architecture patterns for integration problems

Generic services for AST, static semantics and runtime handling are available such that the spec modules of a specific logical composition structure need not be built up from scratch (fig. 7.b). So, the components of fig. 7.a are now instantiations of these generic templates.

We have argued in section 3 that we need *views* for *tools* and *unparsers* which translate logical documents into representations (fig. 7.c). The step to get a suitable view from a logical document interface is a kind of customization and involves coarsening, abstracting, and partially hiding information. This specific view has to be realized using the services of the underlying logical document interface.

Fig. 7.d describes the pattern for specifying an *integration* situation. It is similar to the pattern for tool customization of fig. 7.c and defines an appropriate *view* which customizes, coarsens, and partially hides the underlying graph structure. Only that information necessary for integration is presented (in a suitable form) in the view. Of course, the integration view may again be based on underlying views (of tools, for different perspectives) or there may be a summation interface as in fig. 7.a. For any integration situation there may be another integration view.

Finally, 7.e is the pattern to specify *integration* and *coupling* itself. It applies to logical document and representation document coupling, but especially to horizontal and vertical integration. In the first case, the views are the corresponding tool view on the logical document and a suitable representation view on the representation document. So, both can be interpreted as integration views for unparser/parser coupling. The relation between both layers in fig. 7.c is to realize the corresponding coupling/integration. In the second case, the situation of fig. 7.e applies to integrating/coupling different classes of logical documents for horizontal or vertical integration.

We see that the *relations* between gragra modules introduced in fig. 7 have quite *different semantics* (inverse relation are given in parentheses): They express abstraction or information hiding (realization), coarsening or restriction (detailing, mapping into a wider context), summation (projection), enrichment (factorization), customization (aggregation), instantiation (creating generic templates), and coupling (realizing integration). These semantical relations may be realized using quite different implementation concepts (functional, object-based, object-oriented, events/eventhandlers etc.). To summarize, fig. 7 corresponds to problems (1) to (5) of section 2, problems (6) to (8) are not studied in detail yet.

5. Coupling Graph Specifications

Within section 3 we presented a new concept how to specify *relationships between* different gragra specification modules. Modules dealing with related graph structures may be glued together by means of additionally defined correspondences such that changes of one graph instance are propagated to the other one. Within the running example, one graph, the graphical architecture representation, played about the role of an *abstract view* onto the other graph structure, the textual architecture representation (cf. fig. 1.b). In general, both graph structures may belong to the same level of abstraction or they may even be two different views of a third common underlying graph structure (cf. fig. 1.a). In the sequel, we will discuss how *correspondences between gragra modules* allow us to realize all patterns of coupling discussed in section 2 and 3, and how the new concepts are related to import/export relationships or inheritance as well as aggregation relationships of software design languages (programming-in-the-large languages).

We start our discussion with the simplest case, the *a posteriori integration* or *direct coupling* of two gragra modules which export node classes, node types etc. for defining graph schemata as well as productions for defining graph modifying operations at their interface. Let us assume that the graphical and the textual software design language of section 3 were developed and specified independently of each other. Each module offers a set of operations at its interface which are used by some tools for generating, modifying, and analyzing its graph instances. Furthermore, the interfaces exports also knowledge about the internal graph structure of these graph instances as a prerequisite for writing graph rewrite rules within client modules and for handling fine-grained relationships between node instances of different graphs (cf. section 3).

Abstracting from the concrete example we get the following situation involving two specification modules and one client for each of them (see fig. 8): Each client i uses either the export operations of module i or modifies directly the visible part of its internally used graph structure. In both cases, all resulting changes of the visible graph part of one specified graph class have to be translated into changes of the visible graph part of the related graph class. In section 3, we have already sketched the usage of *correspondence definitions* between graph schema components as well as between productions of both involved specifications for this purpose. In order to be able to write these correspondence definitions between two already existing modules, we need *access to all necessary details* of related graph structures through their interfaces. These interfaces have to be extended appropriately, if necessary details are not yet exported (as node class, node type, and edge type declarations).

The definition of *correspondences* is *restricted by rules* like: (1) Superclasses of related subclasses/node types as well as source and target classes of related edge type definitions must be related, too. (2) Related node patterns within left and right sides of related productions have related node type or class inscriptions. (3) Related productions possess identical formal parameter lists (the exchange of parameter types by related types is permitted).

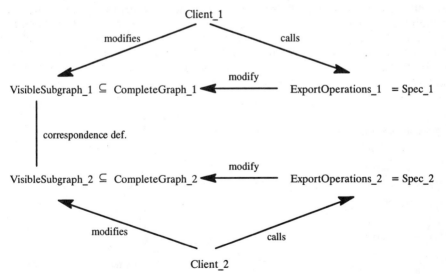

Fig. 8: A posteriori coupling/integration of related gragra specification modules

To summarize, a posteriori integration of two modules is currently possible *without modifying* them as long as these specifications export sufficient knowledge about their internal graph structures. This situation *violates the principle of data abstraction*. However, achieving fine-grained integration of two graph data types without modifying their already existing realizations and without exploiting any knowledge about their internal organization seems to be a contradiction in itself.

Another *drawback* of correspondence definitions is that they are a very high-level declarative specification approach and, therefore, not directly executable. A given correspondence definition is even independent from the fact whether propagation is needed in one direction only or in both directions (horizontal/vertical integration of section 2), or whether propagation should be realized as a batch process or as a continuously active incrementally working process. Currently, we are able to deal with correspondence definitions which involve productions, where the left hand side is a subgraph of its right hand side. In this case, a rather simple pattern matching and propagation algorithm is sufficient to keep related graph instances in a consistent state [16, 17, 18]. The development of algorithms, which are able to deal with correspondences between more general forms of productions, is part of ongoing graph grammar parsing research activities [24].

Until now, we considered the case of direct coupling independent modules with already existing implementations for all interface operations. A related but still more complex problem is the definition of a *new abstract gragra specification module* on top of an already existing more concrete gragra specification module (cf. patterns of fig. 7.c, 7.d, and 7.e). Again, correspondence definitions are used to translate updates of the abstract graph structure into updates of the underlying concrete graph structure and vice-versa. But in addition to the previously discussed situation we have to specify the *effects of new abstract graph operations* (see below).

Fig. 9 contains a diagram of the new situation including the generalization, where more than one abstract specification module is built on top of their common interface to a given concrete specification module. In order to be able to handle this scenario properly, including the well-known *update view problem*, two correspondence definitions must be provided which tie both abstract graph structures to their underlying concrete graph structure. In this case, it seems to be reasonable to view the CorrespondenceDef_i as an essential part of the implementation of the AbstractSpec_i, such that we have access to all internal details of the more AbstractGraph_i and to omit even the distinction between the whole graph structure and its visible part on the abstract level.

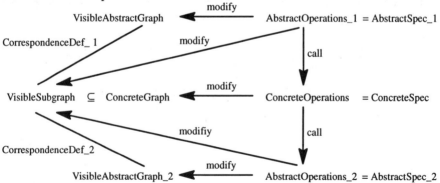

Fig. 9: Building abstract views on top of concrete gragra specification modules

Having established the framework of fig. 9 offers now the opportunity to specify abstract graph operations in three different ways:

1) An abstract graph operation is implemented by *calling* already existing *concrete graph operations*. These concrete graph operations modify the visible as well as invisible part of the concrete graph instance. Afterwards, the given correspondence definition translates all modifications of the visible concrete graph instance back into corresponding modifications of the abstract graph instances.
2) The abstract graph operation *modifies directly* the *visible part* of the *concrete graph instance* (which is defined by the interface of the concrete specification and, therefore, the same for all abstract specifications). Again, the additionally existing correspondence definition is used to propagate modifications of the visible concrete graph instance back into the corresponding abstract instances.
3) The abstract graph operation *works directly on* the new *abstract graph structure*. In this case, all abstract graph modifications are propagated the other way round from the abstract to the concrete layer.

The new concept contains various commonly used *design patterns* supported by programming-in-the-large languages *as special cases*. It should be evident that the architectural patterns of section 4 in fig 7.c., 7.d, and 7.e are more concrete examples of the abstract situation depicted in fig. 9. Furthermore, it is easy to see how *traditional data abstraction* may be realized by hiding any knowledge about concrete (abstract) graph structures in fig. 9. In this case, any direct access to the internal details of concrete graph instances is not possible and abstract operations must be realized by calling sequences of concrete operations.

From a theoretical point of view it should even be possible to model aggregation and inheritance in a similar way by following the lines of another paper [5]. *Aggregation*, i.e. the case where one graph instance contains copies of elsewhere defined graph instances as subgraphs, must be modelled as a situation where the aggregate graph is a view onto more than one related graph instance. In this case, correspondence definitions are partial mappings from an aggregate to its elements (subgraphs). In the simplest case, sometimes called *association*, all element graphs of an aggregate graph are instances of the same specification module. But in general, aggregates may be defined which contain instances of more than one type, thereby supporting integration of previously distinct graphs into a single whole (cf. related architectural pattern of fig. 7.d).

Inheritance or better *specialization* and its inverse relationship, *generalization*, may be seen as a special case of abstraction, where the interface and graph structure of the abstract gragra module is just a subset of the interface and graph structure of the more specific gragra module (cf. fig. 7.a and fig. 7.b). Using this point of view, it seems to be possible to model inheritance relationships by interpreting fig. 9 as follows: (1) Redefined operations of an abstract gragra module call their redefinitions of a more specific gragra module (2) redefined or added operations of the more specific gragra module modify their own graph instances, and (3) correspondence definitions translate modifications of subclass instances into corresponding modifications of (virtually existing) superclass instances.

It is subject for future research to find out whether these rather vague ideas of how aggregation and inheritance may be modelled are of any practical relevance and how they might be integrated into a forthcoming concept of hierarchical graphs which has to support aggregation and inheritance on the level of whole graphs and not just on the level of nodes and edges.

6 Summary

Our background and experience in building complex integrated environments and, especially, our goal to mechanize their realizations based on gragra specifications were the main motivation for studying *gragra specification and composition problems* within this paper (see section 2). These problems arise as soon as we are trying to produce complete gragra specifications of integrated environments which manipulate very complex configurations of related documents, i.e. complexes of graphs with mutual fine-grained dependencies. These specifications must be structured and built by writing first separate gragra modules (subspecifications) with well-defined interfaces and by interconnecting these modules afterwards.

Motivated by the detailed example of section 3, definition of *gragra module views and coupling of views* for different purposes were identified as a key concept. Afterwards, typical architectural patterns which are more less related to the definition of views and view coupling were presented and discussed in section 4.

Abstracting from specific architectural patterns and conceptual relationships between modules in section 5, we were able to suggest a rather general concept for defining and interconnecting gragra modules. It is based on *correspondence definitions* which connect *interfaces of modules* or, more precisely, their exported declarations. Correspondences between graph schema declarations (for node classes, node types, and edge types) establish relationships between graph structures (static graph properties), whereas correspondences between productions define relationships between graph modifying operations (dynamic graph properties).

References

1. Ehrig H., Engels G.: Towards a Module Concept for Graph Transformation Systems, Technical Report 93-34, Department of Computer Science, Leiden University, Netherlands (1993)
2. Engels G., Lewerentz C., Schäfer W.: Graph Grammar Engineering – A Software Specification Method, in: Proc. 3rd Int. Workshop on Graph Grammars and Their Application to Computer Science, LNCS 153, 186-201, Berlin: Springer-Verlag (1987)
3. Engels G., Lewerentz C., Nagl M., Schäfer W., Schürr A.: Building Integrated Software Development Environments, Part I: Tools Specification, in: ACM Transactions on Software Engineering and Methodology 1, 2, 135-167 (1992)
4. Engels G.: Graphs as Central Data Structures in a Software Development Environment (in German), Dissertation Univ. Osnabrück, VDI Fortschrittsberichte 62, Düsseldorf: VDI-Verlag (1986)
5. Ehrich H.D., Saake G., Sernadas A.: Concepts of Object–Orientation, Proc. 2nd IS/KI Workshop, Ulm 1992
6. Eversheim W.,Weck M., Michaeli W.,Nagl M., Spaniol O.: The SUKITS Project – An a posteriori Approach to Integrate CIM-Components, Proc. Ann. GI Conf., informatik aktuell, 494-504 (1992)
7. Claus N., Ehrig H., Rozenberg G. (Eds.): Proc. 1st Int. Workshop on Graph Grammars and Their Application to Computer Science, LNCS 73 (1979), Ehrig H., Nagl M., Rozenberg G. (Eds.): Proc. 2nd Workshop, LNCS 153 (1983), Ehrig H., Nagl M., Rozenfeld A., Rozenberg G. (Eds.): Proc. 3rd Workshop, LNCS 291 (1987), Ehrig H., Kreowski H. J., Rozenberg G. (Eds.): Proc. 4th Workshop, LNCS 532 (1991), Ehrig H., Engels G., Rozenberg G.: (Eds.): Proc. 5th Workshop, this volume
8. Nagl M.: (Ed.): Building Tightly Integrated Software Development Environments – The IPSEN Approach, to appear in LNCS.
9. Janning Th.: Integration of Languages and Tools for Requirements Engineering and Programming in the Large (in German), Dissertation RWTH Aachen, Wiesbaden: Deutscher Universitätsverlag (1992)
10. Klein P., Lacour J., Nagl M., Schmidt V.: Restructuring Client/Server Applications: An example from Business Administration (in German), Proc. Online '95, vol. VI, C630.01-25 (1995)
11. Kohring Ch.: Tightly Integrating Editing, Analysis, and Execution of Requirements Specifications (in German), Ph.D. Thesis, RWTH Aachen (1995), to appear
12. Kreowski H.J.: Graph Grammars for Software Specification and Programming: An Eulology in the Praise of GRACE. in: Proc. Colloquium on Graph Transformations and its Application in Computer Science,

Technical Report B-14, Departament de Ciencies Matematiques i Informatica, Universitat de les Illes Balears, 27-28 (1994)
13. Kiesel N., Schwartz J., Westfechtel B.: Object and Process Management for the Integration of Heterogeneous CIM Components, GI-Jahrestagung 1992, 484-493, Berlin: Springer-Verlag (1992)
14. Kiesel N., Schürr A., Westfechtel B.: GRAS, A Graph-oriented (Software) Engineering Database System, Information Systems 20, 1, 21-51 (1995)
15. Lewerentz C.: Extended Programming in the Large in a Software Development Environment, in Proc. 3rd ACM Symp. on Practical Software Development Environments, Software Engineering Notes 13, 5, 173-182 (1988)
16. Lefering M.: Tools to Support Life Cycle Integration, in: Proc. 6th Software Engineering Environments Conference 1993 (SEE 93), 2-16, Los Alamitos: IEEE Computer Society Press (1993)
17. Lefering M.: Fine-Grained Integrators in Software Development Environments (in German) Ph.D.Thesis, RWTH Aachen, (1994), Aachen: Shaker Verlag
18. Lefering M.: Development of Incremental Integration Tools Using Formal Specifications, Technical Report AIB-94-2, Fachgruppe Informatik, Aachen University of Technology, Germany (1994)
19. Nagl M.: Graph-Grammars: Theory, Applications, and Implementation (in German), Braunschweig: Vieweg-Verlag (1979)
20. Nagl M.: Graph Technology Applied to a Software Project, in Rozenberg, Salomaa (Eds): *The Book of L*, 303-322, Berlin: Springer-Verlag (1985)
21. Nagl M.: Methodological Programming-in-the-Large (in German), Berlin: Springer-Verlag (1990)
22. Nagl M.: Uniform Modelling in Graph Grammar Specifications, in Proc. Dagstuhl Seminar 9301 on Graph Transformations in Computer Science, LNCS 776, Berlin: Springer-Verlag (1994).
23. Nagl M., Westfechtel B.: A Universal Component for the Administration in Distributed and Integrated Development Environments, Technical Report AIB 94-8, RWTH Aachen (1994)
24. Rekers J., Schürr A.: A Parsing Algorithm for Context-Sensitive Graph Grammars, Technical Report 95-05, Departement of Computer Science, Leiden University, Netherlands (1995)
25. Schäfer W.: An Integrated Software Development Environment: Concepts, Design, and Implementation (in German), Dissertation Univ. Osnabrück, VDI Fortschrittsberichte 57, Düsseldorf: VDI-Verlag (1986)
26. Schürr A.: PROGRES: A VHL Language Based on Graph Grammars, in: Proc. 4th Int. Workshop on Graph Grammars and Their Application to Computer Science, LNCS 532, 641-659, Berlin: Springer-Verlag (1991)
27. Schürr A.: Operational Specifications with Programmed Graph Rewriting Systems (in German), Ph.D. Thesis, Wiesbaden: Deutscher Universitätsverlag (1991)
28. Schürr A.: Specification of Graph Translators with Triple Graph Grammars, in: E.W. Mayr (Ed.): Proc. 20th Int. Workshop on Graph-Theoretic Concepts in Computer Science (WG '94), Herrsching, Germany, LNCS 903, Berlin: Springer-Verlag (1995)
29. Proc. International Software Process Workshops, since 1984, IEEE Computer Society Press
30. Westfechtel B.: Revision and Consistency Control in an Integrated Software Development Environment (in German), Ph.D. Thesis, Informatik Fachberichte 280, Berlin: Springer-Verlag (1991)
31. Zündorf A.: Programmed Graph Rewriting Systems: Implementation and Usage (in German); Ph.D. Thesis; Aachen University of Technology (1995)

Using Attributed Flow Graph Parsing to Recognize Clichés in Programs

Linda Mary Wills

School of Electrical and Computer Engineering
Georgia Institute of Technology
Atlanta, Georgia 30332-0250
linda@ee.gatech.edu
http://www.ee.gatech.edu/users/linda/

Abstract. This paper presents a graph parsing approach to recognizing common, stereotypical computational structures, called clichés, in computer programs. Recognition is a powerful technique for efficiently reconstructing useful design information from existing software. We use a flow graph formalism, which is closely related to hypergraph formalisms, to represent programs and clichés and we use attributed flow graph parsing to automate recognition. The formalism includes mechanisms for tolerating variations in programs due to structure sharing (a common optimization in which a structural component is used to play more than one functional role). The formalism has also been designed to capture aggregation relationships on graph edges, which is used to encode aggregate data structure clichés and the abstract operations on them. A chart parsing algorithm is used to solve the problem of determining which clichés in a given cliché library are in a given program.

1 Program Recognition

An experienced programmer can often reconstruct much of the hierarchy of a program's design by recognizing commonly used data structures and algorithms in it and reasoning about how they typically implement higher-level abstractions. We call these commonly used computational structures *clichés* [21]. Examples of clichés are algorithmic computations, such as list enumeration, binary search, and event-driven simulation, and common data structures, such as priority queue and hash table. Since clichés have well-known properties and behaviors, the process of recognizing clichés, which we refer to as *program recognition*, provides an efficient means of reconstructing and understanding a program's design. It bypasses complex reasoning about how behaviors and properties arise from certain combinations of language primitives.

Several researchers have shown the feasibility and usefulness of automating recognition, most recently [9, 10, 12, 13, 20, 25, 26]. A primary motivation for automating recognition is to facilitate tasks requiring program understanding, such as maintaining, debugging, and reusing software.

We have developed an experimental recognition system, called GRASPR ("GRAph-based System for Program Recognition") [26], to automate program recognition.

Given a program and a library of clichés, GRASPR finds all instances of the clichés in a program. It can generate multiple views of a program as well as near-miss recognitions of clichés. It can also recognize clichés in programs even if they are surrounded by or interleaved with unfamiliar code.

The concept of programming clichés is a pre-theoretic notion with no precise formalization. The cliché library is a collection of *standard, frequently used* algorithms and data structures, encoded in terms of their data and control flow constraints. Finding all instances of a cliché in a program means finding program structures that satisfy these constraints; it does not mean recognizing any arbitrary program structure that has the same functionality as the cliché. For example, GRASPR will recognize common implementations of hash tables that are captured in the given library, not all possible program structures that can be viewed as hash tables.

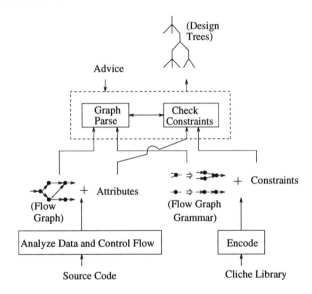

Fig. 1. GRASPR's architecture.

GRASPR uses a graph parsing approach to automating program recognition, shown in Figure 1. It uses data and control flow analysis to represent a program as a restricted form of directed acyclic graph, called a *flow graph* [1, 26], which is annotated with attributes. Nodes in the flow graph represent functions, ports on nodes represent inputs and outputs of the functions, edges connect ports and denote dataflow, and attributes capture additional information, such as recursion, control flow and data aggregation. The cliché library is encoded as an attributed graph grammar, whose rules impose constraints on the attributes of flow graphs matching the rules' right-hand sides. An example constraint is to require that a node have the same "control environment" attribute as another node (i.e., the nodes "co-occur"), which means that the two nodes represent operations that are performed under the same control conditions (e.g., within the same branch of a conditional).

The grammar rules capture implementation relationships between the clichés. Recognition is achieved by parsing the flow graph representing the program in accordance with the grammar. Attribute constraint checking and evaluation are interleaved with the flow graph parsing process. The control flow attributes are themselves represented and manipulated as graphs, since they encode a type of structural information; this allows well-defined graph operations to be used in attribute evaluation and constraint checking.

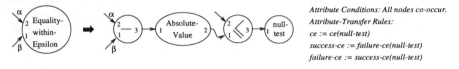

Fig. 2. A flow graph grammar rule encoding the Equality-within-Epsilon cliché.

Figure 2 shows an example of a flow graph grammar rule encoding a simple cliché: testing whether two numbers are within some "epsilon" of each other. The right-hand side is a typical flow graph. The rule for Equality-within-Epsilon constrains all the nodes that match its right-hand side to co-occur. The attribute-transfer rules specify how to synthesize the left-hand side node attributes from the attributes of the flow graph matching the right-hand side. In this example, the null-test is a terminal representing the primitive conditional test of a boolean value. In addition to a control environment (ce), it has a *success-ce* (resp. *failure-ce*) attribute representing the control environment of operations that are performed when the conditional test succeeds (resp. fails).

The control environment attribute of a node indicates under which conditions the operation represented by the node is executed, relative to when other operations in the program are executed. Control environments form a partial order imposed by the control structure of the program (i.e., its branches, iterations, and recursive calls). Other typical constraints on nodes are that they are within the control environment of the success branch of a conditional test, or that they are in the same control environment as a recursive function call. Similar constraints are also imposed on edges, restricting the control environments in which they carry dataflow. (Attribute constraints and transfer rules are stated informally in Fig. 2; [26] gives a formal description of the attribute language.)

Parsing yields a hierarchical description of a plausible design of the program in the form of derivation trees specifying the clichés found and their relationships to each other. In general, GRASPR generates a forest of design trees for a given program. These provide views of the program on multiple levels of abstraction.

Automating program recognition is difficult due to the wide range of possible variations among programs. An instance of a cliché may appear in a wide variety of forms in the text of a program, depending on the constructs or programming language chosen to express their data and control flow. The effort to encode these variations in a cliché library or to perform canonicalizing transformations to the *program text* tends to limit the variability and complexity of the structures that can be recognized. Also, delocalized clichés pose a serious problem.

Our graph-based approach overcomes these problems by shifting the representation of programs and clichés from *text* to a *flow graph*. GRASPR is able to overcome many of the difficulties of syntactic variation and noncontiguousness which hinder recognition. The flow graph representation abstracts away the syntactic features of the code, exposing the program's algorithmic structure. It concisely captures the data and control flow of programs, while suppressing details of the language in which they are expressed. Also, many clichés that are delocalized in the program text are much more localized in the flow graph representation. The flow graph formalism also provides a firm, mathematical basis for a well-defined, algorithmic recognition process that is more robust and easier to analyze and evaluate than previous, more ad hoc approaches.

The next two sections of this paper describe flow graphs and flow graph grammars more formally. This is followed by a formulation of the program recognition problem as flow graph parsing. It focuses on how variations in programs and clichés due to structure sharing are handled. It also describes how aggregation relationships are encoded and how variations in the way aggregate data structures are organized and nested is handled. The chart flow graph parser is then briefly described, followed by an evaluation of its efficiency and of the expressiveness of the flow graph formalism for representing programs and clichés.

2 Flow Graphs

We formulate the program recognition problem in terms of solving a parsing problem for flow graphs. To do this, we are building upon the flow graph formalisms of Brotsky [1] and Lutz [15]. A *flow graph* is an attributed, directed, acyclic graph, whose nodes have *ports* – entry and exit points for edges. Flow graphs have the following properties and restrictions:

1. Each node has a *type* which is taken from a vocabulary of node types.
2. Each node has two disjoint tuples of ports, called its *inputs* and *outputs*. Each port has a *type*, taken from a vocabulary of port types. All nodes of the same type have the same number and type of ports in their input and output port tuples. The size of the input (resp. output) port tuple of a node is called the *input* (resp. *output*) *arity* of the node.
3. A node's inputs (or outputs) may be empty, in which case the node is called a *source* (or *sink*, respectively).
4. Edges do not merely adjoin nodes, but rather edges adjoin ports on nodes. (This is important in representing programs, where input and output ports represent distinct inputs and outputs of functions.) All edges run from an output port on one node to an input port on another node. The ports connected by an edge must have the same port type.
5. More than one edge may adjoin the same port. Edges entering the same input (resp. output) port are called *fan-in* (resp. *fan-out*) *edges*.

6. Ports need not have edges adjoining them. Any input (or output) port in a flow graph that does not have an edge running into (or out of) it is called an *input* (or *output*) of that graph.
7. Each flow graph has a vocabulary of attributes, which is partitioned into two disjoint sets of node attributes and edge attributes. Each attribute has a (possibly infinite) set of possible values.

Notions of flow graphs and flow diagrams have appeared frequently in the literature for more than twenty years. However, our specific type of flow graph was first defined by Brotsky [1], drawing upon the earlier work on *web grammars* [4, 16, 17, 19, 23]. Wills [22, 25] extended Brotsky's definition so that flow graphs can include sinks and sources, fan-in and fan-out edges, and attributes.

Our flow graph formalism is related to that of Lutz, which in turn is equivalent to *hypergraph* formalisms with *hyperedge replacement* [8]. More specifically, Lutz's "flowgraphs" are a special type of our flow graph. (They derived from research on *plex* languages [7].) In addition to nodes, ports, and edges, they contain *tie-points*, which are intermediate points through which ports are connected to each other. Since each port is connected to exactly one tie-point, fan-in and fan-out are not captured to the same level of granularity as is captured by our flow graphs. For example, they cannot express the situation in which an output port p_1 fans out to input ports p_3 and p_4, while output port p_2 is only connected to p_4. Hypergraphs are analogous to Lutz's flowgraphs: nodes in a hypergraph correspond to tie-points and hyperedges correspond to flowgraph nodes.

3 Flow Graph Grammars

A *flow graph grammar* is a set of rewriting rules (or productions), each specifying how a node in a flow graph can be replaced by a particular sub-flow graph. (A flow graph H is a *sub-flow graph* of a flow graph F if and only if H's nodes are a subset of F's nodes, and H's edges are the subset of F's edges that connect only those ports found on nodes of H.)

In addition, the flow graph grammar may be attributed: Each rule specifies how to compute attribute values from the attributes of nodes and edges in the rule. Each rule also imposes constraints on the attributes of the rule's nodes. Every flow graph in the language of an attributed grammar has attribute values that satisfy the constraints of the rules generating the flow graph.

More precisely, a flow graph grammar G has four parts: two disjoint sets N and T of node types, called non-terminals and terminals, respectively, a set P of *productions*, and a set S of distinguished non-terminal types, called the *start types* of G.

Each production in P consists of the following parts:
1. A flow graph L, called the *left-hand side*, containing a single node having a non-terminal type.
2. A flow graph R, called the *right-hand side*, containing nodes of non-terminal or terminal types.

3. A binary *embedding relation* C that specifies the correspondence between the ports of L and R.
4. A set of *attribute conditions* that impose constraints (in the form of relations) on the attribute values of nodes and edges in R.
5. A set of *attribute transfer rules*, each of which specifies the value of an attribute of L's node in terms of the attributes of the nodes and edges in R.
6. A partial or total *node ordering* which constrains the order in which nodes in R are matched (e.g., match the most salient node types first).

3.1 Embedding Relation

A binary embedding relation C relates each left-hand side port to a *tuple* of right-hand side and left-hand side port *sets*, where the position in the tuple is significant and the size of the tuple is ≥ 1.

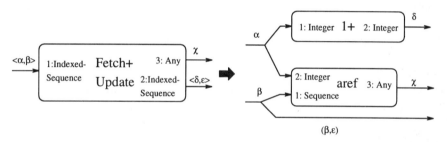

Fig. 3. A grammar rule encoding a clichéd operation on an aggregate data structure. (Attribute conditions and transfer rules are not shown.)

For example, Figure 3 illustrates a graph grammar rule that encodes the cliché Fetch+Update. This is a common implementation of the cliché Stack-Pop in which the Stack is implemented as an Indexed-Sequence (which has two parts: a base sequence and an integer index pointing to the "top" element). Fetch+Update accesses the base sequence and increments the index. This is encoded in the right-hand side of the rule for Fetch+Update. The embedding relation specifies how the inputs and outputs of the right-hand side are aggregated to form the Indexed-Sequences that are the input and an output of the Fetch+Update operation. The embedding relation is shown pictorially in the rule by corresponding Greek letters, where each Greek letter denotes a set of right- or left-hand side ports (e.g., α denotes the set of ports $\{1+_1, \text{aref}_2\}$).

A left-hand side port l_i and a right-hand side port or another left-hand side port p_j are said to "correspond" if $(l_i, t_j) \in C$ and p_j is a member of a set in the tuple t_j. Intuitively, the nonsingleton sets denote fan-in or fan-out of the right-hand side inputs and outputs; the tuples represent aggregation relationships. When a left-hand side port l_1 corresponds with another left-hand side port l_2, the rule is said to contain a *straight-through*. These are used in representing clichéd operations in which some of the input data is not acted upon, but passes directly to the output, as is the case with the base sequence part of the Indexed-Sequence input and output of Fetch+Update. Further details of how the embedding relation is restricted are given in [26].

4 Partial Program Recognition as Flow Graph Parsing

The *subgraph parsing problem* for flow graphs is: Given a flow graph F and a context-free flow graph grammar G, find all parses of all sub-flow graphs of F that are in the language of G.

The *program recognition problem* of determining which clichés in a given library are in a given program (and their locations) is formulated as a subgraph parsing problem: Given a flow graph F representing the program's dataflow and a cliché library encoded as a flow graph grammar G, solve the subgraph parsing problem on F and G. This formulates *partial* program recognition as well as recognition of the program as a whole. A cliché instance may be surrounded by or interleaved with unfamiliar code, but if it is localized in a sub-flow graph of the program's flow graph, it will be recognized by subgraph parsing.

To solve the subgraph parsing problem, GRASPR uses a graph parser which has evolved from Earley's string parsing algorithm [5] and string chart parsing. It incorporates four key improvements:

1. generalization of string parsing to flow graph parsing (Brotsky [1], Lutz [15]);
2. generalization of the control strategy to allow flexibility in the rule-invocation and search strategies used (Kay [11], Thompson [24], Lutz [15], Wills [27]);
3. extension of the grammar formalism to handle variation in graphs due to *structure-sharing* (Lutz [15], Wills [25, 26]), which is useful in dealing with variation due to common function-sharing optimizations; and
4. extension of the grammar formalism to capture *aggregation* relationships (Wills [26]) between single inputs or outputs of a left-hand side node and a tuple of inputs or outputs of a right-hand side sub-flow graph.

The fourth improvement is used to express the relationships between the inputs and outputs of an abstract operation on aggregate data structures and aggregates of the inputs and outputs of the lower-level operations that make up its concrete implementation. This is used, for example, in encoding the Fetch+Update cliché shown in Figure 3.

Other parsers have been developed which are related to our first two extensions described above. Bunke and Haller [2] and Peng, et al. [18] have both developed a parser for plex grammars which are generalizations of Earley's algorithm similar to Brotsky's. Wittenburg, et al. [28] give a unification-based, bottom-up chart parser which is similar to Lutz's and our chart parser. Like our grammar rules, Wittenburg's rules constrain the order in which right-hand side nodes are matched. However, in Wittenburg's rules this is always a strict (total) ordering. This creates fewer partial analyses, which is advantageous in terms of efficiency, but is a drawback in terms of generating partial results when the graph contains unrecognizable sections. We allow partial orderings to allow more partial matches to be attempted. Specifying node orderings and their strictness on a per-rule basis allows us to selectively expend more resources to increase the recognition power for promising partial recognition candidates.

Efficient parsers for *control* flow graphs have also been developed and applied to problems in global dataflow analysis [6] and program restructuring [14]. These parsers are able to take advantage of the fact that their reduction rules have the finite Church-Rosser property, which gives rise to a linear-time parsing process. These parsers also aim for a single full parse of the entire control flow graph. GRASPR, on the other hand, finds all recognizable subgraphs in order to deal with programs that are not constructed completely of clichés. This partial recognition process is inherently more expensive, but more flexible in dealing with code that contains novel or buggy structures.

4.1 Share-Equivalence

Following Lutz [15], we expand the language of a flow graph grammar to include all flow graphs derivable not only from a start type of the flow graph grammar, but also from flow graphs that are "share-equivalent" to a sentential form[1] of the grammar. The notion of share-equivalence is used to deal with variation due to *structure-sharing*: in a structure-sharing flow graph, a node plays the role of more than one node of the same type by generating output that fans out or by receiving input that fans in.

Share-equivalence is defined in terms of a binary relation *collapses* (denoted $\triangleleft$) on flow graphs. Flow graph F_1 collapses flow graph F_2 if and only if there are two nodes n_1 and n_2 of the same node type t in F_2, having input arity I and output arity O, such that all of these conditions hold:
1. Either one or both of the following are true:
    (a) $\forall i = 1...I$, the ith input port of n_1 is connected to the same set of output ports as the ith input port of n_2.
    (b) $\forall j = 1...O$, the jth output port of n_1 is connected to the same set of input ports as the jth output port of n_2.
2. F_1 can be created from F_2 by replacing n_1 and n_2 with a new node n_3 of type t with the ith input (resp., output) of n_3 connected to the union of the ports connected to the ith inputs (resp., outputs) of n_1 and n_2. We call this operation "zipping up" $F2$, and its inverse "unzipping."
3. The attribute values of n_1 and n_2 can be "combined." This is done by applying an *attribute combination function*, which is defined for each attribute, to the attribute values of n_1 and n_2. The attribute combination functions may be partial functions. If the function is not defined for n_1 and n_2's attributes, then the attribute values cannot be combined (and F_1 does not collapse F_2).

The reflexive, symmetric, transitive closure of collapses, $\triangleleft^*$, defines the equivalence relation *share-equivalent*. The *directly derives* relation ($\Rightarrow$) between flow graphs is redefined as follows. A flow graph F_1 directly derives another flow graph F_2 if and only if either F_2 can be produced by applying a grammar rule to F_1, $F_1 \triangleleft F_2$, or $F_2 \triangleleft F_1$.

---
[1] A *sentential form* of a graph grammar is any flow graph that is derivable from a start type of the grammar by the application of zero or more productions of the grammar.

As in string grammars, the reflexive, transitive closure of $\Rightarrow$, is the *derives* relation ($\Rightarrow^*$). The language of a flow graph grammar G (denoted $L(G)$) is the set of all flow graphs, whose nodes are of terminal type and which can be derived from a start type of G. Thus, the notion of a language of a flow graph grammar G has been extended to include flow graphs that are generated by a series of not only production rule applications but also zip-up and unzipping transformations. Since a zip-up or unzipping step can happen anywhere in the derivation sequence, the language of a graph grammar G in this extended formalism is a superset of the set of flow graphs share-equivalent to flow graphs in the "core" language of G in the unextended formalism.

Both generators and parsers for the language of a flow graph grammar can interleave zipping and unzipping transformation steps with their usual expansion and reduction steps. The parser used by the program recognition system reported here simulates the introduction of these transformations into its reduction sequence, as is described in Section 5.1.

4.2 Aggregation-Equivalence

Grammar rules in our flow graph formalism specify how a non-terminal node can be rewritten as a particular grouping of terminal and non-terminal nodes (in the form of a flow graph). We now extend it to also specify how a single input or output of a non-terminal node can correspond to an *aggregation* of the inputs or outputs of a flow graph to which the non-terminal node is rewritten. We define an additional, *aggregation-equivalence* relation to relate flow graphs that differ only in how they aggregate port types. The language of the flow graph grammar now includes all flow graphs aggregation-equivalent to flow graphs derivable from a start type of the grammar.

A simple way to capture the aggregation of port types into fewer, more abstract port types is to use special *Make* and *Spread* nodes. A Make node represents an aggregate type constructor: the tuple of its input port types compose its single output port type. A Spread node represents aggregate type selectors: its input port type is decomposed into its tuple of output port types.

Aggregation-equivalence captures the equivalence between flow graphs that differ solely in the way port types are aggregated within the graphs, i.e., in the order and nesting of aggregation (aggregation is commutative and associative) and whether there is any aggregation at all.

We define the reflexive, symmetric, transitive relation *aggregation-equivalent* as follows. A flow graph F_1 is aggregation-equivalent to another F_2 (denoted $F_1 \equiv_A F_2$) if and only if there exists a flow graph F_3, such that F_1 and F_2 can each be transformed to a flow graph isomorphic to F_3, using a (possibly empty) sequence of the following transformations: the permutation of part port tuples in Spreads and Makes, the flattening of compositions of Spread (resp. Make) nodes into a single Spread (resp. Make) node, the replacement of any composition of corresponding Spread and Make pairs (in either order) with the equivalent edges, and the removal of any Spread node whose input is an input of the flow graph or any Make node whose output is an output of the flow graph. We call the first type of transformation the *permutation* transformation. The rest of the transformations are *aggregation-removal* transformations and their inverses are called *aggregation-introduction* transformations.

A generator or parser for the language of a flow graph grammar may perform the permutation, aggregation-introduction and aggregation-removal transformations as steps in their derivation or reduction sequence. Because there are many possible orderings in which to apply the transformations and because doing this efficiently involves an extension to the embedding relation of the graph grammar formalism, it is important to discuss how such a recognizer is constructed.

One way a recognizer for the language can work, given an input flow graph F, is in two stages. The first would apply some sequence of the permutation, aggregation-removal and aggregation-introduction transformations to F to produce a flow graph F', while the second would apply a recognizer for the core language to F'. A flow graph F would be recognized if a sequence of transformations is found which yields a new flow graph F' that is accepted by a recognizer for the core language. Unfortunately, the first stage could involve a great deal of search to find the appropriate transformation sequence.

A more promising approach is to divide up the stages differently so that no choices need to be made. In the first stage, only aggregation-removal transformations that work "downward" by creating less-aggregated flow graphs are applied until a minimally-aggregated flow graph is obtained. (Note that in the minimally-aggregated flow graph, there are residual Make and Spread nodes only if there are terminal nodes that have aggregate port types.) Then in the second stage, the aggregation-introduction and permutation transformations are interleaved with the reduction actions of the recognizer for the core language. The idea is that the grammar rules can provide guidance as to what to aggregate and how to organize the aggregation so that the flow graph will be recognizable as a member of the core language. This approach is taken by our parser.

5 Chart Parsing Algorithm

We have developed a chart parsing algorithm for solving the subgraph parsing problem for flow graphs. Chart parsers maintain a database, called a *chart*, of partial and complete analyses of the input. The elements in the chart are called *items*. (In string chart parsing, they are called "edges." Lutz [15] calls them "patches.") An item might be either complete or partial. Complete items represent the recognition of some terminal or non-terminal in the grammar. Partial items represent a partial recognition of a non-terminal.

The basic operation of a chart parser is to create new items by combining a partial item with a complete one. This is called the *fundamental event*. If there is a partial item that needs a non-terminal A at a particular location and if there is a complete item for non-terminal A at that location, then the partial item can be *extended* with the complete item. During extension, a copy of the partial item is created and augmented. This results in a new item which is added to the chart. Items are never removed from and duplicate items are never added to the chart. This avoids redoing work and guarantees termination.

The parser continually generates items, conceptually in parallel, but to implement the algorithm on a sequential machine, we use an agenda to queue up the items to be added to the chart. The agenda makes it easy to control which items are added to the chart and when they are added. This explicit control can be used to enforce a particular rule invocation strategy or search strategy. In fact, the parser has several "control knobs" that can be set to achieve a desired control strategy or to implement focusing heuristics [27]. These include parameters like bottom-up or top-down rule invocation, the criterion used to determine whether one item can extend another, the ordering in which right-hand side nodes are matched, and the ordering of attribute condition checking in general.

The chart parser also has chart monitors that trigger on opportunities to create additional, new views of selected sub-flow graphs. These alternative views can be used, for example, to canonicalize certain sub-flow graphs or fix and resume unsuccessful matches. They can also be used as question-triggering patterns to elicit advice from external agents. An important use of additional monitors is in performing "zip-up" steps in parsing share-equivalent flow graphs.

The basic chart parsing algorithm is as follows. During initialization, add complete items to the agenda for each input graph node. When using a top-down rule invocation strategy, add empty items for each rule that derives a start type of the grammar. (An "empty" item is a partial item that needs complete items for *all* of its rule's right-hand side constituents.) Then, until the agenda is empty, continually pull an item X from the agenda and if X is not a member of the chart, do the following:
- Add X to the chart.
- If X is a *complete* item and X's attribute conditions are satisfied, then for each partial item P in the chart that is extendable by X, make a new item extending P with X and put it on the agenda. If a *bottom-up* rule invocation strategy is being used, add empty items for rules that need X to get started
  - i.e., add empty items for each rule that has a minimal node (as defined by the rule's node ordering) of the same type as the node derived in X.
- If X is a *partial* item, then for each complete item C in the chart that can extend X, make a new item extending X with C and put it on the agenda. If using a *top-down* rule invocation strategy, add a new empty item to the agenda for each rule that derives a non-terminal needed by the partial item.
- Apply the tests and operations of the additional monitors to the item. For example, for each complete item X whose constraints are satisfied, the zip-up monitor determines whether there are items that can zip up with X. If so, it performs the zip-ups and adds the results to the agenda.

5.1 Recognizing Share-Equivalent Flow Graphs

A parser for a structure-sharing flow graph grammar must interleave zipping and unzipping transformation steps with the usual reduction steps. Our chart parser *simulates* this introduction in two ways. First, the grammar is made to maximally share by canonicalizing all right-hand sides and by allowing sub-derivations to be shared. Second, the input graph is made to maximally share by using a "zip-up" monitor. All items that represent nodes that are collapsible are merged into a new item representing the result of the zip up.

5.2 Recognizing Aggregation-Equivalent Flow Graphs

Recall from Section 4.2 that a recognizer for the flow graph grammar's language must interleave permutation, aggregation-introduction, and aggregation-removal steps into the reduction sequence. During recognition, Spread and Make nodes must be "inserted" whenever an isomorphic occurrence of a right-hand side is reduced to a left-hand side non-terminal with aggregate ports. The Spread and Make nodes serve to bundle up the edges surrounding the non-terminal node. The recognition process must also "simplify" any composition of Makes and Spreads that results from aggregation-introduction steps. These actions are simulated by our flow graph chart parser working with a bottom-up invocation strategy.

In particular, items keep track of where the right-hand side is found, using a set of *location pointers*, which indicate which edges correspond to the inputs and outputs of the right-hand side of the item's rule. To represent the addition of a Make or Spread, the location pointers are placed in tuples, which are nested in tree structures. The nested tuples reflect the organization of the aggregation of the edges to which they refer. An element of the tuple can be either another tuple or a set of location pointers. (A set of more than one location pointer represents fan-in or fan-out.) When items are combined, their location pointers are compared to see if they represent a Make/Spread composition that simplifies correctly. The corresponding parts of the tuples are compared. If both parts are tuples, they are compared recursively. If both are sets, the sets must have a non-empty intersection for the comparison to succeed. If one is a set and the other a tuple, the comparison fails.

This is a brief description; a fuller account of how the parser deals with structure-sharing and aggregation is given in [26], including how residual Spreads and Makes are handled and how straight-throughs are matched.

6 Efficiency

We are studying the graph parsing approach by experimenting with two real-world simulator programs, written in Common Lisp by parallel processing researchers [3]. These programs are in the 500 to 1000 line range. The largest program recognized by any other existing recognition system is a 300-line database program recognized by CPU[13]. All other systems work with "toy programs" on the order of tens of lines.

We empirically and analytically studied the computational cost of GRASPR's parsing algorithm with respect to the simulator programs. GRASPR is performing a constrained search for matches of clichés – for each rule of the grammar, the parser is searching for a way to match each right-hand side node to an instance of the node's type in the input graph. This is inherently exponential. In fact, the subgraph parsing problem for flow graphs is NP-complete [26], so it is unlikely that there is a subgraph parsing algorithm that is not worst-case exponential.

However, in the practical application of graph parsing to recognizing *complete* instances of clichés, constraints are strong enough to prevent exponential behavior in practice. The three key constraints that come into play are: 1) constraints on node types, which correspond to function types and are highly varied, reducing ambiguity; 2) edge connection constraints, which represent dataflow dependencies and tend to be sparse, so fewer pairs of incorrect matches between nodes satisfy these constraints; and 3) co-occurrence constraints, which are a class of control flow constraints that are especially powerful in reducing ambiguity in recognizing clichéd operations on aggregate data structures.

As we increase the recognition power of **GRASPR** to make it generate more *partial* recognitions of clichés, we lose the advantage of strong constraint pruning. What is most expensive for **GRASPR** to do is the task of near-miss recognition of clichés – recognizing all possible *partial* (as well as complete) instances of clichés. This task is useful in robustly dealing with buggy programs, learning new clichés, and eliciting advice. Fortunately, the complexity of near-miss recognition can be controlled by using grammar indexing and flow graph partitioning advice and controlling the application order of constraints [26, 27].

7 Expressiveness

The flow graph representation is able to suppress many common forms of program variation which hinder recognition. In particular, the flow graph formalism and graph parser enable **GRASPR** to be robust under many types of variation, including *syntactic* (e.g., differences in binding or control constructs chosen and in statement ordering), *organizational* (e.g., differences in procedural modularization and data structure nesting), and *implementational* variation (e.g., differences in algorithms chosen to achieve common abstract operations). It is also robust under variation due to delocalization, unfamiliar code, and common function-sharing optimizations.

We have used flow graph grammars to concisely encode algorithmic and data aggregation clichés whose constraints are primarily based on data and control flow. Our cliché library contains a core set of general-purpose, "utility" clichés, along with a set of clichés from the domain of sequential simulation. These were acquired by manually extracting and generalizing commonly occurring patterns in a corpus of example programs and by speaking with the designers of the simulator programs to codify their experience with typical algorithms and data structures in the simulation domain. We also gathered clichés from textbooks in general computer programming as well as the area of simulation. Clichés in our library include algorithmic computations, such as list enumeration, binary search, instruction-fetch, decode, and execute and event-driven simulation, as well as common data structures, such as priority queue and hash table. These are encoded in approximately 200 graph grammar rules. The library's coverage is by no means absolute. However, it demonstrates the kinds of algorithms and data structures that can be expressed within our graph grammar formalism.

GRASPR recognizes structured programs and clichés containing conditionals, loops with any number of exits, recursion, aggregate data structures, and simple side effects due to variable assignments. Except for CPU [13], existing recognition systems cannot handle aggregate data structure clichés and a majority do not handle recursion. Side effects to mutable data structures still present an open problem in program recognition; a possible solution may be to interleave the recognition of stereotypical aliasing patterns with dataflow analysis [26].

8 Future Directions

One reason we developed a parsing algorithm with flexible control is that we wanted to complement our purely code-driven recognition with other design recovery techniques based on information from other sources, such as comments, identifier names, documentation, specifications, and testing suites. The parser's flexibility allows GRASPR to accept advice and heuristic guidance from external agents. In the future, such a hybrid system will allow us to explore the interactive potential of the chart parser and extensions for incremental analysis to which the chart parser lends itself. Another interesting future direction is the application of GRASPR to multiple tasks that require program understanding. This will help us determine which constraints various tasks place on the recognition process and representational formalism and what new types of control strategies are needed. Finally, our empirical studies have been limited to a few programs with a cliché library that co-evolved with GRASPR. More empirical studies are needed to expand and refine the cliché library, identify more classes of variation that can be tolerated, refine our understanding of the parser's performance in the context of practical reverse engineering problems, and evaluate the ability of the existing system to recognize clichés in new programs.

Acknowledgments: I am grateful to Chuck Rich, Richard Waters, David McAllester, and Peter Szolovits for guidance and insights. The comments from anonymous reviewers were also valuable. The research described here was conducted at the Artificial Intelligence Laboratory of the Massachusetts Institute of Technology. Support for the laboratory's research has been provided in part by the following organizations: National Science Foundation under grants IRI-8616644 and CCR-898273, Advanced Research Projects Agency of the Department of Defense under Naval Research contract N00014-88-K-0487, IBM Corporation, NYNEX Corporation, and Siemens Corporation.

References

1. D. Brotsky. An algorithm for parsing flow graphs. Technical Report 704, MIT Artificial Intelligence Lab., March 1984. Master's thesis.
2. H. Bunke and B. Haller. A parser for context free plex grammars. In M. Nagl, editor, *15th Int. Workshop on Graph-Theoretic Concepts in Computer Science*, pages 136–150. Springer-Verlag, June 1989. LNCS, Vol. 411.
3. W. Dally, et al. The J-Machine: A fine-grain concurrent computer. In *Int. Fed. of Info. Processing Societies*, 1989.
4. P. Della-Vigna and C. Ghezzi. Context-free graph grammars. *Information and Control*, 37(2):207–233, 1978.
5. J. Earley. An efficient context-free parsing algorithm. *Comm. of the ACM*, 13(2):94–102, 1970.

6. R. Farrow, K. Kennedy, and L. Zucconi. Graph grammars and global program data flow analysis. In *Proc. 17th Annual IEEE Symposium on Foundations of Computer Science*, pages 42–56, Houston, Texas, 1976.
7. J. Feder. Plex languages. *Information Sciences Journal*, 3:225–241, 1971.
8. A. Habel. *Hyperedge Replacement: Grammars and Languages*. Springer-Verlag, New York, 1992. Lecture Notes in Computer Science Series, Vol. 643.
9. J. Hartman. Automatic control understanding for natural programs. Technical Report AI 91-161, University of Texas at Austin, 1991. PhD thesis.
10. W. L. Johnson. *Intention-Based Diagnosis of Novice Programming Errors*. Morgan Kaufmann Publishers, Inc., Los Altos, CA, 1986.
11. M. Kay. Algorithm schemata and data structures in syntactic processing. In B. Grosz, K. Sparck-Jones, and B. Webber, editors, *Readings in Natural Language Processing*, pages 35–70. Morgan Kaufmann Publishers, Inc., Los Altos, CA, 1986.
12. V. Kozaczynski and J.Q. Ning. Automated program understanding by concept recognition. *Automated Software Engineering*, 1(1):61–78, March 1994.
13. S. Letovsky. Plan analysis of programs. RR 662, Yale University, Dec. 1988.
14. U. Lichtblau. Decompilation of control structures by means of graph transformation. In H. Ehrig, editor, *LNCS, Vol. 185*, pages 284–297. Springer-Verlag, 1985.
15. R. Lutz. Chart parsing of flowgraphs. In *Proc. 11th Int. Joint Conf. Artificial Intelligence*, pages 116–121, Detroit, Michigan, 1989.
16. U. G. Montanari. Separable graphs, planar graphs, and web grammars. *Information and Control*, 16(3):243–267, March 1970.
17. T. Pavlidis. Linear and context-free graph grammars. *Journal of the ACM*, 19(1):11–23, January 1972.
18. K. Peng, T. Yamamoto, and Y. Aoki. A new parsing algorithm for plex grammars. *Pattern Recognition*, 23(3-4):393–402, 1990.
19. J. L. Pfaltz and A. Rosenfeld. Web grammars. In *Proc. 1st Int. Joint Conf. Artificial Intelligence*, pages 609–619, Washington, D.C., September 1969.
20. A. Quilici. Memory-based approach to recognizing programming plans. *Comm. of the ACM*, 37(5):84–93, May 1994.
21. C. Rich and R. C. Waters. *The Programmer's Apprentice*. Addison-Wesley, Reading, MA and ACM Press, Baltimore, MD, 1990.
22. C. Rich and L. M. Wills. Recognizing a program's design: A graph-parsing approach. *IEEE Software*, 7(1):82–89, January 1990.
23. A. Rosenfeld and D. Milgram. Web automata and web grammars. In B. Meltzer and D. Michie, editors, *Machine Intelligence 7*, pages 307–324. John Wiley and Sons, New York, 1972.
24. H. Thompson. Chart parsing and rule schemata in GPSG. In *Proc. 19th Annual Meeting of the ACL*, Stanford, CA, 1981.
25. L. Wills. Automated program recognition: A feasibility demonstration. *Artificial Intelligence*, 45(1-2):113–172, 1990.
26. L. Wills. Automated program recognition by graph parsing. Technical Report 1358, MIT Artificial Intelligence Lab., July 1992. PhD Thesis.
27. L. Wills. Flexible control for program recognition. In *Proc. 1st Working Conference on Reverse Engineering*, Baltimore, MD, May 1993.
28. K. Wittenburg, L. Weitzman, and J. Talley. Unification-based grammars and tabular parsing for graphical languages. TR ACT-OODS-208-91, MCC, 1991.

Reconfiguration Graph Grammar for Massively Parallel, Fault Tolerant Computers

M. D. Derk* L. S. DeBrunner
School of School of
Computer Science Electrical Engineering

University of Oklahoma
Norman, Oklahoma 73019
(405) 325-4852
FAX: (405) 325-7066
email:molisa@vela.ecn.uoknor.edu

Abstract. Reconfiguration for fault tolerance is the process of excluding faulty processors from an array of interconnected processors and including spare processors to take their place. Reconfiguration Graph Grammar (RGG) is introduced as a model supporting the design and analysis of these reconfiguration algorithms. A formal description is given, as well as several theorems, with proofs, concerning the properties of RGG that make it well suited for modeling reconfiguration. An example RGG-based reconfiguration algorithm is described and demonstrated.

1 Introduction

In massively parallel computers consisting of large numbers of small, interconnected processors, the likelihood that one or more of these processors is faulty due to manufacturing defects, or that it (they) will fail during operation is fairly high. Fault tolerance can be achieved by including spare processors in the design. Should one or more processors fail, reconfiguration is performed which substitutes spare processors for the faulty ones by activating certain communication lines and deactivating others within the processor array. Reconfiguration algorithms should be fast while making efficient use of available spares. This process has been studied at length [1-3], yet, to our knowledge, graph grammars have never been applied to reconfiguration before this study.

Reconfiguration Graph Grammar (RGG) is introduced as a model supporting the design and analysis of these reconfiguration algorithms. Processor arrays

---
* Also affiliated with Oklahoma City University, Dept. of Computer Science, Oklahoma City, Oklahoma, U.S.A. 73106

are represented using graphs, with the nodes representing processors and the edges representing active communication lines between them. An RGG production represents a change in the communication lines which performs the necessary reconfiguration. The addition of edges between nodes indicates the activation of communication lines, and the removal of edges indicates their deactivation.

2 Description

Reconfiguration Graph Grammars differ from most graph grammars in several respects. First, the start graph represents the entire processor array in its fault-free state. Second, the left and right hand sides of all productions have the same number of nodes and edges. Third, RGGs use inheritance functions, similar to the inheritance function defined by Bailey and Cuny for Aggregate Rewriting graph grammars [4-7], to define the edges connecting the right hand side of a production to the rest of the graph.

RGGs use two sets of node labels, alphabetic and numeric. The numeric labels indicate the logical coordinates of the processors within a processor array. In a fault-free processor array, the logical and physical coordinates are identical, but reconfiguration alters the logical coordinates of certain nodes so that faulty processors are not used. The alphabetic node labels indicate the reconfiguration states of the processors, which change according to the reconfiguring productions applied in the neighborhood of the processor. These states determine which production, if any, may be applied to compensate for the failure of a particular processor.

3 Definition

The formal definition of Reconfiguration Graph Grammars is as follows:

RGG = (S, A, a, N, ß, P), where

- S is a finite starting graph representing the initial, non-faulty configuration of the processor array where S=(V,E), a set of nodes V representing processors, and a set of edges E representing the active communication lines between them,

- A is a finite, non-empty set of alphabetic node labels, indicating the possible reconfiguration states of the processors (nodes) V,

- a is a mapping from V to A,

N   is a finite, non-empty set of numeric node labels corresponding to the logical coordinates of the active processors of S, such that $|N| \leq |V|$,

ß   is an injective partial mapping from V to N.

P   is a finite, non-empty set of productions of the form $P = (\sigma, \tau, \phi)$, where

    σ is the left-hand side, a subgraph whose nodes are labeled with symbols of N,

    τ is the right-hand side, a subgraph whose nodes are labeled with symbols of N,

    ϕ is the inheritance function, $\phi: \tau \to \sigma$, a partial, injective function that indicates the nodes of τ that will inherit the connecting edges of the nodes of σ.

The rewriting of an RGG on a graph H occurs as follows:

1. Rewriting occurs every time there is a match between a subgraph of H and the left-hand side of a rewriting production (σ). Rewriting must occur for every match. A match is defined to be present between σ and a subgraph of H if the subgraph is isomorphic to σ with respect to its numeric labels (which are represented in a generalized fashion in the production).

2. The subgraph of H that is isomorphic to σ is removed, and all edges connecting it to the rest of H (referred to here as H-σ) are removed.

3. τ replaces σ. Edges are added between the nodes of τ and H-σ according to the inheritance function of the production, ϕ. ϕ is expressed in the form $\phi(n_\tau) = n_\sigma$, where each node in τ with a particular numeric label n_τ is connected to the same nodes in H-σ that were connected to the node in σ with numeric label n_σ.

There are many types of reconfiguration, and Reconfiguration Graph Grammars have been defined to be a general reconfiguration tool. However, if reconfiguration is for fault tolerance, certain restrictions must apply. The first restriction ensures that some nodes represent spare processors. The other two restrictions ensure that the graph after any production is isomorphic to the graph before the production. These last two restrictions are necessary for the corresponding processor array to be functionally equivalent before and after reconfiguration.

1. |N| is strictly less than |V|. Nodes without a numeric label have degree zero, and represent spare and/or faulty processors. Nodes with a numeric label have degree greater than zero and represent the active processors. This applies to S, to the right hand side of every production, and by consequence, every member of L.

2. In every production, σ and τ have the same number of nodes and edges, and use the same subset of numeric labels. Therefore, the set of numeric labels for any member of L is the same as for S. Also, τ preserves the adjacencies of σ with respect to the numeric labels of the nodes.

3. In every production, for the inheritance function, $\Phi(n_\tau) = n_\sigma$, $n_\tau = n_\sigma$. That is, every node of τ is connected to the nodes in H-σ that were connected to the node in σ with the same numeric label.

4 Properties and Theorems

There are certain properties of RGG graph grammars that should be mentioned at this point. First, there is no distinction between terminal and nonterminal symbols. Each production application results in a member of the generated language. Since every node is unique, then the left and right sides of a production, though isomorphic with respect to numeric labels, are two distinct graphs. Since ß is injective, no numeric label ever appears on more than one node. This property is necessary because the numeric labels correspond to the logical coordinates of the represented processor array. The productions will be presented in a generalized fashion, with numeric labels indicated by variables, in order to save space and to represent productions that can serve for similar processor arrays of any size.

We have proven the following theorems concerning this graph grammar.

Theorem 1. *In a language L generated by an RGG, if* $H \rightarrow I$, *then* $|V_H| = |V_I|$, *and* $|E_H| = |E_I|$.

Proof: Let H be a member of L. Each production of R replaces σ, a subgraph of H, with τ, producing I. Since σ and τ have the same number of nodes by definition, then $|V_H| = |V_I|$.

Since σ and τ have the same number of edges, we need only show that the number of edges connecting σ and H-σ is the same as the number of edges connecting H-σ to τ to produce I. Since nodes with no numeric labels have degree zero, there are only connecting edges between H-σ and those nodes in σ or τ with numeric labels. Since the subsets of numeric labels used in σ and τ are the same, and because the mapping is injective, every node in τ with a numeric label corresponds to exactly one node in σ with the same numeric label, and vice versa. Since $n_τ = n_σ$ in the inheritance function, every node in τ with a numeric label will be connected to the same nodes in H-σ as its counterpart in σ with the same numeric label. Therefore, the number of connecting edges will be the same, and $|E_H| = |E_I|$.

Corollary 1(a). *RGGs preserve the cardinality of the node and edge sets.*
Corollary 1(b). *Every member of L, including S, has the same number of nodes and edges.*

Theorem 2. *Every member of L, the language generated by an RGG, is isomorphic to every other member, including S, with respect to the numeric node labels.*

Proof: Since every member of L has the same number of nodes and edges, we only need to show that adjacency is preserved under the productions of R with the specified restrictions. The adjacency of the nodes of σ is preserved in τ under restriction #2. The adjacencies of σ to H-σ are preserved in the connections of H-σ to τ because $n_τ = n_σ$ in the inheritance production under restriction #3. Since nodes with no numeric label are of degree zero, adjacencies are preserved for all nodes under the productions of R.

Corollary 2(a). *For any given arrangement of numeric labels on a member of L, only one edge arrangement is allowed.*

Theorem 3. *The language L generated by an RGG is a finite language.*

Proof: Every member of L has the same number of nodes and edges. Since the edge arrangement of any H ∈ L is uniquely determined by the arrangement of numeric labels, then H is uniquely determined. The set of numeric labels is finite, and identical for every H ∈ L. The number of possible arrangements of the set of numeric labels over the set of nodes is finite, therefore L must have a finite number of members.

Theorem 4. *Let the supergraph of a language L produced by an RGG be defined as the smallest graph for which every member of L is a subgraph. Then, the needed physical interconnections to implement the algorithm derived from the graph grammar are given by the edges of the supergraph of L.*

Proof: Since every H ∈ L is a subgraph of the supergraph, every edge of H is also an edge of the supergraph. Therefore, every active communication line represented by an edge of H has a corresponding physical interconnection represented by an edge of the supergraph.

5 Example

An example will show how a reconfiguration graph grammar is applied. Figure 1 shows a processor array with one column of spare processors on the right side.

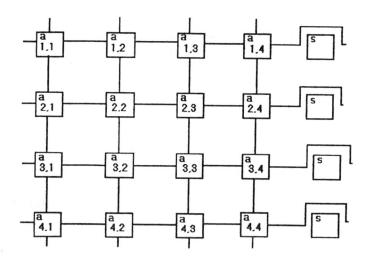

Fig. 1. Fault-free processor array

The letters indicate the reconfiguration states. The spares have a reconfiguration state of "s", and the rest have an initial reconfiguration state of "a". Figures 2-8 show the productions for a reconfiguration graph grammar that defines the reconfiguration procedure for any series of faults that occur in this processor array or any similar array with one column of spares.

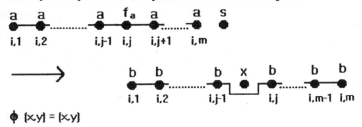

φ [x,y] = [x,y]

Fig. 2. Production #1

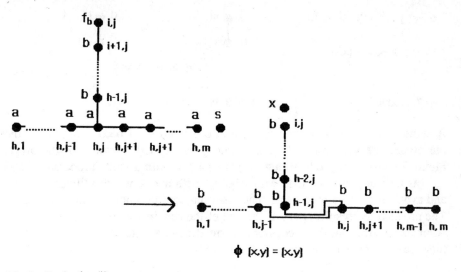

Fig. 3. Production #2

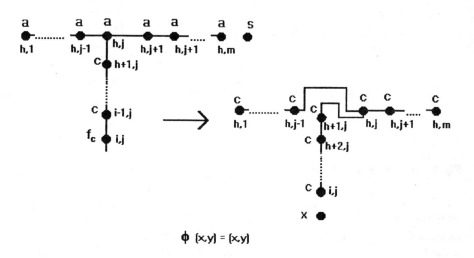

Fig. 4. Production #3

Fig. 5. Production #4

Fig. 6. Production #5

c •⎯1,j ⟶ d •⎯1,j

Φ {x,y} = {x,y}

d •⎯i,j   d •⎯i,j
  | ⟶ |
c •⎯i+1,j   d •⎯i+1,j

Φ {x,y} = {x,y}

Fig. 7 Production #6   **Fig. 8** Production #7

Assume at this point that the processor at coordinates 2,4 becomes faulty. Since the processor's state was "a", production #1 is applied to row 2 of the array. Figure 2 shows that production #1 replaces a row with a faulty processor of state "a" (as indicated by the node with label f_a) with a row which utilizes the spare and avoids the faulty processor. The resulting configuration is shown in Fig. 9. The production uses the spare at the end of the row in which the fault occurred. After the production, the processors in row 2 have a state of "b", indicating that they are on a row without an available spare.

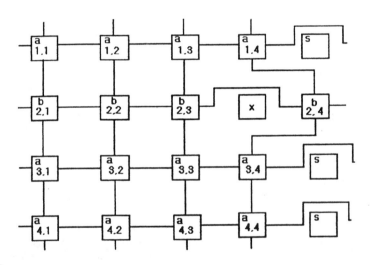

Fig. 9. Array after one fault in row 2

Next, assume that the processor at coordinates 2,1 becomes faulty. Since its state was "b", production #2 is applied. Figure 3 shows that production #2 changes the column in which the fault was found from the faulty node downward to the first row with an available spare. We know the first row with an available spare will have processors with a state of "a". This row is also replaced by the production, so that its spare can be utilized. The production shows that the numeric labels of the column plus the right side of the row are altered so that the spare is used and the faulty node is avoided. The resulting configuration is shown in Fig. 10. In this case, the row immediately below the row with the fault was used, since it had an available spare. Since the spare at the end of row 3 has now been used, the processors in that row have a state of "b".

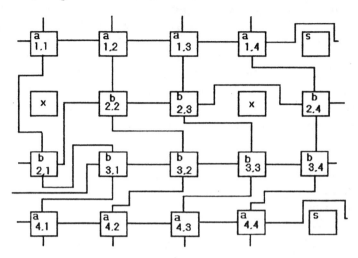

Fig. 10. Array after second fault in row 2

Production #3 is applied when the bottom row of processors has a state of "b". Then, repeated applications of production #4 assure that all processors below the lowest spare have a state of "c". Then, should a processor with a state of "c" fail, production #5 must be applied, which utilizes a spare in a row above rather than in a row below (since no spares are available below). In this way, all spares can be used for any and all faults in the array, a quality which is rare among reconfiguration methods. When the top row of processors has a state of "c", this indicates that all spares have been used. Productions #6 and #7 then change the state of every processor to "d". A failure of a processor with a state of "d" is a fatal failure to the system. However, this reconfiguration method is unique in that the system "knows" in advance that the next failure with be fatal.

6. Conclusions

Using Reconfiguration Graph Grammars to study reconfiguration has yielded several new results, as listed below.

1. RGG-based algorithms are efficient, and can be distributively implemented by the array processors without reliance on a host processor. To our knowledge, the RGG model is the only model than

yields efficient algorithms (O(N) complexity) that can make use of <u>every</u> available spare processor[8].

2. In some processor arrays, the array is divided into "local areas", and reconfiguration for the failed processors in that area is limited to the use of spares only in that area. This approach is used to limit the time taken by reconfiguration. The drawback is that all of the spares in one area may be used, preventing further reconfiguration, while adjacent areas have unused spares. The RGG model of reconfiguration provides the means to combine local areas should this situation arise[9]. This combination is possible by altering the reconfiguration states of the processors. Our model is the only one that provides for area combination.

3. A dangerous processor, as defined in [2], is one whose failure will result in the fatal failure of the system because no reconfiguration is possible if it should fail, due to the reconfigurations that have already occurred. The RGG model is the only reconfiguration model that can identify dangerous processors before they fail. Advance knowledge of the presence of dangerous processors, and their locations, is very valuable for system managers.

4. A direct, logical relationship to the physical implementation of the processor array is provided by the supergraph of the language. The RGG model is the only reconfiguration model, to our knowledge, that provides such a direct relationship between the algorithm and the hardware requirements.

We believe that by using a model based on the theoretically sound framework of graph grammars, additional properties of reconfiguration for fault tolerance can be discovered.

References

1. R. Negrini, M.G. Sami, R. Stefanelli: <u>Fault Tolerance Through Reconfiguration in VLSI and WSI Arrays</u>. MIT Press, Cambridge, Mass., 1989.

2. C. Chen, A. Feng, T. Kikuno, K. Torrii: Reconfiguration algorithm for fault tolerant arrays with minimum number of dangerous processors. <u>21st International Symposium on Fault-Tolerant Computing, Montreal, Quebec,</u>

Canada, 25-27 June, 1991. IEEE Computer Society Press, Los Alamitos, CA. pp. 452-459.

3. M. Chean, J. A. B. Fortes: A taxonomy of reconfiguration techniques for fault-tolerant processor arrays. Computer, vol. 23, no. 1, January, 1990, pp. 55-69.

4. D. A. Bailey, J. E. Cuny: Graph grammar based specification of interconnection structures for massively parallel computation. Graph grammars and their application to computer science, 3rd international workshop, Warrenton, VA, USA, Dec. 1986 Proceedings. Lecture Notes in Computer Science #291. Springer-Verlag, Berlin, Heidelberg, 1987. pp. 73-85.

5. D. A. Bailey, J. E. Cuny: An approach to programming process interconnection structures: aggregate rewriting graph grammars. PARLE Parallel Architectures and Languages Europe, Vol. II: Parallel Languages. Eindhoven, The Netherlands, June 15-19, 1987 Proceedings. LNCS #259 Springer-Verlag, Berlin, Heidelberg, 1987. pp. 112-123.

6. D. A. Bailey, J. E. Cuny: Visual extensions to parallel programming languages. Languages and Compilers for Parallel Computing. ed. by Gelernter, David, et al. MIT Press, Cambridge, MA, 1990. p. 17-36.

7. D. A. Bailey, J. E. Cuny, C. D. Fisher: Programming with very large graphs. Graph Grammars and their Application to Computer Science, 4th International Workshop, Bremen, Germany, March 1990 Proceedings. LNCS #532. Springer-Verlag, Berlin, Heidelberg, 1991. pp. 84-97.

8. M. Derk, L. DeBrunner: Dynamic reconfiguration for fault tolerance with complete use of spares. Parallel and Distributed Computing and Systems. Proceedings of the Sixth IASTED/ISMM International Conference, Washington, D.C., Oct. 3-5, 1994 IASTED, 1994, pp. 331-334.

9. M. Derk, L. DeBrunner: Dynamic reconfiguration for fault tolerance for critical, real-time processor arrays. Conference Record: Twenty-eighth Asilomar Conference on Signals, Systems and Computers, 1994. Vol. 2. Pacific Grove, CA, Maple Press, 1994, pp. 1058-1062.

The Use of Tree Transducers to Compute Translations Between Graph Algebras

Frank Drewes*
Universität Bremen, Fachbereich 3, D–28334 Bremen (Germany)
E-mail: drewes@informatik.uni-bremen.de

Abstract. The power of top-down, bottom-up, and tree-to-graph-to-tree transducers (tgt transducers) to compute translations from hyperedge-replacement algebras into edge-replacement algebras is investigated. Compositions of top-down and bottom-up tree transducers are too weak if the operations in the target algebra are powerful enough to define all series-parallel graphs, 2-trees, or related types of graphs. Tgt transducers are shown to be more powerful. These are able to compute translations into ER algebras whose operations are so-called *2tree** operations, which are generalizations of the well-known operations to generate 2-trees.

1 Introduction

Using the notion of hyperedge replacement a hypergraph H may be understood as an operation on hypergraphs (see [3]). If H contains hyperedges $e_1, \ldots, e_n$ it represents an operation that takes n argument hypergraphs $H_1, \ldots, H_n$ and yields the hypergraph obtained by the substitution of H_i for e_i ($1 \leq i \leq n$). Roughly speaking, this means that a term over these *hyperedge-replacement operations* corresponds to a derivation tree in a hyperedge-replacement grammar, where operation symbols correspond to right-hand sides of production rules. Every finite set of such hyperedge-replacement operations yields an algebra called a *hyperedge-replacement algebra*, or HR algebra, for short. Thus, the values of terms in HR algebras are hypergraphs. A case of special interest is provided by *edge-replacement algebras* (ER algebras), in which only the use of edge-replacement operations is allowed (and hence only graphs are defined).

In this paper we consider the question whether certain sorts of tree transducers are able to compute *translations* from arbitrary HR algebras into the simpler ER algebras. Suppose an ER algebra B defining a set of graphs of interest is given. Then we ask whether, for every HR algebra A, there is a transduction of the terms of A into those of B which yields a term with the same value as the input term whenever such a term of B exists (and is undefined otherwise). For logical transductions such questions were studied by Courcelle in [3]. We prove both a negative and a positive result. As for the negative result, sufficient conditions on the power of ER operations in B are formulated such that (compositions of) top-down and bottom-up tree transducers cannot compute translations from

---
*Supported by COMPUGRAPH II, ESPRIT Basic Research Working Group 7183.

arbitrary HR algebras into B. The conditions are not very restrictive. In particular, the natural ER algebras for the definition of series-parallel graphs and 2-trees are covered by the result.

The positive result is obtained for *tree-to-graph-to-tree transductions* (tgt transductions). These were introduced by Engelfriet and Vogler [9], who showed that deterministic total tgt transductions compare to macro tree transductions. It is shown here that linear tgt transducers without sharing can compute translations from arbitrary HR algebras into ER algebras whose operations are so-called *2tree* operations*. The class of graphs that can be defined by these operations properly contains the class of all 2-trees and is one of those that are not successfully handled by top-down and bottom-up tree transducers.

This is a short version of [4]; in particular, all proofs are omitted.

2 Hypergraphs, hypergraph operations, and tree transducers

We recall briefly the notational conventions and definitions used.

The natural extensions of a mapping f to sets and sequences are denoted by f, too. If s is a sequence the set of all its members is denoted by $[s]$ and the ith component is $s(i)$. If $f: X \to Y^*$ is a mapping yielding sequences we write $f(x,i)$ for $f(x)(i)$. The notation $s \cdot s'$ means concatenation of (finite) sequences.

Let L be a set of *hyperedge labels* and let P be a set of *port labels*. A *hypergraph* with hyperedge labels in L and port labels in P is a tuple $H = (V, E, att, lab, port)$, where V is the finite set of *nodes*, E is the finite set of *hyperedges*, $att: E \to V^*$ is the *attachment* of hyperedges, $lab: E \to L$ is the *labelling* of hyperedges, and $port: P \to V$ is an injective mapping, called the *port labelling* of H. As usual, we also write V_H, E_H, att_H, lab_H, $port_H$, and P_H for the components V, E, att, lab, $port$, and the set P of port labels of a hypergraph H. A node in the range of *port* is a *port*[1], and the set of all ports of H is denoted by $ports(H)$.

As usual, we consider an unlabelled hypergraph (that is, a hypergraph without hyperedge labels) as a hypergraph all of whose hyperedges are labelled with some distinguished label meaning "no label". For $e \in E$ we call $|att_H(e)|$ the *type* of e in H, denoted by $type_H(e)$. By $F(H)$ we denote the portless hypergraph obtained from H by removing all port labels. A *graph* is a hypergraph all of whose hyperedges are edges, that is, have type 2.

Call a hypergraph H with $P_H = \{1, \ldots, k\}$ for some $k \in \mathbb{N}$ a k-*hypergraph*. Let H be a hypergraph with pairwise distinct $e_1, \ldots, e_n \in E_H$, and let H_i be a $type_H(e_i)$-hypergraph for $i = 1, \ldots, n$. Then $H[e_1/H_1 \cdots e_n/H_n]$, the *replacement of $e_1, \ldots, e_n$ with $H_1, \ldots, H_n$* yields the hypergraph obtained from the disjoint union of H and $F(H_1), \ldots, F(H_n)$ by removing $e_1, \ldots, e_n$ and identifying $att_H(e_i, j)$ with $port_{H_i}(j)$ for $i = 1, \ldots, n$ and $j = 1, \ldots, type_H(e_i)$. The resulting hypergraph keeps the port labelling of H.

---
[1] See [8] for this terminology.

Using hyperedge replacement, every hypergraph H as above can be seen as a (partial) n-ary operation on (isomorphism classes of) hypergraphs (see [3]) by defining $H(H_1, \ldots, H_n) = H[e_1/H_1 \cdots e_n/H_n]$. For simplicity, we assume that an arbitrary, but fixed order on the hyperedges is chosen. The hyperedges $e_1, \ldots, e_n$ are then called the *virtual* ones of this operation. Algebras whose operations are such *hyperedge-replacement operations* are called *HR algebras*; *ER algebras* are those whose operations are *edge-replacement operations*— HR operations given by 2-graphs (that is, graphs with port labels 1 and 2). By $val(t)$ we denote the value of a term. In general, if A is any algebra we denote by $T(A, n)$ the set of all terms in A with variables $x_1, \ldots, x_n$, where $T(A) = T(A, 0)$. The mapping val extends to $T(A, n)$ in an obvious way, yielding an n-ary *derived operation* for every term $t \in T(A, n)$. It is defined by $val(t)(H_1, \ldots, H_n) = val(t[H_1, \ldots, H_n])$ (where $t[H_1, \ldots, H_n]$ denotes the substitution of the constants $H_1, \ldots, H_n$ for $x_1, \ldots, x_n$ in t).

In this paper, whenever derived operations of some algebra are considered we restrict our attention to those given by linear and nondeleting terms which do not consist of a single variable. (A term $t \in T(A, n)$ is linear and nondeleting if every variable x_i ($1 \leq i \leq n$) occurs exactly once in it.) Hyperedge replacement has the nice property that the set of HR operations is closed under this type of derived operations, that is, $val(t)$ is an HR operation. It is obtained by performing all the replacements in t, which yields an equivalent term of the form $H(x_{i_1}, \ldots, x_{i_n})$. Now, the HR operation is obtained by reordering the virtual hyperedges $e_1, \ldots, e_n$ of H as $e_{i_1}, \ldots, e_{i_n}$.

Finally, let us recall the different types of tree transducers considered. A *top-down tree transducer* [5, 1, 7] is actually a special term rewrite system. If A and B are algebras, a top-down tree transducer $td :: T(A) \to T(B)$ consists of a set of *states* including an *initial state* γ_0, where the set of states must not intersect with the sets of operation symbols in A and B, and a set of *rules*. Every rule has the form $\gamma(f(x_1, \ldots, x_n)) \to t$, where γ is a state, f is an operation symbol in A, and t is a term containing operation symbols from B and subterms of the form $\gamma'(x_i)$, where γ' is a state and $i \in \{1, \ldots, n\}$. The *transduction* defined by td (which is also denoted by td) is given by $td(t) = \{t' \in T(B) \mid \gamma_0(t) \to^* t'\}$ for all $t \in T(A)$, where $\to^*$ is the derivation relation of the term rewrite system given by the rules of td. A top-down tree transducer is *linear* if no variable occurs twice in the right-hand side of any rule, and *nondeleting* if every variable in the left-hand side of a rule occurs also in the right-hand side of that rule.

Bottom-up tree transducers [5, 1, 7] and the computed transductions are defined in a similar way. Here, the initial state is replaced with a set of *final states* and the rules are of the form $f(\gamma_1(x_1), \ldots, \gamma_n(x_n)) \to \gamma(t)$, where f is an operation symbol of A, $\gamma, \gamma_1, \ldots, \gamma_n$ are states, and $t \in T(B, n)$. For a bottom-up tree transducer $bu :: T(A) \to T(B)$, $bu(t)$ is defined to be the set of all terms $t' \in T(B)$ such that $t \to^* \gamma(t')$ for some final state γ. By TB^* we denote the set of all transductions $\tau_1 \circ \cdots \circ \tau_n$ composed of finitely many top-down and bottom-up tree transductions (see also [6, 8]).

To define *tree-to-graph-to-tree transductions* (tgt transductions, see [9]) we

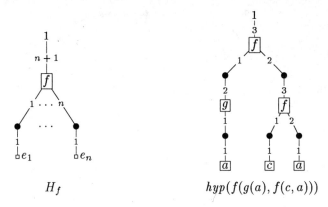

Figure 1: The HR operation H_f and the term hypergraph for $f(g(a), f(c,a))$. Ports are denoted by their labels; ordinary nodes are dots. The virtual hyperedges are unlabelled because their labels are irrelevant, anyway.

need the notion of *term hypergraphs*. For an operation symbol f of arity n let H_f be the 1-hypergraph depicted in Figure 1 on the left. For every term $t = f(t_1, \ldots, t_n)$ the *term hypergraph* $hyp(t)$ is the hypergraph H given by $H = H_f[e_1/hyp(t_1) \cdots e_n/hyp(t_n)]$. Its *root*, denoted by $root(H)$, is the node $port_H(1)$. An example showing the term hypergraph which represents the term $f(g(a), f(c,a))$ is depicted in Figure 1.

Let A, B be algebras. A *tgt transducer* $td :: \mathrm{T}(A) \to \mathrm{T}(B)$ is a top-down tree transducer $td :: \mathrm{T}(A) \to \mathrm{T}(C)$ for some HR algebra C such that for all $t \in \mathrm{T}(A)$ and $t' \in td(t)$ we have $val(t') \in hyp(\mathrm{T}(B))$. The *tgt transduction* τ computed by td is then given by $\tau(t) = hyp^{-1}(val(td(t)))$ for all $t \in \mathrm{T}(A)$.

A tgt transduction is said to be linear if it is computed by a linear top-down tree transducer. Notice that being nondeleting is not a property of tgt transducers that is worthwhile to be mentioned. This is due to the fact that there are binary HR operations for which the empty graph is the neutral element. Using this it is easy to modify a given tgt transducer in such a way that it becomes nondeleting. It must be noted that, apart from the less important fact that Engelfriet and Vogler define only the total deterministic version of their tgt transducers, tgt transductions as introduced here differ from those by Engelfriet and Vogler in one important respect: Those of [9] are allowed to produce representations of terms in which equal subterms are shared. Clearly, this means that the tgt transductions defined here are a subclass of those by Engelfriet and Vogler.

A tree transduction $\tau :: \mathrm{T}(A) \to \mathrm{T}(B)$ is a *translation* if $\tau(t) \neq \emptyset$ exactly when there exists some $t' \in \mathrm{T}(B)$ with $val(t') = val(t)$, and for all $t' \in \tau(t)$ we have $val(t') = val(t)$.

Figure 2: The ER operation K_3 to define 2-trees, and a 2-tree.

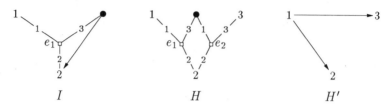

Figure 3: Additional operations for 2-trees.

3 The insufficiency of TB^* transductions

In this section we work with unlabelled hypergraphs only. It is shown that transductions in TB^* are not very powerful for what concerns the computation of translations. Roughly speaking, one can say they are too weak to perform translations into ER algebras whenever there is an operation a 2-subgraph[2] of which is 2-connected. We first consider as an example the set of all *oriented 2-trees*. This is the class of graphs defined by the ER operations K_3 and $\twoheadrightarrow$, where K_3 is the 2-graph given in Figure 2 on the left (with virtual edges e_1, e_2, and e_3) and $\twoheadrightarrow$ is the constant given by the unlabelled 2-graph with nodes v_1 and v_2 and a single edge e, where $att_\twoheadrightarrow(e) = v_1 v_2$. Let B be the ER algebra containing these two operations and look at 2-trees of the form indicated in Figure 2 on the right. In B 2-trees like this are represented by terms of the form $K_3(\twoheadrightarrow, \twoheadrightarrow, K_3(\twoheadrightarrow, \twoheadrightarrow, \ldots K_3(\twoheadrightarrow, \twoheadrightarrow, \twoheadrightarrow) \ldots))$, and no other terms in $T(B)$ yield the same 2-trees. Therefore, a translation into B-terms for some HR algebra A would have to compute, for every $t \in T(A)$ for which $val(t)$ is a 2-tree like the one in the figure, exactly this term. To see that this is impossible to achieve in general, define A by enriching B with the operations I, H, and H' given in Figure 3, with virtual hyperedges e_1 in I, e_1, e_2 in H, and none in H'. Intuitively, the operation I produces the "last" edge of the 2-tree in Figure 2 and a hyperedge to remember the ports as well as the new node. Substituting H for this hyperedge we generate a node somewhere in the middle of the 2-tree. The first argument of H becomes the part to the left of this node and the second argument gives the part to the right. Now the sort of 2-trees in Figure 2 can be represented by a term of the form $I(t_{(n)})$, where $t_{(n)}$ is the fully balanced term of height n over H and H'. Clearly, the corresponding term in B that defines the same graph is the term $\underbrace{K_3(\twoheadrightarrow, \twoheadrightarrow, \ldots K_3(\twoheadrightarrow, \twoheadrightarrow, \twoheadrightarrow) \ldots)}_{2^n \text{ times}}$. It is well-known (and follows

---

[2] A *2-subgraph* of a 2-graph is a subgraph which is itself a 2-graph, that is, contains both ports of the supergraph.

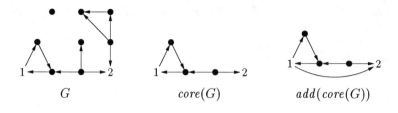

Figure 4: A port connected 2-graph G, $core(G)$, and $add(core(G))$

directly from the form of their rules) that for every top-down or bottom-up tree transducer the height of the output terms is at most linear in the height of the corresponding input terms. Consequently, the same holds for the class TB^*, so the translation is no member of TB^*.

In the following we use results on 2-connected graphs by MacLane [11], Hopcroft and Tarjan [10], and Vogler [12] to make the example into a general result. A 2-graph G is called *substantial* if $|V_G| > 2$ or $|E_G| > 1$. Edges e, e' in a graph G are *parallel* if $[att_G(e)] = [att_G(e')]$. For every (unlabelled) 2-graph G we denote by $add(G)$ the 2-graph G' obtained from G by adding (disjointly) a new unlabelled edge e_0 with $att_{G'}(e_0) = port_G(1 \cdot 2)$.

A vv'-*path* is an alternating sequence $v_1 e_1 \cdots v_n$ of nodes $v_i \in V_H$ and edges $e_i \in E_H$ such that $v_1 = v$, $v_n = v'$, and for $i = 1, \ldots, n-1$ we have $\{v_i, v_{i+1}\} = [att_H(e_i)]$. We say that the path *avoids* all nodes and edges but $v_1, \ldots, v_n$ and $e_1, \ldots, e_{n-1}$. Two vv'-paths $v_1 e_1 \cdots v_m$ and $v'_1 e'_1 \cdots v'_n$ are *openly disjoint* if $\{v_2, \ldots, v_{m-1}\} \cap \{v'_2, \ldots, v'_{n-1}\} = \emptyset$ and, in case $m = n = 2$, $e_1 \neq e'_1$. In a 2-graph G a $1 \cdot 2$-path is a $port_G(1) port_G(2)$-path; G is *port connected* if there is a $1 \cdot 2$-path in G.

For $k \in \mathbb{N}$, a graph G is k-*connected* if $|V_G| \geq k$ and for all nodes $v, v' \in V_G$ ($v \neq v'$) there are at least k pairwise openly disjoint vv'-paths in G. A 2-graph G is *almost* k-*connected* if $add(G)$ is k-connected. The intuition behind the latter notion is that the ports of a 2-graph belong together because they stem from the top-most operation of any term defining the graph and should thus not be separable from each other. One can show that, if G is a port connected 2-graph, then there is a uniquely determined maximal 2-subgraph $core(G)$ of G such that $core(G)$ is almost 2-connected. Figure 4 shows a port connected 2-graph G together with $core(G)$ and $add(core(G))$.

We are now able to state the main result of this section.

Theorem 3.1 *Let B be an ER algebra containing at least one nullary operation and one unary derived operation given by a graph G with one virtual edge e. If some $1 \cdot 2$-path in G avoids e, $[att_G(e)] \neq ports(G)$, and $e \in E_{core(G)}$, then there is an HR algebra A such that there is no translation $\tau :: T(A) \to T(B)$ in TB^*.*

We can apply the theorem to the example discussed in connection with Figure 2. As G we choose $val(K_3(\twoheadrightarrow, \twoheadrightarrow, x_1))$, which yields the operation given by the graph in the left-hand side of Figure 2 with virtual edge $e = e_3$. Obviously, this operation satisfies the requirements.

As another example, consider the graph G in Figure 4. If B contains G as a unary derived operation, where the virtual edge is one of those in the left triangle, then Theorem 3.1 applies since these edges are avoided by a 1·2-path, do not connect the ports directly, and are part of $core(G)$.

Let us see what the main steps to prove Theorem 3.1 are. First, one can show that $core(G)$ can be computed on terms over ER operations, using a transduction in TB^*. This is mainly because one can show that $core$ distributes over edge replacement.

Lemma 3.2 *Let B be an ER algebra. There is an ER algebra C and there is a transduction $\tau' :: T(B) \to T(C)$ in TB^* such that for all $t \in T(B)$, $\tau'(t) \neq \emptyset$ if and only if $val(t)$ is port connected, and for all $t' \in \tau'(t)$ we have $val(t') = core(val(t))$.*

Next, we need some results about almost 2-connected graphs. These were obtained by MacLane ([11], see also Vogler [12], whose presentation is much closer to the one used here) and Hopcroft and Tarjan [10].

A *string* is an unlabelled 2-graph G such that $V_G = \{v_1, \ldots, v_n\}$ for some $n \geq 2$, $port_G(1\cdot 2) = v_1 v_n$, and $E_G = \{e_1, \ldots, e_{n-1}\}$ with $[att_G(e_i)] = \{v_i, v_{i+1}\}$ for $i = 1, \ldots, n$. A *bond* is an unlabelled 2-graph G with $|V_G| = 2$ and $|E_G| \geq 2$. A *block* is an almost 3-connected 2-graph G such that $add(G)$ does not contain parallel edges.

An ER operation given by a bond, a substantial string, or a block is called a *bsb operation*. A term over bsb operations is a *bsb term*. A bsb term is *reduced* if it contains no subterm of the form $G(t_1, \ldots, t_{i-1}, G'(\cdots), t_{i+1}, \ldots, t_n)$ where both G and G' are bonds or both are strings.

The mentioned result by MacLane states that all reduced bsb terms t, t' with $val(t) = val(t')$ are equal up to permutation and reversal of virtual hyperedges and interchange of ports. In particular, all reduced bsb terms whose value is the same graph G have the same height, and thus all bsb terms whose value is G have at least the height of an arbitrarily chosen reduced bsb term with this value. In addition, Hopcroft and Tarjan [10] showed that for every almost 2-connected substantial 2-graph G there is some bsb term t (and also a reduced one) with $val(t) = G$. This result can be used to obtain an easy proof of the following lemma.

Lemma 3.3 *For every ER algebra C such that $val(T(C))$ is a set of almost 2-connected graphs a translation into bsb terms can be computed by a transduction $\tau'' \in TB^*$.*

Based on Lemmas 3.2 and 3.3 one can prove Theorem 3.1 along the lines of the example presented in the beginning of the section (Figures 2 and 3). Suppose a translation $\tau :: T(A) \to T(B)$ exists for all HR algebras A. Then the composition $\tau'' \circ \tau' \circ \tau$ (τ followed by τ' followed by τ'') yields for every term $t \in T(A)$ such that $val(t)$ is port connected a bsb term whose value is

$core(val(t))$. If G is as in the theorem, it is not too hard to show that there is a reduced bsb term of height $\geq n$ defining $core(G^n(H))$ for $n \in \mathbb{N}$ (using the already mentioned fact that $core$ distributes over edge replacement). Here, H is some port connected nullary derived operation of B, which must exist because G without its virtual edge is port connected and there is a nullary operation in B. By MacLane's theorem we thus get that all bsb terms defining $G^n(H)$ are of at least linear height. We can now define A along the lines of our example, using G and H instead of $K_3(\twoheadrightarrow, \twoheadrightarrow, x_1)$ and $\twoheadrightarrow$, so that fully balanced terms of logarithmic height define the graphs $G^n(H)$. Hence, $\tau'' \circ \tau' \circ \tau \notin TB^*$ as it involves an exponential increase of the height of terms, and thus $\tau \notin TB^*$.

4 Computing translations for 2tree* graphs by tgt transducers

In this section we introduce the so-called *2tree* graphs*—a superclass of the class of all (oriented) 2-trees—for which tgt transducers can compute translations. Whereas 2-trees consist solely of triangles one may use complete graphs of arbitrary size to build a *2tree** graph.

In the following we work with unlabelled graphs and hypergraphs without multiple edges, that is, for all distinct edges $e, e' \in E_G$ we have $att_G(e) \neq att_G(e')$. Therefore, we drop in the following the components att and lab, assuming that hyperedges are given as sequences of nodes (in other words, $e = att_G(e)$ for all $e \in E_G$).

Definition 4.1 (*2tree graph and algebra)** *(1) For all natural numbers $n \geq 3$ we let K_n be the (complete) unlabelled 2-graph $(V, E, port)$ without multiple edges given by $V = \{v_1, \ldots, v_n\}$, $E = \{v_i v_j \mid 1 \leq i < j \leq n\}$, $port(1) = v_1$, and $port(2) = v_n$. We view K_n as an ER operation of arity $k(n) = (n^2 - n)/2$ (that is, all edges are considered to be virtual).*
*(2) A *2tree** operation is an operation in $\{K_n \mid n \geq 3\} \cup \{\twoheadrightarrow\}$, a *2tree** algebra is an ER algebra whose operations are *2tree** operations, and a *2tree** term is a term over *2tree** operations. The set $2TREE^*$ of *2tree** graphs is the set of all graphs defined by *2tree** terms without variables. If G is a *2tree** graph then the edge $s(G) = port_G(1 \cdot 2)$ is called the source of G.*

As a convention, we let the first argument of K_n be the one to replace the source, that is, the edge $v_1 v_n$. Thus, the source of the first argument graph becomes the source of the value of an expression like $K_n(G_1, \ldots, G_{k(n)})$ (where $G_1, \ldots, G_{k(n)}$ are *2tree** graphs). The main theorem of this section is the one below.

Theorem 4.2 *For all HR algebras A and all *2tree** algebras B one can effectively construct a linear tgt transducer computing a translation $\tau :: T(A) \to T(B)$.*

In order to prove this theorem it is convenient to consider instead of HR operations another type of operations closely related to the operations on graphs

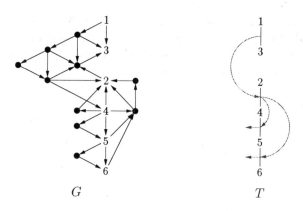

Figure 5: A 2-tree and its tree-like structure.

invented by Bauderon and Courcelle [2]. The two operations used here are those variants considered by Engelfriet in [8].

Let H, H' be hypergraphs without multiple hyperedges. For every set P of ports we define $res_P(H)$, called the *port restriction* of H by P, as $res_P(H) = (V_H, E_H, port_H|_P)$.

Furthermore, let H_1 and H_2 be hypergraphs without multiple hyperedges. Then the *composition* $H_1 \mathbin{\!/\mkern-5mu/\!} H_2$ of H_1 and H_2 is the hypergraph obtained from the disjoint union of H_1 and H_2 by identifying equally labelled ports and all multiple hyperedges this produces.

The operations res_P (for arbitrary P), $\mathbin{\!/\mkern-5mu/\!}$, and all unlabelled hypergraphs H without multiple hyperedges (viewed as constants) are called *RC operations*. Terms over these operations are *RC terms* and algebras whose operations are RC operations are *RC algebras*.

One can show that linear nondeleting top-down tree transducers can compute translations between HR and RC algebras in both directions. As it is well-known that linear nondeleting top-down tree transductions are closed under composition this means it does not matter whether we work with RC or HR terms in tgt transductions. Thus, we have to show how to compute from an RC term denoting a *2tree** graph another RC term which denotes the corresponding term hypergraph over $\{K_n \mid n \geq 3\} \cup \{\rightarrow\}$.

The basic idea is quite simple. We mainly have to consider the composition operation since this is the more complex one. Look at the graph G in Figure 5. This one is a *2tree** graph (and in fact a 2-tree) up to the port labelling. Suppose G is composed of the parts to the left and to the right of the ports. Intuitively, both parts are again collections of *2tree** graphs, and the tree-like structure of G (shown in Figure 5 on the right) is the composition of the tree structures of its parts. In order to be able to decide whether the parts fit together it suffices to remember (or guess and verify), just as in Figure 5, the structure imposed on the edges between ports. To make this precise, however, one needs to be sure that the considered terms have the two properties defined next.

If Π is a property of graphs let us say an RC term t has property Π if all subterms of t define graphs with property Π. We call a graph *isolation free* if it does not contain isolated nodes[3]. (So, an RC term is isolation free if all of its subterms define isolation free graphs.) A graph G is *stable* if G and $G|_{ports(G)}$ are isolation free.[4] An RC term $t \in T(A)$ is *saturated* if for all its subterms of the form $t_1 /\!/ t_2$, where $H = val(t_1 /\!/ t_2)$ and $H_i = val(t_i)$, we have $H_i = H|_{V_{H_i}}$. The lemma below states that stable and saturated terms can be computed by linear nondeleting top-down tree transducers.

Lemma 4.3 *For every RC term there is an equivalent stable and saturated RC term. Furthermore, for every RC algebra A a translation into stable and saturated RC terms is computed by a linear nondeleting top-down tree transducer.*

Thus, by the closure of linear nondeleting top-down transductions under composition we can restrict our attention to stable and saturated terms. In order to exploit the idea sketched above it is not very convenient to work with term hypergraphs since, intuitively, one would prefer the edges of a *2tree** graph to be in one-to-one correspondence with the nodes of the tree representing it. Therefore, we define a special kind of representations of *2tree** graphs as tree-like hypergraphs. (For this, recall that F removes the port labels of a hypergraph.)

Definition 4.4 (*2tree *representation*)** *The set $\mathcal{R}$ of 2tree* representations is the set of unlabelled hypergraphs T with $P_T = \{1 \cdot 2\}$ defined inductively as follows. The hypergraph $\bullet = (\{v\}, \emptyset, port\colon \{1 \cdot 2\} \to \{v\})$ is a 2tree* representation with $rval(\bullet) = \to$. For all $n \in \mathbb{N}$, $n \geq 3$, and $T_1, \ldots, T_{k(n)} \in \mathcal{R}$ the hypergraph $T = h_n\langle T_1, \ldots, T_{k(n)}\rangle$ obtained from the disjoint union of T_1, $F(T_2), \ldots, F(T_{k(n)})$ by adding the hyperedge $e = port_{T_1}(1 \cdot 2) \cdots port_{T_{k(n)}}(1 \cdot 2)$, is a 2tree* representation with $rval(T) = K_n(rval(T_1), \ldots, rval(T_{k(n)}))$.*

The root of a 2tree representation T is its unique port.*

In Figure 6 a *2tree** representation T and its value $rval(T)$ are depicted. Notice that the port label of the root is $1 \cdot 2$, not 1. It should be clear that $root(T)$ is the unique node v such that for all $e \in E_T$, $v \in [e]$ implies $v = e(1)$. We may thus use the notation $root(T)$ for all hypergraphs T with $F(T) \in F(\mathcal{R})$. For such T we denote by $\tilde{T}$ the *2tree** representation $(V_T, E_T, port\colon \{1 \cdot 2\} \to V_T)$ with $port(1 \cdot 2) = root(T)$. A similar remark as above holds for graphs H with $F(H) \in F(2TREE^*)$, to which we extend the notation $s(H)$. For these graphs $s(H) = vv'$, where v and v' are the unique nodes such that $v \in [e]$ implies $v = e(1)$ and $v' \in [e]$ implies $v' = e(2)$, for all $e \in E_H$.

In a *2tree** representation T with $rval(T) = H$ every node represents a unique edge in H (up to automorphisms, of course) via the so-called *associated mapping* $ed\colon V_T \to E_H$: For $T = \bullet$ there is only one such mapping. For $T = h_n\langle T_1, \ldots, T_{k(n)}\rangle$, if ed_i is the mapping associated with T_i we let $ed(v) = ed_i(v)$ for all $v \in V_{T_i}$. Clearly, ed is a bijection that maps $root(T)$ to $s(rval(T))$.

---
[3] A node v in a graph G is *isolated* if it is not incident with any edge.
[4] For $V \subseteq V_G$ the induced subgraph $G|_V$ is obtained from G by removing all nodes not in V, together with their incident edges.

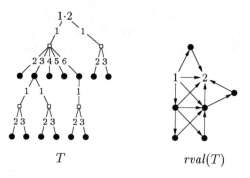

Figure 6: A *2tree** representation and its value.

One can show that terms defining *2tree** representations can be transformed into terms defining equivalent term hypergraphs, using a linear nondeleting top-down tree transducer. Therefore, in order to prove Theorem 4.2 it suffices to show how *2tree** representations can be computed. For this, if *2tree** representations shall be constructed using the operations of port restriction and composition the components must be allowed to have more ports than only the root. Therefore, we now define generalized *2tree** representations and, accordingly, generalized *2tree** graphs.

Definition 4.5 (*generalized 2tree representation* **and** *graph*)** *A generalized 2tree** representation is a hypergraph T such that P_T is a set of sequences of length two, $F(T) \in F(\mathcal{R})$, and $root(T) \in ports(T)$. In this case we let* $labels(T) = \bigcup_{p \in P_T}[p]$ *and* $rlabels(T) = [port_T^{-1}(root(T))]$. *The set of all generalized 2tree** representations is denoted by $\mathcal{R}_g$.*

*The set $2TREE_g^*$ of all generalized 2tree** graphs is the set of all stable graphs H such that $F(H) \in F(2TREE^*)$ and $[s(H)] \subseteq ports(H)$.*

We are now going to extend *rval* to generalized *2tree** representations T in a consistent way, such that the label of a port v in T yields the labels of the two nodes incident with $ed(v)$. Obviously, this is not always possible because the port labelling of T may be inconsistent.

Definition 4.6 (*consistency*) *Let $T \in \mathcal{R}_g$ and $H = rval(\tilde{T})$ with associated mapping $ed\colon V_{\tilde{T}} \to E_H$. Then T is consistent if the following hold.*

(C1) *If $port_T(q) = v$, $port_T(q') = v'$, and $q(i) = q'(j)$ for some $q, q' \in P_T$ and some $i, j \in \{1, 2\}$ then $ed(v, i) = ed(v', j)$.*

(C2) *Conversely, if $port_T(q) = v$, $port_T(q') = v'$, and $ed(v, i) = ed(v', j)$ for some $q, q' \in P_T$ and some $i, j \in \{1, 2\}$ then $q(i) = q'(j)$.*

(C3) *For all $v \in V_T$, if there are $q_i \in P_T$ and $j_i \in \{1, 2\}$ ($i = 1, 2$) such that $ed(port_T(q_i), j_i) = ed(v, i)$ for $i = 1, 2$, then $v \in ports(T)$.*

If $T \in \mathcal{R}_g$ is consistent we let $rval(T) = (V_H, E_H, port)$ with associated mapping $ed\colon V_T \to E_H$, where H and ed are as above and $port\colon labels(T) \to V_H$ is given by $port(q(i)) = ed(port_T(q), i)$ for $q \in P_T$ and $i = 1, 2$.

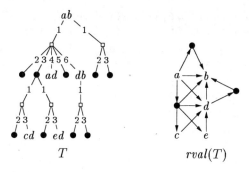

Figure 7: Some consistent $T \in \mathcal{R}_g$ and $rval(T)$.

Note that the extension of $rval: \mathcal{R} \to \mathit{2TREE}^*$ provided by Definition 4.6 is consistent. Figure 7 shows some consistent $T \in \mathcal{R}_g$ (obtained from the $\mathit{2tree}^*$ representation in Figure 6) together with the corresponding generalized $\mathit{2tree}^*$ graph $rval(T)$. It can be shown that for every graph G, $G \in \mathit{2TREE}_g^*$ if and only if $G = rval(T)$ for some consistent $T \in \mathcal{R}_g$ (which is uniquely determined if it exists). Notice that this would not be possible without the restriction that all graphs in $\mathit{2TREE}_g^*$ be stable, since $rval(T)$ is stable by definition.

The following lemma provides a condition which makes it easy to verify whether the composition $T_1 \parallel T_2$ of consistent $T_1, T_2 \in \mathcal{R}_g$ is a consistent element of $\mathcal{R}_g$.[5]

Lemma 4.7 *Let $T_1, T_2 \in \mathcal{R}_g$ be consistent, with $P_{T_1} \cap P_{T_2} = port_{T_2}^{-1}(root(T_2))$. Then we have $T_1 \parallel T_2 \in \mathcal{R}_g$, and $T_1 \parallel T_2$ is consistent if and only if $labels(T_1) \cap labels(T_2) = rlabels(T_2)$.*

For all $T \in \mathcal{R}_g$ and $v, v' \in V_T$ let $v \succeq_T v'$ denote the fact that v' is a descendant of v in T, that is, v lies on the simple path leading from $root(T)$ to v'. (As usual, a path is called *simple* if it does not contain any node twice.) Using this order on the nodes of $T \in \mathcal{R}_g$ we now define the so-called *abstraction* of T. Abstractions will provide the states of the transducer to be constructed.

Definition 4.8 (*abstraction*) *Let $T \in \mathcal{R}_g$ be consistent. The binary relation $\leadsto_T$ on P_T such that for all $q, q' \in P_T$ we have $q \leadsto_T q'$ if $port_T(q) \succeq_T port_T(q')$ is called the* abstraction *of T. A relation $\leadsto$ which is of the form $\leadsto_T$ for some $T \in \mathcal{R}_g$ is called an* abstraction; *in this case $P_{\leadsto}$ denotes P_T, $labels(\leadsto) = labels(T)$, and $rlabels(\leadsto) = rlabels(T)$. The root $\rho(\leadsto)$ is the unique $q \in P_{\leadsto}$ such that $q \leadsto q'$ for all $q' \in P_{\leadsto}$.*

As an example, let the graph G in Figure 5 be obtained as $G_1 \parallel G_2$, where G_1 is the left part and G_2 is the right (the latter containing only the ports labelled $2, 4, 5, 6$). Then G_1 has the form $G_1 = rval(T_1) \parallel rval(T_2) \parallel rval(T_3)$ for suitable $T_1, T_2, T_3 \in \mathcal{R}_g$ and the corresponding abstractions are $\{(1 \cdot 3, 1 \cdot 3), (1 \cdot 3, 4 \cdot 2),$

---

[5]The author wishes to thank the referee upon whose suggestion this lemma is based. The corresponding lemma in the submitted version was unnecessarily complicated and less useful.

$(4\cdot2, 4\cdot2)\}$, $\{(4\cdot5, 4\cdot5)\}$, and $\{(5\cdot6, 5\cdot6)\}$. Similarly, G_2 gives rise to the single abstraction $\{(4\cdot2, 4\cdot2), (4\cdot2, 4\cdot5), (4\cdot5, 4\cdot5), (4\cdot2, 5\cdot6), (5\cdot6, 5\cdot6)\}$.

For abstractions $\leadsto_1$, $\leadsto_2$, based on Lemma 4.7 we let $\leadsto_1 * \leadsto_2$ be defined if $P_{\leadsto_1} \cap P_{\leadsto_2} = \{\rho(\leadsto_2)\}$ and $labels(\leadsto_1) \cap labels(\leadsto_2) = rlabels(\leadsto_2)$. In this case we define $\leadsto_1 * \leadsto_2 = (\leadsto_1 \cup \leadsto_2)^*$.

In general, a subterm of an RC term t with $val(t) \in \mathit{2TREE}^*$ defines a graph which is a disjoint union of $rval(T_1), \ldots, rval(T_k)$ for consistent $T_1, \ldots, T_k \in \mathcal{R}_g$, rather than being of the simple form $rval(T)$. (For instance, in the example above we had $G_1 = rval(T_1) \mathbin{/\!\!/} rval(T_2) \mathbin{/\!\!/} rval(T_3)$.) For this reason it is necessary to consider sets of abstractions rather than single abstractions. An *abstraction set* is a set $\mathcal{A}$ of abstractions such that $P_{\leadsto_1} \cap P_{\leadsto_2} = \emptyset$ for all distinct $\leadsto_1, \leadsto_2 \in \mathcal{A}$ (notice that $labels(\leadsto_1)$ and $labels(\leadsto_2)$ may nevertheless intersect).

The two definitions below say when and how abstraction sets can be restricted and composed.

Definition 4.9 (*P-restriction*) *Let P be a set of port labels and let $\mathcal{A}$ be an abstraction set. For all $\leadsto \in \mathcal{A}$ let $\leadsto'$ be given by $P_{\leadsto'} = P_{\leadsto} \cap P^2$ and, for all $q, q' \in P_{\leadsto'}$, $q \leadsto' q'$ if $q \leadsto q'$. Then $\mathcal{A}$ P-restricts to $\{\leadsto' \mid \leadsto \in \mathcal{A}\}$ if*

- *for all $\leadsto \in \mathcal{A}$ it holds that $rlabels(\leadsto) \subseteq P$,*
- *for all $\leadsto \in \mathcal{A}$ we have $P \cap labels(\leadsto) = labels(\leadsto')$, and*
- *for all $\leadsto_1, \leadsto_2 \in \mathcal{A}$, $\leadsto_1 \neq \leadsto_2$, we have $labels(\leadsto_1) \cap labels(\leadsto_2) \subseteq P$.*

The first condition ensures that all the $\leadsto'$ are abstractions and are pairwise distinct since $P_{\leadsto_1} \cap P_{\leadsto_2} = \emptyset$ for distinct $\leadsto_1, \leadsto_2 \in \mathcal{A}$. The two remaining conditions are needed because, if $\mathcal{A}$ is the set of abstractions of consistent $T_1, \ldots, T_n \in \mathcal{R}_g$ the correctness proof for the transducer given below requires the equalities

$$\begin{aligned} & rval(res_{P^2}(T_1)) \mathbin{/\!\!/} \cdots \mathbin{/\!\!/} rval(res_{P^2}(T_n)) \\ = {} & res_P(rval(T_1)) \mathbin{/\!\!/} \cdots \mathbin{/\!\!/} res_P(rval(T_n)) \\ = {} & res_P(rval(T_1) \mathbin{/\!\!/} \cdots \mathbin{/\!\!/} rval(T_n)). \end{aligned}$$

In order to get the first equality we must prevent a situation where, for instance, $P_{T_i} = \{p \cdot p', q \cdot q'\}$ and $P = \{p, p', q\}$. Then q would get lost in $rval(res_{P^2}(T_i))$ but not in $res_P(rval(T_i))$. The third condition makes sure that nodes are identified in $res_P(rval(T_i)) \mathbin{/\!\!/} res_P(rval(T_j))$ if they are identified in $rval(T_i) \mathbin{/\!\!/} rval(T_j)$, which is obviously needed to make the second equality hold.

Definition 4.10 (*composition of abstraction sets*) *Let $\mathcal{A}$, $\mathcal{A}_1$, and $\mathcal{A}_2$ be abstraction sets. $\mathcal{A}_1$ and $\mathcal{A}_2$ compose to $\mathcal{A}$ if the elements of $\mathcal{A}_1$ and $\mathcal{A}_2$ can be arranged into sequences*[6] *$(\leadsto_{1,1} \cdots \leadsto_{1,k_1}), \ldots, (\leadsto_{l,1} \cdots \leadsto_{l,k_l})$ such that for $i = 1, \ldots, l$ the abstraction $\leadsto_i = (\cdots (\leadsto_{i,1} * \leadsto_{i,2}) * \cdots \leadsto_{i,k_i})$ is defined and $\{\leadsto_1, \ldots, \leadsto_l\} = \mathcal{A}$.*

---

[6] Here, an abstraction appears twice if it occurs in both of $\mathcal{A}_1$ and $\mathcal{A}_2$.

As an example, the reader might perhaps wish to compose the abstraction sets discussed with Definition 4.8, and figure out the required order. We can now define the top-down tree transducer td that computes (terms denoting) $2tree^*$ representations from stable saturated input terms. For this, let A be an RC algebra. We let td contain as states all abstraction sets over port labels that occur in A, where the abstraction set $\{\leadsto_0\}$ with $P_{\leadsto_0} = \{1\cdot2\}$ is the initial state. (We can assume that the port labels 1 and 2 occur in A as otherwise the domain cannot contain any $2tree^*$ graph.) The rules of td are the following ones.

Rules for constants. Let H be a constant of A such that there are consistent $T_1, \ldots, T_k \in \mathcal{R}_g$ with $rval(T_1) \mathbin{/\!\!/} \cdots \mathbin{/\!\!/} rval(T_k) = H$ and $P_{T_i} \cap P_{T_j} = \emptyset$ for $1 \leq i < j \leq k$. Then td contains the rule $\{\leadsto_{T_1}, \ldots, \leadsto_{T_k}\}(H) \to T_1 \mathbin{/\!\!/} \cdots \mathbin{/\!\!/} T_k$.

Rules for port restrictions. For every operation res_P in A and every state $\mathcal{A}$ such that $\mathcal{A}$ P-restricts to $\mathcal{A}'$, td contains the rule $\mathcal{A}'(res_P(x_1)) \to res_{P^2}(\mathcal{A}(x_1))$.

Rules for compositions. For all states $\mathcal{A}$, $\mathcal{A}'$, and $\mathcal{A}''$, where $\mathcal{A}'$ and $\mathcal{A}''$ compose to $\mathcal{A}$ we include the rule $\mathcal{A}(x_1 \mathbin{/\!\!/} x_2) \to \mathcal{A}'(x_1) \mathbin{/\!\!/} \mathcal{A}''(x_2)$.

It can be shown that this transducer yields the desired transduction, as stated in the lemma below.

Lemma 4.11 *Let $t \in T(A)$ be a stable and saturated term. Then $td(t) \neq \emptyset$ if and only if $val(t) \in 2TREE^*$, and for all $s \in td(t)$ with $T = val(s)$ we have $T \in \mathcal{R}$ and $rval(T) = val(t)$.*

The proof requires saturatedness of t since otherwise there could occur compositions of subgraphs that form a $2tree^*$ graph together, yet in one of the components an edge between ports is missing, which makes the transducer "reject" the input although it should not. As mentioned before, by composing td with an appropriate transducer that transforms (terms defining) $2tree^*$ representations into (terms defining) equivalent term hypergraphs a tgt transducer is obtained which computes the translation whose existence is asserted in Theorem 4.2. The constructions are all effective (which is now easier to see than in [4], thanks to the improved Lemma 4.7). Unfortunately, it is not obvious how to extend the result to other sorts of graphs like series-parallel graphs or even partial 2-trees. Rather simple generalizations which are not hard to obtain are those which allow, in addition to the constant $\to$, a number of labelled variants $\to_a$ (where a is the label of the edge) and constants $_a\!\leftarrow$ pointing backwards, that is, from the second port towards the first. Furthermore, using the results obtained by Engelfriet in [6] one can easily show that all these translations can be performed by deterministic tgt transducers if we allow the use of regular look-ahead (cf. [9] for the definition of tgt transducers with regular look-ahead).

Acknowledgement. I am grateful to the referees for their helpful comments. Special thanks are due to one of them, who wrote a remarkably thorough report without a single useless or unclear statement. I also thank Renate Klempien-Hinrichs, who read the final manuscript and (still) found some errors.

References

[1] Brenda S. Baker. Composition of top-down and bottom-up tree transductions. *Information and Control* 41, 186–213, 1979.

[2] Michel Bauderon, Bruno Courcelle. Graph expressions and graph rewriting. *Mathematical Systems Theory* 20, 83–127, 1987.

[3] Bruno Courcelle. The monadic second–order logic of graphs V: on closing the gap between definability and recognizability. *Theoretical Computer Science* 80, 153–202, 1991.

[4] Frank Drewes. The use of tree transducers to compute translations between graph algebras. Report 8/94, Univ. Bremen, 1994.

[5] Joost Engelfriet. Bottom-up and top-down tree transformations — a comparison. *Mathematical Systems Theory* 9(3), 198–231, 1975.

[6] Joost Engelfriet. On tree transducers for partial functions. *Information Processing Letters* 7, 170–172, 1978.

[7] Joost Engelfriet. Some open questions and recent results on tree transducers and tree languages. In R.V. Book, editor, *Formal Language Theory: Perspectives and Open Problems*, 241–286. Academic Press, New York, 1980.

[8] Joost Engelfriet. Graph grammars and tree transducers. In Proc. CAAP 94, *Lecture Notes in Computer Science* 787, 15–37, 1994.

[9] Joost Engelfriet, Heiko Vogler. The translation power of top-down tree-to-graph transducers. *Journal of Computer and System Sciences* 49, 258–305, 1994.

[10] J. E. Hopcroft, R. E. Tarjan. Dividing a graph into triconnected components. *SIAM Journal on Computing* 2(3), 135–158, 1973.

[11] S. MacLane. A structural characterization of planar combinatorial graphs. *Duke Mathematical Journal* 3, 460–472, 1937.

[12] Walter Vogler. Recognizing edge replacement graph languages in cubic time. In H. Ehrig, Hans-Jörg Kreowski, G. Rozenberg, editors, Proc. Fourth Intl. Workshop on Graph Grammars and Their Application to Comp. Sci., *Lecture Notes in Computer Science* 532, 676–687, 1991.

The Bounded Degree Problem for Non-Obstructing eNCE Graph Grammars[*]

K. Skodinis[1] and E. Wanke[2]

[1] University of Passau, 94032 Passau, Germany
e-mail: skodinis@fmi.uni-passau.de
[2] University of Düsseldorf, 40225 Düsseldorf, Germany
e-mail: wanke@cs.uni-duesseldorf.de

Abstract. A graph grammar is called non-obstructing if each graph G derivable from the axiom can derive a terminal graph. In this paper, the bounded degree problem for non-obstructing eNCE graph grammars is proved to be in the complexity class NL.

1 Introduction and summary

In this paper, so-called *edge label neighborhood controlled* (eNCE) graph grammars are analyzed. Such graph grammars are studied by several authors in various respects; see, for example, [Bra87, ELW90, Kau85, Kau87]. In eNCE graph grammars, all nodes and additionally all edges of the graphs are associated with labels. There are terminal and nonterminal node labels as well as terminal and nonterminal edge labels. Thus, the graphs involved have terminal and nonterminal nodes and edges. An eNCE graph grammar is specified by a set of *productions*. A production is a triple $p = (A, H, D)$, where A is a nonterminal label, H is a *labeled graph*, and D is an *embedding relation*. A is called the *left-hand side* and H together with D the *right-hand side* of p.

A production (A, H, D) can only be applied to a node u of a graph G, if u is labeled by A. A derivation step is performed by substituting u by H taking into account the embedding relation D as follows. Node u and its incident edges are removed from G. A copy of H is inserted into G. An edge labeled m is established between a node w from the copy of H and a former neighbor v of u labeled by a which was connected to u by an edge labeled by l if and only if the tuple (a, l, m, w) is contained in the embedding relation D.

The nodes in H can be treated separately, whereas all equally labeled nodes in the former neighborhood of u which are connected to u by equally labeled edges are treated identically. The language $L(\mathcal{G})$ of an eNCE graph grammar $\mathcal{G}$ is the set of all terminal labeled graphs derivable from a start graph S, the *axiom* of the grammar. By a simple extention of eNCE graph grammars we obtain edNCE graph grammars, which allow the edges to be directed. Thus edNCE graph grammars can generate directed graphs. In ordinary NCE graph grammars

---

[*] The work of the first author was supported by the German Research Association (DFG) grant Br-835-3/2

[JR80a, JR80b], the graphs have no edge labels. That is, the embedding relations D consist of pairs (a, w), where a is a node label and w a node of H. Hence in NLC graph grammars all equally labeled nodes in the copy of H are treated identically. Here, the embedding relations consist of pairs (a, b), where a and b are two node labels.

In this paper, we analyze *degree problems* for eNCE graph grammars defined as follows. The degree of a node u in a graph is the number of edges incident to u. The degree $deg(G)$ of a graph G is the maximal degree of its nodes. The *bounded degree problem* for an eNCE graph language $L(\mathcal{G})$ is the question whether or not there is an integer k such that for all graphs G in $L(\mathcal{G})$, $deg(G) < k$. Janssens et al [JRW86] have shown that the bounded degree problem is decidable for NLC graph grammars. The proof is obtained by a reduction to the finiteness problem for ETOL systems which has recently been shown to be PSPACE-hard[3]; see [MR93]. We will substantially improve this result.

Analyzing eNCE graph grammars seems to be much harder than analyzing NCE or NLC graph grammars. The property that makes eNCE graph grammars difficult to analyze is the existence of so-called *blocking edges* (nonterminal edges incident to two terminal nodes). Since only nonterminal nodes and their incident edges disappear in a rewriting step, graphs containing a blocking edge and all graphs derivable from them have at least one nonterminal edge, and thus are not in the language of the grammar. Blocking edges can be used to generate graph languages consisting of graphs with 2^n nodes for all $n \geq 1$. Such languages cannot be generated by eNCE graph grammars without blocking edges.

A simple condition that implies the non-blocking property for eNCE graph grammars is the assumption that for each graph G derivable from the axiom of $\mathcal{G}$, the graph G can derive into a terminal graph. An eNCE graph grammar which fulfills this property is called *non-obstructing*. A non-obstructing eNCE graph grammar is, clearly, always non-blocking.

The main result of this paper is that the bounded degree problem for non-obstructing eNCE graph grammars belongs to the complexity class NL[4]. This result can be easily extended to non-obstructing edENCE graph grammars and implies the existence of efficient parallel algorithms for the bounded degree problem for non-obstructing edNCE graph grammars. Our result carries over to non-obstructing NCE, non-obstructing NLC, and non-obstructing hyperedge-replacement systems; see [Hab92] for a definition of hyperedge replacement systems.

A non-blocking eNCE, edNCE, NCE, NLC, and hyperedge replacement system $\mathcal{G}$ can easily be transformed in linear time into a non-obstructing system $\mathcal{G}'$ such that $L(\mathcal{G}) = L(\mathcal{G}')$. Thus, for all these graph grammars, the bounded degree problem can be solved in polynomial time. On the other hand, the bounded degree problem is at least as hard as the emptiness problem and thus

---

[3] PSPACE is the class of languages recognizable in polynomial space by deterministic Turing machines.

[4] NL is the class of languages recognizable by nondeterministic Turing machines with logarithmic work-space.

P-complete[5] for non-blocking eNCE, NCE, NLC, and hyperedge replacement systems and, moreover, by the results from [SW94] undecidable, DEXPTIME-hard[6], and PSPACE-hard for general eNCE, confluent or boundary eNCE, and linear eNCE graph grammars, respectively.

2 Preliminaries

We define eNCE graph grammars in a sequence of definitions.

Definition 1 (Graphs). Let Σ and Γ be two finite sets of labels (node and edge labels, respectively). A *node/edge labeled graph* over Σ, Γ is a system $G = (V, E, \Phi)$, where

1. V is a finite set of nodes,
2. $\Phi : V \to \Sigma$ is a *node labeling* that associates with each node u a node label $\Phi(u)$,
3. $E \subseteq \{\{u,v\} \mid u,v \in V, u \neq v\} \times \Gamma$ is a finite set of labeled edges, i.e., each edge $e = (\{u,v\}, l)$ consists of two distinct nodes u, v and an edge label l.

A node or an edge labeled by $a \in \Sigma$ or $l \in \Gamma$ is called an *a-node* or *l-edge*, respectively.

We deal with undirected node and edge labeled graphs over Σ, Γ which we simply call *graphs*. Although the definition allows *multiple edges* (differently labeled edges between the same pair of nodes), we will assume that for each pair of nodes u, v there is at most one edge $(\{u,v\}, l)$. For the analysis in this paper, this is not a restriction but more a convenient way to keep the proofs less technical.

Next we define the composition of two graphs G and H by replacing a node u from G by H. This composition mechanism is used in derivation steps of eNCE graph grammars.

Definition 2 (Substitutions). An *embedding relation* for a graph H over Σ, Γ is a set
$$D \subseteq \Sigma \times \Gamma \times \Gamma \times V_H,$$
i.e., each tuple (a, l_1, l_2, w) from an embedding relation consists of a node label a, two edge labels l_1, l_2, and a node w of H.

Let G and H be two graphs over Σ, Γ. Let D be an embedding relation for H, and let u be a node from G. The graph $G[u/_D H]$ is obtained by replacing node u by H with respect to D as follows.

1. Let J be the disjoint union of G without node u and its incident edges and an isomorphic copy of H.

---

[5] P is the class of languages recognizable in polynomial time by deterministic Turing machines.
[6] DEXPTIME is the class of languages recognizable in exponential ($= 2^{\text{poly}(n)}$) time by deterministic Turing machines.

2. For each edge $(\{v,u\}, l_1)$ from G, add an edge $(\{v,w\}, l_2)$ to J if and only if $(\Phi(v), l_1, l_2, w) \in D$.
3. The resulting graph is $G[u/_D H]$.

Intuitively speaking, the substitution of a node u in G by a graph H is controlled by the embedding relation D as follows. Let $N(u) = \{v \mid (\{u,v\}, l) \in E_G,$ for some $l\}$ be the *node neighborhood* of u in G. If (a, l_1, l_2, w) is a tuple in the embedding relation D then node w from H will be connected to an a-node v from $N(u)$ by an l_2-edge if and only if the a-node v was previously connected to u by an l_1-edge.

Fig. 1 shows an example of such a substitution.

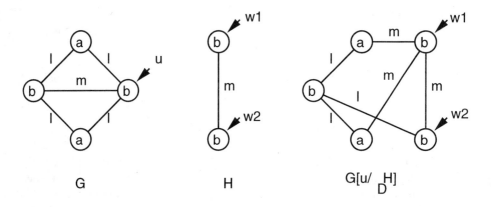

Fig. 1. Three graphs G, H, and $G[u/_D H]$ for $D = \{(a, l, m, w_1), (b, m, l, w_2)\}$.

Since we do not consider multiple edges, we assume that $(a, l, l_1, w), (a, l, l_2, w) \in D \Rightarrow l_1 = l_2$. We continue with the definition of eNCE graph grammars.

Definition 3 (eNCE graph grammars). An *eNCE (edge label neighborhood controlled embedding)* graph grammar is a system $\mathcal{G} = (\Sigma, \Delta, \Gamma, \Omega, S, P)$, where

1. $\Sigma, \Delta \subseteq \Sigma, \Sigma - \Delta, \Gamma, \Omega \subseteq \Gamma$, and $\Gamma - \Omega$ are finite sets of *node labels, terminal node labels, nonterminal node labels, edge labels, terminal edge labels,* and *nonterminal edge labels*, respectively,
2. S is a graph over Σ, Γ, the *axiom* of $\mathcal{G}$, and
3. P is a finite set of *productions*. Each *production* is a triple (A, G, C) where
   (a) A is a nonterminal node label from $\Sigma - \Delta$,
   (b) G is a graph over Σ, Γ,
   (c) C is an embedding relation for G.
   A is called the *left hand side* and (G, C) is called the *right hand side* of the production.

The definition of eNCE graph grammars divides the node and edge labels into terminal and nonterminal labels. A node (edge) labeled by a terminal or nonterminal label is called a *terminal* or *nonterminal node (edge)*, respectively. A graph is called *terminal* if all its nodes and edges are terminal.

The next definition shows how eNCE graph grammars define sets of graphs by derivations.

Definition 4 (Derivations and languages). Let $\mathcal{G}$ be an eNCE graph grammar and G, H be two graphs over Σ, Γ. We say that G *directly derives* H in $\mathcal{G}$, denoted by

$$G \Longrightarrow_{\mathcal{G}} H$$

if and only if G has some A-node u, $\mathcal{G}$ has a production (A, J, F), and $H = G[u/_F J]$.

We say that G *derives* H in $\mathcal{G}$, if

$$G \stackrel{*}{\Longrightarrow}_{\mathcal{G}} H,$$

where $\stackrel{*}{\Longrightarrow}_{\mathcal{G}}$ is the transitive and reflexive closure of $\Longrightarrow_{\mathcal{G}}$.

The *language* $L(\mathcal{G})$ of an eNCE graph grammar $\mathcal{G}$ is the set of all terminal graphs derivable from the axiom of $\mathcal{G}$.

We intend to analyze the complexity of problems using eNCE graph grammars as input. Thus, we have to define the size of an eNCE graph grammar. For our complexity analysis, the size of a grammar is just the size of the string that you get when writing down the grammar in the usual way. This implies that the size of a single node, edge, or label is logarithmic in the number of all nodes, all edges, and all labels, respectively.

We analyze so-called *degree problems* which are formally defined as follows.

Definition 5 (Degree of nodes and graphs). The degree $deg(u)$ of a node u in a graph $G = (V, E)$ is the number of edges incident to u, i.e.,

$$deg(u) := |\{(\{u, v\}, l) \in E\}|.$$

The degree $deg(G)$ of G is the maximum degree of the nodes of G, i.e.,

$$deg(G) := \max_{u \in V} deg(u).$$

In this paper, we consider the following degree problem.

Name: BOUNDED DEGREE PROBLEM

Instance: An eNCE graph grammar $\mathcal{G}$.

Question: Is there an integer k such that $deg(G) \leq k$ for all graphs G from $L(\mathcal{G})$?

3 The bounded degree problem

Lower bounds for the BOUNDED DEGREE PROBLEM can simply be shown with the emptiness problem. The *emptiness problem* for eNCE graph grammars is the question whether or not the language of a given eNCE graph grammar is empty. In [SW94], it is shown that the emptiness problem for eNCE, confluent/boundary eNCE, and linear eNCE graph grammars are undecidable, DEXPTIME-complete, and PSPACE-complete, respectively.

The BOUNDED DEGREE PROBLEM for an eNCE graph grammar $\mathcal{G}$ is at least as hard, with respect to log-space reductions, as the emptiness problem for $\mathcal{G}$. This is shown by the following simple modification of $\mathcal{G}$. Extend each terminal graph at the right hand side of a production by two nodes and one edge as in graph G of Fig. 2, where a, A, l are new labels, a, l are terminal, and A is nonterminal. Extend the set of productions of $\mathcal{G}$ by (A, H, D) and (A, J, F), where H and J are defined as in Fig. 2, $D = \{(a, l, l, u), (a, l, l, v)\}$, and $F = \{(a, l, l, w)\}$. The language of the modified eNCE graph grammar is of unbounded degree (G and productions (A, H, D) and (A, J, F) generate all star-graphs) if and only if the language of the original eNCE graph grammar is not empty.

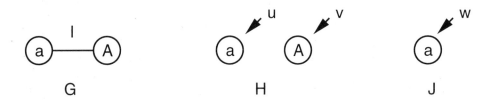

Fig. 2.

The same effect is obtained by simply extending only the axiom of $\mathcal{G}$ as above, but note that such a modification possibly would transform a linear eNCE graph grammar into a boundary eNCE graph grammar.

As stated in the introduction, the property which makes eNCE graph grammars difficult to analyze is the existence of so-called *blocking edges*. These are nonterminal edges incident to two terminal nodes. During a substitution step only nonterminal nodes and their incident edges can disappear. Thus, graphs containing a blocking edge and all graphs derivable from them have at least one nonterminal edge, and are not in the language of the grammar.

A second blocking mechanism—which is much less powerful than the one of blocking edges—is given by the possibility that certain nonterminal labels A either have no productions or all graphs G from the productions (A, G, C) derive always graphs having further nonterminal A-nodes. This situation can be eliminated in linear time by removing certain productions. Nevertheless, even this simple blocking mechanism makes the emptiness problem for non-blocking eNCE graph grammars P-complete and thus the BOUNDED DEGREE PROBLEM P-hard.

To determine some upper bounds, we show that the BOUNDED DEGREE PROBLEM can be solved very efficiently (nondeterministically in logarithmic space) if there is no blocking mechanism. This implies that the BOUNDED DEGREE PROBLEM is at most as difficult as eliminating the blocking mechanisms. We consider *non-obstructing* eNCE graph grammars.

Definition 6 (non-obstructing eNCE graph grammars). An eNCE graph grammar ist *non-obstructing*, if each graph derivable from its axiom can derive a terminal graph.

Each non-obstructing eNCE graph grammar is, obviously, free of any blocking mechanism.

To continue, we need the following notions.

Definition 7 (Descendant and ancestor relations). Let $G \Rightarrow H$ be a direct derivation step performed by applying a production (A, J, F) to a node u of G. The nodes in H which are from J are called *direct descendants* of u, whereas u is called a *direct ancestor* of each node from J.

The reflexive and transitive closure of the direct descendant and direct ancestor relation is simply called *descendant relation* and *ancestor relation*, respectively.

Definition 8 (Subgraph derivations). We say a graph H is *subgraph derivable* from a graph G, denoted by $G \stackrel{*}{\Rightarrow}_\subseteq H$, if G can derive a graph J such that H is a subgraph of J.

The definition of subgraph derivations is not restricted to derivations of terminal graphs. Given as input an eNCE graph grammar $\mathcal{G}$ and graphs G, H, where H has a number of nodes bounded by some constant k, then it is always nondeterministically decidable in logarithmic work-space whether or not H is subgraph derivable from G. This can be done as follows. Let $v_1, \ldots, v_k$ be the nodes of H and let $a_1, \ldots, a_k$ be variables each of which may store some node. Initially, each a_i is assigned to some node from G. Two variables a_i and a_j may store the same node. The intuition is that v_i is a descendant of the node stored in a_i. Let $induce(G, \{a_1, \ldots, a_k\})$ be the subgraph of G induced by the nodes $\{a_1, \ldots, a_k\}$. That is, $induce(G, \{a_1, \ldots, a_k\})$ has vertex set $\{a_1, \ldots, a_k\}$ and all edges from G between two nodes from $\{a_1, \ldots, a_k\}$. We guess step by step a subgraph derivation by substituting a node u from $induce(G, \{a_1, \ldots, a_k\})$ by some J with respect to a production $(lab(u), J, F)$. If $a_i = u$ for some i then variable a_i is assigned to some node from J. Finally, we verify whether

1. for each a_i the labels $\Phi(a_i)$ and $\Phi(v_i)$ are equal,
2. all a_i are pairwise distinct, and
3. for each edge $(\{v_i, v_j\}, l)$ in H there is an edge $(\{a_i, a_j\}, l)$ in $induce(G, \{a_1, \ldots, a_k\})$.

During this guessing process, only a constant number of nodes and edges need to be stored. The guessing process is also possible in polynomial space if we allow more than one edge between two nodes. Then we do not have to consider the complete induced subgraph $induce(G, \{a_1, \ldots, a_k\})$ but only some subgraph of $induce(G, \{a_1, \ldots, a_k\})$ with a bounded number of edges.

Definition 9 (Edge-preserving derivations). Let G be a graph with two nodes u, v labeled by A and B, respectively, and connected by an l-edge. We say that there is an *edge-preserving* (A, l, B)-*derivation* in an eNCE graph grammar $\mathcal{G}$ if G can derive a graph H which has a terminal labeled edge incident to a terminal labeled descendant of u and a terminal labeled descendant of v.

Given as input an eNCE graph grammar $\mathcal{G}$, it is obviously nondeterministically decidable in logarithmic work-space whether there is an edge-preserving (A, l, B)-derivation. This can be done as above by starting with a graph G consisting of one A-node u_1, one B node u_2, and one l-edge between u_1 and u_2. The variables a_1 and a_2 are initialized by u_1 and u_2. The guessing process is finished when a_1 and a_2 are terminal and $induce(G, \{a_1, a_2\})$ contains a terminal edge between a_1 and a_2.

Lemma 10. *Let G be a graph with $k+1$ nodes $u, v_1, \ldots, v_k$ and k edges $(\{u, v_1\}, l)$, $\ldots, (\{u, v_k\}, l)$ for some edge label l, where u is labeled by A and all $v_1, \ldots, v_k$ are labeled by B. If there is an edge-preserving (A, l, B)-derivation in an eNCE graph grammar $\mathcal{G}$, then G derives a terminal graph H in $\mathcal{G}$ with $deg(H) \geq k$.*

Proof. The idea is to apply the productions of an edge-preserving (A, l, B)-derivation simultaneously to all v_i's.

We uniquely name each node by some string over some alphabet. Let H be a graph with some node w. If w is substituted by some graph J with respect to some embedding relation F, then the nodes in $H[w/_F J]$ from J get new names, where all other nodes in $H[u/_F J]$ from H keep their old names. If y is the name of w and x is the name of some node w' of J, then yx is the name of the copy of w' in $H[w/_F J]$. Here yx means the concatenation of the strings y and x.

The nodes $u, v_1, \ldots, v_n$ in G initially get the names $y, y_1, \ldots, y_k$, respectively. Consider any edge-preserving (A, l, B)-derivation of a graph with two nodes u, v_1 labeled by A, B, named by y, y_1, respectively, and connected by some l-edge. A derivation $G \stackrel{*}{\Longrightarrow}_\mathcal{G} H$ with $deg(H) \geq k$ can be generated with the help of the edge-preserving (A, l, B)-derivation above as follows. Whenever a node with name yx for some x is substituted in the edge-preserving (A, l, B)-derivation substitute the node with name yx in the derivation from G. Whenever a node with name $y_1 x$ for some x is substituted in the edge-preserving (A, l, B)-derivation substitute all nodes with names $y_j x$ for $j = 1, \ldots, n$ in the derivation from G one after the other. Clearly, the resulting derivation generates a graph H with $deg(H) \geq k$.

Our main result of this section is based on the following lemma.

Lemma 11. *The language of a non-obstructing eNCE graph grammar $\mathcal{G}$ is of unbounded degree if and only if*

1. G_1 *is subgraph derivable from the axiom of $\mathcal{G}$ and H_1 is subgraph derivable from G_1, where G_1 and H_1 are defined as in Fig. 3 for some labels A, B, l, such that*
   (a) *node w_1 is a descendant of node u_1,*
   (b) *node w_2 and node w'_2 are descendants of node u_2, and*
   (c) *there is an edge-preserving (A, l, B)-derivation in $\mathcal{G}$,*

   or

2. G_2 *is subgraph derivable from the axiom of $\mathcal{G}$ and H_2 is subgraph derivable from G_2, where G_2 and H_2 are defined as in Fig. 3 for some labels A, B, C, l, m, such that*
   (a) *node w_1 is a descendant of node u_1,*
   (b) *node w_2 and node w'_2 are descendants of node u_2,*
   (c) *node w_3 is a descendant of node u_3, and*
   (d) *there is an edge-preserving (A, l, B)-derivation in $\mathcal{G}$.*

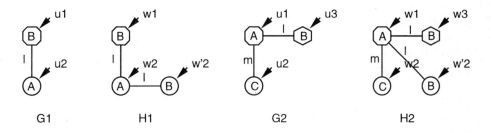

Fig. 3. The descendant relation with respect to the nodes in G_1 and G_2 is emphasized by different node shapes.

Proof. To prove the "if" case, we have to show that for each integer k the grammar $\mathcal{G}$ can derive a graph of a degree higher than k. By the assumptions we know that there is a subgraph derivation from the axiom to graph G_1 or G_2 from Fig. 3, and then to graph H_1 or H_2, respectively, from Fig. 3. This implies that there is a subgraph derivation from the axiom into a graph with an A-node and an arbitrary number of B-nodes all adjacent to the A-node by l-edges. Such a subgraph derivation can explicitly be generated using the idea in the proof of Lemma 10. The statement of Lemma 10 now implies that there is a derivation of a graph H with an arbitrary degree.

To prove the "only if" case, let k be some arbitrarily large integer. Consider any derivation

$$G_1 \Rightarrow G_2 \Rightarrow \cdots \Rightarrow G_n$$

of $\mathcal{G}$ such that G_1 is the axiom of $\mathcal{G}$ which is assumed to be a single node and G_n is a terminal graph with a node of degree at least k.

For every node u of graph G_n let $(u)^i$ be the ancestor of u in graph G_i. Note that each node u of G_n has exactly one ancestor in each G_i. Two different nodes of G_n may have the same ancestor in some G_i. For example, since G_1 consists of one single node, all $(u)^1$ for all nodes u of G_n are equal.

Let $u_{\max}$ be the node of G_n which has degree at least k. Let $v_1, \ldots, v_k$ be k neighbors of $u_{\max}$ in G_n. The number of neighbors of $u_{\max}$ corresponds to the degree of $u_{\max}$, because we do not consider multiple edges. We distinguish between three types of direct derivation steps. A direct derivation step $G_i \Rightarrow G_{i+1}$ for $1 \leq i < n$ is

of *type-0*, if $|\{(v_1)^{i+1}, \ldots, (v_k)^{i+1}\}| = |\{(v_1)^i, \ldots, (v_k)^i\}|$
of *type-1*, if $|\{(v_1)^{i+1}, \ldots, (v_k)^{i+1}\}| > |\{(v_1)^i, \ldots, (v_k)^i\}|$ and the substituted node is $(u_{\max})^i$, and
of *type-2*, if $|\{(v_1)^{i+1}, \ldots, (v_k)^{i+1}\}| > |\{(v_1)^i, \ldots, (v_k)^i\}|$ and the substituted node is some $(v_j)^i$ for some j.

Only type-1 and type-2 derivation steps contribute to the degree of the $(u_{\max})^i$'s for $i = 1, \ldots, n$. Let $c_{\max}$ be the maximum number of nodes of the graphs of the productions of $\mathcal{G}$. For each direct derivation step $G_i \Rightarrow G_{i+1}$, we have $deg(G_{i+1}) \leq deg(G_i) + c_{\max}$. That is, altogether we have at least $k/c_{\max}$ derivation steps of type-1 and type-2. Since the degree k is arbitrarily large it follows that the number of type-1 derivation or the number of type-2 derivation steps is arbitrarily large.

Let w_1, w_2 be two distinct nodes of G_n. As mentioned above, we know that $(w_1)^1 = (w_2)^1$, because G_1 has only one node. The *latest common ancestor index* of w_1, w_2, denoted by $index(w_1, w_2)$, is the highest index i, $1 \leq i \leq n$, such that the ancestor $(w_1)^i$ of w_1 is the ancestor $(w_2)^i$ of w_2. We also say that node w_1 and w_2 are *separated in the ith derivation step* if i is the latest common ancestor index of w_1 and w_2, i.e., if $(w_1)^i = (w_2)^i$ and $(w_1)^{i+1} \neq (w_2)^{i+1}$. Fig. 4 shows an example of the ancestor relations.

Assume the derivation

$$G_1 \Rightarrow G_2 \Rightarrow \cdots \Rightarrow G_n$$

has an arbitrary number K_1 of type-1 derivation steps. Let $S \subseteq \{1, \ldots, n-1\}$ initially be the set of all indices i such that $G_i \Rightarrow G_{i+1}$ is a type-1 derivation step. Then $|S| = K_1$.

First determine the node label which is most frequently used by the nodes $(u_{\max})^i$ for $i \in S$. Let, for example, A be this node label, then remove all indices i from S for which $(u_{\max})^i$ is not labeled by A. So the index set S now specifies at least

$$\frac{K_1}{|\Sigma|}$$

indices i such that node $(u_{\max})^i$ has label A.

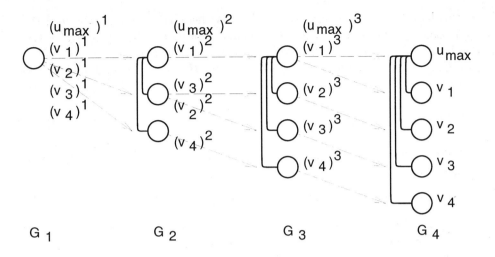

Fig. 4. The figure shows four graphs G_1, G_2, G_3, G_4 of some possible derivation. Usually, the graphs may have more nodes than just the ancestors of $u_{\max}, v_1, \ldots, v_k$. The dotted arrows indicate the descendent relation. The latest common ancestor index of v_2 and v_3 is 2.

Let $N := \{v_1, \ldots, v_k\}$ and v_j be some node from N with the least index $index(v_j, u_{\max}) \in S$, i.e., for all $v_{j'} \in N$ either $index(v_{j'}, u_{\max}) \geq index(v_j, u_{\max})$ or $index(v_{j'}, u_{\max}) \notin S$. Then determine the node/edge label pair which is most frequently used by the nodes $(v_j)^i$ for $i \in S$ and $i > index(v_j, u_{\max})$ and by the edges which connect the $(v_j)^i$'s with the $(u_{\max})^i$'s. Let B_j, l_j be this label pair, then remove all indices $i > index(v_j, u_{\max})$ from S for which $(v_j)^i$ is either not labeled by B_j or not adjacent to $(u_{\max})^i$ by an l_j-edge. Now the index set S contains at least

$$\frac{K_1}{|\Sigma|^2 \cdot |\Gamma|}$$

indices i such that all $(u_{\max})^i$ are equally labeled and for the selected v_j all $(v_j)^i$, $i > index(v_j, u_{\max})$, are equally labeled and adjacent to $(u_{\max})^i$ by equally labeled edges. Then remove all nodes $v_{j'}$ from N for which

$$index(v_{j'}, u_{\max}) = index(v_j, u_{\max})$$

and do the same procedure for the next node v_j from N with the least index $index(v_j, u_{\max}) \in S$. Obviously node v_j is always in the updated set N.

When $|\Sigma| \cdot |\Gamma| + 1$ iterations of the procedure are accomplished then the index set S contains at least

$$\frac{K_1}{|\Sigma|^{|\Sigma| \cdot |\Gamma| + 2} \cdot |\Gamma|^{|\Sigma| \cdot |\Gamma| + 1}}$$

indices i such that all $(u_{\max})^i$ are equally labeled and for each selected $v_j \in N$ all $(v_j)^i$, $i > index(v_j, u_{\max})$, are equally labeled and adjacent to $(u_{\max})^i$ by equally labeled edges.

If the final set S has more than $|\Sigma| \cdot |\Gamma|$ indices, then we have obtained the situation stated in the first case of the lemma. For there exist labels A, B, l, two nodes v_{j_1}, v_{j_2}, and two indices $i_1, i_2 \in S$, with $i_1 = index(v_{j_1}, u_{max}) < i_2 = index(v_{j_2}, u_{max})$, such that the graphs G_1 and H_1 in Fig. 3 can be defined as follows:

1. $u_1 = (v_{j_1})^{i_1+1}$, $u_2 = (u_{max})^{i_1+1}$,
2. $w_1 = (v_{j_1})^{i_2+1}$, $w_2 = u_{max}^{i_2+1}$ and $w_2' = v_{j_2}^{i_2+1}$.

Now assume that we have an arbitrarily large number K_2 of type-2 derivation steps. For every index j, with $v_j \in N$ we define

$$S_j = \{i \mid i = index(v_j, v_{j'}), G_i \Longrightarrow G_{i+1} \text{ is a type 2 derivation step}, v_{j'} \in N\}.$$

Let S initially be the set S_t from $\{S_j \mid j = 1, \ldots, k\}$ having the highest cardinality. Per construction of S there exists a node v_t from N, such that for each index $i \in S$ there exists a node $v_j \in N$ with $i = index(v_j, v_t)$ and $v_j \neq v_t$. Since at most c_{max} node pairs $v_j, v_{j'}$ can be separated in the same type-2 derivation step, it is not difficult to verify that $|S| \geq log_{c_{max}}(K_2)$ holds.

Let us continue with the assumption that number K_2 of type-2 derivation steps is arbitrarily large. Let v_t be the special node, such that each of the selected type-2 derivation step separates some node v_j, $1 \leq j \leq k$, $j \neq t$, from v_t.

Now, we continue quite similarly as in the first part of the proof. We determine the node label which is most frequently used by the nodes $(u_{max})^i$ for $i \in S$. Let A be this node label, then remove all indices i from S for which $(u_{max})^i$ is not labeled by A. The index set S now specifies at least

$$\frac{log_{c_{max}}(K_2)}{|\Sigma|}$$

graphs in which each $(u_{max})^i$ has label A.

Then determine the node/edge label pair which is most frequently used by the nodes $(v_t)^i$ for $i \in S$ and the edges which connect the $(v_t)^i$s to the $(u_{max})^i$s. Let C, m be this label pair, then remove all indices i from S for which $(v_t)^i$ is either not labeled by C or not adjacent to $(u_{max})^i$ by some m-edge. The index set S now specifies at least

$$\frac{log_{c_{max}}(K_2)}{|\Sigma|^2 \cdot |\Gamma|}$$

graphs such that all $(u_{max})^i$, $i \in S$, are equally labeled and all $(v_t)^i$, $i \in S$, are equally labeled and adjacent to $(u_{max})^i$ by equally labeled edges.

Let $N := \{v_1, \ldots, v_k\}$ and v_j be some node from N with the least index $index(v_j, v_t) \in S$. Then determine the node/edge label pair which is most frequently used by the nodes $(v_j)^i$ for $i \in S$ and $i > index(v_j, v_t)$ and by the edges which connect the $(v_j)^i$s with the $(u_{max})^i$s. Let B_j, l_j be this label pair, then remove all indices $i > index(v_j, v_t)$ from S for which $(v_j)^i$ is either not labeled

by B_j or not adjacent to $(u_{\max})^i$ by some l_j-edge. Then remove all nodes $v_{j'}$ from N for which
$$index(v_{j'}, v_t) = index(v_j, v_t)$$
and do the same procedure for the next node v_j from N with the least index $index(v_j, v_t) \in S$.

When $|\Sigma| \cdot |\Gamma| + 1$ iterations are accomplished then the index set S contains at least
$$\frac{log_{c_{max}}(K_2)}{|\Sigma|^{|\Sigma| \cdot |\Gamma| + 3} \cdot |\Gamma|^{|\Sigma| \cdot |\Gamma| + 2}}$$
indices i such that all $(u_{\max})^i$ are equally labeled, all $(v_t)^i$, $i \in S$, are equally labeled and adjacent to $(u_{\max})^i$ by equally labeled edges, and for each selected $v_j \in N$ all $(v_j)^i$, $i > index(v_j, v_t)$, are equally labeled and adjacent to $(u_{\max})^i$ by equally labeled edges.

If the final set S has more than $|\Sigma| \cdot |\Gamma|$ indices, then we have obtained the situation stated in the second case of the Lemma. For there exist labels A, B, C, l, m, two nodes v_{j_1}, v_{j_2} from N, and indices $i_1, i_2 \in S$, with $i_1 = index(v_{j_1}, v_t) < i_2 = index(v_{j_2}, v_t)$, such that the graphs G_2 and H_2 in Fig. 3 can be defined as follows:

1. $u_1 = (u_{max})^{i_1+1}$, $u_2 = (v_t)^{i_1+1}$, $u_3 = (v_{j_1})^{i_1+1}$,
2. $w_1 = (u_{max})^{i_2+1}$, $w_2 = (v_t)^{i_2+1}$, $w'_2 = (v_{j_2})^{i_2+1}$ und $w_3 = v_{j_1}^{i_2+1}$.

Since either K_1 or K_2 is unbounded the assumptions are satisfied either for the first part or the second part of the proof.

The properties stated in the lemma above can obviously be nondeterministically verified in logarithmic work-space. This and the fact that NL is closed under complementation [Imm87] implies the following theorem.

Theorem 12. *The BOUNDED DEGREE PROBLEM for non-obstructing eNCE graph grammars belongs to the complexity class NL.*

Since each non-blocking eNCE graph grammar can be transformed into a non-obstructing eNCE graph grammar using polynomial time, it follows that the BOUNDED DEGREE PROBLEM for non-blocking eNCE graph grammars belongs to the complexity class P. Furthermore, since each confluent eNCE graph grammar can be transformed in exponential time into a non-blocking eNCE graph grammar, see [SW94], it follows that the BOUNDED DEGREE PROBLEM for confluent eNCE graph grammars belongs to the complexity class DEXPTIME. An upper bound for the BOUNDED DEGREE PROBLEM for linear eNCE graph grammars is PSPACE. We omit the technical details of the proof in this abstract.

4 Acknowledgment

We are grateful for many fruitful suggestions and comments by some anonymous referee of an earlier version of this paper.

References

[Bra87] F.J. Brandenburg. On partially ordered graph grammars. In H. Ehrig, M. Nagl, A. Rosenfeld, and G. Rozenberg, editors, *Proceedings of Graph-Grammars and Their Application to Computer Science*, volume 291 of *LNCS*, pages 99–111. Springer-Verlag, 1987.

[ELW90] J. Engelfriet, G. Leih, and E. Welzl. Boundary graph grammars with dynamic edge relabeling. *Journal of Computer and System Sciences*, 40:307–345, 1990.

[Hab92] A. Habel. *Hyperedge Replacement: Grammars and Languages*, volume 643 of *LNCS*. Springer-Verlag, 1992.

[Imm87] N. Immerman. Languages that capture complexity classes. *SIAM Journal on Computing*, 16(4):760–778, 1987.

[JR80a] D. Janssens and G. Rozenberg. On the structure of node label controlled graph languages. *Information Science*, 20:191–216, 1980.

[JR80b] D. Janssens and G. Rozenberg. Restrictions, extensions, and variations of NLC grammars. *Information Science*, 20:217–244, 1980.

[JRW86] D. Janssens, G. Rozenberg, and E. Welzl. The bounded degree problem for NLC graph grammars is decidable. *Journal of Computer and System Sciences*, 33:415–422, 1986.

[Kau85] M. Kaul. Syntaxanalyse von Graphen bei Präzedenz-Graphgrammatiken. Dissertation, Universität Osnabrück, Osnabrück, Germany, 1985.

[Kau87] M. Kaul. Practical applications of precedence graph grammars. In H. Ehrig, M. Nagl, A. Rosenfeld, and G. Rozenberg, editors, *Proceedings of Graph-Grammars and Their Application to Computer Science*, volume 291 of *LNCS*, pages 326–342. Springer-Verlag, 1987.

[MR93] A. Monti and A. Roncato. On the computational complexity of some decidable problems for EOL systems and BSTA. Technical Report 12/93, University of Pisa, I-56125 Pisa,Italy, 1993.

[SW94] K. Skodinis and E. Wanke. The Complexity of Emptiness Problems of eNCE Graph Languages. *LNCS*, 903:180–192, 1994, to appear in *Journal of Computer and System Sciences*.

Process Specification and Verification

Klaus Barthelmann

Johannes Gutenberg-Universität, Institut für Informatik,
Staudinger Weg 9, Postfach 3980, D-55099 Mainz, Germany,
barthel@informatik.mathematik.uni-mainz.de

Summary. Graph grammars provide a very convenient specification tool for distributed systems of processes. This paper addresses the problem how properties of such specifications can be proven. It shows a connection between algebraic graph rewrite rules and temporal (trace) logic via the graph expressions of [2]. Statements concerning the global behavior can be checked by local reasoning.

1. Introduction

A distributed system is a network of processes which exchange messages over channels. If channels are buffered and shared by several processes, message passing is asynchronous. A typical example for this kind of network are actors [1]. For synchronous message passing, each channel connects exactly two processes and these processes must engage simultaneously in any communication. Synchronous message passing is the underlying mechanism in process algebras (for example, CCS [8]) and small programming languages (for example, CSP [6]). A common feature of all these different approaches is that a process nondeterministically chooses one among all possible communications and, depending on which path was taken, transforms itself into a process net. We are interested in a graph grammar semantics for synchronous communication which lends itself to the verification of properties of a process net. The properties can be checked by looking at the rewrite rules. The method is very similar to the equational reasoning known from functional programming. The graph grammar semantics can be extended to cope with higher order process calculi (like the π-calculus [9] or CHOCS [12]), where messages in turn may contain channels. The verification method, however, allows only very limited use of the rich possibilities of these powerful calculi.

The state of a distributed system is modeled as an edge-labeled hypergraph. The labels are taken from an algebra (also called magma) over some signature Σ. To model the behavior in time, these hypergraphs can be rewritten in two ways:

- Form a category of labeled hypergraphs by choosing a suitable notion of morphism [11]. Then, use the well-known double-pushout approach.
  A morphism in this category consists of a hypergraph morphism and a Σ-algebra morphism (see Definitions 2.5 and 2.6 below). Therefore, it is possible to do some pattern matching with the labels when rewrite rules are applied.

- According to [2], a labeled hypergraph is the value of a term over a suitable algebra (an extension of the Σ-algebra for labels). Then, ordinary term rewriting modulo equivalence will do the job (see Definitions 2.8 and 2.9 below).

A straightforward adaption of the proofs given in [2] shows that both approaches remain equivalent in our more general setting. We recommend pushouts for specification and terms for verification.

One hyperedge replacement rule is able to transform a process into a process net. Before that, however, two processes communicate and, therefore, rewrite rules cannot be applied to them independently. Both rules must be amalgamated [3]. The mechanism is the same as the one used in the Grammars for Distributed Systems [4] and is completely symmetric. In our setting, the amalgamation of two rewrite rules requires that the most general unifier of the message labels is computed. In practice, however, a sender and a receiver of the message will be distinguished, where the sender's message label contains no variables; unification reduces to pattern matching.

The most interesting aspect of a process (net) is its observable behavior. That are the messages exchanged with the outside world via external channels. To describe it, a suitable logic is needed. There is no canonic choice; however, a temporal logic dealing with traces [13] seems reasonable. It turns out that the formulas which characterize subprocesses compose in correspondence to the structure of the graph written as a term. This observation indicates a connection to compatible and inductive graph properties (see [5] for further references). A superficial comparison suggests that they are weaker than temporal logic because they express invariants only, that is, the properties hold for every graph in the derivation process. On the other hand, by the use of monadic second order logic, the invariants can be stronger than in (first order) temporal logic.

Seen the other way round, a graph rewrite system serves as a model for the trace logic. In comparison to the one adopted in [10] it has the advantage to handle the dynamic evolution of a network more elegantly. Furthermore, it allows to "derive" characterizing formulas in a certain way.

The paper is organized as follows. After a rather long preliminaries section, we show how the evolution of a process net is modeled by graph rewrite rules. We also present the main topic of this paper, namely the connection between the specification and temporal logic. The last section suggests some generalizations and lists open problems.

2. Preliminaries

The following four definitions (signature, Σ-algebra, term algebra, directed hypergraph with sources) are standard and are given here for convenience only.

Definition 2.1 (Signature). *A signature Σ consists of a set S of sorts, a set $\mathcal{F}$ of symbols, and two mappings $\alpha: \mathcal{F} \to S^*$ and $\varrho: \mathcal{F} \to S$ which together give the profile $\alpha(f)\varrho(f)$ of $f \in \mathcal{F}$.*

Definition 2.2 (Category of Σ-algebras). *An algebra A over a signature Σ consists of a set $A_s \neq \emptyset$ for each $s \in S$ and a mapping $f_A: A_{s_1} \times \ldots \times A_{s_n} \to A_s$ for each $f \in \mathcal{F}$ with profile $s_1 \ldots s_n s$.*

A morphism $\varphi: A \to A'$ of Σ-algebras consists of a mapping $\varphi_s: A_s \to A'_s$ for each $s \in S$ such that $\varphi_s(f_A a_1 \ldots a_n) = f_{A'} \varphi_{s_1}(a_1) \ldots \varphi_{s_n}(a_n)$ for each $f \in \mathcal{F}$ with profile $s_1 \ldots s_n s$ and elements $a_i \in A_{s_i}$, $1 \leq i \leq n$.

An important special case of a Σ-algebra is the term algebra defined next. We require that it exists (that is, the signature Σ must be *sensible*) and consider only *reachable* Σ-algebras (that is, epimorphic images of $\mathcal{T}_s(\Sigma, \emptyset)$) in the sequel.

Definition 2.3 (Term algebra over Σ). *Let $\mathcal{X}$ be a set of variables together with a mapping $\varrho: \mathcal{X} \to S$. We denote by $\mathcal{T}(\Sigma, \mathcal{X})$ the algebra of Σ-terms. $\mathcal{T}_s(\Sigma, \mathcal{X})$, for each $s \in S$, is the smallest set containing $\{x \in \mathcal{X} \mid \varrho(x) = s\}$ and*

$$\{ft_1 \ldots t_n \mid f \in \mathcal{F}, \varrho(f) = s, \alpha(f) = s_1 \ldots s_n, t_i \in \mathcal{T}_{s_i}(\Sigma, \mathcal{X}), 1 \leq i \leq n\}$$

as subsets. Furthermore, for every $f \in \mathcal{F}$, $f_{\mathcal{T}(\Sigma, \mathcal{X})}: (t_1, \ldots, t_n) \mapsto ft_1 \ldots t_n$.

2.1 Σ-labeled Graphs

Definition 2.4 (Category of hypergraphs). *A (directed) hypergraph G (with sources) consists of a set V of vertices, a set E of hyperedges, a vertex mapping $v: E \to V^*$, and a source list $\mathbf{s} \in V^*$. We call $|\mathbf{s}|$ the rank of G. Whenever it is convenient, we view $\mathbf{s}$ as a mapping $\{1, \ldots, |\mathbf{s}|\} \to V$.*

A morphism $\varphi: G \to G'$ of hypergraphs consists of two mappings $\varphi_V: V \to V'$ and $\varphi_E: E \to E'$ such that $v'(\varphi_E(e)) = \varphi_V^*(v(e))$ for all $e \in E$, and $\varphi_V^*(\mathbf{s}) = \mathbf{s}'$.

Hypergraphs of rank n, for arbitrary $n \in \mathbb{N}$, and their morphisms form a cocomplete category $\mathcal{H}_n$ and the discrete hypergraph with n distinct sources (and no hyperedges or additional vertices) is the initial object. Forgetting the sources gives a functor $F_n: \mathcal{H}_n \to \mathcal{H}_0$ for every $n \in \mathbb{N}$, which is not cocontinuous, however.

The following definition entails labeled hypergraphs as the special case, where all function symbols $f \in S$ are constants, that is, $\alpha(f)$ is empty.

Definition 2.5 (Category of Σ-labeled graphs). *A Σ-labeled graph H, which has its hyperedges labeled with elements from a Σ-algebra, consists of*

- *$V, E, v, \mathbf{s}$ which form a hypergraph G,*
- *A_s ($s \in S$), f_A ($f \in \mathcal{F}$) which form a Σ-algebra A, and*
- *a label mapping $l: E \to \biguplus_{s \in S} A_s$.*

A morphism $\varphi\colon H \to H'$ of Σ-labeled graphs consists of mappings $\varphi_V\colon V \to V'$, $\varphi_E\colon E \to E'$ which form a hypergraph morphism and a Σ-algebra morphism $\varphi_A\colon A \to A'$ such that $\varphi_A(l(e)) = l'(\varphi_E(e))$ for all $e \in E$.

Σ-labeled graphs of rank n and their morphisms form a category Σ-$\mathcal{G}_n$. Since colimits are formed for the Σ-algebra and hypergraph component separately, it has all the pushouts required below. The source-forgetting functor F_n from above can be extended to $F_n\colon \Sigma$-$\mathcal{G}_n \to \Sigma$-$\mathcal{G}_0$.

We will be interested only in the isomorphism class of a given Σ-labeled graph. All the definitions given so far and the following notion of rewriting are invariant under isomorphisms.

Definition 2.6 (Rewriting). *A (Σ-labeled graph) rewrite rule is a pair $\langle L, R\rangle$ of Σ-labeled graphs L, R with the same rank $n \geq 0$ and Σ-algebra $\mathcal{T}(\Sigma, \mathcal{X})$. Equivalently, it can be given as the two unique morphisms $L \leftarrow N \to R$ in Σ-$\mathcal{G}_n$, where the algebra components are identities, and N is the discrete graph with rank n and Σ-algebra $\mathcal{T}(\Sigma, \mathcal{X})$. A ($\Sigma$-labeled graph) rewrite system P is a finite set of (Σ-labeled graph) rewrite rules. The induced (Σ-labeled graph) rewrite relation $\xrightarrow[P]{}$ on Σ-$\mathcal{G}_m$ is defined as follows: $G \xrightarrow[P]{} H$ if there are a rewrite rule $\langle L, R\rangle \in P$, an object C (the context) and morphisms l', r' in Σ-$\mathcal{G}_m$, and morphisms l, r, d in Σ-$\mathcal{G}_0$ such that all morphisms are injective on edges and the following diagram in Σ-$\mathcal{G}_0$ consists of two pushouts:*

$$\begin{array}{ccccc} L & \longleftarrow & N & \longrightarrow & R \\ {\scriptstyle l}\downarrow & & {\scriptstyle d}\downarrow & & {\scriptstyle r}\downarrow \\ G & \xleftarrow{\,l'\,} & C & \xrightarrow{\,r'\,} & H \end{array}$$

The horizontal lines come from category Σ-$\mathcal{G}_n$ and Σ-$\mathcal{G}_m$ by applying F_n and F_m, respectively.

It follows that G, C and H draw their labels from the same Σ-algebra and the Σ-algebra components of l' and r' are the identities. The vertical morphisms l, d, r share their algebra component. It should be clear also, that this rewrite relation is undecidable in general.

2.2 Graph Expressions

We need a special signature and an algebra.

Definition 2.7 (Hypergraph signature). *The hypergraph signature consists of sorts $n \in \mathbf{N}$ and the following symbols: $n \in \mathbf{N}$ of profile n, $\oplus_{m,n}$ of profile $mn(m+n)$ for every $m, n \in \mathbf{N}$, $\theta_{\delta,n}$ of profile nn for every $n \in \mathbf{N}$, where δ is an equivalence relation on $\{1, \ldots, n\}$, and $\sigma_{f,m,n}$ of profile nm for every $m, n \in \mathbf{N}$, where $f\colon \{1, \ldots, m\} \to \{1, \ldots, n\}$ is a mapping.*

Definition 2.8 (Σ-labeled graph expression). *Let Σ-Γ be an extension of a signature Σ by the hypergraph signature together with new symbols (called constructors) $\langle s, n \rangle \in S \times \mathbf{N}$ of profile sn. A Σ-labeled graph expression is a term $g \in \mathcal{T}_n(\Sigma\text{-}\Gamma, \mathcal{X})$ of sort $n \in \mathbf{N}$. It is evaluated in the following Σ-Γ-algebra based on a Σ-algebra A:*

- *Sort $n \in \mathbf{N}$ is interpreted as the class of hypergraphs of rank n up to isomorphism.*
- *Constant $n \in \mathbf{N}$ denotes the discrete hypergraph with n vertices.*
- *$\oplus_{m,n} GH$ denotes the disjoint union of two hypergraphs G, H with ranks m and n, respectively, plus a reintroduction of common variables in the labels of G and H (which become different through disjoint union).*
- *$\theta_{\delta,n} G$ denotes source fusion in hypergraph G with rank n induced by δ. That is, two sources in positions i and j are identified if $\langle i, j \rangle \in \theta$.*
- *$\sigma_{f,m,n} G$ denotes source renaming in hypergraph G with rank n through f. That is, the result has source list $\mathbf{s} \circ f$.*
- *Given an element $a \in A_s$ of sort s, $\langle s, n \rangle a$ denotes the hypergraph with one hyperedge labeled a.*

Remark 2.1. It is easy to handle hierarchical graphs in this framework (we only need constructors of the form $\langle n, n \rangle$, for example, where $n \in \mathbf{N}$), but we will not consider this possibility here.

From now on, we omit the signatures of the operators and write $\oplus$ in infix notation. Anticipating our later example, the expression $\sigma_f \theta_\delta (\langle \pi, 2 \rangle \text{store} \oplus \langle \pi, 2 \rangle \text{save}(m))$, where π is the result sort of function symbols store (nullary) and save (unary), δ is generated by the pair $\langle 2, 3 \rangle$, and f maps $1 \mapsto 1, 2 \mapsto 4$, evaluates to the graph shown below.

Hyperedges are drawn as polygons and vertices as bullets.

Definition 2.9 (Rewriting). *A term rewrite rule is a pair of terms $\langle l, r \rangle \in \mathcal{T}_n(\Sigma\text{-}\Gamma, \mathcal{X}) \times \mathcal{T}_n(\Sigma\text{-}\Gamma, \mathcal{X})$ of the same sort $n \in \mathbf{N}$. A term rewrite system P is a finite set of term rewrite rules. The induced* term rewrite relation $\xrightarrow[P]{}$ *on Σ-labeled graphs is defined as follows: $G \xrightarrow[P]{} H$ if there are a rewrite rule $\langle l, r \rangle \in P$, a graph expression c (the context), a position p in c and a substitution σ for the variables in l and r such that $c[p := \sigma(l)]$ evaluates to G and $c[p := \sigma(r)]$ evaluates to H.*

Proposition 2.1. *1. If the Σ-algebra of labels is reachable then every graph is the value of a graph expression. 2. The rewrite relations in Definitions 2.6 and 2.9 coincide.*

Proof. (Sketch) Obviously, the graph structure is independent from the labeling. Therefore, the hard part of the proof was done in [2] already. The common algebra component of the vertical arrow in Definition 2.6 corresponds to the substitution σ in Definition 2.9. □

2.3 Temporal Trace Logic

Definition 2.10 (Traces). *A given signature Σ is extended by a sort $\langle s \rangle$ for finite sequences of values of sort s, called* traces. *We need a constant symbol $\langle \rangle$ of profile $\langle s \rangle$ denoting the empty sequence and a binary operator : of profile $s\langle s \rangle\langle s \rangle$ to put one term in front of a trace. Further function and predicate symbols may be added to make the logic more expressive. For convenience, we will use the binary operator · of profile $\langle s \rangle\langle s \rangle\langle s \rangle$ to concatenate two traces and the binary relation $\preceq$ stating that the first trace is a prefix of the second.*

To arrive at (linear) temporal logic, we extend the predicate logic over Σ (including traces): Besides the usual logical connectives ($\neg$, $\wedge$, $\vee$, $\rightarrow$, $\leftrightarrow$), equality (=) and quantifiers ($\forall$, $\exists$), we use the modalities $\mathcal{U}$ for "(strong) until", $\square$ for "always", $\Diamond$ for "sometimes", and $\bigcirc$ for "next" to talk about the logical validity of formulas at various points in time. Time is represented by the natural numbers. Furthermore, we distinguish a new set of *local* variables from the usual (now *global*) variables. Local variables may change their value over time whereas global variables remain constant. To save us irrelevant considerations and shorten the formulas, local variables for traces must be *monotonous*: The value at one time is a prefix of all the values at a later time.

Depending on the local variables, the validity of a formula can change from one moment to another. The values below, of course, are taken from an interpretation of predicate logic over Σ (including traces).

Definition 2.11 (Satisfaction). *An infinite sequence $w_0 w_1 w_2 \ldots$ of assignments, mapping local variables to values, satisfies a formula*

- *φ of predicate logic, iff φ is satisfied in the usual sense using w_0 for the local variables.*
- *$\varphi \mathcal{U} \psi$, iff there is an $i \in \mathbb{N}$ such that $w_i w_{i+1} \ldots$ satisfies ψ and $w_j w_{j+1} \ldots$ satisfy φ for all $0 \leq j < i$.*
- *$\square \varphi$, iff $w_i w_{i+1} \ldots$ satisfy φ for all $i \in \mathbb{N}$. $\square \varphi$ is equivalent to false $\mathcal{U} \varphi$.*
- *$\Diamond \varphi$, iff there is an $i \in \mathbb{N}$ such that $w_i w_{i+1} \ldots$ satisfies φ. $\Diamond \varphi$ is equivalent to true $\mathcal{U} \varphi$ as well as $\neg \square \neg \varphi$.*
- *$\bigcirc \varphi$, iff $w_1 w_2 \ldots$ satisfies φ.*

The remaining logical connectives are handled as usual.

Remark 2.2. All these temporal connectives are special cases of the so-called strict strong until operator $\bigcirc(\varphi \mathcal{U} \psi)$, which in fact we will use often. For a more elaborate definition of the sematics of (linear) temporal logic the reader is referred to [7].

3. The Operational Model

We represent a process by a labeled hyperedge (where the label is taken from a suitable Σ-algebra and encodes the process state) and (external) channels by (source) vertices. Messages are modeled by ordinary edges labeled with terms over the same Σ-algebra and attached to the corresponding channels. They form a chain, where the most recent message lies closest to the channel vertex. (Processes and messages could be distinguished by their outermost function symbol, but see the last section.)

Example 3.1. The three rewrite rules in Fig. 3.1 specify a buffer. The buffer

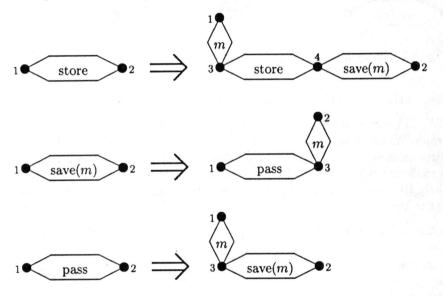

Fig. 3.1. Specification of a buffer. The numbers 3 and 4 on the right-hand sides are for later use; they do not denote sources.

receives its input at its left-hand side, that is, source 1. It is represented by the label store. m is a variable for a message. A new process save(m) is created to hold the message and output it later at the right-hand side, that is, source 2. Further messages are simply handed through by the process pass.

The amalgamation of rewrite rules mentioned in the introduction works as follows. We form the pushout of the left-hand and right-hand sides separately along the respective gluing graph shown below.

The left-hand gluing graph is simply a channel vertex. The right-hand gluing graph consists of one message hyperedge and adjacent vertices. The hyperedge is labeled with a variable for messages.

Example 3.2. Continuing our example, the amalgamation of rules 2 and 3 from Fig. 3.1 is shown in Fig. 3.2. The message variable m is the most general

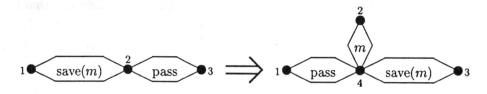

Fig. 3.2. Amalgamated rewrite rule. The number 4 does not denote a source.

unifier of the (distinct!) variables m in the original productions.

We are now able to define the behavior of a process net represented by a graph. While we see process states and communications "from inside", the environment has nothing better than observing the messages appearing on external channels. This is why we want to assign a trace to each channel, namely the sequence of messages that has been passed over it up to a given point in time.

Definition 3.1 (Derivation). *A derivation is a sequence of graphs $G_0 \to G_1 \to G_2 \to \ldots$, where, in each step, G_{i+1} is derived from G_i by applying an arbitrary number of (possibly amalgamated) rewrite rules in parallel. (Of course, the rewrite rules must be applicable in this way.) Furthermore, application must be weakly fair, that is, we are forced to apply a rewrite rule if otherwise it would be applicable to a graph G_i and all the following. As a special case, to apply no rewrite rule at all is allowed only finitely often in a row unless there are no applicable rewrite rules. (This is called* finite stuttering.*)*

A derivation represents a collection of sequences of traces in the following way. There is a trace for each source vertex. The value of the ith trace at time $t \in \mathbb{N}$ is the sequence of labels on the longest path starting from the ith source vertex in intermediate graph G_t and using only message edges.

A derivation represents a collection of sequences of traces under finite stuttering if there is a repetition function $r\colon \mathbb{N} \to \mathbb{N}$ which turns each element $t_0 t_1 t_2 \ldots$ into the element $\underbrace{t_0 \ldots t_0}_{r(0)+1} \underbrace{t_1 \ldots t_1}_{r(1)+1} \underbrace{t_2 \ldots t_2}_{r(2)+1} \ldots$ (exactly) represented by the derivation. ("Invisible actions" in the derivation may be left out.)

Example 3.3. Assuming the rewrite rules from the previous example, a typical derivation is depicted in Fig. 3.3. Rewrite rules 1, 1, 2 and the amalgamation from Fig. 3.2 are applied, in this order. The derivation represents the following two sequences of traces: $\langle\rangle, \langle a\rangle, \langle a,b\rangle, \langle a,b\rangle, \langle a,b\rangle, \ldots$ (for source 1) and $\langle\rangle, \langle\rangle, \langle\rangle, \langle a\rangle, \langle a\rangle, \ldots$ (for source 2). Under finite stuttering, the last two traces may be collapsed into one. Please note that a and b are values, not variables.

The next step is to apply Definition 2.11 to derivations. We assume that our formulas obey two restrictions:

- Among the free variables, the local ones are traces and the global ones are not.
- Variables bound by a universal quantifier are always global.

A pleasant side effect of this requirement is that we need a syntactic distinction between local and global variables only for those which are bound by an existential quantifier. We extend this convention further and take such variables as local unless any subformula marks them as global.

Definition 3.2 (Correctness, Preciseness). *Assume that graph G with rank n whose labels contain m free variables represents a process net. (If $m \neq 0$ then G is the right-hand side of a rewrite rule.) A formula with n free local variables and m free global variables matching those of G is*

- *correct for G if it is satisfied by every collection of sequences of traces which are represented by a derivation starting with G.*
- *precise for G if every collection of sequences of traces satisfying it is represented by a derivation starting with G.*
- *correct/precise under finite stuttering if the collections of sequences of traces above are represented under finite stuttering.*

It is clear that finite stuttering weakens the notion of preciseness but gives a stronger notion of correctness.

Now we are going to construct a formula to characterize a process net. It will contain one free local variable x_i for the ith source vertex, and one bound local variable x_j for each other vertex of the graph. These variables range over traces. (The small numbers in Figures 1–3 correspond to them!) If the graph is part of a rewrite rule and its labels contain additional variables, these will be free and global in the formula (avoiding name clashes).

Proposition 3.1. *If formulas φ and ψ are correct/precise (under finite stuttering) for the graphs represented by the expressions g and h, respectively, then*

1. $\Box \bigwedge_{1 \leq i \leq n} x_i = \langle\rangle$
2. $\varphi \wedge \psi$, *where the variables in ψ are renamed according to the sources in h,*
3. $\varphi \sigma$
4. $\varphi \wedge \Box \bigwedge_{\langle i,j \rangle \in \delta} x_i = x_j$

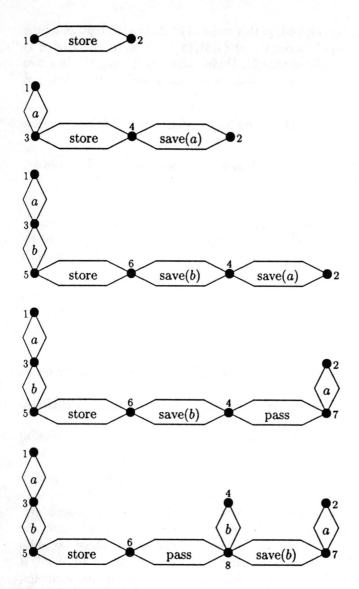

Fig. 3.3. Sample derivation sequence. Only the numbers 1 and 2 denote sources, all the others are for later use.

5. $\exists x_{j_1} \ldots \exists x_{j_{n-r}} \varphi[x_{i_1} := x_{f^{-1}(i_1)}, \ldots, x_{i_r} := x_{f^{-1}(i_r)}] \wedge \Box \bigwedge_{f(i)=f(j)} x_i = x_j,$

where $\{i_1, \ldots, i_r\}$ is the image of f, $\{j_1, \ldots, j_{n-r}\}$ its complement, and $f^{-1}(i)$ picks one of those j for which $f(j) = i$,

6. $\exists y (\Box x_i = t : y \wedge \varphi[x_i := y])$

is correct/precise (under finite stuttering) for the graph represented by

1. $n \in \mathbf{N}$
2. $g \oplus h$, where g and h have no (global) variables in common,
3. $g\sigma$, where σ is a substitution on (global) variables,
4. $\theta_\delta g$
5. $\sigma_f g$
6. g with one message labeled t at source i

respectively.

Proof. (Sketch) The variant with finite stuttering always follows easily from the exact case. Let us look at the items in turn.

1. is clear since $N \to N \to \ldots$ is the only derivation, where N is the value of n.
2. To prove correctness, consider a derivation $G_0 \oplus H_0 \to G_1 \oplus H_1 \to \ldots$, where G_0 and H_0 are the values of g and h, respectively. Since the derivation is weakly fair it can be split into two derivations $G_0 \to G_1 \to \ldots$ and $H_0 \to H_1 \to \ldots$, for which φ and ψ are presumed correct. Therefore, $\varphi \wedge \psi$ holds for the original derivation.

    For preciseness, consider a collection of sequences of traces satisfying $\varphi \wedge \psi$. There exist derivations $G_0 \to G_1 \to \ldots$ and $H_0 \to H_1 \to \ldots$ realizing subsets of these collections. Their parallel combination $G_0 \oplus H_0 \to G_1 \oplus H_1 \to \ldots$ realizes the whole collection.
3. by definition.
4. To prove correctness, consider a derivation $G_0 \to G_1 \to \ldots$, where G_0 is the value of $\theta_\delta g$. Clearly, φ is correct for it. We can prove $\Box \bigwedge_{\langle i,j \rangle \in \delta} x_i = x_j$ by induction on the number of derivation steps. It holds in the beginning. And it remains valid since any change of an x_i requires the same change for all x_j, $\langle i, j \rangle \in \delta$, through amalgamated productions.

    For preciseness, consider a collection of sequences of traces satisfying φ and $\Box \bigwedge_{\langle i,j \rangle \in \delta} x_i = x_j$. It is realized by a derivation on the value of g. Rewrite rules can be amalgamated as necessary to give a realizing derivation on the value of $\theta_\delta g$.
5. is clear since derivations do not depend on the naming of sources.
6. Correctness is clear since the message t can be "factored out" of any derivation starting at the given graph.

    For preciseness, note that for every collection of sequences of traces satisfying $\exists y(\Box x_i = t : y \wedge \varphi[x_i := y])$ there is a sequence of traces for y satisfying $\varphi[x_i := y]$. The new collection, where x_i is replaced with y, is realized by a derivation. We only have to add message t back to source y in each intermediate graph to get a realizing derivation for the original collection. □

Of course, there is no way to generate correct and/or precise formulas for the processes $\langle s, n \rangle a$ automatically; these have to be invented. The following proposition allows to check them afterwards.

Proposition 3.2. *If there is a complete list of hyperedge replacement rules*

$$\langle s,n \rangle a \Rightarrow g_1 \mid \ldots \mid g_k$$

for $\langle s,n \rangle a$ (where $g_1, \ldots, g_k$ are graph expressions of sort n) and the formulas $\varphi_1, \ldots, \varphi_k$ are correct (under finite stuttering) for $g_1, \ldots, g_k$, respectively, then φ is correct (under finite stuttering) for $\langle s,n \rangle a$ if

$$\bigwedge_{1 \leq i \leq n} x_i = \langle\rangle \wedge \bigcirc \bigvee_{1 \leq j \leq k} \left(\left(\bigwedge_{1 \leq i \leq n} x_i = \langle\rangle \right) \mathcal{U}\, \exists y_1 \ldots \exists y_{m_j}\, \varphi_j \right) \to \varphi,$$

where $y_1, \ldots, y_{m_j}$ are the free global variables of φ_j not free in φ. Similarly, if the $\varphi_1, \ldots, \varphi_k$ are precise (under finite stuttering) for $g_1, \ldots, g_k$, respectively, then φ is precise (under finite stuttering) for $\langle s,n \rangle a$ if

$$\varphi \to \bigwedge_{1 \leq i \leq n} x_i = \langle\rangle \wedge \bigcirc \bigvee_{1 \leq j \leq k} \left(\left(\bigwedge_{1 \leq i \leq n} x_i = \langle\rangle \right) \mathcal{U}\, \exists y_1 \ldots \exists y_{m_j}\, \varphi_j \right),$$

where $y_1, \ldots, y_{m_j}$ are the free global variables of φ_j not free in φ.

Proof. We prove that

$$\bigwedge_{1 \leq i \leq n} x_i = \langle\rangle \wedge \bigcirc \bigvee_{1 \leq j \leq k} \left(\left(\bigwedge_{1 \leq i \leq n} x_i = \langle\rangle \right) \mathcal{U}\, \exists y_1 \ldots \exists y_{m_j}\, \varphi_j \right)$$

is correct for $\langle s,n \rangle a$. Consider a collection of sequences of traces realized by a derivation $G_0 \to G_1 \to \ldots$ starting with the value of $\langle s,n \rangle a$. There exists $t > 0$ such that $G_0 = \ldots = G_{t-1}$ and G_t is the result of applying the jth rewrite rule, $1 \leq j \leq k$. Therefore, G_t is the value of $g_j \sigma$ under a substitution σ which eliminates all free variables of g_j. (The first part of Proposition 2.1 guarantees the existence of σ.) We already know that the collection of sequences of traces satisfies $\bigwedge_{1 \leq i \leq n} x_i = \langle\rangle$ up to time t. The subsequences of traces starting at t satisfy $\exists y_1 \ldots \exists y_{m_j}\, \varphi_j$ since φ_j is correct for the subsequence starting at G_t.

The preciseness of

$$\bigwedge_{1 \leq i \leq n} x_i = \langle\rangle \wedge \bigcirc \bigvee_{1 \leq j \leq k} \left(\left(\bigwedge_{1 \leq i \leq n} x_i = \langle\rangle \right) \mathcal{U}\, \exists y_1 \ldots \exists y_{m_j}\, \varphi_j \right)$$

is proven analogously. Finite stuttering adds nothing new. □

It is interesting that the "converse" of the last proposition holds, too.

Proposition 3.3. *If there is a complete list of hyperedge replacement rules*

$$\langle s,n \rangle a \Rightarrow g_1 \mid \ldots \mid g_k$$

for $\langle s,n\rangle a$ (where $g_1, \ldots, g_k$ are graph expressions of sort n) and the formulas $\varphi_1, \ldots, \varphi_k$ are precise (under finite stuttering) for $g_1, \ldots, g_k$, respectively, then φ is correct (under finite stuttering) for $\langle s,n\rangle a$ only if

$$\bigwedge_{1\leq i\leq n} x_i = \langle\rangle \wedge \bigcirc \bigvee_{1\leq j\leq k} \left(\left(\bigwedge_{1\leq i\leq n} x_i = \langle\rangle\right) \mathcal{U} \exists y_1 \ldots \exists y_{m_j} \varphi_j\right) \to \varphi,$$

where $y_1, \ldots, y_{m_j}$ are the free global variables of φ_j not free in φ. Similarly, if the $\varphi_1, \ldots, \varphi_k$ are correct (under finite stuttering) for $g_1, \ldots, g_k$, respectively, then φ is precise (under finite stuttering) for $\langle s,n\rangle a$ only if

$$\varphi \to \bigwedge_{1\leq i\leq n} x_i = \langle\rangle \wedge \bigcirc \bigvee_{1\leq j\leq k} \left(\left(\bigwedge_{1\leq i\leq n} x_i = \langle\rangle\right) \mathcal{U} \exists y_1 \ldots \exists y_{m_j} \varphi_j\right),$$

where $y_1, \ldots, y_{m_j}$ are the free global variables of φ_j not free in φ.

Proof. We only prove the case in which φ is correct, the other cases are similar. Consider a collection of sequences of traces satisfying

$$\bigwedge_{1\leq i\leq n} x_i = \langle\rangle \wedge \bigcirc \bigvee_{1\leq j\leq k} \left(\left(\bigwedge_{1\leq i\leq n} x_i = \langle\rangle\right) \mathcal{U} \exists y_1 \ldots \exists y_{m_j} \varphi_j\right).$$

Then there exist $t > 0$ and $1 \leq j \leq k$ such that $\langle\rangle, \ldots, \langle\rangle$ of length t is a prefix of each sequence of traces and the subsequences of traces starting at t satisfy $\exists y_1 \ldots \exists y_{m_j} \varphi_j$. Since φ_j is precise for the subsequence of traces starting at t there is a derivation $G_t \to G_{t+1} \to \ldots$ starting with the value of $g_j\sigma$ (where σ puts in suitable values for $y_1, \ldots, y_{m_j}$) which realizes them. We may choose the value of $\langle s,n\rangle a$ for $G_0, \ldots, G_{t-1}$ and apply the jth rewrite rule to realize the step $G_{t-1} \to G_t$. □

Example 3.4. Let us apply Propositions 3.1 and 3.2 to our running example. To make sure that the rewrite rules in Fig. 3.1 specify a buffer, we would like to prove $\Box x_2 \preceq x_1$ correct. This is $\varphi_{\text{store}}(x_1, x_2)$ already. We still have to find formulas $\varphi_{\text{save}(m)}(x_1, x_2; m)$ and $\varphi_{\text{store}}(x_1, x_2)$ and prove

$$x_1 = \langle\rangle \wedge x_2 = \langle\rangle \wedge \bigcirc((x_1 = \langle\rangle \wedge x_2 = \langle\rangle) \mathcal{U} \psi_1(x_1, x_2)) \to \varphi_{\text{store}}(x_1, x_2)$$

$$x_1 = \langle\rangle \wedge x_2 = \langle\rangle \wedge \bigcirc((x_1 = \langle\rangle \wedge x_2 = \langle\rangle) \mathcal{U} \psi_2(x_1, x_2; m)) \to \varphi_{\text{save}(m)}(x_1, x_2; m)$$

$$x_1 = \langle\rangle \wedge x_2 = \langle\rangle \wedge \bigcirc((x_1 = \langle\rangle \wedge x_2 = \langle\rangle) \mathcal{U} \psi_3(x_1, x_2)) \to \varphi_{\text{pass}}(x_1, x_2)$$

where

$$\psi_1(x_1, x_2) \equiv \exists m \exists x_3 \exists x_4 \, (\Box x_1 = m : x_3 \wedge$$
$$\varphi_{\text{store}}(x_3, x_4) \wedge \varphi_{\text{save}(m)}(x_4, x_2; m))$$
$$\psi_2(x_1, x_2; m) \equiv \exists x_3 \, (\Box x_2 = m : x_3 \wedge \varphi_{\text{pass}}(x_1, x_3))$$
$$\psi_3(x_1, x_2) \equiv \exists m \exists x_3 \, (\Box x_1 = m : x_3 \wedge \varphi_{\text{save}(m)}(x_3, x_2; m))$$

The reader is invited to check that the following formulas do the job:

$$\Box((x_1 = \langle\rangle \wedge x_2 = \langle\rangle) \vee x_2 = m : x_1 \vee \exists y \exists n \, (x_1 = y \cdot \langle n\rangle \wedge x_2 = m : y))$$

for $\varphi_{\text{save}(m)}(x_1, x_2; m)$ and

$$\Box(x_1 = x_2 \vee \exists m \, x_1 = x_2 \cdot \langle m\rangle)$$

for $\varphi_{\text{pass}}(x_1, x_2)$. They describe how we intuitively expect these processes to behave. Since all the formulas describe invariants, they happen to remain correct under finite stuttering.

But what is a precise formula for $\varphi_{\text{store}}(x_1, x_2)$? Since our intuitive understanding tells us that the time required to pass a message from the left-hand side to the right-hand side cannot be less than the number of elements stored in the buffer before, we will not find such a formula. However, if we state that traces cannot grow arbitrarily, we easily get a precise formula under finite stuttering:

$$x_1 = \langle\rangle \wedge x_2 = \langle\rangle \wedge$$
$$\Box(x_2 \preceq x_1 \wedge \forall y \, (x_1 = y \, \mathcal{U} \, \exists m \, x_1 = y \cdot \langle m\rangle) \wedge \forall y \, (x_2 = y \, \mathcal{U} \, \exists m \, x_2 = y \cdot \langle m\rangle))$$

$\varphi_{\text{save}(m)}(x_1, x_2; m)$ and $\varphi_{\text{pass}}(x_1, x_2)$ must be extended in a similar way. The formula states some kind of additional "liveness" property: The process net is always able to reach a state where $x_2 = x_1$.

4. Conclusion and Further Work

We claimed that our operational model can be extended to allow the passing of channels in messages. Let us observe from Proposition 3.1 that a message edge with label m

behaves like a process hyperedge of rank 2 characterized by the formula $\Box x_1 = m : x_2$! Therefore, if we extend amalgamation to certain processes with more than two adjacent channels, the effect will be the same as if these channels were transferred in the message. For example, if the two rewrite rules

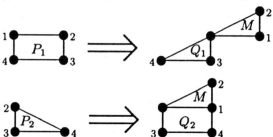

are amalgamated along process M then channel 1 is passed from P_1 to P_2 along channel 2.

An idea for verification suggests itself in this case: We must be able to characterize the amalgamation processes (like M above) by formulas. Further experiences are needed to show whether this is possible in practical examples.

But work has to be done in the basic calculus as well. The most important task is to find a (partial or infinitary) axiomatization of the logic. Although reasoning about traces is easier than reasoning about processes, it is still error-prone without a formal system. Systems tailored to special rewriting systems can also make things clearer. This paper is meant as a starting point to show that graph grammars are also well-suited for verification, a point that seems to have been neglected in the literature so far.

Acknowledgement. The author would like to thank the referees for their comments.

References

1. G. Agha: Concurrent Object-oriented Programming. Communications of the ACM **33**(3), 125–141 (1990)
2. M. Bauderon, B. Courcelle: Graph Expressions and Graph Rewritings. Mathematical Systems Theory **20**, 83–127 (1987)
3. P. Böhm, H.-R. Fonio, A. Habel: Amalgamation of Graph Transformations: A Synchronization Mechanism. Journal of Computer and System Sciences **34**, 377–408 (1987)
4. P. Degano, U. Montanari: A Model for Distributed Systems Based on Graph Rewriting. Journal of the ACM **34**(2), 411-449 (1987)
5. A. Habel, H.-J. Kreowski, C. Lautemann: A comparison of compatible, finite, and inductive graph properties. Theoretical Computer Science **110**, 145–168 (1993)
6. C.A.R. Hoare: Communicating Sequential Processes. Prentice Hall 1985
7. F. Kröger: Temporal Logic of Programs. Springer 1987
8. R. Milner: Communication and Concurrency. Prentice Hall 1989
9. R. Milner, J. Parrow, D. Walker: A Calculus of Mobile Processes I & II. Information and Control **100**, 1–77 (1992)
10. V. Nguyen, A. Demers, D. Gries, S. Owicki: Behavior: a Temporal Approach to Process Modelling. In: R. Parikh (Ed.): Logics of Programs. Lecture Notes in Computer Science 193 (1985)
11. G. Schied: Über Graphgrammatiken, eine Spezifikationsmethode für Programmiersprachen und verteilte Regelsysteme. Dissertation. Arbeitsberichte des IMMD 25, 2, Universität Erlangen-Nürnberg 1992
12. B. Thomsen: Plain CHOCS A second generation calculus for higher order processes. Acta Informatica **30**, 1–59 (1993)
13. J. Widom, D. Gries, F. B. Schneider: Trace-Based Network Proof Systems: Expressiveness and Completeness. ACM Transactions on Programming Languages and Systems **14**(3), 396–416 (1992)

An Event Structure Semantics for Graph Grammars with Parallel Productions[*]

A. Corradini[1], H. Ehrig[2], M. Löwe[2], U. Montanari[1], and F. Rossi[1]

[1] Università di Pisa, Dipartimento di Informatica, Corso Italia 40, 56125 Pisa, Italy
({andrea,ugo,rossi}@di.unipi.it)
[2] Technische Universität Berlin, Fachbereich 13 Informatik, Franklinstraße 28/29, 10587 Berlin, Germany ({ehrig,loewe}@cs.tu-berlin.de)

Abstract. We propose a truly concurrent semantics for graph grammars, based on event structures, that generalizes to arbitrary consuming grammars (i.e., such that each production deletes some items) the semantics presented in [4] for the subclass of safe grammars. Also, parallel derivations are explicitly considered, instead of sequential ones only as in [4]. The "domain" and the "event structure" of a grammar are introduced independently, and one main result shows that they are strongly related, since the domain is the domain of finite configurations of the event structure. Another important result provides an abstract characterization of when two (parallel) graph derivations should be considered as equivalent from a true-concurrency perspective.

1 Introduction

Graph grammars are a powerful formalism for the specification of concurrent and distributed systems. In this respect, they are far more expressive than, for example, Petri nets [15]. In fact, on the one hand they allow for a more structured description of states (which are graphs, instead of sets of tokens); on the other hand they allow one to specify context-dependent operations, where part of the state is read but not consumed (see [7] for more about the relationship between Petri nets and graph grammars).

Most formalisms for the specification of concurrent systems are equipped with *truly-concurrent* semantics, where concurrency is regarded as a primitive notion. In such semantics, a single computation is usually represented by some partial ordering of events modelling causality: Thus two events are not related by the causal ordering if and only if they are concurrent. Moreover, the collection of all computations of a system can be represented by an algebraic structure including both a causal ordering and some representation of the branching (choice) points of the computations. For example, *event structures* [19] are widely accepted as a truly-concurrent semantic model for systems manifesting concurrency and non-determinism.

---
[*] Research partially supported by the COMPUGRAPH Basic Research Esprit Working Group n. 7183

The algebraic theory of graph grammars comprises many results concerning parallelism and concurrency (see [11, 12, 8]), but for a long time it lacked a truly concurrent semantics able to represent both the intrinsic parallelism of derivations and the non-determinism caused by the application of conflicting productions. Recent works in this direction include [16] and [4]. In [16], Schied proposes an event structure semantics for graph grammars. He considers *concrete* graph derivations only, and gets rid of the non-determinism of direct derivations by using a clever algorithm for naming the items (arcs or nodes) generated by a production. Technically, he obtains the event structure of a graph grammar by associating a trace language with it, and by applying general results [1] relating Mazurkiewicz traces and event structures.

In [4] we presented an event structure semantics for a subclass of graph grammars, called *safe* grammars, where *abstract* derivations (i.e., isomorphism classes of derivations) are considered. Thus we leave the application of a production non-deterministic, but we identify in the semantics all the derivations which differ only for the names of the newly generated items. Moreover, the formal technique used to obtain the event structure of a grammar is original and very different from that of [16]. In fact we first define the *category of abstract derivations* and the *category of derivation traces* of a grammar, which are structures interesting in itself; then we show that with a simple comma category construction we get immediately the domain of configurations of an event structure. It is worth stressing that both [16] and [4] restrict to *consuming* grammars, i.e., grammars where each production consumes something.

In this paper we go beyond [4], generalizing its results in two directions. The first one is mainly technical: We consider also *parallel* derivations, and not only sequential ones. The second extension is more substantial: We show that the event structure semantics can be extended to consuming grammars which are not safe. *Safe grammars* are grammars where each graph reachable from the start graph has no automorphisms different from the identity. For such grammars, abstract derivations are defined simply as isomorphism classes of derivations, and it turns out that sequential composition of abstract derivations is well-defined, yielding a category. This is false in general for non-safe grammars [3].

Since the category of abstract derivations is a milestone for the event structure semantics we are looking for, we propose a finer definition of abstract derivation: Two derivations are *abstraction equivalent* if they are not only isomorphic, but also, the two isomorphisms relating the starting and the ending graphs, respectively, are *standard*. Standard isomorphisms [2] are a fixed family of isomorphisms, one between each pair of isomorphic graphs, satisfying suitable properties. Thanks to this additional constraint, equivalence classes of derivations can be composed sequentially

Next we introduce the well-known *shift equivalence* [11, 12, 8] that relates derivations which differ only because independent productions may be applied in different orders. The union of the shift and abstraction equivalences yields the *(concatenable) truly concurrent* equivalence, whose equivalence classes are the *(concatenable) derivation traces*. A fundamental result shows that given any two derivations in a derivation trace of a consuming grammar, there is a unique bijective mapping between their sets of production applications, which satisfies suitable consistency conditions. This result also provides an abstract characterization of when two (parallel) graph derivations should be considered as equivalent from a true-concurrency perspective.

We introduce then category $\mathbf{Tr}(\mathcal{G})$, having abstract graphs as objects and derivation traces as arrows, and the comma category of objects of $\mathbf{Tr}(\mathcal{G})$ under the start graph G_0, denoted $(G_0 \downarrow \mathbf{Tr}(\mathcal{G}))$. While in the case of safe grammars this comma category turns out to be a partial order, in the case of arbitrary consuming grammars it is in general only a preorder. The *domain* of a consuming grammar, $\mathbf{Dom}(\mathcal{G})$, is thus defined as the partial order generated by the preorder $(G_0 \downarrow \mathbf{Tr}(\mathcal{G}))$.

At this point we introduce the event structure $\mathbf{ES}(\mathcal{G})$ of a consuming grammar. Intuitively, an event of $\mathbf{ES}(\mathcal{G})$ is determined by a specific direct derivation α belonging to a derivation ρ that begins from the start graph of $\mathcal{G}$, together with all its causes, i.e., with all the preceeding direct derivations of ρ that caused α, either directly or indirectly. More precisely, such an event will be defined as a set of derivations of $\mathcal{G}$, namely the set of all derivations having (a copy of) α as the last step, containing (a copy of) all its causes, in any order consistent with the causality relation, and possibly containing other direct derivations that are independent from α. Given this definition of events, the causality and conflict relations are easily defined, yielding an event structure.

The main result of the paper shows that there is a bijective mapping between the set of finite configurations of $\mathbf{ES}(\mathcal{G})$ and the elements of $\mathbf{Dom}(\mathcal{G})$. Thus by well-known results [19] these two structures are conceptually equivalent, in the sense that one can be recovered from the other: They constitute the truly-concurrent semantics that we propose for a grammar. As a corollary, we get that $\mathbf{Dom}(\mathcal{G})$ enjoys suitable algebraic properties: More precisely, it is a prime algebraic, finitely coherent, finitary complete lower semilattice, where all elements are finite. Thus we indirectly recover (in a simplified way) the result that was explicitly proved in [4] for safe grammars.

The paper is organized as follows. Section 2 recalls some definitions concerning the double-pushout approach to graph grammars [8] and the abstraction equivalence on derivations [3]. Section 3 introduces the shift and the truly concurrent equivalences, and the derivation traces, presenting an important characterization of these equivalences. Moreover, the category of derivation traces $\mathbf{Tr}(\mathcal{G})$ is defined. In Section 4, we define first the domain $\mathbf{Dom}(\mathcal{G})$ of a grammar, via a suitable comma category construction, and then, independently, the prime event structure of a grammar, $\mathbf{ES}(\mathcal{G})$: The last results of the paper explore the tight relationships between these two structures. Finally, Section 5 summarizes the paper and hints at some directions for future investigation.

Because of space limitation, the proofs have been omitted. A copy of the paper with proofs can be requested to the first author. The reader is assumed to have some familiarity with the basics of the double-pushout approach to graph grammars [8] and of category theory [13].

2 Graph grammars in the double-pushout approach

We introduce here the basic definitions of the algebraic theory of graph grammars, following the double-pushout approach first introduced in [10]. We also present a precise definition of an "abstraction equivalence" relating all graph derivations that only differ for representation details.

In this paper we stick to colored graphs: although most of the definition and results are formulated in pure categorical terms, some specific properties of the category of graphs are sometimes exploited.

Definition 1 (colored graphs). Given two fixed alphabets Ω_E and Ω_N for edge and node labels, respectively, a *(colored) graph (over (Ω_E, Ω_N))* is a tuple $G = \langle E, N, s, t, l_E, l_N \rangle$, where E is a finite set of *edges*, N is a finite set of *nodes*, $s, t : E \to N$ are the *source* and *target* functions, and $l_E : E \to \Omega_E$ and $l_N : N \to \Omega_N$ are the *edge* and the *node labeling* functions, respectively.

A *graph morphism* $f : G \to G'$ is a pair $f = \langle f_E : E \to E', f_N : N \to N' \rangle$ of functions which preserve sources, targets, and labels, i.e., such that $f_N \circ t = t' \circ f_E$, $f_N \circ s = s' \circ f_E$, $l'_N \circ f_N = l_N$, and $l'_E \circ f_E = l_E$; it is an *isomorphism* if both f_E and f_N are bijections. We write $G \cong H$ if there is an isomorphism from G to H; moreover, an *abstract graph* $[G]$ is an isomorphism class of graphs, i.e., $[G] = \{H \mid H \cong G\}$. An *automorphism* of G is an isomorphism $\phi : G \to G$; it is *non-trivial* if $\phi \neq id_G$. The category having colored graphs as objects and graph morphisms as arrow is called **Graph**. □

Definition 2 (graph productions, graph grammars). A *graph production* $(L \xleftarrow{l} K \xrightarrow{r} R)$ is a pair of injective graph morphisms $l : K \to L$ and $r : K \to R$; it is *consuming* (or *consumptive*, as in [4]) if morphism l is not an isomorphism. The graphs L, K, and R are called the *left-hand side*, the *interface*, and the *right-hand side* of the production, respectively. A *graph grammar* $\mathcal{G}$ is a tuple $\mathcal{G} = \langle G_0, P, \pi \rangle$, where G_0 is the *start graph*, P is a set of *production names*, and π maps each production name in P to a graph production. We shall write $p : (L \xleftarrow{l} K \xrightarrow{r} R)$ if $\pi(p) = (L \xleftarrow{l} K \xrightarrow{r} R)$. Grammar $\mathcal{G}$ is called *consuming* if all its productions are consuming. □

Definition 3 (parallel productions). A *parallel production* (over a given graph grammar $\mathcal{G}$) has the form $\langle (p_1, in^1), \ldots, (p_k, in^k) \rangle : (L \xleftarrow{l} K \xrightarrow{r} R)$ (see Figure 1 (a)), where $k \geq 0$, $p_i : (L_i \xleftarrow{l_i} K_i \xrightarrow{r_i} R_i)$ is a production of $\mathcal{G}$ for each $i \in \underline{k}$,[3] L is a coproduct object of the graphs in $\langle L_1, \ldots, L_k \rangle$, and similarly R and K are coproduct objects of $\langle R_1, \ldots, R_k \rangle$ and $\langle K_1, \ldots, K_k \rangle$, respectively. Moreover, l and r are uniquely determined, using the universal property of coproducts, by the families of arrows $\{l_i\}_{i \in \underline{k}}$ and $\{r_i\}_{i \in \underline{k}}$, respectively. Finally, for each $i \in \underline{k}$, in^i denotes the triple of injections $\langle in_L^i : L_i \to L, in_K^i : K_i \to K, in_R^i : R_i \to R \rangle$. A parallel production like the one above is *proper* if $k > 1$; the *empty* production is the (only) parallel production with $k = 0$, having the empty graph $\emptyset$ as left- and right-hand sides and as interface. □

---
[3] For each $n \in \mathbb{N}$, by $\underline{n}$ we denote the set $\{1, 2, \ldots, n\}$ (thus $\underline{0} = \emptyset$).

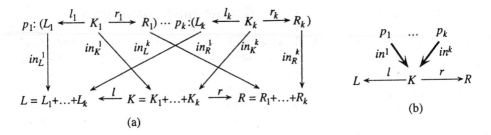

Fig. 1. (a) The parallel production $\langle (p_1, in^1), \ldots, (p_k, in^k)\rangle : (L \xleftarrow{l} K \xrightarrow{r} R)$. (b) Its compact representation.

In the rest of the paper we will often denote the parallel production of Figure 1(a) simply as $p_1 + \ldots + p_k : (L \xleftarrow{l} K \xrightarrow{r} R)$; note however that the "+" operator is not assumed to be commutative. We will also use the more compact drawing of Figure 1(b) to depict the same parallel production. Furthermore, we will freely identify a production p of $\mathcal{G}$ with the parallel production $\langle (p, \langle id_L, id_K, id_R\rangle)\rangle$; thus, by default productions will be parallel in the rest of the paper.

We assume that the reader is familiar with the categorical constructions of pushout and pushout complement, which are the formal basis of the definition of direct derivation (see [8]).

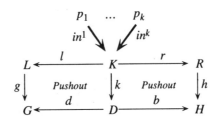

Fig. 2. (Parallel) direct derivation as double-pushout construction.

Definition 4 (direct derivation). Given a graph G, a (parallel) production $q = p_1 + \ldots + p_k : (L \xleftarrow{l} K \xrightarrow{r} R)$, and an *occurrence* (i.e., a graph morphism) $g : L \to G$, a *(parallel) direct derivation from G to H using q (based on g)* exists if and only if the diagram in Figure 2 can be constructed, where both squares are required to be pushouts in **Graph**. In this case, D is called the *context graph*, and we write $G \Rightarrow_{q,g} H$, or also $G \Rightarrow_q H$; only seldom we shall write $G \xRightarrow{\langle g,k,h,b,d\rangle}_q H$, indicating explicitly all the morphisms of the double-pushout. If $q = \emptyset$ (i.e., it is the empty production), then $G \Rightarrow_\emptyset H$ is an *empty direct derivation*. □

We refer the reader to [8] for an analysis of the conditions (called the *gluing conditions*) that guarantee the existence of the left pushout of Figure 2, given morphisms l and g. It is worth stressing that since pushouts—like all universal constructions in categories—are defined only up to isomorphisms, given a production q and an occurrence morphism g there are in general infinitely many different, although isomorphic, direct derivations using q and based on g.

The next fact shows that the application of the empty production to a graph produces an isomorphic graph.

Fact 5 (empty direct derivations). *If $G \stackrel{\langle g,k,h,b,d \rangle}{\Rightarrow}_\emptyset H$ is an empty direct derivation, then morphisms g, k, and h are necessarily the only morphisms from the empty graph (since $\emptyset$ is initial in **Graph**), while b and d must be isomorphisms. Morphism $b \circ d^{-1} : G \to H$ is called the isomorphism induced by the empty direct derivation. Moreover, for any pair of isomorphic graphs $G \cong H$, there is one empty direct derivation $G \stackrel{\langle \emptyset, \emptyset, \emptyset, b, d \rangle}{\Rightarrow}_\emptyset H$ for each triple $\langle D, d : D \to G, b : D \to H \rangle$, where d and b are isomorphisms.* □

Definition 6 (derivations). A *(parallel) derivation (over $\mathcal{G}$)* is either a graph G (called an *identity derivation*, and denoted by $G : G \Rightarrow^* G$), or a sequence of (parallel) direct derivations $\rho = \{G_{i-1} \Rightarrow_{q_i} G_i\}_{i \in \underline{n}}$ such that $q_i = p_{i_1} + \ldots + p_{i k_i}$ is a (parallel) production over $\mathcal{G}$ for all $i \in \underline{n}$ (as in Figure 3). In the last case, the derivation is written $\rho : G_0 \Rightarrow^*_\mathcal{G} G_n$ or simply $\rho : G_0 \Rightarrow^* G_n$. If $\rho : G \Rightarrow^* H$ is a (possibly identity) derivation, then graphs G and H are called the *starting* and the *ending graph* of ρ, and will be denoted by $\sigma(\rho)$ and $\tau(\rho)$, respectively. The *length* of a derivation ρ, denoted by $|\rho|$, is the number of direct derivations in ρ, if it is not an identity, and 0 otherwise. The *order* of ρ, denoted by $\#\rho$, is the total number of elementary productions used in ρ, i.e., $\#\rho = \sum_{i=1}^n k_i$; moreover, $prod_\rho : \underline{\#\rho} \to P$ is the function returning for each j the name of the j-th production applied in ρ—formally, $prod_\rho(j) = p_{rs}$ if $j = (\sum_{i=1}^r k_i) + s$. The *sequential composition* of two derivations ρ and ρ' is defined if and only if $\tau(\rho) = \sigma(\rho')$; in this case it is denoted $\rho ; \rho' : \sigma(\rho) \Rightarrow^* \tau(\rho')$, and it is the diagram obtained by identifying $\tau(\rho)$ with $\sigma(\rho')$ (thus if $\rho : G \Rightarrow^* H$, then $G ; \rho = \rho = \rho ; H$, where G and H are the identity derivations). □

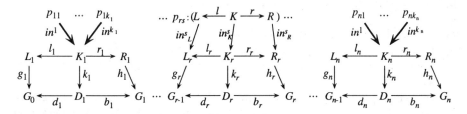

Fig. 3. A (parallel) derivation, with explicit drawing of the s-th production of the r-th direct derivation.

In the theory of the algebraic approach to graph grammars, it is natural to reason in terms of *abstract graphs* and *abstract derivations*, considering as equivalent graphs or derivations, respectively, which only differ for representation dependent details. The problem of defining a reasonable notion of equivalence on derivations is the main topic of [3]. We showed there that such an equivalence should satisfy a number of requirements, motivated by the desire of lifting to the abstract setting some relevant results which hold on concrete derivations. Here we just recall the last equivalence we introduced in [3] (called there equivalence $\equiv_3$ and here *abstraction equivalence*). Before that, we have to introduce some definitions about permutations, and the notion of *standard isomorphisms* [2], on which the mentioned equivalence is based.

Definition 7 (permutations). Recall that $\underline{n}$ denotes the set $\{1, 2, \ldots, n\}$. A *permutation* on $\underline{n}$ is a bijective mapping $\Pi : \underline{n} \to \underline{n}$. The *identity permutation* on $\underline{n}$ is the identity function, and will be denoted by Π_{id}^n. The *composition* of two permutations Π_1 and Π_2 on $\underline{n}$ is their composition as functions, $\Pi_1 \circ \Pi_2$, while the *concatenation* of two permutations Π_1 on $\underline{n_1}$ and Π_2 on $\underline{n_2}$ is denoted by $\Pi_1 \mid \Pi_2$, and it is the permutation on $\underline{n_1 + n_2}$ defined as

$$\Pi_1 \mid \Pi_2(x) = \begin{cases} \Pi_1(x) & \text{if } 1 \leq x \leq n_1 \\ \Pi_2(x - n_1) + n_1 & \text{if } n_1 < x \leq n_2 \end{cases}$$

Concatenation and composition of permutations are clearly associative. □

Definition 9 (abstraction equivalence of graph derivations). Let S be an in category **Graph** is a family of isomorphisms indexed by pairs of isomorphic graphs (i.e., $S = \{S_{G'}^G \mid G \cong G'\}$), satisfying the following conditions for each G, G' and G'' in **Graph**:

(S-iso) $S_{G'}^G : G \to G'$; (S-id) $S_G^G = id_G$; (S-comp) $S_{G'}^{G''} \circ S_{G''}^G = S_{G'}^G$.

A family of standard isomorphisms certainly exists by the axiom of choice. □

Definition 9 (abstraction equivalence of graph derivations). Let S be an arbitrary but fixed family of standard isomorphisms of category **Graph**. Moreover, let $\rho : G_0 \Rightarrow^* G_n$ and $\rho' : G_0' \Rightarrow^* G_{n'}'$ be two derivations as depicted in Figure 4, and suppose that $q_i = p_{i1} + \ldots + p_{ik_i}$ for each $i \in \underline{n}$, and $q_j' = p_{j1}' + \ldots + p_{jk_j'}'$ for each $j \in \underline{n'}$. Then they are *abstraction equivalent* if

1. $n = n'$, i.e., they have the same length;
2. for each $i \in \underline{n}$, $k_i = k_i'$; i.e., the number of productions applied in parallel at each direct derivation is the same; therefore $\#\rho = \#\rho'$, i.e., they have the same order;
3. there exists a family of permutations $\{\Pi_1, \ldots, \Pi_n\}$ such that Π_i is a permutation of $\underline{k_i}$, and $p_{ij} = p_{i\Pi_i(j)}'$ for each $i \in \underline{n}$ and $j \in \underline{k_i}$; thus the productions applied in corresponding direct derivations must be the same, up to a permutation;

4. there exists a family of isomorphisms $\{\phi_{X_i} : X_i \to X'_i \mid X \in \{L, K, R, G, D\}, i \in \underline{n}\} \cup \{\Phi_{G_0}\}$ between corresponding graphs of the parallel productions used in the two derivations such that (1) the isomorphisms relating the starting and ending graphs are standard, i.e., $\phi_{G_0} = S_{G'_0}^{G_0}$ and $\phi_{G_n} = S_{G'_n}^{G_n}$; (2) the resulting diagram commutes; and (3) all triangles formed by the injections from the productions must commute, taking into account the corresponding permutations.

We say that ρ and ρ' are equivalent via Π (written $\rho \equiv_\Pi^{abs} \rho'$) if $\Pi = \Pi_1 \mid \ldots \mid \Pi_n$ is the permutation on $\#\rho$ obtained by concatenating the permutations in $\{\Pi_i\}_{i \in \underline{n}}$. Furthermore, let relation $\equiv^{abs}$ on derivations be defined as $\rho \equiv^{abs} \rho'$ iff $\rho \equiv_\Pi^{abs} \rho'$ for some permutation Π: it will be called *abstraction equivalence*. In fact, from the definition (using the fact that standard isomorphisms are closed with respect to composition) it follows that $\rho \equiv_{\Pi_{id}^{\#\rho}}^{abs} \rho$, that $\rho \equiv_\Pi^{abs} \rho'$ implies $\rho' \equiv_{\Pi^{-1}}^{abs} \rho$, and that if $\rho \equiv_\Pi^{abs} \rho'$ and $\rho' \equiv_{\Pi'}^{abs} \rho''$, then $\rho \equiv_{\Pi' \circ \Pi}^{abs} \rho''$, thus $\equiv^{abs}$ is an equivalence relation. The equivalence classes of derivations with respect to $\equiv^{abs}$ are called *abstract derivations*. □

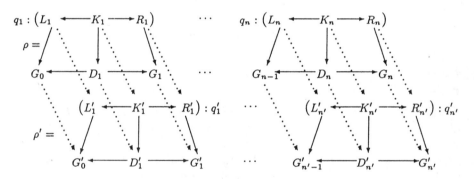

Fig. 4. Abstraction equivalence of derivations (the component productions of the parallel productions q_i and q'_i have not been depicted).

A relevant property of the equivalence on derivations just introduced is that their sequential composition is well-defined, thanks to the use of standard isomorphisms at the extremes (see [3, 5]).

3 From abstract derivations to derivation traces

From a truly concurrent perspective two derivations should be considered as equivalent when they apply the same productions to the "same" subgraph of a certain graph, although the order in which the productions are applied may be different: in particular, they can be at different direct derivations. This basic idea is formalized in the literature through the *shift equivalence* on derivations, introduced below. The shift equivalence is based on the possibility of sequentializing a parallel direct derivation (the *analysis* construction) and on the inverse construction (*synthesis*), which is possible only in the case of *sequential independence*. For the proof of Proposition 11, see for example [11, 9, 5].

Definition 10 (sequential independence). Given a two-step derivation $G \Rightarrow_{q'} X \Rightarrow_{q''} H$ (as in Figure 5), it is *sequential independent* iff there exist graph morphisms $f : L_2 \to X_2$ and $t : R_1 \to Y_1$ such that $j \circ f = s$ and $i \circ t = r$. □

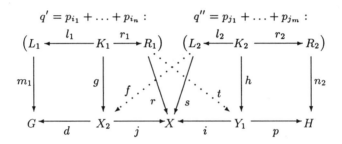

Fig. 5. A sequential independent derivation.

Proposition 11 (analysis and synthesis). *Let $\rho : G \Rightarrow_q H$ be a parallel direct derivation using $q = p_1 + \ldots + p_k : (L \xleftarrow{l} K \xrightarrow{r} R)$. Then for each partition $\langle I = \{i_1, \ldots, i_n\}, J = \{j_1, \ldots, j_m\}\rangle$ of $\underline{k}$ (i.e., $I \cup J = \underline{k}$ and $I \cap J = \emptyset$) there is a constructive way—in general non-deterministic—to obtain a sequential independent derivation $\rho' : G \Rightarrow_{q'} X \Rightarrow_{q''} H$, called an* analysis *of ρ, where $q' = p_{i_1} + \ldots + p_{i_n}$, and $q'' = p_{j_1} + \ldots + p_{j_m}$ as in Figure 5. If ρ and ρ' are as above, we shall write $\rho \equiv_\Pi^{an} \rho'$, where Π is the permutation on $\underline{k}$ defined as $\Pi(i_x) = x$ for $x \in \underline{n}$, and $\Pi(j_x) = x + n$ for $x \in \underline{m}$.*

Conversely, let $\rho : G \Rightarrow_{q'} X \Rightarrow_{q''} H$ be a sequential independent derivation. Then there is a constructive way to obtain a parallel direct derivation $\rho' = G \Rightarrow_{q'+q''} H$, called a synthesis *of ρ. In this case, we shall write $\rho \equiv_\Pi^{syn} \rho'$, where $\Pi = \Pi_{id}^{\#\rho}$.* □

Note that unlike the usual definition of analysis and synthesis, we explicitly keep track of the permutation of the applied productions induced by the constructions.

Informally, two derivations are shift equivalent if one can be obtained from the other by repeated applications of the analysis and synthesis constructions. In the next definition we emphasize the fact that the sets of productions applied in two shift equivalent derivations are related by a permutation, which is constructed inductively starting from the permutations introduced by analysis and synthesis.

Definition 12 (shift equivalence). Derivations ρ and ρ' are *shift equivalent via permutation* Π, written $\rho \equiv_\Pi^{sh} \rho'$, if this can be deduced by the following inference rules:

$$(SH\text{-}id) \; \frac{}{\rho \equiv_{\Pi_{id}^{\#\rho}}^{sh} \rho} \qquad (SH\text{-}\emptyset) \; \frac{b \circ d^{-1} = id_G}{G \equiv_\emptyset^{sh} G \stackrel{\langle \emptyset, \emptyset, \emptyset, b, d \rangle}{\Rightarrow_\emptyset} G} \qquad (SH\text{-}an) \; \frac{\rho \equiv_\Pi^{an} \rho'}{\rho \equiv_\Pi^{sh} \rho'}$$

$$(SH\text{-}syn) \; \frac{\rho \equiv_\Pi^{syn} \rho'}{\rho \equiv_\Pi^{sh} \rho'} \qquad (SH\text{-}sym) \; \frac{\rho \equiv_\Pi^{sh} \rho'}{\rho' \equiv_{\Pi^{-1}}^{sh} \rho} \qquad (SH\text{-}trans) \; \frac{\rho \equiv_\Pi^{sh} \rho', \; \rho' \equiv_{\Pi'}^{sh} \rho''}{\rho \equiv_{\Pi' \circ \Pi}^{sh} \rho''}$$

$$(SH\text{-}comp) \; \frac{\rho_1 \equiv_{\Pi_1}^{sh} \rho'_1, \; \rho_2 \equiv_{\Pi_2}^{sh} \rho'_2, \; \tau(\rho_1) = \sigma(\rho_2)}{\rho_1 ; \rho_2 \equiv_{\Pi_1 | \Pi_2}^{sh} \rho'_1 ; \rho'_2}$$

Note that by *(SH-∅)* an empty direct derivation is shift equivalent to the identity derivation G if and only if the induced isomorphism is the identity. The *shift equivalence* is the equivalence relation $\equiv^{sh}$ defined as $\rho \equiv^{sh} \rho'$ iff $\rho \equiv_\Pi^{sh} \rho'$ for some permutation Π. □

Despite the unusual definition, it is easy to check that the shift equivalence just introduced is the same as in [11, 8, 3]. From the definitions of the shift equivalence and of the analysis and synthesis constructions, it follows that $\rho \equiv^{sh} \rho'$ implies that ρ and ρ' have the same order and the same starting and ending graphs (i.e., $\#\rho = \#\rho'$, $\sigma(\rho) = \sigma(\rho')$, and $\tau(\rho) = \tau(\rho')$; by the way, this guarantees that rule *(SH-comp)* is well defined). Thus equivalence $\equiv^{sh}$ does not subsume abstraction equivalence, since, for example, it cannot relate derivations starting from different but isomorphic graphs. Therefore we introduce a further equivalence, obtained simply as the union of $\equiv^{abs}$ and $\equiv^{sh}$, and we call it *truly-concurrent* (or *tc-*) *equivalence*, since in our view it correctly equates all derivations which are not distinguishable from a true concurrency perspective. A small variation of this equivalence is introduced as well, called *ctc-equivalence*, where the first "c" stays for "concatenable". Equivalence classes of (c)tc-equivalent derivations are called *(concatenable) derivation traces*, a name borrowed from [16].

Definition 13 (truly-concurrent equivalences and derivation traces). Derivations ρ and ρ' are *ctc-equivalent via permutation* Π, written $\rho \equiv_\Pi^c \rho'$, if this can be deduced by the following inference rules:

$$(CTC\text{-}abs) \; \frac{\rho \equiv_\Pi^{abs} \rho'}{\rho \equiv_\Pi^c \rho'} \qquad (CTC\text{-}sh) \; \frac{\rho \equiv_\Pi^{sh} \rho'}{\rho \equiv_\Pi^c \rho'} \qquad (CTC\text{-}trans) \; \frac{\rho \equiv_\Pi^c \rho', \; \rho' \equiv_{\Pi'}^c \rho''}{\rho \equiv_{\Pi' \circ \Pi}^c \rho''}$$

Equivalence $\equiv^c$, defined as $\rho \equiv^c \rho'$ iff $\rho \equiv_\Pi^c \rho'$ for some permutation Π, is called the *concatenable truly concurrent (ctc-) equivalence*. Equivalence classes of derivations with respect to $\equiv^c$ are denoted as $[\rho]$ and are called *concatenable derivation traces*. A *derivation trace* is an equivalence class of derivations with respect to the *truly-concurrent (tc-) equivalence* $\equiv$ defined by the following rules:

$$(TC\text{-}ctc) \; \frac{\rho \equiv_\Pi^c \rho'}{\rho \equiv_\Pi \rho'} \qquad (TC\text{-}iso) \; \frac{\rho \equiv_\Pi \rho', \; \alpha \text{ empty direct derivation}}{\rho \equiv_\Pi \rho' ; \alpha} \qquad \square$$

By Fact 5, a derivation trace contains all derivations which are ctc-equivalent up to an isomorphism for the target graph. Unlike in [4], we had to distinguish here between *concatenable* derivation traces—whose name is justified by the next definition—and derivation traces *tout court*, that will be used in the next section for the event structure semantics; indeed, it is easy to verify that the two notions coincide for the restricted class of derivations considered in [4], i.e., for derivations of *safe* grammars beginning in the start graph.[4] The next definition introduces an interesting category, called the *category of derivation traces* in [4].

Definition 14 (category of concatenable derivation traces). The *category of concatenable derivation traces* of a grammar $\mathcal{G}$, denoted by $\mathbf{Tr}(\mathcal{G})$, is the category having abstract graphs as objects, and concatenable derivation traces as arrows. In particular, if $\rho : G \Rightarrow^*_\mathcal{G} H$, then $[\rho]$ is an arrow from $[G]$ to $[H]$. The identity arrows are ctc-equivalence classes of identity derivations, and the composition of arrows $[\rho] : [G] \to [H]$ and $[\rho'] : [H] \to [X]$ is defined as $[\rho\,;\rho''] : [G] \to [X]$, where ρ'' is a derivation in $[\rho']$ such that $\tau(\rho) = \sigma(\rho'')$. □

Category $\mathbf{Tr}(\mathcal{G})$ is well defined because so is the sequential composition of arrows: In fact, if $\rho_1 \equiv^c \rho_2$, $\rho'_1 \equiv^c \rho'_2$, $\tau(\rho_1) = \sigma(\rho'_1)$ and $\tau(\rho_2) = \sigma(\rho'_2)$, then $\rho_1\,;\rho'_1 \equiv^c \rho_2\,;\rho'_2$ (hence the attribution "concatenable"). For a deeper analysis of the properties of equivalence $\equiv^c$, as well as for counterexamples hinting why sequential composition is not meaningful for (non-concatenable) derivation traces, we refer the reader to [3].

The main result of this section will state that any two tc-equivalent derivations are related by a unique "consistent" permutation among the applied productions. We shall derive this from a more general result, that, in a sense, characterizes an invariant of a derivation trace.

Definition 15 (consistent four-tuples and five-tuples). If $\rho : G_0 \Rightarrow^* G_n$ is the derivation depicted in Figure 3, let $\approx_\rho$ be the smallest equivalence relation on $\bigcup_{i=0}^n Items(G_i)$[5] containing relation $\sim_\rho$, defined as

$$x \sim_\rho y \Leftrightarrow \exists r \in \underline{n}\,.\,x \in Items(G_{r-1}) \land y \in Items(G_r) \land$$
$$\land \exists z \in Items(D_r)\,.\,d_r(z) = x \land b_r(z) = y.$$

Denote by $Items(\rho)$ the set of equivalence classes of $\approx_\rho$, and by $[x]_\rho$ the class containing item x.[6] Then a four-tuple $\langle \rho, h_\sigma, f, \rho' \rangle$ is *consistent* if ρ and ρ' are derivations, $h_\sigma : \sigma(\rho) \to \sigma(\rho')$ is a graph isomorphism between their starting graphs, $f : \#\rho \to \#\rho'$ is an injective function such that $prod_\rho = prod_{\rho'} \circ f$, and there exists a total function $\xi : Items(\rho) \to Items(\rho')$ satisfying the following conditions:

---

[4] A grammar is safe if each graph reachable through a derivation from the start graph has no non-trivial automorphisms.

[5] $Items(G)$ is the union of nodes and edges of G; for simplicity, we assume that all involved sets are disjoint.

[6] The skilled reader may have recognized that this construction builds a colimit of the bottom row of ρ, although we ignored its graphical structure. A related, explicit colimit construction is proposed in [6].

- $\forall x \in \mathit{Items}(\sigma(\rho)) . \xi([x]_\rho) = [h_\sigma(x)]_{\rho'}$, i.e., ξ must be consistent with isomorphism h_σ;
- for each $j \in \#\rho$, let r and s be determined by $j = (\sum_{i=1}^{r} k_i) + s$ (i.e., the j-th production of ρ is the s-th production of its r-th parallel direct derivation), and similarly let s' and r' satisfy $f(j) = \left(\sum_{i=1}^{r'} k'_i\right) + s'$. Then for each item x "consumed" by production $\mathit{prod}_\rho(j) : \left(L \stackrel{l}{\leftarrow} K \stackrel{r}{\rightarrow} R\right)$, i.e., $x \in L - l(K)$, it must hold $\xi([g_r(in_L^s(x))]_\rho) = [g'_{r'}(in_L^{s'}(x))]_{\rho'}$. In words, ξ must relate the items consumed by corresponding production applications (according to f);
- A similar condition must hold for the items "created" by corresponding production applications. Using the above notations, for each $x \in R - r(K)$, $\xi([h_r(in_R^s(x))]_\rho) = [h'_{r'}(in_R^{s'}(x))]_{\rho'}$.

Similarly, say that the five-tuple $\langle \rho, h_\sigma, f, h_\tau, \rho' \rangle$ is *consistent* if the "underlying" four-tuple $\langle \rho, h_\sigma, f, \rho' \rangle$ is consistent, f is a bijection, $h_\tau : \tau(\rho) \to \tau(\rho')$ is an isomorphism relating the ending graphs, and the function ξ is an isomorphism consistent with h_τ as well (i.e., for each item $x \in \tau(\rho)$, $\xi([x]_\rho) = [h_\tau(x)]_{\rho'}$). □

Lemma 16 (unique function between productions). *Let ρ and ρ' be two derivations of a consuming grammar $\mathcal{G}$, and suppose that $\langle \rho, h, f, \rho' \rangle$ and $\langle \rho, h, f', \rho' \rangle$ are consistent four-tuples. Then $f = f'$. The corresponding statement holds for consistent five-tuples as well.* □

This result, whose proof can be found in the full version of the paper, strongly relies on the assumption that productions are consuming, and it generalizes Lemma 30 of [4] to arbitrary parallel (consuming) derivations: thus it implies that no "autoconcurrency" is possible for consuming productions. If instead grammar $\mathcal{G}$ contains a non-consuming production p (i.e., $p : (L \stackrel{l}{\leftarrow} K \stackrel{r}{\rightarrow} R)$ and l is an isomorphism), then it is easy to find a counterexample to Lemma 16. Consider indeed the parallel production $p+p$, and the occurrence morphism $g \stackrel{def}{=} [id_L, id_L] : L + L \to L$: Since l is an isomorphism, the gluing conditions are satisfied, and thus there is a graph H such that $\rho : L \Rightarrow_{\langle p+p, g \rangle} H$.[7] Then it is easy to check that both $\langle \rho, id_L, \Pi_{id}^2, \rho \rangle$ and $\langle \rho, id_L, \Pi_\times^2, \rho \rangle$ are consistent four-tuples, where $\Pi_\times^2(1) = 2$ and $\Pi_\times^2(2) = 1$.

The next theorem provides an abstract characterization of when two (parallel) graph derivations should be considered as equivalent from a truly concurrent perspective.

Theorem 17 (characterization of ctc- and tc-equivalence). *Let S be the fixed family of standard isomorphisms, and $\mathcal{G}$ be any graph grammar.*

1. *Two derivations ρ and ρ' of $\mathcal{G}$ are ctc-equivalent if and only if there is a permutation Π such that the five-tuple $\langle \rho, S_{\sigma(\rho')}^{\sigma(\rho)}, \Pi, S_{\tau(\rho')}^{\tau(\rho)}, \rho' \rangle$ is consistent.*

---

[7] Clearly, if l is not surjective, then the occurrence morphism g would not satisfy the identification condition. Thus this counterexample does not apply if p is consuming.

2. Similarly, ρ and ρ' are tc-equivalent if and only if there is a permutation Π such that the four-tuple $\langle \rho, S^{\sigma(\rho)}_{\sigma(\rho')}, \Pi, \rho' \rangle$ is consistent. □

The (omitted) proof of the theorem exploits, for the *only if* parts, a structural induction based on the inference rules defining the ctc- and tc-equivalence. For the *if part*, it shows that given a consistent five-tuple, $\langle \rho, S^{\sigma(\rho)}_{\sigma(\rho')}, \Pi, S^{\tau(\rho)}_{\tau(\rho')}, \rho' \rangle$, derivations ρ and ρ' can be transformed into equivalent sequential derivations ρ_1 and ρ'_1 related by the identity permutation; at this point, by exploiting the function $\xi : Items(\rho) \to Items(\rho')$, one determines a family of isomorphisms relating the graphs in ρ_1 and ρ'_1 that satisfy the conditions of Definition 9.

Corollary 18 (unique permutation among tc-equivalent derivations).
Let ρ and ρ' be two derivations of a consuming grammar. If $\rho \equiv \rho'$, then there is a unique permutation Π on $\#\rho$ such that $\rho \equiv_\Pi \rho'$.

A fortiori, the same holds for equivalence relations $\equiv^c$, $\equiv^{sh}$, and $\equiv^{abs}$.

Proof. The statement follows from Theorem 17 and from Lemma 16. □

4 An event structure semantics for consuming grammars

In this section we provide an event structure semantics for consuming graph grammars. In [4], a similar semantics for the subclass of safe grammars has been obtained via a standard comma category construction on category $\mathbf{Tr}(\mathcal{G})$. We showed that the "category of objects under the starting graph $[G_0]$" of a safe grammar is a partial order (called the "domain" of the grammar) satisfying nice algebraic properties—it is a prime algebraic, finitely coherent, finitary complete lower semilattice, where all elements are finite. As such, by well-known results [19] it is the domain of finite configurations of a uniquely determined event structure, which is taken as the truly concurrent semantics of the grammar.

For consuming, non-safe grammars, things are slightly more complex, because the same comma category construction does not yield a partial order, but just a preorder, as we shall see below. Fortunately, the obvious partial order induced by the preorder turns out to have the desired algebraic structure, thus it indirectly determines an event structure. However, unlike in [4] where we explicitly prove the algebraic properties of the domain, we get here to the main result by presenting first an explicit construction of the event structure, and then by showing that its finite configurations are one-to-one with the element of the domain. This presentation has the advantage of giving a more explicit understanding of the event structure, and of simplifying slightly the proofs.

First we recall the basic definitions of event structure and of its configurations, and the comma category construction we need.

Definition 19 (prime event structures and configurations [19]). A *prime event structure* (or simply *event structure*) $\mathcal{E}$ is a triple $\mathcal{E} = \langle E, \leq, \# \rangle$ where:

- E is a set of events.

- $\leq\ \subseteq E \times E$ is a partial order relation which satisfies the *axiom of finite causes*, i.e., $\{e' \mid e' \leq e\}$ is finite for all $e \in E$. Relation $\leq$ is called the *causal dependency relation*.
- $\# \subseteq E \times E$ is a binary, symmetric, irreflexive relation which is *hereditary*, i.e., if $e \# e_1$ and $e_1 \leq e_2$, then $e \# e_2$. Relation $\#$ is called the *conflict relation*.

A *configuration* C of $\mathcal{E}$ is a subset of events $C \subseteq E$ that is *conflict-free* (i.e., for all $e, e' \in C$, $\neg(e \# e')$) and *left-closed* (i.e., for all $e, e' \in E$, $e' \leq e$ and $e \in C$ implies $e' \in C$). The *domain of configurations* of $\mathcal{E}$, denoted $\mathcal{L}(\mathcal{E})$, is the partial order $\mathcal{L}(\mathcal{E}) = \langle \mathcal{C}, \subseteq \rangle$, where $\mathcal{C}$ is the set of all configurations of $\mathcal{E}$, and $\subseteq$ is the inclusion relation. The *domain of finite configurations* of $\mathcal{E}$, denoted $\mathcal{FL}(\mathcal{E})$, is the partial order of all *finite* configurations of $\mathcal{E}$ ordered by inclusion. □

Definition 20 (comma category [13]). Let **C** be a category, and let x be an object of **C**. Then the *comma category* $(x \downarrow \mathbf{C})$ (also called the *category of objects under x in* **C**) is the category defined as follows. The objects of $(x \downarrow \mathbf{C})$ are pairs $\langle f, y \rangle$ where $f : x \to y$ is an arrow in **C**. Furthermore, $g : \langle f, y \rangle \to \langle k, z \rangle$ is an arrow of $(x \downarrow \mathbf{C})$ if $g : y \to z$ is an arrow of **C** and $f \,;\, g = k$. □

The *pre-domain* of a consuming graph grammar $\mathcal{G}$ is defined as the category of objects under the start graph in its category of concatenable derivation traces. We will show below that the resulting category is a pre-order: Then the *domain* of $\mathcal{G}$ is defined as the partial order induced by the pre-domain. In the following, we will use $\delta, \delta', \delta_1, \ldots$ to range over concatenable derivation traces, and $\eta, \eta', \eta_1, \ldots$ to range over derivation traces.

Definition 21 (pre-domain and domain of consuming grammars). Let $\mathcal{G} = \langle G_0, P, \pi \rangle$ be a consuming graph grammar, and let $\mathbf{Tr}(\mathcal{G})$ be its category of concatenable derivation traces (see Definition 14). The *pre-domain of $\mathcal{G}$*, $\mathbf{PreDom}(\mathcal{G})$, is defined as $([G_0] \downarrow \mathbf{Tr}(\mathcal{G}))$, i.e., it is the category of objects under $[G_0]$ in $\mathbf{Tr}(\mathcal{G})$.

Let $\sim$ be the equivalence relation on objects of $\mathbf{PreDom}(\mathcal{G})$ defined as $x \sim y$ iff there are two arrows $f : x \to y$ and $g : y \to x$. The *domain of $\mathcal{G}$*, $\mathbf{Dom}(\mathcal{G})$, is the partial order of $\sim$-equivalence classes of objects of $\mathbf{PreDom}(\mathcal{G})$ with the ordering given by $[x]_\sim \leq [y]_\sim$ iff there is an arrow $f : x \to y$ in $\mathbf{PreDom}(\mathcal{G})$. □

Lemma 22 (structure of $\mathbf{PreDom}(\mathcal{G})$ and $\mathbf{Dom}(\mathcal{G})$). *Let $\mathcal{G}$ be a consuming graph grammar. Then category $\mathbf{PreDom}(\mathcal{G})$ is a pre-order. Moreover, the elements of the partial order $\mathbf{Dom}(\mathcal{G})$ are one-to-one with derivation traces starting from $[G_0]$, and the partial ordering is given by $\eta \leq \eta'$ iff there are concatenable derivation traces $\delta \subseteq \eta$ and δ' such that $\delta \,;\, \delta' \subseteq \eta'$.* □

The last result exploits the fact that (c)tc-equivalent derivations of a consuming grammar are related by a unique permutation, as stated in Corollary 18. We give now an explicit definition of the event structure of a consuming graph grammar: we shall relate it with the domain just introduced later on.

Definition 23 (event structure of a consuming graph grammar). Let $\mathcal{G} = \langle G_0, P, \pi \rangle$ be a consuming graph grammar. A *pre-event* for $\mathcal{G}$ is a pair $\langle \rho, \alpha \rangle$, where ρ is a derivation starting from a graph isomorphic to G_0, and α is a direct derivation using a single production in P such that $\rho; \alpha$ is defined (i.e., $\tau(\rho) = \sigma(\alpha)$). Let $\approx$ be the equivalence relation on pre-events induced by the following relation:

- $\langle \rho, \alpha \rangle \simeq \langle \rho', \alpha' \rangle$ if $\rho \equiv^c \rho'$ and $\alpha \equiv^{abs} \alpha'$;
- $\langle \rho; \rho', \alpha \rangle \simeq \langle \rho, \alpha' \rangle$ if $\rho'; \alpha \equiv^{sh}_{\Pi_\times} \alpha'; \rho''$ for some derivation ρ'', where $\Pi_\times$ is the permutation on $\#\rho' + 1$ such that $\Pi_\times(x) = 1$ if $x = \#\rho' + 1$ and $\Pi_\times(x) = x + 1$ otherwise.

If $\langle \rho, \alpha \rangle$ is a pre-event, let ε^ρ_α be the set of derivations defined by the following rules:

$$\frac{}{\rho; \alpha \in \varepsilon^\rho_\alpha}, \quad \frac{\rho'; \alpha' \in \varepsilon^\rho_\alpha, \langle \rho', \alpha' \rangle \approx \langle \rho'', \alpha'' \rangle}{\rho''; \alpha'' \in \varepsilon^\rho_\alpha}$$

$$\frac{\rho' \in \varepsilon^\rho_\alpha, \alpha' \text{ empty direct derivation}, \tau(\rho') = \sigma(\alpha')}{\rho'; \alpha' \in \varepsilon^\rho_\alpha}$$

An *event* ε for $\mathcal{G}$ is a set of derivations $\varepsilon = \varepsilon^\rho_\alpha$ for some pre-event $\langle \rho, \alpha \rangle$. For each derivation ρ, let $Ev(\rho)$ be the set of events $Ev(\rho) = \{\varepsilon \mid \rho' \in \varepsilon \text{ with } \rho' \text{ prefix of } \rho\}$, and for each event ε let $Der(\varepsilon) = \{\rho \mid \varepsilon \in Ev(\rho)\}$.

Then the *event structure of grammar* $\mathcal{G}$, denoted $\mathbf{ES}(\mathcal{G})$, is the triple $\mathbf{ES}(\mathcal{G}) = \langle E, \leq, \# \rangle$, where E is the set of all events for $\mathcal{G}$, $\varepsilon \leq \varepsilon'$ if $Der(\varepsilon') \subseteq Der(\varepsilon)$, and $\varepsilon \# \varepsilon'$ if $Der(\varepsilon) \cap Der(\varepsilon') = \emptyset$. □

It is worth explaining informally the last definition. Conceptually, an event ε^ρ_α of a consuming grammar $\mathcal{G}$ is determined by the application of a production to a graph reachable from the start graph of $\mathcal{G}$ (i.e., by the direct derivation α), together with the history that generated the graph items needed by that production application. Clearly, isomorphic production applications or different linearizations of the same history should determine the same event. Therefore an event is defined as a set of derivations of $\mathcal{G}$, namely the set of all derivations having (a copy of) α as the last step, containing (a copy of) all its causes in any order consistent with the causality relation, and possibly containing additional direct derivations that are independent from α (they can be switched with α via the shift equivalence).

Given this definition of events, the causality and conflict relations are easily defined. In fact, considering for each event ε the set $Der(\varepsilon)$ of all the derivations that performed ε at some point, we have that two events are in conflict if there is no derivation that can perform both, and they are causally related if each derivation that performs one also performs the other. The proof that $\mathbf{ES}(\mathcal{G})$ is a well-defined event structure, (i.e., that $\leq$ is a partial ordering satisfying the axiom of finite causes, and that $\#$ is symmetric, irreflexive and hereditary) can be found in the full paper.

We are now ready to state the main result of this section, namely, that the domain of a grammar is exactly the domain of finite configurations of the associated event structure.

Theorem 24 (domain and event structures). *The domain of a consuming graph grammar $\mathcal{G}$, $\mathbf{Dom}(\mathcal{G})$, is isomorphic to the domain of finite configurations of its event structure $\mathbf{ES}(\mathcal{G})$.* □

From this result it follows as an immediate consequence the main result proved in [4] for the subclass of safe grammars.

Corollary 25 (algebraic structure of the domain of a grammar). *The domain of a consuming grammar $\mathcal{G}$, $\mathbf{Dom}(\mathcal{G})$, is a a prime algebraic, finitely coherent, finitary complete lower semilattice, where all elements are finite.*

Proof. The statement follows from Theorem 24 because such algebraic properties uniquely characterize the domain of finite configurations of a prime event structure (see [19, 4]). □

5 Conclusions and Future Work

We presented an event structure semantics for graph grammars that generalizes to arbitrary consuming grammars (i.e., such that each production deletes some items) the semantics presented in [4] for the subclass of safe grammars. We introduced various equivalences on parallel graph derivations, and (concatenable) graph derivations were defined as equivalence classes of the "truly concurrent" equivalence on derivations: An important result gives a precise characterization of this equivalence. The *domain* of a grammar is defined as the partial order induced by a comma category construction on the category of derivation traces, while the *event structure* of a grammar is defined using techniques which are reminiscent of those used for example in [16] or in [1] for traces. It is worth stressing that those techniques (like the "transition systems with independence" in [18]) were not directly applicable to abstract derivations of graph grammars. One main result of the paper shows that the domain of a grammar is isomorphic to the domain of finite configurations of its event structure.

A direction for future investigation is the application of the proposed event structure semantics to other rewriting formalisms based on the double-pushout construction, like the High-Level Replacement Systems [9]. Also, since P/T Petri nets are a strict subclass of graph grammars, as shown in [4], the event structure semantics we propose can be applied to P/T nets as well. The resulting semantics should be strongly related to that proposed in [14], and we intend to make explicit these relationships. Finally, for what concerns the requirement that grammar are consuming, we showed by a simple counterexample that one of the fundamental results of the paper does not hold for non-consuming grammars. As a further line of investigation, we intend to study other semantic models for concurrent systems able to provide a truly concurrent semantics for non-consuming graph grammars as well.

References

1. M.A. Bednarczyk, *Categories of asynchronous systems*, Ph.D. Thesis, University of Sussex, Report no. 1/88, 1988.
2. A. Corradini, H. Ehrig, M. Löwe, U. Montanari and F. Rossi, *Note on Standard Representation of Graphs and Graph Derivations*, in [17], 104–118. (Also as Technical Report "Bericht-Nr. 92-25", Technische Universität Berlin, Fachbereich 20 Informatik, 1992.)
3. A. Corradini, H. Ehrig, M. Löwe, U. Montanari and F. Rossi, *Abstract Graph Derivations in the Double Pushout Approach*, in [17], 86–103.
4. A. Corradini, H. Ehrig, M. Löwe, U. Montanari and F. Rossi, *An Event Structure Semantics for Safe Graph Grammars*, in *Programming Concepts, Methods and Calculi*, E.-R. Olderog (ed.), IFIP Transactions A-56, North-Holland, 1994, 423–444.
5. A. Corradini, H. Ehrig, M. Löwe, U. Montanari and F. Rossi, *Algebraic Approach to Graph Transformation II: Models of Computation in the Double Pushout Approach*, submitted, 1995.
6. A. Corradini, U. Montanari and F. Rossi, *Graph Processes*, to appear in Fundamenta Informaticae, 1995.
7. A. Corradini, *Concurrent computing: From Petri nets to graph grammars*, in Proceedings SEGRAGRA '95, Electronic Notes in Theoretical Computer Science, Volume 2, URL: http://www.elsevier.nl/locate/entcs/volume2.html.
8. H. Ehrig, *Tutorial introduction to the algebraic approach of graph-grammars*, LNCS 291, Springer-Verlag, 1987, 3–14.
9. H. Ehrig, A. Habel, H.-J. Kreowski, F. Parisi-Presicce, *Parallelism and Concurrency in High-Level Replacement Systems*, in Mathematical Structures in Computer Science, **1**, 1991, 361–404.
10. H. Ehrig, M. Pfender, H.J. Schneider, *Graph-grammars: an algebraic approach*, Proc, IEEE Conf. on Automata and Switching Theory, 1973, 167–180.
11. H.-J. Kreowski, *Manipulationen von Graphmanipulationen*, Ph.D. Thesis, Technische Universität Berlin, 1977.
12. H.-J. Kreowski, *Is parallelism already concurrency? Part 1: Derivations in graph grammars*, LNCS 291, Springer-Verlag, 1987, 343–360.
13. S. Mac Lane, *Categories for the working mathematician*, Springer Verlag, 1971.
14. J. Meseguer, U. Montanari, and V. Sassone, *On the semantics of Petri Nets*, in Proceedings CONCUR '92, LNCS 630, 286–301.
15. W. Reisig, *Petri Nets: An Introduction*, EATCS Monographs on Theoretical Computer Science, Springer-Verlag, 1985.
16. G. Schied, *On relating Rewriting Systems and Graph Grammars to Event Structures*, in [17], 326–340.
17. H.J. Schneider and H. Ehrig (Eds.), *Proceedings of the Dagstuhl Seminar 9301 on Graph Transformations in Computer Science*, LNCS 776, Springer-Verlag, 1994.
18. V. Sassone, M. Nielsen and G. Winskel, *A classification of models for concurrency*, in Proceedings CONCUR '93, LNCS 715, Springer-Verlag, 1993, 82–96.
19. G. Winskel, *An Introduction to Event Structures*, in Linear Time, Branching Time and Partial Order in Logics and Models for Concurrency, LNCS 354, Springer-Verlag, 1989, 364–397.

Synchronized Composition of Graph Grammar Productions

Andrea Corradini and Francesca Rossi

Università di Pisa
Dipartimento di Informatica
Corso Italia 40, 56125 Pisa, Italy
E-mail: {andrea,rossi}@di.unipi.it

Abstract. In the framework of the double-pushout approach to graph grammars, we propose a new notion of parallel composition, called *synchronized composition*, of production applications. Our aim is to allow (pairs of) productions which are possibly not parallel independent, to be applied in parallel, provided that they are not mutually exclusive. The notion of synchronized composition we propose is not comparable to amalgamation [BFH87], since two productions which are amalgable are not necessarily synchronizable, and viceversa. Our different idea of which productions should be applicable in parallel comes from our view of graph grammars as a generalization of Petri nets. The definitions and constructions we use are a conservative extension of the ones used in the classical theory of parallelism in the algebraic approach to graph grammars. Moreover, they can be the basis for the development of a more general concept of canonical derivations, and also for a generalization of the recently developed partial order and event structure semantics for graph grammars.

1 Introduction

According to the double-pushout approach [Ehr87], a graph grammar is a collection of productions, each of which consists of a left-hand side and of a right-hand side graph, plus an interface graph included in both. A production, when applied to a graph, has the effect of replacing an occurrence of its left-hand side with its right-hand side, while preserving an occurrence of its interface.

Graph grammars were originally thought of as an extension of string grammars to deal with more complex structures, and therefore they were used and studied assuming a sequential environment where only one production at a time could be applied. However, quite early the research extended also to the possibility of applying two or more productions simultaneously, and thus a theory of parallelism in graph grammars was developed [Kre77, Kre87, Ehr87]. One of the basic ideas (and results) of such theory is that productions can be applied in parallel whenever they can also be applied in sequence in any order yielding the "same" result. Then they are said to be *parallel independent*. Two production applications in a sequence are instead said to be *sequentially independent* whenever their order of application can be reversed without changing the resulting graph.

More concretely, the parallel independence of two production applications means that the items needed by both productions (that is, that are in the left-hand sides) are all preserved by (that is, they are in the interface of) both. This in fact assures that, in some sense, such productions do not interfere with each other, since none of them prevents the application of the other one. Or, in other words, that there is no conflict between the two productions.

However, there are many situations, even outside the field of graph grammars, where the possibility of applying in parallel some productions (or, in general, rewrite rules) which are not parallel independent, and which therefore have some conflict, seems reasonable and very convenient. Consider in fact the general situation in which there is a resource which is only read by an agent A_1 and consumed by another agent A_2. Then one would say that these agents have some conflicting behavior on the common resource. However, they could be safely executed in parallel, with the resulting situation in which the resource has been deleted. Notice that, in general, this can be done only if we are sure that the two agents, when executed in parallel, act exactly at the same time (think for example at two agents sharing two resources, where each agent consumes exactly one resource, which is read by the other one). In other words, these agents need to be *synchronized*. Notice also that the resulting effect is the same as executing first agent A_1 and then A_2, while executing A_2 first (i.e., consuming the resource first) makes the execution of A_1 not possible anymore. Thus agents A_1 and A_2 manifest a form of *asymmetric conflict*, since they are serializable only in one order.

This does not mean that any two agents can be run in parallel. In fact, two agents which both delete a common resource cannot be executed in parallel even in this less restrictive world. However, there could be agents which are serializable only in some orders (as agents A_1 and A_2 described above), or even not serializable at all, that could be allowed to run in parallel.

Situations of the type just described, and the need to be able to express the parallel execution of this kind of agents, occur in many contexts. For example, in non-orthogonal Term Graph Rewriting Systems (TGRs) [KKSdV93], or also in Elementary Net Systems (ENSs) with activator and inhibitor arcs [JK91]. As in graph grammars, also the classical development of such systems allows one to run in parallel only agents (specifically, rewrite rules in TGRSs, and events in ENSs) which do not have any conflict.

Our idea of which productions should be applicable in parallel comes from our view of graph grammars as a generalization of Petri nets [Rei85]. In Petri nets, two transitions are mutually exclusive if they consume the same item. Being graph grammar states more structured than Petri nets states (graphs instead of multisets), the corresponding notion is slightly more involved. In fact, we also have to make sure that no node deleted by one production is used by the other one to attach a new arc (such nodes will be called *connection nodes*).

To embed our notion of parallelism within the existing theory of graph grammars, we propose a new definition of *synchronizable* direct derivations, and then we define their *synchronized composition* by using categorical diagrams. This same operation can be extended from two to any number of production applications, and it can be proved to be associative (that is, the order in which the compositions are accumulated does not matter). We also show that it is a conservative extension of the usual parallel composition [Ehr87], since it is equivalent to it when the considered production applications are parallel independent.

Other extensions of the notions of parallel and sequential independence have been proposed in the literature. In particular, the notions of *weak parallel independence* and *weak sequential independence*, developed for the single-pushout approach in [LD94, EHKRW95], are strongly related to our notion of synchronized composition. More precisely, two weakly parallel independent productions are always synchronizable, while two synchronizable productions are weakly parallel independent only if they are serializable in some order.

The notion of *amalgamation* defined in [BFH87] allows for the parallel application of two productions which "agree" on shared items. That is, for every shared item, either they both preserve it or they both delete it. It is easy to see that our synchronized composition is incomparable to this: there are amalgable productions which are not synchronizable, and viceversa. The reason of this discrepancy lies in the fact that, as noted above, we see graph grammars as a generalization of Petri nets, where two transitions wanting to consume the same token are mutually exclusive, and therefore not concurrent. From the viewpoint of [BFH87], instead, two such transitions can be considered as partial specifications of a bigger transition, obtained as their amalgamation. Then the two partial specifications must clearly be "consistent", i.e., they must agree on common items.

The notion of *sequential composition* [Ehr87] provides instead a generalization of sequential independence: even if two consecutive production applications are not sequentially independent, they can nevertheless be composed to obtain a unique production application (via their *concurrent production*) yielding the same final graph. We claim that our synchronized composition produces the same result as the sequential composition of [Ehr87] when the considered production applications are serializable in at least one order. However, the two constructions are uncomparable in general. In fact, on the one hand there are pairs of consecutive productions which are not synchronizable but have a corresponding sequential composition; on the other hand, there are synchronizable productions that cannot be serialized in any order, and thus their synchronized composition cannot be obtained as a concurrent production. Another important difference between synchronized composition and amalgamation or sequential composition is that our composition is completely deterministic, while the other two are parametric w.r.t. a subrule or a subgraph.

Starting from the new notion of synchronized composition, the whole theory of concurrency in graph grammars could be suitably redone in order to express the amount of concurrency in single derivations or in the whole given grammar. As predictable, given the same grammar, the new theory will show at least as much concurrency as the old one. In particular, a new kind of canonical derivations can be defined, based on what we call *serialization equivalence* (according to which two derivations are equivalent if one can be obtained from the other one by serializing some parallel steps and/or synchronizing some other steps) instead of on the classical *shift equivalence*.

The same can be done also with the semantics of graph grammars based on partial orders [MR92] or on event structures [CELMR94b, CELMR95b]. However, in these cases new structures have to be used in order to correctly express the amount of concurrency. In fact, as shown in [JK91] for Elementary Net Systems with inhibitor arcs (which can be shown to be a subcase of graph grammars), and also in [JK93] in a more general setting, a partial order is not sufficient to express concurrency when simultaneity does not imply serializability in both orders. Instead, *composets* (which basically consist of a partial order expressing causality, plus a binary relation expressing weak causality) can be fruitfully used.

Since our notion of parallelism is able to express agents' synchronization (as discussed above), we are confident that it could be useful in expressing the behaviour of agents of those concurrent languages, like CCS [Mil80], which employ synchronized compositions.

2 Derivations with synchronization

Following the intuitive ideas discussed in the previous section, we present here the technical definitions of the synchronized composition of two or more direct derivations. We assume that the reader is familiar with the algebraic theory of graph grammars, and in particular with the double-pushout approach (a standard reference is [Ehr87], see also [CELMR95b] for some basic defintions). We assume also that a category of (colored) graphs **Graph** is fixed. The next proposition recalls the necessary and sufficient conditions for the existence of pushout complement objects in category **Graph**.

Proposition 1 (gluing conditions). *Let $l : K \to L$ and $g : L \to G$ be two morphisms in* **Graph**, *with l injective. Then there exists a pushout complement of $\langle l, g \rangle$ (i.e., an object D and two arrows $k : K \to D$ and $d : D \to G$) such that the resulting square is a pushout) if and only if the following conditions are satisfied:*

[*Dangling condition*] *No arc $e \in Arcs(G) - g(Arcs(L))$ is incident with any node in $g(Nodes(L) - l(Nodes(K)))$;*

[*Identification condition*] *There is no $x, y \in Nodes(L) \cup Arcs(L)$ such that $g(x) = g(y)$ and $y \notin l(Nodes(K) \cup Arcs(K))$.*

In this case we say that $\langle l, g \rangle$ satisfy the **gluing conditions**, *and since l is injective then D is unique up to isomorphism.* □

Definition 2 (redexes). A *redex* in a graph G is a pair $\Delta = \langle p, g \rangle$, where $p = (L \xleftarrow{l} K \xrightarrow{r} R)$ is an *injective* production (that is, both l and r are injective), and $g : L \to G$ is an occurrence morphism such that $\langle l, g \rangle$ satisfies the gluing conditions. □

Definition 3 (synchronizable redexes – set-theoretic). Let $p_1 = (L_1 \xleftarrow{l_1} K_1 \xrightarrow{r_1} R_1)$ and $p_2 = (L_2 \xleftarrow{l_2} K_2 \xrightarrow{r_2} R_2)$ be two productions, and let $\Delta_1 = \langle p_1, g_1 \rangle$ and $\Delta_2 = \langle p_2, g_2 \rangle$ be two redexes in the same graph G. Also, for $i = 1, 2$, let I_i be the subset of nodes of K_i defined as $I_i = \{n \in Nodes(K_i) \mid \exists a \in Arcs(R_i - r_i(K_i))$ such that $r_i(n) = source(a)$ or $r_i(n) = target(a)\}$. Nodes in I_i are called **connection nodes** for p_i.

Then Δ_1 and Δ_2 are **synchronizable** if the following conditions are satisfied:

(Set-Synchr-1) $g_1(L_1) \cap g_2(L_2) \subseteq g_1(l_1(K_1)) \cup g_2(l_2(K_2))$
(Set-Synchr-2) for $i, j \in \{1, 2\}$ and $i \neq j$, $g_i(l_i(I_i)) \cap g_j(L_j) \subseteq g_j(l_j(K_j))$.

In the following we will write Set-Synchr(Δ_1, Δ_2) whenever Δ_1 and Δ_2 satisfy the above conditions. □

In words, two redexes are synchronizable whenever the items of G which are needed by both productions are preserved by at least one of them, and also no connection node for one production is deleted by the other one. Thus, in particular, two redexes are *not* synchronizable if there is an item in G which is deleted by both productions.

If two redexes are *parallel independent*, then it can be shown that they are synchronizable.

Proposition 4 (parallel independence implies synchronizability). *Let Δ_1, Δ_2 be two redexes in a graph G as in Definition 3. If they are parallel independent [Ehr87], then they are synchronizable.*

Proof: Parallel independence is characterized in [Ehr87] by the set-theoretic condition $g_1(L_1) \cap g_2(L_2) \subseteq g_1(l_1(K_1)) \cap g_2(l_2(K_2))$. Clearly, this implies condition Set-Synchr-1. For condition Set-Synchr-2, we have $g_i(l_i(I_i)) \cap g_j(L_j) \subseteq g_1(L_1) \cap g_2(L_2) \subseteq g_1(l_1(K_1)) \cap g_2(l_2(K_2)) \subseteq g_j(l_j(K_j))$. □

The same property (that is, redex synchronizability) can also be given in categorical terms. First we need the following lemma.

Lemma 5 (decomposition of injective graph morphisms). *Let $f: A \to B$ be an injective graph morphism. Then there exists, up to isomorphism, a unique graph I minimal with respect to graph inclusion and a unique graph C such that the following diagram is a pushout where all morphisms are injective:*

$$\begin{array}{ccc} I & \longrightarrow & C \\ \downarrow & PO & \downarrow \\ A & \xrightarrow{f} & B \end{array}$$

Considering injective morphisms as inclusions, I is the discrete graph containing all nodes of A (and thus of B) to which some arcs of $B - f(A)$ are attached.

Proof: Let $I = \{n \in Nodes(A) \mid \exists a \in Arcs(B - f(A))$ such that $f(n) = source(a)$ or $f(n) = target(a)\}$, and let $I \to A$ be the obvious inclusion. Furthermore, let C be the subgraph of B containing all nodes that are either in I, or are not in $f(A)$, and all the arcs in $B - f(A)$; let $C \to B$ be the obvious inclusion, and $I \to C$ be the restriction of f to I. Then it is easy to show that the resulting diagram is a pushout.

Now, suppose by absurd that there exists a similar diagram where I has fewer nodes, with injective morphism $g: I \to A$. Then there exists at least one node n in $A - g(I)$ such that an arc a in $B - f(A)$ is attached to $f(n)$. But then the dangling condition of Proposition 1 is not satisfied by $\langle g, f \rangle$, and therefore there can be no C making the square a pushout. □

Definition 6 (synchronizable redexes – category-theoretic). Two redexes $\Delta_1 = \langle p_1, g_1 \rangle$ and $\Delta_2 = \langle p_2, g_2 \rangle$ in G are *synchronizable* if the following two conditions are satisfied:

(*Cat-Synchr-1*) In the following diagram, let $L_1 \cap_G L_2$ be the pullback of g_1 and g_2, with the induced morphism $g': L_1 \cap_G L_2 \to G$. Let also $K_1 \cap_G K_2$ be the pullback of $l_1; g_1$ and $l_2; g_2$. Finally, let $K_1 \xrightarrow{in_1} K_1 \cup_G K_2 \xleftarrow{in_2} K_2$ be the pushout of $K_1 \leftarrow K_1 \cap_G K_2 \to K_2$, and let $k': K_1 \cup_G K_2 \to G$ be the induced morphism. Then there must exist an arrow $h: L_1 \cap_G L_2 \to K_1 \cup_G K_2$ such that $h; k' = g'$.

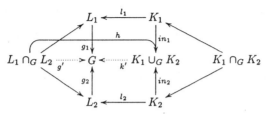

(*Cat-Synchr-2*) In the following diagram, let I_1 and C_1 be the graph uniquely determined by Lemma 5 since r_1 is injective. Also, let $I_1 \cap_G L_2$ be the pullback of $i_1; l_1; g_1$ and g_2. Then there must exist an arrow $j_1: I_1 \cap_G L_2 \to K_2$ such that $j_1; l_2 = I_1 \cap_G L_2 \to L_2$. The symmetric condition with 1 and 2 exchanged must hold as well.

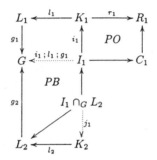

We will write *Cat-Synchr*(Δ_1, Δ_2) whenever Δ_1 and Δ_2 satisfy the above conditions. □

Intuitively,

- $g'(L_1 \cap_G L_2)$ contains all the items of G which are needed by both p_1 and p_2;
- (the image in G of) $K_1 \cap_G K_2$ contains the items needed by p_1 and p_2 *and* preserved by both;
- $K_1 \cup_G K_2$ is the union of K_1 and K_2 where the items in $K_1 \cap_G K_2$ appear only once;
- $I_1 \cap_G L_2$ contains all connection nodes of p_1 that are needed by p_2.

We show now that the two conditions for synchronizability of redexes are equivalent.

Theorem 7. *The two conditions of Definition 3 are equivalent to the corresponding conditions of Definition 6. As a consequence, for any two redexes $\Delta_1 = \langle p_1, g_1 \rangle$ and $\Delta_2 = \langle p_2, g_2 \rangle$, Set-Synchr($\Delta_1, \Delta_2$) holds iff Cat-Synchr($\Delta_1, \Delta_2$) holds.*

Proof: For the first conditions of the two definitions, let us assume, without loss of generality, that $L_1 \cap_G L_2$ is the graph having as nodes and arcs $\{\langle z, v \rangle \mid g_1(z) = g_2(v)\}$. We first show that ¬*Set-Synchr-1* implies ¬*Cat-Synchr-1*, i.e., there is no morphism h such that $h; k' = g'$. In fact, ¬*Set-Synchr-1* implies that there exists two items (arcs or nodes) $x \in L_1$ and $y \in L_2$ such that $g_1(x) = g_2(y)$, $x \notin l_1(K_1)$, and $y \notin l_2(K_2)$. Thus $\langle x, y \rangle$ is an item of $L_1 \cap_G L_2$, and $g'(\langle x, y \rangle) = g_1(x)$. Now since in_1 and in_2 are jointly surjective, and $l_i; g_i = in_i; k'$ for $i = 1, 2$, it follows that $g_1(x)$ cannot be in the image of k'. Thus no $h: L_1 \cap_G L_2 \to K_1 \cup_G K_2$ can satisfy $h; k' = g'$.

Now we show that *Set-Synchr-1* implies *Cat-Synchr-1*, since it implies that the following morphism $\underline{h}: L_1 \cap_G L_2 \to K_1 \cup_G K_2$ is well defined, and that it satisfies $\underline{h}; k' = g'$:

$$\underline{h}(\langle x, y \rangle) = \begin{cases} in_1(z) & \text{if } z \in K_1 \text{ and } l_1(z) = x, \\ in_2(v) & \text{if } v \in K_2 \text{ and } l_2(v) = y. \end{cases}$$

Let us show that for all $\langle x, y \rangle$ in $L_1 \cap_g L_2$, either $x \in l_1(K_1)$, or $y \in l_2(K_2)$, or both. In fact, by *Set-Synchr-1* it follows that either $g_1(x)(= g_2(y)) \in g_1(l_1(K_1))$, or $g_2(y)(= g_1(x)) \in g_2(l_2(K_2))$. Suppose that $g_1(x) \in g_1(l_1(K_1))$ (the other case is similar), and in particular that $g_1(x) = g_1(l_1(z))$; then either $l_1(z) = x$ (thus $x \in l_1(K_1)$, and we are done), or $l_1(z) \neq x$, and in this case by the identification condition (that is satisfied by Δ_1 since it is a redex) there exists $z' \in K_1$ such that $l_1(z') = x$, thus also in this case $x \in l_1(K_1)$. It remains to show that if $x = l_1(z)$ and $y = l_2(v)$, then $in_1(z) = in_2(v)$ (thus $\underline{h}$ is well defined). In fact, in this case $\langle z, v \rangle$ is in $K_1 \cap_G K_2$, and by the property of the pushout $K_1 \cup_G K_2$, this implies $in_1(z) = in_2(v)$.

To conclude the proof it remains to show that *Set-Synchr-2* $\Leftrightarrow$ *Cat-Synchr-2*. This follows easily by the characterization of graph I_i in *Cat-Synchr-2* (for $i \in \{1, 2\}$), since by Lemma 5 it is exactly the set of connection nodes of p_i. □

Thanks to this result, in the rest of the paper we will write sometimes $Synchr(\Delta_1, \Delta_2)$ to say that Δ_1 and Δ_2 are synchronizable, indicating by *Synchr-1* and *Synchr-2* the two conditions of Definitions 3 or 6.

Example: For simplicity reasons, in all the examples of this paper we will deal with discrete graphs, that is, graphs with only nodes and no arc (thus condition *Synchr-2* is trivially satisfied). Consider the two productions in the following diagram, where the arrows are just inclusions.

$$p_1: \quad \bullet^a \bullet^b \, (L_1) \xleftarrow{l_1} \bullet^a(K_1) \xrightarrow{r_1} \bullet^a \bullet^c \, (R_1)$$

$$p_2: \quad \bullet^a \bullet^b \, (L_2) \xleftarrow{l_1} \bullet^b(K_2) \xrightarrow{r_1} \bullet^b \bullet^d \, (R_2)$$

Here, both p_1 and p_2 are applicable to a (discrete) graph whenever one can find a node labelled a and another node labelled b. When applied, p_1 removes the node labelled b, preserves that labelled a, and generates a new node labelled c. On the other hand, p_2 removes the node labelled a, preserves that labelled b, and generated a new node labelled d. Consider now a graph g containing just two nodes, one labelled a and another one labelled b. It is easy to see that $\langle p_1, id \rangle$ and $\langle p_2, id \rangle$ are two redexes in G. Also, $g_1(l_1) \cap g_2(l_2) = id(L_1) \cap id(L_2) = \{a, b\} \subseteq g_1(l_1(K_1)) \cup g_2(l_2(K_2)) = \{a\} \cup \{b\} = \{a, b\}$ (where we have identified the node labels with the nodes themselves), thus the two redexed satisfy condition *Set-Synchr-1*, and since *Set-Synchr-2* trivially holds, they are synchronizable according to Definition 3.

Let us now see how this same conclusion can be reached via the category-theoretic description of Definition 6. Following the series of constructions described in condition *Cat-Synchr-1*, we first get the pullback of g_1 and g_2, $L_1 \cap_G L_2$, as in the following diagram.

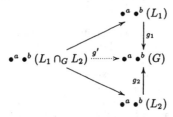

Then we build the pullback $K_1 \cap_G K_2$ of $l_1; g_1$ and $l_2; g_2$, thus getting the following diagram.

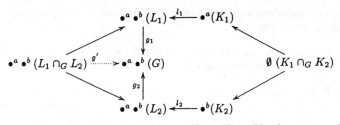

Finally, we get the pushout $K_1 \cup_G K_2$ of $K_1 \leftarrow K_1 \cap_G K_2 \to K_2$, thus getting the following diagram.

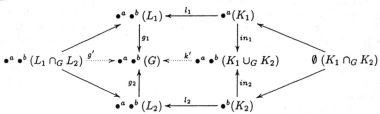

Now, it is easy to see that in our diagram there exists an arrow $h : L_1 \cap_G L_2 \to K_1 \cup_G K_2$ such that $h ; k' = g'$: it is just the identity arrow. Thus the two redexes are synchronizable also according to Definition 6. □

It is worth stressing that although in the example there is only one arrow h satisfying $h ; k' = g'$ (thus it coincides with $\underline{h}$ of Theorem 7), in general this is not true. For example, if $p_1 = (\bullet^a\bullet^a \leftarrow \bullet^a\bullet^a \to \bullet^a\bullet^a), p_2 = (\bullet^a \leftarrow \emptyset \to \emptyset)$, and G has only one node, then there are four different morphisms from $L_1 \cap_G L_2$ to $K_1 \cup_G K_2$ satisfying that condition.

Definition 8 (synchronized composition of two redexes). Let $\Delta_1 = \langle p_1, g_1 \rangle$ and $\Delta_2 = \langle p_2, g_2 \rangle$ be two synchronizable redexes in graph G. Then their **synchronized composition** is the redex $\Delta = \Delta_1 \parallel \Delta_2$ defined as follows. Let $L_1 \cap_G L_2$, $K_1 \cap_G K_2$, and $K_1 \cup_G K_2$ be as in Definition 6, and $\underline{h}$ as in Theorem 7; let $L_1 \cup_G L_2$ be the pushout of $L_1 \leftarrow (L_1 \cap_G L_2) \to L_2$, and let $R_1 \cup_G R_2$ be the pushout of $R_1 \leftarrow (K_1 \cap_G K_2) \to R_2$, as in the following diagrams.

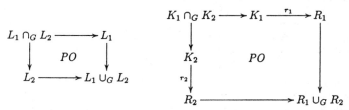

Let K be the pushout complement object of arrows $\underline{h}$ and $l^* : K_1 \cap_G K_2 \to L_1 \cap_G L_2$ which is the mediating arrow between the two pullback objects; similarly, let R be the pushout complement object of the resulting morphism $k^* : K \to K_1 \cup_G K_2$ and $K_1 \cup_G K_2 \to R_1 \cup_G R_2$, as in the following diagram.

$$
\begin{array}{ccc}
K_1 \cap_G K_2 \xrightarrow{l^*} L_1 \cap_G L_2 & \qquad & K \xrightarrow{k^*} K_1 \cup_G K_2 \\
\downarrow \quad \text{PO Compl.} \quad \downarrow \underline{h} & & {\scriptstyle r}\downarrow \quad \text{PO Compl.} \quad \downarrow {\scriptstyle \underline{r}} \\
K \xrightarrow{k^*} K_1 \cup_G K_2 & & R \longrightarrow R_1 \cup_G R_2
\end{array}
$$

Then, the synchronized redex $\Delta = \Delta_1 \parallel \Delta_2$ is obtained by taking $\Delta = \langle p, g \rangle$, where $p = (L_1 \cup_G L_2 \xleftarrow{l} K \xrightarrow{r} R)$ (l is the composition of k^* and the mediating arrow $K_1 \cup_G K_2 \to L_1 \cup_G L_2$), and $g : L_1 \cup_G L_2 \to G$ is the unique arrow induced from g_1 and g_2 via the pushout construction. □

Theorem 9 (synchronized composition is well defined). *The synchronized composition of two redexes is well defined.*

Proof : We have to show that (1) the gluing conditions are satisfied in the two pushout complement constructions defining graphs K and R, and (2) Δ is a redex in G, i.e., it satisfies the gluing conditions of Proposition 1.

We start showing that the identification condition is satisfied by $\langle l^*, \underline{h} \rangle$. Without loss of generality, we assume that $L_1 \cap_G L_2$ is the graph having as nodes and arcs $\{\langle x, y \rangle \mid g_1(x) = g_2(y)\}$, and similarly $K_1 \cap_G K_2$ consists of items $\{\langle z, w \rangle \mid g_1(l_1(z)) = g_2(l_2(w))\}$; then $l^*(\langle z, w \rangle) = \langle l_1(z), l_2(w) \rangle$.

Now suppose that there are two distinct items $\langle x, y \rangle \neq \langle x', y' \rangle$ of $L_1 \cap_G L_2$ such that $\underline{h}(\langle x, y \rangle) = \underline{h}(\langle x', y' \rangle)$: we must show that there is an item $\langle z, w \rangle$ in $K_1 \cap_G K_2$ such that $l^*(\langle z, w \rangle) = \langle x, y \rangle$. Suppose that $x \neq x'$ (a similar argument holds if $y \neq y'$); then clearly $g_1(x) = g_1(x')$, and since Δ_1 satisfies the gluing conditions by hypothesis, there exist z, z' in K_1 such that $l_1(z) = x$ and $l_1(z') = x'$. Thus by definition of $\underline{h}$, $in_1(z) = in_1(z')$, which implies (since $K_1 \cup_G K_2$ is a pushout) that there exist w, w' in K_2 (possibly equal) such that $\langle z, w \rangle$ and $\langle z', w' \rangle$ are in $K_1 \cap_G K_2$. Since $g_2(l_2(w)) = g_1(l_1(z)) = g_1(x) = g_2(y)$, either $l_2(w) = y$, and thus $l^*(\langle z, w \rangle) = \langle x, y \rangle$ and we are done, or $l_2(w) = y'' \neq y$. In this last case, by the identification condition of redex Δ_2 there is a w'' in K_2 with $l_2(w'') = y''$, and thus necessarily $\langle z, w'' \rangle$ is in $K_1 \cap_G K_2$, with $l^*(\langle z, w'' \rangle) = \langle l_1(z), l_2(w'') \rangle = \langle x, y \rangle$.

For the dangling condition, suppose that there is a node $n \in \underline{h}(L_1 \cap_G L_2)$ incident to an arc $e \in (K_1 \cup_G K_2) - \underline{h}(L_1 \cap_G L_2)$. Then e must be either in K_1 or in K_2: Suppose that $e = in_1(e_1)$ for e_1 in K_1. Since $e \notin \underline{h}(L_1 \cap_G L_2)$, it follows that there is no e_2 in L_2 such that $g_2(e_2) = g_1(l_1(e_1))$; moreover, from $n \in \underline{h}(L_1 \cap_G L_2)$ it follows that there is $\langle x, y \rangle$ in $L_1 \cap_G L_2$ and x' in K_1 with $n = in_1(x')$ and $l_1(x') = x$, and $g_2(y)$ incident to $g_1(l_1(e_1))$. Since Δ_2 satisfies the dangling condition, y must be in $l_2(K_2)$, i.e., there is a y' in K_2 with $l_2(y') = y$; thus $l^*(\langle x', y' \rangle) = \langle x, y \rangle$, as desired.

For the pushout complement defining graph R, the identification condition is trivially satisfied because the mediating morphism $\underline{r} : K_1 \cup_G K_2 \to R_1 \cup_G R_2$ is injective. For the dangling condition, suppose that there is an edge e of $R_1 \cup_G R_2$ not in the image of $\underline{r}$, but incident to a node $n = \underline{r}(n')$. This means that e is created either by p_1 or by p_2, and that it is incident to a node preserved by that production. But then condition *Synchr-2* ensures that the corresponding node is preserved by the other production as well, thus it is in $K_1 \cap_G K_2$ and in K as well. Therefore $n' \in k^*(K)$, as required.

It remains to show that Δ is a redex. That is, that morphisms $l : K \to L_1 \cup_G L_2$ and $g : L_1 \cup_G L_2 \to G$ satisfy the gluing conditions. Suppose that $g(x) = g(y)$ and $x \neq y$. It can be shown that since $x \neq y$, then x and y must both belong to the image in $L_1 \cup_G L_2$ of either L_1 or L_2 but not of both. Then the identification condition follows from the corresponding conditions for Δ_1 and Δ_2, which hold by assumption. For the dangling condition, suppose that arc a is in $G - g(L_1 \cup_G L_2)$ and that $g(n)$ is connected to a for some node n in $L_1 \cup_G L_2$. Then the same node n must be either in L_1 or in L_2. Now the corresponding dangling conditions for Δ_1 and Δ_2 can be used, and considering also the way K is defined it can be shown that n is in $l(K)$. □

Example : Let us consider the same two redexes of the previous example, which have been shown to be synchronizable, and let us follow Definition 8 to get their synchronized composition. First we need to compute the pushout $L_1 \cup_G L_2$ $L_1 \leftarrow (L_1 \cap_G L_2) \to L_2$, and also the pushout $R_1 \cup_G R_2$ of $R_1 \leftarrow (K_1 \cap_G K_2) \to R_2$, as in the following diagrams (where we use some graphs that have been computed in the previous example).

$$\begin{array}{ccc} \bullet^a \; \bullet^b \; (L_1 \cap_G L_2) & \longrightarrow & \bullet^a \; \bullet^b \; (L_1) \\ \downarrow & \text{PO} & \downarrow \\ \bullet^a \; \bullet^b \; (L_2) & \longrightarrow & \bullet^a \; \bullet^b \; (L_1 \cup_G L_2) \end{array}$$

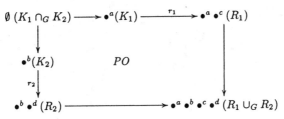

Then, we need the pushout complement object, K, of h and $l^* : K_1 \cap_G K_2 \to L_1 \cap_G L_2$, and also the pushout complement object, R, of k^* and $(K_1 \cup_G K_2 \to R_1 \cup_G R_2)$, as in the following diagrams.

$$\emptyset\ (K_1 \cap_G K_2) \xrightarrow{l^*} \bullet^a \bullet^b\ (L_1 \cap_G L_2) \qquad \emptyset\ (K) \xrightarrow{k^*} \bullet^a \bullet^b\ (K_1 \cup_G K_2)$$

with *PO Compl.* and h on the left; *PO Compl.* and r_1 on the right, giving

$$\emptyset(K) \longrightarrow \bullet^a \bullet^b\ (K_1 \cup_G K_2) \qquad \bullet^c \bullet^d\ (R) \longrightarrow \bullet^a \bullet^b \bullet^c \bullet^d\ (R_1 \cup_G R_2)$$

Now we are ready to produce the synchronized redex, $\Delta = \langle p_1, g_1 \rangle \parallel \langle p_2, g_2, \rangle$, which is our case is given by the following production

$$p: \quad \bullet^a \bullet^b\ (L_1 \cup_G L_2) \xleftarrow{l} \emptyset\ (K) \xrightarrow{r} \bullet^c \bullet^d\ (R)$$

and by the occurrence morphism $g : \bullet^a \bullet^b\ (L_1 \cup_G L_2) \to \bullet^a \bullet^b\ (G)$. Thus the synchronized redex would delete both node a and node b and generate nodes c and d. □

The synchronized composition of two redexes can also be described as the union of the two redexes, followed by the deletion of some items. In fact, the production $(p_1 \cup_G p_2) : L_1 \cup_G L_2 \leftarrow K_1 \cup_G K_2 \to R_1 \cup_G R_2$ can be seen as the "gluing" of p_1 and p_2 along the common subrule $L_1 \cap_G L_2 \xleftarrow{l^*} K_1 \cap_G K_2 \xrightarrow{id} K_1 \cap_G K_2$. Then arrow $l^* : K_1 \cap_G K_2 \to L_1 \cap_G L_2$, which is the inclusion of the items *preserved* by both productions in the collection of items *needed* by both productions, is used in two pushout complement constructions to remove from $K_1 \cup_G K_2$ and $R_1 \cup_G R_2$, respectively, the items which are preserved by one of the productions and deleted by the other one.

The next proposition states precisely in which sense the synchronized composition of redexes is a conservative extension of the classical parallel composition.

Proposition 10 (synchronized vs. parallel composition). *Let $\Delta_1 = \langle p_1, g_1 \rangle$ and $\Delta_2 = \langle p_2, g_2 \rangle$ be two parallel independent redexes in a graph G, let $p_1 + p_2$ be the parallel composition of p_1 and p_2 obtained as their disjoint union, and let $[g_1, g_2] : L_1 + L_2 \to G$ be the induced occurrence morphism. Then $G \Rightarrow_{\langle p_1+p_2,[g_1,g_2]\rangle} H$ iff $G \Rightarrow_{\langle p_1,g_1\rangle \parallel \langle p_2,g_2\rangle} H$.*

Proof: Since Δ_1 and Δ_2 are parallel independent, they are synchronizable (by Proposition 4). Moreover, morphism $l^* : K_1 \cap_G K_2 \to L_1 \cap_G L_2$ is an isomorphism and thus the two pushout complement constructions of Definition 8 do not delete anything. Therefore in this case the synchronized production is (up to isomorphism) simply $p_1 \cup_G p_2 : (L_1 \cup_G L_2 \xleftarrow{l_1 \cup_G l_2} K_1 \cup_G K_2 \xrightarrow{r_1 \cup_G r_2} R_1 \cup_G R_2)$. Now, to prove the statement, we can show that the pushout complement objects D and D' in diagrams (a) and (b) are isomorphic, and similarly that the pushout objects H and H' in diagrams (d) and (e) are isomorphic. This follows from the observation that diagrams (c) and (f) are pushouts, that the composition of two pushouts is again a pushout (thus (c)+(d) and (f)+(e) are pushouts), and that pushout complement objects D and D' are unique up to isomorphism because morphisms $l_1 + l_2$ and $l_1 \cup_G l_2$ are injective. □

$$
\begin{array}{ccc}
L_1 + L_2 \xleftarrow{l_1+l_2} K_1 + K_2 & L_1 \cup_G L_2 \xleftarrow{l_1 \cup_G l_2} K_1 \cup_G K_2 & L_1 + L_2 \xleftarrow{l_1+l_2} K_1 + K_2 \\
{\scriptstyle [g_1,g_2]}\downarrow \quad PO\ Compl. \quad\downarrow & g\downarrow \quad PO\ Compl. \quad\downarrow & \downarrow \quad PO \quad\downarrow \\
G \xleftarrow{} D & G \xleftarrow{} D' & L_1 \cup_G L_2 \xleftarrow{l_1 \cup_G l_2} K_1 \cup_G K_2 \\
(a) & (b) & (c)
\end{array}
$$

$$
\begin{array}{ccc}
K_1 + K_2 \xrightarrow{r_1+r_2} R_1 + R_2 & K_1 \cup_G K_2 \xrightarrow{r_1 \cup_G r_2} R_1 \cup_G R_2 & K_1 + K_2 \xrightarrow{r_1+r_2} R_1 + R_2 \\
\downarrow \quad PO \quad\downarrow & \downarrow \quad PO \quad\downarrow & \downarrow \quad PO \quad\downarrow \\
D \xrightarrow{} H & D' \xrightarrow{} H' & K_1 \cup_G K_2 \xrightarrow{r_1 \cup_G r_2} R_1 \cup_G R_2 \\
(d) & (e) & (f)
\end{array}
$$

Although for parallel independent redexes the synchronized composition has the same effect of the parallel one, it enjoys an additional property that may be useful in some cases. Namely, if the occurrence morphisms g_1 in Δ_1 and g_2 in Δ_2 are both injective, then also the occurrence morphism in $\Delta_1 \parallel \Delta_2$ is injective. On the contrary, in the same situation, the occurrence morphism $[g_1, g_2]$ of the parallel composition is not necessarily injective.

Since our construction of the synchronized composition of two redexes gives priority to deletion over preservation of items, and this kind of priority is an intrinsic feature of the single-pushout approach [Löw93], it can be argued that such a composition would be more naturally expressed in the single-pushout approach. In fact, the *asynchronous parallel transformation* defined in [Löw90] seems to have the same effect of our synchronized composition, and is actually a much easier construction. However, and apart from the inconsistency of the terminology which deserves a deeper analysis, the construction in [Löw90] is applicable also if two redexes delete the same item, or if one deletes a connection node of the other one: This situation has in our view no clear computational interpretation, as discussed in the introduction.

Proposition 11 (associativity of synchronized composition). *Let Δ_1, Δ_2, and Δ_3 be three redexes in a graph G. Suppose that Δ_1 and Δ_2 are synchronizable, and that $\Delta = (\Delta_1 \parallel \Delta_2)$ is their synchronized composition. Then Δ_3 is synchronizable with Δ if and only if it is synchronizable with both Δ_1 and Δ_2. As a consequence, the synchronized composition construction can be iterated, and it is associative (up to isomorphisms).*

Proof: The statement follows by the observation that, if each pair of redexes in $\{\Delta_1, \Delta_2, \Delta_3\}$ is synchronizable, then any item of G can be deleted by at most one of the redexes, and any connection node for one of the redexes must be preserved by all the others. Since the synchronized composition of two redexes deletes what is deleted by one of them and preserves the connection nodes of both, it is still synchronizable with the third redex.

Therefore, both $(\Delta_1 \parallel \Delta_2) \parallel \Delta_3$ and $\Delta_1 \parallel (\Delta_2 \parallel \Delta_3)$ can be constructed, because the '$\parallel$' operator is applied only to synchronizable redexes. The fact that the two resulting redexes are isomorphic follows by the uniqueness of colimits up to isomorphisms. □

An easy consequence of the last result is that synchronized composition is well-defined for any finite set $\underline{\Delta}$ of mutually synchronizable redexes, i.e., such that for all $\Delta, \Delta' \in \underline{\Delta}$, Δ and Δ' are synchronizable.

Definition 12 (derivations with synchronization). A **synchronized direct derivation** is a direct derivation (defined in the usual way as a double pushout construction) via the synchronized composition of a set of redexes. A **derivation with synchronizations** is a derivation (i.e., a sequence of direct derivations) where each direct derivation is synchronized.

□

For a fixed graph grammar $\mathcal{G}$ we denote by **Der** the collection of its (sequential) derivations, by **PDer** the collection of its parallel derivations (i.e., derivations where at each direct derivation a parallel production is applied), and by **SDer** the collection of all its derivation with synchronizations.

Fact 13 (strict inclusion of sets of derivations). *For a grammar $\mathcal{G}$, there are natural mappings j : **Der** $\to$ **PDer** and i : **PDer** $\to$ **SDer**. Mapping j simply regards every sequential direct derivation as a parallel one. In turn, mapping i regards every parallel direct derivation as a synchronized one, which is possible by Proposition 10. Mappings i and j are injective, but in general they are not surjective. Moreover, both of them preserve both source and target graphs of a derivation.* □

3 Serialization of synchronized direct derivations

It is well known in the algebraic approach to graph grammars that a parallel direct derivation (i.e., a direct derivation via a parallel production obtained as coproduct of two productions) can be sequentialized, via a suitable construction called *analysis*, as a derivation consisting of the consecutive application of the two productions: this can be done in both orders, and the resulting graphs are the same. Moreover, this construction induces a relation on derivations, whose reflexive and transitive closure, called *shift-equivalence*, is the basis of the classical theory of concurrency in algebraic graph grammars [Kre87].

Given instead two synchronizable redexes in a graph G, although by Definition 8 we can build their synchronized composition and we can apply it to G in a synchronized direct derivation, in general it is not possible to obtain the same resulting graph by applying first one redex and then the other. In fact, the following stronger condition should be satisfied.

Definition 14 (serializability of redexes). Let $\Delta_1 = \langle p_1, g_1 \rangle$ and $\Delta_2 = \langle p_2, g_2 \rangle$ be two redexes in the same graph G. Then they are **serializable in the order** Δ_1, Δ_2 (written $\Delta_1 \nearrow \Delta_2$) if the following two conditions are satisfied:

(Ser-1) $g_2(L_2) \cap g_1(L_1) \subseteq g_1(l_1(K_1))$
(Ser-2) $g_1(l_1(I_1)) \cap g_2(L_2) \subseteq g_2(l_2(K_2))$

where I_1 is defined as in Definition 3. □

In words, two redexes are serializable whenever the items needed by both redexes are preserved by the first one, and the connection nodes of the first one are not deleted by the second one.

A purely categorical formulation of serializability is possible as well. Notice that serializability in some order implies synchronizability, and serializability in both orders is equivalent to parallel independence.

Theorem 15 (serialization of synchronized direct derivations). *Let $\Delta_1 = \langle p_1, g_1 \rangle$ and $\Delta_2 = \langle p_2, g_2 \rangle$ be two serializable redexes in G (i.e., such that $\Delta_1 \nearrow \Delta_2$). Suppose moreover that $G \Rightarrow_{\Delta_1 \| \Delta_2} H$ is the corresponding synchronized direct derivation, and that $G \Rightarrow_{\Delta_1} H_1$. Then there exists a graph morphism $g_2' : L_2 \to H_1$ such that $H_1 \Rightarrow_{\langle p_2, g_2' \rangle} H'$, and H' is isomorphic to H. In this case we say that derivation $G \Rightarrow_{\Delta_1} H_1 \Rightarrow_{\langle p_2, g_2' \rangle} H'$ is a **serialization** of derivation $G \Rightarrow_{\Delta_1 \| \Delta_2} H$.*

Proof (sketch): The condition of serializability guarantees that if $G \Rightarrow_{\Delta_1} H_1$, then the occurrence morphism $g_2 : L_2 \to G$ factorizes through the context graph D_1 as $g_2 : L_2 \to D_1 \to G$; thus g_2' can be defined as the arrow $L_2 \to D_1 \to H_1$. The fact that $\langle p_2, g_2' \rangle$ is a redex of H_1 follows by the fact that Δ_2 is a redex of G, and that the connection nodes of p_1 are not deleted by Δ_2. Then the construction of the direct derivation $H_1 \Rightarrow_{\langle p_2, g_2' \rangle} H'$ and the proof that $H' \cong H$ are similar to that of the analysis construction [Kre77, CELMR95a]. □

It is worth stressing that the notion of serializability introduced above is can be seen as the equivalent of that of *weak parallel independence*, defined in [EHKRW95], for the double-pushout approach. Also, the last theorem can be seen as a rephrasing, for the double pushout approach, of the "weak parallelism" theorem in [EHKRW95].

It can be shown that the serialization construction has an inverse (like the *synthesis* construction is an inverse of the *analysis* in the classical theory of parallelism), which is applicable to a sequence of two direct derivations $G \Rightarrow_{\Delta_1} X \Rightarrow_{\Delta_2} H$, provided that they satisfy a condition weaker than, but similar to, *sequential independence* [Ehr87]. More precisely, the property is as follows.

Definition 16 (weak sequential independence). Consider the derivation $G \Rightarrow_{\Delta_1} X \Rightarrow_{\Delta_2} H$, where $\Delta_1 = \langle p_1, g_1 \rangle$, $\Delta_2 = \langle p_2, g_2 \rangle$, $p_1 = (L_1 \xleftarrow{l_1} K_1 \xrightarrow{r_1} R_1)$ and $p_2 = (L_2 \xleftarrow{l_2} K_2 \xrightarrow{r_2} R_2)$, as in the following diagram.

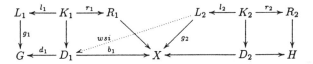

Then Δ_1 and Δ_2 are **weakly sequential independent** if and only if there exists a morphism $wsi : L_2 \to D_1$ such that

(wsi-1) $wsi; b_1 = g_2$;
(wsi-2) $wsi; d_1$ and l_2 satisfy the dangling condition. □

Informally, being weakly sequential independent means that what is needed by the second redex is left unchanged by the first one (that is, either preserved or not even mentioned), and also that the second redex can be applied to G. Note that this condition is strictly weaker than the condition giving sequential independence, which consists of *(wsi-1)* and also requires the existence of an arrow from R_1 to D_2 which makes the diagram commute.

If two redexes are weakly sequential independent, it is possible to show that they are synchronizable, and that the result of applying to G their synchronized composition is the same (up to isomorphism) as that obtained by applying them in the given sequence. More precisely, we have the following.

Theorem 17. *Consider the derivation* $G \Rightarrow_{\Delta_1} X \Rightarrow_{\Delta_2} H$, *where* $\Delta_1 = \langle p_1, g_1 \rangle$, $\Delta_2 = \langle p_2, g_2 \rangle$, $p_1 = (L_1 \xleftarrow{l_1} K_1 \xrightarrow{r_1} R_1)$ *and* $p_2 = (L_2 \xleftarrow{l_2} K_2 \xrightarrow{r_2} R_2)$. *Assume also that* Δ_1 *and* Δ_2 *are weakly sequential independent. Then*

1. Δ_1 *and* $\Delta'_2 = \langle p_2, wsi; d_1 \rangle$ *are serializable, i.e.* $\Delta_1 \nearrow \Delta'_2$, *and thus they are synchronizable;*
2. *if* $G \Rightarrow_{\Delta_1 \| \Delta'_2} H'$, *then* H *and* H' *are isomorphic.*

Proof (sketch): Point 1 follows by the observation that the connection nodes of Δ_1 cannot be deleted by Δ_2, otherwise the dangling condition would not be satisfied by g_2 and l_2. For point 2, the construction of the synchronized direct derivation mimics the synthesis construction in [Kre77, CELMR95a]. □

We think that this inverse construction is strictly related to the *sequential composition* of productions as defined in [Ehr87], and that the resulting *concurrent production* is essentially our synchronized production in the case of weak sequential independence.

The serialization of synchronized direct derivations induces a relation on derivations with synchronization, reminiscent of the classical shift-equivalence. We restrict our considerations to the subclass of *safe* graph grammars, to avoid the problems concerned with automorphisms discussed in depth in [CELMR94a]. In fact, the next definition would not be correct for non-safe grammars, since it would identify too many derivations.

A graph grammar $\mathcal{G}$ with initial graph G_0 is **safe** if every graph H reachable from G_0 via a derivation with synchronizations has no automorphisms different from the identity. For the rest of this section we refer to a given safe graph grammar $\mathcal{G}$.

Definition 18 (serialization equivalence). Given two derivations with synchronizations $\rho = G \Rightarrow_{\Delta_1 \| \Delta_2} H$ and $\rho' = G' \Rightarrow_{\Delta'_1} X \Rightarrow_{\Delta'_2} H'$, we write $\rho \approx \rho'$ if ρ' is obtained (up to isomorphism) from ρ via the serialization construction. Moreover, we denote by $\equiv_{ser}$ the closure of $\approx$ with respect to reflexivity, symmetry, transitivity, and sequential composition of derivations. □

For a given safe grammar $\mathcal{G}$, let $i : \mathbf{PDer} \to \mathbf{SDer}$ be the natural mapping from parallel derivations to derivations with synchronizations introduced in Fact 13. Moreover, let $\equiv_{shift}$ be the classical shift equivalence on parallel derivations. Then the following proposition can be easily proved.

Proposition 19. (serialization vs. shift equivalence) *Let $\rho, \rho' \in \mathbf{PDer}$. Then $\rho \equiv_{shift} \rho'$ if and only if $i(\rho) \equiv_{ser} i(\rho')$. That is, relation $\equiv_{ser}$ is a conservative extension of $\equiv_{shift}$.* □

It is well-known that shift-equivalence classes of parallel derivations have standard representatives, called *canonical derivations*. On the basis of the results presented in [JK91, JK93], we claim that this fact can be generalized to derivations with synchronizations. A canonical derivation with synchronization will manifest, in general, more parallelism than the canonical parallel derivation belonging to the same equivalence class.

4 Conclusions and Future Work

We introduced a notion of "synchronized composition" of redexes in graph grammars, following the algebraic, double-pushout approach. This property extends the classical parallel composition of productions, in the sense that there are redexes which are synchronizable but are not parallel independent. Synchronizable redexes can be applied sequentially (producing the same effect of their synchronized composition) only if they satisfy the stronger condition of serializability, which is asymmetric. This serialization construction induces an equivalence on derivations with synchronizations which is a conservative extension of the classical shift equivalence for parallel derivations.

We plan to investigate the existence of standard representatives for equivalences classes of derivations up to the serialization equivalence, which should play a role analogous to that of canonical derivations for classes of derivations up to shift equivalence.

The proposed synchronization mechanism has potential applications in the specification of concurrent or distributed systems exhibiting some sort of asymmetric conflicts, like non-orthogonal Term Graph Rewriting systems [KKSdV93] and Elementary Net Systems with activator and inhibitor arcs [JK91].

Other topics that we intend to pursue include a comparison of our approach with that of [JK91, JK93], with the intent of providing a faithful partial order semantics to graph grammars with synchronized productions. Also, we plan to study the possibility of using it to model the synchronization of CCS agents [Mil80] using graph grammars.

Acknowledgements

This research has been partially supported by the COMPUGRAPH Basic Research Esprit Working Group n. 7183.

References

[BFH87]   P. Böehm, H. Fonio and A. Habel. Amalgamation of graph transformations: a synchronization mechanism. In *JCSS 34* (1987), 307–408.

[CELMR94a] A. Corradini, H. Ehrig, M. Löwe, U. Montanari and F. Rossi. Abstract Graph Derivations in the Double-Pushout Approach. In *Proceedings Dagstuhl Seminar on Graph Trasformations in Computer Science*. LNCS 776, Springer-Verlag, 1994, 86–103.

[CELMR94b] A. Corradini, H. Ehrig, M. Löwe, U. Montanari and F. Rossi. An event structure semantics for safe graph grammars. In *Programming Concepts, Methods and Calculi*, E.-R. Olderog ed., IFIP Transactions A-56, North Holland, 1994, 423–444.

[CELMR95a] A. Corradini, H. Ehrig, M. Löwe, U. Montanari and F. Rossi. Algebraic approach to graph transformation II: models of computations in the double pushout approach. Submitted.

[CELMR95b] A. Corradini, H. Ehrig, M. Löwe, U. Montanari and F. Rossi. An event structure semantics for graph grammars with parallel productions. In this volume.

[Ehr87]   H. Ehrig. Tutorial introduction to the algebraic approach of graph-grammars. In *Proceedings 3rd International Workshop on Graph Grammars and their Application to Computer Science*, LNCS 291, Springer-Verlag, 1987, 3–14.

[EHKRW95]   H. Ehrig and R. Heckel and M. Korff and L. Ribeiro and A. Wagner. Algebraic Approach to Graph Transformation I: Tutorial Introduction and Single-Pushout Approach. Submitted for the *Handbook of Graph Grammars*, 1995.

[JK91]   R. Janicki and M. Koutny. Invariant semantics of nets with inhibitor arcs. In *Proc. CONCUR*. Springer-Verlag, LNCS 527, 1991.

[JK93]   R. Janicki and M. Koutny. Structure of concurrency. *Theoretical Computer Science*, 112:5–52, 1993.

[KKSdV93]   J.R. Kennaway, J.W. Klop, M.R. Sleep, and F.J. de Vries. Event structures and orthogonal term graph rewriting. In M.J. Plasmejier M.R. Sleep and M.C.J.D. van Eekelen, editors, *Term Graph Rewriting*. Wiley Professional Computing, 1993.

[Kre77]   H.-J. Kreowski. Manipulation von Graph Transformationen. Ph.D. Thesis, Technische Universität Berlin, 1977.

[Kre87]   H.-J. Kreowski. Is parallelism already concurrency? Part 1: Derivations in graph grammars. In *Proceedings of the 3rd International Workshop on Graph-Grammars and Their Application to Computer Science*. LNCS 291, Springer-Verlag, 1987, 343–360.

[Löw90]   M. Löwe. Extended algebraic graph transformation. Ph.D. Thesis, Technische Universität Berlin, 1990.

[Löw93]   M. Löwe. Algebraic approach to single-pushout graph transformation. In *Theoret. Comput. Sci.* 109 (1993) 181–224.

[LD94]   M. Löwe and J. Dingel. Parallelism in the Single-Pushout Graph Rewriting. In *Proceedings Dagstuhl Seminar on Graph Trasformations in Computer Science*, LNCS 776, Springer-Verlag, 1994.

[Mil80]   R. Milner. A Calculus of Communicating Systems. In LNCS 92, 1980.

[MR92]   U. Montanari and F. Rossi. Graph grammars as context-dependent rewriting systems. In *Proceedings CAAP92*. Springer-Verlag, LNCS, 1992.

[Rei85]   W. Reisig. Petri Nets: An Introduction. *EACTS Monographs on Theoretical Computer Science*, Springer-Verlag, 1985.

The Decomposition of ESM Computations

D. Janssens

University of Antwerp - UIA
Universiteitsplein 1
B-2610 Wilrijk, Belgium
email: dmjans@uia.ua.ac.be

Abstract. The work presented continues the exploration of the value of ESM systems as a model of concurrent computation. In [4] and [5] a true concurrent semantics for these systems is presented, where concurrent histories are represented by process objects, called computation structures. In this paper it is shown that, if one refines the notion of a computation structure in a suitable way, then each computation structure can be decomposed according to an arbitrary partitioning of the system into subsets of rules. As a result the semantics is compositional with respect to the union of sets of rules. An application to some variants of Petri nets is sketched.

1 Introduction

ESM systems are a type of graph rewriting systems that originates from a graph grammar model of actor systems, called Actor Grammars (see, e.g., [4],[5]. The basic idea is that system states are modeled by graphs, computations by graph rewriting processes and systems by sets of graph rewriting productions. The latter suggests a natural and simple way to compose systems P_1 and P_2, yielding a larger system P: let P be the union of P_1 and P_2. It will be shown that the process semantics proposed for ESM systems is compositional with respect to this operation. Obviously this research is in line with a significant body of existing results about compositional semantics for concurrent systems, e.g. in the framework of CCS [7] and CSP [2], or Petri boxes [3]. However in most of those cases one uses system descriptions, such as process terms or nets, that are rather different from sets of graph rewriting productions. This different structure is reflected in the way these systems are composed, and therefore the compositionality results obtained for them are not readily applicable to graph rewriting systems.

The semantics of an ESM system is a set of computation structures describing all possible (finite) rewriting processes. Composition on the semantic level will be defined via the composition of individual computation structures: the composition of the semantics of two systems P_1 and P_2 is obtained by the elementwise composition of computation structures of P_1 with computation structures of P_2. The essential step in obtaining the desired compositionality result is to show that each computation structure of the composed system $P_1 \cup P_2$ can be decomposed into two computation structures (one of P_1, the other of P_2). This requires the replacement of the original notion of a computation structure by a slightly more complex one, called conditional computation structure.

In the first sections the basic notions of ESM graph rewriting and computation structures are introduced: in Section 3 the very simple case of the rewriting of discrete graphs is considered, and in Section 4 this is generalized to ESM rewriting. In Section 5 the composition of computation structures is introduced, and in Section 6 the notion of a conditional computation structure is explained. Section 7 contains the basic composition and decomposition results for computation structures, leading to the compositional property of the semantics, as well as a small application to Petri nets: it is demonstrated that the ideas presented yield compositional process semantics for various kinds of nets, such as nets without self-concurrency or nets with bounded capacities.

2 Preliminaries

In this section we introduce some basic terminology to be used in the paper.

(1) Throughout the paper, Σ and Δ denote arbitrary but fixed alphabets. Σ is used for node labels and Δ for edge labels.
(2) For a set V, $Edges(V)$ denotes the set $V \times \Delta \times V$. For a relation $R \subseteq V \times V$, R^+ denotes the transitive closure of R. R is a *partial order* if R is antireflexive, transitive and antisymmetric. For a partially ordered set (V, R) the sets of minimal and maximal nodes of V with respect to R are denoted by $Min(V, R)$ and $Max(V, R)$, respectively. A *cut* of R is a maximal subset K of V such that, for each $x, y \in K$, $(x, y) \notin R$ and $(y, x) \notin R$.
(3) Let V, W be sets such that $W \subseteq V$. For a relation $R \subseteq V \times V$, $R|_W$ denotes the restriction of R to W: $R|_W = R \cap (W \times W)$. Similarly, for $E \subseteq Edges(V)$, $E|_W = E \cap Edges(W)$ and for a function $\lambda : V \to \Sigma$, $\lambda|_W$ is the restriction of λ to W.
(4) Let V, W be sets and let $f : V \to W$ be an injective function. Then $f(V)$ denotes the range of f. For a relation $R \subseteq V \times W$, $f(R)$ is the relation $\{(f(x), f(y)) \mid (x, y) \in R\}$. Similarly, for $E \subseteq Edges(V)$, $f(E) = \{(f(x), \delta, f(y)) \mid (x, \delta, y) \in E\}$, and for a function $\lambda : V \to \Sigma$, $f(\lambda)$ is the function $\lambda' : f(V) \to \Sigma$ such that $\lambda' \circ f = \lambda$. We use similar notations for the inverse relation f^{-1} of f.
(5) The notations from (3) and (4), as well as $\cup$ for set union and $\subseteq$ for the subset relation, are extended in the obvious componentwise way to structures consisting of a carrier set V and various functions and relations; e.g., if $S = (V, E, \lambda, R)$ then $f(S) = (f(V), f(E), f(\lambda), f(R))$ and, for $W \subseteq V$, $S|_W = (W, E|_W, \lambda|_W, R|_W)$. Also, $S_1 \subseteq S_2$ denotes the fact that S_1 is a substructure of S_2, and a structure S_1 is an *induced* substructure of S_2 if $S_1 \subseteq S_2$ and $S_2|_{V_1} = S_1$.
(6) A (Σ, Δ)-*graph* is a system $g = (V, E, \lambda)$ where V is a finite set (called the *set of nodes of g*), $E \subseteq Edges(V)$ (called the *set of edges* of g), and λ is a function from V into Σ (called the *node–labeling function of g*). For a (Σ, Δ)-graph g, its components are denoted by V_g, E_g and λ_g, respectively.

(7) A *graph morphism from g into h* is an injective function $f : V_g \to V_h$ such that $f(g) \subseteq h$. So we consider only injective graph morphisms that preserve labels. f is a *graph isomorphism* if its inverse is a graph morphism from h into g. For (Σ, Δ)-graphs g and h, g is a *subgraph of h* if $g \subseteq h$.

3 Rewriting Discrete Graphs

In this section graph rewriting systems for discrete node labeled graphs are considered; we call them node rewriting (shortly NR) systems. The main aim is not to investigate these systems as such, but to set up a framework for the description of their behaviour. This framework will be extended to ESM graph rewriting in Section 4. For a more extensive introduction of ESM graph rewriting we refer to [4] or [5]. The framework consists of the following elements.

- The notion of a NR computation structure is used to represent a rewriting process. A NR computation structure is a set of (labeled) nodes equipped with a partial order R^c; the nodes represent the nodes occurring in the rewriting process and the partial order represents the causal relationhip between them. More specifically: a node x is a direct predecessor of a node y if and only if there is an atomic rewrite (i.e., an application of a production) in which x is removed and y is created. NR computation structures are compared using injective structure preserving functions on nodes, called NR morphisms.
- A NR production is a NR computation structure of a restricted kind: it is a NR computation structure that corresponds to a rewriting process consisting of just one atomic rewrite. In such a rewriting process each node of the initial configuration is a direct predecessor of each node of the final configuration.
- A NR system is simply a set of NR productions.
- A set of NR computations is associated to a system using the notion of a covering: a covering specifies how a particular NR computation structure may be built from production occurrences.

Definition 3.1. *(1) A NR computation structure is a 3-tuple (V, λ, R^c) where V is a finite set, $\lambda : V \to \Sigma$ is a function and $R^c \subseteq V \times V$ is a partial order.*
(2) Let $C_1 = (V_1, \lambda_1, R_1^c)$ and $C_2 = (V_2, \lambda_2, R_2^c)$ be NR computation structures. A NR morphism from C_1 into C_2 is an injective function $f : V_1 \to V_2$ such that $f(C_1) \subseteq C_2$.

For a NR computation structure C, its components are denoted by V_C, λ_C, R_C^c, respectively. We write $Min(C)$ and $Max(C)$ instead of $Min(V_C, R_C^c)$ and $Max(V_C, R_C^c)$.

Definition 3.2. *(1) A NR production is a NR computation structure such that there exist $V_1, V_2 \subseteq V$ for which $V_1 \neq \emptyset$, $V_2 \neq \emptyset$, $V_1 \cup V_2 = V$, $V_1 \cap V_2 = \emptyset$ and $R^c = V_1 \times V_2$.*
(2) A NR system is a set of NR productions.

(3) Let C be a NR computation structure and let P be a NR system. A P-covering of C is an indexed set $(\pi_i, f_i)_I$ of pairs such that
 (3.1) for each $i \in I$, $\pi_i \in P$ and $f_i : \pi_i \to C$ is a NR-morphism,
 (3.2) for each $x \in V_C$, there exists at most one $i \in I$ such that $x \in f_i(Min(\pi_i))$ and at most one $i \in I$ such that $x \in f_i(Max(\pi_i))$,
 (3.3) Let $R^c_{cov} = \bigcup_{i \in I} f_i(R^c_{\pi_i})$. Then $R^c_C = (R^c_{cov})^+$
(4) The set of NR computation structures of a NR system P is the set of all NR computation structures C such that there exists a P-covering of C.

Informally, each f_i determines an occurrence of a production π_i in C. The set $f_i(Min(\pi_i))$ is the set of nodes that are replaced by this occurrence, and $f_i(Max(\pi_i))$ is the set of nodes that are created by it. (3.2) expresses the fact that each node of C is removed by at most one production occurrence, and that each node of C is created by at most one production occurrence. (3.3) expresses the fact that the causal relationships in C are exactly those specified by the production occurrences.

Example 3.1. Let $\Sigma = \{a, b, c, d\}$ and let π_1, π_2 be the NR productions depicted in Figure 1. Then the structure C depicted in Figure 2 is a NR computation structure (the arrows implied by the fact that R^c_C is transitive are omitted). C has a $\{\pi_1, \pi_2\}$-covering consisting of one occurrence of π_1 and one occurrence of π_2. C represents a rewriting process where first π_1 is applied to replace the nodes 2,3 by the nodes 5,6, and then π_2 is applied to replace the nodes 6,4 by the nodes 7,8,9.

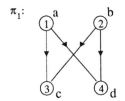 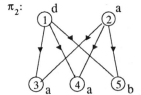

Fig. 1.

It is well-known that a Petri net may be viewed as a rewriting system for discrete node-labeled graphs, and hence, as a NR system: a marking may be represented by a discrete graph where nodes represent tokens and node labels represent places. A transition with input places p_1, p_2 and output places p_3, p_4 corresponds to a graph grammar production where the left-hand side is the graph $\overset{p_1}{\bullet} \overset{p_2}{\bullet}$ and the right-hand side is the graph $\overset{p_3}{\bullet} \overset{p_4}{\bullet}$. On the other hand, each NR system may be viewed in this way as a representation of a Petri net. Since, in graph grammars, one usually allows the concurrent rewriting of arbitrary nonoverlapping occurrences of left-hand sides of productions we consider Petri

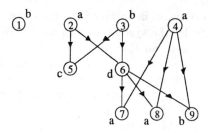

Fig. 2.

nets where self-concurrency of transitions is allowed. In Section 7 it is discussed how one can remove this self-concurrency if desired.

Remark. A pair $(C, (\pi_i, f_i)_I)$ consisting of a NR computation structure and a covering is very similar to a Petri net process (see, e.g., [8]). In fact a process of a net N is a mapping p from an occurrence net into N, and p is comparable to the union of the mappings f_i^{-1}.

When a rewriting process is represented in this way by a partially ordered structure (V, R^c) where the partial order R^c represents causal relationships, then the cuts of R^c are the intermediate configurations that may occur in the rewriting processes corresponding to this structure. (In a similar way the slices of a Petri net process p correspond to the markings occurring in the runs described by p.) In fact, for a computation structure C, each cut of R^c_C yields a sequential decomposition of C. Formally, one easily verifies that the following holds.

Proposition 3.1 *Let P be a NR system, let C be a NR computation structure, let $(\pi_i, f_i)_I$ be a P-covering of C and let K be a cut of R^c_C. Let K_{before} and K_{after} be the sets defined by*

$$K_{before} = K \cup \{x \mid \text{ there exists a } k \in K \text{ such that } (x, k) \in R^c_C \},$$

$$K_{after} = K \cup \{x \mid \text{ there exists a } k \in K \text{ such that } (k, x) \in R^c_C \}.$$

Let $I' = \{i \mid f_i(V_{\pi_i}) \subseteq K_{before}\}$ and $I'' = \{i \mid f_i(V_{\pi_i}) \subseteq K_{after}\}$. Then the substructures of C induced by K_{before} and K_{after} are NR computation structures, and $(\pi_i, f_i)_{I'}$ and $(\pi_i, f_i)_{I''}$ are P-coverings of them.

In particular, the cuts consisting of the minimal and the maximal nodes with respect to R^c may be interpreted as the initial and final configurations of the rewriting process represented by C.

4 From Node Rewriting systems to ESM systems

The aim of this section is to demonstrate that the framework for node rewriting systems and their behaviour introduced in Section 3 may be extended to ESM graph rewriting. The extension is obtained in two steps: on the one hand, the datastructures rewritten by the system are generalized from discrete graphs to arbitrary (Σ, Δ)-graphs and, on the other hand, a suitable embedding mechanism is introduced. The embedding mechanism is first explained in a rather informal way, and then it is demonstrated in which way the notions of Definitions 3.1 and 3.2 may be adapted to ESM graph rewriting.

Consider a rewriting process transforming a graph g into a graph h. The embedding mechanism should specify a way to infer a set of edges E_{gh} of h from edges of g. An easy way to do this is to specify a relation $\to \subseteq Edges(V_g) \times Edges(V_h)$ and to let E_{gh} be the set

$$\{e' \in Edges(h) \mid \text{there exists an } e \in E_g \text{ such that } e \to e'\}.$$

In ESM graph rewriting the relation $\to$ is constructed from two auxiliary relations, $R^s \subseteq (V_g \times \Delta) \times (V_h \times \Delta)$ and $R^t \subseteq V_g \times V_h$ by requiring that $(x, \gamma, y) \to (u, \delta, w)$ if and only if $((x, \gamma), (u, \delta)) \in R^s$ and $(y, w) \in R^t$. Hence, informally speaking, R^s and R^t transfer the source part and the target part of edges, respectively.

The notion of a NR computation structure C is adapted in the following way to take into account the edges and the embedding mechanism:

- a set E_C of edges is added, and
- each computation structure is equipped with pair of relations R^s, R^t, which together specify the embedding mechanism. R^s and R^t are locally generated: each production π is equipped with relations R^s_π and R^t_π, and the global relations R^s, R^t are generated by combining the occurrences of these local relations. As a result, one may assume that edges are transferred only between nodes that are causally related, i.e. one may assume that $((x, \gamma), (y, \delta)) \in R^s$ implies $(x, y) \in R^c$ and $(x, y) \in R^t$ implies $(x, y) \in R^c$. For technical reasons it is convenient to assume that R^s and R^t are transitive, and to combine R^s and R^t with the causal relation R^c into a so-called transfer relation.

Definition 4.1. *(1) Let V be a set. A transfer relation on V is a 3-tuple $R = (R^c, R^s, R^t)$ such that $R^c \subseteq V \times V$, $R^t \subseteq V \times V$, $R^s \subseteq ((V \times \Delta) \times (V \times \Delta))$, $((x, \gamma), (y, \delta)) \in R^s$ implies $(x, y) \in R^c$, and $R^t \subseteq R^c$. The relation $\xrightarrow{R} \subseteq Edges(V) \times Edges(V)$ is the transitive and reflexive closure of the relation $\xrightarrow{R,1}$, defined by: $(x, \gamma, y) \xrightarrow{R,1} (u, \delta, w)$ if $((x, \gamma), (u, \delta)) \in R^s$ and $y = w$, or $(y, w) \in R^t$ and $(x, \gamma) = (u, \delta)$.*

(2) An ESM computation structure is a 4-tuple (V, E, λ, R) such that (V, E, λ) is a (Σ, Δ)-graph, R is a transfer relation on V such that R^c is a partial order, R^s and R^t are transitive and, and for each $e \in E$ and $e' \in Edges(V)$, $e \xrightarrow{R} e'$ implies that $e' \in E$.

The transfer relation $(\emptyset, \emptyset, \emptyset)$ is denoted by $R_\emptyset$. If $e \xrightarrow{R} e'$ then we say that e' is *transferred from e via R*.

Example 4.1. Let $\Sigma = \{a, b, c, d\}$ and $\Delta = \{\alpha, \beta\}$. Then the structure C depicted in Figure 3 is an ESM computation structure (the straight arrows depict R_C^c, the curved edges depict elements of E_C, and it is assumed that $R_C^s = \{((3, \alpha), (6, \beta))\}$, $R_C^t = \{(6, 7)\}$). One has $(3, \alpha, 4) \xrightarrow{R_C} (6, \beta, 4)$ and $(5, \alpha, 6) \xrightarrow{R_C} (5, \alpha, 7)$.

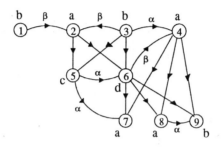

Fig. 3.

ESM computation structures are compared using injective structure preserving functions on nodes, called CS morphisms. Formally, one has the following counterpart of (2) of Definition 3.1.

Definition 4.2. Let $C_1 = (V_1, E_1, \lambda_1, R_1^c)$ and $C_2 = (V_2, E_2, \lambda_2, R_2^c)$ be ESM computation structures. A CS morphism from C_1 into C_2 is an injective function $f : V_1 \to V_2$ such that $f(C_1) \subseteq C_2$.

For an ESM computation structure C, its components are denoted by V_C, E_C, λ_C, R_C, respectively. We will often simply write "computation structure" instead of "ESM computation structure", and we write $Min(C)$ and $Max(C)$ instead of $Min(V_C, R_C^c)$ and $Max(V_C, R_C^c)$, respectively. For a graph (V, E, λ) and a transfer relation R on V such that $(R^c)^+$ is a partial order, $CS(V, E, \lambda, R)$ denotes the smallest computation structure containing (V, E, λ, R). Hence one has $CS(V, E, \lambda, R) = (V, E', \lambda, R')$, where $E' = \{e' \in Edges(V) \mid$ there exists an $e \in E$ such that $e \xrightarrow{R} e'\}$ and $R' = R^+$ (the transitive closure is taken componentwise). Also note that, for a computation structure C, each induced substructure of C is a computation structure.

An ESM production differs from a NR production in two ways: one uses an ESM computation structure instead of a NR computation structure, and one equips it with what is usually called the left-hand side of a production. This "left-hand side" is a graph D which is added as an extra component; D specifies where the production may be applied. It is assumed that the set of nodes of D is the set of minimal nodes of the production, i.e, the nodes of D correspond to the nodes that are replaced when the production is applied.

The notion of a covering of a computation structure C is extended by requiring that

- C is generated by the production occurrences together with the initial graph: the transfer relation of C is obtained by combining the transfer relations of all the production occurrences, and each edge of C is either transferred from an edge of the initial graph via the transfer relation, or it is transferred from an edge created in one of the production occurrences.
- a production (C_i, D_i) can only be applied to a graph if its left-hand side is present; i.e., an occurrence f_i of (C_i, D_i) in C is valid only if $f_i(D_i)$ is a subgraph of C.

Formally, one has the following.

Definition 4.3. *(1) An ESM production is a pair (C, D) where C is a computation structure such that (V_C, λ_C, R_C^c) is a NR production and D is a graph such that $V_D = Min(C)$.*
(2) An ESM system is a set of ESM productions.
(3) Let P be an ESM system and let C be a computation structure. A P-covering of C is an indexed set $(\pi_i, f_i)_I$ of pairs such that
  (3.1) for each $i \in I$, $\pi_i = (C_i, D_i) \in P$ and $f_i : C_i \to C$ is a CS morphism,
  (3.2) for each $x \in V_C$, there exists at most one $i \in I$ such that $x \in f_i(Max(C_i))$ and at most one $i \in I$ such that $x \in f_i(Min(C_i))$,
  (3.3) let $E_{Min} = E_C \cap Edges(Min(C))$, $E_{cov} = \bigcup_{i \in I} f_i(E_{C_i})$ and $R_{cov} = \bigcup_{i \in I} f_i(R_{C_i})$. Then $C = CS(V_C, E_{cov} \cup E_{Min}, \lambda_C, R_{cov})$, and
  (3.4) for each $i \in I$, $f_i(D_i) \subseteq C$.
(4) The set of computation structures of P is the set of all computation structures C such that there exists a P-covering of C.

Condition (3.3) of Definition 4.3 expresses the fact that C is obtained by applying the computation structure $C_{cov} = \bigcup_{i \in I} f_i(C_i)$, i.e. the computation structure generated by the production occurrences $f_i(C_i)$, to the graph $g = (Min(C), E_{Min}, \lambda_C|_{Min(C)})$, which is the initial graph of C (see also Proposition 4.1). Condition (3.4), on the other hand, expresses the fact that a production may only be applied if its left-hand side is present. The notions of Definition 4.3 are illustrated in Example 4.2.

Example 4.2. Let $\pi_1 = (C_1, D_1)$, $\pi_2 = (C_2, D_2)$ be the ESM productions depicted in Figure 4, where it is assumed that $R_{C_1}^s = \{((2, \alpha), (4, \beta))\}$, $R_{C_1}^t = \emptyset$, $R_{C_2}^s = \emptyset$ and $R_{C_2}^t = \{(1, 3)\}$. Then the computation structure C of Example 4.1 has a $\{\pi_1, \pi_2\}$-covering consisting of an occurrence f_1 of π_1 and an occurrence f_2 of π_2: f_1 maps the nodes $1,2,3,4$ of C_1 onto the nodes $2,3,5,6$ of C, respectively, and f_2 maps the nodes $1,2,3,4,5$ of C_2 onto the nodes $6,4,7,8,9$ of C, respectively.

Remark. A graph g may be identified with a computation structure where the causal relation R^c is empty (no rewriting occurs), so g corresponds to the computation structure $(V_g, E_g, \lambda_g, R_\emptyset)$.

It is easily verified that the counterpart of Proposition 3.1 holds for ESM computation structures (instead of NR computation structures).

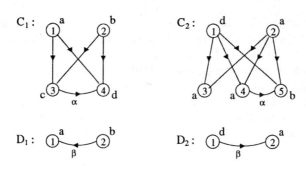

Fig. 4.

Proposition 4.1. *Let P be an ESM system, let C be a computation structure, let $(\pi_i, f_i)_I$ be a P-covering of C and let K be a cut of R_C^c. Let K_{before} and K_{after} be the sets defined by*

$$K_{before} = K \cup \{x \mid \text{ there exists a } k \in K \text{ such that } (x,k) \in R_C^c \},$$

$$K_{after} = K \cup \{x \mid \text{ there exists a } k \in K \text{ such that } (k,x) \in R_C^c \}.$$

Let $I' = \{i \mid f_i(V_{\pi_i}) \subseteq K_{before}\}$ and $I'' = \{i \mid f_i(V_{\pi_i}) \subseteq K_{after}\}$. Then the substructures of C induced by K_{before} and K_{after} are computation structures, and $(\pi_i, f_i)_{I'}$ and $(\pi_i, f_i)_{I''}$ are P-coverings of them.

Hence the substructures of a computation structure C induced by the cuts of the causal relation R_C^c may be interpreted as intermediate configurations occuring in the rewriting processes that correspond to C. Note that these substructures are (Σ, Δ)-graphs by the remark following Example 4.2. In particular, the subgraphs of C induced by $Min(C)$ and $Max(C)$ are the ititial graph and the result graph of these rewriting processes, and $Min(C)$ and $Max(C)$ may be viewed as the sets of input nodes and output nodes of C, respectively.

5 The Composition of Computation Structures

In [5] and [6] a composition operation on computation structures is introduced that models the way rewriting processes interact with each other. This composition operation consists in the gluing of computation structures over a common part, or, formally, of a pushout construction in the category CS of computation structures and CS-morphisms. The composition operation allows both sequential and parallel composition of rewriting processes as special cases.

In the definition of the composition of two computation structures C_1 and C_2, one needs the notion of a (C_1, C_2)-interaction. An interaction specifies the way in which nodes of C_1 and C_2 are glued together; it consists of a discrete computation stucture together with CS-morphisms into C_1 and C_2. Our main interest in considering the composition of rewriting processes $Proc_1$ and $Proc_2$

is in the case where they interact only through their input and output nodes: some of the output nodes of $Proc_1$ are used as input for $Proc_2$, and vice versa. If one takes into account only this type of interaction, then one may say that the only nodes of a rewriting process that are visible to other processes (or available for interaction) are its input and output nodes. These are represented in a computation structure by the minimal and the maximal nodes of the causal relation, respectively. The special case where C_1 and C_2 interact only via their input and output nodes is called an *external* interaction.

Definition 5.1. *Let C_1 and C_2 be computation structures. A (C_1, C_2)-interaction is a 3-tuple $J = (C_J, j_1, j_2)$ where C_J is a computation structure such that $E_J = \emptyset$, $R_J = R_\emptyset$, and $j_1 : C_J \to C_1$, $j_2 : C_J \to C_2$ are CS-morphisms such that the relation $(j_1^{-1}(R_1^c) \cup j_2^{-1}(R_2^c))^+$ is a partial order. J is an external (C_1, C_2)-interaction if, for each $v \in V_J$, either $j_1(v) \in Max(C_1)$ and $j_2(v) \in Min(C_2)$, or $j_1(v) \in Min(C_1)$ and $j_2(v) \in Max(C_2)$.*

The composition of computation structures is defined as follows. Let CS be the category where computation structures are objects and CS-morphisms are arrows.

Definition 5.2. *Let C_1, C_2 be computation structures and let $J = (C_J, j_1, j_2)$ be a (C_1, C_2)-interaction. The composition of C_1 and C_2 over J, denoted by $C_1 {}^J\square\, C_2$, is the set*

$$\{(C, c_1, c_2) \mid \text{the diagram } \begin{array}{c} C_J \xrightarrow{j_2} C_2 \\ j_1 \downarrow \quad \downarrow c_2 \\ C_1 \xrightarrow{c_1} C \end{array} \text{ is a pushout in } CS\}.$$

The situation is illustrated in Figure 5.

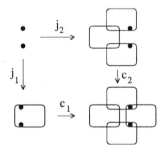

Fig. 5.

Note that the composition of computation structures C_1 and C_2 over an external interaction generally models a *two-way* interaction: output of C_1 is used as input of C_2 and vice versa. One may say that C_1 and C_2 are composed sequentially if

$j_2 \circ j_1^{-1}$ is an isomorphism from the result graph of C_1 onto the initial graph of C_2. Such a situation occurs in Proposition 4.1.

It follows from the properties of pushouts that C is unique up to an isomorphism. Moreover, the pushout from Definition 5.2 always exists; this follows from the next result, which gives a concrete construction for the composition of computation structures. As one may expect, it is based on the construction of pushouts for sets and its proof is a straightforward verification.

Proposition 5.1. *Let* $C_1 = (V_1, E_1, \lambda_1, R_1)$, $C_2 = (V_2, E_2, \lambda_2, R_2)$ *be computation structures and let* $J = (C_J, j_1, j_2)$ *be a* (C_1, C_2)*-interaction. Let* V,
$c_1 : V_1 \to V$ *and* $c_2 : V_2 \to V$ *be such that*
$$\begin{array}{ccc} V_J & \xrightarrow{j_2} & V_2 \\ {}_{j_1}\downarrow & & \downarrow{}_{c_2} \\ V_1 & \xrightarrow{c_1} & V \end{array}$$
is a pushout in sets, and let $C = CS(c_1(C_1) \cup c_2(C_2))$. *Then* $(C, c_1, c_2) \in C_1 {}^J\square\, C_2$.

6 Conditional Computation Structures

In order to obtain a decomposition result for ESM computations that corresponds to the general case of the composition of computation structures (Definition 5.2), it is essential to consider parts of a P-covering $(\pi_i, f_i)_I$ of a computation structure C, and to relate these parts to substructures of C. The parts considered are obtained by restricting I to a subset I'. It turns out that such a part of a covering corresponds to a *pair* of substructures of C, rather than to one single substructure. Such pairs will be called conditional computation structures, and they will be viewed as a generalization of the original notion of a computation structure.

Let, for some $i \in I$, $\pi_i = (C_i, D_i)$. We investigate separately the relationship between the two components C_i and D_i of π_i, and C. On the one hand, C_i may be interpreted as a part of C that is *generated* when π_i is applied, replacing the nodes $f_i(Min(C_i))$ by the nodes $f_i(Max(C_i))$: C_i specifies the nodes and edges created, and the transfer relation between the nodes involved. On the other hand, D_i expresses a *condition* on the context in which π_i may be applied: f_i is a valid occurrence only if the graph $f_i(D_i)$ is a subgraph of C. One may generalize these observations from a single occurrence to the set of occurrences corresponding to I'. On the one hand, the structure *generated* by those occurrences is

$$C' = CS(\bigcup_{i \in I'} f_i(C_i)).$$

On the other hand, one cannot assume that $f_i(D_i) \subseteq C'$, because in general C' is not an induced substructure of C. Hence in general $(\pi_i, f_i)_{I'}$ is not a P-covering of C'. However, one has $f_i(D_i) \subseteq D'$, where D' is the substructure of C induced by $\bigcup_{i \in I'} f_i(V_{C_i})$. Thus one may associate D' with $(\pi_i, f_i)_{I'}$, and again interpret D' as a *condition* on the context in which the production occurrences described by this part of the covering may occur: if that context contains D' as a substructure, then all the left-hand sides are present.

The situation is illustrated by Example 4.2: the edge $(6, \beta, 4)$ does not belong to any of the occurrences of C_1 or C_2, and neither is it inferred from an edge of those occurrences. Hence $(6, \beta, 4)$ is not an edge of $CS(\bigcup_{i \in I} f_i(C_i))$ $(let\, I' = I)$. However, $(6, \beta, 4)$ corresponds to the edge of D_2, and thus its presence is required, because otherwise π_2 would not be applicable here. Evidently, $(6, \beta, 4)$ is an edge of the substructure of C induced by $\bigcup_{i \in I} f_i(V_{C_i})$, i.e. the substructure induced by the nodes 2,3, ..., 9.

We conclude that a P-covering corresponds to a *pair* of computation structures (such as (C', D')), where the first component describes the structure that is generated, and the second component expresses a condition on the context. Such a pair can be viewed as a natural generalization of a production, where the first and the second component correspond to the right-hand side and the left-hand side of a production, respectively. Formally, one has the following.

Definition 6.1. *Let C be a computation structure. A C-condition is a computation structure D such that $V_D = V_C$ and $C \subseteq D$. A conditional computation structure is a pair (C, D) where C is a computation structure and D is a C-condition.*

Informally, to say that a conditional computation structure (C, D) belongs to the semantics of an ESM system P means that productions from P may generate C on the condition that C occurs in context D. To express this formally, the notion of a P-covering is extended to conditional computations in the following way.

Definition 6.2. *Let P be an ESM system.*

(1) Let (C, D) be a conditional computation structure. A P-covering of (C, D) is an indexed set $(\pi_i, f_i)_I$ of pairs such that
  (1.1) for each $i \in I$, $\pi_i = (C_i, D_i) \in P$ and $f_i : C_i \to C$ is a CS morphism,
  (1.2) for each $x \in V_C$, there exists at most one $i \in I$ such that $x \in f_i(Max(C_i))$ and at most one $i \in I$ such that $x \in f_i(Min(C_i))$,
  (1.3) $C = CS(C_{cov})$, where $C_{cov} = \bigcup_{i \in I} f_i(C_i)$, and
  (1.4) for each $i \in I$, $f_i(D_i) \subseteq D$.
(2) The set of conditional computation structures of P is the set of all conditional computation structures (C, D) such that there exists a P-covering of (C, D).

Remark. A computation structure C has a P-covering ((3) of Definition 4.3) if and only if there exists a conditional computation structure (C_{cov}, C) such that $C = CS(C_{cov} \cup C|_{Min(C)})$ and (C_{cov}, C) has a P-covering. In this way each computation structure that has a P-covering may be viewed as a conditional computation structure.

Since a (Σ, Δ)-graph g corresponds to the computation structure $(V_g, E_g, \lambda_g, R_\emptyset)$, and this computation has a P-covering (the empty one), g corresponds to the conditional computation structure $((V_g, \emptyset, \lambda_g, R_\emptyset), (V_g, E_g, \lambda_g, R_\emptyset))$. Observe that this is consistent with the intuitive interpretation of conditional computations: g corresponds to a trivial rewriting that occurs in the context g. Also note that an ESM production (C, D) may be viewed as a conditional computation structure (C, D'), where $D' = CS(C \cup D)$. Hence in this approach system states, systems and semantics are all represented by the same kind of objects, which facilitates the investigation of relationships between these three notions. Note that such a situation does not occur in most other approaches to graph rewriting and graph processes (e.g. [1]), where one first introduces a notion of a graph transformation (the application of a production to a graph), using as basic objects graphs representing system states, and then one considers derivation sequences and equivalence classes of derivation sequences.

To compose two conditional computation structures (C_1, D_1) and (C_2, D_2) over a (C_1, C_2)-interface J, one composes C_1 and C_2 and then equips the resulting computation structure C with a C-condition D such that D is consistent with D_1 and D_2; i.e. D_1 and D_2 correspond to induced substructures of D. The composition of conditional computation structures is formally defined as follows.

Definition 6.3. *Let (C_1, D_1), (C_2, D_2) be conditional computation structures and let $J = (C_J, j_1, j_2)$ be a (C_1, C_2)-interaction. The composition of (C_1, D_1) and (C_2, D_2) over J, denoted by $(C_1, D_1) {}^J\square\, (C_2, D_2)$, is the set*

$$\{((C, D), c_1, c_2) \mid (C, c_1, c_2) \in C_1 {}^J\square\, C_2,\ D_1 = c_1^{-1}(D|_{W_1})\ \text{and}\ D_2 = c_2^{-1}(D|_{W_2})\},$$

where $W_1 = c_1(V_{C_1})$ and $W_2 = c_2(V_{C_2})$.

Note that, in general, the elements of $(C_1, D_1) {}^J\square\, (C_2, D_2)$ are not isomorphic to each other because D_1, D_2 do not uniquely determine D (even for a fixed C).

7 The Decomposition of ESM Computations

In this section the main result of the paper is shown. It states that, for an ESM system P and a conditional computation (C, D) which is built from production occurrences in the way specified by a P-covering $cov = (\pi_i, f_i)_I$, each partition of I into parts I', I'' corresponds to conditional computations (C', D'), (C'', D'') such that (C, D) is the composition of (C', D') and (C'', D''), and the restrictions cov', cov'' of cov to I' and I'' are P-coverings of (C', D') and (C'', D''), respectively. The result generalizes Proposition 4.1, where only the sequential composition of computation structures is considered, to the general case of composition defined in Section 5. This allows one, e.g., to decompose a sequential computation into a number of nonsequential computations. At the end of the section, it is demonstrated how the results presented may be used to obtain a compositional semantics for ESM systems. It is also shown that they may be applied to the special case of NR systems to yield a compositional semantics for various types of Petri nets.

Theorem 7.1 *Let (C, D) be a conditional computation structure and let $(\pi_i, f_i)_I$ be a P-covering of (C, D). For each $i \in I$ let $\pi_i = (C_i, D_i)$ and $C_i = (V_i, E_i, \lambda_i, R_i)$. Let $I', I'' \subseteq I$ such that $I' \cap I'' = \emptyset$ and $I' \cup I'' = I$. Let $W', W'' \subseteq V_C$ such that $\bigcup_{I'} f_i(V_i) \subseteq W'$, $\bigcup_{I''} f_i(V_i) \subseteq W''$ and $W' \cup W'' = V_C$. Let $c' : W' \to V_C$ and $c'' : W'' \to V_C$ be inclusions.*

Then there exist conditional computation structures (C', D'), (C'', D'') and an external (C', C'')-interaction J such that $((C, D), c', c'') \in (C', D')^J \square (C'', D'')$, and $(\pi_i, f_i)_{I'}$, $(\pi_i, f_i)_{I''}$ are P-coverings of (C', D') and (C'', D''), respectively.

Proof. Let $E_{cov'}, E_{cov''}, R_{cov'}, R_{cov''}, C', C'', D', D''$ and C_J be defined by

$$E_{cov'} = \bigcup_{I'} f_i(E_i), \quad E_{cov''} = \bigcup_{I''} f_i(E_i),$$
$$R_{cov'} = \bigcup_{I'} f_i(R_i), \quad R_{cov''} = \bigcup_{I''} f_i(R_i),$$
$$C' = CS(W', E_{cov'}, \lambda_{W'}, R_{cov'}),$$
$$C'' = CS(W'', E_{cov''}, \lambda_{W''}, R_{cov''}),$$
$$D' = D|_{W'}, \quad D'' = D|_{W''},$$
$$C_J = (W' \cap W'', \emptyset, \lambda|_{W' \cap W''}, R_\emptyset).$$

Furthermore, let $j' : C_J \to C'$ and $j'' : C_J \to C''$ be inclusions. Then it follows from $I' \cap I'' = \emptyset$ that $J = (C_J, c', c'')$ is an external (C', C'')-interaction. It follows from the fact that $(\pi_i, f_i)_I$ is a P-covering of (C, D) that $C = CS(C' \cup C'')$ and hence, by Proposition 5.1, that $(C, c', c'') \in C' \square C''$. It follows from Definition 6.3 that

$$(C, D) \in (C', D')^J \square (C'', D'').$$

It remains to show that $(\pi_i, f_i)_{I'}$ and $(\pi_i, f_i)_{I''}$ are P-coverings of (C', D') and (C'', D''), respectively. We consider only $(\pi_i, f_i)_{I'}$ and (C', D'); the proof for $(\pi_i, f_i)_{I''}$ and (C'', D'') is analogous. It is easily verified that the definition of C' implies that (1.1), (1.2) and (1.3) of Definition 6.2 are satisfied for $(\pi_i, f_i)_{I'}$ and (C', D'). On the other hand, (1.4) also holds because D' is an induced substructure of D and $f_i(D_i) \subseteq D$ for each $i \in I$. This completes the proof. □

Next it is shown that the converse of Theorem 7.1 also holds: the composition of conditional computation structures that have a P-covering yields conditional computation structures that have a P-covering.

Theorem 7.2. *Let (C', D''), (C'', D'') be conditional computation structures such that there exist P-coverings cov', cov'' of (C', D'') and (C'', D''), respectively. Let J be an external (C', C'')-interaction and let $((C, D), c', c'') \in (C', D')^J \square (C'', D'')$. Then there exists a P-covering of (C, D).*

Proof. Let $cov' = (\pi_i, f_i)_{I'}$, $cov'' = (\nu_i, g_i)_{I''}$ and let I be the disjoint union of I' and I''. Let $cov = (\rho_i, h_i)_I$ be defined by

$$\begin{cases} \rho_i = \pi_i \text{ and } h_i = c' \circ f_i, & \text{if } i \in I' \\ \rho_i = \nu_i \text{ and } h_i = c'' \circ g_i, & \text{if } i \in I'' \end{cases}$$

Then, for each $i \in I$, $h_i : C_i \to C$ is a CS morphism, so (1.1) of Definition 6.2 is satisfied for cov and (C, D). It follows from the fact that J is external that (1.2) is also satisfied. Furthermore, it follows from Definition 6.3 that $c'(D') \subseteq D$ and $c''(D'') \subseteq D$, which implies that (1.4) holds, and it follows from Proposition 5.1 that (1.3) holds. Thus one may conclude that cov is a P-covering of (C, D). □

As a first application of the decomposition result, it is demonstrated that the semantics corresponding to the conditional computation structures is compositional with respect to the union of ESM systems. Such compositionality results are quite important since they provide a first step towards a modular design of ESM sytems. Formally, for an ESM system P, let

$$Sem(P) = \{(C, D) \mid (C, D) \text{ is a conditional computation structure} \\ \text{such that there exists a } P\text{-covering of } (C, D)\}$$

and let the composition of conditional computation structures be extended to sets as follows. Let X, Y be sets of conditional computation structures. Then the set $X \square Y$ is defined by

$$X \square Y = \{(C, D) \mid \text{there exist } (C_1, D_1) \in X, (C_2, D_2) \in Y, \text{ a } (C_1, C_2)\text{-interaction } J \text{ and functions } c_1, c_2 \text{ such that } ((C, D), c_1, c_2) \in (C_1, D_1) {}^J\square (C_2, D_2)\}.$$

Then one has the following result.

Proposition 7.1. *Let P_1, P_2 be ESM systems. Then*

$$Sem(P_1 \cup P_2) = Sem(P_1) \square Sem(P_2).$$

Proof. To show the inclusion $Sem(P_1 \cup P_2) \subseteq Sem(P_1) \square Sem(P_2)$, let $(C, D) \in Sem(P_1 \cup P_2)$. Then there exists a $P_1 \cup P_2$-covering $(\pi_i, f_i)_I$ of (C, D). One may now apply Theorem 7.1 to decompose (C, D) into an element of $Sem(P_1)$ and an element of $Sem(P_2)$: let, in Theorem 7.1, $I' = \{i \mid \pi_i \in P_1\}$ and $I'' = \{i \mid \pi_i \in P_2\}$. The inclusion $Sem(P_1) \square Sem(P_2) \subseteq Sem(P_1 \cup P_2)$ follows from Theorem 7.2. □

In the last part of this section an application of the framework for the description of ESM systems to Petri nets is outlined. As explained in Section 3, Petri nets may be viewed as NR systems. Applying the ideas of Section 6 to NR systems, we consider conditional computation structures of NR systems. Clearly, a conditional computation structure of a NR system is a pair of NR computation structures; i.e., one may omit the set of edges (it is always empty) and the relations R^s and R^t (which only serve to infer edges). Proposition 7.1 could now be applied to to obtain a compositional semantics for NR systems, and hence, for Petri nets, but it is easily verified that the semantics consisting of NR computation structures ((4) of Definition 3.2) already has this compositional property. So one may wonder what is gained in this context by introducing conditional computation structures, as opposed to computation structures or processes in the usual sense. It turns out, however, that the notion of a conditional computation structure is useful if one wants to consider Petri nets with a more restricted behaviour, e.g. when one is interested only in computations that do not exhibit self-concurrency, or in computations where the number of tokens in a given place does not exceed a certain bound. The first case, the elimination of self-concurrency, corresponds to considering only computations in which, for each pair of occurrences o_1, o_2 of a transition t, either o_1 precedes o_2 in the causal order, or vice versa. Translating this into the framework of NR sytems, this amounts to considering a restricted form of coverings: in (1) of Definition 6.2, one may add the restriction that, for each $\pi \in P$ and each $i, j \in I$ such that $i \neq j$ and $\pi = \pi_i = \pi_j$, there exist $x \in V_{C_i}$ and $y \in V_{C_j}$ such that either $(f_i(x), f_j(y)) \in R_D^c$ or $(f_j(y), f_i(x)) \in R_D^c$. The second case, where e.g. place p has capacity k, amounts to considering only conditional computation structures (C, D) where, for each set of k different p-labelled nodes, at least two of these nodes are ordered under R_D^c. In both cases the results of Theorem 7.1 and 7.2 still hold for the restricted semantics, and hence one obtains again a compositional semantics. The situation is illustrated in Example 7.1.

Example 7.1. *Let P_1, P_2 and P be the Petri nets of Figure 6. Obviously, P is the net obtained by gluing P_1 and P_2 over the places a and b. Then proc of Figure 7 depicts a conditional computation structure of P that may be decomposed into the conditional computation structures $proc_1$ and $proc_2$, which are conditional computation structures of P_1 and P_2, respectively. The first components of $proc_1$ and $proc_2$ are drawn in full lines, the dotted lines represent the parts of*

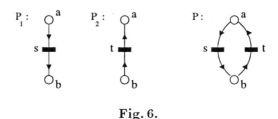

Fig. 6.

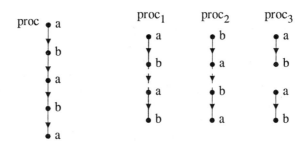

Fig. 7.

the causal relations that do belong to the second components of $proc_1$ and $proc_2$, but not to their first components. If one assumes that place a has capacity 1, then $proc_1$ is a conditional computation structure of P_1, but that is not the case for $proc_3$.

References

1. A. Corradini, H.Ehrig, M.Löwe, U.Montanari and F.Rossi, Abstract Graph derivations in the Double-Pushout Approach, in *Graph-Transformations in Computer Science*, Lecture Notes in Computer Science, vol. 776, Springer-Verlag, Berlin, 1994, 86-103.
2. C. Hoare, *Communicating Sequential Processes*, Prentice-Hall, 1985.
3. M. Koutny, J. Esparza and E. Best, Operational semantics for the Petri box calculus, in *CONCUR '94*, Lecture Notes in Computer Science, vol. 836, Springer-Verlag, Berlin, 1994, 210-225.
4. D. Janssens, M.Lens and G.Rozenberg, Computation Graphs for Actor Grammars, *Journal of Computer and System Sciences*, **46** (1993), 60-90.
5. D. Janssens, ESM Systems and the Composition of Their Computations, in *Graph-Transformations in Computer Science*, Lecture Notes in Computer Science, vol. 776, Springer-Verlag, Berlin, 1994, 203-217.
6. D. Janssens and T. Mens, Abstract semantics for ESM systems, *Fundamenta Informaticae*, to appear (also available as: UIA report 95-04, dept. of Math. and Comp. Sci., University of Antwerp).
7. R. Milner, *A Calculus of Communicating Systems*, Lecture Notes in Computer Science, vol. 92, Springer-Verlag, Berlin, 1980.
8. W. Reisig, *Petri Nets*, EATCS Monographs on Theoretical Computer Science, Springer-Verlag, Berlin, 1985.

Formal Relationship between Graph Grammars and Petri Nets *

Martin Korff and Leila Ribeiro

TU Berlin

Abstract. The main aim of this paper is to analyze the similarities and differences between graph grammars and Petri nets. Particularly we compare the high-level versions of nets and grammars, namely AHL nets and attributed graph grammars and the corresponding flattening constructions to PT nets and labeled graph grammars, and show that these flattenings are compatible (syntactically and semantically) with the translations of nets to grammars. Thus Petri nets are equivalent to their graph grammar translations. Due to the fact that graphs correspond to sets of tokens, Petri nets can syntactically be considered as special (simple) graph grammars. In their basic semantics however this is not precisely true: tokens without individuality are in contrast to corresponding vertices showing individuality. (Reachability and derivation) Trees have been chosen as a common semantical domain.

1 Introduction

Petri nets have been used since the early 70's as a formalism to describe concurrent systems. The main reasons for the success of Petri nets is that they rely on a simple concept of states and transitions and that they provide a graphical representation of the system, what makes the understanding of the model and its behavior easier even for non specialists. Besides the great acceptance in the industry, there is a solid theory of concurrency behind Petri nets. Although most of the theory of nets is developed for very basic net models like condition/event and place/transition nets, these are not suitable for the complete description of large systems. The main reason for this is that Petri nets stress on the description of the flow of informations (tokens) in a system, and not on the representation of the informations themselves, that this representation has then to be coded graphically using places and/or transitions leading to very large and complex nets. This problem gave raise to definition of high-level nets: nets in which tokens have some internal structure (see [JR91] for a survey). For example, algebraic high-level (AHL) nets [EGH92, PER93] extend the classical place/transition (PT) nets by their ability to deal with tokens carrying algebraic data type structure. Formally, the relationship of AHL nets to the well-investigated world of PT nets is provided by a corresponding flattening construction [EPR94].

Graph grammars have originated from Chomsky grammars. Using graph grammars for (concurrent) system specifications means to provide a set of rules by which

---

* This paper was partially supported by a CNPq-grant for Leila Ribeiro and by the ESPRIT Basic Research working group # 7183 "Computing by Graph Transformation (COMPUGRAPH II)".

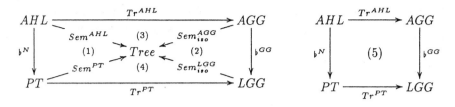

Fig. 1. Compatibility Results

a given initial object, i.e. a (labeled) graph, may be transformed. In fact there is a long series of works concerned about the relation between Petri nets and graph grammars. [Wil79, Kre81, Rei81, Sch93]. See [Sch94] for a survey. In this paper we follow the 'minimal way' of modeling Petri nets by algebraic graph grammars as given in [CEL+94a] (see diagram (4))—although we focus the single pushout approach to graph grammars [Löw93]. Summarizing Petri nets appear as a very simple kind of graph grammars. In this paper we show that this does not only hold for simple Petri (P/T) nets but also for the (high-level) variants: A (high-level) Petri net has

(1) no direct references between tokens (no edges). Using high-level tokens still does not allow to directly relate tokens. That means that for example a database would have to be represented in an AHL net by a single token carrying all the structure of the database. This is not adequate because then only one access to this database can be made at each time (tokens can't be shared). Of course one could code references into data types and let the database be distributed into many tokens having cross references coded into their internal structure. But then it would be very difficult to keep the integrity of the database and assure uniqueness of references during the execution of transitions. Moreover, this kind of coding of references into data types leads very quickly to a high complexity of the model.

(2) no individuality of tokens. It is a basic assumption of 'classical' Petri nets [Rei85] that the tokens don't have individuality, i.e. the tokens that are on one place are indistinguishable. As a consequence, at the semantical level of Petri nets we usually only deal with the number of tokens that are in each place at a time (markings). Contrastingly, for graph grammar derivations, it makes a difference if one vertex or another is chosen by a rule (via a match).

(3) no read-only access of tokens. Each time a token is accessed by a transition this token must be consumed. Similarly, a graph rule may access vertices and edges when it is applied to a graph. Such a vertex may either be deleted or, contrastingly, preserved. Preservation is essential in order to connect (glue) new elements to the already existing structure.

Analogously to the case of nets, using labeled graphs leads to complex models of states. A higher level of representation is obtained in graph grammars by using attributed graphs [LKW93], that are graphs provided with algebraic data types (attributes). This allows to use variables (and terms) in the rules and therefore reduces considerably the number of rules that are necessary to describe a system. As a second result of this paper, we give a flattening construction that transforms an attributed graph grammar into a labeled graph grammar (i.e. graph grammars using attributed resp. labeled graphs). In Section 3 we show that this construction is compatible with the semantics of graph grammars in terms of derivations (see diagram (2)).

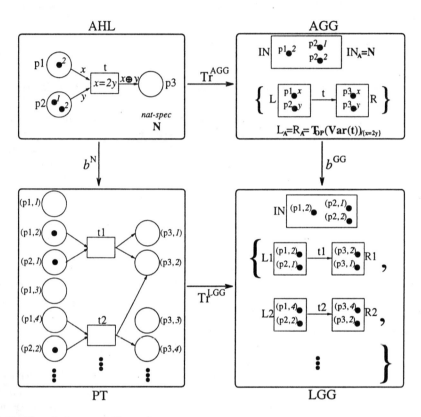

Fig. 2. Translations and Flattenings

The relationship between AHL nets and attributed graph grammars is analogous to the one of PT nets and labeled graph grammars: Although AHL nets are a higher-level formalism to describe systems, their translation to attributed graph grammars yields again grammars over graphs without edges. In Sect. 4 we show that this translation is compatible with semantics (see diagram (3)).

Finally we show that flattenings and translations are also compatible, i.e. the (syntax) square (5) commutes (up to isomorphism).

Figure 2 shows an example of the translation and flattening constructions. Each component, as well as each construction, will be explained in corresponding sections of this paper.

Basic Notions

We assume that the reader is familiar with basic notions of Petri nets [Rei85], graph grammars [EKL91] and algebraic specifications [EM85]. Besides we introduce the following conventions.

Functions are partial by default. For a set X we denote by X^* the finite sequences over X. X^ω denotes the infinite sequences, and $X^\infty = X^* \cup X^\omega$. The <u>length</u> of a sequence $\sigma \in X^\infty$ is denoted by $|\sigma| \in \mathbb{N} \cup \{\omega\}$. σ may be considered as a partial function $\sigma : \mathbb{N} \to X$ with domain $dom(\sigma) = \{i \in \mathbb{N} | i \leq |\sigma|\}$. Function application shall be denoted by $\sigma[i]$ or σ_i. $\rho \in X^\infty$ is a <u>prefix</u> of σ, written $\rho \leq \sigma$, if $|\rho| \leq |\sigma|$ and $\rho(i) = \sigma(i)$ for all $i \in dom(\rho)$. For a total function $f : X \to Y$, $f^* : X^* \to Y^*$ denotes its corresponding free extension. The append operation on sequences is denoted by $\bullet$. A tree T (over X) is a subset of X^∞ where $\lambda \in T$ and T is closed under prefixes. A morphism $f : T1 \to T2$ is a monotonic function that satisfies $f(\lambda) = \lambda$. The set of all trees is denoted by **Tree**.

The free abelian group $A^\oplus$ over a set A is given by the set of all functions $\underline{a} : A \to \mathbb{Z}$ with finite support (which means that the set $\{a \in A | \underline{a}(a) \neq 0\}$ is finite). The elements $\underline{a}$ in $A^\oplus$ can be considered as linear combinations $\underline{a} = \sum_{i=1}^{n} c_i a_i$ where $a_i \in A, c_i = \underline{a}(a_i)$. The zero element, sum and difference operators of abelian groups are represented by $\underline{0}$, $\oplus$ and $\ominus$ resp. The total ordering $\leq$ on $\mathbb{Z}$ induces a partial order $\leq_{ab}$ on $A^\oplus$ defined for $\underline{a}, \underline{b} \in A^\oplus$ by $\underline{a} \leq_{ab} \underline{b}$ if $\underline{a}(a) \leq \underline{b}(a)$ for all $a \in A$. A morphism between free abelian groups $f^\oplus : A^\oplus \to B^\oplus$ is a homomorphic function induced by $f : A \to B$.

Alg(Sig) denotes the category of all Sig-algebras and Sig-homomorphisms. **GAlg** is the category of generalized algebras [EBO91, PER93]. The objects of this category are algebras w.r.t. some algebraic specification, and the morphisms are composed of a specification morphism and a homomorphism. U denotes the forgetful functor from **Alg(Sig)** (or **GAlg**) to **Set** yielding the disjoint union of carrier sets (and homomorphisms).

2 Petri Nets

In this paper we use the definition of PT nets based on abelian groups [EGH92].[2] In this way we obtain a tight analogy to the AHL nets in the sense of [EPR94, RP94] which will be presented afterwards. Despite the more general definition, we will emphasize on restricted classes of positive (PT and AHL) nets, i.e. nets that have (i) positive pre- and post-conditions (as this is the standard case of place/transition nets), (ii) no isomorphic transitions and (iii) no isolated transition (that are transitions t where $pre(t) = post(t) = \underline{0}$). Note that if the pre- and post-conditions of all transitions are positive, only positive markings will be reachable (by the definition of enabling and because the initial marking is always positive).[3] This restriction tributes to translating nets to grammars in Sect. 4.

---

[2] In Petri nets terminology it is usual to call 'nets' together with an initial marking as 'net systems'. In this paper we use these two names as synonyms.

[3] Condition (i) assures that positive PT nets can also be seen as PT nets in the sense of [MM90].

Definition 2.1 A <u>place/transition (PT) net</u> is a tuple $N = (P, T, pre, post, M, m)$ where P, T and M are sets of places, transitions and tokens respectively, $pre, post: T \to P^{\oplus}$ are functions representing the pre- and post-conditions of transitions to switch, and $m: M \to P$ represents the (initial) marking of the net N. The net N is called <u>positive</u> if for all $t \in T$ (i) $\underline{0} \leq_{ab} pre(t), post(t)$, (ii) for all $t1, t2 \in T$, $(pre(t1) = pre(t2)$ and $post(t1) = post(t2))$ implies $t1 = t2$, (iii) $pre(t) \neq \underline{0}$ or $post(t) \neq \underline{0}$. The set of all positive PT nets is denoted by **PT**. ∗

Definition 2.2 A <u>marking</u> of a net N is an element of $P^{\oplus}$.[4] A transition $t \in T$ is <u>enabled by a marking</u> $m1$, written $m1[t\rangle$ if $pre(t) \leq_{ab} m1$. In this case t may <u>switch</u> and the <u>successor marking</u> $m2$ is given by $m2 = m1 \ominus pre(t) \oplus post(t)$. $m1[t\rangle m2$ is called a <u>step</u> of N. The set of all steps of N is denoted by $PTSteps_N$. A sequence $\sigma \in PT\overline{Steps}_N^{\infty}$ is called a <u>switching sequence</u> if $\sigma = \lambda$ or $\sigma \neq \lambda$ and $m_1 = ab(m)$, $\sigma_i = m_i[t_i\rangle m_{i+1}$ for all $i \in dom(\sigma)$. The set of all switching sequences of N, denoted by $Sem^{PT}(N) = SwiPTSteps_N^{\infty}$, is the (tree) <u>semantics</u> of N. ∗

Example 1. A PT net PT is shown in the left lower corner of Fig. 2. Places, transitions and tokens of PT as drawn as usual, i.e. as circles, squares and black dots resp. The functions pre and $post$ are indicated by the incoming and outcoming arrows of each transition (multiplicities are indicated by labeling arcs with numbers, label 1 is omitted). The function m assigns to each black dot the place it lies on. ∗

Algebraic high-level nets (AHL nets) combine PT nets with algebraic specifications [EGH92, PER93, RP94], thus joining suitable formalisms to specify data flow and data types respectively. Similar approaches can be found in [Jen92, Rei91, GL81].

Definition 2.3 An <u>algebraic high-level (AHL) net</u> is a tuple $AN = (P, T, pre, post, M, m, X, SPEC, A, cond)$ where P, T and M are sets of places, transitions and tokens respectively, $SPEC = (S, OP, E)$ is an algebraic specification [EM85], A is a generalized $SPEC$-algebra, $X = (X_s)_{s \in S}$ is a family of sets of variables, $m: M \to (U(A) \times P)$ represents the (initial) marking of the net N, $cond: T \to \mathcal{P}(Eqns((S, OP), X))$ $pre, post: T \to U((T_{OP}(X)) \times P)^{\oplus}$ are functions representing the algebraic, and the pre- and post-conditions of transitions to switch. The net N is called <u>positive</u> if (i) $\underline{0} \leq_{ab} pre(t), post(t)$, (ii) for all $t1, t2 \in T$, $(pre(t1) = pre(t2)$, $post(t1) = post(t2)$ and $T_{OP}(Var(t1))_{/cond(t1)} = T_{OP}(Var(t2))_{/cond(t2)})$ implies $t1 = t2$, (iii) $pre(t) \neq \underline{0}$ or $post(t) \neq \underline{0}$ and (iv) $Var(t)_s \cap Var(t)_{s'} = \emptyset$, for all $t \in T_{AN}, s \in S_{AN}$.[5] The set of all positive AHL nets is denoted by **AHL**. ∗

Definition 2.4 A <u>marking</u> of an AHL net is an element of $(U(A) \times P)^{\oplus}$. A transition t is <u>enabled under an assignment</u> $asg: Var(t) \to A$ <u>by a marking</u> m [6], written $m[(t, asg)\rangle$, if (i) the necessary tokens are present, i.e. $\overline{asg}^{\oplus}(pre(t)) \leq_{ab} m$

---
[4] The initial marking $m: M \to P$ can be uniquely translated to a marking $ab(m) \in P^{\oplus}$ by $ab(m) = \sum_{x \in M} m(x)$.
[5] Condition (iv) is necessary in order to obtain a suitable category of AHL nets in which morphisms are compatible with switching of transitions, yielding a flattening functor to the category of PT nets.
[6] $Var(t)$ is the S-indexed family of variables occurring in $pre(t), post(t)$ and/or $cond(t)$. Thus for all $s \in S$, $Var(t)_s \subseteq X_s$. $Var(t)$ is called the family of local variables for the transition t.

and [7] (ii) all equations in $cond(t)$ do hold under asg, i.e. $A \models \overline{asg}^{\oplus}(cond(t))$. An enabled transition t can **switch**; then the **successor marking** m' defined by $m' = m \ominus \overline{asg}^{\oplus}(pre(t)) \oplus \overline{asg}^{\oplus}(post(t))$, giving raise to a **step** $m[(t, asg)\rangle m'$. The set of all steps of an AHL net AN is denoted by $AHLSteps_{AN}$. A sequence $\sigma \in AHLSteps_{AN}^{\infty}$ is called **switching sequence** iff $\sigma = \lambda$ or $\sigma \neq \lambda$ and $m_1 = ab(m)$, $\sigma_i = m_i[(t_i, asg_i)\rangle m_{i+1}$ for all $i \in dom(\sigma)$. The set of all switching sequences of N, denoted by $Sem^{AHL}(AN) = SwiAHLSteps_{AN}^{\infty}$, is the (tree) **semantics** of AN. ∗

Example 2. An AHL net AHL is shown in the left upper corner of Fig. 2. Places, transitions and tokens are represented as in PT nets. The (algebraic) specification is just indicated by its name ($nat-spec$ for the specification of natural numbers) as well as the algebra (IN for the algebra of natural numbers). The numbers connected to the tokens describe their values according to the algebra IN. The algebraic condition of t, namely $x = 2y$, indicates e.g. that, given the initial marking, only the switching using the tokens $\bullet^2$ of $p1$ and $\bullet^1$ of $p2$ is possible (in this case, $x = 2$ and $y = 1$, what makes $x = 2y$ true). The result of such a switching would put two tokens ($\bullet^2$ and $\bullet^1$) on $p3$. ∗

In [EPR94] a flattening construction was introduced that associates to each AHL net a corresponding PT net which is compatible with the switching of transitions.

Definition 2.5 The **flattening** function $\flat^N: \mathbf{AHL} \to \mathbf{PT}$ is defined, for all AHL nets $AN = (P, T, pre, post, M, m, X, SPEC, A, cond)$ by $\flat^N(AN) = (P', T', pre', post', M', m')$ where $T' = \{(t, asg) | t \in T, asg: Var(t) \to A$ and $A \models cond(t)$ under $asg\}$, $P' = U(A) \times P$, $pre', post': T' \to P'^{\oplus}$, $pre'(t, asg) = \overline{asg}^{\oplus}(pre(t))$, $post'(t, asg) = \overline{asg}^{\oplus}(post(t))$, $M' = M$ and $m' = m$. ∗

Example 3. The flattening of AHL in Fig. 2 generates the PT net PT containing an infinite number of places ($\{p1, p2, p3\} \times$ IN) and transitions (there is one transition for each possible assignment of values in IN to x and y that makes $x = 2y$ true). ∗

This flattening function is compatible with the semantics.

Theorem 2.6 The flattening function $\flat^N$ is compatible with the semantics of nets, i.e. $Sem^{AHL} = \flat^N \circ Sem^{PT}$. ∗

Proof: Let $AN = (P, T, pre, post, M, m, X, SPEC, A, cond) \in \mathbf{AHL}$, $N' = \flat^N(AN) = (P', T', pre', post', M', m')$. We prove that for each step (switching of transition) of AN, there is a corresponding step of N', and vice-versa.
(1) $\forall s1 = m1[t, asg\rangle m2 \in AHLSteps_{AN}$ implies $s1 \in PTSteps_{N'}$: As $m1, m2 \in (U(A) \times P)^{\oplus}$, then $m1, m2 \in P'$. As $t \in T$, $asg: Var(t) \to A$ and $A \models cond(t)$ under asg, then $t' = (t, asg) \in T'$. Thus we just have to show that t' is enabled under $m1$ and that the successor marking when it switches is $m2$. By definition, $pre'(t, asg) = \overline{asg}^{\oplus}(pre(t))$ and $post'(t, asg) = \overline{asg}^{\oplus}(post(t))$. As these were exactly the definitions used in the definition of enabling and successor marking for AHL-nets (Def. 2.4), t' is enabled under $m1$ and the corresponding successor marking is $m2$. Thus we conclude that $m1[t'\rangle m2 \in PTSteps_{N'}$.

---
[7] $\overline{asg}^{\oplus}: (U(T_{OP}(Var(t)) \times P)^{\oplus} \to (U(A) \times P)^{\oplus}$ is the free extension of $asg: Var(t) \to A$.

(2) $\forall s1 = m1[t, asg\rangle m2 \in PTSteps_{N'}$ implies $s1 \in AHLSteps_{AN}$: Anal. to (1).

(3) By construction of $SwiAHLSteps_{AN}^{\infty}$ and $SwiPTSteps_{N'}^{\infty}$, based on single steps, and the fact that $m' = m$, we conclude that $SwiAHLSteps_{AN}^{\infty} = SwiPTSteps_{N'}^{\infty}$. □

3 Graph Grammars

According to [Löw93], a graph transformation is a single pushout in a category of graphs and partial morphisms. The graphs we use here are labeled and attributed on vertices only [LKW93].

Definition 3.1 An <u>Lgraph</u> $G = (G_V, G_E, G_L, s^G, t^G, l^G)$ consists of a set of <u>vertices</u> G_V, a set of <u>edges</u> G_E, a set of <u>labels</u> G_L together with <u>source</u>, <u>target</u>, and label functions $s^G, t^G: G_E \to G_V$ and $l^G: G_V \to G_L$ respectively. An <u>Lmorphism</u> $f: G \to H$ is a tuple $f = (f_V, f_E)$ consisting of functions $f_V: G_V \to H_V$ and $f_E: G_E \to H_E$ such that $s^H(f_E(e)) = f_V(s^G(e))$ and $t^H(f_E(e)) = f_V(t^G(e))$ and $l^G(v) = l^H(f_V(v))$ for all $e \in dom(f_E)$ and all $v \in dom(f_V)$.

An <u>Agraph</u> $G = (G_V, G_E, G_L, G_A, s^G, t^G, l^G, a^G)$ consists of an Lgraph $G = (G_V, G_E, G_L, s^G, t^G, l^G)$ together with an algebra $G_A \in \mathbf{Alg(Sig)}$ and an attribution function $a^G: G_V \to U(G_A)$. An <u>Amorphism</u> $f: G \to H$ between Agraphs G and H is a tuple $f = (f_V, f_E, f_A)$ consisting of an Lmorphism (f_V, f_E) and a homomorphism $f_A \in \mathbf{Alg(Sig)}$ such that $U(f_A)(a^G(v)) = a^H(f_V(v))$ for all $v \in dom(f_V)$.

∗

A main result is that this leads to cocomplete categories [Löw93, LKW93] on which the following definition of grammars and derivations fundamentally rely.

Theorem 3.2 Lgraphs and Lmorphisms and Agraphs and Amorphisms form categories denoted by **LG** and **AG** respectively.[8] $L: \mathbf{AG} \to \mathbf{LG}$ denotes the obvious functor forgetting the A-component. By fixing an algebra Alg, we obtain subcategories $\mathbf{AG^{Alg}}$ of **AG** containing as objects all $A \in \mathbf{AG}$ such that $A_A = Alg$ and all morphisms $f: A \to B \in \mathbf{AG}$ where f_A is the identity on Alg. The categories **LG**, **AG**, and $\mathbf{AG^{Alg}}$ are (finitely) cocomplete. ∗

Definition 3.3 An <u>Lrule</u> $r: L \to R$ is an injective, non-isomorphic Lmorphism.[9] An <u>Arule</u> $r: L \to R$ is an Amorphism such that $L(r)$ is an Lrule and r_A is the identity on a quotient of the term algebra $T_{OP}(X)$ with variables in X. An <u>Lgraph grammar</u>, short LGG, $GG = (P, G)$ consists of a set of Lrules P and an Lgraph G in **LG**. An <u>Agraph grammar</u>, short AGG, $GG' = (P', G')$ consists of a a countable set of Arules P', and a finite Agraph G' in **AG**. ∗

The intuition of a rule $r: L \to R$ is that all objects in $L - DOM(r)$ shall be deleted, all objects in $R - r(L)$ shall be added and $DOM(r)$ resp. $r(L)$ provides the gluing context. Sometimes the application of the rule shall be constrained w.r.t. the

---
[8] Agraphs implicitly depends on Sig and L whereas Lgraphs depend only on L.
[9] i.e. a rule always deletes or generates something.

attribution, like e.g. $eq(x, 4) = True$. In this case we 'specify' the attribution algebra of the rule equationally, i.e. the required equations are syntactically added to the rule implying that the terms annotating the rule shall be interpreted into the corresponding quotient term algebra.

$$\begin{array}{ccc} L_d & \xrightarrow{r_d} & R_d \\ m_d \downarrow & & \downarrow m_d^\bullet \\ IN_d & \xrightarrow{r_d^\bullet} & OUT_d \end{array}$$

Definition 3.4 An <u>Lmatch</u> $m: L \to IN$ of r into an Lgraph IN is a total Lmorphism where m_V is injective. An <u>Amatch</u> $m: L \to IN$ of r into an Agraph IN is an Amorphism where $L(m)$ is an Lmatch. Given an LGG $GG = (P.G)$, an <u>Lderivation step</u> d of an Lgraph IN_d, called <u>input</u> of d, with Lrule $r_d \in P$ at Lmatch m_d is a pushout of m_d and r_d in **LG** as on the left. It is denoted by $IN_d[d\rangle OUT_d$. OUT_d is called the <u>output graph</u>. The morphisms $r_d^\bullet: IN_d \to OUT_d$ and $m_d^\bullet: R_d \to OUT_d$ are called the <u>co-rule</u> and the <u>co-match</u> respectively. The class of all Lderivation steps in GG is denoted by $LSteps_{GG}$.

A sequence $\sigma \in LSteps_{GG}^\infty$ is called a <u>sequential derivation of GG</u> also denoted by $G[\sigma\rangle_{GG}$, if either $\sigma = \lambda$ or $\sigma \neq \lambda$ and $IN_{\sigma[1]} = G$, $OUT_{\sigma[i]} = IN_{\sigma[i+1]}$ for all $i \in dom(\sigma)$. If σ is finite we additionally define the output graph as the output graph of the last step, i.e. $OUT_\sigma = OUT_{\sigma[|\sigma|]}$. A finite sequential derivation $\sigma \in LSteps_{GG}^*$ is then also denoted by $IN_\sigma[\sigma\rangle OUT_\sigma$. The sets of finite, infinite, and arbitrary sequential derivations of GG are denoted by $LSDer_{GG}^*$, $LSDer_{GG}^\omega$, $LSDer_{GG}^\infty$ respectively. The <u>semantics</u> $Sem^{LGG}(GG)$ is the tree $T = LSDer_{GG}^\infty$. Given an AGG $GG' = (P', G')$, an <u>Aderivation step</u> d' of an Agraph $IN_{d'}$ where $IN_{d'A} = G'_A$, with Arule $r_{d'} \in P'$ at Amatch $m_{d'}$ is a pushout of $m_{d'}$ and $r_{d'}$ in **AG** such that $r_{d'A}^\bullet = id_{G'_A}$. The other notions are defined analogously to the ones for LGras. ∗

Example 4. The upper and lower right corners of Fig. 2 show an attributed (AGG) and a labeled (LGG) graph grammar resp. Each graph is drawn within a box. Vertices are drawn as black dots, labels at the upper left and attributes at the upper right sides of vertices (clearly LGG has no attributes). There are no edges. The label sets of AGG and LGG are P and $P \times U(A)$ respectively, where P are the places and A is the algebra of AHL. Consider the rule $t: L \to R$ of AGG. The left-hand side consists of two vertices having as labels $p1$ and $p2$ and as attributes x and y (x and y are elements of the carrier sets of the quotient term algebra $T_{OP}(Var(t))_{/\{x=2y\}}$). The only possible match m of L into IN (that have the algebra of natural numbers as attribute algebra) is $m({}^{p1}\!\bullet^x) = {}^{p1}\!\bullet^2$ and $m({}^{p2}\!\bullet^y) = {}^{p2}\!\bullet^1$ because labels must be identically mapped and attributes must be homomorphically mapped ($x \mapsto 2$ and $y \mapsto 2$ does not yield a homomorphism). The result of applying t at match m would be a graph consisting of two vertices: ${}^{p3}\!\bullet^2$ and ${}^{p3}\!\bullet^1$. ∗

Similarly to the flattening construction of AHL nets, we define a flattening construction from AGGs to LGGs. But first we define a functor from $\mathbf{AG^A}$ to $\mathbf{LG}$ relating attributed and labeled graphs and their corresponding morphisms (rules). The idea of this functor is to code the algebra of attributes into labels. Based on this functor, we define the flattening of attributed graph grammars into labeled graph grammars.

Definition 3.5 Given an algebra A, the <u>flattening functor</u> $\flat_A^G: \mathbf{AG^A} \to \mathbf{LG}$ is defined for each graph $G \in \mathbf{AG^A}$ and each morphism $f: G \to H \in \mathbf{AG^A}$ by $\flat_A^G(G) = (G_V, G_E, L_L, s^G, t^G, l^L)$ where $L_L = G_L \times U(G_A)$ and $l^L = (l^G, a^G)$ and $\flat_A^G(f) = (f_V, f_E, l_L)$ where $l_L = (f_L, U(f_A))$.

Let S be a set, $G1, G2$ be LGraphs and $f: G1 \to G2$ be an Lmorphism. Then for $i = 1, 2$ we define $Gi \otimes S = (G_{iV} \times S, G_{iE}, G_{iL}, (s^{Gi}, id_S), (t_{Gi}, id_S), l^{GiS})$ where for all $(v, s) \in Gi \times S, l^{GiS}(v, s) = l^{Gi}(v); f \otimes S = ((f_V, id_S), f_E, f_L): G1 \otimes S \to G2 \otimes S$. Obviously, $Gi \otimes S$ are LGraphs and $f \otimes S$ is an Lmorphism.

Let $GG = (P, G)$ be an attributed graph grammar and $A = G_A$ be the attributing algebra of G. Then the <u>flattened LGG</u> $\flat^{GG}(GG) = (P', G')$ of AG is defined by $G' = \flat_A^G(G)$ and $P' = \{\flat_A^G(r_d^\bullet) \otimes \{d\} | \exists d \in ASteps_{AG}$ such that $r_{dA}^\bullet = id_A, L(r_d^\bullet) = L(r_d)\}$.

∗

Example 5. The (labeled) grammar LGG in Fig. 2 is (up to isomorphism) the flattening of the attributed grammar AGG. Formally, each vertex • of a rule in LGG is represented by a pair $(\bullet, d)$ where d is the derivation of AGG that gave raise to the corresponding rule of LGG (these pairs are obtained by the $\otimes$ construction and are, for readability reasons, not explicitly represented in the picture). The grammar LGG contains one rule for each possible homomorphism from L_A to $\mathbb{N}$.

∗

Proposition 3.6 $\flat_A^G$ and $\flat^{GG}$ are well-defined. Moreover $\flat_A^G$ is an isomorphism between categories.

∗

Proof: Let $L = G_L \times U(G_A)$.

(1a) By construction $\flat_A^G(G)$ is an Lgraph. Now we have to show that for each morphism $f: G \to H \in \mathbf{AG^A}$, $\flat_A^G(f)$ is a morphism of Lgraphs, i.e. the component l_L must be the identity and f must be compatible with source, target and label functions. By definition, $l_L = (f_L, U(f_A))$. f_L and f_A are the identities on G_L and G_A resp. because f is a $\mathbf{AG^A}$-morphism. As U is a functor, $U(f_A)$ is the identity of $U(G_A)$. Thus we conclude that l_L is the identity of $G_L \times U(G_A)$. The compatibility with the source, target and label functions follow from the fact that f is an $\mathbf{AG^A}$-morphism.

(1b) $\flat_A^G$ obviously preserves identities. The preservation of composition follows from the fact that the composition of identities is again an identity.

(1c) Let $F: \mathbf{LG} \to \mathbf{AG^A}$ be defined for all $I \in \mathbf{LG}, g: I \to J \in \mathbf{LG}$ as follows: $F(I) = (I_V, I_E, L, A, s^I, t^I, l^A, a^A)$ where for all $v \in I_V$ such that $l^I(v) = (l, a)$, $l^A(v) = l$ and $a^A(v) = a$; $F(g) = (g_V, g_E, id_L, id_A)$. Analogously to $\flat_A^G$, F is also a functor. The facts that $\flat_A^G \circ F = id_{\mathbf{LG}}$ and $F \circ \flat_A^G = id_{\mathbf{AG^A}}$ follow directly from the definition of these functors.

(2) We have to show that $\flat^{GG}(GG)$ is an LGG grammar, that is the initial graph and all the rules must have the same label set L. Let $L = G_L \times U(G_A)$. Then, by construction G' is labeled over L. As all the rules in P' must stem from morphisms in $\mathbf{AG^A}$. Thus these rules have also L as label set. □

Now we show the compatibility of the flattening of graph grammars with their semantics.

Theorem 3.7 Let $GG = (P, G)$ be an AGG and $\flat^{GG}(GG) = (P', G')$ be its flattened LGG. Then $Sem^{AGG}(GG) \cong Sem^{LGG} \circ \flat^{GG}(GG)$. ∗

Proof: As $G' = \flat^G_{G_A}$ (by definition of $\flat^{GG}$) and the fact that derivations in $LSDer^\infty_{\flat GG(GG)}$ and $ASDer^\infty_{GG}$ are constructed in the same way, it suffices to show that $d \in ASteps_{GG} \iff d' \in LSteps_{\flat GG(GG)}$ where d' is the flattening of the derivation step d as it will be described below.

1. Assume $d \in ASteps_{GG}$ as illustrated in (1). Then we have to find corresponding rule and match such that the diagram (2) becomes a derivation $d' \in LSteps_{\flat GG(GG)}$.

$$
\begin{array}{ccc}
L_d \xrightarrow{r_d} R & & \flat^G_A(L^A) \otimes \{dr\} \xrightarrow{\flat^G_A(r^A) \otimes \{dr\}} \flat^G_A(R^A) \otimes \{dr\} \\
m_d \downarrow \quad (1) \quad \downarrow m^\bullet_d & \flat^G_A(m^A) \downarrow \quad (2) \quad \downarrow \flat^G_A(m^{A\bullet}) \\
IN_d \xrightarrow{r^\bullet_d} OUT_d & & \flat^G_A(IN_d) \xrightarrow{\flat^G_A(r^\bullet_d)} \flat^G_A(OUT_d)
\end{array}
$$

   (a) Rule: Let $m = (id_V, id_E, id_L, m_{dA})$ where $L_d = (V, E, L, A)$. Then m is a match for r_d (it is total and injective). Therefore the derivation dr illustrated in (3) where $r^A = (r_{dV}, r_{dE}, r_{dL}, id_{G_A})$ is also in $ASteps_{GG}$ (as $d \in ASteps_{GG}$, the image of m_{dA} must be the algebra G_A). As $L(m) = id_{L(L_d)}$, $L(r^A) = L(r_d)$ and thus by the definition of $\flat^{GG}$, $\flat^G_A(r^A) \otimes \{dr\} \in P'$.

   (b) Match: Let $\overline{m^A} = (m_{dV}, m_{dE}, m_{dL}, id_A): L^A \to IN_d$. Then (4) commutes. Let $\flat^G_A(\overline{m^A}): \flat^G_A(L^A) \otimes \{dr\} \to \flat^G_A(IN_d)$ be defined as $\flat^G_A(m^A) = (x, m_{dE}, (m_{dL}, id_A))$ where for all $(v, d) \in \flat^G_A(L^A) \otimes \{dr\}_V$, $x(v, d) = m_{dV}(v)$. As m_{dV} is injective and $\{dr\}$ is a singleton, x is also injective and thus $\flat^G_A(m^A)$ is an Lmatch.

   (c) Derivation step: As (1) and (3) are pushouts in **AG** and (4) commutes, (5) is also a pushout (decomposition of pushouts). As all morphisms in (5) have the identity on A as algebra component, they are also $\mathbf{AG^A}$-morphisms and (5) is obviously also pushout in $\mathbf{AG^A}$. Let $u1, u2$ be LG morphisms such that $\flat^G_A(m^A) \circ u1 = \overline{\flat^G_A(m^A)}$ and $\flat^G_A(m^{A\bullet}) \circ u2 = \overline{\flat^G_A(m^{A\bullet})}$. Then we have that (6) is a pushout (because (5) is a pushout and $\flat^G_A$ is iso) and (7) is a pushout because $u1$ and $u2$ are isomorphisms and $\flat^G_A(r^A) \cong \flat^G_A(r^A) \otimes \{dr\}$ (by the definition of $\otimes$ and the fact that $\{dr\}$ is a singleton). Thus by composition of pushouts (2), called d', is also a pushout and a derivation in $LSteps_{\flat GG(GG)}$.

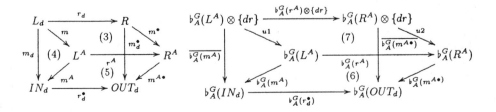

2. Assume $d' \in LSteps_{\flat GG}(GG)$ be a derivation step with rule $r_{d'}: \flat_A^G(L^A) \otimes \{dr\} \to \flat_A^G(R^A) \otimes \{dr\}$, match $\overline{m_{d'}}: \flat_A^G(L^A) \otimes \{dr\} \to IN_{d'}$ and output graph $OUT_{d'}$.

   (a) Rule: As $r_{d'} \in P'$ then by construction of P', $r_{dr} \in P$.

   (b) Match: Let $m = F(\flat_A^G(m_{d'})) \circ m_{dr}$ (see Prop. 3.6 for the definition of the functor F). As $m_{d'} \cong \overline{m_{d'}}$, $\overline{m_{d'}}$ and m_{dr} are total and injective (they are matches), and F is an iso, $m: L^A \to F(IN_{d'})$ is a match.

   (c) Derivation step: Similarly to case 1.c, by using decomposition of pushouts it can be shown that the pushout object OUT of rule r_{dr} and match m is isomorphic to $F(OUT_{d'})$, and thus that the corresponding derivation step is in $ASteps_{GG}$.

□

In contrast to the semantics of Petri nets, the sequential derivations presented above are quite concrete: if there are two (isomorphic) choices of matches of a rule L to a graph IN, this gives raise to two different sequential derivations. In the case of Petri nets, the choice of which tokens are consumed is not considered in the semantics. Thus in order to compare graph grammars and Petri nets we have to define a different semantics either for Petri nets or for graph grammars. In this paper we define 'isomorphic derivations' and use them as a basis for a semantic of graph grammars in which matches and derivations are just considered up to isomorphism.

Definition 3.8 Let $GG = (P, G)$ be an LGG. Then we define a congruence relation $\sim$ on $LSteps_{GG}$: for all $s, s' \in LSteps_{GG}$, $s \sim s'$ if $r_s = r_{s'}$ and $m_s \cong m_{s'}$. Let $isoLSteps_{GG} = LSteps_{GG}/_\approx$ where $\approx$ is the equivalence relation induced by $\sim$. For all $[d] \in isoLSteps_{GG}$, let $IN_{[d]} = \{IN_s | s \in [d]\}$ and $OUT_{[d]} = \{OUT_s | s \in [d]\}$. The iso-semantics of GG are all sequential derivations of isoSteps of GG, denoted by $Sem_{iso}^{LGG}(GG) = isoLSDer_{GG}^\infty$, i.e. the set (tree) consisting of all $\sigma \in isoLSteps_{GG}^\infty$ such that $\sigma = \lambda$ or $\sigma \neq \lambda$ and $G \in IN_{\sigma[1]}$, $OUT_{\sigma[i]} = IN_{\sigma[i+1]}$ for all $i \in dom(\sigma)$. Analogously these notions are defined for attributed graph grammars. ∗

Note that if $s \sim s'$ then $r_s = r_{s'}$, $IN_s \cong IN_{s'}$ and $OUT_s \cong OUT_{s'}$. Now we define the relationship between these two graph grammar's semantics.

Proposition 3.9 For all GG, there is a tree morphism $f: Sem^{LGG}(GG) \to Sem_{iso}^{LGG}(GG)$ defined by $f(\lambda) = \lambda$, $f(s \bullet \sigma) = [s] \bullet f(\sigma)$ for all $s \bullet \sigma \in LSDer_{GG}^\infty$. ∗

Proof: f is obviously a monotonic function and thus a tree morphism. □

Theorem 3.10 The iso-semantics of graph grammars is compatible with the flattening, i.e. $Sem_{iso}^{AGG}(GG) \cong Sem_{iso}^{LGG} \circ \flat^{GG}(GG)$. ∗

Proof: Let $i: Sem^{AGG}(GG) \to Sem^{LGG}(\flat^{GG}(GG))$ be the isomorphism described in Theorem 3.7, and $f: Sem^{LGG}(\flat^{GG}(GG)) \to Sem_{iso}^{LGG}(\flat^{GG}(GG))$ be the tree morphism described in Prop. 3.9. Then we define $u: Sem_{iso}^{AGG}(GG) \to Sem_{iso}^{LGG}(\flat^G(GG))$ as $u(\lambda) = \lambda$, for all $[s] \in isoASteps_{GG}$, $u([s]) = f \circ i(s)$ and for all $[s] \bullet \sigma \in isoASDer_{GG}^\infty$, $u([s] \bullet \sigma) = u([s]) \bullet u(\sigma)$. Based on the fact that i is an isomorphism and on the definition of f, it can easily be shown that u is also an isomorphism. □

4 Translating Petri Nets into Graph Grammars

The way in which we see the relationship between PT nets and graph grammars here is similar to the approach presented in [CEL+94a]: a PT net $N = (P, T, pre, post, M, m)$ is translated into an LGG $Tr^{LGG}(N) = (SR, G)$ over the label set $L = P$ (the labels of the LGG are the places of the PT net), $G = (M, \emptyset, P, \emptyset, \emptyset, m)$ is the initial graph, consisting only of vertices (in M) and the labeling function given by m. SR is a set of rules corresponding to the transitions in T, i.e. $SR = \{r_t : L_t \to R_t | t \in T, L_t$ and R_t are discrete graphs[10] labeled by places corresponding to $pre(t)$ and $post(t)$ resp.$\}$. An element of $P^\oplus$ can be considered as a function $p : P \to \mathbb{Z}$ with finite support. So the graph L_t corresponding to $pre(t)$ can be defined as $L_t = (V, \emptyset, P, \emptyset, \emptyset, l)$ where $V = \{(p, i) \in P \times \mathbb{Z} | 0 < i \leq pre(t)(p)\}$ and $l(p, i) = p$. The left-hand side of a rule $r_t \in SR$ represents the tokens that must be present for a transition to switch; the right-hand side of r_t represents those which are created by the switching of t, and r_t is the empty graph morphism. For an example, see lower part of Fig. 2 ($Tr^{LGG}: PT \to LGG$).

The translation in Def. 4.3 of AHL nets into AGGs correspondingly generalizes that of PT nets into LGGs: vertices become attributes (corresponding to structured tokens). The attribute algebra for a rule is obtained from the environment of the corresponding transition of the net.

As the translation of nets into graph grammars always results into discrete graphs (graphs without edges), in the following we consider only this kind of graphs, i.e. *LSteps* and *ASteps* are the sets of derivation steps involving discrete graphs only. Before we define the translation of AHL nets into AGGs, we define how markings of an AHL net are translated into attributed graphs and vice versa.

Definition 4.1 Let $m \in (U(A) \times P)^\oplus$ be a positive marking of an AHL net AN, $h : A \to A'$ be a homomorphism and $G = (G_V, \emptyset, L, A, \emptyset, \emptyset, l^G, a^G)$ be an attributed graph.

(1) The **translation** of m to a graph with attribute algebra A' is defined by $AG_h(m) = (AG_V, \emptyset, P, A', \emptyset, \emptyset, l^{AG}, a^{AG})$ where $AG_V = \{(a', p, i) \in U(A') \times P \times \mathbb{N} | 0 < i \leq m(a, p)$ and $(U(h))(a) = a'\}$, and for all $(a', p, i) \in AG_V$, $l^{AG}(a', p, i) = p$ and $a^{AG}(a', p, i) = a'$.

(2) The **translation** of G to a marking of AN is defined as follows $M(G) = \sum_{v \in G_V} (a^G(v), l^G(v))$. 　＊

Proposition 4.2 Let $m \in (U(A) \times P)^\oplus$ be a positive marking of an AHL net AN, G be a graph with attribute algebra A. Then $M(AG_{id_A}(m)) = m$ and $AG_{id_A}(M(G)) \cong G$. 　＊

Proof: (1) By definition of AG we have that $AG_{id_A}(m) = (AG_V, \emptyset, P, A, \emptyset, \emptyset, l^{AG}, a^{AG})$ where $AG_V = \{(a, p, i) | 0 < i \leq m(a, p), a \in U(A), p \in P\}$, $l^{AG}(a, p, i) = p$ and $a^{AG}(a, p, i) = a$. By definition of M, $M(AG_{id_A}(m)) = \sum_{(a,p,i) \in AG_V} (a^{AG}(a, p, i), l^{AG}(a, p, i)) = \sum_{(a,p,i) \in AG_V} (a, p) = m$.

(2) The set of vertices G_V can be partitioned into pairwise disjoint subsets $G_V^{a,p} = \{v | a^G(v) = a$ and $l^G(v) = p\}$. By the well ordering axiom of set theory we may

---

[10] Graphs without edges.

assume that the enumeration of vertices in $G_V^{a,p}$ is implementing the construction AG. Hence $u_V : G_V \to AG_{id_A}(M(G))_V$ is a bijection which, by construction, is compatible with labeling and attribute functions. □

Definition 4.3 Let $AN = (P, T, pre, post, M, m, X, (Sig, E), A, cond)$ be an AHL net. Then the <u>translated AGG</u> $Tr^{AGG}(AN) = TA$ is defined as $TA = (TA_P, TA_G)$ over Sig where $TA_G = AG_{id_A}(ab(m))$ and $TA_P = \{r_t : L_t \to R_t | t \in T, L_t = AG_{nat}(pre(t)), R_t = AG_{nat}(post(t))$ and $r_t = (\emptyset, \emptyset, id_P, id_{A'})$ where $nat : T_{OP}(Var(t)) \to T_{OP}(Var(t))_{/cond(t)} = A'$ is the natural homomorphism obtained by the factorization of $T_{OP}(Var(t))$ over $cond(t)\}$. ∗

Example 6. The translation of AHL into the attributed graph grammar AGG is shown in the upper part of Fig. 2. The initial marking $ini = (2, p1) \oplus (1, p2) \oplus (2, p2)$ is translated to the graph IN, where a vertex $^p\bullet^n$ corresponds formally to a triple $(n, p, 1)$ labeled with p and attributed with n. That is, places become labels and values become attributes. Transition t becomes a rule $t : L \to R$ having as attribute algebra the quotient term-algebra obtained by adding the equation $x = 2y$ to $nat - spec$. The vertex $^{p1}\bullet^x$ of L is formally described by the triple $([x], p1, 1)$ labeled with $p1$ and attributed with $[x]$, that is, with the equivalence class of x in $T_{OP}(Var(t))_{/\{x=2y\}}$. ∗

Proposition 4.4 Tr^{AGG} is well-defined. ∗

Proof: It is obvious by the construction of $Tr^{AGG}(AN)$ that TA_G is an attributed graph and that all rules in TA_P are attributed rules (the algebra and label components are always identities by definition). Thus $Tr^{AGG}(AN)$ is an AGG. □

The following theorem states the semantical compatibility of the translation of AHL nets into AGGs.

Theorem 4.5 Given an AHL net AN and its translated AGG $Tr^{AGG}(AN) = TA$. Then $Sem^{AHL}(AN) \cong Sem^{AGG}(TA)$. ∗

Proof: Let $AN = (P, T, pre, post, I, ini, X, (Sig, E), A, cond)$ and $TA = (TA_P, TA_G)$. Then we have to show that
 (1) $AG_{id_A}(ab(ini)) = TA_G$ and $M(TA_G) = ab(ini)$. (2) There is an isomorphism $u : AHLSteps_{AN} \to isoASteps_{Tr^{AGG}(AN)}$. (3) For all $s_i = (m_i[t_i, asg_i\rangle m'_i) \in AHLSteps_{AN}$, $i = 1..2$, $m'_1 = m2 \Leftrightarrow OUT_{u(s1)} = IN_{u(s2)}$.
 (1) Holds by definition of Tr^{AGG}.
 (2) Let $s = m[t, asg\rangle m' \in AHLSteps_{AN}$. Then $u(s) = [d]$ where d is defined as follows
  (2a) Rule: $r_t : L_t \to R_t$ as defined in Def. 4.3.
  (2b) Match: Since t is enabled under m using assignment asg we have that $\overline{asg}^\oplus(pre(t)) \leq_{ab} m$ (def. of enabling). This implies that there is a match $m_{asg} : AG_{nat}(pre(t)) \to AG_{id_A}(m)$ where $m_{asg} = (m_v, \emptyset, id_P, \overline{asg})$ and for all $(a, p, i) \in L_{tV}$, $m_V(a, p, i) = (U(\overline{asg})(a), p, i)$.
  (2c) Derivation step: Using rule r_t and match m_{asg}, we obtain a derivation step in $ASteps_{TA}$ as a pushout of r_t and m_{asg} in $\mathbf{AG^A}$, i.e. we obtain a graph

OUT and morphisms $r_t^\bullet : AG_{id_A}(m) \to OUT$ and $m_{asg}^\bullet : R_t \to OUT$ where $r_{tA}^\bullet = id_A$. As r_t is a 'translated transition', it is defined as $r_t = (\emptyset, \emptyset, id_P, id_A)$. By the construction of pushouts in **AG** (see [LKW93]), the graph OUT can be defined as $OUT = (OUT_V, \emptyset, P, A, \emptyset, \emptyset, l^{OUT}, a^{OUT})$ where $OUT_V = (AG_{id_A}(m)_V - m_{asgV}^\bullet(L_tV)) \uplus m_{asgV}^\bullet(R_tV)$ and for all $v \in OUT_V$, if $v \in AG_{id_A}(m)_V$ then $l^{OUT}(v) = l^{AG(m)v}(v)$, $a^{OUT}(v) = a^{AG(m)v}(v)$, and if $v \in m_{asgV}^\bullet(R_tV)$ then $l^{OUT}(v) = l^{R_tv}(v)$, $a^{OUT}(v) = a^{R_tv}(v)$. By definition of the translation of graphs to markings,

$M(OUT) = \sum_{v \in OUT_V}(a^{OUT}(v), l^{OUT}(v))$
$= \sum_{v \in (AG_{id_A}(m)_V - m_{argV}(L_tV)) \uplus m_{argV}^\bullet(R_tV)}(a^{OUT}(v), l^{OUT}(v))$
$= \sum_{v \in (AG_{id_A}(m)_V - m_{argV}(L_tV))}(a^{AG(m)v}(v), l^{AG(m)v}(v)) \oplus \sum_{v \in m_{argV}^\bullet(R_tV)}(a^{R_tv}(v), l^{R_tv}(v))$
$= \sum_{v \in AG_{id_A}(m)_V}(a^{AG(m)v}(v), l^{AG(m)v}(v)) \ominus \sum_{v \in m_{argV}(L_tV)}(a^{L_tv}(v), l^{L_tv}(v)) \oplus$
$\quad \sum_{v \in m_{argV}^\bullet(R_tV)}(a^{R_tv}(v), l^{R_tv}(v))$
$= m \ominus \overline{asg}^\otimes(pre(t)) \oplus \overline{asg}^\otimes(post(t))$
$= m'$

As $M(OUT) = m'$ then by Prop. 4.2 $AG_{id_A}(m') \cong OUT$ and therefore $AG_{id_A}(m')$ is also a pushout object of r_t and m_{asg}. Thus we define $u(s) = [d]$ (see below).

$$\begin{array}{ccc} L_t & \xrightarrow{r_t} & R_t \\ {\scriptstyle m_{arg}}\downarrow & d & \downarrow{\scriptstyle m_{arg}^\bullet} \\ AG_{id_A}(m) & \xrightarrow{r_t^\bullet} & AG_{id_A}(m') \end{array}$$

Let $[d] \in isoASteps_{TA}$. Then $u^{-1} : isoASteps_{TA} \to AHLSteps_{AN}$ is defined as follows: $u^{-1}([d]) = m[t, asg\rangle m'$ where t is the transition of T whose translation is r_t (it is unique because there are no isomorphic transitions in T), $asg = m_{asgA} \circ \overline{nat}$ where $\overline{nat} : Var(t) \to T_{OP}(Var(t))_{/cond(t)}$ is the unique extension of the natural homomorphism nat. The fact that there is a total and injective morphism (match) from L_t to IN_d assures that $\overline{asg}^\otimes(pre(t)) \leq_{ab} M(IN_d)$. The switching of t under marking $M(IN_d) = m$ and assignment asg gives raise to the successor marking $m' = m \ominus \overline{asg}^\otimes(pre(t)) \oplus \overline{asg}^\otimes(post(t)) = M(OUT)$ (see above).

By definition of u and u^{-1}, we conclude that $u \circ u^{-1} = id_{isoASteps_{TA}}$ and $u^{-1} \circ u = id_{AHLSteps_{AN}}$.

(3) Let $s_i = (m_i[t_i, asg_i\rangle m'_i) \in AHLSteps_{AN}$, $i = 1..2$, and $m'_1 = m2$. Then by Prop. 4.2 $G(m'_1) \cong G(m_2)$. By definition of u and the fact that $G(m'_1) \in OUT_{u(s_1)}$ and $G(m_2) \in IN_{u(s_2)}$ we have $OUT_{u(s_1)} = \{G | G \cong G(m'_1)\} = \{G | G \cong G(m_2)\} = IN_{u(s_2)}$.

Let $[d], [d'] \in isoASteps_{TA}$, $OUT_{[d]} = IN_{[d']}$, $[d] = u(m_1[t1, asg1\rangle m'_1)$ and $[d'] = u(m_2[t1, asg1\rangle m'_2)$. This implies that $OUT_d \cong IN_{d'}$ and thus $M(OUT_d) = M(IN_{d'})$ (by Prop. 4.2). By definition of u we have $OUT_d = AG_{id_A}(m'_1)$ and $IN_{d'} = AG_{id_A}(m_2)$. As $M(AG_{id_A}(m)) = m$ we conclude that $m'_1 = m_2$. □

Moreover the flattening of AHL nets into PT nets and of AGGs into LGGs, and the translations PT nets→ LGGs and AHL nets → AGGs are compatible syntactically and semantically as stated in the following theorem.

Theorem 4.6 Given an AHL net system AN, then the following diagrams representing syntactical (1) and semantical (2) aspects commute (up to isomorphism), where $GG = \flat^{GG}(Tr^{AGG}(AN)) \cong Tr^{LGG}(\flat^N(AN))$ ∗

$$\begin{array}{ccc} AN & \xrightarrow{Tr^{AGG}} & Tr^{AGG}(AN) \\ \flat^N \downarrow & (1) & \downarrow \flat^{GG} \\ \flat^N(AN) & \xrightarrow{Tr^{LGG}} & GG \end{array} \qquad \begin{array}{ccc} Sem^{AHL}(AN) & \longrightarrow & Sem^{AGG}_{iso}(Tr^{AGG}(AN)) \\ \downarrow & (2) & \downarrow \\ Sem^{PT}(\flat^N(AN)) & \longrightarrow & Sem^{LGG}_{iso}(GG) \end{array}$$

Proof: Commutativity of (1) is based on: (i) the set of places of $\flat^N(AN) = U(A) \times P$ is isomorphic to the label set of $\flat^{GG}(Tr^{AGG}(AN)) = P \times U(A)$, (ii) whenever there exists a homomorphism $asg: T_{OP}(Var(t))_{/cond(t)} \to A$ then $A \models cond(t)$ under asg, i.e. the matches of $Tr^{AGG}(AN)$ that lead to rules in $\flat^{GG}(Tr^{AGG}(AN))$ correspond to the assignments that enable a transition in AN, leading to transitions in $\flat^N(AN)$ and corresponding rules in $Tr^{LGG}(\flat(AN))$, (iii) the translation between markings and graphs, AGraphs and LGraphs, and AHL-markings and PT-markings always preserve the number of involved tokens/vertices.

Commutativity of (2) follows from the commutativity of diagrams (1) to (4) in Fig. 1, given respectively by Theorems 2.6, 3.10, 4.5, and from the fact that the compatibility of translation of place/transition nets into labeled graph grammars can be shown analogously to the compatibility shown in Theorem 4.5. □

5 Concluding Remarks

In order to reveal the formal relationship between graph grammars and Petri nets both high-level nets and grammars were mapped into their corresponding low-level variants. These intuitive flattenings were formally justified by showing their semantical compatibility. Thereafter, we gave a direct translation of both high-level and low-level versions of nets into grammars. Again the constructions were justified by their compatibility w.r.t. the previously defined flattenings. In fact this not only applied to the syntactical but also to the semantical level. However this last step required to replace the usual graph grammar semantics by a more abstract one.

It remains an open question whether a translation of nets into graph grammars could also be achieved for a more concrete semantics for Petri nets (respecting the individuality of tokens — see e.g. [MMS94]).

In a preliminary version of this paper PT nets and AHL nets were defined as categories rather than sets. Correspondingly we dealt with semantics and flattening functors rather than functions. The more general compatibility result (see diagram (1) in the introduction) there was here restricted to the object part in order to make the construction more comprehensible. A full version is in preparation.

References

[CEL+94a] A. Corradini, H. Ehrig, M. Löwe, U. Montanari, and F. Rossi, *An event structure semantics for safe graph grammars*, To appear in Proc. of the IFIP Working Conference PROCOMET'94, 1994.

[EBO91] H. Ehrig, M. Baldamus, and F. Orejas, *New concepts for amalgamation and extension in the framework of specification logics*, Proc. ADT-Workshop Durdan (Durdan), 1991, LNCS 655, pp. 199 – 221.

[EGH92] H. Ehrig, M. Große-Rhode, and A. Heise, *Specification techniques for concurrent and distributed systems*, Tech. Report 92/5, Technical University of Berlin, jan. 1992, Invited paper for 2nd Maghr. Conference on Software Engineering and Artificial Intelligence, Tunis,1992.

[EKL91] H. Ehrig, M. Korff, and M. Löwe, *Tutorial introduction to the algebraic approach of graph grammars based on double and single pushouts*, 4th Int. Workshop on Graph Grammars and their Application to Computer Science, Springer, 1991, LNCS 532, pp. 24–37.

[EM85] H. Ehrig and B. Mahr, *Fundamentals of algebraic specifications 1: Equations and initial semantics*, EACTS Monographs on Theoretical Computer Science, vol. 6, Springer, Berlin, 1985.

[EPR94] H. Ehrig, J. Padberg, and L. Ribeiro, *Algebraic high-level nets: Petri nets revisited*, Recent Trends in Data Type Specification (Caldes de Malavella, Spain), Springer, 1994, LNCS 785, pp. 188–206.

[GL81] H.J. Genrich and K. Lautenbach, *System modelling with high-level Petri nets*, TCS **13** (1981), 109–136.

[Jen92] K. Jensen, *Coloured Petri nets. basic concepts, analysis methods and practical use*, Springer, Berlin, 1992.

[JR91] K. Jensen and G. Rozenberg (eds.), *High-level Petri nets: theory and application*, Springer, 1991.

[Kre81] H.-J. Kreowski, *A comparison between Petri nets and graph grammars*, Springer, 1981, LNCS 100, pp. 306–317

[LKW93] M. Löwe, M. Korff, and A. Wagner, *An algebraic framework for the transformation of attributed graphs*, Term Graph Rewriting: Theory and Practice, John Wiley & Sons Ltd, 1993, pp. 185–199.

[Löw93] M. Löwe, *Algebraic approach to single-pushout graph transformation*, TCS **109** (1993), 181–224.

[MM90] J. Meseguer and U. Montanari, *Petri nets are monoids*, Information and Computation **88** (1990), no. 2, 105–155.

[MMS94] J. Meseguer, U. Montanari, and V. Sassone, *On the model of computation of place/transition Petri nets*, ATPN'94, Springer, 1994, LNCS 815, pp. 16–38.

[PER93] J. Padberg, H. Ehrig, and L. Ribeiro, *Algebraic high-level net transformation systems*, Tech. Report 93-12, Technical University of Berlin, 1993, Revised Verion accepted for Mathematical Structures in Computer Science.

[Rei81] W. Reisig, *A graph grammar representation of nonsequential processes*, Springer, 1981, LNCS 100, pp. 318–325

[Rei85] W. Reisig, *Petri nets*, Springer, 1985.

[Rei91] W. Reisig, *Petri nets and algebraic specifications*, TCS **80** (1991), 1–34.

[RP94] L. Ribeiro and J. Padberg, *Algebraic high-level nets and transformations with initial markings*, Tech. Report 94/7, Technical University of Berlin, 1994.

[Sch93] H.-J. Schneider, *On categorical graph grammars integrating structural transformations and operations on labels*. TCS **109** (1993), pp. 257–274.

[Sch94] H.-J. Schneider, *Graph grammars as a tool to define the behaviour of process systems: from Petri nets to Linda*. In Proc. 5th Int. Workshop on Graph Grammars and their Application to Computer Science, 1994.

[Wil79] J.C. Wileden, *Relationships between graph grammars and the design and analysis of concurrent software*, Springer, 1979, LNCS 73, pp 456–463.

Hierarchically Distributed Graph Transformation*

Gabriele Taentzer

Computer Science Department, Technical University of Berlin,
Franklinstr. 28/29, Sekr. FR 6-1, D-10587 Berlin,
e-mail: gabi@cs.tu-berlin.de

Abstract. Hierarchically distributed graph transformation offers means to model different aspects of open distributed systems very intuitively in a graphical way. The distribution topology as well as local object structures are represented graphically. Distributed actions such as local actions, network activities, communication and synchronization can be described homogeneously using the same method: graph transformation. This new approach to graph transformation follows the lines of algebraic and categorical graph grammars and fits into the framework of double-pushout high-level replacement systems.

Keywords: Graph transformation, distributed systems, communication, synchronization

1 Introduction

Graphical representations are an obvious means to describe different aspects of systems. Modeling distributed and concurrent systems graphs are often used to describe the *topological structure* of the system. The graphical structure shows then which parts are involved and what are the ways of communication. Graph transformations can be used conveniently to model dynamic changes of the system structure. For example, the distribution of some local parts is rearranged or communication channels are created or deleted. Local states are typically coded in some specification or programming text or not considered. This idea is followed, for example, in [3], by Δ-grammars in [9], in [13] and by actor graph grammars in [11].

Graphs can be used also to model complex *object relations* inside of local parts of a system as they arise, for example, in database systems (entity-relationship models described in [2]) or software process modeling (development graphs as used in [12] or project flow graphs in [10]). Graph transformations are useful then on these lower levels to specify changes of object relations.

The possibility to allow local actions to run concurrently can be modeled by distributed graph transformation following the algebraic approaches in [5], [8]

---

* This work has been partly supported by the ESPRIT Working Group 7183 "Computing by Graph Transformation (COMPUGRAPH II)"

and [15]. Here, some restricted types of network structures are allowed which can be changed by special operations, namely SPLIT and JOIN. SPLIT splits a graph into two or more local graphs with an interface graph between each two where the connection to the local graphs is described by graph morphisms. A local graph is not allowed to be split again, i.e. hierarchical network structures cannot be modeled. JOIN joins a distributed graph of such a kind again to one graph.

An approach which combines graph transformation on the network and on the local level is the work of categorical graph grammars in [14]. This kind of graph grammar allows a flexible change of network structures and a description of local actions for example by graph transformation. Communication is not expressed by means of graphs and graph morphisms, i.e. object identity has to be coded into names, for example.

With hierarchically distributed graph transformation the advantages of the categorical and algebraic graph grammars for modeling distributed systems are combined. It is possible to handle complex network structures, especially hierarchical ones. Communication and synchronization can be modeled by interface graphs which are connected to their local graphs by graph morphisms, i.e. object identities are described by graph morphisms between different local states.

This paper is organized as follows: In the next section the modeling of distributed systems based on graph transformations is discussed. A simplified example of distributed software development serves as illustration. All main features are introduced in that section. In section 3 the formal description of hierarchically distributed graph transformation is presented following the double-pushout approach to graph transformation ([4]). It is shown that this new approach fits into the framework of HLR-systems introduced in [6]. For this section the reader is supposed to be familiar with basic notions of category theory as they are presented, f. ex. in [1]. A reader not interested in the formal description of hierarchically distributed graph transformation can skip section 3 and gets a good impression of hierarchically distributed graph transformation anyway.

2 Distributed Systems Modeled by Graph Transformation

In this section the main features for modeling of distributed systems by graph transformations are presented. Distributed states are modeled by graphs and state transitions, i.e. local actions, network activities or communication are modeled by graph transformation.

Example 1 (Distributed software development). Several people developing a software system in parallel have to cooperate with each other. As an support *development graphs* (introduced in [12]) or *project flow graphs* (presented in [10]) can be used which describe different states of software development. These kinds of graphs have been introduced in order to assure software quality and possibly speed up the project. The developers are allowed to concurrently work on the software. Every *development step* is described by a graph transformation, in the following. All further examples are closely related to project flow graphs and their developments, although they are simplified in some minor points.

2.1 Distributed States

Usually a state of a system can be described by a graph where the nodes represent objects and the edges relations between them. If the state is a distributed one it can be reflected by several graphs where each of them shows a local state. In the following we call graphs describing a local state *local graphs*. The objects and relations in such a local state are called local, too.

Example 2 (Local software development graph). In so-called local development graphs dependencies between different units or, more concrete, documents, input and output relations of development tools and revisions of documents are modeled. Such a graph stores about the same information as a revision and configuration management system together with a tool like "make". The local graph in figure 1 contains three nodes of type "doc" modeling documents, two "tool"-nodes modeling two different development tools, i.e. editors, compilers, etc. and two "rev"-nodes standing for two revisions of documents. Furthermore, input and output relations are described by edges from "doc"-nodes to "tool"-nodes and vice versa. They are drawn as solid arrows. Edges drawn as dashed arrows model dependencies between different documents. A "doc"-node with a "rev"-node at its lower right part represents a document which has a revision. Notice that there is not an arrow which explicitly shows this relation. Internally the relation is modeled by an edge as all other relations, too.

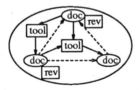

Fig. 1. A local development graph

Different local systems are usually connected by some kind of network. They interact with each other by some interface. The interfaces as well as the local systems are allowed to change their states. This means that we have not only local states but also interface states and, moreover, network states. In the network states the current distribution structure of the whole system is stored. It changes if, for example, new local systems are added or connections are changed.

Modeling the network structure of local systems by a *network graph* its nodes represent the local systems. These nodes are called *network nodes*. The edges of the network graph describe relations between local systems. These *network edges* can model some kind of links that have to be hold consistent.

The whole distributed state is described by a so-called *hierarchically distributed graph*, short HD-graph, which consists of a network graph where each network node is equipped with a local graph representing the current state of its local system. Each network edge is equipped with a total graph morphism describing the relation between two local states. These total graph morphisms are the essential basis for interaction of local systems. They are used to describe which local objects and relations correspond to each other in different local parts. Local graphs which are the target of such a graph morphism are called *target graphs*. Analogously, the sources of graph morphisms are called *source graphs*.

Example 3 (Development graph). Considering the development of a big software system it has to be determined first which groups and, more concrete, which persons have to do which parts. Such a distribution which is usually rather hierarchical can be modeled by a network graph. In our example a "big project" in the state described in example 2 is modeled by a so-called node which contains the local development graph in figure 1. The "big project" is distributed in those portions which are handled by one developer each. Thus, the network graph in figure 2 contains two "developer"-nodes which are connected with the "big project"-node by a network edge. It models how a developer part fits into the whole project shown by corresponding layout of local graphs.

Both developers should cooperate via some interface modeled by an "interface"-node and two network edges from this node to the "developer"-nodes. The local development graphs in the "developer"-nodes show the local states in each case which are parts of the "big project"-graph. The interface between the developers should contain those objects and relations which can be used or should be handled by both. These can be some kind of prereleases of produced documents which should be forwarded to other developers. All objects and relations which do not have a correspondence in the interface are considered as hidden for other developers, i.e. they do not have access.

Furthermore, the developers can use different views on their development parts, namely the "semantical view" and the "operational view" modeled by three so-called nodes and their connections to the corresponding developer parts.

The semantical view shows all documents as well as revisions and their interdependencies. The operational view is restricted to the connections between tools and documents.

In [5] and [8] distributed graphs with exactly two local graphs and one interface graph, i.e. a source graph which describes common parts are considered. This notion is extended in [15] to an arbitrary number of local graphs where an interface graph has to be established between each two. In both approaches it is not possible to define *distribution hierarchies*. This means, for example, that there can be an interface between interface graphs.

Usually a hierarchy means some kind of tree or, more generally, a directed acyclic graph. In our case of hierarchically distribution it is not an essential step to allow arbitrary graphs since loops or cycles model a distribution hierarchy, too, but in an abbreviated notation. For example, a network described by a

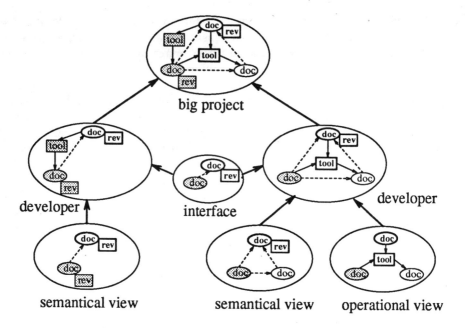

Fig. 2. An HD-graph for distributed software development

graph that consists of a node with an associated loop can also be modeled by one node and its copy with an edge in between.[2]

2.2 Local Actions

In the graph grammar field actions are usually described by *graph productions* and modeled by *graph transformation*. In this paper we use the double-pushout approach to graph transformation which characterizes some kind of cautious rewriting of graphs. This means that for a production and its matching part in some graph the following conditions have to be satisfied. Context edges are not allowed to dangle, i.e. a node which is connected to a context node has to be preserved, and two items (nodes or edges) are not allowed to be identified if at least one of them should be deleted. These conditions are combined in the well-known *gluing condition*. If the gluing condition is satisfied for a production and a match of its left hand side in the current graph a new graph is derived by deleting this occurrence and adding the right hand side of the production. (More details to this kind of graph transformation can be found in [4], etc.)

Conceptually local actions are described by *HD-graph productions* which consist of a *local production* describing the local action and a *network production* which is identical here, since the network graph is actually not changed. It is transformed by an identical production which preserves that network node where the local action took place.

---

[2] The property "hierarchical" does not belong to the levels of abstraction which have been invented to describe this kind of graph transformation. There are just two abstraction levels, the network level and the local level.

Such an HD-graph production can be applied to an HD-graph if the local action is somehow compatible with the context where it takes place. An action on a local graph which is a target graph is not permitted if it destroys the reference structure to other local graphs, i.e. a source graph cannot be mapped totally to its target graphs any more.

Applying an HD-graph production which describes a local action, first the network production is applied to the current network graph. This just means that the matching of the only network node is replaced identically. The local production is applied to that local graph which is equipped with the matched network node. The local production can be applied if the gluing condition is satisfied for this local graph transformation. After the application the matched network node is equipped with a new graph, the transformed one.

Example 4 (Local development steps). Typical development steps are the introduction of new documents, merging documents, storing new revisions, merging revisions, introducing new tools and therefore changing processing relations or changing dependency relations. These actions do not change the network structure. Development steps of that kind are considered in the next section. Here, we consider for example the introduction of a new tool as a local development step done by the "big boss" in the state of the "big project". The HD-graph production *new tool* in figure 3 describes this local development step where a tool requiring one input and one output document is introduced.

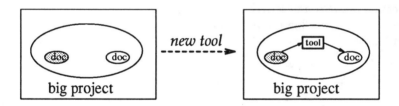

Fig. 3. Introduction of a new tool

Since there is nothing deleted in this production it can be applied to the development graph in figure 2. Otherwise a reference to an object or relation in the "big project" would be destroyed. Such destructions have to be arranged with the developers and, thus, are not local. A similar action as modeled in figure 3 cannot be performed by a developer since it has to be reported to the "big boss", i.e. all objects and relations belonging to the "developer"-node have to be totally mapped into the "big project".

2.3 Network Administration

Modeling distributed systems by graph transformations means usually the description of network activities. Changes of the network topology can be described by transformation of the network graph. The deletion and creation of network nodes and edges has to be done very cautious to avoid inconsistencies. Deletions may not destroy the reference structure. Creations have to fit into the reference structure, for example, a node in a new interface graph has to have a correspondent in all the local graphs it is interface of.

If such conditions are satisfied the network actions can be done concurrently, i.e. connections can be changed, new local parts can be inserted and other local parts (possibly with connections) can be deleted. Moreover, connections can be deleted and created, too.

Altogether a network activity is also described by an HD-graph production where its network production describes the changes of the network topology. For each network node which is preserved an identical production has to be applied to its local state graph. Each network node (edge) which is deleted or created is equipped with a local graph (morphism) on the left- or right-hand side of the HD-graph production, resp. The HD-graph production can be applied if the conditions for deletion and creation described above are satisfied. Moreover, the gluing condition has to be satisfied for the network part of the application.

Example 5 (Network activities). In a software development process network activities can be the introduction of a new developer or a new interface, the establishment or changing of connections between developers and the "big boss", etc. In figure 4 the HD-graph production *change connection* is shown where the connection between a developer and a "big project" is relaxed which releases the developer from this "big brother is watching you"-situation modeled in example 3. This means that the local state of the developer is no longer a part of that of the "big project".

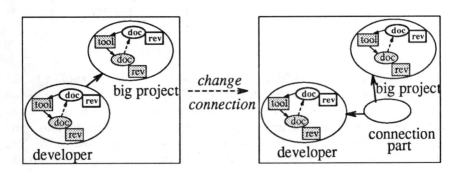

Fig. 4. Changing a connection

Applying this production to the HD-graph in figure 2 the left "developer"-graph is then connected to the "big project"-graph by an empty "interface"-graph. The whole relation between the corresponding local graphs is completely described in the HD-graph production and can be deleted.

2.4 Communication and Synchronization

Synchronization of different local actions is modeled by such actions on network nodes connected with each other. In this case an action on a source graph has to be a subaction in some sense of those on connected target graphs. This means more or less that the same actions on local items are allowed as in their adjoined local graphs. But in the source graph items may be deleted whereas some of their images in the corresponding target graph are preserved. Furthermore, a local item may be inserted in some source graph if its correspondents are preserved in all connected target graphs.

Synchronization is modeled by an HD-graph production where the network production is an identical production similarly to local actions. But here the whole network part which is more than one node is identically replaced. All nodes where a part of the synchronized action should be performed have to be described in the network production. Moreover, all network edges which should be used are identically replaced. For each local part a local production, which models that action part of the synchronized action that should be performed there, should be applied. As before, the application of an HD-graph production should not destroy the reference structure. For each local production application the gluing condition has to be satisfied.

Asynchronous communication consists of at least two concurrent actions. They can be performed by using some interface graphs (as in the examples) for the description of channels or common memory. First one local part puts something into the channel or memory and then the other gets it. Considering the interface as an independent local part the other local parts have to synchronize themselves with the interface. Thus, HD-graph productions modeling synchronization can be used here, too.

Example 6 (Communication between developers). Synchronous communication between two developers can be described by an "interface"-graph and its connections to the "developer"-graphs. In figure 5 HD-graph production *sync new revision* is shown where a new revision of a document created by one developer is established in the interface and also in the current state of the other developer.

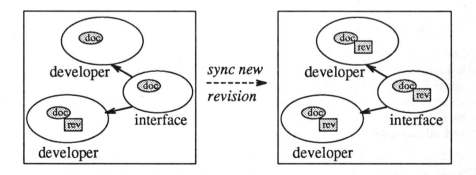

Fig. 5. Synchronous releasing of a new revision

The application of HD-graph production *sync new revision* to the HD-graph in figure 2 leads for example to the slightly modified HD-graph where new "rev"-nodes are established in the "interface"- and in the right "developer"-graphs. Additionally, the new "rev"-node in the "interface"-graph is mapped to its pendants in the "developer"-graph.

An asynchronous releasing of a new revision may be modeled by HD-graph production *async new revision* in figure 6. This action has not to be synchronized with actions of other developers. "New revision" just means the introduction of a new revision in the interface which has a link to the original revision where the developer worked on.

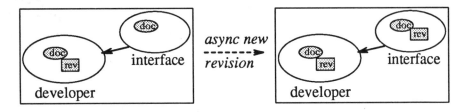

Fig. 6. Asynchronous releasing of a new revision

HD-graph production *async new revision* cannot be applied to the HD-graph in figure 2 since a new "rev"-node in the "interface"-graph cannot be mapped totally into the other "developer"-graph. But such a partial mapping of the "interface"-graph into the "developer"-graph should be possible for asynchronous communication. This can be achieved by a pair of total mappings of an additional "interface"-graph into the original "interface"- and the "developer"-graph. This additional interface graph would contain those objects and relations which should be mapped in any case, i.e. cannot be treated asynchronously. Applying an HD-graph production similar to *change connection* to the HD-graph in figure 2 an HD-graph as indicated in figure 7 can be achieved. This HD-graph can be transformed with HD-graph production *async new revision* yielding the HD-graph in figure 7 with the difference that a new "rev"-node is established in the left "interface"-graph. Additionally, this new node is mapped to the corresponding old "rev"-node in the left "developer"-graph.

2.5 Distributed Systems

Starting with some network topology and local initializations the initial state of a distributed system can be described by an HD-graph. Local actions, network activities, some kind of communicating actions and, moreover, also mixtures of these actions are allowed as state transitions. The whole distributed system is described by a so-called *HD-graph grammar* consisting of the starting HD-graph and all HD-graph productions modeling distributed actions.

Example 7 (Distributed software development system). A distributed software development system can be described by an HD-graph grammar consisting of

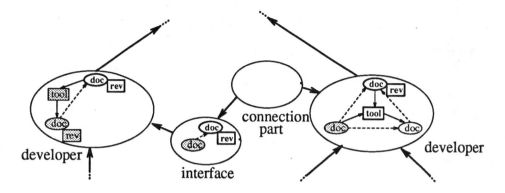

Fig. 7. Section of development graph with changed developer connection

a start graph which could be the HD-graph depicted in figure 2 and a set of HD-graph productions modeling distributed development steps also allowed to be performed concurrently. This set may contain the HD-graph productions *new tool, change connection, sync new revision, async new revision*, etc. As an example, a distributed development step may be the parallel application of *new tool* and *async new revision* to the state described in figure 2.

3 Formal Description of Distribution Concepts

The distribution concepts presented in the previous section are formally described in the framework of the double-pushout approach to graph transformation. This approach is comprehensively described in [4] for directed and labeled graphs. The double-pushout approach has been generalized to so-called high-level replacement systems in [6] where the main results of the graph grammar theory are abstracted to arbitrary objects and morphisms. In the following we show that HD-graph transformation fits into the framework of HLR-systems.

Definition 1. Let **GRAPH** be the category of labeled graphs and (total) graph morphisms which are label preserving. Given a graph G [3] of **GRAPH** which is called **network graph** a functor $\hat{G} : G \to$ **GRAPH** is called **hierarchically distributed graph (HD-graph)**[4].

A local graph is denoted $\hat{G}(i)$ for a network node $i \in G^N$. Given a network edge $e \in G^E$ the local graph $\hat{G}(s(e))$ is an interface graph or source graph, $\hat{G}(t(e))$ is called target graph and $\hat{G}(e)$ is a total graph morphism.

Example 8 (HD-graph). In figure 2 an HD-graph $\hat{G} : G \to$ **GRAPH** is shown. Graph G consists of all big ellipses with solidly drawn edges in between. Let d_1 be the left "developer"-node, its local graph which is drawn inside the ellipse is

---

[3] G^N describes the set of nodes and G^E the set of edges of G. s and t are the source and target mappings between G^E and G^N.

[4] Graph G and its induced small category are identified.

called $\hat{G}(d_1)$. Correspondingly the local graph of node b labeled by "big project" is called $\hat{G}(b)$. The network edge e with source node d_1 and target node b is equipped with a graph morphism $\hat{G}(e) : \hat{G}(d_1) \to \hat{G}(b)$ coded into the fill and frame styles of local nodes, i.e. nodes which are mapped to each other are filled and framed alike. Local edges are mapped in a structure compatible way.

Definition 2. Given two HD-graphs $\hat{G}$ and $\hat{H}$ an **HD-graph morphism** $\hat{f} = (\eta, f) : \hat{G} \to \hat{H}$ is a natural transformation $\eta : \hat{G} \to \hat{H} \circ f = (\hat{f}_i)_{i \in G^N}$ in **GRAPH** with $f : G \to H$ being a graph morphism of **GRAPH**, called **network morphism**.

If f is injective $\hat{f}$ is called **n-injective**. If moreover, all $\hat{f}_i$ with $i \in G^N$ are injective, $\hat{f}$ is called **injective**, too.

An HD-graph morphism $\hat{f}$ assigns to each node $i \in G^N$ a graph morphism $\hat{f}_i : \hat{G}(i) \to \hat{H} \circ f(i)$, called **local graph morphism** such that $\forall e \in G^E$ where $s(e) = i$ and $t(e) = j$ the diagram in figure 8 commutes. More concretely, given two local graphs $\hat{G}(i)$ and $\hat{G}(j)$ which are connected by $\hat{G}(e)$ and these graphs are mapped by local graph morphisms $\hat{f}_i$ and $\hat{f}_j$ to graphs $\hat{H} \circ f(i)$ and $\hat{H} \circ f(j)$ the connection $\hat{G}(e)$ has to be adapted leading to $\hat{H} \circ f(e)$.

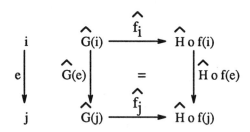

Fig. 8. HD-graph morphism

Example 9 (HD-graph morphism). HD-graph morphisms are shown in figures 3, 5 and 6. In all figures the network morphisms are not explicitly shown but given by the layout of the source and target graphs. Nodes and edges at corresponding places are mapped to each others.

Definition 3. The **composition** $\hat{f} = (\eta_f, f) : \hat{G} \to \hat{H}$ of two HD-graph morphisms $\hat{g} = (\eta_g, g) : \hat{G} \to \hat{K}$ and $\hat{h} = (\eta_h, h) : \hat{K} \to \hat{H}$ is defined by $f = h \circ g$ and $\eta_f = \eta_h \circ \eta_g : \hat{G} \to \hat{K} \circ g \to \hat{H} \circ h \circ g$.

Proposition 4. *All HD-graphs and HD-graph morphisms as defined above form a category* **DISTR(GRAPH)**.

In the following we do not consider the existence of pushouts in the category **DISTR(GRAPH)** in general, but concentrate on those which can be constructed component wise. Therefore, we need the following pushout conditions.

Definition 5. Two n-injective HD-graph morphisms $\hat{a} : \hat{A} \to \hat{C}$ and $\hat{b} : \hat{A} \to \hat{B}$ satisfy the **pushout conditions** (1) and (2) if (1) $\forall e \in C^E - a(A^E) : \exists y \in A^N$ with $a(y) = s(e)$ implies $\hat{b}_y$ is bijective and (2) $\forall e \in B^E - b(A^E) : \exists y \in A^N$ with $b(y) = s(e)$ implies $\hat{a}_y$ is bijective.

Proposition 6. *Given two n-injective HD-graph morphisms $\hat{a} : \hat{A} \to \hat{C}$ and $\hat{b} : \hat{A} \to \hat{B}$ which satisfy the pushout conditions (1) and (2) the pushout of $\hat{a}$ and $\hat{b}$ in **DISTR(GRAPH)** exists and can be constructed componentwise.*

Proof sketch: Let D with $c : C \to D$ and $d : B \to D$ be the pushout of a and b. (For pushouts in the category **GRAPH** compare [4], etc.) The pushout graph $\hat{D} : D \to \mathbf{GRAPH}$ is constructed in the following way:[5]

$$\hat{D}(x) := \begin{cases} PO(\hat{a}_y, \hat{b}_y) = PO_x & , \text{ if } \exists y \in A^N \text{ with } c \circ a(y) = x \\ \hat{C}(z) & , \text{ if } \exists z \in C^N - a(A^N) \text{ with } c(z) = x \\ \hat{B}(v) & , \text{ if } \exists v \in B^N - b(A^N) \text{ with } d(v) = x \\ IND(PO_{s(x)}, PO_{t(x)}) & , \text{ if } \exists e \in A^E \text{ with } c \circ a(e) = x \\ \hat{c}_{t(x)} \circ \hat{C}(e) \circ \hat{c}_{s(x)}^{-1} & , \text{ if } \exists e \in C^E - a(A^E) \text{ with } c(e) = x \\ \hat{d}_{t(x)} \circ \hat{B}(e) \circ \hat{d}_{s(x)}^{-1} & , \text{ if } \exists e \in B^E - b(A^E) \text{ with } d(e) = x \end{cases}$$

where $IND(PO_{s(x)}, PO_{t(x)})$ is the induced morphism from pushout graph $\hat{D}(s(x))$ to pushout graph $\hat{D}(t(x))$ with $\hat{a}_{t(e)} \circ \hat{A}(e) = \hat{C}(a(e)) \circ \hat{a}_{s(e)}$ and $\hat{b}_{t(e)} \circ \hat{A}(e) = \hat{B}(b(e)) \circ \hat{b}_{s(e)}$. Pushout morphisms $\hat{c} : \hat{C} \to \hat{D}$ and $\hat{d} : \hat{B} \to \hat{D}$ are defined as follows:

$$\hat{c}_x := \begin{cases} id_{\hat{C}(x)} & , \text{ if } x \in C^N - a(A^N) \\ \text{PO-morphism of } PO_{c(x)} & , \text{ otherwise} \end{cases}$$

$$\hat{d}_x := \begin{cases} id_{\hat{B}(x)} & , \text{ if } x \in B^N - b(A^N) \\ \text{PO-morphism of } PO_{d(x)} & , \text{ otherwise} \end{cases}$$

It is straightforward to show that $\hat{D}$ is an HD-graph (using the pushout conditions) and $\hat{c}$ as well as $\hat{d}$ are HD-graph morphisms. Now, the pushout properties have to be shown. Commutativity follows directly from the construction. For the universal property the induced morphism $\hat{u}$ consists of the induced morphisms for the underlying pushouts in **GRAPH** where they exist. Otherwise it is defined suitable to the comparing HD-graph morphisms. Well-definedness of $\hat{u}$ follows from the universal property of local pushouts and the pushout conditions. The universal property can be obtained directly from the definition of $\hat{u}$. □

A pushout as described above can be constructed in the following steps. First the pushout on network morphisms is created. For all network nodes in A which have images in B and C the pushout on their local graph morphisms is constructed. All other local graphs of $\hat{B}$ and $\hat{C}$ are carried over to $\hat{D}$ unchanged. A network edge in D which has a preimage in A is equipped with the induced

---

[5] $PO(\hat{a}_y, \hat{b}_y)$ is the pushout graph of the pushout of $\hat{a}_y$ and $\hat{b}_y$

morphism between its source and target pushout graph. All other local graph morphisms of $\hat{B}$ and $\hat{C}$ are adapted to their new target graphs. The source graphs have to be mapped structure equivalent according to the pushout conditions.

Definition 7. An **HD-graph production** $\hat{p} = (\hat{L} \xleftarrow{\hat{l}} \hat{I} \xrightarrow{\hat{r}} \hat{R})$ consists of HD-graphs $\hat{L}$, $\hat{R}$ and $\hat{I}$, called left- and right-hand side and intermediate HD-graphs and two n-injective HD-graph morphisms $\hat{l}$ and $\hat{r}$.

If all $\hat{p}_x = (\hat{L}(l(x)) \xleftarrow{\hat{l}_x} \hat{I}(x) \xrightarrow{\hat{r}_x} \hat{R}(r(x)))$ for $x \in I^N$ are left-injective, i.e. all $\hat{l}_x$ are injective, $\hat{p}$ is called *left-injective*, too.

Example 10 (HD-graph production). The HD-graph morphisms in figures 3, 4, 5 and 6 are all left injective HD-graph productions if they are interpreted in the following way. Except figure 4 the figures mentioned show the right morphism $\hat{I} \xrightarrow{\hat{r}} \hat{R}$ of an HD-graph production. For each of these examples the left morphism $\hat{I} \xrightarrow{\hat{l}} \hat{L}$ is the identity $id_{\hat{I}}$. As usually in graph transformation this means that nothing is deleted. In figure 4 HD-graphs $\hat{L}$ and $\hat{R}$ are given explicitly. Graph I can be constructed by all nodes and edges drawn alike, i.e. the "developer"- and the "big project"-nodes. All local graph morphisms of this example are identities. All local and HD-graph morphisms shown or constructed are injective.

Definition 8. Given an HD-graph production $\hat{p} = (\hat{L} \xleftarrow{\hat{l}} \hat{I} \xrightarrow{\hat{r}} \hat{R})$ an n-injective HD-graph morphism $\hat{m} : \hat{L} \to \hat{G}$ is called **HD-match of** $\hat{p}$ if the following **distributed gluing condition** is satisfied:

1. m satisfies the *gluing condition* wrt. p
2. $\forall x \in I^N : \hat{m}_{l(x)}$ satisfies the *gluing condition* wrt. $\hat{p}_x = (\hat{L}(l(x)) \xleftarrow{\hat{l}_x} \hat{I}_x \xrightarrow{\hat{r}_x} \hat{R}(r(x)))$ (compare [4], etc.)
3. connection condition: (a) $\forall e \in G^E - m((L^E) - l(I^E))$ with $t(e) = m \circ l(y)$ for $y \in I^N$:

$$\hat{G}(e)(\hat{G}(s(e)) - \hat{m}_{l(x)}(\hat{L}(l(x)) - \hat{l}_x(\hat{I}(x)))) \subseteq \hat{G}(t(e)) - \hat{m}_{l(y)}(\hat{L}(l(y)) - \hat{l}_y(\hat{I}(y)))$$

where $s(e) = m \circ l(x)$, $x \in I^N$ or $\hat{L}(l(x))$, $\hat{I}(x)$, $\hat{m}_{l(x)}$ and $\hat{l}_x$ are empty
(b) $\forall e \in G^E - m(L^E) : \exists y \in I^N$ with $m \circ l(y) = s(e)$ implies $\hat{l}_y$ and $\hat{r}_y$ are bijective.
4. network condition: (a) $\forall x \in L^N - l(I^N)$ and $\forall x \in l(I^N)$ where $\exists e \in L^E - l(I^E)$ with $s(e) = x : \hat{m}_x$ is bijective
(b) $\forall e \in R^E - r(I^E) : \exists y \in I^N$ with $r(y) = s(e)$ implies $\hat{m}_{l(y)}$ is bijective.

The connection condition means the following: (a) An action on a target graph is not permitted to delete local items where some of their pendants in connected source graphs are not deleted. (b) A local action on a source graph is not allowed to extend this graph since new items would not have images in connected target graphs. Furthermore, such an action is not allowed to delete local items without deleting references to items in target graphs. Lastly, local objects in a source graph are not allowed to be glued together since this is not reflected in the connected target graphs.

The network condition is interpreted as follows: (a) The deletion of network nodes can be done if its local state graph is deleted as a whole in the same production, i.e. if the current local state corresponds with that in the production. If a network edge should be deleted its source graph has to correspond bijectively with that of the production. (b) The other way around, if a new connection from an existing source graph should be established this graph has to be structural equivalent with its correspondent given in the production.

The conditions above are shortly discussed within examples 4 and 6.

Definition 9. Given an HD-graph production $\hat{p} = (\hat{L} \xleftarrow{\hat{l}} \hat{I} \xrightarrow{\hat{r}} \hat{R})$ and an HD-match $\hat{m} : \hat{L} \to \hat{G}$ the following HD-graph $\hat{C} : C \to \mathbf{GRAPH}$ is called **HD-context graph** of $\hat{p}$ and $\hat{m}$. Let $C = G - m(L - l(I))$ be the context graph of p and m in **GRAPH**.

$$\hat{C}(x) := \begin{cases} \hat{G}(x) - \hat{m}_{l(y)}(\hat{L}(l(y)) - \hat{l}_y(\hat{I}(y))) \,, & \text{if } \exists y \in I^N \text{ with } m \circ l(y) = x \\ \hat{G}(x) & , \text{ if } x \in G^N - m(L^N) \\ \hat{g}_{t(x)}^{-1} \circ \hat{G}(x) \circ \hat{g}_{s(x)} & , \text{ if } x \in G^E - m(L^E - l(I^E)) \end{cases}$$

where HD-graph morphisms $\hat{g} : \hat{C} \to \hat{G}$ and $\hat{c} : \hat{I} \to \hat{C}$ are defined as follows. Let $c = m \circ l$ and $g = id_{G|_C}$ HD-graph morphism $\hat{c}$ is defined by $\hat{c}_y = \hat{m}_{l(y)} \circ \hat{l}_y$, $\forall y \in I^N$.

$$\hat{g}_x := \begin{cases} id_{\hat{G}(x)/\hat{C}(x)} \,, & \text{if } x \in m(l(I^N)) \\ id_{\hat{G}(x)} & , \text{ otherwise} \,, \forall x \in C^N \end{cases}$$

Remark: $\forall y \in I^N$ the PO-complement $(\hat{C}(c(y)), \hat{g}_{c(y)}, \hat{c}_y)$ of $\hat{m}_{l(y)}$ and $\hat{l}_y$ in **GRAPH** is constructed.

Proposition 10 (Applicability of HD-graph productions). *Given an HD-graph production $\hat{p} = (\hat{L} \xleftarrow{\hat{l}} \hat{I} \xrightarrow{\hat{r}} \hat{R})$ and an HD-match $\hat{m} : \hat{L} \to \hat{G}$ there are an HD-context graph $\hat{C}$ as well as HD-graph morphisms $\hat{g}$ and $\hat{c}$ as defined above such that $(\hat{G}, \hat{m}, \hat{g})$ is the pushout of $\hat{c}$ and $\hat{l}$ in $\mathbf{DISTR(GRAPH)}$. Furthermore, the pushout of $\hat{c}$ and $\hat{r}$ exists.*

Proof sketch: $\hat{C}$ is an HD-graph since the gluing condition for all local transformations and the connection condition (a) are satisfied. Clearly, $\hat{g}$ and $\hat{c}$ are HD-graph morphisms. Next we construct the pushout of $\hat{l}$ and $\hat{c}$ (which is possible since the distributed gluing conditions contain the gluing condition for m and p and the pushout conditions which are part of network condition (a) and connection condition (b)) and have to show that the resulting pushout graph $\hat{X}$ is isomorphic to $\hat{G}$. According to pushout properties there is an HD-graph morphism $\hat{u} : \hat{X} \to \hat{G}$. Vice versa, a suitable HD-graph morphism $\hat{w} : \hat{G} \to \hat{X}$ can be defined using the local induced morphisms. $\hat{w}$ is well defined according to pushout complement properties in category **GRAPH** and network condition (a). □

Definition 11. Given an HD-graph production $\hat{p} = (\hat{L} \xleftarrow{\hat{l}} \hat{I} \xrightarrow{\hat{r}} \hat{R})$ and an HD-graph match $\hat{m} : \hat{L} \to \hat{G}$ an **HD-graph transformation** $\hat{G} \Longrightarrow_{hd} \hat{H}$ via $\hat{p}$ and $\hat{m}$, short $\hat{G} \overset{\hat{p},\hat{m}}{\Longrightarrow}_{hd} \hat{H}$, from an HD-graph $\hat{G}$ to an HD-graph $\hat{H}$ is given by the two pushout diagrams (1) and (2) in the category **DISTR(GRAPH)** shown in figure 9.

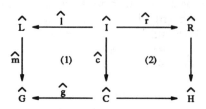

Fig. 9. HD-graph transformation

An **HD-graph transformation sequence** $\hat{G} \overset{P}{\Longrightarrow}^{*}_{hd} \hat{H}$ is a sequence of $n \geq 0$ HD-graph transformations $\hat{G} = \hat{G}_0 \Longrightarrow_{hd} \hat{G}_1 \Longrightarrow_{hd} \ldots \Longrightarrow_{hd} \hat{G}_n = \hat{H}$ via HD-graph productions of a set P. $\hat{H}$ is also called **HD-derivable** from $\hat{G}$ by P.

Proposition 12 (Uniqueness of HD-graph transformation). *Given a left-injective HD-graph production $\hat{p} = (\hat{L} \xleftarrow{\hat{l}} \hat{I} \xrightarrow{\hat{r}} \hat{R})$ and an HD-match $\hat{m} : \hat{L} \to \hat{G}$ the HD-graph transformation $G \overset{\hat{p},\hat{m}}{\Longrightarrow}_{hd} H$ is unique up to isomorphism.*

Proof sketch: First, we have to show that $\hat{C}$ together with $\hat{c}$ and $\hat{g}$ as defined in 9 are unique. This is done similar to the proof of uniqueness of pushout complements in **GRAPH**. Together with the fact that pushouts are unique up to isomorphisms we can state that $G \overset{\hat{p},\hat{m}}{\Longrightarrow}_{hd} H$ is unique up to isomorphism. □

Definition 13. Given two HD-graph productions $\hat{p}_1 = (\hat{L}_1 \xleftarrow{\hat{l}_1} \hat{I}_1 \xrightarrow{\hat{r}_1} \hat{R}_1)$ and $\hat{p}_2 = (\hat{L}_2 \xleftarrow{\hat{l}_2} \hat{I}_2 \xrightarrow{\hat{r}_2} \hat{R}_2)$ the HD-graph production $\hat{p}_1 + \hat{p}_2 = (\hat{L}_1 + \hat{L}_2 \xleftarrow{\hat{l}_1+\hat{l}_2} \hat{I}_1 + \hat{I}_2 \xrightarrow{\hat{r}_1+\hat{r}_2} \hat{R}_1 + \hat{R}_2)$ defined by disjoint union of graphs and morphisms is called **parallel (HD-graph) production** of $\hat{p}_1$ and $\hat{p}_2$.

Definition 14. An **HD-graph grammar** $HDGG = (\hat{S}, P)$ is given by an HD-graph $\hat{S}$, called the **start graph**, and a set of HD-graph productions P.[6]

Let P^+ be the smallest extension of the set P including all parallel HD-graph productions $\hat{p}_1 + \hat{p}_2$ for $\hat{p}_1, \hat{p}_2 \in P^+$. The **observation set** $O(HDGG) = \{\hat{G} \mid \hat{S} \overset{P^+}{\Longrightarrow}^{*}_{hd} \hat{G}\}$ of $HDGG$ consists of all HD-graphs $\hat{G}$ HD-derivable from $\hat{S}$ by P^+.

---

[6] Since we are not interested in classical language aspects here, we do not distinguish terminal and nonterminal graphs.

Since all matches have to be n-injective parallel productions can be applied to different local graphs only, i.e. the matches of the original HD-graph productions are not allowed to overlap. Thus, it seems to be possible to show that given an HD-graph transformation $\hat{G} \overset{\hat{p}_1 + \hat{p}_2}{\Longrightarrow}_{hd} \hat{H}$ where $\hat{p}_1$ and $\hat{p}_2$ are applied in parallel there is a corresponding HD-graph transformation sequence $\hat{G} \overset{\hat{p}_1}{\Longrightarrow}_{hd} \hat{X} \overset{\hat{p}_2}{\Longrightarrow}_{hd} \hat{H}$ applying the original HD-graph productions $\hat{p}_1$ and $\hat{p}_2$ sequentially in some order.(See also the analysis construction of the parallelism theorem in [6].) This means that the observation set $O(HDGG)$ is "closed under parallelism", i.e. there is not a graph in $O(HDGG)$ which can only be derived by applying at least one parallel production.

4 Conclusion and Open Problems

In this paper hierarchically distributed graph transformation has been introduced to offer the possibility of modeling the main aspects of open distributed systems, such as distributed software development. This approach allows arbitrary distribution topologies, especially hierarchical ones, which can be handled dynamically. The internal structures of local parts can be modeled in a graphical way, too, and consistent copies are indicated by graph morphisms between local graphs. By means of graph transformation the topological structure as well as local object structures can be manipulated in an integrated way, i.e. graph transformation is performed on both levels of description.

It might be useful to increase the number of description levels to capture additional aspects of system modeling. For example in our running example of distributed software development, the documents, revisions and tools described by local nodes could be refined to graphs again showing their internal structures. Moreover, it might be useful to structure the local parts according to access rights. This would yield some kind of encapsulated components with well-defined interfaces. Such a concept for modular systems emphasizing encapsulation is described in [16]. It builds up on HD-graph transformation.

On the theoretical side HD-graph grammars are shown to fit into the framework of HLR-systems. Further results for HD-graph transformations such as independence or embedding results can be easily achieved if so-called HLR-conditions can be proven. These condition are mainly based on the existence of pushouts in the given category. In this paper we concentrated on that kind of pushouts in **DISTR(GRAPH)** being built component wise because it reflects best the distribution of local actions. The investigation of other kinds of pushouts and the cocompleteness of **DISTR(GRAPH)** altogether will be further work.

Allowing pushouts on HD-graph morphisms which need not to be n-injective the parallel execution of actions on one local graph can be modeled. This feature has to be used to simulate the special operations SPLIT and JOIN of distributed graph transformation in the algebraic approach by HD-graph transformation ([5]).

Acknowledgment: I thank Annika Wagner and the referees for their valuable comments on this paper.

References

1. J. Adamek, H. Herrlich, and G. Strecker. *Abstract and Concerte Categories*. Series in Pure and Applied Mathematics. John Wiley and Sons, 1990.
2. I. Classen, M. Löwe, S. Wasserroth, and J. Wortmann. Static and dynamic semantics of entity-relationship models based on algebraic methods. to appear in proc. IFIP-Congress and GI-Fachgespräche, Hamburg, 1994.
3. P. Degano and U. Montanari. A model of distributed systems based on graph rewriting. *Journal of the ACM*, 34(2):411–449, 1987.
4. H. Ehrig. Introduction to the algebraic theory of graph grammars. In V. Claus, H. Ehrig, and G. Rozenberg, editors, *1st Graph Grammar Workshop, Lecture Notes in Computer Science 73*, pages 1–69. Springer Verlag, 1979.
5. H. Ehrig, P. Boehm, U. Hummert, and M. Löwe. Distributed parallelism of graph transformation. In *13th Int. Workshop on Graph Theoretic Concepts in Computer Science, LNCS 314*, pages 1–19, Berlin, 1988. Springer Verlag.
6. H. Ehrig, A. Habel, H.-J. Kreowski, and F. Parisi-Presicce. From graph grammars to High Level Replacement Systems. In Ehrig et al. [7], pages 269–291. Lecture Notes in Computer Science 532.
7. H. Ehrig, H.-J. Kreowski, and G. Rozenberg, editors. *4th International Workshop on Graph Grammars and Their Application to Computer Science*. Springer Verlag, 1991. Lecture Notes in Computer Science 532.
8. H. Ehrig and M. Löwe. Parallel and distributed derivations in the single pushout approach. *TCS*, 109:123 – 143, 1993.
9. S.M. Kaplan, J.P. Loyall, and S.K. Goering. Specifying concurrent languages and systems with Δ-grammars. In Ehrig et al. [7], pages 475–489. Lecture Notes in Computer Science 532.
10. D. Kips and G. Heidenreich. Project flow graphs – a meta-model to support quality assurance in software-engineering. to appear in proc. of IEPM'95, 1995.
11. M. Korff. Single pushout transformations of equationally defined graph structures with applications to actor systems. In *Proc. Graph Grammar Workshop Dagstuhl 93*, pages 234–247. Springer Verlag, 1994. Lecture Notes in Computer Science 776.
12. P. Pepper and M. Wirsing. KORSO: A methodology for the development of correct software. to be published in LNCS, 1995.
13. G. Schied. *Über Graphgrammatiken, eine Spezifikationsmethode für Programmiersprachen und verteilte Regelsysteme*. Arbeitsberichte des Institus für mathematische Maschinen und Datenverarbeitung (Informatik), University of Erlangen, 1992.
14. H.-J. Schneider. On categorical graph grammars integrating structural transformation and operations on labels. *TCS*, 109:257 – 274, 1993.
15. G. Taentzer. Towards synchronous and asynchronous graph transformations. accepted for special issue of Fundamenta Informaticae, 1995.
16. G. Taentzer and A. Schürr. DIEGO, another step towards a module concept for graph transformation systems. to appear in proc. of SEGRAGRA'95 " Graph Rewriting and Computation", published in Electronic Notes of TCS, 1995.

On Edge Addition Rewrite Systems And their Relevance to Program Analysis

Uwe Aßmann

INRIA Rocquencourt
Domaine de Voluceau, BP 105, 78153 Le Chesnay Cedex, France
Uwe.Assmann@inria.fr[*]

Abstract. In this paper we define a special class of graph rewrite systems for program analysis: *edge addition rewrite systems* (EARS). EARS can be applied to distributive data-flow frameworks over finite lattices [Hec77] [RSH94], as well as many other program analysis problems. We also present some techniques for optimized evaluation of EARS. They show that EARS are very well suited for generating efficient program analyzers.

1 Introduction

This section presents three programm analysis problems, specified by graph rewrite systems. Although the problems are — for the purpose of demonstration — somewhat simplified, they represent realistic tasks in state-of-the-art program optimizers. First we state some preliminaries.

A directed Σ-*Graph* $G = (N, E, \Sigma = \Sigma_N \cup \Sigma_E, l_N, m_N, A_N)$ consists of the following. N is a finite set of *nodes*. E, the *edges*, is a subset of the cross product $(N \times N \times \Sigma_E)$. Nodes and edges are labeled with labels from two finite sets $\Sigma_N(\Sigma_E)$. $l_N : N \mapsto \Sigma_N$, $m_N : (N \times N \times \Sigma_E) \mapsto \Sigma_E$ provide the labels for nodes and edges. Because $E \subseteq (N \times N \times \Sigma_E)$, between two nodes several edges are allowed, however, their labels must be distinct. The set $N_t = \{n \in N | l_N(n) = t\}$ of all nodes with the same label forms a *node domain*. The set $E_t = \{e \in E | m_N(e) = t\}$ of all edges with the same label forms an *edge domain (relation)*. Furthermore, $A_N = \{f | f : N \mapsto A_i, i \in I\!\!N\}$ is a set of functions which annotate each node with a set of attribute values of some domains A_i. The set of all Σ-graphs over a set of nodes N, $(\mathcal{L}_N^{\Sigma}, \cup, \subseteq)$, forms a finite lattice with Σ-graph-union $\cup$ and Σ-graph-inclusion $\subseteq$. $\cup$ and $\subseteq$ are standard extensions of set union and inclusion over edges, extended to Σ-graphs.

In the given figures we annotate a node by its label, its attributes, and an identification number, in order to identify the same nodes on left and right hand sides. It is important for the given examples that edges are added to a graph only when their source and target nodes are not already linked by an edge of the same label.

---

[*] This work has been supported by Esprit project No. 5399 COMPARE. It has been done while the author was at Universität Karlsruhe IPD, Vincenz-Prießnitz-Str. 3, 76128 Karlsruhe, Germany

Basic block graph construction Our first example is the construction of the basic block graph during control flow analysis. Simply speaking, a basic block is a sequence of intermediate code instructions that is terminated by a jump instruction. The terminating jumps induce a relation among the basic blocks, the *basic block graph*. This graph is constructed during control flow analysis, starting from the basic blocks and the jump instructions.

This process can be described by the graph rewrite system BB in Figure 1. Intermediate code instructions are marked by label I, basic blocks by label B. BB adds edges with labels `blocks-succ` and `blocks-pred` which denote the successor and predecessor relation of the basic blocks. BB-1 finds all successors of blocks that are terminated by an unconditional jump. BB-2 and BB-3 do the same for conditional jumps, each for a different jump label.

May data-flow analysis Our next example is the problem of *reaching definitions*. This is a classical bitparallel data-flow analysis [Hec77]. It computes the relation of definitions and uses of variables[2]. Its result are *definition-use-chains* that link definitions of variables to their uses. It is a prerequisite for many other program analyses.

The graph rewrite system RD (Figure 2) constructs a bipartite graph between basic blocks (B) and definitions (D). Several edge labels model the information: if an edge with label DEF is drawn from a basic block node to a definition node, the definition occurs in the basic block. PRESERVED models the definitions that are not killed during the execution of a basic block, i.e. those whose variables are not redefined. RD-IN and RD-OUT model the definitions that reach the beginning and the end of a basic block. Thus if an edge with label RD-IN (RD-OUT) is drawn from a basic block node to a definition node the definition reaches the beginning (end) of the block.

RD-1 and RD-2 model the flow functions DEF and PRESERVED from the data-flow framework *reaching definitions*. Rule RD-3 models the union operation in the lattice of the data-flow framework by combining the values of the basic block graph predecessors (edge `blocks-pred`). A fixpoint evaluation over the rules of RD on a finite set of basic blocks and definitions gives a unique graph which is the reaching definition information.

Equivalence classes (value numbering) The next problem is the construction of equivalence classes on intermediate code expression trees. We would like to identify those expression trees which are syntactically identical. Within a single basic block this procedure is called *value numbering* [Hec77] because within basic blocks structurally equivalent expressions denote the same value.

Figure 3 contains essential parts of the specification. Expression nodes are marked by an E and fall into two categories. Expressions such as IntC (integer constants) do not have children and form leafs of expression trees, while expressions such as Plus have children and form non-leaf expressions. We say that an

---

[2] A *definition* is a program point where some variable is defined (assigned).

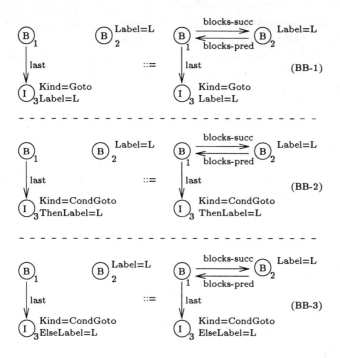

Fig. 1. Graph rewrite system BB: Basic block graph construction during control flow analysis

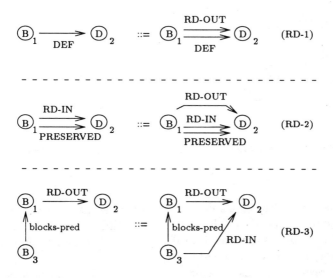

Fig. 2. Graph rewrite system RD for reaching definitions (data-flow analysis)

expression is simple-equivalent to another (`simple-eq`) if it has the same expression subtype and the same attributes. This is expressed for leaf expressions by rules such as EQ-1, and for non-leaf expressions in rules such as EQ-2. A specification must provide such rules for all expression subtypes.

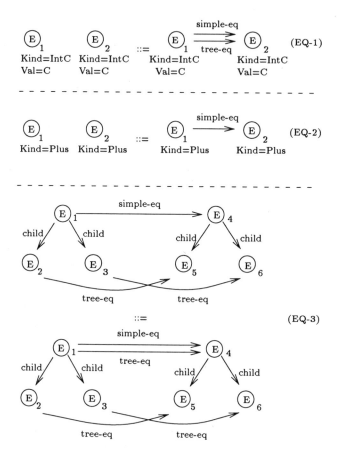

Fig. 3. Graph rewrite system EQ for tree equivalence classes (value numbering)

Two expressions are tree-equivalent (`tree-eq`) if either they are leaf nodes of a tree and simple-equivalent (also expressed by EQ-1), or they are simple-equivalent and all subtrees are tree equivalent (EQ-3). We have to provide rules such as EQ-3 for all subtypes of non-leaf expressions.

While these rewrite systems have different forms of rules, they have one thing in common: they only add edges to the axiom, and thus enlarge graphs. This gives rise to the definition of edge addition rewrite systems.

2 Edge addition rewrite systems

Definition 1. (Edge addition rewrite system)

An *edge addition rewrite system* (EARS) $\mathcal{E} = (S, Z)$ consists of a set of edge addition rules S and a Σ-graph Z (*axiom*). An *edge addition rule* $r = (L, R) \in S$, $L = (N_L, E_L, \ldots) \in \mathcal{L}_{N_L}^{\Sigma}$, $R = (N_R, E_R, \ldots) \in \mathcal{L}_{N_R}^{\Sigma}$ does not delete nodes and edges, and only adds edges to the axiom ($N_L = N_R, E_L \subset E_R, E_L \neq E_R$). L is called *left hand side (rule test)*, R is called *right hand side (rule addition)*.

r is called *recursive* if it adds an edge of label $l \in \Sigma_E$ in its right hand side and also tests an edge with label l in its left hand side. $\mathcal{E}$ is *indirectly recursive* if it adds an edge with label l in one rule and tests an edge with label l in another rule. $\mathcal{E}$ is *recursive* if it contains a recursive rule or is indirectly recursive. If a rule tests (adds) an edge with label l, we also say that l is tested (added). The set of all edge addition rewrite systems is called $\mathcal{L}_{EARS}$.

Definition 2. (Rule application)

Let be $\mathcal{E} = (S, Z) \in \mathcal{L}_{EARS}$, $r = (L = (N, E_L, l_N, m_N, A_N), R = (N, E_R, l_N, m_N, A_N)) \in S$. Let *attributes* : $N \mapsto 2^{A_N}$ deliver all attributes of a node. We say that r is *applicable* to a Σ-graph $G = (N_G, E_G, \ldots)$ if the following conditions hold:

1. there exists a graph morphism g_L from L to a subgraph $g_L(L) \subset G$. g need not be an isomorphism.
2. $\forall n \in N, a \in attributes(n) : a \in attributes(g_L(n))$, i.e. all attribute values of nodes in L are equal to those of corresponding nodes in $g_L(L)$.
3. there is no graph morphism g_R which maps R to G such that $g_L(L) \subset g_R(R)$, i.e. not all added edges of r are already in G.

If r is applicable, r derives from G the Σ-graph $H = (N_G, E_H, \Sigma, l_{N_G}, m_{N_G}, A_{N_G})$ (*direct derivation* $G \to H$) by performing the following:

1. r adds those of its edges to G, which are not already in G: $E_H = E_G \cup \{(g_L(n_1), g_L(n_2), l) | (n_1, n_2, l) \in E_R \backslash E_L\}$. E_G and E_H are sets, i.e. after the application of r there will still be only one edge of label l between $g_L(n_1)$ and $g_L(n_2)$, so that H is a well-formed Σ-graph.

2.1 Termination and strong confluence

EARS have the following important property:

Theorem 3. *(Termination and strong confluence)* If $\mathcal{E} = (S, Z) \in \mathcal{L}_{EARS}$, and Z is finite, $\mathcal{E}$ terminates and is strongly confluent.

Proof. Starting from Z, a derivation $d = G \xrightarrow{*} H$ of $\mathcal{E}$ forms an ascending chain in $(\mathcal{L}_{N_Z}^{\Sigma}, \cup, \subset)$ because the rules of $\mathcal{E}$ only add edges. Due to rule addition action (1) edges are added only *once*. Thus every direct derivation leads to another Σ-graph in $(\mathcal{L}_{N_Z}^{\Sigma}, \cup, \subset)$. Because $(\mathcal{L}_{N_Z}^{\Sigma}, \cup, \subset)$ is finite, the chain is noetherian. Thus $\mathcal{E}$ terminates on Z.

EARS are also strongly confluent: if $\to^\lambda$ is the reflexive closure of $\to$ then $\forall G, H_1, H_2 \in \mathcal{L}_{N_Z}^\Sigma : H_1 \leftarrow G \to H_2 \Longrightarrow \exists I \in \mathcal{L}_{N_Z}^\Sigma : H_1 \to^\lambda I \leftarrow^\lambda H_2$. A direct derivation does not delete nodes or edges of the graph. Thus it never destroys rule application conditions (1) and (2) for alternative direct derivations. However, rule application condition (3) may be destroyed for other direct derivations, if they add the same edges. Let in the following be d_1, d_2 be two direct derivations with rules $r_1 = (L_1, R_1), r_2 = (L_2, R_2) \in S$, $edges(G)$ the set of edges of a Σ-graph G, and $\setminus$ the Σ-graph difference.

1) $edges(g(R_1\setminus L_1)) \cap edges(g(R_2\setminus L_2)) = \emptyset$, i.e. the rules do not add edges competitively (Figure 4 left). Then d_1 does not influence application condition (3) for d_2 and vice versa. Thus d_1 and d_2 must be parallel independent [EK76].

2) $edges(g(R_1\setminus L_1)) = edges(g(R_2\setminus L_2))$, i.e. the rules add the same edges competively (Figure 4 middle). Then d_1 prevents the application of d_2 and vice versa because of condition (3). However, the pair of derivations still fulfils the condition of strong confluence because $H_1 \leftarrow_{r_1} G \to_{r_2} H_2 \Longrightarrow H_1 = H_2$.

3) $edges(g(R_1\setminus L_1)) \subset edges(g(R_2\setminus L_2))$, i.e. d_1 adds a subset of the edges of d_2 (Figure 4 right). Then d_2 prevents application of r_1; however, $r_2 \circ r_1$ yields the same result as r_2. Thus these derivations do not destroy strong confluence.

4) $edges(g(R_1\setminus L_1)) \supset edges(g(R_2\setminus L_2))$. This is symmetrical to case 3).

Thus, for all pairs of derivations, a direct common reduct can be found, and we have strong confluence.

For a certain subclass of EARS (such as RD) termination and strong confluence coincides with the existence of a fixpoint in distributive data-flow frameworks over finite lattices [MR90b] [Aß95a].

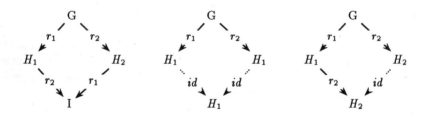

Fig. 4. Non-competitive and competitive direct direct derivations

2.2 Mapping to Datalog

EARS can be mapped to Datalog with binary predicates and to relational algebra [CGT89] [Aß95a]. We only give a sketch of the proof idea. The axiom of an EARS is mapped to Datalog facts. Edges in the axiom map to binary facts. Node identifiers, node labels, and attributes of nodes are represented by n-ary facts. The rules of an EARS are mapped to Datalog rules. The added edges of an EARS rule correspond to the left hand side predicates of Datalog rules, while the tested edges and nodes correspond to the rule body. Nodes of rules form common variables between the predicates that stem from the corresponding in- and out-going edges.

There is one minor difference. EARS add their added edges directly to the graph, while Datalog does not change its extensional database and only infers intensional knowledge. Nevertheless, strong confluence of EARS and fixpoint semantics of Datalog are related. Thus the evaluation schemes for Datalog can be used for EARS, e.g. *naive evaluation* and other optimized evaluation techniques.

3 Optimized evaluation of EARS

Our aim is to represent the graphs as nodes with neighbor sets, not as tuples of relations. This is how a program analyzer represents the analysis information in order to apply efficient graph algorithms. It would be convenient if these efficient algorithms along with their data representation could be generated automatically from an EARS specification. This section will show that this is possible for the problem classes of our examples. First we define a characteristic feature of an EARS, its *order*. In the following a *source node* is a node that has in-degree 0.

Definition 4. (Order of an EARS) Let be $\mathcal{E} = (S, Z) \in \mathcal{L}_{EARS}$. Let $sig(L)$ be the multiset of the source nodes labels in a left hand side L. $sig(L)$ is called *signature* of L. Then the order k of $\mathcal{E}$ is the maximum of the cardinalities of sig over all left hand sides:

$$k := \max_{(L,R) \in S} |sig(L)|$$

We call an EARS of order k an EARS(k).

The order of RD is 1, because each rule has one source node. The order of BB is 2, because each rule has two source nodes. The order of EQ is 2, because EQ-1 and EQ-2 have two isolates on their left hand sides.

3.1 Evaluation on directed graphs

Using the order of an EARS we can give an algorithm which evaluates them efficiently. The idea is the following. Because EARS are strongly confluent, we may interchange rule tests arbitrarily. Because we deal with directed graphs, we can test a redex by pattern matching starting at its source nodes. Thus we need only loop over all node domains of the sources of the left hand sides and, starting at these nodes, look up neighbor nodes for redexes. If all labels of source nodes of two left hand sides coincide we can overlay their rule tests. If there are other rules which cannot be overlaid we have to repeat the process for them separately.

This idea results in the generic algorithm in Figure 5. It is generic because loops marked by **forall-const** rely only on parameters of the rules of the EARS,

Input: $\mathcal{E} = (S, Z)$, order k. Rule classes V. Path covers $P(L)$
Output: Added relations
fix ← FALSE;
while fix = FALSE **do** (1)
   fix ← TRUE;
   forall-const $v \in V$ **do** (2)
      forall $(x_1, \ldots, x_{k_v}) \in (N_1 \times \ldots \times N_{k_v})$ **do** (3)
         /* tests of rules with overlapping signature */
         forall-const $r \in v$ **do** (4)
            /* Test rule r with sources $x_1, \ldots, x_{k_v}$ */
            fix ← fix or RULETEST$(r, x_1, \ldots, x_{k_v})$;
end
/* Rule test for all redexes that can be reached from source nodes */
procedure RULETEST$(r = (L, R), x_1, \ldots, x_{k_v})$ **return** BOOLEAN
begin
   fix ← TRUE;
   /* Cross product of all paths P_1 to $P_p \in P(L)$ by traversal of */
   /* the neighbor sets, starting from source nodes $x_1, \ldots, x_{r_{k_v}}$ */
   forall $((y_{1_1}, \ldots, y_{1_l}), \ldots, (y_{p_1}, \ldots, y_{p_l})) \in (P_1 \times \ldots \times P_p)$ **do** (5)
      /* Test all rule edges by neighbor set tests */
      forall-const $(n_1, n_2, a) \in E_L, n_1 \in N_i, n_2 \in N_j$ **do** (6)
         if $y_j \notin y_i.a$ **then**
            continue (5); /* No redex here */
      /* Redex found. Add all edges of the rule, if they are not there */
      forall-const $(n_1, n_2, a) \in E_R \backslash E_L, n_1 \in N_i, n_2 \in N_j$ **do** (7)
         if y_j NOT $\in y_i.a$ **then**
            $y_i.a \leftarrow y_i.a \cup y_j$; fix ← FALSE;
      end
   end
   return fix;
end

Fig. 5. Generic algorithm ORDER: Evaluation starting from source node domains

not of the manipulated graph. For a concrete algorithm these loops can be unrolled. Thus the cost for the loop in the concrete algorithm is constant. The algorithm consists of two parts. First ORDER enumerates all permutations of source nodes of redexes. Its complexity corresponds directly to the order k of an EARS. Then, for one single permutation of source nodes, RULETEST finds all reachable redexes. It does so by traversing neighbor sets. Finally, if the EARS is recursive, we have to embed this algorithm into a fixpoint loop (loop (1), *fixpoint evaluation*).

In order to apply the algorithm the rule set S has to be partitioned into a set of equivalence classes V. A rule $r_1 = (L_1, R_1)$ is equivalent to a rule $r_2 = (L_2, R_2)$, if $sig(L_1) \subseteq sig(L_2)$. This expresses whether the tests for the rules can be done together in loop (3). In the algorithm, k_v is the maximal number of elements in a signature in an equivalence class v. k_v is always smaller

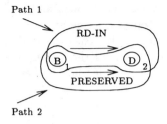

Fig. 6. An edge-disjoint path cover for RD-2

or equal to order k.

For the test of a single rule we compute for each left hand side L an *edge-disjoint path cover* $P(L), P : \mathcal{L}_{N_L}^{\Sigma} \mapsto 2^{\mathcal{L}_{N_L}^{\Sigma}}$. It covers a left hand side by a set of paths which intersect each other exactly at their end points (Figure 6). RULETEST uses the path cover to find all redexes for a given set of source nodes: it computes the cross product over all paths and all nodes which are reachable under the paths. In other words, RULETEST performs a nested loop join over the path problems of the paths (loop (5)). $P(L)$ should be chosen such that the cost for loop (5) minimizes.

Theorem 5. *(Order theorem) Let be $\mathcal{E} = (S, G) \in \mathcal{L}_{EARS}$ with order k. Let l the length of the longest path of a path cover over all left hand sides, p the maximum number of paths in a path cover, n the maximal number of nodes in a node domain with an arbitrary label, e the maximum out-degree of a node concerning an arbitrary edge label. The addition of an edge shall be performed in constant time.*

Then, if $\mathcal{E}$ is non-recursive, a concrete algorithm generated from ORDER can be evaluated in $O(n^k e^{lp})$. If $\mathcal{E}$ is recursive, it can be evaluated with $O(n^{k+2} e^{lp})$.

Proof. First, a concrete algorithm generated from ORDER terminates. In each round of the fixpoint loop the algorithm adds at least one edge, otherwise the fixpoint loop stops. It also terminates when the complete graph is reached.

Then we have to show that a concrete algorithm enumerates all redexes in the manipulated graph. Because we work on directed graphs, all nodes and edges of a redex can be reached from the set of its source nodes. A procedure which is generated from RULETEST enumerates, starting from a permutation of source nodes, all path problems of the edge disjoint path cover. Thus RULETEST finds all possible redexes for a fixed set of source nodes. Because loop (3) of ORDER enumerates all possible permutations of source nodes, all redexes are found in the manipulated graph.

(1: Complexity of enumeration of source nodes) The order k of $\mathcal{E}$ yields the maximal number of source nodes over all left hand sides. Loops (2) and (4) perform on all rules $r \in S$, using the partitioning V ($O(|S|)$). Because both loops only depend on S they can be unrolled and neglected for the complexity of the concrete algorithm. Loop (3) enumerates the cross product of all source node domains of a rule group $v \in V$. Thus it has complexity $O(n^k)$ because k was chosen maximally. Thus we get for the loops (2), (3), and (4) $O(n^k)$.

(2: Complexity of enumeration redex nodes) Let $r = (R, L)$ be a rule, $P(L)$ an edge disjoint path cover of L, l the length of the longest path in $P(L)$, and p the number of paths in $P(L)$. Loop (5) enumerates the cross product over all paths of the path cover $P(L)$. The path problems are solved by walking over the neighbor sets. Thus the following costs result:

$$O\left(\prod_{\substack{K=(N_K,E_K,\ldots)\\ \in P(L)}} \prod_{f \in E_K} e_{m_{N_K}(f)}\right) = O\left(\prod_{K \in P(L)} e^l\right) = O(e^{lp})$$

(3: Complexity of rule test/addition) If m is the maximal number of tested or added edges in a rule, loops (6) and (7) have complexity $O(m)$. Both loops can be unrolled and do not contribute to the complexity of a concrete algorithm.

(4: Fixpoint evaluation) Then a fixpoint is only reached if the EARS is non-recursive. Otherwise, we must apply the algorithm until the rules can no longer be applied. In the worst case every fixpoint iteration only adds one edge. If a the number of different labels of added edges this costs $O(a|N|^2) = O(an^2|\Sigma_N|^2)$. Because a and $|\Sigma_N|$ are constant for a fixed EARS we can neglect them and have $O(n^2)$.

Loop	Task	Generic algorithm	Concrete algorithm
1	Fixpunkt evaluation	$O(an^2|\Sigma_N|^2)$	$O(n^2)$
2,3,4	Enumeration source nodes	$O(|S|n^k)$	$O(n^k)$
5,6,7	Rule test	$O(me^{lp})$	$O(e^{lp})$
	Total	$O(m|S|n^{k+2}e^{lp}a|\Sigma_N|^2)$	$O(n^{k+2}e^{lp})$

Table 1. The complexity of ORDER

Table 1 summarizes these results. The cost formulas contain variables which depend on the rules *and* on the manipulated graph. For a fixed rule set only polynomial algorithms result because k, l, and p are constant. Often they are rather small. Additionally, in program analysis many graphs tend to be sparse, so e is rather small. It may even be bound by a constant, in which case the cost reduces to $O(n^k)/O(n^{k+2})$. Thus for many cases linear, quadratic and cubic algorithms in the maximum cardinality of a node domain result.

A concrete algorithm for RD (Figure 7) has complexity $O(|N_B|^3 e^4)$, because $k = 1, l = 2$ (in RD-3) and $p = 2$ (in RD-2). All rule tests can be overlayed into one rule group (see algorithm in Figure 7). If we represent the edges of the graph as bitvectors, loops (3), (4), and (6) become bitvector operations. Thus the order theorem results in a standard dataflow analysis method, round-robin iteration over bitvectors [Hec77].

Input: Blocks, Definitions. Relations DEF, PRESERVED, blocks-pred
Output: Relations RD-IN, RD-OUT

```
fix ← FALSE;
while fix = FALSE do (1)
    fix ← TRUE;
    forall x₁ ∈ Blocks do (2)
        forall x₃ ∈ x₁.DEF do /* rule test RD-1 */ (3)
            if not x₃ ∈ x₁.RD-OUT do
                x₁.RD-OUT ← x₁.RD-OUT ∪ x₃; fix ← FALSE;
            end
        forall x₃ ∈ x₁.PRESERVED do /* rule test RD-2 */ (4)
            if not x₃ ∈ x₁.RD-OUT do
                x₁.RD-OUT ← x₁.RD-OUT ∪ x₃; fix ← FALSE;
            end
        forall x₂ ∈ x₁.blocks-pred do /* rule test RD-3 */ (5)
            forall x₃ ∈ x₂.RD-OUT do (6)
                if not x₃ ∈ x₁.RD-IN do
                    x₁.RD-IN ← x₁.RD-IN ∪ x₃; fix ← FALSE;
                end
    end
end
```

Fig. 7. Concrete algorithm for RD, generated from ORDER. RULETEST is inlined and all generic loops are expanded

A concrete algorithm for BB has complexity $O(|N_B|^2 e^2)$ ($k = 2, l = 1, p = 2$). Because each basic block has exactly one last instruction, the complexity can be reduced to $O(|N_B|^2)$. All rule tests can be overlayed into one rule group.

Also for EQ all rule tests can be overlayed into one rule group. A concrete algorithm has complexity $O(|N_E|^3 e^8)$, because $k = 1$, $l = 2, p = 4$ (in EQ-3). However, the out-degree of expressions concerning edge label `child` is always bound by a constant, and the subgraph with edge label `simple-eq` is a partition on the expressions. Thus we can expect e to be rather small, or even of constant size, and the algorithm reduces to $O(|N_E|^3)$.

3.2 Index structures

We can reduce the complexity of a concrete ORDER algorithm by the use of index structures (dictionaries). Let in the following be an index structure a single- or multi-valued mapping from node attributes to nodes, in the sense of the relational data model.

Theorem 6. *(Index) Consider a non-recursive $\mathcal{E} = (S, Z) \in \mathcal{L}_{EARS}$ with order k. Let j be number of attribute equality tests on source nodes. Assume that the lookup and insert in an index structure over the attributes of the equality tests can be done in constant time.*

Then a concrete algorithm for $\mathcal{E}$, generated from ORDER, may be evaluated in $O(n^q e^{lp}), k - j \leq q \leq k$.

Proof. For a test on attribute equality of source nodes of label t_1 and t_2, we have to compare all nodes N_{t_1} and N_{t_2} pairwise. In a concrete algorithm this is done by the cross product over the source node domains (Figure 5, loop (3)). Instead performing a cross product between the node domains, we can use an index structure on N_{t_1} or N_{t_2} to lookup matching neighbor nodes. For the corresponding rule the cross product has to be performed only over $k - 1$ source node domains. However, the order of $\mathcal{E}$ is not necessarily reduced, because other rules of the same rule equivalence class need not be affected. If j indices are applied, the order of $\mathcal{E}$ reduces at most by j. As additional effort we have to build index structures for each relevant node domain. This, however, can be done in linear time if the insert in the index needs constant time.

We use the index structure to implement 'virtual' edges between the node domains. The mechanism is equivalent to the standard mechanism in the relational model for speeding up queries by index structures [Ull89]. The type of the index structure depends on the relationship of the attributes to the nodes. If an attribute is a *key* (in the sense of the relational model), the index is a single-valued mapping and a single-valued index can be used. Otherwise a multi-valued index must be used.

The index theorem can be applied to our example BB. All its rules have two sources in their left hand sides, thus we would have complexity $O(|N_B|^2)$. However, we can construct an index from block labels to blocks and use it in a single pass over the blocks to look up all successors/predecessors. Block labels are unique for a block. Hence the index domain is a key, and we can use a simple array with constant access and insert time. Then the standard 2-pass algorithm for basic block graph construction results $(O(|N_B|))$. It is clear that the index theorem can be applied in the same way to other identification problems, e.g. construction of call graphs.

The index theorem can also be applied to our example EQ. For rules EQ-1 and EQ-2 we can construct an index from attributes to expressions and use it in a single pass over the expressions to look up all others which are simple-equivalent $(O(|N_E|))$. Because the index domain is a product domain, the attributes are multi-valued keys to the expressions, and their domain is not dense. Hence a hash table may be used as index structure. This shows that the index theorem can be used for all analysis algorithms which construct equivalence classes, e.g. grouping of types into subtypes or simple aliasing.

3.3 Avoiding fixpoint evaluation

Some recursive EARS do not need fixpoint evaluation, if we perform the rule application in a certain order.

Definition 7. (2-level-rule) Let be $\mathcal{E} = (S, Z) \in \mathcal{L}_{EARS}$. Consider $r = (L, R) \in S$ with acyclic L and one source node label t. Let a be an edge label of an edge from L which has the following features: source and target nodes of edges of label a have label t, if all edges of label a are removed from L, L is partitioned into two non-connected subgraphs, edges of label a only connect these subgraphs, and their sources are found in one subgraph.

Then a is called the *carrier-graph label* of r, the two non-connected subgraphs are called *levels*, and r is called a *2-level-rule*. If the edge addition is done between the levels of a 2-level-rule, parallel to the carrier-graph edges, or if it is done in the level which contains the source nodes of the carrier-graph edges, we call the rule *upward-adding*.

In rule RD-3 the edge of label `blocks-pred` forms the carrier-graph label. If all edges with this label are removed from the left hand side of the rule, it partitions into two non-connected components. RD-3 is upward-adding because the added edge is added parallel to the carrier-graph. RD-1 and RD-2 do not have carrier-graph labels because edges have different source and target node labels.

Definition 8. (Carrier-graph) Let be $\mathcal{E} = (S, Z = (N, E, \Sigma, l_N, m_N, A_N))$ a recursive EARS(1) with source node label t. All left hand sides of rules of $\mathcal{E}$ are acyclic. Let a be the carrier-graph label of a 2-level-rule of $\mathcal{E}$. a is tested, but not added. All 2-level-rules of $\mathcal{E}$ contain the same carrier-graph label a. One of the 2-level-rules takes part in the recursion of $\mathcal{E}$. Then $(N_t \subseteq N, E_a \subseteq (N_t, N_t, \{a\}) \subseteq E, \{t\} \cup \{a\}, l_N, m_N, A_N)$ is called the *carrier-graph* of $\mathcal{E}$.

BB does not have a carrier-graph because it is non-recursive.

In RD there is one 2-level-rule, RD-3. `blocks-pred` is only tested, and with source and target node label B. RD-3 adds the edge label `RD-OUT` which takes part in the recursion of RD. Thus all nodes of label B and the relation `blocks-pred` form the carrier-graph. This is the predecessor relation of the basic block graph.

EQ has one upward-adding 2-level-rule, EQ-3. Its carrier-graph label is `child`. With the index theorem EQ can be regarded as EARS(1), because virtual edges are drawn between the source nodes in EQ-1 and EQ-2. Thus the carrier-graph of EQ consists of the intermediate code expression trees.

Theorem 9. *(Avoiding fixpoint evaluation) Consider $\mathcal{E} = (S, Z) \in \mathcal{L}_{EARS}$ recursive, with order 1, and acyclic carrier-graph. Let all 2-level-rules $r \in S$ be upward-adding. Then the outer fixpoint loop (1) in an algorithm generated by ORDER can be avoided.*

Proof. First we consider the case that $\mathcal{E}$ has only one rule, a 2-level-rule r. Assume that the edge addition is done between the levels of r, parallel to the carrier-graph. In an algorithm generated by ORDER for an EARS(1) we walk over the node domain of the left hand side source nodes. This is the node domain of the carrier-graph. The source of an added edge belongs to the same level of r as the source of the carrier-graph edge, the target belongs to the other level. Thus new redexes are only created towards the root of the carrier-graph. We can apply r while walking the carrier-graph bottom-up. At the top of the carrier-graph r cannot be applied anymore and the fixpoint is reached. Furthermore, if the edge addition is done within the level of the source node of the added edge, the arguing stays the same.

If $\mathcal{E}$ has several rules, the same technique can be applied. However, we have to take into account dependencies among the rules. If $r_1 \in S$ adds an edge which is tested by $r_2 \in S$ r_1 must be executed before r_2 at a certain point in the up-walk of the carrier-graph. This is possible, even if recursive dependencies among the added and tested edges exist. In the worst case we have to apply all rules twice at a certain point in the up-walk of the carrier-graph. Thus the outer fixpoint loop (1) of algorithm ORDER can be avoided.

The theorem affects EQ. The relation child is acyclic. Thus, if we apply index theorem and fixpoint avoidance the standard algorithm for value numbering results: the fixpoint evaluation is avoided by a bottom-up visit of the expression trees; the sharing of equivalent expression subtrees is achieved by a lookup in a hash table.

The theorem explains why data-flow analysis on acyclic basic block graphs can be performed in a single bottom-up traversal. Hybrid global data-flow analysis uses this fact systematically [MR90a]. This method first constructs an acyclic condensation of a cyclic flow graph and then divides the data-flow problem into a local and a global part. The local problems are solved on each component with fixpoint iteration because the components are cyclic. The global problem, however, works on the acyclic condensed graph and can be solved in a single pass. Furthermore, it is clear that if the carrier-graph is cyclic, but reducible, interval analysis or other elimination methods can be applied to find the fixpoint [RP86].

4 Conclusion

In this paper we have defined *edge addition rewrite systems* and have shown that they can be used to abstractly specify program analysis problems. We also have presented an efficient evaluation scheme which can even be improved for EARS with special features. With these techniques efficient algorithms can be derived systematically for many program analysis problems. Thus EARS are very appropriate for use in a generator of program analyzers.

In [Aß95a] EARS are applied to other program analysis problems which have not been mentioned here. This is possible because EARS are related to Datalog and relational algebra. Up to now most of these problems have been solved by ad-hoc algorithms. Also, if node addition and deletion is allowed, program optimizations may be specified. Thus EARS form — together with more general graph rewrite systems — a novel uniform framework for program optimization. This has been demonstrated in the Esprit project COMPARE by the optimizer generator OPTIMIX [Aß95b].

We thank the COMPARE group at the University of Karlsruhe for their inspiring atmosphere, especially T. Müller, H. Emmelmann, J. Vollmer and C.-T. Buhl. We also thank the reviewers who provided valuable comments.

References

[Aß95a] Uwe Aßmann. *Generierung von Programmoptimierungen mit Graphersetzungssystemen*. PhD thesis, Universität Karlsruhe, Kaiserstr. 12, 76128 Karlsruhe, Germany, July 1995.

[Aß95b] Uwe Aßmann. Optimix Language Report. Technical Report 31, Universität Karlsruhe, 1995.

[CGT89] S. Ceri, G. Gottlob, and L. Tanca. *Logic Programming and Databases*. Springer Verlag, 1989.

[EK76] H. Ehrig and H.-J. Kreowski. Parallelism of manipulations in multidimensional information structures. In *Proc. Mathematical Foundations of Computer Science*, volume 45 of *Lecture Notes in Computer Science*, pages 284–293. Springer Verlag, 1976.

[Hec77] M. S. Hecht. *Flow Analysis of Computer Programs*. Elsevier North-Holland, 1977.

[MR90a] T. J. Marlowe and B. G. Ryder. An Efficient Hybrid Algorithm for Incremental Data Flow Analysis. In *ACM Symp. on Principles on Programming Languages*, pages 184–196, 1990.

[MR90b] T. J. Marlowe and B. G. Ryder. Properties of Data Flow Frameworks. *Acta Informatica*, 28:121 161, 1990.

[RP86] B. G. Ryder and M. C. Paull. Elimination algorithms for data flow analysis. *ACM Computing Surveys*, 18(3):277–316, September 1986.

[RSH94] Thomas Reps, Mooly Sagiv, and Susan Horwitz. Interprocedural Dataflow Analysis via Graph Reachability. Technical Report 94-14, Datalogisk Institut, University of Copenhagen, April 1994.

[Ull89] J. D. Ullman. *Principles of Database and Knowledge Base Systems*. Computer Science Press, Stanford University, 1989. volumes 1 and 2.

Graph Automata for Linear Graph Languages

F.J. Brandenburg and K. Skodinis

University of Passau,
94032 Passau, Germany
e-mail: {brandenb, skodinis}@fmi.uni-passau.de

Abstract. We introduce graph automata as devices for the recognition of linear graph languages. A graph automaton is the canonical extension of a finite state automaton recognizing a set of connected labeled graphs. It consists of a finite state control and a collection of heads, which search the input graph. In a move the graph automaton reads a new subgraph, checks some consistency conditions, changes states and moves some of its heads beyond the read subgraph. It proceeds such that the set of currently visited edges is an edge-separator between the visited and the yet undiscovered part of the input graph. Hence, the graph automaton realizes a graph searching strategy. Our main result states that finite graph automata recognize exactly the set of graph languages generated by connected linear NCE graph grammars.

1 Introduction

The theory of graph languages is based on generative devices, i.e., on graph grammars. A graph grammar consists of a finite set of productions, which are used to grow graphs or hypergraphs by repeated replacements of nodes or hyperedges. There are various types of graph grammars introduced so far in literature. They differ mainly in the form of the productions and in the embedding mechanisms, see, for example [Cou90, EKR91, Eng89]. However, the dual to generative devices is still missing. There is no systematic approach on recognizing devices for graph languages. There are no graph automata, which fit to the major classes of graph grammars. This is a gap in the theory of graph languages. Here we do a first step to fill this gap. In an early paper, Rosenfeld and Milgram [RM72] have introduced web automata, which however have the computational power of Turing machines. Similarly, the approaches by Wu and Rosenfeld [WR79a, WR79b] and Remila [Rem94] on cellular automata have a high computational power in the range of linear bounded automata and have not been studied as duals to important types of graph grammars.

We consider linear graph languages. These are generated by linear NCE graph grammars. Linear graph grammars are special node replacement systems and are investigated in detail in [EL89]. They can be seen as the extension of linear context-free grammars from strings to graphs. The productions of an NCE graph grammar are triples (A, R, C), where A is a nonterminal node label, R is a nonempty graph of the right-hand side and C is the embedding relation.

C establishes edges between the nodes of R and the former neighbours of the replaced node. The language of a graph grammar consists of the set of all terminal labeled graphs derivable from the axiom. Such a graph grammar is linear, if the right-hand side graphs have at most one nonterminal node. The theory of graph grammars and their languages has been explored in great detail and depth. We refer the reader to e.g. [Bra95, Cou90, EKR91, Eng89, EL89, Nag79, RW86].

In this paper we introduce graph automata. They continue the line of finite automata and tree automata. These machines operate on strings and trees. Graph automata work on connected labeled graphs. A graph automaton is a multihead automaton with a finite set of states and a finite set of instructions. It uses its heads for a systematic search of the input graph. In a move, some heads are advanced and read a pre-defined subgraph of the remaining input graph. The automaton scans this subgraph for consistency and changes states. It is inherently nondeterministic. A graph automaton may choose among several next states. This can be made deterministic by the power set construction. However, it must choose the proper subgraph to be read, particularly at the start. A graph automaton accepts if after a sequence of consistent moves the input graph is completely scanned and it has reached a final state.

Although a graph automaton has a finite number of states and finitely many instructions, the number of heads is unbounded. It depends on the size and the structure of the graph given to the input and is bounded from below by the edge-search number [BS91]. The heads are placed on nodes and edges, where they guard nodes and clear edges. At any time, the set of currently visited edges is an edge-separator between the visited and the yet undiscovered part of the input graph. In a move, the border of separating edges continuously moves beyond the read subgraph. Thus already cleared nodes and edges cannot be recontaminated. This means a monotone search strategy, graph searching without recontamination [BS91, LaP93, MHGJP88]. Hence, graph automata are plans for monotone search strategies on graphs. The search strategies are special. They are given by a finite set of instructions and can be executed by nondeterministic finite state machines.

Our main result states that graph automata are equivalent to linear graph grammars and recognize exactly the class of connected linear graph languages.

2 Basic notions

We assume that the reader is familiar with the basic notions from graph theory and from the theory of node-replacement graph grammars. We deal with undirected, connected, node labeled graphs. The approach can be extended to directed graphs with node and edge labels.

Definition 1. Let Σ be a finite alphabet. A *graph* $g = (V, E, m)$ is a simple, undirected, node labeled graph and consists of a finite set of nodes V, a set of undirected edges E without self-loops and multiple edges, and a node labeling function $m : V \to \Sigma$. Let $g = (V(g), E(g), m(g))$.

Our graph automata impose a direction on the edges $e = \{u, v\}$, such that u is the "old" node and v is the "new" node. They slide along e from u to v and so they define edge-separators. Directed edges from u to v are written as pairs $e = (u, v)$.

Definition 2. Let $g = (V, E, m)$ be a graph and let $U \subseteq V$ be a subset of the nodes of g. The *subgraph induced* by U is $g|_U = (U, D, n)$, where $D = \{\{u, v\} \in E \mid u, v \in U\}$ and $n(v) = m(v)$ for every node $u \in U$. Let $h = g|_U$. The *complementary graph* $g - h$ is the subgraph induced by $V - U$. Thus $g - h = g|_{V-U}$. The *edge-separator* of an induced subgraph $h = g|_U$ is the set of edges between h and its complementary graph, $sep(g, h) = \{(u, v) \mid u \in U, v \in V - U\}$. These edges are directed from U to $V - U$. Their sources in U are called *ports*, i.e., $port(g, h) = \{u \in U \mid (u, v) \in sep(g, h)\}$. The endnodes in $V - U$ are the *neighbours* of h, $neigh(g, h) = \{v \in V - U \mid \{u, v\} \in E \text{ and } u \in U\}$. For these notions we may identify an induced subgraph $g|_U$ with its defining set of nodes U.

The sets of separator edges and ports can easily be updated, if a subgraph h is extended by another disjoint induced subgraph. Moreover, the separator edges and the ports uniquely determine the induced subgraph of a connected graph.

Lemma 3. *Let h and h' be node disjoint induced subgraphs of a graph g, i.e., $h = g|_U$ and $h' = g|_{U'}$ with $U \cap U' = \emptyset$. Then the edge-separator and the ports of the subgraph $g|_{h \cup h'}$ induced by $U \cup U'$ are*
$sep(g, h \cup h') = sep(g, h) \cup sep(g, h') - \{\{u, u'\} \mid u \in U, u' \in U'\}$ *and*
$port(g, h \cup h') = port(g, h) \cup port(g, h') - \{u \in V(g) \mid \{u, v\} \in E(g) \text{ implies } v \in U \cup U'\}$.

Lemma 4. *Let g be a connected graph and $h = g|_U$ an induced subgraph of g. A node v of g is in U iff v is connected with a port $u \in port(g, h)$ by a path which contains no separating edge from $sep(g, h)$. v belongs to the complementary graph $g - h$ iff v is connected to some port by a path whose last edge is the only separator edge of that path.*
These characterizations of h and $g - h$ can be generalized to an inside test such that $v \in U$ iff all paths from a port have an even number of separating edges.

Note that lemma 4 holds if g is connected. Only in this case the separator edges uniquely describe subgraph h and its complementary subgraph $g - h$. Therefore we consider connected graphs. In every computation step our graph automaton places its edge-heads on the separator edges between the already visited part h and the not yet visited part $g - h$ of the connected input graph g. The graph automaton does not need to store all nodes and edges of h. It is sufficient to place its edge-heads on the separator edges betwenn h and $g - h$.

Next we review graph grammars from the NCE family of node replacement systems, see, for example, [Cou90, EKR91, Eng89, RW86].

Definition 5. A *graph grammar* is a tuple $GG = (N, T, P, S)$, where N is the alphabet of nonterminal node labels, T is the alphabet of terminal node labels, $S \in N - T$ is the axiom and P is a finite set of productions of the form $p = (A, R, C)$, where $A \in N$ is the node label of the left-hand side, the right-hand side R is a nonempty graph and the connection relation C consists of pairs (a, w) with $a \in (N \cup T)$ and $w \in V(R)$.

According to their labels we speak of terminal and nonterminal nodes.
For a node $w \in V(R)$ let $C^{-1}(w) = \{a \mid a \in N \cup T, (a, w) \in C\}$ denote the set of labels in the connection relation of w.

There is a natural and well-established graphic notation for the productions, which we shall use throughout. For $p = (A, R, C)$ draw the right-hand side graph R with the nonterminal nodes as unit size squares, terminal nodes as points and (whenever possible) straight line edges. For the left-hand side draw a big rectangle with label A around R. Finally, for every connection $(a, w) \in C$ draw a line from the node w of R to an a-labeled point outside the big rectangle. This concepts helps in understanding the application of productions and the (visual) definition of derivations, see [Bra95, Hic94].

A derivation step means replacing a node v with label A by the right-hand side R and establishing connections between the neighbours of v and the nodes of R as specified by C. The language $L(GG)$ generated by GG consists of all terminal graphs that can be derived from the axiom S.

Instead of a formal definition we give an example for the generation of chains of the form $a^n b^n c^n$, $n \geq 1$. By productions p_3, a new a-, b- and c-node is generated and is connected to its left neighbour with the same label. Using p_2 the first a- and b-nodes are connected and the final production p_4 connects the last b- and c-nodes.

Example 1. Let $GG = (N, T, P, S)$ be a graph grammar with $N = \{S, A\}$, $T = \{a, b, c\}$ and $P = \{p_1, p_2, p_3, p_4\}$. The productions of GG are shown in Fig. 1.

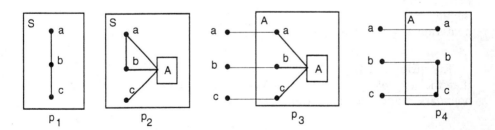

Fig. 1. The productions of GG

A derivation is illustrated in Fig. 2.

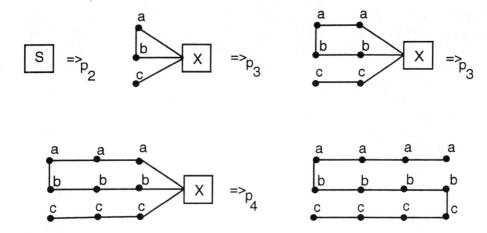

Fig. 2. A derivation of GG

Forthcoming we restrict ourselves to linear graph languages and consider linear graph grammars in normal form for their generation. The normal form is important for the construction of an equivalent graph automaton. Graph automata have a built-in check for the conditions set by the normal form.

Definition 6. A graph grammar $GG = (N, T, P, S)$ is *linear*, if the right-hand side of every production has at most one nonterminal node. Then the connection relations consist of pairs (a, w) with terminal node labels $a \in T$.
GG is *connected*, if all graphs of $L(GG)$ are connected.
GG is *in normal form*, if GG is chain-free, context consistent and neighbourhood preserving. Thus for every production (A, R, C), R does not solely consist of a nonterminal node, there is a context describing function $c : N \to P(T)$ with $c(A) = \{a \in T \mid S \Rightarrow^* g$ and $v \in V(g)$ with $m(g)(v) = A$ implies that there is a neighbour $u \in neigh(g, v)$ with $m(g)(u) = a\}$, and for every application of some production (A, R, C) to some node v in g with $S \Rightarrow^* g$ such that $g \Rightarrow g'$ we have $neigh(g, v) = neigh(g', R)$. Hence, c records the labels of the neighbours of the nonterminal vertex v and edges do not go lost by a rewriting step.

Engelfriet and Leih [EL89] and Rozenberg and Welzl [RW86] have shown that the normal form is no restriction for the generative power of linear and boundary graph grammars, respectively.

Lemma 7. *It is decidable whether or not a linear graph grammar is connected. For every linear graph grammar GG there is a linear graph grammar GG' in normal form generating all connected graphs of $L(GG)$, i.e., $L(GG') = \{g \in L(GG) \mid g$ is connected $\}$.*

3 Graph Automata

Before we give a formal definition of a graph automaton let's take a look at finite state automata and tree automata and explore some analogies. Recall the way a nondeterministic finite automaton A works, and suppose that it makes no λ-moves. In each move A reads a new symbol of the input string and changes states. The change of states is determined by the transition function. The automaton accepts, if the input string is completely scanned and it has reached a final state. If A is a one-way automaton, then it reads every letter exactly once. It uses its read-head to separate the already scanned part of the input string from the part yet to be visited. The read-head plays the role of a separator.

Next reconsider a bottom-up tree automaton, see [GS84]. Initially, it marks each leaf of an input tree with a final state. In each computation step a new node v with label a will be marked with some state s, if its sons $v_1, v_2, \ldots, v_n$ were marked with the states $s_1, s_2, \ldots, s_n$, respectively, and there is an instruction $\delta(s_1, s_2, \ldots, s_n, a) \to s$. The bottom-up tree automaton accepts an input tree t, if the root of t is marked with an initial state. Again, the roots of the already marked subtrees are a separator. The scenario for top-down tree automata is similar, now starting at the root and finishing, when all leaves are marked by a final state.

A graph automaton GA consists of a finite state control and a collection of edge-heads. The number of edge-heads is unbounded and depends on the edge-search number of the given input graph [BS91]. The edge-heads mark the current separator edges. The end nodes of the separator edges are the ports of the subgraph visited so far.

In each step, the graph automaton reads a new subgraph according to some instruction and checks its compatibility. The compatibility is described by augmented graphs with node labels with three components. It includes a dynamic check for connectivity and neighbourhood preservation. Then the automaton changes states and moves its edge-heads beyond the read subgraph such that they occupy the new separator edges. Graph automata are nondeterministic in several respects. There is a choice of the next state and of the subgraph to be read and there must be a proper initialization. Otherwise the computation will fail. This is made such that a graph automaton can reconstruct the derivations of its associated connected linear graph grammar.

An augmented graph h over the base-alphabet T is a node labeled graph $h = (V, E, m)$ such that $m : V \to T \times P(T) \times \{0,1\}$. For $i = 1, 2, 3$ let m_i denote the projection of m onto the i-th component. If $m(v) = (a, X, j)$, then $m_1(v) = a$ is the ordinary node label, $m_2(v) = X$ describes a set of node labels for neighbours of v and $m_3(v) = j$ indicates the existence or nonexistence of other edges. A graph g and an augmented graph h are taken as isomorphic, if they are isomorphic on the ordinary node labels. Moreover, we identify a graph and its isomorphic copy.

Now we are ready to define graph automata and their computations.

Definition 8. A *graph automaton* $GA = (Q, T, \delta, q_o, F)$ consists of a finite set of states Q, the alphabet T of node labels, the start state $q_o \in Q$, the set of final states $F \subseteq Q$, and the transition function $\delta : Q \times \Gamma \times P(T) \to P(Q)$, where Γ is a finite set of augmented graphs. Each such tuple $(q, \gamma, Y) \to q'$ is an instruction of GA.

Let $g = (V, E, m)$ be a connected graph, which is an input to GA. A *configuration* of GA on g is a pair $K = (q, h)$, where $q \in Q$ is the current state and $h = g|_U$ is an induced subgraph of g.

Since an induced subgraph $h = g|_U$ of a connected graph g is completely determined by h, by the set of vertices U, by the edge-separators $S = sep(g, h)$ and by the set of ports $R = port(g, h)$ we may replace h by any of U, S, or R. Furthermore, $S = \emptyset$ iff $R = \emptyset$ iff $h = \emptyset$ or $h = g$.

An instruction $(q, \gamma, Y) \to q'$ of a graph automaton defines a *computation step* on configurations $K \vdash K'$. Let $K = (q, k)$ and let $\gamma = (W, D, m)$ be an augmented graph, such that its projection onto the first component is $\gamma_1 = (W, D, m_1)$. Then the instruction $(q, \gamma, Y) \to q'$ is *applicable* to K if the following holds: There is an induced subgraph of $g - k$, which is isomorphic to γ_1. The isomorphic copy of $\gamma_1 = (W, D, m_1)$ is new and is read in this computation step. Let $m(w) = (a, X, j)$ be the augmented node label of a new node $w \in W$.

1. For every node label b, $b \in X$ iff there is a port $u \in port(g, k)$ with label b and an edge $\{u, w\}$ between u und w and for every port $u' \in port(g, k)$ with label b there is an edge $e = \{u', w\}$ from u' to w.
   Hence, if $e = \{u, w\}$ is an edge between a new node $w \in W$ and some port $u \in port(g, k)$, then $m(g)(u) \in X$.
2. Moreover, $j = 0$ iff $\{w, w'\} \in E(g)$ implies $w' \in port(g, h) \cup W$. I.e., w is directly connected only with "old" nodes.
3. Finally, for the third component of the instruction, $b \in Y$ implies that there is a port $u \in port(g, k)$ with label b and an edge between u and some new node $z \in V(g) - (V(k) \cup W)$, and this holds for all b-labeled ports. Conversely, if $e = \{u, z\}$ is an edge from a port u to some new node $z \in V(g) - (V(k) \cup W)$, then e is registered in Y by the label $m(u)$.

If the ports, $port(g, k)$ and the new set of nodes W satisfies the conditions set by X, j and Y, then it was legal to read the subgraph induced by W and the instruction can be executed. Then $K \vdash K'$ where $K' = (q', k')$ and $h' = k \cup \gamma_1$. As usual, let $K \vdash^* K'$ denote the transitive closure of $\vdash$ such that $K \vdash^* K'$ describes a computation of a graph automaton from configuration K to configuration K'.

The graph automaton halts, if there is no applicable instruction or if it has deactivated all its edge-heads. Then it may accept. GA starts in the initial state q_o with all its edge-heads deactivated. Thus, the *language* accepted by final state and deactivation of all edge-heads is $L(GA) = \{g \mid g \text{ is connected}, (q_0, \emptyset) \vdash^*_{(q,g)}$ and $q \in F\}$.

The operational view to a computation step $K \vdash K'$ by the instruction $(q, \gamma, Y) \to q'$ with $\gamma = (W, D, m)$ is as follows. The graph automaton is in state q. Its edge-heads visit the set of separating edges in direction from h to the complementary graph $g - k$. The node-heads guard the ports $port(g, k)$. Then GA reads an isomorphic copy $g|_W$ of γ_1 in the rest graph $g - k$. If there exist ports, then every node $w \in W$ is connected to some port by a path whose last and only edge is a separator edge. Thus some heads search the copy of γ_1 starting from separator edges. Furthermore, if $m(w) = (a, X, j)$ is the augmented node label of some node $w \in W$ and $X \neq \emptyset$, then the isomorphic copy of w is directly connected to all ports u whose label is in X, and for each such label $b \in X$ there exists such a port and such an edge. From the viewpoint of a port u with label b, every separator edge $e = (u, w)$ is registered by the augmented label b at w. If there is such a port, then all ports u with this label are treated the same and are put into one class. If $j = 0$, then all edges incident with w either are separator edges or are edges from γ_1. Otherwise, if $j = 1$, there is at least one edge from w to a yet unvisited node of g. The graph automaton will clear the edges of the read subgraph and the separator edges between the ports of $port(g, h)$ and the nodes from the read subgraph γ_1. Some edge-heads will be advanced to the edges between the nodes of γ_1 and the rest graph. The graph automaton moves some node-heads to those new nodes w of γ_1, whose augmented label $m(w) = (a, X, j)$ has $j = 1$. These nodes become new ports. It removes node-heads from those old ports, whose incident edges are all cleared. These are exactly the ports $u \in port(g, h)$ whose label $m(u)$ is not recorded in the third component Y of the instruction. If all these checks succeed, then the graph automaton can excecute the instruction and it enters the next state q'.

Example 2. Let $GA = (Q, T, \delta, q_0, F)$ be a graph automaton with $Q = \{q_0, q_1, q_2, f\}$, $T = \{a, b, c\}$, $F = \{f\}$ and the instructions $(q_0, \gamma_1, \emptyset) \to \{f\}$, $(q_0, \gamma_2, \emptyset) \to \{q_1\}$, $(q_1, \gamma_3, \emptyset) \to \{q_1\}$, and $(q_1, \gamma_4, \emptyset) \to \{f\}$, where $\gamma_i = (U_i, E_i, m_i)$ for $1 \leq i \leq 4$, are the augmented graphs given below in Fig. 3.

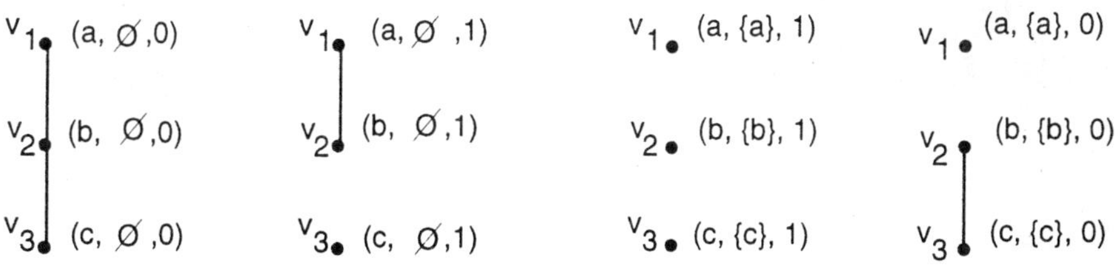

v_1 (a, ∅, 0) v_1 (a, ∅, 1) v_1 (a, {a}, 1) v_1 (a, {a}, 0)

v_2 (b, ∅, 0) v_2 (b, ∅, 1) v_2 (b, {b}, 1) v_2 (b, {b}, 0)

v_3 (c, ∅, 0) v_3 (c, ∅, 1) v_3 (c, {c}, 1) v_3 (c, {c}, 0)

Fig. 3. The augmented graphs $\gamma_1, \gamma_2, \gamma_3,$ and γ_4

On an input $a^n b^n c^n$ with $n \geq 1$, which should be drawn like an "S", GA

can only apply the second instruction. It must read the first a-node and the first b-node in the upper left corner of the "S", which are connected by an edge and it can read any c-node. However, if it does not read the first c-node in the lower corner of the "S", it will run into an error. If $n \geq 3$ and it picks a c-node in the middle, then after the next move, there are two c-nodes, which are ports, and each of these c-ports must be connected by the other c-nodes. Similarly, if GA picks the last c-node that is connected to the b-node, there will be two c nodes, which are ports and which must be connected to the other c-nodes. If the proper nodes are read in the first step, then the graph automaton works deterministically and sweeps over the "S" from left to right, reading the next a-, b-, and c-nodes in its next move.

Observe, that the automaton GA is the associate of the linear graph grammar GG generating $\{a^n b^n c^n \mid n \geq 1\}$. A computation of the automaton is shown in Fig. 4.

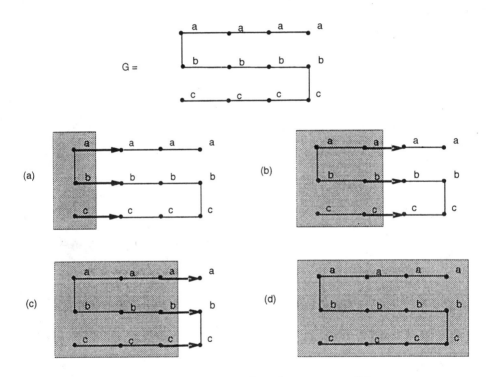

Fig. 4. A computation of graph automaton GA

Lemma 9. *Let $GA = (Q, T, \delta, q_0, F)$ be a graph automaton and let $(q, h) \vdash (g', h')$ be a computation step by some instruction $(q, \gamma, Y) \to q'$. Then GA deactivates all its edge-heads if and only if $Y = \emptyset$ and $j = 0$ for every node $w \in V(\gamma)$ with augmented node label (a, X, j).*

We are going to simplify the GA. A graph automaton $GA = (Q, T, \delta, q_0, F)$ is quasi-deterministic, if $|\delta(p, \gamma, Y)| = 1$.

Using the power set construction known for finite state automata we obtain:

Lemma 10. *For every graph automaton $GA = (Q, T, \delta, q_0, F)$ there exists a quasi-deterministic graph automaton $GA' = (Q', T', \delta', q_0', F')$, such that $L(GA) = L(GA')$.*

Although GA' has no choice for its next state, it is not deterministic in the classical sense. From the example above it is easy to see that a graph automaton must be initialized properly. This is the first place for nondeterminism. Also, there may be a choice, which subgraph should be read. This can easily be seen from a cycle with a-labeled nodes.

For an illustration see our example with chains of the form $a^n b^n c^n$. In the first step the graph automaton must choose the nodes with labels a and b, which are connected. On the a's and b's it proceeds deterministically, and picks the next neighbours. However, it may choose any c but the last, if $n > 0$. But if it picks a c-node in the middle, there will be two c-ports after one step. Each of them must be connected to the next c-node, but these edges do not exist. Hence GA would fail and run into an error.

For our main result on the equivalence of graph automata and connected linear graph grammars the following observation is useful.

Definition 11. Let $p = (A, R, C)$ be a production of a linear graph grammar GG in normal form. The augmented graph $\gamma(p)$ associated with p is the subgraph of R induced by the terminal nodes W of R, where these nodes have augmented labels $m(w) = (a, C^{-1}(v), j)$ with $a = m(R)(w)$ and $j = 1$ iff the nonterminal node v of R is directly connected to w by an edge $\{v, w\}$.
Moreover, the pair $(\gamma(p), C^{-1}(v))$ completely characterizes the production p, where v is the nontermial node of R. Conversely, if GG is in normal form, then $(\gamma(p), C^{-1}(v))$ can be constructed from p and the context describing function. If R is a terminal graph, then $j = 0$ for all augmented node labels and $Y = \emptyset$, where $Y = C^{-1}(v)$, and conversely. This one-to-one correspondence is the key to the equivalence of linear graph grammars and graph automata.

Theorem 12. *For every connected linear graph grammar GG in normal form there is a graph automaton GA such that $L(GG) = L(GA)$.*

Proof. For $GG = (N, T, P, S)$ construct the associate graph automaton $GA = (Q, T, \delta, q_0, F)$.
Let $Q = N \cup \{f\}$, where f is a new state. Define $F = \{f\}$ and $q_0 = S$.
Every production $p = (A, R, C)$ of GG is transformed one-to-one into an associate instruction $(A, \gamma(p), Y) \to A'$, such that $\gamma(p)$ is the augmented graph associated with p and A' is the label of the single nonterminal node label v' of

R or $A' = f$, if R is a terminal graph. Let $Y = C^{-1}(v') = \{a \in T \mid (a, v') \in C\}$ and $Y = \emptyset$, if v' does not exist.

It remains to prove by induction that derivations of GG translate one-to-one into computations of GA, and vice-versa. Let $S \Rightarrow^* h \Rightarrow h' \Rightarrow^* g$ be a derivation of some graph $g \in L(GG)$, where $h \Rightarrow h'$ is obtained by applying a production (A, R, C) to the nonterminal node v. Let U be the set of terminal nodes of h and v its nonterminal node. Let W be the set of terminal nodes of R, which is nonempty, and let v' be its nonterminal node, which may not exist. Then $V(h) = U \cup \{v\}$ and $V(h') = U \cup W \cup \{v'\}$ or $V(h') = U \cup W$, if h' is terminal. Let $K_0 \vdash^* K \vdash K' \vdash^* K_f$ be the associated computation of GA on g, such that $K = (A, k)$ is associated with h iff the following invariant holds.

(*) The graphs h and k coincide on the terminal nodes, i.e. $k = h|_U$, where U is the set of terminal nodes of h. If h is not terminal, then $V(h) = U \cup \{v\}$, where v is the nonterminal node of h. Then $m(v) = A$; otherwise $A = f$. Moreover, the neighbours of the nonterminal node coincide with the ports of k, $\{u \in V(h) \mid \{u, v\} \in E(h)\} = port(g, k)$. Finally, the labels of these nodes are stored both in the context describing set $c(A)$ of the nonterminal A and in the second components X of the augmented node labels $m(w) = (a, X, j)$ of an applicable instruction.

This invariant holds for S and K_0, since all relevant sets are empty.
Suppose that (*) holds for h and $K = (A, k)$. If the production $p = (A, R, C)$ is applied to the nonterminal node v of h such that $h \Rightarrow h'$, then $K \vdash K'$ by the instruction $(A, \gamma(p), Y) \to A'$ and h' and K' are associated. To see this, first observe that $(A, \gamma(p), Y) \to A'$ is applicable to K. Since GG is chain-free, W and $R|_W$ are nonempty. Hence, $\gamma(p)$ is nonempty. Let $W = V(\gamma(p))$ up to isomorphism and augmented labels.

For $w \in W$ let $m(w) = (a, X, j)$ be its augmented label. Then $b \in X$ iff there is a port $u \in U$ with $m(u) = b$ and $\{u, w\} \in E(g)$ iff there is an edge $\{u, v\} \in E(h)$ and $(b, w) \in C$. This holds for all nodes $u \in U$ with label b, which are connected to the nonterminal node v, and these are the ports of k with label b.
Moreover, for $w \in W$ there is an edge $\{w, v'\} \in E(R)$ iff $j = 1$ in the augmented label $m(w) = (a, X, j)$.
Finally,

R has a nonterminal node v' iff there is an edge $\{u, v'\} \in E(h')$ with $u \in U$
iff $\{u, v\} \in E(h)$ and $(m(u), v') \in C$
iff $u \in port(g, k)$ and $m(u) \in Y$.

Now, the application of $(A, \gamma(p), Y) \to A'$ to K yields $K' = (A', k')$, where

$V(k') = V(k) \cup W$, the nodes have the proper terminal labels, and

$E(k') = E(k)$
$\cup\ E(R|_W)$
$\cup\ \{\{u,w\} \mid u \in port(g,k), m(w) = (a,X,j) \text{ and } m(u) \in X\}.$

This coincides with the set of edges of h' between terminal nodes,

$E(h'|_{U \cup W} = E(h|_U)$
$\cup\ E(R|_W)$
$\cup\ \{\{u,w\} \mid u,w \in W, \{u,v\} \in E(h) \text{ and } (m(u),w) \in C\}.$

Finally,

$port(g,k') = \{u \in port(g,k) \mid m(u) \in Y\} \cup \{w \in W \mid m(w) = (a,X,j) \text{ and } j = 1\}$
$= \{u \in V(h') \mid \{u,v'\} \in E(h')\}.$

Hence, the invariant (*) holds for h' and K'.
Conversely, by the same reasoning, if $K \vdash K'$ by the instruction $(A, \gamma(p), Y) \to A'$ and (*) holds for K and h, then $h \Rightarrow h'$ by the application of $p = (A, R, C)$ to the nonterminal node v of h, and (*) holds for K' and h'.
If (*) holds for a terminal graph g and a configuration $K = (A, k)$, then $A = f$ and $k = g$. Hence, $L(GG) = L(GA)$.

Example 3. Consider the connected linear graph grammar GG in normal form of example 1 which generates the language $a^n b^n c^n$. The equivalent graph automaton $GA = (Q, T, \delta, q_0, \{f\})$ associated with GG is the graph automaton GA of example 2.

For the converse simulation there is again a one-to-one transformation from instructions to productions. Here, the application conditions for the instructions are translated into the context describing function for the productions. This prevents a misuse of productions. Moreover, disconnectivity must be excluded.

Theorem 13. *For every graph automaton GA there is a linear graph grammar GG such that $L(GA) = L(GG)$. Moreover, GG is connected and in normal form.*

Proof. Suppose that $GA = (Q, T, \delta, q_0, F)$ has only a single final state f, which is reached only at termination, when the edge-heads are deactivated. Furthermore, instructions are excluded, which would disconnect a graph, when they are translated into productions. This happens, when there are no edges to ports, i.e. if for some instruction $(q, \gamma, Y) \to q'$ with $q' \neq t$, the set Y and all sets X, where (a, X, j) is the augmented label of the nodes of γ, are empty. Such instructions are deleted from GA.

Let $GG = (N, T, P, S)$. The set of nonterminals $N \subseteq Q \times P(T)$ consists of pairs of states and sets of terminal node labels and is constructed with the productions. The second components are the context describing function. Let $S = (q_0, \emptyset)$. For every instruction $(q, \gamma, Y) \to q'$ of GA there is an associate

production $p = (A, R, C)$ of GG. Let $\gamma = (W, E, m)$ be an augmented graph with $m(w) = (a, X, j)$ for every node w.
Define
$$X(\gamma) = \{a \in T \mid a \in X \text{ and } w \in W\} \text{ and}$$
$$Z(\gamma) = \{a \in T \mid m(w) = (a, X, 1) \text{ for } w \in W\}.$$
$X(\gamma)$ collects all node labels stored in the second components of the augmented node labels of the nodes of γ. $Z(\gamma)$ collects all node labels, whose third component is set to one.
Then $A = (q, X(\gamma) \cup Y)$.
If $Z(\gamma) \cup Y \neq \emptyset$, then
$$R = (W', E', m') \text{ with}$$
$W' = W \cup \{v'\}$ for some new nonterminal node v' and terminal nodes W,
$m'(w) = m_1(w) \in T$ for $w \in W$,
$m'(v') = (q', Z(\gamma) \cup Y)$, and
$E' = E \cup \{\{w, v'\} \mid w \in W \text{ and } m(w) = (a, X, j) \text{ with } j = 1\}$.
If $Z(\gamma) \cup Y = \emptyset$ and $q' = f$ is the final state, then there is no nonterminal node v' and $V = V'$ and $E = E'$.
Let $C = \{(b, w) \mid w \in W, m(w) = (a, X, j) \text{ and } b \in X\} \cup \{(b, v') \mid b \in Y\}$.

By construction, GG is chain-free, neighbourhood preserving and context consistent, where the context describing function is the projection onto the second components of the nonterminals. Moreover, GG is connected, since the instance for disconnectivity is deleted from GA. It remains to prove that computations of GA correspond one-to-one to derivations of GG. This follows along the lines of the proof of Theorem 1.

Let $K_0 \vdash^* K \vdash K' \vdash^* K_f$ be a computation of GA on some connected graph g with $K \vdash K'$ by the instruction $(q, \gamma, Y) \to q'$. Then there is an associated derivation $S \Rightarrow^* h \Rightarrow h' \Rightarrow^* g$ in GG with $h \Rightarrow h'$ by the production $p = (A, R, C)$, and conversely. If the invariant (*) holds for K and h, then it does so for K' and h'.
Hence L(GA) = L(GG).

Example 4. Let GA be the automaton of example 2. The equivalent linear graph grammar GG associated with GA is shown in Fig.5. The graph language of GG is exactly $\{a^n b^n c^n \mid n \geq 1\}$.

Combining these results we obtain the equivalence of connected linear graph grammars and graph automata.

Theorem 14. *For graph languages of connected graphs the following are equivalent:*
(1) L is generated by a connected linear graph grammar.
(2) L is accepted by a finite graph automaton.

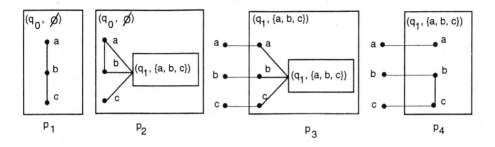

Fig. 5. The linear graph grammar associated with GA

Observe, that finite graph automata are more powerful than finite state automata, when they are used to recognize strings as chains of labeled graphs. Our running example $\{a^n b^n c^n \mid n \geq 1\}$ is a famous witness.

For the complexity, given a linear NCE graph grammar GG the construction of an equivalent graph automaton GA takes polynomial space in the size of GG.

Finally, let's consider the complexity of the membership problem for connected graphs generated by linear NCE graph grammars. The membership problem is known to be in NP, and is in NL for linear graph grammars of bounded degree [EL89]. In fact, the membership problems are complete for these classes. These bounds are easy to see for graph automata. A graph automaton runs in linear time. Using a standard representaton of graphs it can be simulated in nondeterministic polynomial time, which shows the inclusion in NP. If the graph grammars has bounded degree, then the constructed graph automaton has a bounded number of edge-heads. Such automata can be simulated on logarithmic space, which shows the inclusion in NL.

4 Conclusion

Our main result shows that the graph automata introduced in this paper recognize exactly the graph languages of connected linear graph grammars. This is the first step towards a systematic study of various types of graph automata, which closely correspond to the major classes of node-replacement graph grammars. Canonical extensions are pushdown graph automata, which store sets of edge-separators in a stack and correspond to the boundary graph grammars. Another important issue are restrictions to graph automata with a bounded number of edge-heads. These should correspond to graph languages of bounded degree, and finally we shall investigate deterministic graph automata, which may have a choice for the initialization and then proceed in a unique fashion. Last but not least we are interested in types of graph automata which realize general graph searching strategies.

References

[BS91] D. Bienstock, P. Seymour. Monotonicity in graph searching. *J. Algorithms 12 (1991), 239-245.*

[Bra95] F.J. Brandenburg. Designing graph drawings by layout graph grammars. *Proc. Workshop on Graph Drawing 94, LNCS 894 (1995), 416-427.*

[Cou90] B. Courcelle. Graph rewriting: an algebraic and logic approach. *Handbook of Theoretical Computer Science, Elsevier, Amsterdam, (1990) 193-242.*

[EKR91] H. Ehrig, H.J. Kreowski, G. Rozenberg. Proc. 4. Workshop on Graph Grammars and Their Application to Computer Science, *LNCS 532 (1991).*

[Eng89] J. Engelfriet. Context-free NCE graph grammars. *Proc. FCT 89, LNCS 380 (1989), 148-161.*

[EL89] J. Engelfriet, G. Leih. Linear graph grammars: power and complexity. *Inform. Comput. 81 (1989), 88-121.*

[GS84] F. Gécseg, M. Steinby. Tree Automata. Akadémiai Kiadó, Budapest (1984).

[Hic94] T. Hickl. Rechtwinkliges Layout von hierarchisch strukturierten Graphen. *Dissertation, Universität Passau (1994).*

[JRW86] D. Janssens, G. Rozenberg, E. Welzl. The bounded degree problem for NLC graph grammars is decidable. *J. Comput. System Sci. 33 (1986), 415-422.*

[Kau87] M. Kaul. Practical applications of precedence graph grammars. *Proc. 3. Workshop on Graph Grammars and their Application to Computer Science, LNCS 291 (1987), 326-342.*

[LaP93] A.S. LaPaugh. Recontamination does not help to search a graph. *J. Assoc. Comput. Mach. 40 (1993), 224-245.*

[MHGJP88] N. Megiddo. S.L. Hakimi, M. R. Garey, D.S. Johnson, C.H. Papadimitriou. The complexity of searching a graph. *J. Assoc. Comput. Mach. 35 (1988), 18-44.*

[Nag79] M. Nagl. Graph Grammatiken. *Vieweg, Braunschweig (1979).*

[PR69] J.L. Pfaltz, A. Rosenfeld. Web Grammars. *Proc. Joint Intern. Conference on Artificial Intelligence, Washington, D.C., (1969), 609-619.*

[Rem94] E. Remila. Fundamental study - Recognition of graphs by automata. *Theor. Comput. Sci. 136 (1994), 291-332.*

[RM72] A. Rosenfeld, D.L. Milgram. Web automata and web grammars. *Machine Intelligence 7 (1972), 307-324.*

[RW86] G. Rozenberg, E. Welzl. Boundary NLC graph grammars - Basic definitions, normal forms and complexity. *Inform. Control 69 (1986), 136-167.*

[WR79a] A. Wu, R. Rosenfeld. Cellular graph automata I. *Inform. Control 42 (1979), 305-329.*

[WR79b] A. Wu, R. Rosenfeld. Cellular graph automata II. *Inform. Control 42 (1979), 330-353.*

The Obstructions of a Minor-Closed Set of Graphs Defined by Hyperedge Replacement Can Be Constructed

B. COURCELLE and G. SÉNIZERGUES

LaBRI (URA CNRS 1304) Bordeaux I University, 351 Cours de la Libération 33405 Talence - France ** ***

Abstract. : We establish that the finite set of obstructions of a minor-closed set of graphs given by a hyperedge replacement grammar can be effectively constructed. Our proof uses an auxiliary result stating that the system of equations associated with a proper hyperedge replacement grammar has a unique solution.

Introduction

Sets of finite graphs can be finitely specified in several ways among which we shall consider in the present paper

- certain graph grammars and in particular the so-called *hyperedge replacement (HR) grammars,*

- logical formulas and in particular the formulas of *monadic second-order (MS) logic,*

- forbidden configuration and in particular *forbidden minors.*

We refer the reader to Courcelle [Cou93] for a survey. (Terminology is defined in Sections 1 and 4). In particular, every minor closed set of graphs of bounded tree-width can be defined

- either by a HR grammar,
- or by an MS formula,
- or by a finite set of forbidden minors.

A natural question is whether the grammar can be computed from the formula or from the minors and vice-versa. The following results are known :

- from the forbidden minors (also called the *obstructions*) one can construct a HR grammar, an MS formula and an upperbound on the tree-width of the graphs,
- from an MS formula and an upperbound on the tree-width of the graphs one can construct a HR grammar and the obstructions.

We complete the picture by showing the following new result:

-from a HR grammar, one can construct the set of obstructions (theorem (4.7)).

This solves an open problem presented in [Cou92, Cou91b].

** email addresses : courcell,ges@labri.u-bordeaux.fr
*** This work has been supported by the ESPRIT Basic Research Working Group COMPUGRAPH II

Rather than a usable algorithm we give a *computability proof* : our method consists in the enumeration of "*candidate* sets of obstructions" and a *test* that a given set is satisfactory. This test is possible by means of already known decidability results for MS properties of the graphs generated by HR graph grammars, together with a new result stating that certain *proper* HR graph grammars have *unique* solutions when appropriately considered as systems of equations in sets of graphs. This technique is (unfortunately) intractable. We do not hope to obtain from it the set of obstructions for classes of graphs, like that of graphs of tree-width at most 4 (the obstructions are unknown for this class). The paper is organized as follows : Section 1 contains only definitions and can be skipped by the reader knowing graph grammars ; Section 2 deals with proper HR grammars in the general case where they generate hypergraphs ; Section 3 gives a construction of the minor-closure of a HR set of graphs ; Section 4 is devoted to the main result.

1 Hypergraphs and HR hypergraph grammars

We review the basic definitions from [BC87]. As in this paper, we deal with a certain class of oriented hypergraphs. The reader knowing [Cou91a] may skip this section.

1.1 Hypergraphs

The *hypergraphs* we define have labelled hyperedges. The alphabet of hyperedge labels is a *ranked alphabet* A, i.e., an alphabet that is given with a mapping $\tau : A \to \mathbb{N}$ (the integer $\tau(a)$ is called the *type* of a). A *hypergraph over* A *of type* n is a 5-tuple $H = <\mathbf{V}_H, \mathbf{E}_H, \mathbf{lab}_H, \mathbf{vert}_H, \mathbf{src}_H>$ where $\mathbf{V}_H$ is the finite set of vertices, $\mathbf{E}_H$ is the finite set of hyperedges, $\mathbf{lab}_H$ is a mapping $\mathbf{E}_H \to A$ defining the *label* of a hyperedge, $\mathbf{vert}_H$ is a mapping $\mathbf{E}_H \to \mathbf{V}_H^*$, defining the (possibly empty) *sequence of vertices* of a hyperedge, and $\mathbf{src}_H$ is a sequence of vertices of length n. We impose the condition that the length of $\mathbf{vert}_H(e)$ is equal to $\tau(\mathbf{lab}_H(e))$ for all e in $\mathbf{E}_H$. One may also have labels of type 0, labelling hyperedges with no vertex. An element of $\mathbf{src}_H$ is called *a source* of H. The sets $\mathbf{E}_H$ and $\mathbf{V}_H$ are assumed to be finite and disjoint.

We denote by $\mathbf{G}(A)$ the set of all hypergraphs over A, by $\mathbf{G}(A)_n$ the set of those of type n. A hypergraph of type n is also called an n-*hypergraph*.

A *graph* is a hypergraph all hyperedges of which are of type 2. A graph is thus directed (unless otherwise specified).

We consider two isomorphic hypergraphs as equal.

We now define the substitution of a hypergraph for a hyperedge in a hypergraph.

1.2 Substitutions

Let $G \in \mathbf{G}(A)$, let $e \in E_G$; let $H \in \mathbf{G}(A)$ be a hypergraph of type $\tau(e)$. We denote by $G[H/e]$ the result of the *substitution* of H for e in G. This hypergraph can be constructed as follows :

(1) construct a hypergraph G' by deleting e from G (but keep the vertices of e) ;
(2) add to G' an isomorphic copy $\overline{H}$ of H, disjoint from G' ;
(3) fuse the vertex $\text{vert}_G(e, i)$, i.e., the ith element of the sequence $\text{vert}_G(e)$ (that is still a vertex of G'), with the ith source of $\overline{H}$; this is done for all $i = 1, \ldots, \tau(e)$;
(4) the sequence of sources of $G[H/e]$ is the image of the sequence of sources of G' under the identifications induced by step (3).

If $e_1, \ldots, e_l$ are pairwise distinct hyperedges of G, if $H_1, \ldots, H_l$ are hypergraphs of respective types $\tau(e_1), \ldots, \tau(e_l)$, then the substitutions in G of H_1 for $e_1, \ldots, H_l$ for e_l can be done in any order ; the result is the same, and it is denoted by $G[H_1/e_1, \ldots, H_l/e_l]$.

1.3 Hypergraph operations

Let U be another finite ranked alphabet, disjoint from A. We call U the alphabet of *unknowns*. Let $U = \{u_1, u_2, \ldots, u_m\}$. Let $G \in \mathbf{G}(A \cup U)_n$. Let us fix some injective enumeration $e_1, e_2, \ldots, e_l$ of the hyperedges of G having a label in U, let us suppose $label(e_j) = u_{i_j}$ and let $n_i = \tau(e_i)$. The operation

$$\bar{G} : \mathbf{G}(A)_{n_1} \times \ldots \times \mathbf{G}(A)_{n_l} \longrightarrow \mathbf{G}(A)_n$$

is defined by:

$$\bar{G}[H_1, H_2, \ldots, H_m] = G[H_{i_1}/e_1, H_{i_2}/e_2, \ldots, H_{i_l}/e_l]$$

In the sequel we use the notation $G[e_1, e_2, \ldots, e_l]$ to denote a graph G together with some injective enumeration $e_1, e_2, \ldots, e_l$ of its U-labelled hyperedges. The operation

$$\hat{G} : \mathcal{P}(\mathbf{G}(A)_{n_1}) \times \ldots \times \mathcal{P}(\mathbf{G}(A)_{n_l}) \longrightarrow \mathcal{P}(\mathbf{G}(A)_n)$$

is defined by:

$$\hat{G}[L_1, L_2, \ldots, L_m] = \{\bar{G}[H_1, H_2, \ldots, H_m] \mid \forall i \in [1, m], H_i \in L_i\}$$

1.4 HR grammars

A *Hyperedge Replacement* grammar is a 4-tuple $\Gamma = (A, U, Q, Z)$ where A is the finite *terminal* ranked alphabet, U is the finite *nonterminal* ranked alphabet, Q is the finite set of *production rules*, i.e., is a finite set of pairs of the form $(u, D) \in U \times \mathbf{G}(A \cup U)_{\tau(u)}$ usually written $u \to D$, and $Z \in U$ is a non-terminal symbol called the *axiom*.

The *one-step derivation* relation, $\xrightarrow[Q]{}$ is defined by:

$K \xrightarrow[Q]{} H$ iff there exists a hyperedge e in K, the label of which is some u in U, and a production rule (u, D) in P, such that $H = K[D/e]$.

The *derivation* relation $\xrightarrow[Q]{*}$ is the reflexive and transitive closure of $\xrightarrow[Q]{}$.

For every hypergraph $K \in \mathbf{G}(A \cup U)_n$, we define

$$\mathbf{L}(\Gamma, K) := \{H \in \mathbf{G}(A)_n \mid K \xrightarrow[Q]{*} H\},$$

The set of hypergraphs *generated* by Γ is defined by:

$$\mathbf{L}(\Gamma) := \mathbf{L}(\Gamma, Z)$$

A set of graph is HR iff it is equal to the language generated by some HR grammar Γ. We shall simply say a grammar in the sequel.

1.5 Systems of equations in sets of hypergraphs

Let $\Gamma = (A, U, Q, Z)$ be a grammar. Let us suppose:

$$Q = \{ u_i \longrightarrow D_{i,j} \mid 1 \leq i \leq m, 1 \leq j \leq n_i \}$$

The *system of equations* S_Γ associated with Γ is then the set of equations:

$$u_i = \sum_{j=1}^{n_i} D_{i,j}$$

for $1 \leq i \leq m$.

A m-tuple $(L_1, L_2, \ldots, L_m) \in \mathcal{P}(\mathbf{G}(A)_{n_1}) \times \ldots \times \mathcal{P}(\mathbf{G}(A)_{n_m})$ is then a *solution* of S_Γ iff:

$$\forall i \in [1, m], L_i = \bigcup_{j=1}^{n_i} \widehat{D_{i,j}}(L_1, L_2, \ldots, L_m)$$

We denote the formal sum $\sum_{j=1}^{n_i} D_{i,j}$ by t_i and the operation $\sum_{j=1}^{n_i} \widehat{D_{i,j}}$ by $\hat{t}_i$.

Theorem 1.1 *([BC87])* : $(\mathbf{L}(\Gamma, u_1), \ldots, \mathbf{L}(\Gamma, u_m))$ *is the least solution of* S_Γ.

2 Proper hyperedge replacement grammars

We denote by $\mathbf{I}_G$ the set of internal vertices of a hypergraph G i.e., those that are not sources. For every hypergraph G, we let $\|G\|$ be the integer $Card(\mathbf{I}_G) + Card(\mathbf{E}_G)$ and we call it the *size* of G. We say that a graph is *empty* iff $\|G\| = 0$.

There is only one empty 0-hypergraph. An empty n-hypergraph, for $n \geq 1$, has vertices, namely the sources (possibly all identical).

For every hypergraph G, we let $\varepsilon(G)$ denote the empty hypergraph obtained by deleting the hyperedges and the internal vertices of G.

Lemma 2.1 *For all hypergraphs $H, G_1, \ldots, G_l$, if $e_1, \ldots, e_l$ are hyperedges of H of respective types $\tau(G_1), \ldots, \tau(G_l)$ we have :*

$$\| H[G_1/e_1, \ldots, G_l/e_l] \| = \| H[\varepsilon(G_1)/e_1, \ldots, \varepsilon(G_l)/e_l] \| + \| G_1 \| + \ldots + \| G_l \|.$$

Proof : Easy verification from the definitions.

A *unit hypergraph* is a graph with exactly one U-labelled hyperedge, no terminal hyperedge and no isolated internal vertex. (A vertex is *isolated* if it is not in the vertex sequence of any hyperedge). A grammar Γ is *proper* if no righthand side of a rule is unit or empty.

Lemma 2.2 *A proper grammar generates no empty graph.*

Proof : Assume we have a derivation $u \xrightarrow[\Gamma]{*} G$ with G empty. Let $u' \to D$ be a rule with terminal righthand side used in this derivation. If D has a hyperedge or an isolated internal vertex, then G has also a hyperedge or an internal vertex. Hence D should be empty. But Γ has no rule with empty righthand side. Contradiction.

Theorem 2.3 *Let Γ be a proper grammar with nonterminals $u_1, \ldots, u_m$. The m- tuple*

$$(\mathbf{L}(\Gamma, u_1), \ldots, \mathbf{L}(\Gamma, u_m))$$

is the unique m-tuple of sets of nonempty hypergraphs that is a solution of S_Γ.

Proof:
Let $L_i = \mathbf{L}(\Gamma, u_i)$ for $i = 1, \ldots, m$. Hence $(L_i)_{1 \leq i \leq m}$ is the least solution of $\mathbf{S}_\Gamma$ in sets of hypergraphs. The sets L_i contain no empty hypergraph (Lemma (2.2)). Hence $(L_i)_{1 \leq i \leq m}$ is the least solution of $\mathbf{S}_\Gamma$ in sets of nonempty hypergraphs.

Assume $(M_i)_{1 \leq i \leq m}$ is any solution in sets of non empty hypergraphs. We have $L_i \subseteq M_i$ for all i. If we do not have the equalities, we let G be a hypergraph of minimal size in $\bigcup \{M_j - L_j / 1 \leq j \leq m\}$, and $G \in M_i - L_i$.

Since $(M_j)_{1 \leq j \leq m}$ is a solution of $\mathbf{S}_\Gamma$ in nonempty hypergraphs, we have

$$G = D[G_1/e_1, \ldots, G_l/e_l]$$

for some rule of Γ:
$$u \to D[e_1, \ldots, e_l]$$
with label$(e_j) = u_{i_j}$ and $G_j \in M_{i_j}$
for all $j \in [1, l]$. By Lemma (2.1) we have
$$\| G \| = \| H[\varepsilon(G_1), \ldots, \varepsilon(G_l)] \| + \| G_1 \| + \ldots + \| G_l \|$$

Case one : $\| G_j \| < \| G \|$ for all j.
By minimality of $\| G \|$, we have $G_j \in L_{i_j}$ for each j hence $G \in L_i$ but this contradicts the choice of G.

Case two : $\| G_j \| = \| G \|$ for some j.
Since the hypergraphs $G_1, \ldots, G_l$ are all nonempty, we must have $l = 1$ and $\| D[\varepsilon(G_1)/e_1] \| = 0$. Hence D has no terminal hyperedge and no internal isolated vertex. Hence it is unit. But this means that Γ contains the rule $u_i \to D$ which has a unit righthand side. Hence Γ is not proper.

In both cases we get a contradiction. It follows that $M_j - L_j = \emptyset$ for every j hence that
$$(M_1, \ldots, M_m) = (L_1, \ldots, L_m)$$

Theorem 2.4 *(Habel [Hab93, Corollary 1.10, page 77]) : For every grammar Γ one can construct a proper grammar Γ' with the same set of non-terminals such that, for every $u \in U$:*
$$\mathbf{L}(\Gamma', u) = \{G \in \mathbf{L}(\Gamma, u) \mid G \text{ is non empty}\}.$$

Proof : For every $u \in U$ we let
$$EMPTY(u) = \{G \in \mathbf{L}(\Gamma, u) \mid G \text{ is empty}\},$$
$$UNIT(u) = \{G \mid u \xrightarrow[\Gamma]{*} G, G \text{ is a unit graph}\}.$$

These sets are finite and can be computed because the membership problem for an HR set of graphs is decidable ([Hab93, Corollary 1.6, page 75]).

We first construct a grammar Γ'' consisting of all rules of Γ together with the following ones :
all rules $\quad u \to H[H'[E_1/e_1, \ldots, E_l/e_l]/u'] \quad (1)$
where

- $H[u'] \in UNIT(u), u' \to H'$ is a rule of Γ,
- $0 \le l \le$ number of nonterminal hyperedges of H'
- $e_1, \ldots, e_l$ are nonterminal hyperedges of H' with respective labels $w_1, \ldots, w_l$
- $E_i \in EMPTY(w_i)$, for all $i \in [1, l]$
- $H[H'[E_1/e_1, \ldots, E_l/e_l]/u']$ is neither empty nor unit.

Remarks

1 - In a rule of the form (1) fulfilling the above conditions it may happen that :

- $H[u'] = u, u' = u$ (because $u \in UNIT(u)$)
- $\ell = 0$, i.e. the rule is of the form $u \to H[H'/u']$

2 - It may happen that H' is empty, $l = 0$, but $H[H'/u']$ is not empty, as in the example shown on figure 1

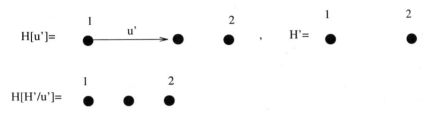

Fig. 1.

3 - It may happen that H' is unit, $l = 0$, but $H[H'/u']$ is not unit, see figure 2

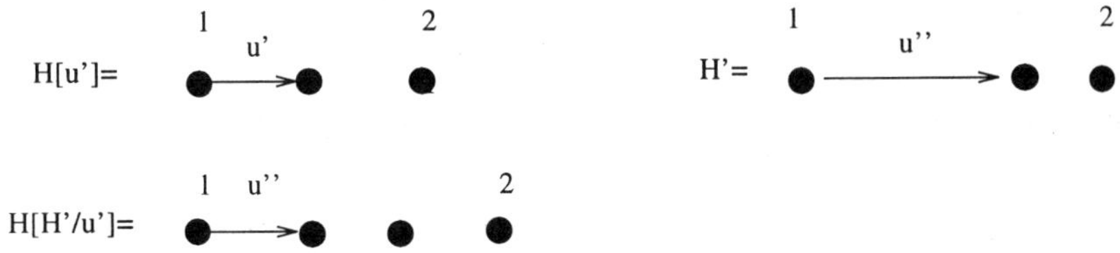

Fig. 2.

For each new rule $u \longrightarrow G$, we have $u \xrightarrow[\Gamma]{+} G$. It follows that

$$\mathbf{L}(\Gamma, u) = \mathbf{L}(\Gamma'', u) \text{ for all } u \in U.$$

We now let Γ' be obtained from Γ'' by the deletion of all rules with an empty or unit righthand side. Hence

$$\mathbf{L}(\Gamma', u) \subseteq \mathbf{L}(\Gamma'', u) \text{ for all } u \in U.$$

Since Γ' is proper $\mathbf{L}(\Gamma', u)$ is contained in the set of nonempty graphs of $\mathbf{L}(\Gamma'', u)$. Let conversely $G \in \mathbf{L}(\Gamma'', u)$ be nonempty. Let

$$d : u \longrightarrow G_1 \longrightarrow G_2 \ldots \longrightarrow G_p = G$$

be a derivation of G in Γ''.

If no rule used in this derivation has an empty or unit righthand side, then $G \in \mathbf{L}(\Gamma', u)$. Otherwise we transform d into another derivation d' of G in Γ'' which does not use rules with empty or unit righthand sides. We do the proof by induction on p (simultaneously for all $u \in U$).

Basis : $p = 1$
Then $u \longrightarrow G_1$ is a rule of Γ'', $G_1 = G$ is non empty and we take $d' = d$.

Inductive step:

Case 1 : $G_1, \ldots, G_{p-1}$ are all unit graphs.
In this case $u \longrightarrow G$ is a rule of Γ'' hence we let $d' : u \longrightarrow G$

Case 2 : G_i (for some $1 < i \leq p-1$) is the first non unit hypergraph in the sequence $G_1, \ldots, G_i$.
Then $G_{i-1} \in UNIT(u)$ and $u \longrightarrow G_i$ is a rule of Γ''.
We take
$$d' : u \longrightarrow G_i \longrightarrow G_{i+1} \longrightarrow \ldots \longrightarrow G_p = G.$$

(Note that, by point 3 of the above remark, we may have $G_{i-1} = H[u']$ and $G_i = H[H'[u'']/u']$ non-unit for some rule $u' \longrightarrow H'[u'']$ with unit righthand side .

Case 3 : G_1 is nonunit.
Then it is nonempty. We have $G_1 = H[e_1, \ldots, e_l]$ and $G = H[G'_1/e_1, \ldots, G'_l/e_l]$ where label$(e_j) = u_{i_j}$ and $u_{i_j} \xrightarrow{*} G'_j$ by some derivations d_j of length at most $p-1$. Let us assume that $G'_1, \ldots, G'_{l_0}$ are empty and the others are not.
By induction, we have derivations
$$d'_j : u_{i_j} \xrightarrow{*} G'_j$$

without unit or empty rules for $j = l_0 + 1, \ldots, l$. We have also in Γ''' a rule of the form $u \longrightarrow H'$ where

$$H' = H[G'_1/e_1, \ldots, G'_{l_0}/e_{l_0}].$$

Hence we can take for d' the derivation starting with $u \longrightarrow H'$ and continued by the derivations $d'_{l_0+1}, \ldots, d'_l$ so as to generate $H'[G'_{l_0+1}/e_{l_0+1}, \ldots, G'_l/e_l] = G$.

Let us illustrate this construction by the following :

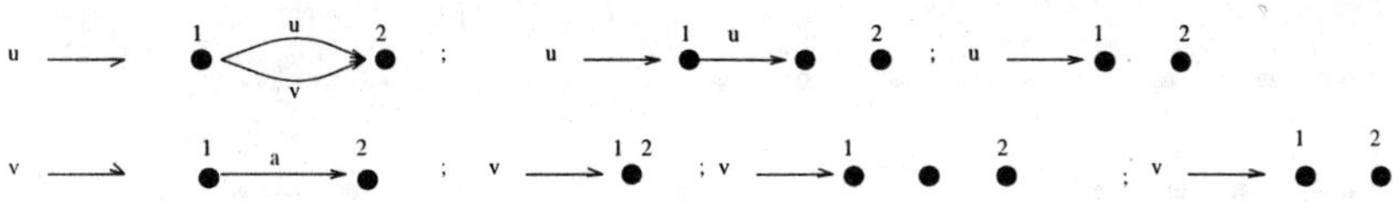

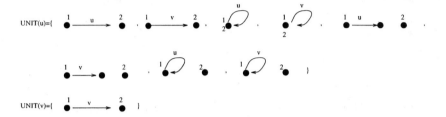

Fig. 3.

UNIT(u)={ [graphs] }

UNIT(v)={ [graph] }

Fig. 4.

Example : $\Gamma = <A, U, Q, Z>$ where $A = \{a\}, U = \{u, v\}, (\tau(u) = \tau(v) = \tau(a) = 2)$, Z=u and Q consists of the rules shown on figure 3

We have :

$$EMPTY(u) = EMPTY(v) = \{\overset{1}{\bullet}\ \overset{2}{\bullet}\ ,\ \overset{12}{\bullet}\}$$

The graphs in UNIT(u), UNIT(v) are shown on figure 4.

Then $\Gamma'' = <A, U, Q'', Z>$ where Q'' consists of the union of Q with the set of rules Q'' shown on figure 5.

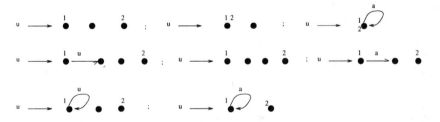

Fig. 5.

Finally, Γ' is shown on figure 6.

3 The minor closure of a HR set of graphs

In this section we let A consist of one symbol of type 2. The elements of $\mathbf{G}(A)$ are thus *directed* graphs (possibly with loops and multiple edges) and sources (we

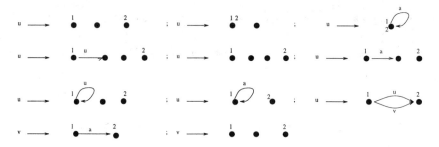

Fig. 6.

abreviate $\mathbf{G}(A)$ as $\mathbf{G}$, $\mathbf{G}(A)_n$ as $\mathbf{G}_n$). We shall use grammars, with nonterminals of arbitrary types, to generate subsets of $\mathbf{G}$.

Minor inclusion is usually defined for *undirected* graphs without sources. The extension to directed graphs with sources is straightforward.

Definition : Minor inclusion

Let $G, H \in \mathbf{G}$. We write $G \trianglelefteq H$ and we say that G *is a minor* of H (or *is included in H as a minor*) iff

1. G and H are of same type
2. G is obtained from a subgraph G' of H by edge contractions.

Since isomorphic graphs are considered as equal, minor inclusion is a partial order (because $G \trianglelefteq H$ and $H \trianglelefteq G$ implies $G = H$). G is a *proper minor* of H if $G \trianglelefteq H$ and $G \neq H$.

Since $G \trianglelefteq H$ implies that G and H are of same type, the sources of H cannot be deleted. By edge contractions several distinct sources can get fused. In the example shown on figure 7, $G \trianglelefteq G' \subseteq H$.

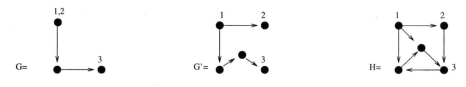

Fig. 7.

For every set of graphs L, we let

$$\trianglelefteq(L) = \{G / G \trianglelefteq H, \text{ for some } H \in L\}$$

and we call it the *minor closure* of L. A set L is *minor closed* iff $L = \trianglelefteq(L)$. We write $G \trianglelefteq' H$ iff G and H are of same type and G can be obtained from H by edge contractions and edge deletions.

It follows that $G \triangleleft' H$ does not hold if, for example
$$G = \overset{1}{\bullet} \overset{2}{\bullet} \bullet \bullet$$
and
$$H = \overset{1}{\bullet} \overset{2}{\bullet} \bullet \bullet \bullet$$

whereas of course $G \triangleleft H$. For every set of graphs L we let

$$\triangleleft'(L) = \{G \mid G \triangleleft' H, \text{ for some } H \in L\}$$

Proposition 3.1 *For every grammar Γ, one can construct a grammar Γ' with same set U of non terminals such that $\mathbf{L}(\Gamma', u) = \triangleleft'(\mathbf{L}(\Gamma, u))$ for every $u \in U$.*

We need first a lemma

Lemma 3.2 *Let $H \in \mathbf{G}(A \cup U)$ with nonterminal hyperedges $e_1, \ldots, e_l$; let $G_1, \ldots, G_l \in \mathbf{G}$ of respective types $\tau(e_1), \ldots, \tau(e_l)$. Then $G \triangleleft' H[G_1/e_1, \ldots, G_l/e_l]$ iff there exist $H' \in \mathbf{G}(A \cup U)$ and $G'_1, \ldots, G'_l \in \mathbf{G}$ such that :*

1. *$G'_i \triangleleft' G_i$ for every $i = 1, \ldots, l$*

2. *H' is obtained from H by contractions and deletions of terminal edges*
3. *$G = H'[G'_1/e_1, \ldots, G'_l/e_l]$.*

Proof : Straightforward.

Proof of Proposition (3.1) : For every rule $u \longrightarrow D$ of Γ, for every D' obtained from D by deletions and contractions of terminal edges, we put $u \longrightarrow D'$ as a rule of Γ'. With the help of lemma 3.2 and by induction on the length of derivation sequences, one can prove that :

1. if $u \xrightarrow[\Gamma]{*} G$, G is terminal and $G' \triangleleft' G$ then $u \xrightarrow[\Gamma']{*} G'$,

2. if $u \xrightarrow[\Gamma']{*} G'$ then there exists $G \in \mathbf{L}(\Gamma, u)$ such that $G' \triangleleft' G$.

Hence $\mathbf{L}(\Gamma', u) = \triangleleft'(\mathbf{L}(\Gamma, u)$. We omit details.

For every set of graphs L, we let
$c(L) = \{G/G \text{ is obtained from a graph in } L \text{ by deletion of some internal isolated vertices}\}$. For every $L \subseteq \mathbf{G}$ we have $L \subseteq c(L)$ and $\triangleleft(L) = c(\triangleleft'(L))$.

Proposition 3.3 *For every grammar Γ with set of nonterminals U, one can construct a grammar Γ' with set of nonterminals $U' \supseteq U$ such that :*

1. *for every $u \in U, \mathbf{L}(\Gamma', u) = c(\mathbf{L}(\Gamma, u))$ (hence $\mathbf{L}(\Gamma', u) = c(\mathbf{L}(\Gamma', u)))$*

2. for every $u \in U' - U$, $\mathbf{L}(\Gamma', u) = c(\mathbf{L}(\Gamma', u))$
3. if $\mathbf{L}(\Gamma, u) = \triangleleft'(\mathbf{L}(\Gamma, u))$ for every $u \in U$ then $\mathbf{L}(\Gamma', u) = \triangleleft(\mathbf{L}(\Gamma', u))$ for every $u \in U' - U$ (hence also for every $u \in U'$).

Sketch of proof: Given Γ, one first constructs a grammar Γ_1 consisting of all rules $u \longrightarrow D_1$ where $D_1 \in c(D)$ for some rule $u \longrightarrow D$ of Γ. Clearly $\mathbf{L}(\Gamma_1, w) \subseteq c(\mathbf{L}(\Gamma, w))$ for all nonterminals w. We would like to have the equality but this is not always the case as shown by the following example (see figure 8), where u is of type 2. The rules of Γ are given on figure 8.

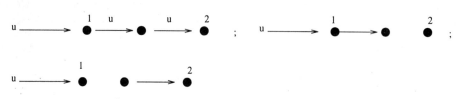

Fig. 8.

No righthand side of a rule has an internal isolated vertex, hence $\Gamma_1 = \Gamma$. However, the graph G_1 (see figure 9) belongs to $\mathbf{L}(\Gamma, u)$, but the graph G_2 (see figure 9) belongs to $c(\mathbf{L}(\Gamma, u)) - \mathbf{L}(\Gamma, u)$.

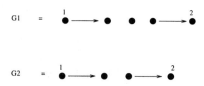

Fig. 9.

The above construction is then insufficient. It works for grammars Γ satisfying the following additional condition:

for every nonterminal u, for every $G \in \mathbf{L}(\Gamma, u)$, there is a derivation of G from u in Γ such that the isolated internal vertices of G come from isolated internal vertices of righthand sides of rules of Γ.

Every grammar Γ can be transformed into an equivalent one satisfying this condition. One needs to introduce new nonterminals. For example, in the case of the above grammar Γ, we add two nonterminals u_1, u_2 of type 1 and the rules shown on figure 10.

The general case is routine. We omit the proof.

Corollary 3.4 *Let Γ be a grammar. One can construct a grammar Γ' such that $\mathbf{L}(\Gamma') = \triangleleft(\mathbf{L}(\Gamma))$ and each set $\mathbf{L}(\Gamma', u)$ (where u is a nonterminal of Γ') is minor closed.*

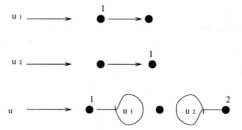

Fig. 10.

Proof : Immediate consequence of Propositions (3.2) and (3.3).

4 Obstructions

The following result is known from Robertson and Seymour [RS88]:

Theorem 4.1 *(Graph Minor Theorem)* : *For every infinite sequence of graphs* $(G_i)_{i \geq 0}$ *all of the same type n, there exists $i < j$ such that $G_i \trianglelefteq G_j$*

Discussion : The Graph Minor Theorem is proved in [RS88] for graphs with pairwise distinct sources. However, Theorem 4.1 follows easily from the special case : consider $(G_i)_{i \geq 0}$ where all graphs are of same type, there is an infinite subsequence $G_{i_1}, G_{i_2}, \ldots, G_{i_n} \ldots$ such that for some empty graph E, $\epsilon(G_{i_j}) = E$ for all j.

The graphs of this subsequence can be considered as having all $\mathbf{Card}(\mathbf{V}_E)$ pairwise distinct sources, hence the result of [RS88] applies and we have $G_{i_j} \trianglelefteq G_{i_{j'}}$, as was to be proved.

Let L be a minor-closed set of graphs, all of same type, say n. We let

$$\Omega(L) = \{G \mid G \text{ has type } n, G \notin L \text{ and every proper minor of } G \text{ is in } L\} \quad (1)$$

Let M be a set of graphs, all of same type, say n. We let

$$\text{FORB}(M) = \{G \mid G \text{ has type n and no minor of } G \text{ is in } M\}$$

so that

$$L = \text{FORB}(\Omega(L)) \quad (2)$$

By Theorem (4.1), $\Omega(L)$ is finite and (2) gives a finitary description of L.

Our aim is to show that one can construct effectively the set $\Omega(\trianglelefteq(\mathbf{L}(\Gamma_0)))$ where Γ_0 is a given grammar. We shall need the notion of tree-width.

Definition *Tree-width.* Let G be a graph. A *tree-decomposition* of G is a pair (T, f) consisting of an undirected tree T, and a mapping $f : \mathbf{V}_T \longrightarrow \mathcal{P}(\mathbf{V}_G)$ such that :

1. $\mathbf{V}_G = \bigcup \{f(i) \mid i \in \mathbf{V}_T\}$,
2. every edge of G has its vertices in $f(i)$ for some i
3. if $i, j, k \in \mathbf{V}_T$, and if j is on the unique cycle-free path in T from i to k, then $f(i) \cap f(k) \subseteq f(j)$,
4. all sources of G are in $f(i)$ for some i in $\mathbf{V}_T$.

The *width* of such a decomposition is defined as

$$\text{Max}\{\text{card}(f(i)) \mid i \in \mathbf{V}_T\} - 1.$$

The *tree-width* of G is the minimum width of a tree-decomposition of G. It is denoted by $\mathbf{twd}(G)$. For a 0-graph, condition (4) is always satisfied in a trivial way. Trees are of tree-width 1, series-parallel graphs are of tree-width 2 (or 1 in degenerate cases), a complete graph with n vertices is of tree-width $n - 1$.

The *tree-width of a set* L of graphs (denoted by $\mathbf{twd}(L)$) is the least upper bound in $\mathbb{N} \cup \{\infty\}$ of $\{\mathbf{twd}(G) \mid G \in L\}$. The set of complete graphs and the set of square grids have infinite tree-width.

Lemma 4.2 ([Cou92]) *Let Γ be a grammar. Let $D[e_1, \ldots, e_l]$ be the righthand side of a rule in Γ.*

1. *If $G_1, \ldots, G_l$ are graphs of respective types $\tau(e_1), \ldots, \tau(e_l)$ and of tree-width at most k, then the graph $D[G_1/e_1, \ldots, G_l/e_l]$ has tree-width at most k, if $k + 1 \geq \mathbf{Card}(\mathbf{V}_D)$.*
2. *For every nonterminal u of Γ, the tree-width of $\mathbf{L}(\Gamma, u)$ is at most k if $k+1 \geq \max \{\mathbf{Card}(\mathbf{V}_D) \mid (u, D)$ is a rule of $\Gamma\}$.*

We shall use Monadic Second-order logic (MS) to describe sets of graphs. We refer the reader to Courcelle [Cou90, Cou92, Cou91a, Cou93] for definitions. (Since we shall not construct explicit formulas, we need not formal definitions). The following fundamental result will be used in the sequel.

Theorem 4.3 ([Cou90]) *The following problem is decidable:* given integers k, n and a MS-formula φ, is it true that,

$$\forall H \in \mathbf{G}_n, \mathbf{twd}(H) \leq k \Longrightarrow (H \models \varphi) \ ?$$

By Corollary (3.4) one can construct a grammar Γ with non-terminals $u_1, \ldots, u_m$ such that

$$\mathrm{L}(\Gamma, u_1) = \trianglelefteq(\mathrm{L}(\Gamma_0)) \tag{3}$$

$$\mathrm{L}(\Gamma, u_i) = \trianglelefteq(\mathrm{L}(\Gamma, u_i)) \text{ for } i = 2, \ldots, m \tag{4}$$

By Theorem(2.4), one can construct a proper grammar Γ' with non-terminals $u_1, \ldots, u_m$, such that

$$\mathrm{L}(\Gamma', u_i) = \mathrm{L}(\Gamma, u_i)) - \text{EMPTY}(u_i) \text{ for } i = 1, \ldots, m \tag{5}$$

In order to give an explicit decidable characterization of the sets of obstructions $(\Omega(L(\Gamma, u_i)))_{1 \leq i \leq m}$, we define the following sets of graphs for every $k, n \in \mathbb{N}$:
$\Lambda(k, n) = \{H \in \mathbf{G}_n | H$ is a n-graph formed from an orientation of the $(k + 1) \times (k + 1)$ square grid augmented with at most n isolated vertices in such a way that the set of isolated vertice is exactly the set of sources $\}$
$\mathrm{TWD}(k, n) = \{H \in \mathbf{G}_n | \mathbf{twd}(H) \leq k\}$
$\mathrm{EMPTY}(n) = \{H \in \mathbf{G}_n | H$ is empty $\}$.

Lemma 4.4 $\forall k, n \in \mathbb{N}, \exists k' \in \mathbb{N}$, such that

$$\mathrm{TWD}(k, n) \subseteq \mathrm{FORB}(\Lambda(k, n)) \subseteq \mathrm{TWD}(k', n) \tag{6}$$

Proof:

1. The undirected $(k + 1) \times (k + 1)$-grid Q_{k+1} has tree-width $k + 1$. It follows that every element in $\Lambda(k, n)$ has tree-width $\geq k + 1$, hence

$$\mathrm{TWD}(k, n) \subseteq \mathrm{FORB}(\Lambda(k, n)).$$

2. Let $k'' = 20^{2(k+1)^5}$, $k' = k'' + n - 1$. It is proved in [RST90] that every undirected 0-graph of tree-width $\geq k''$ contains Q_{k+1} as a minor. Let H be a (directed) n-graph of tree-width $\geq k' + 1$. Let H' be obtained from H by removing the sources (and the edges incident with the sources) and forgetting the orientation.

$$\mathbf{twd}(H') \geq k''$$

(because adding the sources and the corresponding incident edges to H' increases the tree-width by at most n).
Hence $Q_{k+1} \trianglelefteq H'$, which implies that some element of $\Lambda(k, n)$ is a minor of H. This proves that

$$\mathrm{FORB}(\Lambda(k, n)) \subseteq \mathrm{TWD}(k', n).$$

From now on we fix an integer k such that

$$k + 1 \geq \max \{\mathbf{Card}(\mathbf{V}_D) \mid (u, D) \text{ is a rule of } \Gamma\}.$$

Lemma 4.5 Let $(\Omega_1, \ldots, \Omega_m)$ be an m-tuple of finite sets of graphs of respective types $n_1, \ldots, n_m$ where $n_i = \tau(u_i)$ and let k' be some integer such that (6) is true for every pair (k, n_i). Then $(\Omega_1, \ldots, \Omega_m) = (\Omega(L(\Gamma, u_1)), \ldots, \Omega(L(\Gamma, u_m)))$ iff the following conditions hold

1. $\forall H \in \Omega_i, H$ is a minimal element of $\mathbf{G}_{n_i} - L(\Gamma', u_i)$ (with respect to the ordering $\trianglelefteq$)

2. $\text{FORB}(\Omega_i) \cap \text{EMPTY}(n_i) = \text{L}(\Gamma, u_i) \cap \text{EMPTY}(n_i)$
3. the m-tuple $(L'_1, \ldots, L'_m)$ where $L'_i = \text{FORB}(\Omega_i) - \text{EMPTY}(n_i)$ is a solution of $\mathbf{S}_{\Gamma'}$
4. $\text{FORB}(\Omega_i) \subseteq \text{FORB}(\Lambda(k, n_i))$

Proof:

1. Let us suppose $(\Omega_1, \ldots, \Omega_m) = (\Omega(\text{L}(\Gamma, u_1)), \ldots, \Omega(\text{L}(\Gamma, u_m)))$. Point (1) is clear from the definition of an obstruction. Points (2)(3) follow from the facts that $\text{FORB}(\Omega_i) = \text{L}(\Gamma, u_i)$ and equality (5). By lemma(4.2),

$$\text{L}(\Gamma, u_i) \subseteq \text{TWD}(k, n_i)$$

Hence, by lemma(4.4)

$$\text{L}(\Gamma, u_i) \subseteq \text{FORB}(\Lambda(k, n_i))$$

which implies point (4).

2. Let us suppose that $(\Omega_1, \ldots, \Omega_m)$ fulfills conditions (1)(2)(3)(4). By (3) and theorem (2.3)

$$\text{FORB}(\Omega_i) - \text{EMPTY}(k, n_i) = \text{L}(\Gamma', u_i)$$

Together with point (2) it shows that

$$\text{FORB}(\Omega_i) = \text{L}(\Gamma, u_i) = \text{FORB}(\Omega(\text{L}(\Gamma, u_i)))$$

Point (1) allows then to conclude that

$$\Omega_i = \Omega(\text{L}(\Gamma, u_i)).$$

Lemma 4.6 *It is decidable whether a given m-tuple of finite sets of graphs of respective types $n_1, \ldots, n_m$ is equal to $(\Omega(\text{L}(\Gamma, u_1)), \ldots, \Omega(\text{L}(\Gamma, u_m)))$.*

Proof:
It suffices to show that the conjunction of conditions (1) to (4) given in lemma (4.5) is decidable. Condition (1) is decidable (because membership in $\text{L}(\Gamma, u_i)$ is decidable). Condition (2) is decidable, because both sides of the equality are finite sets which can be computed. Condition (4) is equivalent to:

$$\forall H \in \Lambda(k, n_i), \exists K \in \Omega_i, K \trianglelefteq H.$$

This can be tested since $\Lambda(k, n_i), \Omega_i$ are finite sets. Let us now assume that condition (4) is fulfilled. By lemma (4.4) and condition (4)

$$L'_i = \text{FORB}(\Omega_i) - \text{EMPTY}(k, n_i) \subseteq \text{TWD}(k', n_i)$$

from Ω_i one can construct a MS-formula φ_i such that

$$\forall H \in \mathbf{G}_{n_i}, H \models \varphi_i \iff H \in \text{FORB}(\Omega_i) - \text{EMPTY}(k, n_i).$$

From t'_i (the i-th right hand side of the system of equations $S_{\Gamma'}$) and $\varphi_1, \ldots, \varphi_m$, one can construct a MS-formula ψ_i such that

for all $H \in \mathbf{G}_{n_i}, H \models \psi_i \iff$ there exists $H_1, \ldots, H_m$ |

for all j, $H_j \models \varphi_j$ and $\{H\} = \hat{t}'_i(\{H_1\}, \ldots, \{H_m\})$.

Condition (3) is then equivalent to

$$\forall i \in [1, m], \forall H \in \mathrm{TWD}(k', n_i), H \models (\varphi_i \iff \psi_i)$$

which is decidable by Theorem (4.3).

Theorem 4.7 *Let L be a set of graphs defined by a given HR-grammar Γ. One can construct effectively the set of obstructions of the minor-closure of L.*

Proof:
By equality (3), $\trianglelefteq(L) = \trianglelefteq(\mathrm{L}(\Gamma_0)) = \mathrm{L}(\Gamma, u_1)$.
It suffices to enumerate vectors $\mathbf{\Omega} = (\Omega_1, \ldots, \Omega_m)$ and test (by lemma (4.6)) each of them for conditions (1) − (4) of lemma (4.5), until a vector $\mathbf{\Omega}$ fulfilling (1) − (4) is reached. Then $\Omega(L) = \Omega_1$.

The proof of Theorem (4.7) extends easily to HR-grammars generating sets of undirected graphs. This will be done in the full version of this paper.

References

[BC87] M. Bauderon and B. Courcelle. Graph expressions and graph rewritings. *Mathematical System Theory 20*, pages 83–127, 1987.

[Cou90] B. Courcelle. The monadic second-order logic of graphs I: Recognizable sets of finite graphs. *Information and Computation 85*, pages 12–75, 1990.

[Cou91a] B. Courcelle. The monadic second-order logic of graphs V: On closing the gap between definability and recognizability. *Theoretical Computer Science 80*, pages 153–202, 1991.

[Cou91b] B. Courcelle. On constructing obstuction sets of words. *Bulletin of EATCS 44*, pages 178–185, 1991.

[Cou92] B. Courcelle. The monadic second-order logic of graphs III:Tree-decompositions, minors and complexity issues. *RAIRO Informatique Théorique et Applications 26*, pages 257–286, 1992.

[Cou93] B. Courcelle. Graph grammars, monadic second-order logic and the theory of graph minors. *in N.Robertson and P.Seymour eds.,Contemporary Mathematics 147,American Mathematical Society*, pages 565–590, 1993.

[FL89] M. Fellows and M. Langston. An analogue of the Myhill-Nerode theorem and its use in computing finite basis characterizations. *Proceedings 30th Symp. FOCS*, pages 520–525, 1989.

[Hab93] A. Habel. *Hyperedge Replacement:Grammars and languages*. Lectures Notes in Comput.Sc.,vol. 643,Springer, 1993.

[RS88] N. Robertson and P. Seymour. Graph minors xx: Wagner's conjecture. *preprint*, 1988.

[RST90] N. Robertson, P. Seymour, and R. Thomas. Quickly excluding a planar graph. *preprint*, 1990.

Concatenation of Graphs

Joost Engelfriet and Jan Joris Vereijken [*]

Department of Computer Science, Leiden University
P.O.Box 9512, 2300 RA Leiden, The Netherlands
e-mail: engelfri@wi.leidenuniv.nl

Abstract. An operation of concatenation is defined for graphs. Then strings are viewed as expressions denoting graphs, and string languages are interpreted as graph languages. For a class K of string languages, $\text{Int}(K)$ is the class of all graph languages that are interpretations of languages from K. For the class REG of regular languages, $\text{Int}(\text{REG})$ might be called the class of regular graph languages; it equals the class of graph languages generated by linear Hyperedge Replacement Systems. Two characterizations are given of the largest class K' such that $\text{Int}(K') = \text{Int}(K)$.

1 Concatenation and Sum

Context-free graph languages are generated by context-free graph grammars, which are graph replacement systems. One of the most popular types of context-free graph grammar is the Hyperedge Replacement System, or HR grammar (see, e.g., [Hab]). A completely different way of generating graphs (introduced in [BauCou]) is to select a number of graph operations, to generate a set of expressions (built from these operations), and to interpret the expressions as graphs. The set of expressions is generated by a classical context-free grammar generating strings (or a regular tree grammar). It is shown in [BauCou] that, for a particular collection of graph operations, this new way of generating graphs is equivalent with the HR grammar. Other work on the generation of graphs through graph expressions is in, e.g., [Cou, CouER, Dre, Eng].

In this framework we investigate another, natural operation on graphs that was introduced (for "planar nets") in [Hot1] (and which is a simple variation of the graph operations in [BauCou]). Due to its similarity to concatenation of strings, we call it *concatenation of graphs*. Together with the sum operation of graphs (introduced for planar nets in [Hot1] and defined for graphs in [BauCou]) and all constant graphs, a collection of graph operations is obtained that is simpler than the one in [BauCou], but also has the power of the HR grammar (which is our first result).

The elementary laws that are satisfied by concatenation and sum of planar nets, are the basis of the theory of x-categories developed in [Hot1] (also called

---

[*] The present address of this author is Faculty of Mathematics and Computing Science, Eindhoven University of Technology, P.O.Box 513, 5600 MB Eindhoven, The Netherlands, e-mail: janjoris@acm.org

strict monoidal categories, see, e.g., [EhrKKK, Ben]). Free x-categories model the sets of derivation graphs of Chomsky type 0 grammars (see [Hot2, Ben]). Finite automata on such graphs are considered, e.g., in [BosDW]. The idea of using concatenation and sum in graph grammars is from [HotKM], where "logic topological nets" are generated by (parallel) graph grammars.

Let us be more precise. We consider the multi-pointed, directed, edge-labeled hypergraphs of [Hab]. Let Σ be a *typed* (or doubly ranked) *alphabet*, i.e., an alphabet Σ together with a mapping type : $\Sigma \to \mathbf{N} \times \mathbf{N}$, where $\mathbf{N} = \{0, 1, 2, \ldots\}$; thus, two nonnegative integers are associated with every symbol of Σ. A *multi-pointed hypergraph* g over Σ consists of a finite set V of nodes and a finite set E of (hyper)edges; every edge e has a sequence of "sources" $s(e) \in V^*$, a sequence of "targets" $t(e) \in V^*$, and a "label" $l(e)$ in Σ such that type$(l(e)) = (|s(e)|, |t(e)|)$; additionally, g has a designated sequence begin$(g) \in V^*$ of "begin nodes", and a designated sequence end$(g) \in V^*$ of "end nodes". If $|\text{begin}(g)| = m$ and $|\text{end}(g)| = n$, then g is said to be of *type* (m, n) and we write type$(g) = (m, n)$. Similarly, for an edge e of g, we write type(e) to denote type$(l(e))$; thus, by the above requirement, if type$(e) = (m, n)$, then e has m sources and n targets. If all edges of a multi-pointed hypergraph are of type $(1, 1)$, then it is an ordinary directed graph (with edge labels, and with begin nodes and end nodes).

For a typed symbol σ, with type$(\sigma) = (m, n)$, we denote by atom(σ) the multi-pointed hypergraph g of type (m, n) such that $V = \{x_1, \ldots, x_m, y_1, \ldots, y_n\}$, $E = \{e\}$ with $l(e) = \sigma$, and begin$(g) = s(e) = \langle x_1, \ldots, x_m \rangle$, and end$(g) = t(e) = \langle y_1, \ldots, y_n \rangle$.

From now on we will just say graph *instead of multi-pointed hypergraph*. As usual we consider both concrete and abstract graphs, where an abstract graph is an equivalence class of isomorphic concrete graphs. The isomorphisms between graphs g and h are the usual ones, which, additionally, should map begin(g) to begin(h), and end(g) to end(h). In particular, isomorphic graphs have the same type. Our operations are defined on abstract graphs. The set of abstract graphs over a typed alphabet Σ will be denoted GR(Σ), and GR denotes the union of all GR(Σ). A (typed) *graph language* is a subset L of GR(Σ), for some Σ, such that all graphs in L have the same type (m, n), also called the type of L, and denoted by type$(L) = (m, n)$.

We now define the graph operations of concatenation and sum. If g and h are graphs with type$(g) = (k, m)$ and type$(h) = (m, n)$, then their *concatenation* $g \circ h$ is the graph obtained by first taking the disjoint union of g and h, and then identifying the ith end node of g with the ith begin node of h, for every $i \in \{1, \ldots, m\}$; moreover, begin$(g \circ h) = $ begin(g) and end$(g \circ h) = $ end(h), and so type$(g \circ h) = (k, n)$. Note that the concatenation of g and h is defined only when $|\text{end}(g)| = |\text{begin}(h)|$. The *sum* $g \oplus h$ of arbitrary graphs g and h (as defined in [BauCou]) is their disjoint union, with begin$(g \oplus h) = $ begin$(g) \cdot $ begin(h) and end$(g \oplus h) = $ end$(g) \cdot $ end(h), where $\cdot$ denotes the usual concatenation of sequences. Intuitively, concatenation is sequential composition of graphs, and sum is parallel composition of graphs.

Figure 1 shows two (ordinary) graphs, g of type $(2, 3)$ and h of type $(3, 1)$,

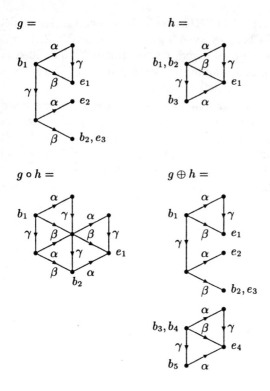

Fig. 1. Two graphs, their concatenation, and their sum.

with their concatenation $g \circ h$ of type $(2,1)$ and their sum $g \oplus h$ of type $(5,4)$. The graphs are drawn in the usual way; the ith begin node is indicated by b_i, and the ith end node by e_i.

These two graph operations have a number of simple properties. First of all, concatenation is associative, i.e., if $|\text{end}(g_1)| = |\text{begin}(g_2)|$ and $|\text{end}(g_2)| = |\text{begin}(g_3)|$, then $(g_1 \circ g_2) \circ g_3 = g_1 \circ (g_2 \circ g_3)$. For every $n \in \mathbf{N}$ there is an *identity* id_n of type (n, n) such that $g \circ \text{id}_n = g$ and $\text{id}_n \circ h = h$ for every g with $|\text{end}(g)| = n$ and h with $|\text{begin}(h)| = n$. In fact, id_n is the (abstract graph corresponding to the) discrete graph with nodes $1, \ldots, n$ and $\text{begin}(\text{id}_n) = \text{end}(\text{id}_n) = \langle 1, \ldots, n \rangle$. Second, sum is associative, with unity id_0 (the empty graph). Also, for every $m, n \in \mathbf{N}$, $\text{id}_{m+n} = \text{id}_m \oplus \text{id}_n$. Finally, concatenation and sum are connected through the following law of *strict monoidality*: if $|\text{end}(g)| = |\text{begin}(g')|$ and $|\text{end}(h)| = |\text{begin}(h')|$, then

$$(g \oplus h) \circ (g' \oplus h') = (g \circ g') \oplus (h \circ h').$$

All these properties together mean that GR is a strict monoidal category (or x-category), see, e.g., [EhrKKK, Hot1, Ben]. The objects of this category are the nonnegative integers, and each (abstract) graph of type (m,n) is a morphism from m to n in this category. Concatenation is the composition of morphisms (but is usually written $h \circ g$ rather than $g \circ h$), and the id_n are the identity morphisms.

Let Δ be the set of operators $\{\circ, \oplus\} \cup \{c_g \mid g \in \mathrm{GR}\}$, where $\circ$ and $\oplus$ denote concatenation and sum of graphs, as discussed above, and c_g is a constant standing for the graph g. A *regular tree grammar* over Δ is an ordinary context-free grammar G such that the nonterminal alphabet of G is typed, and the right-hand side of each production of G is a (well-formed) expression over the operators from Δ and the nonterminals of the grammar (which should be treated as constant operators, with the given type) of the same type as the left-hand side. Obviously, the language $L(G)$ generated by G is a set of expressions over Δ (and it is called a regular tree language, see [GecSte]). But G can also be viewed as a *context-free graph grammar*, generating the graph language $\mathrm{val}(L(G)) = \{\mathrm{val}(e) \mid e \in L(G)\} \subseteq \mathrm{GR}$, where the graph $\mathrm{val}(e)$ is the value of the expression e. Let $\mathrm{Val}(\mathrm{REGT}) = \{\mathrm{val}(L(G)) \mid G$ is a regular tree grammar over $\Delta\}$. Intuitively, $\mathrm{Val}(\mathrm{REGT})$ is the class of "values of regular tree languages" over Δ (where values of expressions are graphs).

Fig. 2. Graphs g and g'.

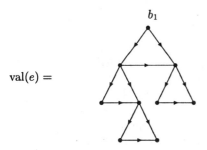

Fig. 3. The value of graph expression e.

As an example, consider the regular tree grammar G_b that has one nonterminal X, with $\mathrm{type}(X) = (1,0)$, and two productions $X \to c_g \circ (X \oplus X)$ and $X \to c_{g'}$, where g is the triangle of type $(1,2)$ with $V = \{x, y, z\}$, $E = \{(x,y), (x,z), (y,z)\}$, $s(u,v) = u$, $t(u,v) = v$, and $l(u,v) = \sigma$ for every edge (u,v), $\mathrm{begin}(g) = \langle x \rangle$ and $\mathrm{end}(g) = \langle y, z \rangle$, and g' is the graph of type $(1,0)$ with one node x, no edges, $\mathrm{begin}(g') = \langle x \rangle$ and $\mathrm{end}(g')$ is the empty sequence $\langle \rangle$. The graphs g and g' are shown in Fig.2. The expression

$$e = c_g \circ (c_g \circ (c_{g'} \oplus c_g \circ (c_{g'} \oplus c_{g'})) \oplus c_g \circ (c_{g'} \oplus c_{g'}))$$

is in $L(G)$; the graph val(e) is shown in Fig.3 (without the edge labels σ). Clearly, val$(L(G_b))$ is the set of all graphs of type $(1,0)$ that are obtained from (directed, rooted) binary trees by connecting each pair of children by an additional edge; the sequence of begin nodes consists of the root of the binary tree. This graph language is therefore in Val(REGT).

Our first result is that generating graph languages in the above way is equivalent to generating them with HR grammars. As observed above, this is a simple variant of the result of [BauCou]. Let HR denote the class of all graph languages generated by HR grammars.

Theorem 1. Val(REGT) = HR.

2 Strings Denote Graphs

Since concatenation of graphs is associative, strings can be viewed as expressions that denote graphs. Thus, as an even simpler variation of the approach with regular tree grammars in Section 1, we can use ordinary string grammars to generate graph languages. More generally, every class K of string languages defines a class Int(K) of graph languages (where Int stands for "interpretation", which is similar to Val in Section 1). Formally, for an alphabet A, an *interpretation of* A is a mapping $h : A \to$ GR that associates a graph with each symbol of A; h is extended to a (partial) function from A^* to GR by

$$h(a_1 a_2 \cdots a_n) = h(a_1) \circ h(a_2) \circ \cdots \circ h(a_n)$$

with $a_i \in A$ for all i, and $n \geq 1$. Note that the extended h is partial because the types of the $h(a_i)$ may not fit; moreover, $h(\lambda)$ is undefined, where λ is the empty string. Thus, the only "technical trouble" is that the concatenation of graphs is typed whereas the concatenation of strings is always possible. For a string language $L \subseteq A^*$, we define, as usual, the set of graphs $h(L) = \{g \in$ GR $\mid g = h(w)$ for some $w \in L\}$; note that $h(L)$ need not be a graph language (in our particular meaning of the term) because not all graphs need have the same type. Now we define

Int$(K) = \{h(L) \mid L \in K, h : A \to$ GR with $L \subseteq A^*, h(L)$ is a graph language$\}$.

In other words, Int(K) consists of all graph languages $h(L)$, where L is any language in K and h is any mapping from the symbols of L to graphs. Intuitively, h determines the interpretation of the symbols, and then the concatenation of those symbols is interpreted as concatenation of the corresponding graphs.

The first class K of interest is the class REG of regular languages. An example of a graph language in Int(REG), of type $(0,0)$, is $h(a(b\cup c)^*d)$ where the graphs $h(a), h(b), h(c)$, and $h(d)$ are shown in Fig.4 (without edge directions and edge

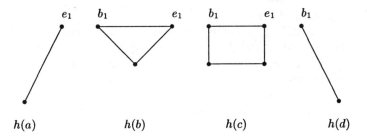

Fig. 4. An interpretation

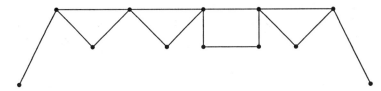

Fig. 5. Graph interpretation of the string $abbcbd$.

labels). The graph $h(abbcbd)$ is shown in Fig.5. Clearly, the graph language $h(a(b \cup c)^*d)$ consists of all "clothes lines" on which triangles and rectangles are hanging to dry. Our second result says that the graph languages that are interpretations of a regular language are precisely those that can be generated by *linear* HR grammars, where 'linear' means that there is at most one nonterminal in each right-hand side of a production of the HR grammar. Let LIN-HR denote the class of graph languages generated by linear HR grammars.

Theorem 2. Int(REG) = LIN-HR.

Proof. (Sketch) It can be shown that the LIN-HR graph languages are generated by the *right-linear* regular tree grammars over Δ, of which all productions are of the form $X \to c_g \circ Y$ or $X \to c_g$ (where X and Y are nonterminals). In fact, if, in particular, all nonterminals have types of the form $(m, 0)$, then the correspondence between such regular tree grammars and linear HR grammars is straightforward. As an example, the graph language of clothes lines is generated by the right-linear regular tree grammar with the productions $S \to c_{h(a)} \circ X$, $X \to c_{h(b)} \circ X$, $X \to c_{h(c)} \circ X$, and $X \to c_{h(d)}$, where $h(a)$, $h(b)$, $h(c)$, $h(d)$ are the graphs in Fig.4, type$(S) = (0,0)$, and type$(X) = (1,0)$.

To show that LIN-HR $\subseteq$ Int(REG), consider a right-linear regular tree grammar G over Δ with set of productions P. Let $R_{\text{der}} \subseteq P^*$ be the set of all production sequences $p_1 \cdots p_n$ that correspond to a successful derivation of G. Clearly R_{der} is regular, and $h(R_{\text{der}}) = \text{val}(L(G))$ where h is the interpretation with

$h(X \to c_g \circ Y) = g$ and $h(X \to c_g) = g$.

To show that Int(REG) $\subseteq$ LIN-HR, consider a regular language $L \subseteq A^*$ and an interpretation h of A. Let R_{def} be the language of all strings $w \in A^*$ such that $h(w)$ is defined; clearly, R_{def} is regular (just check that $|\text{end}(h(a))| = |\text{begin}(h(b))|$ for every substring ab of w). This implies that the language $h(L) = \{c_{h(a_1)} \circ \cdots \circ c_{h(a_n)} \mid a_1 \cdots a_n \in L \cap R_{\text{def}}\}$ can be generated by a right-linear regular tree grammar over Δ (where the types of the nonterminals can be determined because of the restriction to R_{def}). $\square$

Next we consider a second characterization of the class Int(REG), corresponding to the characterization of REG by regular expressions. The operation of graph concatenation is extended to graph languages L and L' in the usual way: if type$(L) = (k,m)$ and type$(L') = (m,n)$, then their *concatenation* is defined by $L \circ L' = \{g \circ g' \mid g \in L, g' \in L'\}$. Then, in the obvious way, the (Kleene) *star* of a graph language is defined by iterated concatenation: for a graph language L with type$(L) = (k,k)$ for some $k \in \mathbf{N}$, $L^* = \bigcup_{n \in \mathbf{N}} L^n$ where $L^n = L \circ \cdots \circ L$ (n times) for $n \geq 1$, and $L^0 = \{\text{id}_k\}$. Also, $L^+ = \bigcup_{n \geq 1} L^n$ is the (Kleene) plus of L. Finally, the *union* $L \cup L'$ of two graph languages L and L' is defined only when type$(L) = $ type(L') (otherwise it would not be a graph language). Thus, the operations of union, concatenation, and star are also typed operations on graph languages (as opposed to the case of string languages for which they are always defined). Let REX$(\cup, \circ, *, \text{SING})$ denote the smallest class of graph languages containing the empty graph language and all singleton graph languages, and closed under the operations union, concatenation, and star. Thus, it is the class of all graph languages that can be denoted by (the usual) regular expressions, where the symbols of the alphabet denote singleton graph languages. As an example, the above graph language of clothes lines is in REX$(\cup, \circ, *, \text{SING})$ because it can be written as $\{h(a)\} \circ (\{h(b)\} \cup \{h(c)\})^* \circ \{h(d)\}$. The next result can be viewed as a variation of a well-known characterization of the rational subsets of a monoid (see, e.g., Proposition III.2.2 of [Ber]).

Theorem 3. Int(REG) = REX$(\cup, \circ, *, \text{SING})$.

Proof. We have to cope with the "technical trouble" of typing, in particular with the empty string. Note that, for a graph language L with type$(L) = (k,k)$, $L^* = L^+ \cup \{\text{id}_k\}$ and $L^+ = L \circ L^*$. This shows that we can replace star by plus, i.e., REX$(\cup, \circ, *, \text{SING}) = $ REX$(\cup, \circ, +, \text{SING})$, the smallest class of graph languages containing the empty graph language and all singleton graph languages, and closed under the operations union, concatenation, and plus.

To show that REX$(\cup, \circ, +, \text{SING}) \subseteq $ Int(REG), it suffices to prove that Int(REG) contains the empty language and all singleton graph languages, and that it is closed under union, concatenation, and plus. Clearly, $h(\emptyset) = \emptyset$ for any interpretation h. Also, if $h(a) = g$, then $h(\{a\}) = \{g\}$. Now let $L_1 \subseteq A_1^*$ and $L_2 \subseteq A_2^*$ be regular languages, and let h_1 and h_2 be interpretations of A_1 and A_2, respectively, such that $h_1(L_1)$ and $h_2(L_2)$ are graph languages in Int(REG). Obviously, by a renaming of symbols, we may assume that A_1 and A_2 are disjoint. Let $h = h_1 \cup h_2$ be the interpretation of $A_1 \cup A_2$ that extends both

h_1 and h_2. It is easy to verify that (with the appropriate conditions on types) $h_1(L_1) \cup h_2(L_2) = h(L_1 \cup L_2)$, $h_1(L_1) \circ h_2(L_2) = h(L_1 \cdot L_2)$, and $h_1(L_1)^+ = h(L_1^+)$, which shows that these graph languages are also in Int(REG).

To show that Int(REG) $\subseteq$ REX($\cup, \circ, +,$ SING), we first note that, since an interpretation is undefined for the empty string, Int(REG) = Int(REG $- \lambda$), where REG $- \lambda = \{L - \{\lambda\} \mid L \in \text{REG}\}$ is the class of all λ-free regular languages. It is well known (and easy to prove) that REG $- \lambda$ is the smallest class of languages containing the empty language and all languages $\{a\}$ where a is a symbol, and closed under the operations union, concatenation, and plus. By induction on this characterization we show that for every language $L \in \text{REG} - \lambda$ and every interpretation h of the alphabet of L, if $h(w)$ is defined for every $w \in L$, and $h(L)$ is a graph language, then $h(L) \in \text{REX}(\cup, \circ, +, \text{SING})$. Note that we can indeed assume that h is defined for all strings in L because, in general, $h(L) = h(L \cap \{w \mid h(w) \text{ is defined}\})$ and the language $\{w \mid h(w) \text{ is defined}\}$ is regular (cf. the proof of Theorem 2). The inductive proof is as follows. If L is empty, then so is $h(L)$. If $L = \{a\}$, then $h(L)$ is a singleton. If $L = L_1 \cup L_2$, then $h(L) = h(L_1) \cup h(L_2)$. Now let $L = L_1 \cdot L_2$ and assume that L_1 and L_2 are nonempty (otherwise L is empty). Since, by assumption, $h(L_1 \cdot L_2)$ is a graph language and $h(w)$ is defined for every $w \in L_1 \cdot L_2$, $h(L_1)$ and $h(L_2)$ are also graph languages; for $h(L_1)$ this is proved as follows: if $w_1, w_1' \in L_1$, then, for any $w_2 \in L_2$, $h(w_1 \cdot w_2) = h(w_1) \circ h(w_2)$ and similarly for w_1', and so $|\text{begin}(h(w_1))| = |\text{begin}(h(w_1 \cdot w_2))| = |\text{begin}(h(w_1' \cdot w_2))| = |\text{begin}(h(w_1'))|$ and $|\text{end}(h(w_1))| = |\text{begin}(h(w_2))| = |\text{end}(h(w_1'))|$. Hence $h(L) = h(L_1 \cdot L_2) = h(L_1) \circ h(L_2)$. Finally, let $L = L_1^+$. Then $h(L_1)$ is a graph language of some type (k, k) by an argument similar to the one above, and $h(L) = h(L_1)^+$. □

By Theorems 2 and 3, LIN-IIR = REX($\cup, \circ, *,$ SING). This suggests that the class LIN-HR of linear HR graph languages might be called the class of "regular" graph languages, because they can be denoted by regular expressions.

The characterization of Theorem 3 still holds after adding the sum operation (extended to graph languages in the usual way). In other words, Int(REG) = REX($\cup, \circ, *, \oplus,$ SING), the smallest class of graph languages containing the empty graph language and all singleton graph languages, and closed under the operations union, concatenation, star, and sum. This is because of the following simple reason.

Lemma 4. *For every class of languages K, if* Int(K) *is closed under concatenation, then it is closed under sum.*

Proof. We first show that if M is in Int(K) then so is $M \oplus \{\text{id}_k\}$ for every k. Let $M = h(L)$ for some $L \in K$ and some interpretation h of the alphabet A of L. Define $h'(a) = h(a) \oplus \text{id}_k$ for every $a \in A$. Then $h'(a_1 \cdots a_n) =$

$(h(a_1) \oplus \text{id}_k) \circ \cdots \circ (h(a_n) \oplus \text{id}_k) = (h(a_1) \circ \cdots \circ h(a_n)) \oplus (\text{id}_k \circ \cdots \circ \text{id}_k)$

because of strict monoidality, and the last expression equals $h(a_1 \cdots a_n) \oplus \text{id}_k$. This implies that $h'(L) = h(L) \oplus \{\text{id}_k\} = M \oplus \{\text{id}_k\}$. Similarly it can be shown that $\{\text{id}_k\} \oplus M$ is in Int(K).

Now, for arbitrary graph languages M and M' with $\text{type}(M) = (m,n)$ and $\text{type}(M') = (m',n')$, $M \oplus M' = (M \circ \{\text{id}_n\}) \oplus (\{\text{id}_{m'}\} \circ M') = (M \oplus \{\text{id}_{m'}\}) \circ (\{\text{id}_n\} \oplus M')$ by strict monoidality. Hence, by the above, and the fact that $\text{Int}(K)$ is closed under $\circ$, $M \oplus M'$ is in $\text{Int}(K)$. □

It turns out that, if we allow $\oplus$ in our regular expressions, then we do not need all singleton graph languages to start with, but only a "small" number of them, with very simple graphs only. In fact, graphs can be decomposed into very simple graphs, using concatenation and sum. Assume that the edge labels of our graphs are taken from a given (possibly infinite) typed alphabet Σ. Recall from Section 1 the definition of the graph $\text{atom}(\sigma)$, with one σ-labeled edge (for $\sigma \in \Sigma$). For $m, n \in \mathbf{N}$, let $I_{m,n}$ be the graph of type (m,n) with one node x, no edges, $\text{begin}(I_{m,n}) = \langle x, \ldots, x \rangle$ (m times), and $\text{end}(I_{m,n}) = \langle x, \ldots, x \rangle$ (n times). Finally, let π_{12} be the graph of type $(2,2)$ with two nodes x and y, no edges, $\text{begin}(\pi_{12}) = \langle x, y \rangle$, and $\text{end}(\pi_{12}) = \langle y, x \rangle$. Recall that id_0 is the empty graph (of type $(0,0)$). Now define the set of *elementary* graphs

$$\text{EL} = \{\text{atom}(\sigma) \mid \sigma \in \Sigma\} \cup \{I_{0,1}, I_{1,0}, I_{1,2}, I_{2,1}, \pi_{12}, \text{id}_0\}.$$

Then every graph (over Σ) can be composed from elementary graphs by the operations $\circ$ and $\oplus$. This yields the following theorem, in which we denote by $\text{REX}(\cup, \circ, *, \oplus, \text{ELSING})$ the smallest class of graph languages containing the empty graph language and all singleton graph languages with a graph from EL as element, and closed under the operations union, concatenation, star, and sum.

Theorem 5. $\text{REX}(\cup, \circ, *, \text{SING}) = \text{REX}(\cup, \circ, *, \oplus, \text{ELSING})$.

3 Characterizations of $\text{Int}(K)$

In this section and the next, we investigate properties of the class of graph languages $\text{Int}(K)$ for arbitrary classes of string languages K. However, to avoid trivialities we will mainly be interested in classes K that are *closed under nsm-mappings*, i.e., nondeterministic sequential machine mappings (where a sequential machine is an ordinary finite automaton that, moreover, at each step outputs one symbol). Equivalently, K is closed under intersection with regular languages and under alphabetical substitutions (where an alphabetical substitution from alphabet A to alphabet B is a relation $\rho \subseteq A \times B$ that is extended to a function from A^* to the finite subsets of B^* by $\rho(a_1 \cdots a_n) = \{b_1 \cdots b_n \mid (a_i, b_i) \in \rho, 1 \leq i \leq n\}$).

In this section we present two characterizations of $\text{Int}(K)$. The first generalizes the one of Theorem 2. To this aim we consider *controlled* linear HR grammars, in the obvious sense. Let G be a linear HR grammar with (finite) set of productions P, and let C be a string language over P (where P is viewed as an alphabet). The graph language generated by G under control C is the set of all terminal graphs g for which there is a derivation $g_0 \Rightarrow_{p_1} g_1 \Rightarrow_{p_2} g_2 \cdots \Rightarrow_{p_n} g_n$ with $g_n = g$, g_0 is the axiom of G, and such that the string $p_1 p_2 \cdots p_n$ is in C.

Of course, $\Rightarrow_p$ denotes a derivation step of G that uses production $p \in P$. Thus, the control language C specifies the sequences of productions that the grammar G is allowed to use in its derivations.

For a class K of string languages, we denote by LIN-HR(K) the class of graph languages generated by linear HR grammars under a control language from K.

Theorem 6. *For every class K that is closed under nsm-mappings,* Int(K) = LIN-HR(K).

Proof. (Sketch) We generalize the proof of Theorem 2, as follows.

LIN-HR(K) $\subseteq$ Int(K). If C is the control language of the grammar G, then $h(C \cap R_{\text{der}}) = \text{val}(L(G))$, where R_{der} is the regular language of successful (uncontrolled) derivations of G. Since K is closed under intersection with regular languages, $C \cap R_{\text{der}} \in K$.

Int(K) $\subseteq$ LIN-HR(K). The language $h(L) = \{c_{h(a_1)} \circ \cdots \circ c_{h(a_n)} \mid a_1 \cdots a_n \in L \cap R_{\text{def}}\}$ can be generated by a right-linear regular tree grammar G over Δ that uses its finite control to generate strings of R_{def}, and has control language $\rho(L)$, where ρ is the alphabetical substitution that substitutes all productions $X \to c_{h(a)} \circ Y$ and $X \to c_{h(a)}$ for a. Since K is closed under alphabetical substitutions, $\rho(L) \in K$. □

The second characterization of Int(K) is through the notion of a "replacement", which generalizes the language theoretic homomorphism of strings. It is based on the well-known notion of replacement of edges by graphs (see Section I.2 of [Hab]). Let g be a graph, let $e_1, \ldots, e_k$ be edges of g, and let $g_1, \ldots, g_k$ be graphs such that type(g_j) = type(e_j) for $1 \leq j \leq k$. Then the replacement of e_i by g_i in g, denoted by $g[e_1/g_1, \ldots, e_k/g_k]$, is obtained from g by first removing edges $e_1, \ldots, e_k$, then adding graphs $g_1, \ldots, g_k$ disjointly, and finally identifying, for all $1 \leq j \leq k$, the ith source of e_j with the ith begin node of g_j (for $1 \leq i \leq |s(e_j)|$), and the ith target of e_j with the ith end node of g_j (for $1 \leq i \leq |t(e_j)|$). The notion of a string homomorphism is now generalized as follows. Let Σ be a typed alphabet. A *replacement* is a type-preserving mapping $\phi : \Sigma \to \text{GR}$; it is extended to a mapping from GR(Σ) to GR by defining, for $g \in \text{GR}(\Sigma)$, $\phi(g) = g[e_1/\phi(l(e_1)), \ldots, e_k/\phi(l(e_k))]$, where $\{e_1, \ldots, e_k\}$ is the edge set of g. Thus, every edge e of g with label $l(e) = \sigma$ is replaced by the graph $\phi(\sigma)$. Replacements are indeed homomorphisms with respect to concatenation: if $g \circ g'$ is defined, then $\phi(g \circ g') = \phi(g) \circ \phi(g')$. This immediately gives the following closure property of Int(K).

Lemma 7. *For every class K, Int(K) is closed under replacements.*

Proof. Clearly, $\phi(h(L)) = h'(L)$, where h' is defined by $h'(a) = \phi(h(a))$ for every $a \in A$. In fact, for $a_1, \ldots, a_n \in A$, $\phi(h(a_1 \cdots a_n)) = \phi(h(a_1) \circ \cdots \circ h(a_n)) = \phi(h(a_1)) \circ \cdots \circ \phi(h(a_n)) = h'(a_1) \circ \cdots \circ h'(a_n) = h'(a_1 \cdots a_n)$. □

We denote the class of all replacements by Repl, and, for a class G of graph languages, we let Repl(G) = $\{\phi(L) \mid \phi \in \text{Repl}, L \in G\}$. The next characterization of Int(K) is based on the fact that every interpretation can be decomposed into an "atomic" interpretation and a replacement. An interpretation $h : A \to$ GR of A is *atomic* if A is a typed alphabet and $h(a) = \text{atom}(a)$ for every $a \in A$ (for the definition of atom(a) see Section 1). By AtInt(K) we denote the set of all $h(L) \in \text{Int}(K)$ such that h is an atomic interpretation.

Theorem 8. *For every class K, Int(K) = Repl(AtInt(K)).*

Proof. One inclusion follows from Lemma 7. For the other inclusion, let $h : A \to$ GR be an interpretation. Make A into a typed alphabet by defining type(a) = type($h(a)$) for every $a \in A$. Let t be the unique atomic interpretation of the typed alphabet A, and let $\phi : A \to$ GR be the replacement defined by $\phi(a) = h(a)$ for every $a \in A$ (i.e., ϕ is h viewed as a replacement). Clearly, for every string $w \in A^*$, $\phi(t(w)) = h(w)$. In fact, if $w = a_1 \cdots a_n$, then $\phi(t(w)) = \phi(t(a_1) \circ \cdots \circ t(a_n)) = \phi(t(a_1)) \circ \cdots \circ \phi(t(a_n)) = \phi(a_1) \circ \cdots \circ \phi(a_n) = h(a_1) \circ \cdots \circ h(a_n) = h(w)$, because $\phi(\text{atom}(a)) = \phi(a)$ for every $a \in A$. □

4 Comparison of Int(K) and Int(K')

From Theorem 2 we know that Int(REG) = LIN-HR. It is not difficult to prove that also Int(LIN) = LIN-HR, where LIN is the (usual) class of languages generated by linear context-free grammars. This suggests that for graph languages the notions 'regular' and 'linear' coincide, as opposed to the string case. Even Int(DB) = LIN-HR, where DB is the class of derivation bounded context-free languages (see, e.g., Section VI.10 of [Sal], where they are called languages of "finite index"). One now wonders when Int(K) = Int(K'), and in particular one wonders how much larger the class K can be made without enlarging the class Int(K).

It is easy to see that for every given class K there is a largest class K' such that Int(K') = Int(K). We will call this the *extension of* K, denoted Ext(K). In fact, Ext(K) = $\bigcup \{K' \mid \text{Int}(K') = \text{Int}(K)\}$. Note that, for arbitrary classes K and K', Int(K) = Int(K') if and only if Ext(K) = Ext(K').

In the next theorem we give a characterization of Ext(K). For a class G of graph languages, let Str(G) denote the class of string languages L such that gr(L) is in G. Here, gr(L) = $\{\text{gr}(w) \mid w \in L\}$, and, for a string $w = a_1 \cdots a_n$ (with $n \geq 1$), gr(w) is the graph of type $(1,1)$ with nodes $1, \ldots, n+1$, an a_i-labeled edge from i to $i+1$ for every $1 \leq i \leq n$, begin node 1, and end node $n+1$. Thus, gr(w) encodes w in the obvious way: it is a path with the symbols of w as edge labels. Note that, in fact, 'gr' is the unique atomic interpretation which is obtained by viewing every symbol as having type $(1,1)$; hence gr(L) $\in$ Int(K) for every $L \in K$, which means that $K \subseteq$ Str(Int(K)).

Theorem 9. *For every class K that is closed under nsm-mappings, Ext(K) = Str(Int(K)).*

Proof. Clearly, if Int(K') = Int(K), then $K' \subseteq$ Str(Int(K')) = Str(Int(K)). Thus, it remains to show that Int(Str(Int(K))) = Int(K). Since $K \subseteq$ Str(Int(K)),

$\mathrm{Int}(K) \subseteq \mathrm{Int}(\mathrm{Str}(\mathrm{Int}(K)))$. For the other inclusion it suffices, by Lemma 7 and Theorem 8, to show that $\mathrm{AtInt}(\mathrm{Str}(\mathrm{Int}(K))) \subseteq \mathrm{Int}(K)$. To prove this, let $L_1 \in K$, let h_1 be an interpretation of the alphabet A of L_1 such that $h_1(L_1) = \mathrm{gr}(L_2)$ for some string language L_2, and let h_2 be an atomic interpretation of the alphabet B of L_2. Thus, $h_1 : A \to \mathrm{GR}(B)$ where each symbol from B has type $(1,1)$. However, for the atomic interpretation h_2 each symbol b from B has another (arbitrary) type that we will denote by $\mathrm{type}(b)$. Note that $h_2(b) = \mathrm{atom}(b)$, where $\mathrm{type}(\mathrm{atom}(b)) = \mathrm{type}(b)$; hence $\mathrm{type}(b) = (|\mathrm{begin}(h_2(b))|, |\mathrm{end}(h_2(b))|)$.

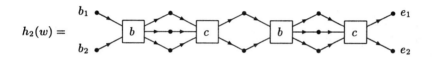

Fig. 6. Graphs $\mathrm{gr}(w)$ and $h_2(w)$ for $w = bcbc$, with $\mathrm{type}(b) = (2,3)$ and $\mathrm{type}(c) = (3,2)$.

We have to construct a language $L \in K$ and an interpretation h such that $h(L) = h_2(L_2)$. For a string $w \in L_2$, the graph $h_2(w)$ can be obtained from the graph $\mathrm{gr}(w)$ (which is an element of $h_1(L_1)$) in an easy way, as follows (see Fig.6). Each node v of $\mathrm{gr}(w)$ has to be replaced by a sequence of nodes $(v,1),(v,2),\ldots,(v,\mu(v))$, where μ stands for "multiplicity". Clearly, $\mu(v)$ is determined by $\mathrm{type}(b)$, where b is the label of an edge e incident with v: if e enters v, then $\mu(v) = |\mathrm{end}(h_2(b))|$, and if e leaves v, then $\mu(v) = |\mathrm{begin}(h_2(b))|$. Every edge e of $\mathrm{gr}(w)$, with source u and target v, should be replaced by an edge e with sources $(u,1),\ldots,(u,\mu(u))$ and targets $(v,1),\ldots,(v,\mu(v))$ (and the same label). In Fig.6, the multiplicity of the nodes of $\mathrm{gr}(w)$ is 2, 3, 2, 3, 2, respectively. The edges of $h_2(w)$ are drawn as squares, with "tentacles" from their sources and to their targets (where we assume that the tentacles are ordered, e.g., from top to bottom).

Based on this idea we change L_1 into L and h_1 into h, as follows. Let A' be the alphabet consisting of all pairs (a,μ) with $a \in A$ and μ is a mapping $V(h_1(a)) \to \mathbf{N}$ where $V(h_1(a))$ is the set of nodes of $h_1(a)$, such that for every edge e of $h_1(a)$ with source u, target v, and label b: $\mathrm{type}(b) = (\mu(u),\mu(v))$, i.e., $\mu(u) = |\mathrm{begin}(h_2(b))|$ and $\mu(v) = |\mathrm{end}(h_2(b))|$. Intuitively, μ is a guess of the multiplicities of the nodes of $h_1(a)$ as they will occur in a graph of $h_1(L_1)$; for nodes that are incident with an edge, this multiplicity is determined by the label of that edge, but for the other nodes (which are necessarily begin or end nodes) their multiplicity will only be clear after concatenation. Let ρ be the alphabetical substitution that substitutes all possible (a, μ) for a. Thus, $\rho(L_1) = \{(a_1, \mu_1) \cdots (a_n, \mu_n) \mid a_1 \cdots a_n \in L_1, (a_j, \mu_j) \in A'\}$. Let R be the regular language over A' that consists of all strings $(a_1, \mu_1) \cdots (a_n, \mu_n)$ such that $h_1(a_1 \cdots a_n)$ is defined, and μ_j of the ith end node of $h_1(a_j)$ equals μ_{j+1} of the ith begin node of $h_1(a_{j+1})$ (for all relevant i and j). This language R checks that the guessed multiplicities are consistent with the identification of nodes when concatenating the graphs $h_1(a_1), \ldots, h_1(a_n)$ (and hence are the correct multiplicities). We now define $L = \rho(L_1) \cap R$; since, by assumption, K is closed under nsm-mappings, L is in K. Finally, for $(a, \mu) \in A'$ we define the graph $h(a, \mu)$ as follows. First, it has all nodes (v, i) where v is a node of $h_1(a)$ and $1 \leq i \leq \mu(v)$. Second, it has the same set of edges as $h_1(a)$ (and they have the same labels), but if edge e has source u and target v in $h_1(a)$, then $s(e) = \langle (u, 1), \ldots, (u, \mu(u)) \rangle$ and $t(e) = \langle (v, 1), \ldots, (v, \mu(v)) \rangle$ in $h(a, \mu)$. Finally, if $h_1(a)$ has the sequence of begin nodes $\langle v_1, v_2, \ldots, v_k \rangle$, then $h(a, \mu)$ has the sequence of begin nodes $\langle (v_1, 1), \ldots, (v_1, \mu(v_1)), \ldots, (v_2, 1), \ldots, (v_2, \mu(v_2)), \ldots, (v_k, 1), \ldots, (v_k, \mu(v_k)) \rangle$, and similarly for the end nodes.

It should now be clear that $h(L) = h_2(L_2)$. □

As a corollary of Theorem 9 we obtain that for arbitrary K and K' (both closed under nsm-mappings), $\text{Int}(K) = \text{Int}(K')$ if and only if $\text{Str}(\text{Int}(K)) = \text{Str}(\text{Int}(K'))$. This means that the graph generating power of K is completely determined by its string generating power (with strings coded as graphs by the mapping gr).

Finally, we show that $\text{Ext}(K)$ is a class of languages that is well known in formal language theory. By $2\text{DGSM}(K)$ we denote the class of images of languages from K under 2dgsm mappings, i.e., the class of all $f(L)$ where f is a 2dgsm mapping and $L \in K$. A 2dgsm (i.e., a two-way deterministic generalized sequential machine) is a deterministic finite automaton that can move in two directions on its input tape (with endmarkers), and outputs a (possibly empty) string at each step. The proof of the next result is obtained by generalizing the proof in [EngHey] that $\text{Str}(\text{LIN-HR})$ equals the class of output languages of 2dgsm mappings.

Lemma 10. *For every infinite class K that is closed under nsm-mappings, $\text{Str}(\text{LIN-HR}(K)) = 2\text{DGSM}(K)$.*

From Theorems 9 and 6, and Lemma 10, we obtain our second characterization of the class $\text{Ext}(K)$.

Theorem 11. *For every infinite class K that is closed under nsm-mappings, $\text{Ext}(K) = 2\text{DGSM}(K)$.*

Corollary 12. *For all infinite classes K and K' that are closed under nsm-mappings, $\text{Int}(K) = \text{Int}(K')$ if and only if $2\text{DGSM}(K) = 2\text{DGSM}(K')$.*

Quite a lot is known about the class $2\text{DGSM}(K)$, see, e.g., [EngRS]. As an example, it equals the class of languages generated by K-controlled ETOL systems of finite index. The trivial fact that $\text{Ext}(\text{Ext}(K)) = \text{Ext}(K)$ corresponds to the known result that $2\text{DGSM}(2\text{DGSM}(K)) = 2\text{DGSM}(K)$; this shows that $\text{Ext}(K)$ is closed under 2dgsm mappings.

Corollary 12 allows us to use known formal language theoretic results for the classes $2\text{DGSM}(K)$ to find out the power of the classes $\text{Int}(K)$. Thus, for $K = \text{REG}$, $\text{Ext}(K)$ is the class $2\text{DGSM}(\text{REG})$ of output languages of 2dgsm mappings. Since it is well known that the class DB of derivation-bounded context-free languages is contained in $2\text{DGSM}(\text{REG})$ (see, e.g., [Raj]), this implies the previously mentioned result that $\text{Int}(\text{LIN}) = \text{Int}(\text{DB}) = \text{Int}(\text{REG})$. Also, since there is a context-free language not in $2\text{DGSM}(\text{REG})$, see [Gre, EngRS], $\text{Int}(\text{REG})$ is properly included in $\text{Int}(\text{CF})$, where CF is the class of context-free languages. We finally note that $\text{Int}(\text{CF})$ is properly included in HR, the class of graph languages generated by HR grammars. The inclusion follows from the fact that context-free grammars generating graph expressions built from concatenation and all constant graphs, can be simulated by HR grammars, by Theorem 1. Properness of the inclusion follows from the more general fact that for any class K, $\text{Int}(K)$ contains graph languages of bounded path-width only. Thus, the set of all binary trees (which is in HR) does not belong to any $\text{Int}(K)$.

5 Conclusion

We have introduced an operation of concatenation of graphs, and we have shown how it can be used for the generation of graph languages. More detailed proofs can be found in [EngVer]. Some remaining problems are the following.

(1) Are there characterizations, comparable to those of Theorems 1 and 2, for the graph languages generated by apex HR grammars? (see, e.g., [EngHL]).

(2) Is there a complete set of equations (including the law of strict monoidality) for the operations $\circ$ and $\oplus$ and a number of constants such as those in EL? This would give a result similar to Theorem 3.10 of [BauCou]. It would characterize $\text{GR}(\Sigma)$ as the free x-category satisfying the equations; such results are shown in [Hot1, Cla] (where $I_{1,0}, I_{1,2}, \pi_{12}$ are called U, D, V, respectively).

(3) Is $\text{Int}(\text{CF})$ the largest class of the form $\text{Int}(K)$ that is included in HR? It is shown in [EngVer] that this is equivalent to the question whether $2\text{DGSM}(\text{CF})$ is equal to the class $\text{Str}(\text{HR})$, which is the class of output languages of deterministic tree-walking transducers (cf. [EngHey]). And is this largest class equal to the class of all HR languages of bounded path-width?

(4) Is it decidable whether an HR language is in LIN-HR ($= \text{Int}(\text{REG})$)? And whether it is in $\text{Int}(\text{CF})$?

(5) Is there another natural concatenation operation on graphs that can be used to characterize the graph languages generated by linear edNCE grammars (which are node replacement graph grammars, see, e.g., [CouER])?

Acknowledgment. We thank the referees for their remarks and suggestions.

References

[BauCou] M.Bauderon, B.Courcelle; Graph expressions and graph rewritings, Math. Syst. Theory 20 (1987), 83-127

[Ben] D.B.Benson; The basic algebraic structures in categories of derivations, Inf. and Control 28 (1975), 1-29

[Ber] J.Berstel; *Transductions and Context-Free Languages*, Teubner, Stuttgart, 1979

[BosDW] F.Bossut, M.Dauchet, B.Warin; A Kleene theorem for a class of planar acyclic graphs, Inf. and Comp. 117 (1995), 251-265

[Cla] V.Claus; Ein Vollständigkeitssatz für Programme und Schaltkreise, Acta Informatica 1 (1971), 64-78

[Cou] B.Courcelle; Graph rewriting: an algebraic and logic approach, in *Handbook of Theoretical Computer Science, Vol.B* (J.van Leeuwen, ed.), Elsevier, 1990, pp.193-242

[CouER] B.Courcelle, J.Engelfriet, G.Rozenberg; Handle-rewriting hypergraph languages, J. of Comp. Syst. Sci. 46 (1993), 218-270

[Dre] F.Drewes; Transducibility - symbolic computation by tree-transductions, University of Bremen, Bericht Nr. 2/93, 1993

[EhrKKK] H.Ehrig, K.-D.Kiermeier, H.-J.Kreowski, W.Kühnel; *Universal Theory of Automata*, Teubner, Stuttgart, 1974

[Eng] J.Engelfriet; Graph grammars and tree transducers, Proc. CAAP'94 (S.Tison, ed.), Lecture Notes in Computer Science 787, Springer-Verlag, Berlin, 1994, pp.15-36

[EngHey] J.Engelfriet, L.M.Heyker; The string generating power of context-free hypergraph grammars, J. of Comp. Syst. Sci. 43 (1991), 328-360

[EngHL] J.Engelfriet, L.M.Heyker, G.Leih; Context-free graph languages of bounded degree are generated by apex graph grammars, Acta Informatica 31 (1994), 341-378

[EngRS] J.Engelfriet, G.Rozenberg, G.Slutzki; Tree transducers, L systems, and two-way machines, J. of Comp. Syst. Sci. 20 (1980), 150-202

[EngVer] J.Engelfriet, J.J.Vereijken; Context-free graph grammars and concatenation of graphs, Report 95-27, Leiden University, September 1995

[GecSte] F.Gécseg, M.Steinby; *Tree Automata*, Akadémiai Kiadó, Budapest, 1984

[Gre] S.Greibach; One-way finite visit automata, Theor. Comput. Sci. 6 (1978), 175-221

[Hab] A.Habel; *Hyperedge Replacement: Grammars and Languages*, Lecture Notes in Computer Science 643, Springer-Verlag, Berlin, 1992

[Hot1] G.Hotz; Eine Algebraisierung des Syntheseproblems von Schaltkreisen, EIK 1 (1965), 185-205, 209-231

[Hot2] G.Hotz; Eindeutigkeit und Mehrdeutigkeit formaler Sprachen, EIK 2 (1966), 235-246

[HotKM] G.Hotz, R.Kolla, P.Molitor; On network algebras and recursive equations, in *Graph-Grammars and Their Application to Computer Science* (H.Ehrig, M.Nagl, G.Rozenberg, A.Rosenfeld, eds.), Lecture Notes in Computer Science 291, Springer-Verlag, Berlin, 1987, pp.250-261

[Raj] V.Rajlich; Absolutely parallel grammars and two-way finite state transducers, J. of Comp. Syst. Sci. 6 (1972), 324-342

[Sal] A.Salomaa; *Formal Languages*, Academic Press, New York, 1973

HRNCE Grammars – A Hypergraph Generating System with an eNCE Way of Rewriting

Changwook Kim and Tae Eui Jeong

School of Computer Science
University of Oklahoma
200 Felgar street, Room 114
Norman, Oklahoma 73019, U.S.A.

Abstract. We introduce a hypergraph-generating system, called HRNCE grammars, which is structurally simple and descriptively powerful, and present their basic properties, in particular their description power and normal forms.

1. Introduction

It is important that any graph grammar model be structurally simple, i.e., easy to use or understand, yet descriptively powerful, i.e., able to describe many interesting graph-theoretical properties. It is also desirable that such a system be flexible so that subclasses with nice features, such as tractable membership complexity or decidability, can be easily defined.

One of the most successful graph grammar models in this sense is the node-label-controlled (NLC) grammars of Janssens and Rozenberg [11], which replace a single node by a graph in a derivation step and the embedding of the newly introduced graph into the existing graph is based on node labels only. They can describe PSPACE-complete languages. Many restrictions and extensions of NLC grammars have been studied in the literature. An extension of NLC grammars that also utilize the edge labels and node identities, called eNCE grammars [8], has been particularly successful. They are as simple as NLC grammars yet possess many nice features of NLC grammars. For example, their linear and separated subclasses have many decidable decision properties and tractable membership complexity.

For grammars generating hypergraphs, there are two well-known models: the context-free hypergraph (CFHG) grammars [2, 10], which replace a hyperedge by a hypergraph through their preidentified gluing points, and an extension of CFHG grammars called the handle-rewriting hypergraph (HH) grammars [4], which replace a handle (i.e., a hyperedge together with its incident nodes) by a hypergraph through an extension of the CFHG rewriting mechanism that can also duplicate or delete the hyperedges surrounding the replaced handle. Both CFHG and HH grammars generate directed hypergraphs. CFHG grammars are structurally simple but are limited in description power, i.e., generate NP languages only. HH grammars are powerful but their rewriting mechanism seems to be rather complicated, in particular because of the duplication and deletion features.

We introduce a new hypergraph-generating system, called HRNCE grammars (hypergraph grammars with an eNCE way of rewriting), generating node- and hyperedge-labeled undirected hypergraphs. An HRNCE grammar replaces a handle by a hypergraph, as in HH grammars, whose nodes are connected to (or contained in) the hyperedges surrounding the replaced handle by using the eNCE rewriting mechanism. HRNCE grammars are as easy to use as eNCE grammars yet can generate all recursively enumerable languages. Furthermore, their subclasses possess many nice features comparable to their eNCE counterparts. We shall present basic properties of HRNCE grammars and their restrictions, emphasizing their normal forms and description power (or complexity). Due to the space limitation, proofs are mostly omitted or very briefly sketched; they will be given in a longer version of the paper.

2. Preliminaries

Let Σ, Γ be alphabets. A *hypergraph* over Σ and Γ is a system $H = (V, E, \phi, \psi)$, where V is a finite set of *nodes*, E is a finite set of *hyperedges* (or simply *edges*), $\phi: V \to \Sigma$ is a *node-labeling function*, and $\psi: E \to \Gamma$ is an *edge-labeling function*. Each edge $e \in E$ consists of a nonempty subset of V, denoted by $V(e)$. Thus, we handle undirected, node- and edge-labeled hypergraphs. There can be multiple edges, i.e., edges e and e' with $V(e) = V(e')$, but isolated nodes are not allowed, i.e., $\bigcup_{e \in E} V(e) = V$. Note that H is a *graph* if each edge consists of one or two nodes. For any hypergraph H, its four components are denoted by V_H, E_H, ϕ_H, and ψ_H.

A hypergraph can be pictorially described by a bipartite graph. Figure 1 shows an example of such a description of a hypergraph with seven nodes (the dots) and three edges (the boxes) together with their labels. The membership of a node in an edge is indicated by a line connecting them.

A node v and an edge e in H are *incident* to each other if $v \in V_H(e)$. Two

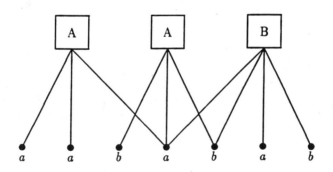

Fig. 1. A hypergraph.

edges e and e' in H are *adjacent* (or *a-adjacent*) if there is a node $v \in V_H(e) \cap V_H(e')$ (with $\phi_H(v) = a$). The *degree* of a node v in H, denoted by $\deg_H(v)$, is the number of its incident edges; the degree of H, denoted by $\deg(H)$, is the maximum degree of its nodes. The *rank* of an edge e in H, denoted by $\text{rank}_H(e)$, is the number of its incident nodes; the rank of H, denoted by $\text{rank}(H)$, is the maximum rank of its edges.

A sequence $e_0, v_1, e_1, \ldots, v_k, e_k$ ($k \geq 1$) is a *path* (between e_0 and e_k) in H if $e_i \in E_H$, $0 \leq i \leq k$, and $v_j \in V_H(e_{j-1}) \cap V_H(e_j)$, $1 \leq j \leq k$. H is a *chain* if it consists of a path and there is no other node, edge or incidence. H is *connected* if there is a path between each pair of its edges. H is the *empty hypergraph*, denoted by Λ, if $E_H = \emptyset$.

The set of all hypergraphs over Σ and Γ is denoted by $HGR_{\Sigma,\Gamma}$. A *hypergraph language* is any subset of $HGR_{\Sigma,\Gamma}$. A hypergraph language is *connected* if it contains connected hypergraphs only and is *degree-bounded* (*rank-bounded*) if the degree (rank) of each of its members is at most k, for some fixed $k \geq 0$.

3. HRNCE Grammars

Existing (hyper)graph grammars rewrite a node, an edge, or a handle in a derivation step. As stated in [4], one can observe that, in general, handle-rewriting grammars are more powerful than node-rewriting grammars, which in turn are more powerful than edge-rewriting grammars. We choose to rewrite a handle in order to maximize description power. There are several ways to connect the newly introduced hypergraph into the neighborhood of the replaced handle: edges to nodes, nodes to edges, or both. One can also create new nodes or edges not in the right-hand side of a production when embedding takes place, as in an HH grammar. Our choice is to simply connect nodes to edges without creating new nodes or edges. However, in order to perform such a connection selectively, we shall fully utilize all available local informations, such as node/edge labels and node identities, as in the eNCE grammar. This combination is simple enough, comparable to the eNCE rewriting mechanism, yet produces power and flexibility as shall be demonstrated in the subsequent sections.

Definition 3.1. A *hypergraph grammar with an eNCE way of rewriting* (*HRNCE grammar*) is a system $G = (\Sigma, \Delta, \Gamma, \Omega, P, Z)$, where

(1) Σ is an alphabet of *node labels*;

(2) Δ ($\subseteq \Sigma$) is the set of *terminal node labels* (the elements in $\Sigma - \Delta$ are *nonterminal node labels*);

(3) Γ is an alphabet of *edge labels*;

(4) Ω ($\subseteq \Gamma$) is the set of *terminal edge labels* (the elements in $\Gamma - \Omega$ are *nonterminal edge labels*);

(5) P is a finite set of *productions*, of the form $\pi = (A, X, C)$, where $A \in \Gamma - \Omega$ (the *left-hand side*), $X \in HGR_{\Sigma,\Gamma}$ (the *right-hand side*), and $C \subseteq V_X \times \Sigma \times \Gamma$ (the *embedding relation*); and

(6) Z ($\in HGR_{\Sigma,\Gamma}$) is the *axiom hypergraph*.

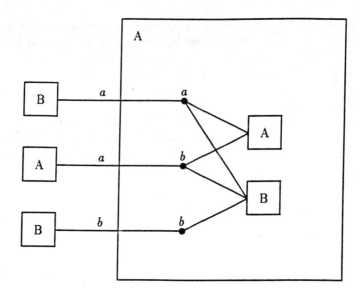

Fig. 2. A pictorial representation of a production rule.

We discuss informally how a production $\pi = $ (A, X, C) is applied to a nonterminal edge e (with $\psi_H(e) = $ A) in a hypergraph $H \in HGR_{\Sigma,\Gamma}$. It is very similar to the eNCE derivation step: First, remove e and all its incident nodes (i.e., a handle as called in [4]) from H. Second, add X (or an isomorphic copy of it) to the resulting hypergraph. Now, for each $v \in V_X$ and each $(v, a, B) \in C$, add v to each edge labeled by B that is a-adjacent to e in H. Such a situation can be well described pictorially as used, e.g., in [8]; we show an example in Fig. 2, where the symbol A located in the left upper corner of the big box represents the left-hand side of π, the hypergraph located inside the big box (with two edges and three nodes) represents the right-hand side of π, and the three lines crossing the big box together with the three edges located outside the big box represent the embedding relation of π, i.e., (x, a, B), (y, a, A), and (z, b, B) if x, y, z are nodes of X read top-down in Fig. 2. If any edge adjacent to e in H contains no node after applying π to e, then it is removed immediately.

As usual, a sequence of direct derivation steps is called a *derivation*. The transitive reflexive closure of $\Rightarrow$ is denoted by $\Rightarrow^*$. A hypergraph $H \in HGR_{\Sigma,\Gamma}$ such that $Z \Rightarrow^* H$ is called a *sentential form* of G. The *language generated by* G, denoted by $L(G)$, is the set $\{H \in HGR_{\Delta,\Omega} \mid Z \Rightarrow^* H\}$.

Example 3.2. Let G $= (\Sigma, \Delta, \Gamma, \Omega, P, Z)$ be an HRNCE grammar such that $\Sigma = \Delta = \{*\}$, $\Gamma = \{A, B, \#\}$, $\Omega = \{\#\}$, $Z = $ $\boxed{A} \overset{*}{\bullet} \boxed{\#}$, and P consists of the four productions given in Fig. 3(a). G generates the set of all "double stars" of the form as shown in Fig. 3(b), where the rank of the center of the left star can grow arbitrarily and the degree of the center of the right star can grow

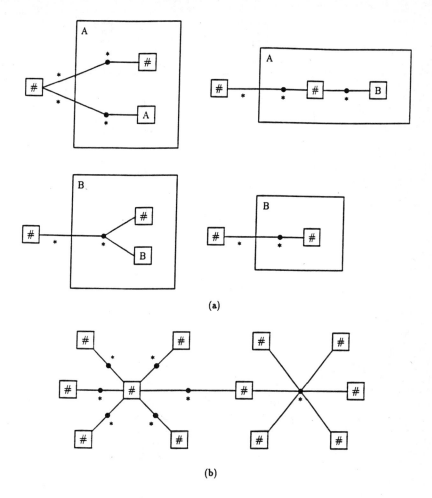

Fig. 3. (a) An HRNCE grammar generating all double stars;
(b) an example of a double star.

arbitrarily. This example shows that an HRNCE grammar can generate rank- and degree-unbounded hypergraphs, while CFHG and HH grammars generate rank-bounded hypergraphs only.

Example 3.3. Let G = (Σ, Δ, Γ, Ω, P, Z) be an HRNCE grammar such that $\Sigma = \Delta = \{*\}$, $\Gamma = \{A, \#\}$, $\Omega = \{*\}$, Z = [A]—*—[#]—*—[A], and P consists of the two productions given in Fig. 4(a). G generates the set of all (duals of) "binary trees" of the form as shown in Fig. 4(b).

HRNCE grammars are not confluent, i.e., derived graphs are dependent upon the order of application of production rules. Two subclasses of confluent (hyper)graph grammars that have been studied much in the literature are the

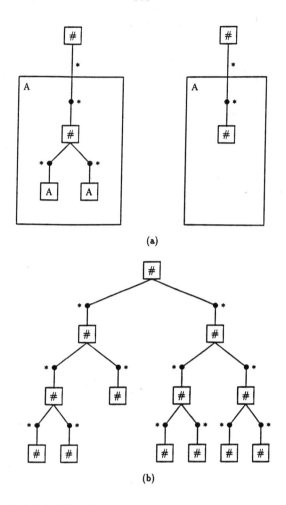

Fig. 4. (a) An HRNCE grammar generating all binary trees;
(b) an example of a binary tree.

separated and linear classes. We define their HRNCE versions, which can be easily seen to be confluent. Observe that the grammar in Example 3.2 is a linear HRNCE grammar and the grammar in Example 3.3 is a separated HRNCE grammar.

Definition 3.4. An HRNCE grammar is a *separated HRNCE (S-HRNCE) grammar* if no two nonterminal edges are adjacent in the axiom and in the right-hand side of any production. It is a *linear HRNCE (Lin-HRNCE) grammar* if there is at most one nonterminal edge in the axiom and in the right-hand side of each production.

4. Description Power

A word $x = a_1 a_2 \cdots a_n$ can be described by a special chain called its *oriented chain*, shown in Fig. 5 and denoted by o-chain(x), where \$ is a special symbol to indicate orientation. For a word language L, o-chain(L) denotes the set $\{\text{o-chain}(x) \mid x \in L\}$.

The following theorem can be proved by using a method similar to the well-known Turing-machine simulation by a phrase-structured grammar. Namely, an HRNCE grammar G can first generate two copies of an arbitrary input string to a Turing machine M in the form of a chain, with one copy of the input string on edges and another on nodes, and simulate M using node labels, adding new edges and nodes if M visits new blank cells. When M enters an accepting state, G can erase unnecessary edges and nodes to produce the oriented chain for the input. G can ensure the correctness of this simulation by keeping each intermediate sentential form in the chain form, i.e., once a disconnected sentential form is derived, no terminal hypergraph can be generated from it.

Theorem 4.1. *For every r.e. language L, we can construct an HRNCE grammar G such that $L(G) = o\text{-}chain(L)$.*

As each HRNCE language is clearly r.e., this implies that HRNCE languages are equivalent to r.e. languages. (This does not mean that every r.e. hypergraph language can be "directly" generated by an HRNCE grammar. For example, the set of all complete graphs cannot be generated by an HRNCE grammar.) This r.e. description power of HRNCE grammars originates from their *edge erasing capability* that can filter out undesired derivations from generating terminal hypergraphs by use of a so-called *blocking edge* or a *blocking node*, similar to the one used in [8]. Note that an HRNCE grammar can remove an edge in a sentential form in two ways:

(1) by an explicit edge-erasing production, called Λ-*production*, whose right-hand side is Λ; and

(2) by an implicit edge-erasing mechanism that removes an edge by removing the incident nodes/surrounding handles in a derivation.

The grammar G in the proof of Theorem 4.1 uses only the erasing mechanism of the second type and a blocking edge. It is also possible to simulate r.e. languages by using HRNCE grammars with the erasing mechanism of the first type only. For this purpose, let us describe a word $x = a_1 a_2 \cdots a_n$ by a *modified oriented chain*, denoted by m-o-chain(x), which is obtained from the oriented chain in Fig. 5 by attaching a $*$-labeled, degree-one node to each edge.

Fig. 5. An oriented chain.

It can be seen that, for each r.e. language L, we can construct an HRNCE grammar G' that uses the erasing of the first type only together with a blocking node such that $L_{conn}(G') = \text{m-o-chain}(L(M))$, where $L_{conn}(G')$ is the set of all connected hypergraphs in $L(G')$. Filtering out undesired derivations by disconnecting the finally generated hypergraphs as discussed here was previously used for NLC grammars simulating context-sensitive grammars [12]. The eNCE grammars use a blocking edge which is permanent once it is created [8]. Our HRNCE model uses a blocking edge and/or a blocking node that can or cannot be removed depending on a particular derivation by its selective erasing capability.

If we eliminate both types of edge erasing from HRNCE grammars, then their languages are in PSPACE. Such a grammar without edge erasing can generate a PSPACE-complete language, by using a blocking node only. See Theorem 4.4 below. This situation is comparable to NLC and eNCE grammars that also characterize the PSPACE languages in a similar way.

Definition 4.2. An *N-HRNCE grammar* is an HRNCE grammar without edge erasing, i.e., it never erases an already created edge except for its rewriting by a nonempty hypergraph.

Lemma 4.3. *For every context-sensitive language L, we can construct an N-HRNCE grammar G such that $L_{conn}(G) = \text{m-o-chain}(L)$.*

Theorem 4.4. *Every N-HRNCE language is in PSPACE. There exists a PSPACE-complete N-HRNCE language.*

The edge erasing property of either type, that an HRNCE grammar ever erases an already created edge to generate a terminal hypergraph, is an undecidable property. In practice, we need a class of HRNCE grammars for which this property is decidable or can be effectively removed, such as S-HRNCE grammars. Sections 5 and 6 investigate their normal forms and complexity.

5. Normal Forms

We shall present some useful normal form results for S-HRNCE grammars. In particular, the edge erasing property can be removed from S-HRNCE grammars and there exist Chomsky and Greibach normal forms for S-HRNCE grammars. (This implies that S-HRNCE languages are in NP.) S-HRNCE languages are context-free in the sense of Courcelle [3], i.e., they are generated by confluent and associative S-HRNCE grammars. The reader can observe that many normal forms discussed in this section are known for B-NLC, B-eNCE and/or B-edNCE grammars and the results stated in this section hold for Lin-HRNCE grammars as well.

The following lemma can be proved by using a standard node contraction technique:

Lemma 5.1. *Every S-HRNCE language can be generated by an S-HRNCE grammar such that the nodes in each nonterminal edge of any sentential form have pairwise distinct labels.*

The *context* of a nonterminal edge e in a hypergraph H, denoted by context$_H(e)$, is the set $\{(a, B) \mid$ an edge e' with $\psi_H(e') = B$ is a-adjacent to $e\}$. An S-HRNCE grammar $G = (\Sigma, \Delta, \Gamma, \Omega, P, Z)$ is *context-consistent* if there is a function $\eta: \Gamma - \Omega \to 2^{\Sigma \times \Omega}$ such that, for each sentential form H in G and for each nonterminal edge e in H, context$_H(e) = \eta(\psi_H(e))$. The context of nonterminal edges created from the right-hand side of a production can be calculated from the context of a replaced hyperedge in each derivation step. Furthermore, the context of a nonterminal edge does not change until it is rewritten. This observation was used in [14] to transform B-NLC grammars into the context-consistent normal form. The following lemma can be proved by using the same technique.

Lemma 5.2. *Every S-HRNCE language can be generated by a context-consistent S-HRNCE grammar.*

A *chain production* is one whose right-hand side contains exactly one edge which is nonterminal. The following lemma can be proved by a method analogous to the case of context-free grammars or eNCE grammars [8].

Lemma 5.3. *Every S-HRNCE language without Λ can be generated by an S-HRNCE grammar without Λ-productions or chain productions.*

An S-HRNCE grammar $G = (\Sigma, \Delta, \Gamma, \Omega, P, Z)$ is *neighborhood preserving* if, for all $H, K \in HGR_{\Sigma,\Gamma}$ such that $Z \Rightarrow^* H \Rightarrow_{(e,\pi)} K$, each edge adjacent to e in H is again adjacent to at least one edge from the right-hand side of π in K. If connection between a nonterminal edge and a terminal edge is broken in future derivation, such a connection can be avoided in the first place by using augmented edge labels. Such a method was used in [14] (for B-NLC grammars) and [9] (for B-edNCE grammars). By a similar method, the following can be proved.

Lemma 5.4. *Every S-HRNCE language without Λ can be generated by a neighborhood-preserving S-HRNCE grammar.*

Now, the following lemma is immediate from Lemma 5.3, that removes Λ-productions, and the neighborhood-preserving property stated in Lemma 5.4.

Lemma 5.5. *Every S-HRNCE language without Λ can be generated by an S-HRNCE grammar without edge erasing.*

An S-HRNCE grammar is *reduced* if it is context-consistent, chain production free, neighborhood-preserving, and edge erasing free. The normal form results stated so far can be combined together by using a standard method.

Theorem 5.6. *Every S-HRNCE language without Λ can be generated by a reduced S-HRNCE grammar.*

An S-HRNCE grammar is in *Chomsky normal form* if the right-hand side of each production consists of either one edge, which is terminal, or two edges, at least one of which is nonterminal. A production whose right-hand side contains two terminal edges or more than two edges can be transformed into a set of productions that must be executed in sequence, as in a context-free grammar. Such a method was used for B-eNCE grammars in [8]. By using a similar method, we have:

Theorem 5.7. *Every S-HRNCE language without Λ can be generated by a reduced S-HRNCE grammar in Chomsky normal form.*

A Lin-HRNCE grammar is a *Lin1-HRNCE grammar* if the right-hand side of each production contains either one or two edges, one of which is terminal. The following theorem is a special case of the previous theorem.

Theorem 5.8. *Every Lin-HRNCE language without Λ can be generated by a reduced Lin1-HRNCE grammar.*

An S-HRNCE grammar is in *Greibach normal form* if the right-hand side of each production contains exactly one terminal edge, and optionally other nonterminal edges. The well-known Greibach normal form transformation for context-free grammars involves two principal transformations on productions: production substitution and recursion removal. It is easy to define such a substitution for S-HRNCE grammars, that combines two derivation steps that can occur consecutively into one. Call a production (A, X, C) in an S-HRNCE grammar recursive if all edges in X are labeled by A. Then, recursions can be removed from S-HRNCE grammars. Now, by using the method for context-free grammars, we can prove:

Theorem 5.9. *Every S-HRNCE language without Λ can be generated by a reduced S-HRNCE grammar in Greibach normal form.*

Our final subject in this section is the context-freeness of S-HRNCE languages, along the definition of context-free rewriting systems given in [3]. The following theorem can be proved in a straightforward way and their proofs are left to the interested reader, referring to Lemma 5.3 in [3] and Lemma 3.2 in [4] for a very similar proof.

Theorem 5.10. *Every S-HRNCE language can be generated by a context-free S-HRNCE grammar (in reduced, Chomsky or Greibach normal form).*

6. Complexity of S-HRNCE Languages

We observed in Section 4 that HRNCE and N-HRNCE languages characterize the r.e. and PSPACE respectively languages. We continue investigation of the description power of S-HRNCE grammars along this line and show that S-HRNCE and Lin-HRNCE languages characterize the NP languages. We also show that certain restricted classes of S-HRNCE and Lin-HRNCE languages characterize the classes LOGCFL and NLOG, respectively.

The existence of reduced Greibach normal form for S-HRNCE grammars (Theorem 5.9) implies that S-HRNCE languages are in NP. The set CB_2 of all graphs with cyclic bandwidth at most two, which is NP-complete, was sometimes used to prove NP-hardness of graph languages, e.g., in [15] for monotone NLC languages and in [14] for B-NLC languages. By using a similar method, we can construct a Lin-HRNCE grammar generating dual(CB_2), the set of all duals of the graphs in CB_2, and other variations of Lin-HRNCE grammars, that yield the following theorem.

Theorem 6.1. *Every S-HRNCE language is in NP. There exists an NP-complete Lin-HRNCE language which satisfies any two of the following conditions: connected; degree-bounded; and rank-bounded.*

However, if all three conditions stated in Theorem 6.1 are imposed, then S-HRNCE and Lin-HRNCE languages characterize the classes LOGCFL and NLOG, respectively (see the following two theorems). This can be proved by using methods similar to the ones for efficient parsing of connected, linear or boundary eNCE languages of bounded degree [1, 6, 7] (the inclusion parts) and the simulation of string languages stated in Lemma 6.4 (the hardness parts).

Theorem 6.2. *Every Lin-HRNCE language which is connected, degree-bounded, and rank-bounded is in NLOG. There exists such a Lin-HRNCE language which is NLOG-complete.*

Theorem 6.3. *Every S-HRNCE language which is connected, degree-bounded, and rank-bounded is in LOGCFL. There exists such an S-HRNCE language which is LOGCFL-complete.*

This situation is similar to the complexity of B-edNCE languages which are in NP: there is an NP-complete Lin-edNCE language satisfying any two of the following conditions: connected; in-degree bounded; and out-degree bounded [1], and every B-edNCE (Lin-edNCE) language satisfying all these three conditions is in LOGCFL (NLOG) [7]. Note also that CFHG languages characterize the NP languages and connected CFHG graph languages of bounded degree are in LOGCFL [13]. In fact, B-edNCE grammars, CFHG grammars, and separated HH grammars generate the same graph languages of bounded degree [5, 9].

Lemma 6.4. *For every context-free (linear) language L, we can construct an S-HRNCE (Lin-HRNCE) grammar G such that $L(G) = o$-chain(L).*

Major decision problems such as equivalence are undecidable for linear grammars. Therefore, a direct consequence of Lemma 6.4 is that these decision problems are also undecidable for Lin-HRNCE grammars. Among others, we shall state the following:

Theorem 6.5. *It is undecidable whether or not $L(G_1) = L(G_2)$ and $L(G_1) \cap L(G_2) = \emptyset$ for Lin-HRNCE grammars G_1 and G_2.*

7. An HRNCE Language Hierarchy

For each $X \in \{$ Lin-HRNCE, S-HRNCE, N-HRNCE, HRNCE $\}$, let $\mathcal{L}(X)$ denote the family of all languages generated by X-grammars.

The language of all duals of binary trees given in Example 3.4 is generated by an S-HRNCE grammar. It can be proved that every Lin-HRNCE language of bounded degree and bounded rank is of bounded cutwidth (with an appropriate definition of the cutwidth of a hypergraph). The language in Example 3.4 is cutwidth-unbounded but is degree- and rank-bounded, and so, it is not a Lin-HRNCE language. Therefore, we have the theorem stated below. A similar method was previously used to separate Lin-eNCE and Apex-eNCE classes [6].

Theorem 7.1. $\mathcal{L}(\text{Lin-HRNCE}) \subsetneq \mathcal{L}(\text{S-HRNCE})$.

Consider the word language $L = \{a^{2^n} \mid n \geq 1\}$. L can be certainly accepted by a linear-bounded automaton. Therefore, by Theorem 4.3, there exists an N-HRNCE grammar G such that $L_{conn}(G) = $ m-o-chain(L). However, it can be easily proved, by using a pumping argument, that there is no S-HRNCE grammar G' such that $L_{conn}(G') = $ m-o-chain(L). Therefore, we have:

Theorem 7.2. $\mathcal{L}(\text{S-HRNCE}) \subsetneq \mathcal{L}(\text{N-HRNCE})$.

Finally, the following separation is immediate from Theorems 4.1 and 4.4 since PSPACE is properly included in the class of r.e. languages.

Theorem 7.3. $\mathcal{L}(\text{N-HRNCE}) \subsetneq \mathcal{L}(\text{HRNCE})$.

8. Discussion

HRNCE grammars are structurally simple, as easy to use as NLC variations of grammars, yet descriptively powerful – they generate all r.e. languages and can simulate string grammars and automata in a very straightforward way. They are also flexible in that subclasses with nice features, such as S-HRNCE and Lin-HRNCE grammars, can be well defined. Known techniques and results can be well extended to these classes without difficulty. There are many interesting research subjects for the HRNCE model not covered in this paper. We shall list a few of them:

(1) There are other properties of HRNCE languages not discussed in this paper, e.g., their combinatorial and closure properties.

(2) There are other subclasses of HRNCE grammars, e.g., neighborhood-uniform and apex subclasses, and extensions of HRNCE grammars, e.g., their directed version obtained via the eNCE to edNCE extension.

(3) There are other language-describing mechanisms for HRNCE grammars, e.g., via squeezing with graph languages or other interesting hypergraph languages.

(4) There may be other tight relations between HRNCE subclasses and traditional complexity classes not discussed in this paper.

(5) Comparision of HRNCE grammars with other (hyper)graph grammars, such as eNCE, CFHG and HH grammars, is certainly an interesting research topic.

References

[1] IJ. J. Aalbersberg, J. Engelfriet and G. Rozenberg, The complexity of regular DNLC graph languages, *J. Comput. System Sci.* **40** (1990), 376–404.

[2] M. Bauderon and B. Courcelle, Graph expressions and graph rewritings, *Math. Systems Theory* **20** (1987), 83–127.

[3] B. Courcelle, An axiomatic definition of context-free rewriting and its application to NLC graph grammars, *Theoret. Comput. Sci.* **55** (1987), 141–181.

[4] B. Courcelle, J. Engelfriet and G. Rozenberg, Handle-rewriting hypergraph grammars, *J. Comput. System Sci.* **46** (1993), 218–270.

[5] J. Engelfriet and L. Heyker, Hypergraph languages of bounded degree, *J. Comput. System Sci.* **48** (1994), 58–89.

[6] J. Engelfriet and G. Leih, Linear graph grammars: power and complexity, *Inform. and Comput.* **81** (1989), 88–121.

[7] J. Engelfriet and G. Leih, Complexity of boundary graph languages, *Theoretical Informatics and Applications* **24** (1990), 267–274.

[8] J. Engelfriet, G. Leih and E. Welzl, Boundary graph grammars with dynamic edge relabeling, *J. Comput. System Sci.* **40** (1990), 307–345.

[9] J. Engelfriet and G. Rozenberg, A comparison of boundary graph grammars and context-free hypergraph grammars, *Inform. and Comput.* **84** (1990), 163–206.

[10] A. Habel and H.-J. Kreowski, Some structural aspects of hypergraph languages generated by hyperedge replacement, *Proc. STACS 87*, Lecture Notes in Computer Science **247** (1987), 207–219.

[11] D. Janssens and G. Rozenberg, On the structure of node-label-controlled graph languages, *Inform. Sci.* **20** (1980), 191–216.

[12] D. Janssens and G. Rozenberg, Restrictions, extensions, and variations of NLC grammars, *Inform. Sci.* **20** (1980), 217–244.

[13] C. Lautemann, The complexity of graph languages generated by hyperedge replacement, *Acta Inform.* **27** (1990), 399–421.

[14] G. Rozenberg and E. Welzl, Boundary NLC graph grammars — Basic definitions, normal forms, and complexity, *Inform. and Control* **69** (1986), 136–167.

[15] Gy. Turán, On the complexity of graph grammars, *Acta Cybernet.* **6** (1983), 271–280.

Node Replacement in Hypergraphs: Simulation of Hyperedge Replacement, and Decidability of Confluence

Renate Klempien-Hinrichs*

Universität Bremen, Fachbereich 3, Postfach 33 04 40, D–28334 Bremen
e-mail: rena@informatik.uni-bremen.de

Abstract. Node replacement in directed labelled hypergraphs as a generalization of node replacement in directed labelled graphs is introduced. This hypergraph grammar approach can simulate hyperedge replacement in hypergraphs. The problem whether a given hNCE-grammar is confluent is rather complex but decidable.

1 Introduction

Node replacement in hypergraphs may be motivated by the close relationship to place or transition refinement in Petri nets. A Petri net problem dealt with in [GG90] is how to connect a refinement net with initial or final concurrency to the environment of the refined transition such that the construction itself does not add a deadlock. The resulting operation on one-safe Petri nets is demonstrated by the following example where the transition r in the net N_1 is expanded into the subnet r' of N_2:

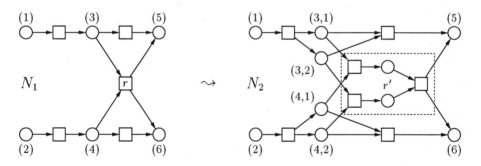

Fig. 1. Transition refinement in a Petri net.

For every preplace p of r in N_1 and every initial transition t in r' there is a new instance of p in N_2 connected to t and inheriting the rest of its connections from p (and analogously for the postplaces of r and the final transitions in r').

---

* Supported by a scholarship of the University of Bremen and the ESPRIT Basic Research Working Group No. 7183: COMPUGRAPH II.

This refinement operation is intriguing from a graph grammatical point of view because it cannot be expressed directly by hyperedge or handle rewriting in hypergraphs ([Hab92], [CER93]) or by node replacement in (bipartite) graphs (for the different approaches, cf. [ER91]). However, it can be modelled naturally by generalizing node replacement with node controlled embedding from edge-labelled directed graphs (edNCE-grammars, introduced as a special case in [Nag76] and studied in e.g. [Bra88], [ER90]) to hypergraphs (hNCE-grammars).

In section 2, this new hNCE-approach is introduced. In section 3, it is shown that if a hypergraph language is generated by a hyperedge replacement grammar then it can be defined by an hNCE-grammar. In section 4, the decidability of confluence for hNCE-grammars is proved. Section 5 contains some concluding remarks.

2 Node replacement in hypergraphs

Let $\Sigma := \Sigma_V \dot\cup \Sigma_E$ be an alphabet of *node* and *hyperedge labels*, respectively. A *hypergraph*[1] $H = (V_H, E_H, m_H)$ over Σ consists of a finite set V_H of *nodes*, a finite set $E_H \subseteq \Sigma_E \times V_H^*$ of *hyperedges*, and a *node labelling* function $m_H: V_H \to \Sigma_V$. The set of all hypergraphs over Σ is denoted by $\mathcal{H}_\Sigma$.

In drawings, dots represent nodes, squares stand for hyperedges, labels are drawn in the vicinity of their objects, and numbered lines from a square to various dots indicate the node sequence of the hyperedge (cf. figure 2).

For every hyperedge $e = (\sigma, v_1 \cdots v_k) \in E_H$, the set of *incident* nodes is $inc_H(e) := \{v_1, \ldots, v_k\}$. For a node set $V' \subseteq V_H$, $H|_{V'} := (V', E', m')$ with $E' := \{e \in E_H \mid inc_H(e) \subseteq V'\}$ and $m'(v) := m_H(v)$ for all $v \in V'$ denotes the *subhypergraph of H induced* by V'.

Fig. 2. A hypergraph G.

A bijective mapping $h: V_G \to V_H$ is called a *(hypergraph) isomorphism* between hypergraphs G and H if $m_H(h(v)) = m_G(v)$ for all $v \in V_G$ and $E_H = \{(\sigma, h(v_1) \cdots h(v_k)) \mid (\sigma, v_1 \cdots v_k) \in E_G\}$.

A *hypergraph grammar with neighbourhood controlled embedding*, for short hNCE-grammar, is a system $HG = (N, T, P, S)$, where the alphabet of *nonterminal* labels $N = N_V \dot\cup N_E$ consists of nonterminal node and hyperedge labels, respectively, the alphabet of *terminal* labels $T = T_V \dot\cup T_E$ consists of terminal node and hyperedge labels, respectively, P is the finite set of *productions*, and $S \in N_V$ is the *initial nonterminal*. In the sequel, let $\Sigma = N \dot\cup T$, $\Sigma_V = N_V \dot\cup T_V$, $\Sigma_E = N_E \dot\cup T_E$.

A *production* $\pi \in P$ is a triple $\pi = (A, K, Emb)$ such that $A \in N_V$ (*left-hand side* of π), $K \in \mathcal{H}_\Sigma$ (*right-hand side* of π), and the *embedding relation* $Emb \subset \Sigma_E \cdot (\{\bullet\} \cup \Sigma_V)^* \times \Sigma_E \cdot (V_K \cup \mathbf{N})^*$ is a finite set of *embedding components*

---

[1] Throughout this paper, *concrete* hypergraphs as defined are used to explain the constructions. To be entirely correct, *abstract* hypergraphs (equivalence classes made up of isomorphic concrete hypergraphs) would be needed to define generated hypergraphs etc.

($ex \triangleright cr$) each of which satisfies the following conditions:[2] (i) $\exists\, i \in \mathbf{N}$: $ex_i = \lozenge$, and (ii) $\forall\, j \in \mathbf{N}$: ($cr_j \in \mathbf{N} \Rightarrow ex_{cr_j} \in \Sigma_V$). The *existence part ex* sifts the hyperedges incident with a replaced node v for those fitting its specification, where the new symbol $\lozenge$ indicates an occurrence of v. With tentacles to some of the nodes thus designated or belonging to the right-hand side, an embedding hyperedge is then constructed following the description in the *creation part cr*.

In a drawing of an embedding component, the leftmost (unlabelled) node corresponds to $\lozenge$, and the one hyperedge to which it is incident represents the existence part which identifies the other nodes incident with that hyperedge. The remaining (unlabelled) nodes stand for those of the right-hand side and are arranged as in its graphical representation. The second hyperedge depicts the creation part. Embedding components with the same existence part will be placed together in one drawing (cf. figure 3).

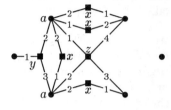

Fig. 3. Five embedding components referring to hypergraph G of figure 2.

Informally, a production $\pi = (A, K, Emb)$ is applied to a nonterminal node v in a hypergraph G with $m_G(v) = A$ as follows. First, v is removed from G together with all incident hyperedges. Next, K is added disjointly to the remainder of G in place of v. Finally, K is embedded in the remainder of G by adding hyperedges which are inferred from hyperedges incident with v in G, and matching embedding components in Emb.

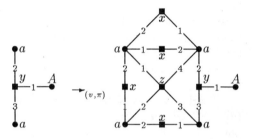

Fig. 4. Application of $\pi = (A, G, Emb)$ to the A-labelled node v (with G as in figure 2 and Emb as in figure 3).

Formally, this is defined as follows. Let $HG = (N, T, P, S)$ be an hNCE-grammar. Let G and H be hypergraphs over Σ, $v \in V_G$, and $\pi = (A, K, Emb)$ in P. G and K are assumed to be disjoint (otherwise, replace K by an isomorphic copy). Then G *directly derives* to H by applying π to v, denoted by $G \rightarrow_{(v,\pi)} H$ or just $G \rightarrow H$, if $m_G(v) = A$ and H is the following hypergraph:

- $V_H = (V_G - \{v\}) \cup V_K$,
- $E_H = (E_G - \{e \in E_G \mid v \in inc_G(e)\}) \cup E_K \cup$
  $\{(cr_1, u_2 \cdots u_l) \mid (ex_1, v_2 \cdots v_k) \in E_G$ and $(ex \triangleright cr) \in Emb$ such that:
  (1) $ex_i = \lozenge$ if $v_i = v$ and $ex_i = m_G(v_i)$ otherwise $(2 \le i \le k)$ and
  (2) $u_j = cr_j$ if $cr_j \in V_K$ and $u_j = v_{cr_j}$ if $cr_j \in \mathbf{N}$ $(2 \le j \le l)\,\}$,
- $m_H(u) = m_G(u)$ if $u \in V_G - \{v\}$ and $m_K(u)$ if $u \in V_K$.

$G \rightarrow_{(v,\pi)} H$ is also called a *derivation step*.

Note that condition (ii) for embedding components ensures that the embedding hyperedges in $E_H - (E_K \cup E_G)$ are well defined, while by condition (i), a

---

[2] $\mathbf{N}$ denotes the set of positive integers; w_i refers to the i-th symbol of a word $w \in A^*$.

hyperedge in E_G can be transformed into an embedding hyperedge only if it is incident with v and thus does not belong to E_H itself.

As usual, $\rightarrow^*$ denotes the reflexive transitive closure of $\rightarrow$, and $\rightarrow^i$ a derivation of i steps. For an hNCE-grammar $HG = (N, T, P, S)$ and $S_\bullet$ designating the *initial* hypergraph consisting of one node which is labelled by the initial nonterminal, $S(HG) := \{G \in \mathcal{H}_\Sigma \mid S_\bullet \rightarrow^* G\}$ is the set of *sentential forms*, and $L(HG) := S(HG) \cap \mathcal{H}_T$ is the *generated language*, i.e. the set of terminally labelled sentential forms.

Remark. An hNCE-grammar $HG = (N, T, P, S)$ is essentially an edNCE-grammar if for all productions $(A, K, Emb) \in P$, K is a graph, i.e. each of its hyperedges is of the form (σ, uv) with $u \neq v$, and all embedding components $(ex \triangleright cr)$ in Emb deal only with ordinary directed edges, i.e. $ex \in \Sigma_E \cdot (\{\blacklozenge\} \cup \Sigma_V)^2$ such that $ex_2 \neq ex_3$, and $cr \in \Sigma_E \cdot (V_K \cup \mathbf{N})^2$ with $\{cr_2, cr_3\} \not\subseteq V_K$ and $\{cr_2, cr_3\} \not\subseteq \mathbf{N}$. Every edNCE-grammar can be converted to this form and thus be perceived as an hNCE-grammar.

An embedding component $(ex \triangleright cr)$ is *truly embedding* if $cr \notin (\Sigma_E \cdot V_K^*) \cup (\Sigma_E \cdot \mathbf{N}^*)$, i.e. the nodes of an embedding hyperedge created by it are neither entirely in the right-hand side of π nor entirely in the remainder of G, so that the embedding hyperedge connects these two parts. An embedding relation Emb (an hNCE-grammar $HG = (N, T, P, S)$) is *truly embedding* if every embedding component $(ex \triangleright cr) \in Emb$ (the embedding relation Emb in every production $(A, K, Emb) \in P$) is truly embedding.

Clearly, every edNCE-grammar induces a truly embedding hNCE-grammar. In the Petri net example above, transitions can be interpreted as nodes, and places with the attached arcs of the flow relation as hyperedges labelled adequately to convey which of the arcs are ingoing (resp. outgoing). Thus, the refinement step from N_1 to N_2 may be seen as the replacement of the node r by the hypergraph r' where the hyperedges $(3,1)$ and $(3,2)$ in N_2 are derived from the hyperedge (3) in N_1, etc. The corresponding embedding relation is truly embedding.

3 Simulation of hyperedge replacement

Hyperedge replacement, a hypergraph rewriting technique thoroughly studied in [Hab92], operates on multiple hypergraphs, i.e. hypergraphs in which two or more distinct hyperedges may have the same label and be incident with the same sequence of nodes. A *multiple hypergraph* over an alphabet Γ is a system $M = (V_M, E_M, att_M, lab_M)$ where V_M is a finite set of *nodes*, E_M is a finite set of *hyperedges*, the function $att_M: E_M \rightarrow V_M^*$ assigns a sequence of *attached nodes* $att_M(e)$ to each $e \in E_M$, and $lab_M: E_M \rightarrow \Gamma$ is a function *labelling* the hyperedges. The set of all multiple hypergraphs over Γ is denoted by $\mathcal{MH}_\Gamma$.

A *hyperedge replacement grammar*, for short HR-grammar, is a system $HR = (\Gamma_N, \Gamma_T, R, Z)$ where Γ_N resp. Γ_T is the alphabet of *nonterminal* resp. *terminal labels* ($\Gamma := \Gamma_N \cup \Gamma_T$), R is the finite set of *rules*, and $Z \in \mathcal{MH}_\Gamma$ is the *axiom*.

A rule $\varrho \in R$ is a triple $\varrho = (A, K, Ext)$ such that $A \in \Gamma_N$ (*left-hand side* of ϱ), $K \in \mathcal{MH}_\Gamma$ (*right-hand side* of ϱ), and $Ext \in V_K^*$ is the sequence of *external* nodes.

Without loss of generality, it can be assumed that every node $v \in V_K$ occurs at most once in Ext (cf. Well-Formedness Theorem in [Hab92] where this property is called *repetition-freeness*), and that HR is *typed*, which insures that a rule can be applied to a hyperedge if and only if the label of the hyperedge equals the left-hand side of the rule (cf. Typification Theorem in [Hab92]).

The replacement of a hyperedge e in a multiple hypergraph M via a rule $\varrho = (lab_M(e), K, Ext)$ yielding a multiple hypergraph N, denoted $M \to_{(e,\varrho)} N$, is roughly executed by first deleting e from M (but not the incident nodes), then adding K disjointly to the remainder M' of M, and finally fusing the i-th external node of K with the i-th attached node of e.

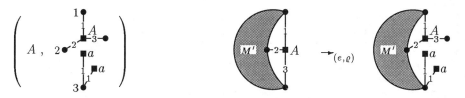

Fig. 5. A rule ϱ (the i-th external node is marked by i in the right-hand side) and its application to a hyperedge e in M.

The reflexive transitive closure of $\to$ is denoted by $\to^*$, and $\to^i$ stands for a derivation of i steps. For an HR-grammar $HR = (\Gamma_N, \Gamma_T, R, Z)$, the set of *sentential forms* is $S(HR) := \{M \in \mathcal{MH}_\Gamma \mid Z \to^* M\}$, and the *generated language* is denoted by $L(HR) := S(HR) \cap \mathcal{MH}_{\Gamma_T}$.

In [ER90], it is shown that HR-grammars and a subtype of edNCE-grammars which are known as boundary edNCE-grammars of bounded nonterminal degree (B-edNCE$_{bntd}$-grammars) have the same (hyper)graph generating power if hypergraphs are understood as bipartite graphs. By contrast, the idea pursued in this section is to simulate hyperedge replacement via node replacement while maintaining the complex structure of hyperedges. Not surprisingly, the constructed hNCE-grammar is also boundary, i.e. cannot have a sentential form in which two distinct nonterminal nodes are linked by a hyperedge. This property guarantees *confluence* which will be investigated in the next section.

Let $\Gamma' = \Gamma \cup \{\#, *\}$ with $*$ the unique hyperedge label; the new labels will be omitted in graphical representations. For $M \in \mathcal{MH}_\Gamma$, its encoding $\varphi(M) \in \mathcal{H}_{\Gamma'}$ is defined by $\varphi(M) = H$ with $V_H := V_M \cup E_M$, $E_H := \{(*, att_M(e) \cdot e) \mid e \in E_M\}$, $m_H(v) := lab_M(v)$ if $v \in E_M$, and $m_H(v) := \#$ otherwise.

Theorem 1. *For every language $L(HR)$ generated by an HR-grammar HR, a truly embedding boundary hNCE-grammar HG with $\varphi(L(HR)) = L(HG)$ can be constructed.*

Lemma 2. *Let $HR = (\Gamma_N, \Gamma_T, R, Z)$ be a (repetition-free, typed) HR-grammar. For every rule $\varrho = (A, K, Ext) \in R$, a production $\pi_\varrho = (A, K_\varrho, Emb_\varrho)$ can be constructed with Emb_ϱ truly embedding such that $M \rightarrow_{(e,\varrho)} N$ if and only if $\varphi(M) \rightarrow_{(e,\pi_\varrho)} \varphi(N)$, where $M, N \in S(HR)$, $e \in E_M$, $lab_M(e) = A$.*

Proof. The idea is to take $\varphi(K)$ except the external nodes of ϱ as the right-hand side of π_ϱ (the external nodes of ϱ cannot be fused with anything in node replacement). For every hyperedge e' of $\varphi(K)$ which is incident with an external node there is one embedding component the existence part of which specifies the encoded version of the hyperedge to be replaced, and the creation part introduces e'. As e' will also be incident with some new node $e \in E_K$ belonging to the right-hand side, the corresponding embedding component is truly embedding. □

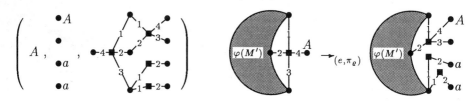

Fig. 6. The production π_ϱ constructed from ϱ as in figure 5, and its application to the node e in $\varphi(M)$.

Proof of the theorem. Let $HR = (\Gamma_N, \Gamma_T, R, Z)$ be an HR-grammar. Then define $HG = (N_V \mathbin{\dot\cup} N_E, T_V \mathbin{\dot\cup} T_E, P, S)$ as $N_V := \Gamma_N \cup \{S\}$, $N_E := \emptyset$, $T_V := \Gamma_T \mathbin{\dot\cup} \{\#\}$, $T_E := \{*\}$, and $P := \{\pi_\varrho \mid \varrho \in R\} \mathbin{\dot\cup} \{(S, \varphi(Z), \emptyset)\}$ with π_ϱ as in lemma 2. For every derivation $Z \rightarrow^k M$ in HR with $M \in L(HR)$ there is a derivation $S_\bullet \rightarrow \varphi(Z) \rightarrow^k G$ in HG such that $G = \varphi(M) \in L(HG)$, and vice versa because the applicability of a rule ϱ to a hyperedge e depends only on the label of e. Every embedding component occurring in HG is truly embedding and therefore HG as well. □

Remark. If HR is not typed, the construction above can be expanded with embedding components creating *blocking* hyperedges whenever a rule would not be applicable to a hyperedge. In general, however, these additional embedding components are not truly embedding.

4 Decidability of confluence

In [Cou87], the notion of context-freeness is generalized from string grammars to more complex types of rewriting systems, such as (hyper)graph grammars. A rewriting system is context-free only if it is *confluent*. For hNCE-grammars, confluence is decidable. In [Kau85] the same result is proved for the restricted case of edNCE-grammars by enumerating and testing all reachable constellations of neighbouring nonterminal nodes. Starting from this idea, the situation turns out to be considerably more complex with hNCE-grammars.

In this section, let $HG = (N, T, P, S)$ be an arbitrary hNCE-grammar.

Let $H \in \mathcal{H}_\Sigma$ be a hypergraph. H is *confluent*[3] in HG (with respect to $U \subseteq V_H$) if for every pair of different nonterminal nodes $v_1, v_2 \in V_H$ ($v_1, v_2 \in U$) and all productions $\pi_k = (A_k, K_k, Emb_k) \in P$ with $m_H(v_k) = A_k$ ($k = 1, 2$) the order in which the productions are applied to the nodes has no influence on the resulting hypergraph (on the subhypergraph of the resulting hypergraphs induced by $U' := (U - \{v_1, v_2\}) \cup V_{K_1} \cup V_{K_2}$). HG is confluent if all its sentential forms are confluent in HG. Figure 7 illustrates the situation.

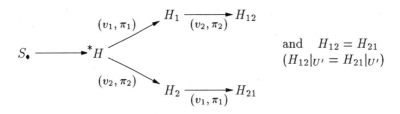

Fig. 7. Defining confluence.

Theorem 3. *For every hNCE-grammar $HG = (N, T, P, S)$, there is a constant $depth_{HG}$ (triply exponential with respect to the size of HG) such that HG is confluent if and only if all sentential forms derived from $S_\bullet$ in up to $depth_{HG}$ steps are confluent in HG.*

Outline of the proof. Certain substructures (called *locales*) of sentential forms G are defined of which only finitely many equivalence classes (so-called *settings*) need to be examined to determine the confluence of G locally. Assuming that HG is not confluent, there is a derivation $S_\bullet \rightarrowtail^n H$ with n minimal such that H is not confluent in HG. This implies that any other sentential form in this derivation is confluent in HG, and therefore the node replacements in the derivation may be reordered to obtain a sequence of "leftmost" subderivations. The sentential form at the end of such a subderivation contains a new setting which cannot be reached in fewer derivation steps and out of which the non-confluent part of H evolves. There is an upper bound on the length of these subderivations. This limit multiplied by the number of settings is an upper bound for n (if n exists) and thus equals $depth_{HG}$.

The locality of (non-)confluence follows from the fact that reversing the order of two consecutive derivation steps can only lead to a different result if the generated embedding hyperedges are not the same as the other way round. This is the case only if the sets of embedding hyperedges caused by some hyperedge incident with both of the rewritten nodes are not equal.

Lemma 4. *Let $H \in S(HG)$. H is confluent in HG if and only if H is confluent in HG with respect to $inc_H(e)$ for all hyperedges $e \in E_H$.*

---

[3] Note that the notion of confluence has a different meaning in the area of term rewriting systems.

The sets of embedding hyperedges caused by some crucial hyperedge need not be disjoint with the hyperedges already present in H, or with the embedding hyperedges caused by a hyperedge which was incident with only one of the rewritten nodes. The following notion takes this into account when reducing H to a critical substructure.

For $H \in \mathcal{H}_\Sigma$ and $V \subseteq V_H$, $N(V)$ denotes the set of nonterminally labelled nodes in V, and the hypergraph $\mathsf{L}(H, V) = (V', E', m')$ is the *locale* of V in H if

- $V' := V \dot\cup \Sigma_V$,
- $E' := \{e \in E_H \mid inc_H(e) \subseteq V\} \dot\cup \{\lambda_{H,V}(e) \mid e \in E_H, inc_H(e) \cap N(V) \neq \emptyset\}$,
  where $\lambda_{H,V}((\sigma, u_1 \cdots u_k)) := (\sigma, \tilde u_1 \cdots \tilde u_k)$
  with $\tilde u_i := u_i$ if $u_i \in V$ and $\tilde u_i := m_H(u_i)$ otherwise $(1 \leq i \leq k)$,
- $m' : V' \to \Sigma_V, m'(u) = m_H(u)$ if $u \in V$ and $m'(u) = u$ otherwise.

V is called the *centre* of $\mathsf{L}(H, V)$ and denoted by $centre(\mathsf{L}(H, V))$.

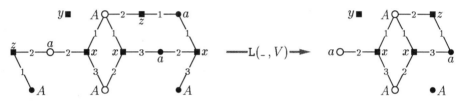

Fig. 8. A hypergraph H, where the white nodes form the set $V \subseteq V_H$, and $\mathsf{L}(H, V)$.

$\mathsf{L}(H, V)$ consists of $H|_V$ supplemented by a simplified version of all other hyperedges incident with a nonterminal node in V. The simplification lies in forgetting the identity of any incident node not in V and retaining only its label. $\mathsf{L}(\_, V)$ can be perceived as a function mapping any hypergraph H containing the nodes in V onto the corresponding locale $\mathsf{L}(H, V)$.

Clearly, $H|_V$ is identical to $\mathsf{L}(H, V)|_V$, where $V \subseteq V_H$. Also, it is not difficult to see that $\mathsf{L}(\mathsf{L}(H, V), U) = \mathsf{L}(H, U)$ if $U \subseteq V \subseteq V_H$, where the first term splits the construction corresponding to the second term into two parts. Another property of a locale is that it can monitor on a limited scope (its centre nodes) the effects of a derivation step which rewrites one of these centre nodes (cf. figure 9).

Lemma 5. *Let $G \in \mathcal{H}_\Sigma$, $U \subseteq V_G$, $u \in U$, and $\pi = (m_G(u), K, Emb)$ a production. Set $W = (U - \{u\}) \dot\cup V_K$. If $G \to_{(u,\pi)} H$ and $\mathsf{L}(G, U) \to_{(u,\pi)} M$, then $\mathsf{L}(H, W) = \mathsf{L}(M, W)$.*

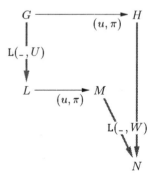

Fig. 9. Locales can monitor a derivation step.

Due to this property, certain locales may be used to detect a violation of the confluence condition.

Corollary 6. Let $H \in S(HG)$ and $e \in E_H$. H is confluent in HG with respect to $inc_H(e)$ if and only if $L(H, inc_H(e))$ is confluent in HG with respect to $inc_H(e)$.

Proof. Let $v_1, v_2 \in inc_H(e)$ and $\pi_i = (m_H(v_i), K_i, Emb_i) \in P$ ($i = 1, 2$). Triple usage of lemma 5 leads to the situation shown in figure 10 for all $i, j \in \{1, 2\}$ with $i \neq j$ and node sets $U_i = (inc_H(e) - \{v_i\}) \cup V_{K_i}$ and $U' = (inc_H(e) - \{v_i, v_j\}) \cup V_{K_i} \cup V_{K_j}$. With $M_{ij}|_{U'} = N_{ij}|_{U'} = H_{ij}|_{U'}$ follows $M_{12}|_{U'} = M_{21}|_{U'}$ if and only if $H_{12}|_{U'} = H_{21}|_{U'}$. □

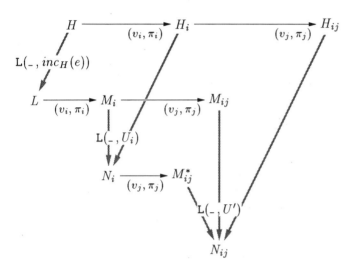

Fig. 10. Monitoring two consecutive derivation steps.

From lemma 4 and corollary 6 one can conclude that HG is not confluent if and only if there are a sentential form $H \in S(HG)$ and a hyperedge $e \in E_H$ such that $L(H, inc_H(e))$ is not confluent in HG with respect to $inc_H(e)$. To enumerate all reachable equivalence classes of locales the centre nodes of which are the nodes incident with some hyperedge in a sentential form, the origin of the hyperedge has to be taken into account. Either it is contained in the right-hand side of a production and thus stems directly from the rewriting of a node, or it is an embedding hyperedge and its creation was caused by another hyperedge with an incident nonterminal node. This classifies the centre nodes of relevant locales.

Locales L_1, L_2 are *equivalent* if a hypergraph isomorphism $h: L_1 \to L_2$ exists which maps $centre(L_1)$ onto $centre(L_2)$ (and thus extends the identity on Σ_V). The equivalence class of a locale L is denoted by $[L]$.

The set of *settings* of HG is given by
$Set(HG) := \{[L(H, \{v\})] \mid H \in S(HG), v \in V_H, m_H(v) \in N_V\} \cup$
$\{[L(H, inc_H(e))] \mid H \in S(HG), e \in E_H, N(inc_H(e)) \neq \emptyset\}$.

Obviously, the size of $Set(HG)$ is bounded only if the number of nodes incident with a hyperedge occurring in a sentential form of HG has a limit. With

$rk(e) := k$ denoting the *rank* of a hyperedge $e = (\sigma, v_1 \cdots v_k)$, define the rank of HG as $rk := \max\{rk(e) \mid e \in E_H, H \in S(HG)\}$. As a hyperedge in a sentential form is either (a copy of) a hyperedge in the right-hand side of a production or created by an embedding component, rk is at most $\max(\{rk(e) \mid e \in E_K, (A, K, Emb) \in P\} \cup \{|cr| - 1 \mid (ex \triangleright cr) \in Emb, (A, K, Emb) \in P\})$.

Lemma 7. *An upper bound for the size[4] of Set(HG) is*

$$s := rk \cdot (\#\Sigma_V)^{rk} \cdot 2^{(rk+1) \cdot (\#\Sigma_E) \cdot (\#\Sigma_V + rk)^{rk}}.$$

Proof. A locale in a setting has between 1 and rk centre nodes and therefore between $\#\Sigma_V + 1$ and $\#\Sigma_V + rk$ nodes in total. There are up to $(\#\Sigma_E) \cdot j^i$ hyperedges of rank i in a hypergraph with j nodes, with $0 \leq i \leq rk$. Therefore the maximal number of hyperedges is bounded by $(rk + 1) \cdot (\#\Sigma_E) \cdot j^{rk}$. With the hyperedges of a particular hypergraph forming a subset of all possible hyperedges, all centre nodes of a locale having some node label, and the number of these nodes varying from 1 to rk, s is a (quite generous) upper bound for the number of distinct settings. □

If HG is not confluent, one derivation of minimal length to yield a non-confluent setting suffices as proof. To find an upper bound for the required number of derivation steps, the given derivation is investigated with the help of the following notions.

For a derivation $G \to^* H$ and nodes $u \in V_G$, $v \in V_H$, u is the *ancestor* of v in G, denoted $u = anc_G(v)$, if $(u = v)$ or $(G \to_{(u,(A,K,Emb))} H$ and $v \in V_K)$ or $(G \to^* G' \to H$ and $u = anc_G(anc_{G'}(v)))$. If $u = anc_G(v)$ for a node $v \in V_H$, then v is a *descendant* of u in H, denoted $v \in desc_H(u)$.

Let $D : S_\bullet = H_0 \to_{(v_0,\pi_0)} H_1 \to_{(v_1,\pi_1)} \ldots H_n = H$ be a derivation with $\pi_i = (A_i, K_i, Emb_i)$ for $0 \leq i < n$ and $e \in E_H$. Then $l = (u_0, \ldots, u_k, e_{k+1}, \ldots, e_n)$ with $u_i \in V_{H_i}$ for $0 \leq i \leq k$ and $e_j \in E_{H_j}$ for $k < j \leq n$ is a *line in the genealogy of e* in D, denoted $l \in gen_D(e)$, if

- $u_i = anc_{H_i}(u_{i+1})$ for $0 \leq i < k$,
- $u_k = v_k$ and $e_{k+1} \in E_{K_{k+1}}$,
- $e_j = e_{j+1}$ or (v_j is incident with e_j and e_{j+1} is an embedding hyperedge constructed from e_j and some $(ex \triangleright cr) \in Emb_j)$ for $k < j < n$,
- $e_n = e$.

For $0 \leq i \leq n$, define the nodes *associated* with $p_i(l)$[5] as $V_{p_i(l)} := \{u_i\}$ if $p_i(l) = u_i \in V_{H_i}$ and $V_{p_i(l)} := inc_{H_i}(e_i)$ if $p_i(l) = e_i \in E_{H_i}$.

Clearly, if all nodes in $V_{p_i(l)}$ have a terminal label then $p_i(l) = e$, and $gen_D(e) \neq \emptyset$ for every derivation $D : S_\bullet \to^* H$ and $e \in E_H$. In general, $gen_D(e)$ contains more than one line, because a hyperedge may simultaneously be generated in more than one way.

---

[4] The size of a set X will be denoted by $\#X$.
[5] $p_i(x) = x_i$ is the projection on the i-th component for a tuple $x = (x_0, \ldots, x_n)$.

For each production $(A, K, Emb) \in P$, let $\prec_K$ be an arbitrary, but fixed total order on V_K. This induces an order on the nodes of any sentential form and thereby the notion of a leftmost derivation: As $S_\bullet$ consists of only one node, the order on it is trivial. For $S_\bullet \rightarrow^* G \rightarrow_{(A,K,Emb)} H$ and $v_1, v_2 \in V_H$, the induced order $\prec_H$ on V_H is defined as $v_1 \prec_H v_2$ for nodes $v_1, v_2 \in V_H$ if $v_1, v_2 \in V_K$ with $v_1 \prec_K v_2$ or $anc_G(v_1) \prec_G anc_G(v_2)$.

A derivation $H_0 \rightarrow_{(v_0, \pi_0)} H_1 \rightarrow_{(v_1, \pi_1)} \ldots H_n$ with $H_0 \in S(HG)$ is *leftmost* if for all $0 < i < n$, $v_i \in V_{H_{i-1}}$ only if $v_{i-1} \prec_{H_{i-1}} v_i$ holds. This property is indicated by writing $H_0 \xrightarrow[lm]{*} H_n$.

Given a derivation of minimal length to produce a non-confluent sentential form, it can be transformed (without altering the length) to pass through a series of pairwise distinct settings each of which contains sufficient information to derive all the following ones.

Lemma 8. *Let* $D : S_\bullet = H_0 \rightarrow_{(v_0,\pi_0)} H_1 \rightarrow_{(v_1,\pi_1)} \ldots H_n = H$ *be a derivation with* n *minimal such that* H *is not confluent in* HG *with respect to* $inc_H(e)$ *for some* $e \in E_H$. *Let* $\mathsf{D} = (v_0, \pi_0) \cdots (v_{n-1}, \pi_{n-1})$ *contain some information of* D.
(1) *A permutation* D' *of* D *defining* $D' : S_\bullet = H'_0 \rightarrow_{(v'_0, \pi'_0)} H'_1 \rightarrow_{(v'_1, \pi'_1)} \ldots H'_n = H$ *with* $gen_{D'}(e)$ *containing a line* $l = (u_0, \ldots, u_k, e_{k+1}, \ldots, e_n)$ *can be constructed such that*
 (i) *for* $0 \le i < n$: $v'_i \in V_{p_i(l)} \Rightarrow V_{H'_i} - V_{p_i(l)} \subseteq V_H$, *and*
 (ii) *for* $0 \le q < r \le n$: $v'_j \notin V_{p_j(l)}$ *for all* $q \le j < r \Rightarrow H'_q \xrightarrow[lm]{r-q} H'_r$.
(2) *For* D' *and* l *as in (1) and* $0 \le i < n$: *If* $v'_i \in V_{p_i(l)}$ *then for all* $S_\bullet \rightarrow^j G$ *with* $j < i$ *and* $U \subseteq V_G$, $\mathsf{L}(G, U)$ *is not equivalent to* $\mathsf{L}(H'_i, V_{p_i(l)})$.

Proof. (1) The condition that every sentential form derived by less than n steps must be confluent guarantees that n derivation steps starting from $S_\bullet$ can be permutated into any sequence yielding the same sentential form (provided that in the new sequence a node will be replaced only if it has been produced by a former step). A derivation fulfilling requirement (i) can be generated by progressing along D, successively putting every pair (v_i, π_i) with $v_i \notin V_{p_i(l)}$ to the point where v_i has just been created. After that, the derivation steps between two replacements of nodes belonging to items in the line only have to be reordered into a leftmost derivation, thus complying with requirement (ii).

(2) Whenever $v'_i \in V_{p_i(l)}$ then in the rest of the derivation D' beginning with H'_i, only descendants of nodes in $V_{p_i(l)}$ will be replaced (condition (i)). Suppose there is a derivation $S_\bullet \rightarrow^j G$ with $j < i$ and $\mathsf{L}(G, U)$ equivalent to $\mathsf{L}(H'_i, V_{p_i(l)})$ for some $U \subseteq V_G$. Then the derivation steps $(v'_i, \pi'_i) \cdots (v'_{n-1}, \pi'_{n-1})$ can be executed analogously on nodes in U and their descendants. Therefore this derivation also results in a non-confluent sentential form, contradicting the minimality of n. □

As the maximal number of settings is already known, the task is now to determine the maximal number of derivation steps connecting two successive settings $[\mathsf{L}(G, U)]$ and $[\mathsf{L}(G', U')]$ in D'. So, let $v'_i \in V_{p_i(l)}$ for some $0 \le i < n$, and set $G := H'_i$, $U := V_{p_i(l)}$. Moreover, let $j > i$ be minimal with $v'_j \in V_{p_j(l)}$ or $j = n$, and set $G' := H'_j$, $U' := V_{p_j(l)}$.

By lemma 8, the subderivation $G \rightarrow^* G'$ consists of one step $G \rightarrow_{(v'_i, \pi'_i)} G''$ and a derivation $G'' \xrightarrow[lm]{*} G'$. In G'', all nodes belonging to U' are already present. This means that the node rewritings in $G'' \xrightarrow[lm]{*} G'$ contribute to $L(G', U')$ only with embedding hyperedges and hyperedges of rank 0. Furthermore, $G'' \xrightarrow[lm]{*} G'$ being leftmost allows to divide it into subderivations each of which starts with rewriting one node u in $U - \{v'_i\}$ or in the right-hand side of π'_i, followed by rewriting only descendants of u. As there are at most rk nodes in U and there is a bound nn for the maximal number of nonterminal nodes in the right-hand side of any production of HG, the maximal length of such a subderivation has to be determined next. This will be done by combining the productions as used in the subderivation into one new production π' (to be applied to u) and restricting it to that information which possibly contributes to $L(G', U')$. It turns out that there are only finitely many of these restricted productions. Finally, the maximal length of the derivation represented by some π' can be determined.

Let P_Σ be the set of all productions in which only symbols from Σ appear, and $\pi_1, \pi_2, \pi_3 \in P_\Sigma$. Then $\pi_1 = (A_1, K_1, Emb_1)$ derives to $\pi_3 = (A_3, K_3, Emb_3)$ by applying $\pi_2 = (A_2, K_2, Emb_2)$ to a node $v \in V_{K_1}$, denoted by $\pi_1 \rightarrow_{(v, \pi_2)} \pi_3$ and called a *production derivation step*, if $m_{K_1}(v) = A_2$, $A_3 = A_1$, $K_1 \rightarrow_{(v, \pi_2)} K_3$, and $Emb_3 = \{(ex \triangleright cr) \in Emb_1 \mid v \text{ does not occur in } cr\} \cup$
$\{(ex \triangleright cr'') \mid (ex \triangleright cr) \in Emb_1, v \text{ occurs in } cr, (ex' \triangleright cr') \in Emb_2,$
$|cr| = |ex'|, cr_1 = ex'_1$, and for $2 \leq i \leq |cr|$:
$$cr_i = v \Rightarrow ex'_i = \blacklozenge,$$
$$cr_i \in V_{K_1} - \{v\} \Rightarrow ex'_i = m_{K_1}(cr_i),$$
$$cr_i \in \mathbf{N} \Rightarrow ex'_i = ex_{cr_i},$$
$|cr'| = |cr''|, cr'_1 = cr''_1$, and for $2 \leq j \leq |cr'|$:
$$cr'_j \in V_{K_2} \Rightarrow cr''_j = cr'_j,$$
$$cr'_j \in \mathbf{N} \Rightarrow cr''_j = cr_{cr'_j} \}$$

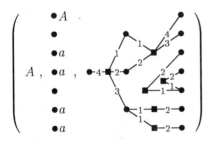

Fig. 11. The production obtained by applying π_ϱ of figure 6 to itself.

The production π_3 is designed such that for all H in $\mathcal{H}_\Sigma$ and π_1, π_2, π_3 in P_Σ with $\pi_1 \rightarrow_{(v, \pi_2)} \pi_3$, $H \rightarrow_{(u, \pi_3)} H_2$ if and only if $H \rightarrow_{(u, \pi_1)} H_1$ and $H_1 \rightarrow_{(v, \pi_2)} H_2$.

Just as "normal" derivation steps, production derivation steps may be iterated to form a *production derivation* $\pi \rightarrow^* \pi'$. Assuming an order on the nodes in the right-hand side of every production, there is also a *leftmost* production derivation $\pi \xrightarrow[lm]{*} \pi'$.

Production derivations tend to result in productions with very large right-hand sides. These have to be reduced to relevant information concerning embedding hyperedges and hyperedges of rank 0.

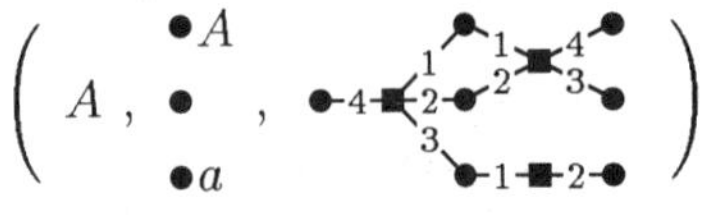

Fig. 12. The locale production of π_ϱ from figure 6.

For a production $\pi = (A, K, Emb)$, the *locale production* of π is $\mathsf{L}(\pi) := (A, K^*, Emb^*)$ with $V_{K^\bullet} = \Sigma_V$, $E_{K^\bullet} = \{e \in E_K \mid inc_K(e) = \emptyset\}$, $m_{K^\bullet}(u) = u$ for all $u \in V_{K^\bullet}$, and $Emb^* = \{(ex \triangleright cr^*) \mid (ex \triangleright cr) \in Emb, |cr^*| = |cr|, cr_i^* = m_K(v)$ if $cr_i = v \in V_K$ and $cr_i^* = cr_i$ otherwise, for $1 \le i \le |cr|\}$.

For all $\pi \in P_\Sigma$, $\pi^* := \mathsf{L}(\pi)$ contains the minimum of information from π such that for all $H \in HG_\Sigma$, if $H \to_{(v,\pi)} H'$ then $H \to_{(v,\pi^*)} H^*$ and $\mathsf{L}(H^*, V_H - \{v\}) = \mathsf{L}(H', V_H - \{v\})$.

Let $\mathbf{e}$ denote the empty derivation of length 0, and f_0 the constant mapping yielding $\mathbf{e}$ for every argument.

Define for every production $\pi = (A, K, Emb) \in P$ and $k \in \mathbf{N}$ sets $LP_k(\pi)$, the elements of which are pairs containing a description of a production derivation and the locale production obtained from the result of this derivation, as follows:

$LP_1(\pi) := \{(f_0, \mathsf{L}(\pi))\}$ and
$LP_{k+1}(\pi) := LP_k(\pi) \cup$
$\{(f, \mathsf{L}(\pi^*)) \mid f : N(V_K) \to \{\mathbf{e}\} \cup \bigcup_{\pi' \in P} p_1(LP_k(\pi'))\}$ describes a leftmost production derivation $\pi \xrightarrow{*}_{lm} \pi^*$ with $f(v) \in \{\mathbf{e}\} \cup p_1(LP_k(\pi'))$ such that $m_K(v)$ is the left-hand side of π' for all $v \in N(V_K)$, and for all $(g, \mathsf{L}(\pi^+)) \in LP_k(\pi) : \mathsf{L}(\pi^+) \ne \mathsf{L}(\pi^*)\}$.

$LP(\pi) := \bigcup_{k \in \mathbf{N}} LP_k(\pi)$.

Intuitively, an f with $(f, \mathsf{L}(\pi_k)) \in LP(\pi_0)$ describes a leftmost production derivation $\pi_0 \to^* \pi_1 \to^* \ldots \pi_k$ such that the subderivation $\pi_{i-1} \to^* \pi_i$ is specified by $f(v_i)$ and starts with replacing v_i by π' if $f(v_i) \in LP(\pi')$, where v_i is the i-th nonterminal node of the right-hand side of π ($1 \le i \le k$).

Lemma 9. *For all $\pi \in P : LP(\pi) = LP_d(\pi)$ if*
$$d := \#P \cdot 2^{(rk+1)^2 \cdot (\#\Sigma_E)^3 \cdot (\#\Sigma_V + rk)^{2 \cdot rk}}.$$

Proof. First, an upper bound for $\#p_2(LP(\pi))$ is determined for all $\pi \in P$: As the left-hand side of every $\mathsf{L}(\pi^*) \in p_2(LP(\pi))$ is inherited from π and the node set of the right-hand side consists of Σ_V, two locale productions in $p_2(LP(\pi))$ differ only with respect to the embedding relations and the hyperedges of rank 0 contained in the right-hand sides. The right-hand side of a locale production comprises at most $\#\Sigma_E$ different hyperedges of rank 0, and the number of embedding components is limited by

$$\sum_{i=1}^{rk}\left((\#\Sigma_E) \cdot (\#\Sigma_V + 1)^i \cdot \sum_{j=0}^{rk}(\#\Sigma_E) \cdot (\#\Sigma_V + i)^j\right)$$

where i and j refer to the ranks of hyperedges specified by the existence and creation part, respectively, of some embedding component. This sum is bounded in its turn by $(rk+1)^2 \cdot (\#\Sigma_E)^2 \cdot (\#\Sigma_V + rk)^{2 \cdot rk}$. As any of the hyperedges and embedding components may or may not appear in a specific locale production,

$$\#p_2(LP(\pi)) \le 2^{(rk+1)^2 \cdot (\#\Sigma_E)^3 \cdot (\#\Sigma_V + rk)^{2 \cdot rk}}$$

for every $\pi \in P$.

Scrutinizing again the definition of $LP_{k+1}(\pi)$, it is easy to see that if for all $\pi \in P : LP_{k+1}(\pi) = LP_k(\pi)$, then for all $j \in \mathbf{N}$ and all $\pi \in P : LP_{k+j}(\pi) = LP_k(\pi)$. (Obviously, $LP(\pi) = LP_k(\pi)$ in this case.) This is not fulfilled only if there is at least one production $\pi \in P$ such that $LP_{k+1}(\pi) \not\subseteq LP_k(\pi)$. Because of the bound on $\#p_2(LP(\pi))$ and the fact that if $x_1, x_2 \in LP(\pi)$ exist with $p_2(x_1) = p_2(x_2)$ then for all $i \in \mathbf{N}$ either $\{x_1, x_2\} \cap LP_i(\pi) = \emptyset$ or $\{x_1, x_2\} \cap LP_i(\pi) = \{x_1, x_2\}$, this means that if $k \geq d$ then for all $\pi \in P : LP_{k+1}(\pi) = LP_k(\pi)$. □

Knowing that the elements in $LP_d(\pi)$ contain all locale productions which can be derived from π, the last question to answer concerns the maximal length of a derivation which is described by some f with $(f, \mathsf{L}(\pi^*)) \in LP_d(\pi)$.

Define $lgth(\mathsf{e}) := 0$ and for f with $(f, \mathsf{L}(\pi^*)) \in LP(\pi)$ for some $\pi \in P$, $lgth(f) := 1 + \sum_{v \in N(V_K)} lgth(f(v))$. This definition reflects the length of a derivation starting with applying π to some node, followed by the rest of the derivation described by f.

Corollary 10. Let $\pi = (A, K, Emb) \in P$ and $(f, \mathsf{L}(\pi^*)) \in LP(\pi)$. Then with d as in lemma 9, $lgth(f) \leq d \cdot nn^d$.

Proof. By lemma 9, $(f, \mathsf{L}(\pi^*)) \in LP_d(\pi)$. Combining this fact with the definition of $lgth$, one gets

$$lgth(f) \leq 1 + nn \cdot \max_{v \in N(V_K)} lgth(f(v)) \leq \left(\sum_{i=0}^{d-1} nn^i\right) + nn^d \cdot lgth(\mathsf{e}) \leq d \cdot nn^d.$$

□

To sum up, the derivation $G \twoheadrightarrow^* G'$ from above has at most $1 + ((rk-1) + nn) \cdot d \cdot nn^d$ steps. Assuming that P contains at least one production (otherwise HG is trivially confluent), this is bounded by $(rk + nn) \cdot d \cdot nn^d$, which means that with s as in lemma 7, the constant $depth_{HG}$ promised in theorem 3 is (bounded by) $s \cdot (rk + nn) \cdot d \cdot nn^d$.

5 Conclusion

As node replacement systems, hNCE-grammars round off the presently known collection of hypergraph grammars. They generalize node replacement in graphs and allow to define any hypergraph language generated by an HR-grammar ([Hab92]), with the defining grammar being confluent (even boundary). The generating power of S-HH-grammars ([CER93]) seems to be included in that of hNCE-grammars. The relationship to unrestricted HH-grammars remains to be investigated. Although the recently introduced HRNCE-grammars ([KJ]) which rewrite hyperhandles use an eNCE-like embedding mechanism, there are major differences with respect to hNCE-grammars (e.g. the generation of rank-unbounded hypergraph languages permitted in the HRNCE framework).

Returning to the area of Petri nets, the example given in the introduction shows that truly embedding hNCE-grammars are suited to describe certain behaviour preserving refinement operations. Finally, noting the close relationship between hierarchical Petri nets as introduced in [Feh93] and the derivation trees of confluent node rewriting graph grammars, it may be interesting to employ confluent hNCE-grammars in the construction of system models.

Acknowledgement. The comments I received from Frank Drewes, Sabine Kuske, and two reviewers helped considerably to improve this paper.

All pictures are drawn with Frank's LaTeX-package for typesetting graphs; if they have unpleasing aspects I am to blame.

References

[Bra88] Franz J. Brandenburg: On polynomial time graph grammars. In: STACS 88 (R. Cori, Martin Wirsing, Eds.), Lecture Notes in Computer Science 294, 227–236, 1988.

[Cou87] Bruno Courcelle: An axiomatic definition of context-free rewriting and its application to NLC graph grammars. Theoretical Computer Science 55, 141–181, 1987.

[CER93] Bruno Courcelle, Joost Engelfriet, Grzegorz Rozenberg: Handle-rewriting hypergraph grammars. Journal of Computer and System Science 46, 218–270, 1993.

[ER90] Joost Engelfriet, Grzegorz Rozenberg: A comparison of boundary graph grammars and context-free hypergraph grammars. Information and Computation 84, 163–206, 1990.

[ER91] Joost Engelfriet, Grzegorz Rozenberg: Graph grammars based on node rewriting: an introduction to NLC graph grammars. In: Graph Grammars and Their Application to Computer Science (Hartmut Ehrig, Hans-Jörg Kreowski, Grzegorz Rozenberg, Eds.), Lecture Notes in Computer Science 532, 12–23, 1991.

[Feh93] Rainer Fehling: A concept of hierarchical Petri nets with building blocks. In: Advances in Petri Nets (Grzegorz Rozenberg, Ed.), Lecture Notes in Computer Science 674, 148–168, 1993.

[GG90] Rob van Glabbeek, Ursula Goltz: Refinement of actions in causality based models. In: Stepwise Refinement of Distributed Systems (Jaco W. de Bakker, Willem Paul de Roever, Grzegorz Rozenberg, Eds.), Lecture Notes in Computer Science 430, 267–300, 1990.

[Hab92] Annegret Habel: Hyperedge Replacement: Grammars and Languages. Lecture Notes in Computer Science 643, 1992.

[Kau85] Manfred Kaul: Syntaxanalyse von Graphen bei Präzedenz-Graph-Grammatiken. Ph.D. thesis, Osnabrück 1985. Report MIP-8610, Passau 1986.

[KJ] Changwook Kim, Tae Eui Jeong: HRNCE grammars – a hypergraph generating system with an eNCE way of rewriting. This volume.

[Nag76] Manfred Nagl: Formal languages of labelled graphs. Computing 16, 113–137, 1976.

Chain-Code Pictures and Collages Generated by Hyperedge Replacement[*]

Jürgen Dassow, Annegret Habel, and Stefan Taubenberger [**]

Abstract. Regular chain-code picture languages can be generated by collage grammars. As a consequence, all undecidability results known for regular chain-code languages can be adapted to picture languages generated by collage grammars.

1 Introduction

The generation and recognition of artificial pictures and patterns are challenging tasks in computer science and other applied areas. In the literature, one encounters quite a variety of syntactic approaches where classes of patterns are described by grammars. Among these approaches are chain-code grammars and collage grammars. Chain-code grammars, as introduced and investigated in [MRW82, SW85, Kim90, DH93], are based on string grammars over the alphabet $\{u, d, l, r, \uparrow, \downarrow\}$ and provide devices for the generation of picture languages by interpreting the symbols u, d, l, r of a string as commands for drawing a unit line segment in the direction up, down, left, and right, respectively, and the symbols $\uparrow, \downarrow$ for lifting the pen up and down. Collage grammars, as introduced and investigated in [HK88, HK91, HKT93, DHKT95, DK95], are based on hyperedge replacement in a geometric environment and provide context-free syntactic devices for the generation of picture languages by joining the parts (sets of points) of a collage.

In this paper, we relate context-free chain-code grammars and collage grammars and show that

1. The two syntactic devices for the generation of picture languages are incomparable, i.e., there exist picture languages which can be generated by collage grammars, but not by a context-free chain-code grammar, and there exist picture languages which can be generated by context-free chain-code grammars, but not by collage grammars.

---

[*] This work was partially supported by the Deutsche Forschungsgemeinschaft and the ESPRIT Basic Research Working Group No. 7183: Computing by Graph Transformation (COMPUGRAPH II).

[**] The author's addresses: Jürgen Dassow: Fakultät für Informatik, Universität Magdeburg, Postfach 4120, 39016 Magdeburg, Germany. E-mail: dassow@cs.uni-magdeburg.de. Annegret Habel: Institut für Informatik, Universität Hildesheim, Marienburger Platz 22, 31141 Hildesheim, Germany. E-mail: habel@informatik.uni-hildesheim.de. Stefan Taubenberger: Tettenbornstraße 26, 28211 Bremen, Germany. E-mail: taube@informatik.uni-bremen.de.

2. Restricted to regular chain-code grammars and a suitable type of linear collage grammars, so-called linear picture grammars, the approaches become comparable, i.e., each regular chain-code grammar can be transformed into an equivalent linear picture grammar, and vice versa.
3. All undecidability results known for regular chain-code picture languages can be carried over to linear collage languages. In particular, it can be shown that it is undecidable whether or not the language of a linear collage grammar contains only (a) pictures with holes, (b) disconnected pictures, or (c) connected patterns.
4. All decidability results known for regular chain-code languages can be carried over to linear picture languages. Unfortunately, up to now only a few questions on geometric properties turn out to be decidable.

The paper is organized as follows. In the sections 2 and 3, the concepts of chain-code picture languages and collage languages are recalled. A comparison of the presented concepts is given in section 4. In particular, it is shown that any regular chain-code picture language can be generated by a collage grammar. In section 5, undecidability results known for regular chain-code languages are adapted to linear collage languages. Finally, in section 6, some decidability results are presented. We conclude with a general discussion of (un-)decidability and possibilities for further research.

2 Chain-Code Picture Languages

In this section, the basic notions and notations concerning pictures, picture description grammars, and chain-code picture languages (cf. [MRW82, DH93]) are recalled and illustrated by an example. For the basic concepts of formal language theory as regular and context-free grammars and languages we refer to Hopcroft, Ullman [HU79].

A picture (in the sense of [MRW82]) consists of a finite set of unit lines in the Euclidean space considered as a square grid. A word over the alphabet $\{u, d, l, r, \uparrow, \downarrow\}$ is a picture description in the sense that it represents a traversal of a picture where the interpretation of the symbols $u, d, l, r, \uparrow, \downarrow$ is:

u	go one unit line up from the current point,
d	go one unit line down from the current point,
l	go one unit line to the left of the current point,
r	go one unit line to the right of the current point,
$\uparrow$	lift up the pen of the plotter, and
$\downarrow$	lift down the pen of the plotter.

A set of picture descriptions forms a picture description language. The interpretation of the words in the picture description language yields a set of pictures, a so-called chain-code picture language.

Assumption 1. *Let $\mathbb{Z}^2$ denote the two-dimensional Euclidean space with integer coordinates. For $z = (x, y) \in \mathbb{Z}^2$, the up-neighbour of z is $u(z) = (x, y+1)$,* the down-neighbour of z is $d(z) = (x, y-1)$, the left-neighbour of z is $l(z) = (x-1, y)$, and the right-neighbour of z is $r(z) = (x+1, y)$. The neighbourhood of z is defined as $N(z) = \{u(z), d(z), l(z), r(z)\}$. For $z \in \mathbb{Z}^2$ and $z' \in N(z)$, $\langle z, z' \rangle$ denotes the (undirected) unit line connecting z and z'.

Definition 2. (pictures) A *picture* is a finite set of unit lines in the grid $\mathbb{Z}^2$. A *drawn picture* is a triple (Q, z, s) where Q is a picture, $z \in \mathbb{Z}^2$ is a point, and $s \in \{\uparrow, \downarrow\}$ gives the state pen-up or pen-down.

Definition 3. (picture description words, grammars, and languages)

1. Let $\pi = \{u, d, l, r, \uparrow, \downarrow\}$ [3]. Every word in π^* is called a *picture description* or a *π-word*, every language over π is called a *picture description language* or a *π-language*, and every Chomsky-grammar generating a π-language is called a *picture description grammar* or a *π-grammar*.
2. The *drawn picture* described by a π-word w, denoted by $dpic(w)$, is defined inductively as follows:
   - If $w = \lambda$, then $dpic(w) = (\emptyset, (0,0), \downarrow)$.
   - If $w = vb$ for some $v \in \pi^*$, $b \in \pi$ and $dpic(v) = (Q, z, s)$, then

   $$dpic(w) = \begin{cases} (Q \cup \{\langle z, b(z) \rangle\}, b(z), \downarrow) & \text{if } s = \downarrow \text{ and } b \in \{u, d, l, r\} \\ (Q, b(z), \uparrow) & \text{if } s = \uparrow \text{ and } b \in \{u, d, l, r\} \\ (Q, z, \downarrow) & \text{if } b = \downarrow \\ (Q, z, \uparrow) & \text{if } b = \uparrow \end{cases}$$

   The *chain-code picture* described by a π-word w, denoted by $pic(w)$, is the picture Q underlying the drawn picture $dpic(w) = (Q, z, s)$. The *chain-code picture language* described by a π-language L, denoted by $pic(L)$, is defined by $pic(L) = \{pic(w) | w \in L\}$.
3. A picture language B is called *regular (context-free, etc.)* if $B = pic(L(G))$ for some regular (context-free, etc.) π-grammar G. The class of all regular and context-free picture languages is denoted by $\mathcal{B}(REG)$ and $\mathcal{B}(CF)$, respectively.

Example 1. Let $stairs = (N, \pi, P, S)$ be the picture description grammar with $N = \{S\}$ and production set $P = \{S \to ruSdr, S \to rurdr\}$. Then $L(stairs) = \{(ru)^n rurdr(dr)^n | n \in \mathbb{N}_0\}$. The pictures in the chain-code picture language $pic(L(stairs))$ look as in figure 1.

---

[3] Note that there are some small differences to the notions used in Dassow and Hinz [DH93]. In their paper, π denotes the picture description alphabet $\{u, d, l, r\}$. The alphabet π, extended by the symbols $\uparrow$ and $\downarrow$, is denoted by $\pi_\uparrow$. Moreover, a picture language described by a regular π-grammar is said to be a *regular* picture language and a picture language described by a regular $\pi_\uparrow$-grammar is said to be a *regular generalized* picture language. In this sense, we consider *regular generalized* picture languages.

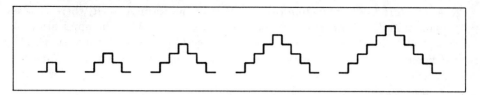

Fig. 1. Some chain-code pictures described by the π-grammar *stairs*.

3 Collage Grammars

In this section, the basic notions and notations concerning collages and collage grammars (cf. [HK91, HKT93]) are recalled and illustrated by an example. For the elementary notions of Euclidean geometry we refer to Coxeter [Cox89].

A collage consists of a set of parts being geometric objects, and a sequence of so-called pin-points. To allow the generation of sets of collages, a collage can be decorated by hyperedges. A hyperedge has a label and an ordered finite set of tentacles, each of which is attached to a point. A hyperedge may be replaced by a decorated collage, if there exists a transformation t from a given set of admissible transformations which maps each pin-point to the corresponding attachment point of the hyperedge. In this case, the result of the replacement is the decorated collage obtained from the original collage by removal of the hyperedge and addition of the transformed image of the replacing collage.

Assumption 4. *Let $\mathbb{R}^2$ denote the Euclidean space of dimension 2. Moreover, let TRANS be an arbitrary, but fixed set of affine transformations $t:\mathbb{R}^2 \to \mathbb{R}^2$.*

Definition 5. (decorated collages, collage grammars and languages)

1. A *collage* in $\mathbb{R}^2$ is a pair $(PART, pin)$ where $PART$ is a finite set of parts, each part being a set of points in $\mathbb{R}^2$, and $pin \in (\mathbb{R}^2)^*$ is a finite sequence of pairwise distinct *pin-points*[4]. Given a set LAB of labels, a *(hyperedge-) decorated collage* in $\mathbb{R}^2$ over LAB is a system $C = (PART, EDGE, att, lab, pin)$, where $(PART, pin)$ is a collage in $\mathbb{R}^2$, $EDGE$ is a finite set of *hyperedges*, and $att: EDGE \to (\mathbb{R}^2)^*$ and $lab: EDGE \to LAB$ are mappings, assigning a sequence of pairwise distinct *attachment-points* as well as a *labeling* to each hyperedge. A decorated collage with empty set of hyperedges and empty mappings att and lab may be seen as a collage. The class of all collages is denoted by $\mathcal{C}$. The class of all decorated collages over LAB is denoted by $\mathcal{C}(LAB)$.

---

[4] For sets A, B, $\mathcal{P}(A)$ denotes the set of the subsets and if A is finite, $|A|$ denotes the cardinality of A. $A-B$ denotes their set-theoretical difference and $A+B$ their disjoint union. Furthermore, A^* denotes the set of all finite sequences over A, including the empty sequence λ.

2. A *(context-free) production* over LAB is of the form $A \to R$ where A is a symbol in LAB and R is a decorated collage over LAB. The application of a production $A \to R$ to a decorated collage C proceeds in three steps: (1) Choose a hyperedge e with label A and a transformation t from $TRANS$ which maps the pin-points of R to the attachment points of e, (2) remove e from C, yielding $C - \{e\}$, and (3) add the transformed image $t(R)$ to the remainder, yielding the decorated collage $C' = C - \{e\} + t(R)$.

   We write $C \Longrightarrow C'$ and say C' is *directly derived from* C, if C' can be obtained from C by the application of a production. We write $C \Longrightarrow^* C'$, and say C' is *derivable from* C, if $C \cong C'$ or there is some finite sequence of decorated collages $C_0, C_1, \ldots, C_k$ with $C_0 = C$ and $C_k = C'$ such that for all i, C_{i+1} is directly derived from C_i. (Thus $\Longrightarrow^*$ is the reflexive and transitive closure of $\Longrightarrow$.)

3. A *(context-free) collage grammar* is a system $CG = (LAB, PROD, START)$ where LAB is a finite set of *labels*, $PROD$ is a finite set of *productions* over LAB, and $START$ is a decorated collage over LAB, called the *axiom*. The *collage language* generated by CG, $L(CG)$ for short, consists of all collages which can be derived from $START$ by applying productions of $PROD$.

4. A collage grammar $CG = (LAB, PROD, START)$ is *linear* if $START$ is a decorated collage with one hyperedge and empty pin-point sequence, and the right-hand side of each production contains at most one hyperedge.

5. A collage language L is called *linear (context-free)* if there is a linear (context-free) collage grammar CG such that $L = L(CG)$. The class of all linear and context-free collage languages is denoted by $\mathcal{L}(LIN)$ and $\mathcal{L}(CF)$, respectively.

Example 2. Consider the collage grammar $spiral = (\{S\}, PROD, START)$ with the axiom and the productions as shown in figure 2. Then the decorated collages derived from the axiom in the first, the second, and the fiftieth step are shown in figure 3.

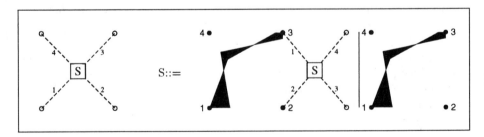

Fig. 2. Axiom and productions of the grammar *spiral*.

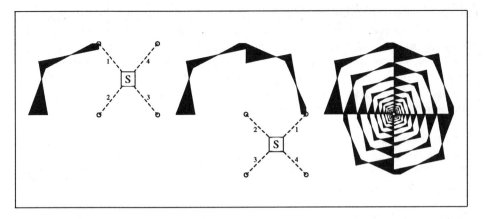

Fig. 3. Collages derived by the grammar *spiral* in the first, second and fiftieth step.

4 Comparison

The two syntactic devices for the generation of picture languages, chain-code picture grammars and collage grammars, are incomparable, i.e., there exist picture languages which can be generated by collage grammars, but not by a context-free chain-code grammar, and there exist picture languages which can be generated by context-free chain-code grammars, but not by collage grammars. Nevertheless, restricting them to regular chain-code grammars and a suitable type of linear collage grammars, so-called linear picture grammars, the approaches become comparable, i.e., each regular chain-code grammar can be transformed into an equivalent linear picture grammar, and vice versa.

A picture in the sense of [MRW82] may be seen as a collage in $I\!\!R^2$ where each part is a unit line and the sequence of pin-points is empty. Vice versa, there are collages in $I\!\!R^2$ consisting of unit lines as well as other types of parts, such as triangles, circuits, etc. Therefore the following lemma is obvious.

Lemma 6. *There is a context-free collage language L such that $L \notin \mathcal{B}(CF)$.*

Proof. Consider the collage language *SPIRAL* generated by the collage grammar given in example 2. Since the generated collages are composed of triangles, the language is not a picture language. □

Lemma 7. *There is a context-free π-language L such that $pic(L) \notin \mathcal{L}(CF)$.*

Proof. Consider the picture description language $STAIRS = \{(ru)^n rurdr(dr)^n \mid n \in I\!\!N_0\}$ generated by the context-free picture description grammar given in example 1. The corresponding chain-code picture language $pic(STAIRS)$ cannot be generated by a collage grammar. (A formal proof can be given by means of a pumping lemma for collage grammars.) □

In the following, we will investigate the relationship between picture description grammars and collage grammars in more detail. For this purpose, we have to restrict the set of admissible affine transformations to a special group of translations and the concept of collage grammars to so-called picture grammars, which generate pictures in the sense of definition 2. In this context, we introduce the notion of a decorated picture, being a decorated collage in which all parts are unit lines in the grid, all hyperedges are attached to points in $\mathbb{Z}^2$, and the sequence of pin-points is a sequence of points in $\mathbb{Z}^2$. Note that these notations are introduced with reference to [MRW82, DH93].

Assumption 8. *In the following, we restrict the set of admissible transformations to the group of all translations generated by the basis translations t_u, t_d, t_l, t_r: $\mathbb{R}^2 \to \mathbb{R}^2$ given by $t_u(x,y) = (x, y+1)$, $t_d(x,y) = (x, y-1)$, $t_l(x,y) = (x-1, y)$, and $t_r(x,y) = (x+1, y)$ for $(x,y) \in \mathbb{R}^2$.*[5]

Definition 9. (decorated pictures, picture grammars and languages)

1. A *decorated picture* is a decorated collage $C = (PART, EDGE, att, lab, pin)$ in which all parts in $PART$ are unit lines in the grid, all hyperedges in $EDGE$ are attached to points in $\mathbb{Z}^2$, i.e. $att(e) \in (\mathbb{Z}^2)^*$ for $e \in EDGE$, and $pin \in (\mathbb{Z}^2)^*$.
2. A *picture grammar* is a collage grammar $CG = (LAB, PROD, START)$ in which $START$ and the right-hand sides of the productions in $PROD$ are decorated pictures. A picture language L is called *linear (context-free)* if there is a linear (context-free) picture grammar CG such that $L = L(CG)$.

Theorem 10. *For every regular π-grammar G, there is a linear picture grammar CG such that $pic(L(G)) = L(CG)$.*

Proof. Let $G = (N, \pi, P, S)$ be a regular π-grammar where all productions of P are of the form $A \to wB$ or $A \to w$ with $A, B \in N$ and $w \in \pi^*$ and $S \in N$. We will construct a linear picture grammar $CG = (LAB, PROD, START)$ with $LAB = \{\uparrow, \downarrow\} \cdot N$ such that $L(CG) = pic(L(G))$. $START$ and $PROD$ are obtained with the help of the translation col that associates with any word $wB \in \pi^* \cdot N$ a decorated picture $col(wB)$ and with any word $w \in \pi^*$ a decorated picture $col(w)$. This is done as follows. If $w \in \pi^*$, $dpic(w) = (Q, z, s)$, and $B \in N$, then $col(wB) = (Q, \{e\}, att, lab, (0,0))$ with $att(e) = z$ and $lab(e) = sB$ and $col(w) = (Q, \emptyset, \emptyset, \emptyset, (0,0))$. Let $START = col(\downarrow S)^{\circ 6}$ and $PROD = \{sA \to col(sx) | A \to x \text{ in } P \text{ and } s \in \{\uparrow, \downarrow\}\}$. Then $L(CG) = pic(L(G))$. This may be seen as follows.

Claim 1. *For $A \in N$, $s \in \{\uparrow, \downarrow\}$, and $w \in \pi^*$, $A \Longrightarrow^* w$ in G implies $col(sA) \Longrightarrow^* col(sw)$ in CG.*

Proof. The argument is an induction on k, the length of a derivation in G.

---

[5] Note that the basis translations t_u and t_r would be sufficient to generate the group.
[6] For a collage C, C° denotes the underlying collage with empty sequence of pin-points, i.e., $C^\circ = (PART_C, EDGE_C, att_C, lab_C, \lambda)$.

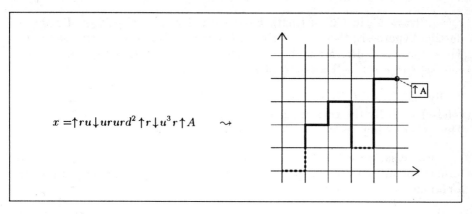

Fig. 4. From a word to a decorated picture.

Induction basis. Suppose $k = 1$. Then $A \Longrightarrow w$ is a direct derivation in G and $A \to w$ is a production in P. By the construction, $sA \to col(sw)$ is a production in $PROD$ and $col(sA) \Longrightarrow col(sw)$ a direct derivation in CG ($s \in \{\uparrow, \downarrow\}$).

Induction step. Suppose $k \geq 2$ and claim 1 holds for all derivations of length less than k. Suppose $A \Longrightarrow w_1 B \Longrightarrow^{k-1} w$ in G is a derivation of length k, where $A, B \in N$, $w_1, w \in \pi^*$. Then there is a word $w_2 \in \pi^*$ and a derivation $B \Longrightarrow^* w_2$ in G of length $k-1$ such that $w = w_1 w_2$. Let $dpic(sw_1) = (Q_1, z_1, s_1)$. By the construction, $sA \to col(sw_1 B)$ is in $PROD$ ($s \in \{\uparrow, \downarrow\}$). By the induction hypothesis, there is a derivation $col(s_1 B) \Longrightarrow^* col(s_1 w_2)$ in CG. Applying the translation by the vector z_1, this derivation can be embedded into the decorated picture $col(sw_1 B)$. Thus, there is a derivation $col(sA) \Longrightarrow col(sw_1 B) \Longrightarrow^* col(sw_1 w_2) = col(sw)$ in CG. □

By claim 1, $S \Longrightarrow^* w \in \pi^*$ in G implies $col(\downarrow S) \Longrightarrow^* col(\downarrow w)$ and $col(\downarrow S)^\circ \Longrightarrow^* col(\downarrow w)^\circ$. By the construction, $START = col(\downarrow S)^\circ$ and $col(\downarrow w)^\circ = pic(w)$. Thus, claim 1 implies that $pic(L(G)) \subseteq L(CG)$. To complete the proof, the following result is needed.

Claim 2. For $A \in N$, $s \in \{\uparrow, \downarrow\}$, and $C \in \mathcal{C}$, $col(sA) \Longrightarrow^* C$ in CG implies $A \Longrightarrow^* w$ in G where $col(sw) = C$.

Proof. The argument is an induction on k, the length of a derivation in CG.

Induction basis. Suppose $k = 1$. Then $col(sA) \Longrightarrow C$ in CG and $sA \to C$ is in $PROD$. By the construction, $A \to w$ is in P for some $w \in \pi^*$ with $col(sw) = C$. Thus, $A \Longrightarrow^* w$ is in G for some $w \in \pi^*$ with $col(sw) = C$.

Induction step. Suppose $k \geq 2$ and claim 2 holds for all derivations of length less than k. Suppose $col(sA) \Longrightarrow C_1 \Longrightarrow^{k-1} C$ in CG is a derivation of length k, where $A \in N$, C_1 is a decorated picture with one hyperedge e, and C is a picture. By the construction, $C_1 = col(sw_1 B)$ for some $A \to w_1 B$ in P. Let $dpic(sw_1) = (Q_1, z_1, s_1)$. Then there is a picture $C_2 \in \mathcal{C}$ and a derivation

$col(s_1B) \Longrightarrow C_2$ in CG of length $k-1$ such that $C = C_1[e/C_2]^7$. By the induction hypothesis, there is a derivation $B \Longrightarrow^* w_2$ in G with $col(s_1w_2) = C_2$. Thus, there is a derivation $A \Longrightarrow w_1B \Longrightarrow^* w_1w_2$ in G with $col(sw_1w_2) = col(sw_1B)[e/col(s_1w_2)] = C_1[e/C_2] = C$. □

By claim 2, $col(\downarrow S) \Longrightarrow^* C \in \mathcal{C}$ in CG implies $S \Longrightarrow^* w \in \pi^*$ in G where $col(\downarrow w) = C$. By the construction, $START = col(\downarrow S)^\circ$ and $col(\downarrow w)^\circ = pic(w)$. Thus, claim 2 implies that $L(CG) \subseteq pic(L(G))$. This completes the proof. □

Vice versa, we will show that every linear picture grammar can be transformed into an equivalent regular chain-code grammar. For this purpose, we use a normal form theorem for picture grammars.

Definition 11. (normalized collage grammars) A collage grammar $CG = (LAB, PROD, START)$ is said to be *normalized* if $START$ is a decorated collage with one hyperedge and no parts and, for all productions $A \to R$ in $PROD$, all decorated collages C in CG, and all hyperedges e in C with label A, there exists at least one transformation $t \in TRANS$ which maps the sequence pin_R of pin-points of R to the sequence $att_C(e)$ of attachment-points of e in C. A normalized collage grammar CG is said to be *k-normalized* for some $k \in \mathbb{N}_0$ if for all right-hand sides R in CG, $|pin_R| = k$, and for all decorated collages C in CG and all $e \in EDGE_C$, $|att_C(e)| = k$.

Theorem 12. *Given a collage grammar CG, we can effectively construct a normalized collage grammar CG' such that $L(CG') = L(CG)$. If, in particular, CG is a picture grammar and $TRANS$ is the group of translations generated by the basis t_u, t_d, t_l, t_r, we can effectively construct a 1-normalized picture grammar CG' such that $L(CG') = L(CG)$.*

Proof. The construction of a 1-normalized grammar is done in three steps.

1. Normalization. Let $CG = (LAB, PROD, START)$ be a collage grammar and $\equiv$ be the equivalence relation on $(\mathbb{R}^2)^*$ defined by $seq \equiv seq'$ if and only if there are transformations $t, t' \in TRANS$ such that $t(seq) = t'(seq')$. Then we construct a normalized collage grammar $CG' = (LAB', PROD', START')$ with $L(CG') = L(CG)$ as follows. Let $LAB' = \{S\} \cup LAB \times PIN$ where S is a new symbol not occurring in LAB and PIN is the set of all pin-point sequences occurring in the grammar. $PROD'$ and $START'$ are obtained with the help of an operation that adds information to each label of a hyperedge: For a decorated collage C over LAB and a mapping $info_C : EDGE_C \to PIN$ with $info_C(e) \equiv att_C(e)$ for $e \in EDGE_C$, called information assignment for C, $(C, info_C)$ denotes the decorated collage $(PART_C, EDGE_C, att_C, lab, pin_C)$ with $lab(e) = (lab_C(e), info_C(e))$ for $e \in EDGE_C$.

---

[7] $C_1[e/C_2]$ denotes the decorated picture obtained from the decorated picture C_1 with hyperedge e by replacing e by C_2. For this purpose, the picture C_2 has to be translated in such a way that the image of the pin-points of C_2 matches the attachment points of e.

Without loss of generality, we may assume that the identity is the only transformation which maps pin_{START} to pin_{START} (Otherwise, we extend the pin-point sequence of $START$ in such a way that the identity becomes the only transformation with this property.) Let $START'$ be the decorated collage $(\emptyset, \{e\}, att, lab, pin_{START})$ with $att_{START'}(e) = pin_{START}$ and $lab_{START'} = S$. Moreover, let $PROD'$ consist of all productions $S \to (START, info_{START})$ where $info_{START}$ is an information assignment for $START$ and the productions of the form $(A, pin_R) \to (R, info_R)$ where $A \to R$ is a production of $PROD$ and $info_R$ is an information assignment for R. Then CG' is normalized and $L(CG') = L(CG)$.

2. *Elimination of productions with empty pin-point sequence.* Let $CG = (LAB, PROD, START)$ be a normalized picture grammar and $TRANS$ be the group of translations generated by the basis t_u, t_d, t_l, t_r. Then we construct a picture grammar $CG' = (LAB', PROD', START)$ with non-empty pin-point sequences as follows. Let $LAB' = LAB \cup \{J\}$ for some new symbol $J \notin N$ and N_λ be the set of symbols $A \in N$ for which there exists a production $A \to R$ with $pin_R = \lambda$. For each symbol $A \in N_\lambda$, replace the non-attached A-labeled hyperedges by $(0,0)$-attached, J-labeled ones. Add the jump productions (J, UP), $(J, DOWN)$, $(J, LEFT)$, $(J, RIGHT)$ where UP, $DOWN$, $LEFT$, and $RIGHT$ are the decorated collages $(\emptyset, \{e\}, att, lab, (0,0))$ with $att(e) = (0,1), (0,-1), (-1,0)$, and $(1,0)$, respectively, and $lab(e) = J$. Finally, replace each production (A, R) with empty pin-point sequence by the production (J, R') where R' is the $(0,0)$-pinned version of R, i.e., $R' = (PART_R, EDGE_R, att_R, lab_R, (0,0))$. Then $L(CG') = L(CG)$.

3. *Elimination of unneccessary pin- and attachment-points.* Let $CG = (LAB, PROD, START)$ be a normalized picture grammar with non-empty pin-point sequences. Then we construct a 1-normalized picture grammar $CG' = (LAB, PROD', START)$ as follows: Replace each production $A \to R$ in $PROD$ by the production $A \to R_1$ where R_1 is the decorated collage obtained from R by forgetting all pin- and attachment-points except the first pin-point and the first attachment-point in a sequence, i.e., $R_1 = (PART_R, EDGE_R, att_1, lab_R, pin_{R,1})$ with $att_1(e) = att_R(e)_1$ for $e \in EDGE$. Then CG' is 1-normalized and $L(CG') = L(CG)$. □

Theorem 13. *For every linear picture grammar CG, there is a regular π-grammar G such that $pic(L(G)) = L(CG)$.*

Proof. Let $CG = (LAB, PROD, START)$ be a linear picture grammar. Without loss of generality, we may assume that CG is 1-normalized and $(0,0)$-attached, i.e., $pin_R = (0,0)$ for all right-hand sides R in CG. We will construct a regular π-grammar $G = (N, \pi, P, S)$ such that $L(CG) = pic(L(G))$. Let $N = LAB$. S and P are obtained with the help of the translation $word$ that associates with any decorated picture C with one hyperedge a word $word(C) \in \pi^* \cdot N$ and with any decorated picture C without hyperedge a word $word(C) \in \pi^*$. This is done as follows. If C is a decorated picture with unit lines $\langle z_1, b_1(z_1)\rangle, \ldots,$

$\langle z_n, b_n(z_n)\rangle$ ($b_1, \ldots, b_n \in \{u, d, l, r\}$) and one hyperedge with attachment point z_{n+1} and label B, then $word(C)$ is a π-word $\uparrow w_0 \downarrow b_1 \uparrow w_1 \ldots w_{n-1} \downarrow b_n \uparrow w_n \downarrow B$ where $w_0 \in \{u, d, l, r\}^*$ is a π-word describing a simple path from $(0,0)$ to the point z_1, w_1 is a π-word describing a path from $b_1(z_1)$ to z_2, $\ldots$, and w_n is a π-word describing a path from $b_n(z_n)$ to z_{n+1}. If C is a decorated picture without hyperedges, then $word(C)$ is constructed analogously, but it ends with $w_{n-1} \downarrow b_n$.

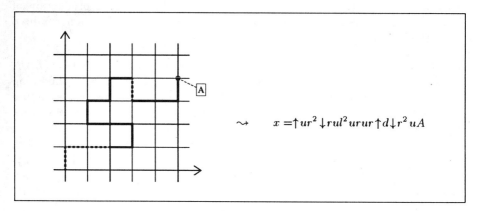

Fig. 5. From a decorated picture to a picture description word.

Now let $S = word(START)$ and $P = \{A \to word(R) | A \to R \text{ in } PROD\}$. Then $L(CG) = pic(L(G))$. This may be seen as follows.

Claim 1. For $A \in LAB$ and $C \in \mathcal{C}$, $col(A) \Longrightarrow^* C$ in CG implies $word(col(A)) \Longrightarrow^* word(C)$ in G.

Proof. The argument is an induction on k, the length of a derivation in CG.

Induction basis. Suppose $k = 1$. Then $col(A) \Longrightarrow C$ in CG and $A \to C$ is in $PROD$. By the construction, $A \to word(C)$ is in P. Thus, $A = word(col(A)) \Longrightarrow^* word(C)$ is in G.

Induction step. Suppose $k \geq 2$ and claim 1 holds for all derivations of length less than k. Suppose $col(A) \Longrightarrow R \Longrightarrow^{k-1} C$ in CG is a derivation of length k, where $A \in LAB$, $A \to R \in PROD$, and $C \in \mathcal{C}$. Since CG is a 1-normalized, linear picture grammar, R is a decorated picture with one hyperedge e with attachment point $att(e) \in \mathbb{Z}^2$ and one pin-point $(0,0)$. The restriction of the derivation $R \Longrightarrow^{k-1} C$ to $col(lab(e))$, the collage induced by the hyperedge $e \in EDGE_R$, yields a derivation $col(lab(e)) \Longrightarrow^{k-1} C(e) \in \mathcal{C}$ such that $R[e/C(e)] = C$. By the induction hypothesis, there is a derivation $word(col(lab(e))) \Longrightarrow^* word(C(e))$ in G. Thus, there is a derivation $word(col(A)) \Longrightarrow word(R) \Longrightarrow^* word(R)[lab(e)/word(C(e))]^8 = word(C)$ in G. □

---

[8] For words w_1, w_2 and a symbol A occurring in w_1, $w_1[A/w_2]$ denotes the word obtained from w_1 by replacing A by w_2.

By claim 1, $col(S) \Longrightarrow^* C \in \mathcal{C}$ in CG implies $word(col(S)) \Longrightarrow^* word(C)$ in G. Moreover, $START = col(S)^\circ \Longrightarrow^* C^\circ \in \mathcal{C}$ in CG and $C^\circ = pic(word(C))$. Thus, claim 1 implies that $L(CG) \subseteq pic(L(G))$.

Claim 2. For $A \in N$ and $w \in \pi^*$, $A \Longrightarrow^* w$ in G implies $col(A) \Longrightarrow^* C$ in CG with $word(C) = w$.

Proof. The argument is an induction on k, the length of a derivation in G.

Induction basis. Suppose $k = 1$. Then $A \Longrightarrow w \in \pi^*$ in G and $A \to w$ is in P for some $A \to R$ in $PROD$ with $word(R) = w$. Then $col(A) \Longrightarrow R$ with $word(R) = w$.

Induction step. Suppose $k \geq 2$ and claim 2 holds for all derivations of length less than k. Suppose $A \Longrightarrow w_1 B \Longrightarrow^{k-1} w$ in G is a derivation of length k, where $A, B \in N$, $w_1, w \in \pi^*$. Then there is a word $w_2 \in \pi^*$ and a derivation $B \Longrightarrow^* w_2$ in G of length $k-1$ such that $w = w_1 w_2$. By the construction, there is a production $A \to R$ in $PROD$ with $word(R) = w_1 B$. By the induction hypothesis, there is a derivation $col(B) \Longrightarrow^* C_2$ in CG with $word(C_2) = w_2$. Embedding of this derivation into the decorated picture R yields a derivation $R \Longrightarrow^* C$ in CG with $word(C) = word(R)[B/word(C_2)] = w_1 w_2 = w$. □

By claim 2, $S \Longrightarrow^* w \in \pi^*$ in G implies $col(S) \Longrightarrow^* C$ in CG with $word(C) = w$. Moreover, $START = col(S)^\circ \Longrightarrow^* C^\circ$ in CG and $C^\circ = pic(word(C)) = pic(w)$. Thus, claim 2 implies that $pic(L(G)) \subseteq L(CG)$. This completes the proof. □

5 Undecidability

All undecidability results known for regular chain-code picture languages can be carried over to linear picture languages. In particular, it can be shown that it is undecidable whether or not the language of a linear picture grammar contains only (a) pictures with holes, (b) disconnected pictures, or (c) connected pictures.

Pictures in the sense of definition 2 may be considered as undirected graphs, where the the set of nodes is given by the set of points of $\mathbb{Z}^2$ belonging to the picture, and the set of edges is given by the set of unit lines of the picture. Therefore, we may ask whether or not the graph $graph(Q)$ of a picture Q satisfies a suitable graph property $PROP$ or not.

Among other properties, we shall consider the following properties of pictures or their graphs. A picture Q is said to be a *simple curve* if each node of $graph(Q)$ has at most degree 2, and is called a *simple closed curve* if all nodes of $graph(Q)$ have degree 2. Q is called *convex* if there is a picture Q' such that $Q \cup Q'$ is a closed simple curve and the intersection of the inner part of $Q \cup Q'$ with any straight line which is parallel to one of the axes is a finite straight line. Analogously, a picture Q is said to be *Eulerian, regular, Hamiltonian, unit-line-colourable by three colours, 2-connected*, if the corresponding graph $graph(Q)$ is Eulerian, regular, Hamiltonian, edge-colourable by three colours, 2-connected, respectively[9].

---

[9] For graph-theoretical concepts we refer to Berge [Ber73].

Theorem 14. *It is undecidable whether or not a linear picture language contains (a) a simple curve, (b) a closed simple curve, (c) an Eulerian picture, (d) an Eulerian cycle, (e) a tree, (f) a regular picture, (g) a Hamiltonian picture, (h) a Hamiltonian cycle, (i) a picture unit-line-colourable by three colours, or (j) a 2-connected picture.*

Proof. The statement follows by theorem 10 since the corresponding problems are undecidable for regular chain-code picture languages (see [DH93], theorem 3.2). □

A picture Q is *connected* if the set $pattern(Q) = \bigcup_{l \in PART} l \subseteq \mathbb{R}^2$ of all points of unit lines is connected, i.e., if each two points $x, y \in pattern(Q)$ are connected by a continous line $L \subseteq pattern(Q)$. If Q is not connected, it is said to be *disconnected*. Moreover, a picture Q is said to possess a *hole*, if there exist paths in $pattern(Q)$ that are not continously deformable into each other.

Theorem 15. *It is undecidable whether or not a linear picture language contains only (a) disconnected pictures or (b) connected pictures.*

Proof. The statement follows from theorem 10 since the corresponding problems are undecidable for regular chain-code picture languages (see [DH93], theorem 4.1). □

Theorem 16. *It is undecidable whether or not a linear picture language contains a picture without holes.*

Proof. The statement follows from theorem 10 since the problem whether or not a picture language contains a tree is undecidable for regular chain-code picture languages described by $\{u, d, l, r\}$-grammars (see [DH93], theorem 3.2) and a connected picture is a tree if and only if the picture does not contain any holes. □

6 Decidability

All decidability results known for regular chain-code picture languages can be carried over to linear picture languages. Unfortunately, up to now only a few questions on geometric properties turn out to be decidable.

Theorem 17. *It is decidable whether or not the language of a linear picture grammar with connected right-hand sides* [10] *contains (a) a rectangle, (b) a convex curve, or (c) only rectangles.*

Proof. The statement follows by theorem 13 since the corresponding problems are decidable for regular chain-code picture languages described by $\{u, d, l, r\}$-grammars (see [DH93], theorem 6.8). □

---

[10] A decorated picture Q is said to be *connected* if each two points $x, y \in pattern(Q) \cup attach(Q)$ are connected by a continous line $L \subseteq pattern(Q)$ where $attach(Q)$ denotes the set of attachment point of hyperedges in Q.

Remark. By theorem 17, serveral problems are decidable for picture languages described by regular $\{u, d, l, r\}$-grammars. But we do not know what is the situation with respect to arbitrary linear picture languages: Is it decidable whether or not a linear picture language contains (a) a rectangle, (b) a convex curve, (c) only rectangles, or (d) only trees?

7 Conclusion

In this paper, we have compared two syntactic devices for the generation of picture languages, chain-code picture grammars and collage grammars, and shown that whenever one regards regular chain-code picture grammars and a suitable type of linear collage grammars, called linear picture grammars, the approaches become comparable. As one consequence, all undecidability results known for regular chain-code picture languages (see [DH93]) can be carried over to linear picture languages and thus to linear collage languages.

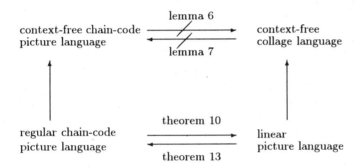

1. In the present paper, it is shown that it is undecidable whether or not the language of a linear picture grammar contains only (a) disconnected pictures or (b) connected pictures. In Drewes, Kreowski [DK95], for fairly different classes of collage grammars (linear, pin-bounded, non-overlapping collage grammars and compact, pin-bounded, non-overlapping collage grammars, respectively) corresponding undecidability results are proved. Due to the non-comparability of the selected grammar classes, those results are not comparable to the results given in this paper.
2. In Dassow, Hinz [DH93], a general method to obtain undecidability results for regular chain-code picture languages is presented. This method is based on the idea to simulate a run of a linear bounded automaton by a regular grammar. This is done in three steps. First, a normal form for linearly bounded automata is presented. Then usual Turing tapes and Turing tapes with defects are introduced such that the existence of a run without defects corresponds to the non-emptiness of the accepted language. Finally, a regular grammar is constructed such that every word generated by the grammar

corresponds to a run of the linearly bounded automaton on a probably defect Turing tape. The picture drawn by a word has a certain property if no defects of the Turing tape occur in the corresponding run, and it fails to have this property if a serious defect occurs in the corresponding run of the linearly bounded automaton.

The same method can be used to obtain further undecidability results for linear collage languages. E.g. it can be shown that it is undecidable whether or not a linear collage language contains a black unit square.

3. In Beauquier [Bea91], it is shown that it is undecidable whether or not a regular picture language contains the contour word of a polyomino[11]. This is done by a reduction of the Halting Problem for Turing machines to the problem of interest. In Robilliard [Rob94], another proof method, based upon the Post Correspondence Problem, for the decidability questions concerning the existence of a tree, a contour of a polyomino, and a connected picture is presented.

4. In Dube [Dub94], it is shown that two questions about iterated function systems (IFS) are undecidable — to test if the attractor intersects a given line segment and to test if the attractor of a given IFS is totally disconnected. The proofs are obtained by reducing the Post Correspondence Problem and by interpreting words as numbers and concatenation operation as composition of affine transformations. In Taubenberger [Tau94], iterated function systems and collage grammars are related. In particular, it is shown that given an iterated function system I and a point set Q, a collage grammar $CG(I, Q)$ can be constructed such that the sequence of patterns generated by I and Q and the sequence of patterns generated by $CG(I, Q)$ coincide. As a consequence, the corresponding questions about collage grammars become undecidable.

5. Recently, Drewes [Dre95] investigated language-theoretic aspects and algorithmic properties of particular classes of context-free collage languages and iterated function systems. In particular, the decidability results shown for "grid" collage languages are in sharp contrast to certain undecidability results known from the literature on picture-generating devices (cf. the work by Dassow and Hinz [DH93], or by Drewes and Kreowski [DK95]).

References

[Bea91]   Danièle Beauquier. An undecidability problem about rational sets and contour words of polyominoes. Information Processing Letters 37, 257–263, 1991.

[Ber73]   Claude Berge. Graphs and Hypergraphs. North Holland, Amsterdam, 1973.

---

[11] A polyomino Q is a finite union of nonoverlapping unit squares such that the interior of Q is a connected set without any hole. The contour of such a polyomino can be represented as a word, called a *contour word*, written on a four letters alphabet $\{u, d, l, r\}$ corresponding to the four directions up, down, left, and right.

[Cox89] H.S.M. Coxeter. Introduction to Geometry. Wiley Classics Library Edition, John Wiley & Sons, Inc., New York, second edition edition, 1989.

[DH93] Jürgen Dassow, Friedhelm Hinz. Decision problems and regular chain code picture languages. Discrete Applied Mathematics 45, 29–49, 1993.

[Dre95] Frank Drewes. Language theoretic and algorithmic properties of collages and pattern living in a grid. Journal of Computer and System Sciences, 1995. To appear.

[DHKT95] Frank Drewes, Annegret Habel, Hans-Jörg Kreowski, Stefan Taubenberger. Generating self-affine fractals by collage grammars. Theoretical Computer Science 145, 159–187, 1995.

[DK95] Frank Drewes, Hans-Jörg Kreowski. (Un-)decidability of geometric properties of pictures generated by collage grammars. Fundamenta Informaticae, 1995. To appear.

[Dub94] Simant Dube. Fractal geometry, turing machines and divide-and-conquer recurrences. RAIRO Theoretical Informatics and Applications 28, 405–423, 1994.

[HK88] Annegret Habel, Hans-Jörg Kreowski. Pretty patterns produced by hyperedge replacement. In H. Göttler, H.J. Schneider, eds., Graph-Theoretic Concepts in Computer Science, Lecture Notes in Computer Science 314, 32–45, 1988.

[HK91] Annegret Habel, Hans-Jörg Kreowski. Collage grammars. In H. Ehrig, H.-J. Kreowski, G. Rozenberg, eds., Graph Grammars and Their Application to Computer Science, Lecture Notes in Computer Science 532, 411–429, 1991.

[HKT93] Annegret Habel, Hans-Jörg Kreowski, Stefan Taubenberger. Collages and patterns generated by hyperedge replacement. Languages of Design 1, 125–145, 1993.

[HU79] John E. Hopcroft, Jeffrey D. Ullman. Introduction to Automata Theory, Languages and Compuation. Addison-Wesley, Reading, Mass., 1979.

[Kim90] Changwook Kim. Complexity and decidability for restricted classes of picture languages. Theoretical Computer Science 73, 295–311, 1990.

[MRW82] Hermann A. Maurer, Grzegorz Rozenberg, Emo Welzl. Using string languages to describe picture languages. Information and Control 54, 155–185, 1982.

[Rob94] Denis Robilliard. Undecidable questions in picture languages. Technical Report IT 262, Université de Lille, 1994.

[SW85] Ivan Hal Sudborough, Emo Welzl. Complexity and decidability for chain code picture languages. Theoretical Computer Science 36, 173–202, 1985.

[Tau94] Stefan Taubenberger. Correct translations of generalized iterated function systems to collage grammars. Technical Report 7/94, Bremen, 1994.

Transformations of Graph Grammars

Francesco Parisi-Presicce

Dipartimento di Scienze dell'Informazione, Universita' di Roma 'La Sapienza',
Via Salaria 113, 00198 Roma (Italy)
e-mail : parisi@dsi.uniroma1.it

Abstract. The notion of multilevel graph representations, where parts of graphs are not visible and the information can be restored via the explicit application of productions, and the corresponding extension of the classical double pushout approach is generalized to the algebraic theory of graph grammars and to the rewriting of these grammars, at the global level by defining High Level Replacement Systems where the productions consist of grammars and grammar morphisms, and at the local level where standard productions are used to rewrite both the initial graph and the productions of a grammar.

1 Introduction

Many approaches to graph generation and transformation have been proposed over the years, some based on the replacement of nodes [2], others on edge replacement [7], some controlled by labels on nodes or arcs [5], others by structural characteristics. All these approaches treat graphs as sets of vertices and edges, all of which are visible at each stage. The representation of graphs is rarely considered and both vertices and edges are always *visible*. In many applications, it is very useful to be able to represent graphs in terms of their particular subgraphs and to hide details of structures needed only in certain conditions; repeated hiding of details gives origin to representations on more than one level of visibility. *Graph representations* consist of a graph together with the set of productions needed to restore the different levels of the original graph, making visible again the information hidden at a different level. Of course, if the set of restoring productions associated to a graph is empty, the development reduces to the algebraic approach to grammars for totally visible graphs. It must be possible to encode the aprropriate order in which the restoring productions must be applied to obtain *only* the intended graph. By removing this constraint, a graph representation may generate several graphs, i.e., it is nothing more than a graph grammar. So the rewriting of graph representations of [11] becomes a special case of the general problem of transforming graph grammars, investigated in this paper.

This paper has also been motivated by the need for a theory of generation and transformation of graph representations. The approach adopted is the classical algebraic approach based on a double pushout construction (the comparison with the equivalent one based on partial functions and single pushout [9] is under investigation). Starting from the idea of representing graphs on different levels of visibility and of rewriting or generating their representations, we define an extension of the algebraic approach to graph grammars, characterizing formally the notions of graph production, derivation, grammar and language. We then define morphisms between graph grammars and investigate properties of the corresponding category of graph grammars and graph grammar morphisms. Following the development of the classical approach, the notion of direct derivation and of gluing condition is formalized. The double pushout approach of [2] becomes a special case where the graph representation is the trivial one consisting of the graph itself.

Among the intended applications of the theory is the possibility to compare formalisms for the representation of the lexical elements, the syntax and the semantics of Visual Languages [6, 8] as well as a framework to give a semantics for compilers of visual languages.

2 Preliminaries

We briefly review the basic notions of the algebraic approach to graph grammars necessary to explain our notation.

Definition 1. Let $C = (C_A, C_N)$ be a pair of sets, called color alphabets. A C–colored graph G is a six-tuple $(G_A, G_N, s, t, m_{aG}, m_{nG})$, consisting of:

- sets G_A and G_N, called the set of arcs and the set of nodes, respectively;
- source and target mappings $s : G_A \to G_N, t : G_A \to G_N$;
- arcs and nodes coloring mappings $m_{aG} : G_A \to C_A, m_{nG} : G_N \to C_N$.

For $e \in G_A$, denote $\{s(e), t(e)\}$ by $end(e)$.
A graph G' is a subgraph of a graph G if $G'_A \subseteq G_A$, $G'_N \subseteq G_N$, and all the mappings s', t', $m_{aG'}$, $m_{nG'}$ are restrictions of the corresponding ones from G.

Definition 2. Given a C-colored graph $G = (G_A, G_N, s, t, m_{aG}, m_{nG})$, and a C'-colored graph $G' = (G'_A, G'_N, s', t', m_{aG'}, m_{nG'})$, a **graph morphism** $f : G \to G'$ is a 4-tuple $(f_{cN} : C_N \to C'_N, f_{cA} : C_A \to C'_A, f_N : G_N \to G'_N, f_A : G_A \to G'_A)$ such that

- $f_N \circ s = s' \circ f_A$; $f_N \circ t = t' \circ f_A$ (the structure is preserved)
- $m_{nG'} \circ f_N = f_{cN} \circ m_{nG}$; $m_{aG'} \circ f_A = f_{aN} \circ m_{aG}$ (the labels uniformly changed).

Given $k = (k_N : C_N \to C'_N, k_A : C_A \to C'_A)$, a *k-based graph morphism* is a graph morphism f such that $f_{cN} = k_N$ and $f_{cA} = k_A$.

Notice that the classical notion of color-preserving graph morphism [2] corresponds to identity–based graph morphisms, where f_{cN} and f_{cA} are the identity.

The idea, common to every theory on Graph Grammars, is to realize, from an initial graph, a sequence of transformations, following rules allowing a subgraph DEL of G to be replaced by another graph ADD, leaving unchanged the subgraph D of G not involved in the deletion of DEL. Every transformation of this kind requires the specification of how ADD must be connected to the graph D (called *embedding*). A graph production p is a pair $(L \leftarrow K \to R)$ of injective graph morphisms, where L, R and K are called the left side, the right side and the interface of the production, respectively. A production specifies that a graph L must be replaced by a graph R, using an interface graph K (whose arcs and nodes are the gluing items) for the embedding.

Definition 3 (Direct Derivation). Given a production $p = (L \leftarrow K \rightarrow R)$, a direct derivation from G_1 to G_2 via p, denoted by $p : G_1 \Rightarrow G_2$ or by $G_1 \Rightarrow_p G_2$, consists of the following double pushout

$$\begin{array}{ccccc} L & \xleftarrow{l} & K & \xrightarrow{r} & R \\ {\scriptstyle g_1}\downarrow & & {\scriptstyle c}\downarrow & & \downarrow{\scriptstyle g_2} \\ G_1 & \xleftarrow{l'} & D & \xrightarrow{r'} & G_2 \end{array}$$

A derivation $G \Rightarrow_P^* H$ is a sequence of direct derivations $G = G_1 \Rightarrow_{p_1} \ldots \Rightarrow_{p_{n-1}} G_n = H$ using productions $p_i \in P$.

Notice that the derivation diagram is symmetric and therefore, if $p : G_1 \Rightarrow G_2$, then $p^{-1} : G_2 \Rightarrow G_1$ where $p^{-1} = (R \leftarrow K \rightarrow L)$.
The applicability of a production $p = (L \leftarrow K \rightarrow R)$ to a graph G_1 is determined by the existence and properties of a total morphism $g1 : L \rightarrow G_1$. The existence of a pushout complement D for a given g1 depends on the following Gluing Conditions being satisfied [2].

Theorem 4. *[Gluing Conditions]* Given $p = (L \leftarrow K \rightarrow R)$ and $g_1 : L \rightarrow G_1$, let

$$ID_{g1} = \{x \in L : \exists y \in L, x \neq y, g_1(x) = g_1(y)\}$$

$$DANG_{g1} = \{n \in L_N : \exists e \in G_{1A} - g_{1A}(L_A) \text{ such that } g_{1N}(n) \in end_{G_1}(e)\}$$

Then the pushout complement D exists **if and only if** $DANG_{g1} \cup ID_{g1} \subseteq l(K)$

This limitation of the traditional double pushout approach can be overcome either by using single pushouts [9] or by using "restricting derivation sequences" [10].

Definition 5. A graph grammar $Gra = ((C,T), PROD, START)$ consists of color alphabets $C = (C_A, C_N)$, terminal alphabets $T = (T_A, T_N)$, included in C, a finite set $PROD$ of productions, and an initial graph $START$. The Graph Language generated is $Lang(Gra) = \{H \mid START \Rightarrow^* H, \text{ and } H \text{ is a } T-\text{colored graph}\}$.

Productions can be compared using triples of graph morphisms, one for each component of a production.

Definition 6. Given a pair of productions $p_1 = (L_1 \leftarrow K_1 \rightarrow R_1)$ and $p_2 = (L_2 \leftarrow K_2 \rightarrow R_2)$, a *production morphism* $f : p_1 \rightarrow p_2$ is a triple $(f_L : L_1 \rightarrow L_2, f_K : K_1 \rightarrow K_2, f_R : R_1 \rightarrow R_2)$ of graph morphisms as in the following commuting diagram

$$\begin{array}{ccccc} L_1 & \xleftarrow{l_1} & K_1 & \xrightarrow{r_1} & R_1 \\ {\scriptstyle f_L}\downarrow & & {\scriptstyle f_K}\downarrow & & \downarrow{\scriptstyle f_R} \\ L_2 & \xleftarrow{l_2} & K_2 & \xrightarrow{r_2} & R_2 \end{array}$$

A production morphism f is *strict* if, in addition, the two squares are pushout diagrams. A production p_1 is a *subproduction* of p_2 (and we write $p_1 \subseteq p_2$) if there is a strict production morphism from p_1 to p_2 where the components are monomorphisms.

3 From Multilevel Graphs to Graph Grammars

In the derivation $p : G_1 \Rightarrow G_2$, if $L = K$, then the application of p *adds* to $G1$ the right side of R via K. The idea behind multilevel graphs is to represent G_2 by G_1, which contains the *interface* K to the hidden part, and by a production p. In turn G_1 could be represented by another graph F by hiding part of it and maintaining the information with a production q which, when applied to F, restores the hidden part. Since a hidden graph can in turn contain interfaces (possibly only their subparts) of other previously hidden graphs, and a graph can simultaneously contain many interfaces, it is possible to obtain a graph represented on various levels of visibility or detail. To realize such a representation, we must associate to a graph G_1 the set of all productions required to expand any interface, restoring the corresponding hidden graph [11]. Furthermore, since a graph might hide an interface of another graph, in general, the application of restoring productions will be realizable without unexpected effects, only in inverse order with respect to the order of their construction. This constraint may be imposed in various way, e.g. structuring the set of restoring productions as a LIFO list, or using not-terminal numbered labels as indices to prevent the productions from being applied in an incorrect order or to an unintended part.

Definition 7. A multilevel graph is a triple $G_M = (G, RP_G, HP_G)$, where

- G is an ordinary graph called base graph,
- RP_G is a sequence of productions of the form $(L \leftarrow L \rightarrow R)$ called restoring productions, which can be used to replace interfaces with the corresponding hidden graphs, and
- HP_G is a sequence of productions of the form $(L \leftarrow R \rightarrow R)$ called hiding productions.

Any multilevel graph of the form $(G, RP_G, \emptyset)$ is said to be in normal form and is simply represented as a pair (G, RP_G).

By moving productions between RP_G and HP_G after applying them to G, we obtain the representations of the original graph at various level of visibility.

Definition 8. An *R-extension* of a multilevel graph $G_M = (G, RP_G, HP_G)$ is any multilevel graph $G'_M = (G', RP'_G, HP'_G)$ obtained as follows

- G' is derived from G by applying (in the imposed order) all the productions of an initial subsequence $R = \{p_1, ..., p_k\}$ of RP_G
- $RP'_G = RP_G - \{p_1, ..., p_k\}$
- $HP'_G = HP_G \cup \{(p_k)^{-1}, ..., (p_1)^{-1}\}$

For the dual notion of an *R-restriction*, $R \subseteq HP_G$, $RP'_G = RP_G \cup \{(p_k)^{-1}, ..., (p_1)^{-1}\}$ and $HP'_G = HP_G - \{p_1, ..., p_k\}$.

Every production of RP_G restores a graph previously hidden by an interface, while any production in HP_G hides graphs contained in G. Extensions of a multilevel graph are considered as temporary representations of a graph in normal form and are used to realize derivations.

As in the classical double-pushout approach, not all occurrence morphisms $L \to G$ allow the application of the production. The Extended Gluing Condition [11] guarantees that any rewriting of multilevel graphs will not create inconsistencies within hidden graphs, or loss of links between them and the associated interfaces.

The notion of morphism between two multilevel graphs establishes a correspondence (graph morphism) between the base graphs and associates to each restoring production of G a restoring production of H in such a way that the order in which they need to be applied to recostruct the hidden part is preserved. Similarly for the hiding productions if the graph is not in normal form [11].

Definition 9. Given $G_M = (G, RP_G, HP_G)$ and $H_M = (H, RP_H, HP_H)$ with
$RP_G = \{p_1, ..., p_n\}$ where $p_i = (L_i \leftarrow L_i \to R_i)$ $i = 1, ..., n$
$RP_H = \{q_1, ..., q_m\}$ where $q_j = (L'_j \leftarrow L'_j \to R'_j)$ $j = 1, ..., m$
$HP_G = \{s_1, ..., s_{n'}\}$ where $s_i = (N_i \leftarrow I_i \to I_i)$ $i = 1, ..., n'$
$HP_H = \{t_1, ..., t_{m'}\}$ where $t_j = (N'_j \leftarrow I'_j \to I'_j)$ $j = 1, ..., m'$
a *multilevel graph morphism* $f_M : G_M \to H_M$ is a triple $(f, \{g_1, ..., g_n\}, \{k_1, ..., k_{n'}\})$ where

1. $f : G \to H$ is a graph morphism
2. with $s : \{1, ..., n\} \to \{1, ..., m\}$, for each $i = 1, ..., n$, $g_i : p_i \to q_{s(i)}$ is a production morphism $(g_{L_i}, g_{L_i}, g_{R_i})$ such that $1 \le i < j \le n$ implies $1 \le s(i) < s(j) \le m$
3. with $s' : \{1, ..., n'\} \to \{1, ..., m'\}$ for each $i = 1, ..., n'$, $k_i : s_i \to t_{s'(i)}$ is a production morphism $(k_{N_i}, k_{I_i}, k_{I_i})$ such that $1 \le i < j \le n'$ implies $1 \le s'(i) < s'(j) \le m'$
4. the morphisms g_i are consistent with respect to the composition of restoring productions
5. the morphisms k_i are consistent with respect to the composition of hiding productions

The second and third conditions require that the correspondence between related restoring (resp. hiding) productions respect the order in which they are intended to be applied. The last two conditions (too cumbersome to be given in full details) require, for example, that if $p_{1R} = p_{2L}$ and $q_{s(1)R} = q_{s(2)L}$ then $g_{1R} = g_{2L}$.

Multilevel graph morphisms do not depend on the level of representation and can migrate from level to level.

Proposition 10. *Given a multilevel graph morphism $f_M : G_M \to H_M$ with $G_M = (G, RP_G, HP_G)$ and $H_M = (H, RP_H, HP_H)$, if G'_M is an R-extension of G and H'_M is the corresponding R'-extension of H, then there exists a unique multilevel graph morphism $f'_M : G'_M \to H'_M$ which extends f_M.*

A generalization of the formalization and the results of this section so far leads us to the definition of the category of Graph Grammars. For simplicity, we assume that the number of productions in each grammar is finite. A multilevel graph in normal form can be seen as a particular type of graph grammar, where the structure of the productions and the appropriate labelling guarantee that only one graph is generated (its 'flattening').

Recall that a graph grammar Gra_1 is a triple $((C_1, T_1), PROD_1, START_1)$ consisting of a pair of alphabets for all the labels and for the terminal labes, respectively, a C_1-colored base graph $START_1$ and a finite set $PROD_1 = \{p_1, \ldots, p_n\}$ of productions.

Definition 11. Given graph grammars Gra_0 and Gra_1 where
$PROD_0 = \{p_1, \ldots, p_n\}$ with $p_i = (L_i \leftarrow K_i \to R_i)$ $i = 1, \ldots, n$ and
$PROD_1 = \{q_1, \ldots, q_m\}$ with $q_j = (L'_j \leftarrow K'_j \to R'_j)$ $j = 1, \ldots, m$
A gra–morphism $F_1 : Gra_0 \to Gra_1$, is a 4-tuple $(l_1, st_1, P_1, \{pr_1(p) : p \in PROD_0\})$ where $l_1 : (C_0, T_0) \to (C_1, T_1)$ is a map between the alphabets, $st_1 : START_0 \to START_1$ is an l_1-based graph morphism, $P_1 : PROD_0 \to PROD_1$ is a function between sets of productions, and $pr_1(p)$ is a strict production morphism from p to $P_1(p)$ for $p \in PROD_0$.

A gra–morphism is a 'relaxed' multilevel graph morphism, where there is no ordering on the productions preserved by the set of production morphisms and no explicit compatibility among the different $pr_1(p)$.

It is possible to extend the result of Proposition 10 to graph grammars and show that the graphs generated by graph–grammars connected by gra–morphisms are related by graph morphisms.

Proposition 12. Given a gra-morphism $F_1 = (l_1, st_1, P_1, \{pr_1(p) : p \in PROD_0\}) : Gra_0 \to Gra_1$ and an l_1-based graph morphism $f : G_0 \to G_1$, if

- $p_0 : G_0 \Rightarrow H_0$ via the occurrence morphism $g_0 : L_0 \to G_0$ in Gra_0
- $p_1 = P_1(p_0) : G_1 \Rightarrow H_1$ via the occurrence morphism $g_1 : L_1 \to G_1$ in Gra_1
- g_0 and g_1 are compatible, i.e., $f \circ g_0 = g_1 \circ pr_1(p_0)_L$

then there exists a unique $h : H_0 \to H_1$ extending f, i.e., such that f and h coincide on the part of G_0 unaffected by the production p_0.

Proof. (Sketch) If we call D_0 and D_1 the pushout complements constructed in the derivations via p_0 and p_1, respectively, it is easy to construct the unique morphism $D_0 \to D_1$ such that $D_0 \to G_0 \to G_1 = D_0 \to D_1 \to G_1$ in the diagram

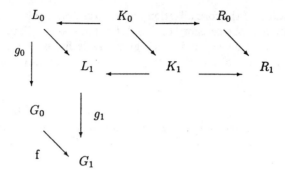

By the injectivity of $D_1 \to G_1$, we also have $K_0 \to D_0 \to D_1 = K_0 \to K_1 \to D_1$. Since the composition of pushout diagrams is again a pushout diagram, the commutative diagram consisting of $K_0 \to L_0 \to G_0 \to G_1$ and $K_0 \to D_0 \to D_1 \to G_1$ is also a pushout. Subtracting from it the first pushout of the derivation via p_0, we obtain the pushout $D_0 \to G_0 \to G_1 = D_0 \to D_1 \to G_1$. By the universal property of pushouts, there is a unique morphism $h : H_0 \to H_1$ so that the appropriate diagrams commute. The restriction of such a morphism to D_0 coincides with the restriction of f.

Unlike the case of multilevel graph grammars where the restoring productions can always be applied to the base graph, if a production p_0 is applicable to $G_0 \in Lang(Gra_0)$, it is not necessary that $pr_1(p_0)$ be applicable to G_1 even if there is a graph morphism $f : G_0 \to G_1$. Hence if $f : G_0 \to G_1$ and $p_0 : G_0 \Rightarrow H_0$, f may not always be extendable to a morphism $H_0 \to H_1$ for some $pr_1(p_0) : G_1 \Rightarrow H_1$. So, the presence of an arbitrary gra–morphism from Gra_0 to Gra_1 is not sufficient to guarantee that any derivation in Gra_0 can be duplicated by the appropriate production in Gra_1. It is possible to relate the derivations if Gra_1 is constructed from Gra_0 in a uniform way, as shown in the next section.

It is easy to see that gra–morphisms can be composed componentwise with the composition being given by $l_2 \circ l_1$, $st_2 \circ st_1$ and $P_2 \circ P_1$ for the first three components and $\{pr_2(pr_1(p)) \circ pr_1(p) : p \in PROD_0\}$ for the fourth one, and that gra–isomorphisms are 4-tuples consisting of an identity, a graph isomorphism, another identity and a set of production isomorphisms. The usual properties are satisfied so that graph grammars and gra–morphisms form a category $GRAgra$.

Theorem 13. *The category GRAgra is finitely cocomplete.*

Proof. For simplicity, we just sketch the construction of pushouts. Given $F_i : Gra_0 \to Gra_i = (l_i, st_i, P_i, \{pr_i(p) : p \in PROD_0\})$ for i=1,2, the pushout object Gra_3 of F_1 and F_2 has as set of labels and initial graph the pushout object of l_1 and l_2 and of st_1 and st_2, respectively. The corresponding components of the induced gra–morphisms are the obvious ones induced by the universal properties. The set of productions is the union of $\{pr_1(p) +_p pr_2(p) : p \in PROD_0\}$ and $(PROD_1 - P_1(PROD_0)) \cup (PROD_2 - P_2(PROD_0))$ where if $p = (L \leftarrow K \to R)$ and $p_i(p) = (L_i \leftarrow K_i \to R_i)$, for i=1,2, then $pr_1(p) +_p pr_2(p) = (L_1 +_L L_2 \leftarrow K_1 +_K K_2 \to R_1 +_R R_2)$. The induced gra–morphisms are the obvious ones from the colimit for the first part and the inclusion for the second part.

4 Transformations of Grammars

There are at least two (uniform) ways of transforming graph grammars : at a global level and at a local level.

On a local level, (some of) the single components of a grammar are modified to obtain the transformed grammar. On a global level, a grammar can be transformed by replacing a specific subgrammar with another grammar, analogous to the situation where a graph is transformed by replacing a subgraph with another one by using a graph production.

4.1 Global Transformation

A transformation system for graph grammars consists of a set of productions of the form $(L \leftarrow K \rightarrow R)$ where now L, K and R are graph grammars and $(L \leftarrow K)$ and $(K \rightarrow R)$ are gra–morphisms.

Rather than duplicating the efforts to prove directly properties such as Local Confluence or Parallelism Theorems, already valid for graph grammars, we use the notion of High Level Replacement System, where sufficient conditions for the validity of certain properties are formalized in an axiomatic way based upon the category from which the objects L, K and R and the corresponding morphisms are defined. In what follows, let M be the class of gra–morphisms $(l, st, P, \{pr(p) : p \in PROD\})$ where l, st and P are injective (or monomorphisms) in their respective categories and each p is a subproduction of $pr(p)$.

The property called Local Confluence is satisfied when a pair of derivations does not overlap in a meaningful way and allows the application of the corresponding productions in either order. We first formalize the notion of independence.

Definition 14. Two direct derivations $p_1 : G \Rightarrow H_1$ and $p_2 : G \Rightarrow H_2$ are called *parallel independent* if there exist morphisms $L_1 \rightarrow C_2$ and $L_2 \rightarrow C_1$ such that $L_1 \rightarrow C_2 \rightarrow G = L_1 \rightarrow G$ and $L_2 \rightarrow C_1 \rightarrow G = L_2 \rightarrow G$ as in the following diagram

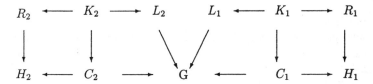

Two direct derivations $p_1 : G \Rightarrow H_1$ and $q : H_1 \Rightarrow H$ are *sequential independent* if $p_1^{-1} : H_1 \Rightarrow G$ and $q : H_1 \Rightarrow H$ are parallel independent.

Parallel and sequential independence formalize the idea of "non meaningful overlap".

Theorem 15. *1. If $p_1 : G \Rightarrow H_1$ and $p_2 : G \Rightarrow H_2$ are parallel independent, then p_2 is applicable to H_1, p_1 is applicable to H_2 and there exists a graph H such that $p_1 : H_2 \Rightarrow H$ and $p_2 : H_1 \Rightarrow H$.*

2. If $p_1 : G \Rightarrow H_1$ and $q : H_1 \Rightarrow H$ are sequential independent, then there exists an H such that $q : G \Rightarrow H$ and $p_1 : H \Rightarrow H_1$.

When two derivations are parallel independent, not only they can be performed in either order but they can be performed simultaneously by applying the coproduct of the corresponding productions. The coproduct of $p_1 = (L_1 \leftarrow K_1 \rightarrow R_1)$ and $p_2 = (L_2 \leftarrow K_2 \rightarrow R_2)$ is defined as $p_1 + p_2 = (L_1 + L_2 \leftarrow K_1 + K_2 \rightarrow R_1 + R_2)$ with the obvious induced morphisms.

Theorem 16. *Given sequentially independent derivations $p_1 : G \Rightarrow H_1$ and $q : H_1 \Rightarrow H$, the coproduct production $p_1 + q$ is applicable to G and $p_1 + q : G \Rightarrow H$.*

For the proofs of these two theorems, we use the notion of High Level Replacement Theorem. As shown in [3, 4], it is sufficient to verify that the base category from which the morphisms of the productions are chosen satisfy properties so as to classify it as HLR0.5, HLR1 or HLR1* category, to be able to conclude that it defines a replacement system based on the double pushout approach which satisfies the Local Confluence (Church–Rosser) property, the Parallelism Theorem or the existence (and uniqueness) of canonical derivation sequences. For convenience, we summarize here the properties which define the different categories.

- An $HLR0$–category is a category CAT with a distinguished class M of morphisms.
- An $HLR0.5$–category is an $HLR0$–category which satisfies
  1. existence of pushouts between a morphism m_1 in M and an arbitrary mor-

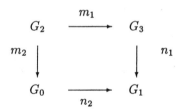

  phism m_2
  2. inheritance of M–morphisms under pushouts, i.e., if $m_1 \in M$ then $n_2 \in M$
  3. existence of binary coproducts such that if $m_1, m_2 \in M$, then their coproduct $m_1 + m_2$ is also in M
- An $HLR0.5^*$–category is an $HLR0.5$–category which, in addition, satisfies
  4. $M \subseteq \{monomorphisms\ in\ CAT\}$
  5. M–pushouts are pullbacks, i.e., if $m_1, m_2 \in M$ in the pushout diagram above, then the diagram is also a pullback
  6. existence of the initial object I in the category CAT with $Id_I \in M$
- An $HLR1$–category is an $HLR0.5$–category satisfying also

7. M–po–pb decomposition property, i.e., in the diagram

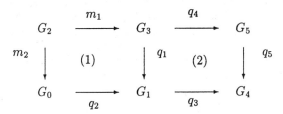

if (1)+(2) is a pushout and (2) is a pullback with morphisms in M, then (1) and (2) separately are pushouts
8. existence and inheritance of M–pullbacks
- An $HLR1^*$-category is an $HLR1$-category which also satisfies properties [4.], [5.] and [6.].

Proposition 17. *Let M be the class of gra–morphisms $(l, st, P, \{pr(p) : p \in PROD\})$ where l, st and P are monomorphisms in their respective categories and each p is a subproduction of $pr(p)$. Then GRAgra is an $HLR0.5$-category*

Proof. (Sketch) By theorem 13, the category Gragra has pushouts for any pair of morphisms. If one of them is in M, then the induced one is also in M using the fact that both SETS and GRAPHS (the categories of sets and of graphs with functions and graph morphisms respectively) with M the set of injective morphisms are $HLR1^*$ and therefore $HLR0.5$ categories. The fact that the induced morphisms are strict is a consequence of a general property of pushouts where, in a diagram such as the one in [7.] above, if (1)+(2) is a pushout and (1) is a pushout, then (2) is also a pushout. Coproducts are constructed componentwise.

As a consequence of this Proposition, we have that if the derivations $Gra_0 \Rightarrow_{p_1} Gra_1 \Rightarrow_{p_2} Gra_2$ are butterfly- independent, then there exists a graph grammar Gra such that $Gra_0 \Rightarrow_{p_2} Gra \Rightarrow_{p_1} Gra_2$. We do not go into details on the precise definition of butterfly–independence [4] since (as we shall see) the category GRAgra is an $HLR1$-category and therefore the notion of butterfly independence coincides with the more intuitive and better known sequential independence. Categories which are $HLR0.5$ also allow the shifting of butterfly-independent derivations so as to maximize left–most parallelism.

Before analyzing the properties to place GRAgra higher in the hierarchy, we need the following closure property.

Proposition 18. *The category GRAgra is closed under pullbacks. Furthermore, if the original morphisms are in M, then the induced morphisms are in M.*

Proof. (Sketch) Given $Gra_i = ((C_i, T_i), PROD_i, START_i)$ for i=1,2,3, and $F_i = (l_i, st_i, P_i, \{pr_i(p) : p \in PROD_i\}) : Gra_i \to Gra_3$, define $Gra_0 = ((C_0, T_0), PROD_0, START_0)$ by letting (C_0, T_0) be the pullback object of l_1 and l_2, $START_0$ the pullback object of st_1 and st_2 and $PROD_0$ the set of productions obtained as the pullback object of $pr_1(p_1)$ and $pr_2(p_2)$ for all p_1 and p_2 such that $P_1(p_1) = P_2(p_2) \in PROD_3$.

The gra–morphism $Gra_0 \to Gra_i$ is the morphism whose components are those induced by the component pullbacks. Since injectivity is inherited under pullbacks in both SETS and GRAPHS, the resulting gra–morphism is in M, again using, as in the previous Proposition, compositional properties of pullback diagrams.

Proposition 19. *With M defined as the set of gra–morphisms $(l, st, P, \{pr(p) : p \in PROD\})$ such that l, st and P are injective and $p \subseteq pr(p)$, GRAgra is an $HLR1$-category*

Proof. (Sketch) By the previous proposition, we need only prove the properties [7.] and [8.]. Once more, exploiting the fact that SETS and GRAPHS are $HLR1$-categories, the diagrams (1) and (2) of [7.], reduced to each of its three components (C, T), $PROD$ and $START$, are separately pushouts in the respective categories. In order to extend the morphisms to gra–morphisms, we need the $\{pr(p) : p \in PROD\}$ which are already defined by (1)+(2) being a pushout

4.2 Local Transformation

As mentioned at the beginning of this section, graph grammars can be transformed at a local level, by using a production to transform the initial graph START and/or each component of the different productions in PROD. A modification of the label sets is considered as a global transformation.

Definition 20. *Given a production $q = (L \leftarrow K \to R)$ to be applied to another production $p_1 = (L_1 \leftarrow K_1 \to R_1)$ via the morphism $h : L \to K_1$, the rewriting of p_1 by q is the production p_2 given by the following diagram where each square is a pushout*

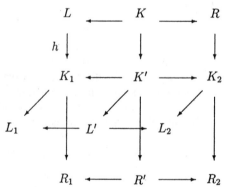

Abusing the notation, we write $q : p_1 \Rightarrow p_2$ via h.

The applicability of q to p_1 depends on the existence of the pushout complements K', L' and R'. For the existence of K', theorem 4 gives necessary and sufficient conditions. For the extendibility of the derivation $q : K_1 \Rightarrow K_2$ to L_1 and R_1, there are results on the embedding of derivation sequence [2] in larger graphs. In this simplified case, we can summarize the conditions in the following theorem.

Theorem 21. *A production $q = (L \leftarrow K \rightarrow R)$ is applicable to $p_1 = (L_1 \leftarrow K_1 \rightarrow R_1)$ via $h : L \rightarrow K_1$ if and only if*

1. $DANG_h \cup ID_h \subseteq l(K)$
2. $DANG_{l_1} \cap h(L) \subseteq h(l(K))$
3. $DANG_{r_1} \cap h(L) \subseteq h(l(K))$

Proof. The first condition is the Gluing Condition relative to the application of q to K_1, which is equivalent to the existence of K'. Given K', the pushout complement L' exists if and only if the morphism $l_1 : K_1 \rightarrow L_1$ satisfies the Gluing Condition relative to $K' \rightarrow K_1$; since l_1 is injective, this is equivalent to requiring that $DANG_{l_1}$ be in the image of $K' \rightarrow K_1$ and condition 2. now follows by construction of K'. Condition 3. is similar.

The rewriting of a production p_1 to a production p_2 via another production q should be reflected in the correspondence between the graphs produced by p_1 and those produced by p_2. The behavior is exactly as expected: if q transforms G_1 into G_2 and the production p_1 into the production p_2, the results of the transformation of G_1 via p_1 and of G_2 via p_2 should be related by the production q.

Theorem 22. *If*

- $q : p_1 \Rightarrow p_2$ via $h : L \rightarrow K_1$,
- $p_1 : G_1 \Rightarrow H_1$ via $g_1 : L_1 \rightarrow G_1$,
- $q : G_1 \Rightarrow G_2$ via $f = g_1 \circ l_1 \circ h : L \rightarrow G_1$ and
- $p_2 : G_2 \Rightarrow H_2$ via $g_2 : L_2 \rightarrow G_2$

then $q : H_1 \Rightarrow H_2$.

Proof. Let K', L' and R' be the contexts constructed in the application of the production q to K_1, L_1 and R_1, respectively. By the construction, the morphisms $K' \rightarrow L'$ and $K' \rightarrow R'$ are injective and the production $p' = (L' \leftarrow K' \rightarrow R')$ is a subproduction of both p_1 and p_2. Since $p' \subseteq p_1$, we have $p' : G_1 \Rightarrow H_1$ via $g'_1 : L' \rightarrow L_1 \rightarrow G_1$; similarly $p' : G_2 \Rightarrow H_2$ via $g'_2 : L' \rightarrow L_2 \rightarrow G_2$. Notice now that the derivations via p' and q of H_1 and G_2, respectively, are parallel independent and thus, by applying the Church-Rosser Theorem for Graph Grammars [2], q is applicable to H_1 and $q : H_1 \Rightarrow H_2$.

Corollary 23. *Let $q : p_1 \Rightarrow p_2$ via $h : L \rightarrow K_1$, $p_1 : G_1 \Rightarrow H_1$ via $g_1 : L_1 \rightarrow G_1$, $q : G_1 \Rightarrow G_2$ and $q : H_1 \Rightarrow H_2$. Then $p_2 : G_2 \Rightarrow H_2$*

Proof. Since $p' \subseteq p_2$, we have $p_2 : G_2 \Rightarrow H_2$ if and only if $p' : G_2 \Rightarrow H_2$; along the lines of the previous theorem and using local confluence, p' is applicable to G_2 and produces H_2.

By repeated application of this corollary, it is possible to relate the languages generated by grammars connected by a gra–morphism. First let us define the rewriting of a graph grammar by a production.

Definition 24. Given $Gra_1 = ((C_1, T_1), START_1, PROD_1 = \{p_1, ..., p_n\})$ and a production $q = (L \leftarrow K \rightarrow R)$, the graph grammar Gra_2 is derived from Gra_1 by q via $h, h_1, \ldots, h_n$, denoted by $(q; h_1, \ldots, h_n) : Gra_1 \Rightarrow Gra_2$, or just $q : Gra_1 \Rightarrow Gra_2$ if the morphisms can be determined from the context, if $Gra_2 = ((C_2, T_2), START_2, PROD_2)$ where

- $(C_1, T_1) = (C_2, T_2)$
- $(q, h) : START_1 \Rightarrow START_2$
- $PROD_2 = \{q_i : (q, h_i) : p_i \Rightarrow q_i, p_i \in PROD_1\}$.

We can now state the main result of this part.

Theorem 25. Let $q : Gra_1 \Rightarrow Gra_2$. Then for each $G \in Lang(Gra_1)$, $q : G \Rightarrow H$ and $H \in Lang(Gra_2)$.

The conclusion of the theorem is denoted by $q : Lang(Gra_1) \preceq Lang(Gra_2)$. Notice that the theorem states that q is applicable **and** the result of the application is a graph in the language generated by (Gra_1). Since a production could be applied to a graph in more than one way by different occurrence morphisms, not every graph in $Lang(Gra_2)$ need be the result of a transformation by q of a graph in $Lang(Gra_1)$.

Notice that if q is a restoring production, i.e., if $L = K$, and $q : G \Rightarrow H$, then there is an obvious morphism from G to H since the context graph of the derivation via q is the graph G itself. This is the basis of the following result mentioned in the previous section.

Proposition 26. Given $Gra_1 = ((C_1, T_1), START_1, PROD_1 = \{p_1, ..., p_n\})$, every restoring production q and tuple $h, h_1, \ldots, h_n$ of graph morphisms induce a unique gra-morphism $F_q : Gra_1 \rightarrow Gra_2$ with second component st_q. Furthermore, for such an F_q, for each $G \in Lang(Gra_1)$ there exists an $H \in Lang(Gra_2)$ and a graph morphism $h : G \rightarrow H$ extending st_q.

Proof. given Gra_1 and q as in 24, define $F_q : Gra_1 \rightarrow Gra_2 = (l_q, st_q, P_q, \{pr_q(p) : p \in PROD_1\})$ as follows:

- $l_q =$ identity
- st_q is the morphism induced by $(q, h) : START_1 \Rightarrow START_2$
- $P_q(p_i) = q_i$ where $(q, h_i) : p_i \Rightarrow q_i$
- $pr_q(p_i) : p_i \rightarrow q_i$ is the morphism induced by the derivation via q

A simple induction on the length of the derivation corresponding to $G \in Lang(Gra_1)$ gives the graph H and the graph morphism $h : G \rightarrow H$ extension of st_q.

The two ways of transforming a graph grammar are not independent. In particular, it is possible to place local transformations in a global framework in an obvious way. If $q : Gra_1 \Rightarrow Gra_2$, then $q : START_1 \Rightarrow START_2$ and $q : p_i \Rightarrow q_i$ for $p_i \in PROD_1$ and $q_i \in PROD_2$. Let $START_0$ be the context graph in the derivation of $START_2$ from $START_1$. Also, for each $q : p_i \Rightarrow q_i$, let p'_i be the subproduction of both p_i and q_i as in the proof of theorem 22. Then by defining Gra_0 as the grammar with the same label set as Gra_1 and Gra_2, with $START_0$ as initial graph and with the set of productions consisting of the productions p'_i, there are obvious gra-morphisms $Gra_0 \to Gra_i$, i=1,2, defining a production of graph grammars.

Not every graph grammar replacement system can be seen as the result of local transformations, even if it is label-preserving, unless it is possible to place limitations in a uniform way on the occurrence morphisms which are allowed to transform a production into another one.

5 Concluding Remarks

We have presented here two approaches to the rewriting of graph grammars where derivations are based on the double pushout approach. The development is a generalization (and a considerable simplification) of the work done on transformations of graphs represented on several levels of visibility, described by productions needed to retrieve the hidden parts [11]. Although the inspiration came from multilevel graph grammars, there are several applications of the idea of transforming graph grammars. Viewing graph grammars as generative devices, a compiler is a procedure which allows, through a correspondence between the two grammars, to transform members of a language into members of the other. In the context-free case, there is no need of a correspondence between the structure of related productions. The transformations discussed in this paper can be used to give semantics to compilers, expecially of visual languages [6, 8]. Viewing productions in a graph grammar as individual steps of a computational model based on state transformation, the work presented could be used to rewrite each atomic transformation using productions at the same level, to model the refinement, the simulation or the evolution of the computational model.

The presentation has been phrased using graphs and graph morphisms, but it is easily extendable, along the lines of [3] to the transformation of High Level Replacement Systems, both at the global and at the local level. With an eye towards applications, the rephrasing of the results obtained in the context of hypergraphs [7] is of particular interest.

Still to be investigated are the properties to classify GRAgra as an HLR2-category, so that the Concurrency Theorem [2] would hold for replacement systems based on it. Under investigation, instead, is the rewriting of graph grammars where the derivation steps use single pushouts.

References

1. Boehm, P., Fonio, H.R., Habel, A.: Amalgamation of Graph Transformations with applications to Synchronization. in TAPSOFT 85, Lect. Notes Comp. Sci. 185, Springer-Verlag (1985) 267-283.
2. Ehrig, H. : Introduction to the Algebraic Theory of Graph Grammars. in Lect. Notes Comp. Sci. 73, Springer-Verlag (1979) 1-69.
3. Ehrig, H., Habel, A., Kreowski, H.-J., Parisi Presicce, F.: ¿From Graph Grammars to High Level Replacement Systems. in Graph Grammars and their applications to Computer Science (H.Ehrig, H.-J.Kreowski, G.Rozenberg, eds.) Lect. Notes Comp. Sci. 532, Springer-Verlag (1991) 269-287.

4. Ehrig, H., Kreowski, H.-J., Taentzer, G.: Canonical Derivations for High Level Replacement Systems. Bericht 6/92, Univ. Bremen, Dec 1992.
5. Engelfriet, J., Rozenberg, G.: Graph Grammars based on Node Rewriting: an Introduction to NLC Graph Grammars. in Graph Grammars and their applications to Computer Science (H.Ehrig, H.-J.Kreowski, G.Rozenberg, eds.) Lect. Notes Comp. Sci. 532, Springer-Verlag (1991) 12-23.
6. Golin, E.J., Reiss, S.P.: The Specification of Visual Language Syntax. in Proc. of the IEEE Workshop on Visual Languages, Rome, Italy, 1989, 105-110.
7. Habel, A.: Hyperedge Replacement: Grammars and Languages. Lect. Notes Comp. Sci. 643, Springer-Verlag (1992)
8. Helm, R., Marriot, K.: Declarative Specification of Visual Languages. in Proc. of the IEEE Workshop on Visual Languages, Skokie, Illinois, USA, 1990, 98-103.
9. Lowe, M.: Extended Algebraic Graph Transformation. Technische Universitat Berlin feb 1991, 180 pages
10. Parisi-Presicce, F. : Single vs. Double pushout derivations of Graph. in Proc. 18th Internat. Wksp. on Graph Theoretic Concepts in Comp. Sci. 1992, Lect. Notes Comp. Sci. 657, Springer-Verlag (1993) 248-262
11. Parisi-Presicce, F., Piersanti, G.: Multilevel Graph Grammars. in Proc. 20th Internat. Wksp. on Graph Theoretic Concepts in Comp. Sci. 1994, Lect. Notes Comp. Sci. 903, Springer-Verlag (1995) 51-64 [extended version: Dip. Matematica Pura ed Applicata, Univ. L'Aquila, Tech.Rep. 33, July 1993]
12. Parisi-Presicce, F., Piersanti, G.: Graph Based Modelling of Visual Languages. in preparation

Drawing Graphs with Attribute Graph Grammars

Gaby Zinßmeister[1] and Carolyn L. McCreary[2]

[1] Wilhelm-Schickard-Institut, Universität Tübingen, Sand 13, 72076 Tübingen, Germany, email: zinssmei@informatik.uni-tuebingen.de
[2] Dept. of Computer Science and Engineering, Auburn University, Auburn, AL, USA, email: mccreary@eng.auburn.edu

Abstract. We address the problem of automatically generating layouts for graphs using graph grammars.

The central idea of our approach to graph layout is viewing layout algorithms as attribute evaluators of attribute graph grammars, thus a layout algorithm is an attribute scheme plus an attribute evaluator. The main advantages are that we make use of the intrinsic structure of graphs to subdivide the layout problem and that different layouts can be specified simply with different attribute schemes for the same graph grammar.

1 Introduction

Attribute graph grammars can be used to automatically construct clear and readable drawings of graphs. We focus on the layout of graphs for visualizing structures for human perception, in contrast, for example, to VLSI design. In this paper we report on node layout and do not discuss the edge routing.

There exist a couple of algorithms specialized to the class of graphs they lay out (see [6]). Some classes of graphs have an intrinsic structure, e.g. trees, flow-graphs, and series-parallel graphs and are therefore well suited for the representation with graph grammars. The structure of other classes, such as directed acyclic graphs (DAGs), can be approximated by a graph grammar. The central idea of our approach to graph layout is viewing layout algorithms as attribute evaluators of attribute graph grammars. Thus a layout algorithm is an attributed grammar plus an attribute evaluator. The advantages are a declarative description of the layout and the easy exchange of different layouts by simply specifying a different attribute scheme.

We present a framework and examples for efficient and easy layout generation for a wide class of graphs. We show that DAGs cannot be generated by a confluent context-free graph grammar and then describe a layout technique that uses a graph grammar that approximates the DAG. Addition work on drawing graphs using graph grammars has been published by Brandenburg [2] and by Minas [14].

2 Graph Grammatical Context

The general idea is to parse a given graph according to a specified graph grammar and generate a derivation tree representation of it. The attribute evaluator takes the derivation tree as input, associates an attribute scheme to specify the node layout, and calculates the x-/y-coordinates of the vertices of the graph as attribute values.

The general idea is illustrated in Fig. 1.

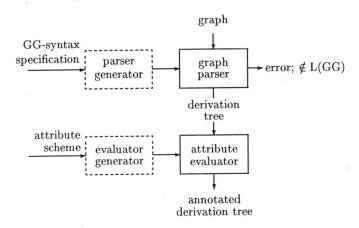

Fig. 1. Framework

The parsing component of the entire solution limits the approach to graph classes describable by polynomial graph grammars, i.e. graph grammars with polynomial membership problem. For precedence graph grammars Kaul [10] presented a parser generator, which generates $O(n^2)$-parsers. Lamshöft [11] has implemented a parser for confluent, ϵ-free edNCE-graph grammars. For some graph classes specific parsers have been developed, eg. McCreary [13] for partial order approximations of DAGs and Lichtblau [12] for flow charts and Valdes [19] for series-parallel graphs. Schütte [17] has defined (node-)attribute graph grammars as a canonical extension of Knuth' attribute string grammars, such that the evaluation mechanisms can be used if a total order for the nodes of the right-hand sides of the productions is given.

We present two example graph decompositions and their corresponding layout attributes. One gives a node layout scheme for trees and is based on a confluent 1-dNCE graph grammar (see [3] for definitions). The other draws DAGs and is based on the application of 2-structures to graphs [7].

The main problem in the practical application of this approach is finding polynomial graph grammars to describe the desired graph class. This is a creative non-automatable task.

3 Layout of Trees

As an example we will elaborate on the layout of arbitrary trees. Several different layout-algorithms have been modelled by attribute graph grammars (see [20]). The graph grammar for trees is defined as follows:
Let $T = (\Sigma_N = \{T, G\}, \Sigma_T = \{t\}, \Sigma_E = \emptyset, P, S = T)$ where Σ_N and Σ_T are the nonterminal- and terminal node alphabets, Σ_E is the (empty) edge label alphabet, P are the productions and S is the start symbol. The productions are given in Fig. 2.

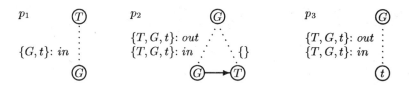

Fig. 2. Productions P of T

G nodes are the 'generic' nodes with which we can add an arbitrary number of children to a treenode, whereas T nodes represent a fully derived subtree. The upper nonterminal nodes are the left hand sides or mother subgraphs, the lower graphs with terminal and nonterminal nodes and solid edges are the right hand sides of the productions or daughter subgraphs. The dotted lines indicate the replacement step. The mark $\{T, G, t\} : in$ means that edges from T-, G- and t-nodes to the mother node, are replaced by edges from the same sources to the daughter graph node specified by the dotted line (*out* denotes the reverse orientation of edges.) The empty set {} at a dotted line denotes that the target node is not connected to the host graph. The terminal alphabet can obviously be extended, but is reduced to one node type for simplicity.

The parser for that grammar is in essence the precedence parser of Kaul where the existing shift-reduce conflicts are solved by shifting. There may be different derivation trees for the same (terminal) tree depending on the choice of the start node for the parsing process and the choice amongst the adjacent nodes to *top(parsestack)* with equal or increasing precedence. This reflects the non-order within siblings of the terminal tree and therefore any derivation tree is suitable for the layout [3].

As an extension to the attribute string grammars of Knuth, one can define attribute graph grammars (see [17]). Synthesized and inherited attributes are associated with the (non)terminals and semantic functions with the production rules. As long as only node attribute values are arguments of the semantic func-

---
[3] If one has ordered trees as input, the graph grammar has to be augmented with different edge labels reflecting parent and left-sibling relations. This graph grammar is an (unambigous) precedence graph grammar.

tions, the attribute evaluators for the graph grammars are the same as in the string case.

The results of the layout algorithms of Moen [15], Reingold/Tilford [16], Carpano [4] and the H-tree algorithm of Ullman [18] for VLSI-Design implemented as attribute graph grammar are shown in Fig. 3.

The main idea in attributing the graph grammar is pushing level information down the derivation tree and synthesizing some relative positions in the next step. Reingold/Tilford put nodes of the same level at a line gathering contours of terminal subtrees. Moen gathers the contours of (terminal) subtrees and joins them trying to save space as can be seen in Fig. 3. Nodes of one level but different subtrees are not necessarily placed on one line. The attribute scheme for the Moen algorithm is shown in Fig. 4 as an example for the elegance of the method. Copying rules are left out. In this figure ↑/↓ denote synthesized/inherited attributes and the functions written in italic are taken from the original algorithm. *layout_leaf* calculates the contour of a single leaf. Contours are represented as polylines. *merge1* calculates the contour around neighbouring subtrees, whereas *merge2* adds up the height of these subtrees. *attach_parent* determines the contour around one subtree by putting its root in the middle above its children. Carpano puts nodes on one level on a concentric circle (dashed circles in Fig. 3). For all layout algorithms the absolute positions are calculated as inherited attributes in the second pass. The attributes are evaluable in two left-to-right passes through the derivation tree.

As stated above the main difficulty consists in finding a polynomial graph grammar for the given graph class. Therefore parsing directed acyclic graphs (DAGs) pose a serious problem. Dags are not describable by a confluent context-free graph grammar[4]. The proof idea is a counting argument, analogous to the one of Brandenburg [1] for the set of all graphs. The number of DAGs having at most n nodes is in $O(c^{n^2/2})$, $c \geq 2$. To represent a monotone derivation sequence for a graph with at most n nodes in a bit pattern we need $O(nlogn)$ bits. So we can derive only $O(2^{n\ logn})$ graphs, much less than the number of DAGs.

4 Parsing and Drawing DAGs

Gill, McCreary et al. [9] report on parser for Hasse graphs that can be used to determine a suitable node layout for arbitrary DAGs. This work is based upon the theory of two structures developed by Ehrenfeucht and Rosenberg [7]. Fig. 5 shows the production rules of the underlying grammar. The productions map a single mother node into a graph that is (a) a set of unconnected nodes (an independent clan, I) or (b) a set of sequentially connected nodes (a linear clan, L) or (c) a Hasse graph that cannot be parsed (a primitive clan, H) or d) a

---

[4] A **context-free** graph grammar has production rules with only one nonterminal node as the left-hand side (see [5] for natural and useful requirements for a graph grammar to be 'context-free'). A graph grammar is **confluent** iff for each sentential form and distinct nonterminal nodes the order of the application of productions at these nodes yield the same graph.

Moen:

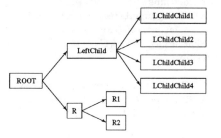

Reingold/Tilford:

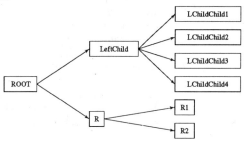

Carpano:

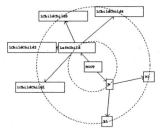

Ullman: (binary tree)

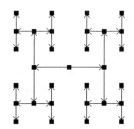

Fig. 3. Different tree layouts

p_1: $\uparrow T$.rootdims　　　　　　:= G.rootdims
　　　$\uparrow T$.contour　　　　　　　:= IF G.contoursubtrees = empty THEN
　　　　　　　　　　　　　　　　　　$layout\_leaf(G$.rootdims$)$
　　　　　　　　　　　　　　　ELSE $attach\_parent(G$.contoursubtrees,G.rootdims,G.sum$)$
　　　$\downarrow G$.heightsubtrees　　　:= G.sum

p_2: $\uparrow G_0$.contoursubtrees　:= IF G_1.contoursubtrees = empty THEN T.contour
　　　　　　　　　　　　　　　ELSE $merge1(G_1$.contoursubtrees,T.contour$)$
　　　$\uparrow G_0$.lastsubtreeheight := T.rootdims.height + 2 * T.rootdims.border
　　　$\uparrow G_0$.rootdims　　　　　:= G_1.rootdims
　　　$\uparrow G_0$.sum　　　　　　　:= IF G_1contoursubtrees = empty THEN
　　　　　　　　　　　　　　　　　　G_0.lastsubtreeheight
　　　　　　　　　　　　　　　ELSE G_0.lastsubtreeheight + G_1.sum +
　　　　　　　　　　　　　　　　　　$merge2(G_1$.contoursubtrees,T.contour$)$
　　　$\downarrow T$.pos.x　　　　　　:= G_0.pos.x + G_1.rootdims.border + G_1.rootdims.width +
　　　　　　　　　　　　　　　　　　PARENTDISTANCE
　　　$\downarrow T$.pos.y　　　　　　:= IF G_1.contoursubtrees = empty THEN
　　　　　　　　　　　　　　　　　　G_0.pos.y + G_1.rootdims.border +
　　　　　　　　　　　　　　　　　　$(G_1$.rootdims.height $-G_1$.heightsubtrees $)/2$
　　　　　　　　　　　　　　　ELSE G_1.lastypos + G_1.lastsubtreeheight +
　　　　　　　　　　　　　　　　　　$merge2(G_1$.contoursubtrees,T.contour$)$
　　　$\uparrow G_0$.lastypos　　　　　:= T.pos.y

p_3: $\uparrow G$.contoursubtrees　　:= empty
　　　$\uparrow G$.lastsubtreeheight　:= 0
　　　$\uparrow G$.sum　　　　　　　　:= 0
　　　$\uparrow G$.rootdims　　　　　　:= t.dims
　　　$\uparrow G$.lastypos　　　　　　:= 0

Fig. 4. Attribute scheme for the Moen layout algorithm

singleton terminal node (t) [7][5]. The daughter graphs are connected to the host graph by the heredity embedding rule, i.e. in-edges to the mother graph are replaced with in-edges to each source of the daughter graph and out-edges from the mother graph are replaced with out-edges from each sink of the daughter graph. Case c) is not describable by a finite set of grammar rules and p7 is a faint suggestion of the inifinite set of rules. Because there exists no confluent context-free graph grammar for DAGs, that is the best we can achieve. The productions build graphs with no transitive edges, but additional edges may be added to the final layout.

Nodes in independent clans from p5 can be placed in adjacent positions within a row and nodes in linear clans from p6 can be placed vertically. However, a clear visualization for primitives from p7 is not easily determined. Since primitives can be arbitrarily large, the parse may not identify subgraphs whose layout is

---

[5] * denotes $\{D, L, I, H, t\}$.

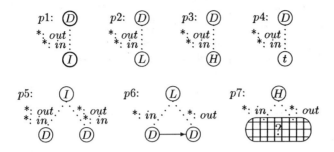

Fig. 5. Productions for DAGs

easily visualized. However, Gill and McCreary augment temporarily the edge set of the original graph to give an approximation that can be parsed into only linear or independent clans. One method of creating an approximation is to decompose the primitive by grouping all the source nodes into an independent clan that is linearly connected to the rest of the primitive or graph tail. The tail may then itself decompose or the augmentation technique may be recursively applied. After decomposing the primitives, the resultant parse tree can be placed in canonical form where no linear (independent) clan has a child which is also linear (independent). This canonical form is a kind of abstract syntax tree, omits all chain productions and reduces multiple applications of either independent clans or linear clans to n-ary instead of binary I- or L-nodes. This bipartite parse tree can be decorated with layout attributes that can be evaluated to produce an aesthetically pleasing drawing, see Fig. 8 on the left side.

Each nonterminal node I or L of the parse tree has two synthesized attributes that represent the width and height of the bounding box enclosing the subtree rooted at this node and two inherited attributes representing the x- and y-coordinate of the midpoint of the bounding box. The width and height attribute of terminal nodes and the x- and y-coordinate of the main root node are external. The attributes can be calculated in a way that is most aesthetically pleasing for a particular application. As an example, Fig. 6 shows the *natural bounding box attribute*. For the independent mother node, I, from production $p5$, the width of I is the sum of the widths of its replacement nodes, C_i, and the height of I is the maximum of the heights of its replacement nodes. Similarly, for the linear mother node, L, from production $p6$, the width of L is the maximum of the widths of its replacement nodes, C_i, and the height of L is the sum of the heights of its replacement nodes. Figure 7 shows how this attribute is applied to a parse tree example to produce the node layout of Fig. 8. Here the external values assigned to the width and height of the terminal nodes is 2. The location of the bounding box for the root places the upper left node at the origin of the graph.

Once the bounding boxes are determined, the nodes are centered within their boxes. The scheme is easily extended to arbitrary directed graphs. When a cycle is encountered, the edge that closes the loop is temporarily reversed. After the

p_5: ↑ I.bbwidth   := $\Sigma_{i=1}^{n} C_i$.bbwidth
     ↑ I.bbheight   := $\max(C_1.\text{bbheight},\ldots,C_n.\text{bbheight})$
     ↓ C_1.x   := I.x - I.bbwidth/2 + C_1.bbwidth/2
     ↓ C_i.x   := C_{i-1}.x + C_{i-1}.bbwidth/2 + C_i.bbwidth/2 $\forall 1 < i \leq n$
     ↓ C_i.y   := I.y $\forall 1 \leq i \leq n$
     where $C_i \in \{L, t\}$

p_6: ↑ L.bbwidth   := $\max(C_1.\text{bbwidth},\ldots,C_n.\text{bbwidth})$
     ↑ L.bbheight   := $\Sigma_{I=1}^{n} C_i$.bbheight
     ↓ C_i.x   := L.x
     ↓ C_1.y   := L.y - L.bbheight/2 + C_1.bbheight/2
     ↓ C_i.y   := C_{i-1}.y + C_{i-1}.bbheight/2 + C_i.bbheight/2 $\forall 1 < i \leq n$
     where $C_i \in \{I, t\}$

Fig. 6. Computing the bounding box attributes

nodes are layed out, the edges can be placed in their proper direction.

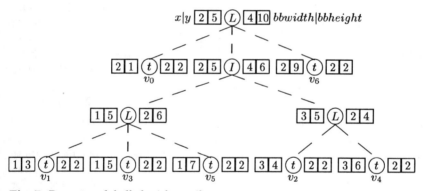

Fig. 7. Parse tree labelled with attributes

In Fig.9 an example of a dag drawing generated by this algorithm, is shown. Nodes are initially placed using the productions and attributes $p1 \ldots p6$. A discussion of the edge routing is beyond the scope of this paper.

5 Final Remarks

The approach of using attribute graph grammars for the layout of graphs offers the possibility to separate the syntactical aspect, i.e. classification and decomposition of graphs from the semantical aspect, i.e. the calculation of x-/y-coordinates for the nodes of the graphs. The implementation of different layout

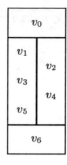

Fig. 8. Node layout corresponding to parse tree of Fig.7

algorithms is easily done by specifying different attribute schemes to the same graph grammar. The attribute evaluators may be generated automatically.

For interactive environments where graphs are not only to be displayed but are manipulated, incremental attribute evaluation algorithms have to be examined, such that only necessary parts of the graph layout have to be recalculated.

References

1. F.-J. Brandenburg. Graph Grammatiken. Script to a lecture, summer semester '89, 1990.
2. F. J. Brandenburg. Layout Graph Grammars: The Placement Approach. In Ehrig et al. [8], pages 144–156.
3. F. J. Brandenburg. The Equivalence of Boundary and Confluent Graph Grammars on Graph Languages of Bounded Degree. In R. V. Book, editor, *Rewriting Techniques and Applications, 4th Int. Conf., Como, Italy*, pages 312–322. Springer, 1991.
4. M.-J. Carpano. Automatic Display of Hierarchized Graphs for Computer-Aided Decsision Analysis. *IEEE Trans. on Systems, Man, and Cybernetics*, SMC-10(11):705–715, Nov. 1980.
5. B. Courcelle. An Axiomatic Definition of Context-Free Rewriting and its Application to NLC-Graph Grammars. *Theoretical Comput. Sci.*, 55(2/3):141–182, 1987.
6. G. Di Battista, P. Eades, R. Tamassia, and I. Tollis. Algorithms For Drawing Graphs: an Annotated Bibliography. Technical report, Brown University, Providence, RI 02912-1910, USA, 1993. Avail. via anonymous ftp from wilma.cs.brown.edu, file /pub/gdbiblio.tex.Z.
7. A. Ehrenfeucht and G. Rozenberg. Theory of 2-Structures, Part I: Clans, Basic Subclasses, and Morphisms. *Theoretical Comput. Sci.*, 70(3):277–303, Feb. 1990.
8. H. Ehrig, H.-J. Kreowski, and G. Rozenberg, editors. *Graph-Grammars and Their Application to Computer Science, 1990.* LNCS 532. Springer, 1991.
9. D. H. Gill, T. J. Smith, T. E. Gerasch, J. V. Warren, C. L. McCreary, and R. E. K. Stirewalt. Spatial - Temporal Anaylsis of Program Dependence Graphs for Useful Parallelism. *Journ. of Parallel and Distr. Comp.*, 19(2):103–118, Oct. 1993.

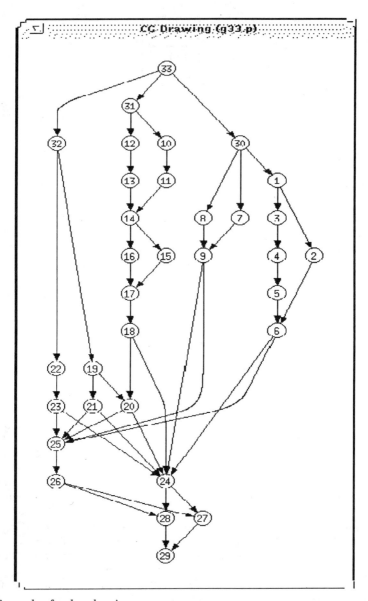

Fig. 9. Example of a dag drawing

10. M. Kaul. Syntaxanalyse von Graphen bei Präzedenz-Graph-Grammatiken. Technical Report MIP-8610, Univ. Passau, 1986. (Dissertation).
11. T. Lamshöft. Ein Parser für Graphgrammatiken. Master's thesis, Univ. Passau, FB Informatik, 1993.
12. U. Lichtblau. Recognizing Rooted Context-Free Flowgraph Languages In Polynomial Time. In Ehrig et al. [8], pages 538–548.
13. C. McCreary. *An Algorithm for Parsing a Graph Grammar.* PhD thesis, Univ. of Colorado, 1987.

14. M. Minas and G. Viehstaedt. Specification of Diagram Editors Providing Layout Adjustment with Minimal Changes. In *IEEE Workshop on Visual Languages, Bergen, Norway*, pages 324–329, Los Alamitos, CA, 1993. IEEE Computer Society Press.
15. S. Moen. Drawing Dynamic Trees. *IEEE Software*, 7(4):21–28, July 1990.
16. E. M. Reingold and J. S. Tilford. Tidier Drawings of Trees. *IEEE Trans. Softw. Eng.*, SE-7(2):223–228, Mar. 1981.
17. A. Schütte. *Spezifikation und Generierung von Übersetzern für Graph-Sprachen durch attributierte Graph-Grammatiken.* Reihe Informatik. EXpress Edition, Berlin, 1987. (Dissertation).
18. J. D. Ullman. *Computational Aspects of VLSI*, chapter 3 Layout Algorithms, pages 80–130. Computer Science Press, 1984.
19. J. Valdes, R. E. Tarjan, and E. L. Lawler. The Recognition of Series Parallel Digraphs. *SIAM J. Computation*, 11(2):298–313, 1982.
20. G. Zinßmeister. Tree Layout by Attribute Graph Grammars. WSI-Fachbericht WSI-93-4, Univ. Tübingen, Wilhelm-Schickard-Institut, 1993.

Graph Pattern Matching in PROGRES

Albert Zündorf[1], AG Softwaretechnik, Fachbereich 17, Uni-GH-Paderborn,
Warburgerstr. 100, D-33098 Paderborn, Germany

e-mail: zuendorf@uni-paderborn.de
phone: ++49/+5251/60-2430   fax: ++49/+5251/60-3530

Abstract

The work reported here is part of the PROGRES (**PRO**grammed Graph **R**ewriting Systems) project. PROGRES is a very high level multi paradigm language for the specification of complex structured data types and their operations. The data structures are modelled as **di**rected, **a**ttributed, **n**ode and **e**dge labeled graphs (diane graphs). The basic programming constructs of PROGRES are graph rewriting rules (productions and tests) and derived relations on nodes (paths and restrictions).

Although graph rewriting systems have successfully been used for specification purposes in many application areas since about 20 years, there was no sufficient tool available for the execution and implementation of graph grammar specifications. Especially, the problem of efficiently searching for a redex for an arbitrary given rewrite rule has been unsolved for a long time. In this paper we propose a new, heuristic, graph based algorithm solving this graph pattern matching problem. This algorithm has been implemented and is used successfully within the PROGRES environment.

1   Introduction

Graph rewriting rules consist of a graph pattern, the so called left-hand side, and a replacement graph, the so called right-hand side. A graph rewriting rule is executed by first matching its left-hand side graph pattern against a subgraph of the current working graph, the so called *redex*, and then replacing this redex by a copy of the right-hand side graph. While the replacement operation itself is quite easy, matching the graph pattern is the central efficiency problem for the execution of graph rewriting rules. In this paper we propose a heuristic algorithm for the execution of graph pattern matching in PROGRES.

In general, a naive algorithm for executing a graph rewriting step has the *time complexity* $O(P * N^L)$ where P is the number of inspected productions, N is the number of nodes in the current work graph, and L is the maximum size of a left-hand side of the system (cf. [BuGlTr 91]). Although being polynomial, this complexity is acceptable at most for a rapid prototype. For a final implementation of e.g. balanced, sorted, binary trees, only insert and delete operations with the time complexity $O(\log N)$ are acceptable.

To achieve this goal we have to consider in more detail the complexity $O(P * N^L)$ for performing a graph rewriting step for the language PROGRES. The factor P results from testing all productions of a graph grammar for applicability. In PROGRES, we use programmed graph rewriting steps, i.e. the user defines which graph rewriting rule (in PROGRES called *production*) may be applied in the next step by means of control structures, (cf. [ZünSchü 92]). So, in PROGRES, P is usually a very small number and in many

---

1. Supported by *Deutsche Forschungsgemeinschaft*.

cases even equal to 1. The factor N^L results from the subgraph matching problem. A naive implementation computes every possible mapping of the L nodes of the left-hand side of a production to the N nodes of the current graph. Then, for every matching, correct adjacency is tested until all required conditions hold. Once a subgraph isomorphic to the left-hand side is found, the replacement step simply has costs proportional to the number of the nodes and edges which have to be removed, created, or redirected. These costs are determined by the modification to be performed and will arise in hand-coded solutions, too. So, the central remaining problem for the execution of PROGRES is the problem of efficiently matching the left-hand side of a fired rule to a subgraph of the current host graph: the *graph pattern matching problem*.

2   The Language PROGRES

The graph pattern matching problem is much more difficult than the corresponding string pattern matching problem. Therefore, the language PROGRES offers various means supporting this task, like edge cardinality assertions, node valued parameters for graph rewriting rules, and additional attribute index structures. Figure 1 shows an (abstract) example of a PROGRES production with a complex left-hand side containing most of the mentioned elements. Its left-hand side describes a graph pattern consisting of five nodes belonging to the classes C, C1, and C2 respectively, and three normal edges of types one and many. Additionally, there are three *path conditions* represented by double line arrows. The path condition Path1 demands that corresponding nodes are connected by a series of -many-> edges of arbitrary length (the * operator computes the transitive closure). Path2 is executed by following -one-> edges as long as possible (due to the { } operator).

While in our implementation edges are bidirectional, paths are computed in forward direction only. Since a path may have a very complex body, in general it is very difficult to compute its inversion. But, note that the declaration of Path2 is preceeded by the keyword static. In our implementation static paths are *materialized*. That means, every pair of nodes in a given graph that is related by such a path is internally connected by an edge of a special type. During graph modifications, an incrementally working two phase lazy algorithm keeps the materialization edges up to date, cf. [KiSchWe 92]. Although materialization needs some bookkeeping and memory overhead, static paths normally can be evaluated much faster than normal ones. Thus, they are preferred for pattern matching. Additionally, since they are materialized by edges, static paths are bidirectional. This can be very useful for pattern matching, as we will see.

Another element of the graph pattern of our example is the *restriction* "def -many->". This restriction is fullfilled if there exists an outgoing edge of type many from the node matched by '3. Like paths, restrictions may be described by complex expressions including iteration and conditional branch operations.

In the declaration part of our example an integer attribute attr is declared for all nodes of class C. In the graph rewriting rule, the *attribute condition* "f('1.attr) > 100" holds if the attribute attr of the node matched by '1 has a value greater than 10 (cf. the definition of f). Attribute conditions may be defined as arbitrary complex expressions including (recursive) function calls, iteration, etc. Nodes of class C2 possess an additional attribute indexattr. The keyword index in the declaration of this attribute indicates that our implementation of the graph data structures have to provide a special indexing mechanism for fast (associative) access to nodes of class C2 via the values of their attribute indexattr. So, on one hand the attribute condition "'2.indexattr = '4.attr" is a restriction on the allowed values of the corre-

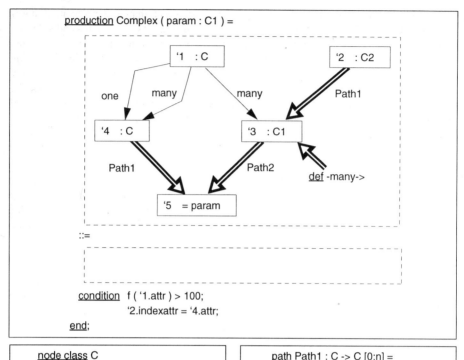

Figure 1: Complex pattern example

sponding attributes of the nodes '2 and '4, but on the other hand it can be used to compute easily a match for '2 via the attr value of '4.

Finally, the inscription " '5 = param" says that the node matched by '5 has to be equal to the parameter param of the production. In general, matching a node with the help of a parameter is the easiest possibility. This mechanism is often used to facilitate the application of several graph rewriting rules (one after the other) at the same 'place' within a graph.

In our implementation we have various basic matching and query operations wich can be used for matching the left-hand side of a given production. The most general operation queries for all nodes of a certain class. This can be used to start pattern matching. Edges can be matched by testing their existence or by traversing them in or against their direction. For paths, several basic query actions are necessary. If not materialized, they can be used in forward direction only. For attribute conditions, efficient access to attribute values is provided. All these basic operations need different computational efforts that may depend on the structure of the current graph. A good pattern matching strategy will try to minimize the overall efforts needed for computing a match for a given production. This is done by regarding the following principles, which are adopted from relational database query languages:

1. The most important principle is to avoid the GetInstance operation which iterates through the set of all nodes of a certain class, since in general, the size of this set is proportional to the size of the current work graph.

2. Several other query operations like traversing an edge or evaluating a path compute sets of nodes, too. We try to keep the size of these sets as small as possible, since we will have to iterate through their elements in order to find a node belonging to a full match. For example, the cardinality assertions [0:1] and [1:1] within the declarations of edge type one and path Path1 assure that traversing a one edge or evaluating Path1 will compute at most one node. Thus, these elements are preferred.

3. We prefer the usage of edges and paths that are declared to be partial using the cardinality assertions [0:1] and [0:n]. Trying to match these elements we may compute an empty set. An empty set as result of a query operation detects a wrong matching attempt. An early termination of a wrong matching attempt avoids the costs of the following query operations for this attempt. This technique is known as first-fail principle in constraint satisfaction systems (cf. [HaraElli 80, Henten 89]).

4. We try to apply restrictions and attribute conditions as early as possible, again in order to terminate wrong matching attempts early. According to the first-fail principle we try to match nodes with many restrictions early.

5. Naturally, we prefer simple query operations like testing or traversing a single edge (perhaps with a single result) over complex operations like evaluating a path which may need several query operations.

6. Finally, we try to use the special efficiency means of PROGRES, like <u>static</u> paths, <u>index</u> attributes, or node valued parameters.

In our example, a good pattern matching strategy would start with '5 using the parameter param. Matching one of the remaining nodes by querying for all instances of the corresponding class in general is very expensive and the Path1 connecting '4 and '5 has the wrong direction. Thus, in the next step we should traverse the materialized Path2 against its direction and thereby compute a set of possible matches for '3. The pattern matching process will proceed by iterating through this set until a full match for the left hand side is determined. Due to the first-fail principle, in the next step we should evaluate the restriction "<u>def</u> -many->". Now we may traverse a many edge against its direction in order to match '1. Iterating through the set of possible matches for '1, we first test the attribute condition "f('1.attr) > 100". In the next step we could use either a one or a many edge to determine '4. Due to its better cardinality assertion we will use the one edge to compute a first match and then we will verify the existence of the parallel many edge. At this point we will check whether Path1 leads from '4 to '5. After matching '4 and applying all its restrictions we can now access candidates for '2 via the <u>index</u> attribute as described by the attribute condition "'2.indexattr = '4.attr". In the last step, the existence of Path1 between '2 and '3 is tested.

3   Modelling all Possible Search Plans

As shown in our example, the language PROGRES offers various means supporting the graph pattern matching problem. Unfortunately, this even complicates the task of (pre-)compiling a graph pattern. Now we will propose a graph based pattern matching optimization algorithm that solves this problem.

The algorithm works in three phases. First, the elements of a graph pattern and their interdependencies are represented by an appropriate *pattern graph*. In the second phase, we construct a so called *operation graph* by adding nodes representing all possible basic query operations for every graph pattern element. For each of these query operation nodes, we add match edges denoting the pattern element(s) it computes and input edges denoting pattern elements which are used as input for this computation. In the third phase, we associate costs with the different query operations and try to choose a sequence of operations with minimal overall costs that computes matches for all graph pattern elements and regards the requirements of its query operations.

Within the PROGRES environment, a production is stored as an abstract syntax tree enriched by context sensitive edges comprising symbol table information (cf. [ELNSS 92]). For our purposes a simpler representation suffices. We reduce the structure of each left-hand side element, i. e. of each node, edge, path condition, restriction, attribute condition and parameter equation, to a single node with a text attribute containing the corresponding specification part, cf. Figure 2. An LSNode (**L**eft **S**ide) within the pattern graph models a normal node of the corresponding left-hand side. An LSEdge node represents a left side edge. SN (**S**ource **N**ode) and TN (**T**arget **N**ode) edges are used to relate a LSEdge to its source and destination LSNode. In the same way paths are modelled by LSPath nodes. Restriction, AttCond, and ParamExpr nodes denote their corresponding LSNode node(s) by TN edges, too.[2] All this information is provided by the analyzer of the PROGRES environment.

An element of a left-hand side can be matched to an element of the current host graph by executing different basic query operations. In the second phase of our algorithm, for every node of the pattern graph in turn we add nodes representing all possible query operations for the corresponding left side element. Such an operation node will be connected to the left side elements it matches by m-edges and by i-edges to the left side nodes it uses as input. We could build these query operation nodes by the means of PROGRES productions applied to the graph of Figure 2.[3] But, in order to avoid the confusions arising by such a bootstrap step we use a slightly modified notation (but with the same semantics).

For an LSNode only the general search operation GetInstance is available, cf. Figure 3, which just queries the current host graph for a node of the corresponding class. Note, that this is a nondeterministic search operation, since in general several nodes of a given class may exist. Within a complex search PROGRES uses backtracking mechanisms to handle such nondeterministic search operations, cf. [Zünd 92]. Applying the rule of Figure 3 to the pattern graph of Figure 2 generates the operation nodes 111, 116, 120, and 124 in Figure 5.

An LSEdge may be matched in three different ways, cf. Figure 4. We may use a GetETarget operation in order to traverse the corresponding edge in its direction, thereby computing a match for the LSEdge and its target LSNode. This operation needs the source LSNode as input. This input node must already be matched by another search operation. If the target of

---

2. An attribute equation with an idexed attribute in some way acts like a path that computes a target node. In this case the source LSNode is related to the AttCond by an SN edge, cf. nodes 10 and 30 in Figure 2.
3. We did so in [Zünd 95].

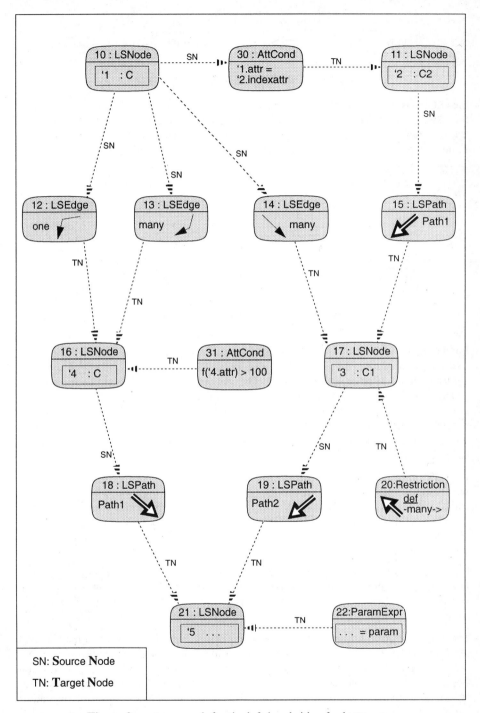

Figure 2: pattern graph for the left-hand side of rule Complex

Figure 3: Search operation for LSNode nodes

the LSEdge is already matched, we can use a GetESource operation to compute a match for the LSEdge and its source node. If both LSNode nodes are already matched a (cheap) TestEdge operation may be used to ensure that the corresponding left side edge exists. E. g. in Figure 5 the rule of Figure 3 generated the operations 100, 101, and 103 for the LSEdge 12.

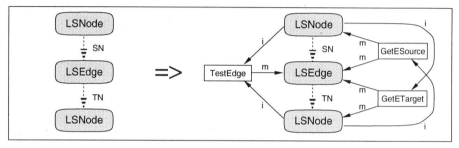

Figure 4: Search operations for LSEdge nodes

For a normal LSPath, like node 18 in Figure 5, only two search operations are generated. A GetPTarget operation, like node 122, is used to traverse the path in its direction using the corresponding source LSNode 16 as input. The cheaper TestPath operation 123 needs source and target node of LSPath 18 as input and thereby computes a match for it. Note, since Path1 is not declared to be static no GetPSource operation is generated for LSPath 18. Such a normal path may not be traversed against ist direction. But, for LSPath 19 such a GetPSource operation 126 is generated, since the corresponding path Path2 is static and thus it may be used like an edge.

For restrictions, like node 20, just TestRestriction operations, like 121, are generated that test wether the corresponding condition holds. In the same way for the attribute conditions 30 and 31 the TestAttCond operations 113 and 114 are generated. Note, that an attribute condition may use several LSNode nodes as input like 113. Since AttCond 30 is an equation using an index attribute we generate an additional GetByIndex operation modelling the index access computing LSNode 11 using the attribute value of LSNode 10.

Last but not least, for the ParamExpr node 22 we generate the GetByParam operation 125, that just assigns the value of param to the corresponding LSNode 21, and the TestParam operation 129, that can be used to verify a match of LSNode 21 computed by some other search operation.

The graph of Figure 5 now represents the set of all possible *search plans* that may be used to match the left hand side of the corresponding rule to a redex of a given host graph. We derive a concrete search plan from this operation graph as follows: In every construction step we first compute the set of all *available* search operations. A search operation is available, if all its input nodes are already matched. Then, the set of available search operations is restricted to the set of usefull operations that compute a match for at least one new left side element. A search operation that is available and usefull is called *enabled*. In every

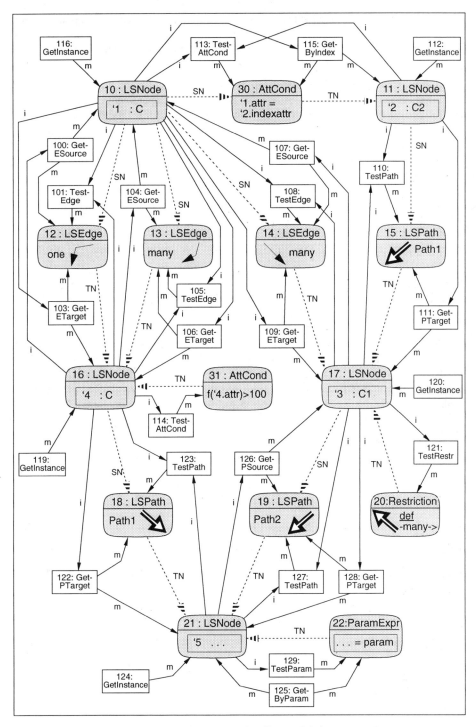

Figure 5 : Operation-Graph für die Produktion Complex

construction step just one enabled search operation is chosen. The chosen search operation computes new left side element(s), maybe enables new search operations and lets some other operations become useless. This construction step is iterated until all elements of a given graph pattern are regarded.

In Figure 5 for the first construction step all GetInstance operations and the GetByParam operation 125 may be chosen, since these operations need no input nodes (there is no incomming i-edge) and they all match 'new' left side elements (reached by the outgoing m-edges). Let's start our search (plan) with operation 116 (in the upper left corner). This computes a match for LSNode 10. This enables the search operations 103, 106, 109, and 115, which are reached from node 10 via i-edges. The also reached operations 101, 105, 108, and 113 are not enabled since they require additional, not yet matched input nodes. So, in the second construction step we may choose the GetETarget operation 103. Via the outgoing m-edges we can mark the left side elements 12 and 16 as matched. Following i-edges from node 16, we can determine further search operations, like 100, 104, 105, 114, 122, and 123. Traversing m-edges from 12 and 16 against their direction we compute now (potentially) useless operations, like 100, 101, 103, and 119. So, the set of available and usefull search operations can be maintained incrementally during search plan construction.

Note, it is always possible to construct a valid search plan for a given left hand side using the algorithm above. Proof: For any construction step we distinguish two cases. 1.) There is an unmatched LSNode. Then, a corresponding GetInstance operation is available (since it needs no input nodes) and this operation is usefull (since its LSNode is unmatched by assumption). 2.) All LSNode nodes are matched. If furthermore all other left side elements are matched, we have constructed a valid search plan and the algorithm terminates. If there exists an unmatched left side element lse by construction of our search operations there exists a operation o that matches lse. Thus, o is usefull and it is available since its input nodes are already matched (since all LSNode nodes are matched by assumption). The algorithm always terminates since in every construction step at least one 'new' left side element is matched and their number is finite.

Note, for a left hand side with l nodes there exist at least l! different search plans. At the beginning at least l GetInstance operations are enabled. In every construction step at most one LSNode is matched (by construction of our search operations), i. e. in step i there are still l-i GetInstance operations enabled.

4 A Cost Model for Search Plans

Our target is to find a good or even optimal search plan from the set of all possible search plans for a given left-hand side. Therefore, in the next step we will introduce a cost model for search plans, that allows us to compare different plans and to choose a good/an optimal one during compilation within the PROGRES environment.

Before we are able to develop a cost model for our search plans, we have to consider their nondeterministic execution in more detail. Let's have a look at the following fragment of a search plan that computes a match for node '1 and '4 and their connecting edges within the rule in Figure 1:

```
   '1 := GetInstance( SC_C )
 & '4 := GetETarget( '1, SC_many )
 & TestEdge( '1, SC_one, '4 )
 & . . .
```

The execution of this operation sequence starts by querying the current host graph for a first node of class C. Using this first match for '1 we traverse an outgoing many edge in order to compute a match for node '2. Now the TestEdge operation is used to verify the existence of an additional one edge between our matches for '1 and '4. This test (probably) may fail. In case of failure backtracking occurs, i. e. the last nondeterministic choice is revisited and another alternative is examined.[4] In our example the system first tries to traverse another many edge. If (perhaps after several trials) there exists no more many edge backtracking proceeds further and the GetInstance operation is revisited. If this (finally) fails, too, the whole pattern matching attempt fails. If after some trials of GetInstance and the corresponding edge traversals a combination of nodes is found, that is connected by an one edge, the search proceeds with the subsequent operations which again may contain choices or cause backtracking steps.

The estimated costs of executing a search plan SP are described by the function CostSum. The costs of executing a search plan are computed by summing up the base costs of executing its search operations multiplied by the estimated necessary number of executions.

$$\text{CostSum}(SP) = \text{CostSum}\Big(Op_1 \& \dots \& Op_l\Big) = \sum_{i=1}^{l} \text{NoOfTrials}_{SP}(i) \times \text{BaseCosts}(Op_i)$$

Figure 6 contains the values of function BaseCosts for our search operations. These values mainly depend on the data structures used to implement the host graph.[5] The description of the estimated number of executions of an operation during the pattern matching process by the function NoOfTrials needs some more efforts. First of all we make some assumptions about the class of graphs we are dealing with: We assume, that a typical graph of our application domain consists of about $G = 20000$ nodes distributed in 40 classes with an average of $n = 500$ node instances. Some additional classes have at most one instance. Traversing an edge of cardinality [0:n] may yield $e = 10$ results, evaluating a path of the same cardinality leads up to $p = 20$ nodes. An attribute equation may hold for $a = 5$ nodes of a given class. We will discuss the evidence of this assumptions at the end of this chapter.

Based on these assumptions we derive the values of the function *NoOfResults* for our search operations as given in Figure 6. A value NoOfResults >= 1 denotes a *get* operation that contains a nondeterministic choice. A *test* operation with NoOfResults < 1 represents a condition that holds only one of 1/NoOfResults times. The product of the NoOfResults of a sequence of search operations gives us the maximal number MaxNoOfResults of combination of nodes considered by this sequence:

$$\text{MaxNoOfResults}_{Op_1 \& \dots \& Op_l}(l) = \prod_{i=1}^{l} \text{NoOfResults}\Big(Op_i\Big)$$

Often only some of the maximal number of combinations have to be examined in order to find a match for a given left-hand side. Thus, we introduce the following function

$$\text{NecessaryNoOfResults}_{SP}(i) = \begin{cases} 1 & \text{if } i = l \\ \max\left(\dfrac{\text{NecessaryNoOfResults}_{SP}(i+1)}{\text{NoOfResults}(Op_{i+1})}, 1\right) & \text{otherwise} \end{cases}$$

---

4. A simple implementation strategy could use nested loops iterating through the results of set valued query operations in order to achieve this behaviour.
5. For operations that evaluate complex expressions like pathes additional assumptions concerning the average costs are made.

	NoOfResults				BaseCosts
Operation	[0:1]	[1:1]	[0:n]	[1:n]	
GetInstance	0,8	1	n = 500	501	1
GetETarget / -ESource	0,8	1	e = 10	11	2
TestEdge	0,0016	0,002	e/n = 0,02	0,022	2
GetPTarget / -Source (Materialized)	0,8	1	p = 20	21	20 / 4
TestPath (Materialized)	0,0016	0,002	p/n = 0,04	0,042	20 / 4
TestRestr	-	-	p/n = 0,04	-	20
GetByIndex	a = 5				4
TestAttCond (Equality test)	0,5				4
	a/n = 0,01				
GetByParam	0,8	1	p = 20	21	1
TestParam	0,0016	0,002	p/n = 0,04	0,042	1

Figure 6 : Key values of the search operations

The function NecessaryNoOfResults reflects that we need (an average of) 1/NoOfResults trials in order to pass a test operation and that a get operation needs only 1/NoOfResults input combinations in order to compute the requested number of result combinations. The maximum function ensures that at least one result is computed.

For some graph patterns and search plans only a small number of nondeterministic trials are necessary in order to compute a full match. These search processes are well described by function NecessaryNoOfResults. But, for some other graph patterns and search plans even the examination of all MaxNoOfResults possible combinations of nondeterministic choices may not suffice to find an overall match. Such search processes are described by MaxNoOfResults. Thus, the following function

$$\text{RealNoOfResults}_{SP}(i) = \min(\text{NecessaryNoOfResults}_{SP}(i), \text{MaxNoOfResults}_{SP}(i))$$

yields a first aproximation of the execution of a search plan SP by combining these two functions using their minimum. Unfortunately, for very restrictive patterns MaxNoOfResults may become quite small (<<1), which leads to a serious systematic error for RealNoOfResults. This error is adjusted by splitting up the function MaxNoOfResults into its parts >= 1 and into a new function SuccessProbability, that models the probability that a given sequence of search operations computes at least one result:

$$\text{MaxNoOfResults}_{SP}(i) = \begin{cases} \text{NoOfResults}(Op_i) & \text{if } i = 1 \\ \max\left(1, \text{MaxNoOfResults}_{SP}(i-1) \times \text{NoOfResults}(Op_i)\right) & \text{otherwise} \end{cases}$$

$$\text{SuccessProbability}_{SP}(i) = \begin{cases} 1 & \text{if } i = 0 \\ \left(\text{SuccessProbability}_{SP}(i-1) \times \min\left(1, \frac{\text{MaxNoOfResults}_{SP}(i-1) \times}{\text{NoOfResults}(Op_i)}\right) \right) & \text{otherwise} \end{cases}$$

This leads to the following modified version of RealNoOfResults:

$$\text{RealNoOfResults}_{SP}(i) = \min\left(\frac{\text{NecessaryNoOfResults}_{SP}(i),}{\text{MaxNoOfResults}_{SP}(i)} \right) \times \text{SuccessProbability}_{SP}(i)$$

For get operations this new version of RealNoOfResults already gives a good approximation of the wanted funtion NoOfTrials estimating the number of executions of an operation within a search plan. For test operations with NoOfResults < 1 we only have to regard, that 1/NoOfResults trials are necessary in order to yield one successful result. This leads to

$$\text{NoOfTrials}_{SP}(i) = \begin{cases} \dfrac{\text{RealNoOfResults}_{SP}(i)}{\text{NoOfResults}(Op_i)} & \text{if NoOfResults}(Op_i) < 1 \\ \text{RealNoOfResults}_{SP}(i) & \text{otherwise} \end{cases}$$

Figure 7 and Figure 8 contain two examples of search plans for the rule Complex of Figure 1 together with the values of the main functions of our cost model.[6] Figure 7 contains a so called *canonical search plan*, which first uses GetInstance operations to match the nodes of the graph pattern and then verifies all remaining left side elements using the corresponding test operations. This strategy always results in very poor search plans, since the GetInstance operations may easily generate a very large number of combinations (cf. column MaxNoOfResults) and in our example the following very restrictive set of test operations actually enforces the examination of all possible alternatives (cf. column NecessaryNoOfResults). The search plan of Figure 7 is the worst case with respect to our cost model, since it uses less restrictive test operations (with high NoOfResults) first. This violates the first-fail principle and results in late elimination of wrong combinations and in very frequent executions of the test operations. Just resorting the test operations using minimal NoOfResults first reduces the over all costs by 75 % to 125.05 E 12 units.

Figure 8 contains an *optimal search plan* for rule complex (with respect to our cost model). Note, that we use get operations with small NoOfResults instead of the GetInstance operations used in Figure 7. Just using all this small sized get operations first and the remaining test operations afterwards already results in search costs of 6014 units. Using the test operations as early as possible (as done in Figure 8) enhances this result by 90 % to 529 units. Within the optimal search plan only a small number of combinations is generated and wrong combinations are eliminated as early as possible.

Provided with our cost model for assessing search plans, we could realize the compiler for graph patterns by just enumerating the set of all possible search plans and then choosing one with minimal costs. But as stated above there exist at least l ! search plans for a rule with l nodes in its left hand side. Enumerating all search plans led to unacceptable compile times (especially for the interactive PROGRES environment). Fortunately, further analysis of our

---

6. Numbers given as x E n are to be read as x times 10 power n.

canonical search plan	NoOf-Results	MaxNoOf-Results	Success-Probability	Necessary-NoOfResults	NoOfTrials
'1 := GetInstance(SC_C)	500	500	1	19,531 E 6	500
& '2 := GetInstance(SC_C2)	500	250 E3	1	9,765 E 9	250 E 3
& '3 := GetInstance(SC_C1)	500	125 E6	1	4,882 E 12	125 E 6
& '4 := GetInstance(SC_C)	500	62,5 E9	1	2,441 E 15	62,5 E 9
& '5 := GetInstance(SC_C1)	500	31,25E12	1	1,22 E 18	31,25E12
& TestAttCond("func('4.attr)>100")	0,5	15,62E12	1	610,351 E 15	31,25E12
& TestRestr('3, "def -many->")	0,04	625 E9	1	24,414 E 15	15,62E12
& TestPath('2, "Path1", '3)	0,04	25 E9	1	976,562 E 12	625 E 9
& TestPath('4, "Path1", '5)	0,04	1 E9	1	39,062 E 12	25 E 9
& TestEdge('1, SC_many, '3)	0,02	20 E6	1	781,25 E 9	1 E 9
& TestEdge('1, SC_many, '4)	0,02	400 E3	1	15,625 E 9	20 E 6
& TestAttCond(" '1.attr='2.indexattr")	0,01	4 E3	1	156,25 E 6	400 E 3
& TestEdge('1, SC_one, '4)	0,002	8	1	312,5 E 3	4 E 3
& TestParam('5, "param")	0,002	1	0,016	625	8
& TestPath('3, "Path2", '5)	0,0016	1	25,6E-6	1	0,016

CostSum = 481,714 E12 units

Figure 7 : Costs of the canonical search plan for rule Complex

optimal search plan	NoOf-Results	MaxNoOf-Results	Success-Probability	Necessary-NoOfResults	NoOfTrials
'5 := GetByParam("param")	1	1	1	1,953 E 3	1
& '3 := GetPSource('5, "Path2")	20	20	1	39,063 E 3	20
& TestRestr('3, "def -many->")	0,04	1	0,8	1,562 E 3	20
& '1 := GetESource('3, SC_many)	10	10	0,8	15,625 E 3	8
& '4 := GetETarget('1, SC_one)	0,8	8	0,8	12,5 E 3	8
& TestEdge('1, SC_many, '4)	0,02	1	0,128	250	6,4
& TestPath('4, "Path1", '5)	0,04	1	5,12E-3	10	0,128
& TestAttCond("func('4.attr)>100")	0,5	1	2,56E-3	5	5,12 E-3
& '2:=GetByIndex('1.attr,SC_indexattr)	5	5	2,56E-3	25	12,8 E-3
& TestPath('2, "Path1", '3)	0,04	1	102,4 E-6	1	12,8 E-3

CostSum = 528,688 units

Figure 8 : Cost of an optimal search plan for rule Complex

cost model showed that a simple greedy algorithm already does the job. Thus, within every construction step of a search plan we just choose an operation with minimal NoOfResults out of the set of enabled search operations.[7]

---

7. Only if a GetInstance operation of cardinality [0:n] or [1:n] has to be chosen, our implementation inspects different alternatives in order to find a good entry point to the corresponding (sub) pattern.

According to our experiences within the PROGRES environment, this optimization algorithm works sufficiently in almost all cases (e. g. it generated the search plan of Figure 8). It avoids the global GetInstance operation whenever possible, it minimizes the number of nodes to be inspected, and it regards the first-fail principle just by using operations with small NoOfResults. Since simple operations like edge traversals normally have smaller NoOfResults than complex operations like path evaluations, the simple operations are preferred as demanded, cf. the optimization principles of section 2.

Now let's reconsider the assumptions made for our cost model like the number of nodes per class and the number of expected results for edge traversals and path evaluations. Further analysis of our algorithm shows, that a quite coarse classification of search operations in (1) global operations like GetInstance, in (2) multiple result get operations evaluating edges and pathes of cardinalities [0:n] or [1:n], in (3) single result get operations belonging to left side elements of cardinalities [0:1] or [1:1], and in test operations of (4) low and (5) high restrictiveness already suffices to yield good search plans by our algorithm. It is quite seldom that a large number of search operations belonging to the same optimal class is enabled at the same time and even if this happens choosing one or another normally makes no significant difference for the quality of the generated search plan. Thus, our algorithm works well for other classes of graphs as long as the classification above holds.

5   Conclusions

This paper proposes a solution to the graph pattern matching problem in executing (programmed) graph rewriting systems. We first introduced a simple graph representation for the graph pattern and the set of all possible search plans. With the help of a sophisticated cost model we derived a simple heuristic optimization algorithm that solves the problem very well, according to our experiences with its implementation within the PROGRES environment. Due to the availability of this algorithm graph rewriting systems not only can be used for the specification but also for the realization of applications.

The proposed algorithm is a significant improvement to other/former systems and approaches that either do not support automatic redex search at all [Beyer 91], use user provided search functions [Gött 88] or restrict themselves to specific classes of graph grammars [Dörr 94b][8].

Our results could be applied to other graph rewriting systems, too. The approach is suitable especially (1) if huge work graphs are considered and only small subgraphs have to be matched (for testing two large graphs for isomorphism, other algorithms are more appropriate, cf. [Gould 88]), (2) if the work graph is stored in a data structure that supports special indexing schemes for efficient access to different graph elements that should be used within the matching task, and (3) if at any time only a small number of graph rewriting rules must be considered (like in PROGRES where we use control structures to direct the choice of an applicable production).

If a large number of rules is considered in every graph rewriting step e.g. the algorithm of [BuGlTr 91] might have advantages. If several rules are considered for execution starting the search with common sub patterns of the rules (as done in [Dörr 94b]) can be an important improvement. We try to enhance our algorithm in this direction. Other ongoing research tries to provide the user with additional means to guide the search process, e. g. by

---

8. For the class of graph grammars introduced in [Dörr 94b] our algorithm yields optimization results comparable to that approach.

assigning priorities to certain pattern elements. This would enable the user to do some fine tuning using additional knowledge of the application that is not available to our optimizer.

So far we have implemented an interpreter and a compiler for PROGRES that generates code for the left-hand side of productions using the proposed algorithm. Thus, we are able to compile and execute PROGRES specifications for rapid prototyping purposes and to generate efficient Modula-2 or C code for PROGRES specifications. The compiler / interpreter is part of the PROGRES environment that also comprises a structure oriented textual / graphical editor, an incremental analyzer, and a graphical browser for the host graph.[9]

References

[Beyer 91]   Beyer M.: GAG: Ein graphischer Editor für algebraische Graphgrammatiksysteme; Masters Thesis, TU Berlin (1991)

[BuGlTr 91]   Bunke H., Glauser T., Tran T.-H.: An efficient implementation of graph grammars based on the RETE matching algorithm; in [EhKrRo 91], 174-189 (1991)

[Dörr 94b]   Dörr H.: Bypass Strong V-Structures and Find an Isomorphic Labelled Subgraph in Linear Time; technical report, Report B-94-08, Freie Universität Berlin (1994)

[EhKrRo 91]   Ehrig H., Kreowski H.-J., Rozenberg G. (Eds.): Proc. 4th Int. Workshop on Graph-Grammars and their Application to Computer Science; LNCS 532, Berlin, Springer-Verlag (1991)

[ELNSS 92]   Engels G., Lewerentz C., Nagl M., Schäfer W., Schürr A.: Experiences in Building Integrated Tools, Part I: Tool Specification; in ACM Transactions on Software Engineering and Methodology, Vol.1, No.2, 135-167, ACM Press (1992)

[Gött 88]   Göttler H.: Graph-Grammatiken in der Softwaretechnik; IFB 178, Springer-Verlag (1988)

[Gould 88]   Gould R.: GRAPH THEORY; The Benjamin Cummings Publishing Company, Menlo Park, California (1988)

[HaraElli 80]   Haralick R. M., Elliot G. L.: Increasing Tree Search Efficiency for Constraint Satisfaction Problems; in Artificial Intelligence, Vol.14, 263-313 (1980)

[Henten 89]   Van Hentenryck P.: Constraint Satisfaction in Logic Programming; MIT Press, London (1989)

[LeReJa 93]   Lee H.-Y., Reid Th. F., Jarzabek St. (Eds): CASE '93 Proc. 6th Int. Workshop on Computer-Aided Software Engineering; IEEE Computer Society Press (1993)

[KiSchWe 92]   Kiesel N., Schürr A., Westfechtel B.: Design and Evaluation of GRAS, a Graph-Oriented Database System for Engineering Applications; in [LeReJa 93], 272-286 (1992)

[SchmiBer 92]   Schmidt G., Berghammer R. (Eds.): Graph-Theoretic Concepts in Computer Science; WG '91, LNCS 570, Berlin, Springer-Verlag (1992)

[Schürr 91]   Schürr A.: Operationales Spezifizieren mit programmierten Graphersetzungssystemen; PhD. Thesis, RWTH Aachen, Wiesbaden, Deutscher Universitäts-Verlag (1991)

[Schürr 91b]   Schürr A.: PROGRES: A VHL-Language Based on Graph Grammars; in [EhKrRo 91], 641-659 (1991)

[Zünd 92]   Zündorf A.: Implementation of the imperative / rule based language PROGRES; technical report, AIB 92-38, RWTH Aachen (1992)

[Zünd 93]   Zündorf A.: A Heuristic for the Subgraph Isomorphism Problem in Executing PROGRES; technical report, AIB 93-5, RWTH Aachen (1993)

[ZünSchü 92]   Zündorf A., Schurr A.: Nondeterministic Control Structures for Graph Rewriting Systems; in [SchmiBer 92], 48-62 (1992)

[Zünd 95]   Zündorf A.: PROgrammierte GRaphErsetzungsSysteme — Spezifikation, Implementierung und Anwendung einer integrierten Entwicklungsumgebung; PhD Thesis, RWTH Aachen (1994)

---

9. The PROGRES environment is available as free software within the world wide web at "http://www-i3.informatik.rwth-aachen.de/research/progres.html".

A Technique for Recognizing Graphs of Bounded Treewidth with Application to Subclasses of Partial 2-Paths

Stefan Arnborg*[1] and Andrzej Proskurowski**[2]

[1] Kungliga Tekniska Högskolan,
NADA, S-1100 44 Stockholm, Sweden
[2] University of Oregon,
Computer Science Department, Eugene, Oregon 97403, USA

Abstract. Regarding members of a class of graphs as values of algebraic expressions allows definition of a congruence such that the given class is the union of some of the equivalence classes. In many cases this congruence has a finite number of equivalence classes. Such a congruence can be used to generate a reduction system that decides the class membership in linear time. However, a congruence for a given problem is often difficult to determine.

We describe a technique that produces an algebra and a congruence relation on its carrier for some classes of graphs. Our technique builds on considering possible representations of the generated graphs as graphs of the desired class. By introduction of a labeling describing the "most parsimonious" such representations, we can work with small labeled graphs instead of large unlabeled ones, and some of the case analysis can be delegated to the algebraic machinery used. The congruence relation is subsequently used to construct a labeled graph reduction system based on a reduction system recognizing a larger class than the one sought.

---

* Supported by the Swedish Research Council for Engineering Sciences.
** Research supported in part by National Science Foundation grants NSF-CCR-9213439 and NSF-INT-9214108.

1 Introduction

Graph reduction is an attractive way to recognize classes of graphs with certain properties and even to find optimal values of certain graph parameters (see [1] and [6]). A terminating graph reduction system consists of a finite set of *rewriting rules* and a finite set of *accepting graphs*. Given any graph G, every sequence of applications of these rewriting rules terminates with an irreducible graph called a *normal form* of G. The system defines a class of graphs such that every normal form of every graph in the class is an accepting graph but no normal form of a graph outside the class is accepting.

A rewriting rule has a left-hand side graph and a right-hand side graph, both with the same set of distinct *source* vertices, and disjoint sets of other (*internal*) vertices. An application of such a rule to graph G means finding a subgraph of G isomorphic to the left-hand side graph of the rule, so that vertices of G that correspond to the internal vertices have only the adjacencies given in the rule, and replacing the subgraph by the right-hand side graph of the rule.

The reason for our interest in graph reduction is two-fold. On the one hand, a large class of terminating graph reduction systems can be implemented as algorithms executing in time proportional to the size of the input graph. On the other hand, there exists an algebraic theory of graph reduction for graphs with bounded treewidth that leads to a construction paradigm for terminating reduction systems (see [1]).

The construction paradigm is intractable in general and this paper deals with a technique that can make it tractable for some classes of graphs with bounded treewidth. The structure of graphs in such a class can be exploited for the purpose of trimming down the sizes of reduction systems and algebras recognizing the class. In this paper we discuss the following "bootstrapping" procedure for computing congruence classes for families of graphs with bounded treewidth. We assume that there is an attribute $\mathcal{A}(x)$ that can be associated with every element x of a graph. This attribute is inherent for the class of graphs we want to recognize and commits the element to play a certain role in a graph of this class. (For instance, an articulation vertex, or an edge between vertices of a minimal separator.) We call the set of such attributes used for a given graph class an *attribution*. We assume further that the structure of the class of graphs on hand allows only a small number of congruence classes among such attributed graphs and that it is easy to determine congruence of two attributed graphs. If the attribution is suitably chosen, it will be easier to prove membership in L using an attributed graph rather than its underlying graph. By using a partial order on congruence classes of attributed graphs the method becomes feasible for some fairly difficult classes L, such as the class of graphs of pathwidth at most two. From the congruence classes and their size-minimal elements one can further construct graph reduction rules and the obstruction set.

We illustrate our technique with its application to graphs with treewidth bounded by 2. We give the congruence classes for acyclic partial 2-paths (also called trees of proper pathwidth 2), relative to a graph algebra of 1-sourced graphs producing the set of connected partial 2-trees. The classes are obtained from an attributed algebra, where attributions are derived from possible embeddings of a given graph in a 2-path.

Our paper is organized as follows. After basic definitions of graphs and algebras in Section 2, we introduce the concept of resource usage, resource use labels and the algebras of labeled graphs for 1-sourced graphs in Section 3. In Section 4, we present algebras on labeled graphs leading to recognition of acyclic partial 2-paths. (Because of the increasing computational intensity – the number of congruence classes – we defer to the full version of the paper applications of our methodology to series-parallel partial 2-paths, and all partial 2-paths.)

2 Basic definitions

In our exposition, we will need concepts dealing with graphs and algebra, especially in combination (graph algebras). To contain the complexity of the relevant algebras we use certain partial orders. In this section, we introduce definitions for these subjects.

2.1 Graphs and partial orders

We consider finite graphs given by sets V and E of vertices and edges, respectively, and the incidence relation $I \subset V \times E$ that is constrained to make every edge incident to exactly two vertices, its *end-vertices*. Two vertices are said to be *adjacent* in G if they are incident to a common edge. A *subgraph* H of a graph G, $H \subseteq G$, is a graph defined by V_H, E_H, I_H that are subsets of the corresponding sets defining G. A subgraph is *induced* by a set $S \subseteq V$ if its vertex set is S and its edge set is the set of edges incident only with vertices in S. A set of m vertices is called a *clique* in G if it induces a complete subgraph of G (the subgraph itself is often called an m-clique and is denoted K_m).

The number of incident edges of a vertex is called the *degree* of the vertex. A sequence of pairwise adjacent degree 2 vertices, starting and ending with degree 1 (*pendant*) vertices, is called a *path* between its extreme vertices. A graph is *connected* if there is a path between any two of its vertices. A set of vertices, $S \subset V$, is a *separator* in a graph G if the removal of S and the edges incident to its vertices disconnects G.

A connected graph with no K_{k+2} subgraph and such that every minimal separator (*wrt.* set inclusion) induces a K_k subgraph is called a *k-tree* [14]. An alternative and more intuitive definition of this class of graphs is given by the construction process: the K_{k+1} clique is a k-tree, and any k-tree with $n > k+1$ vertices can be constructed from a k-tree with $n-1$ vertices by adding a vertex adjacent to a K_k subgraph of that k-tree. Degree k vertices of a k-tree are called *k-leaves*. Subgraphs of k-trees are called *partial k-trees*. A *k-path* is a k-tree with exactly two k-leaves. The class of partial k-trees is exactly the class of graphs of treewidth at most k [19]. Partial graphs of k-paths are said to have *proper pathwidth* at most k.

A *partial order* $\prec$ on domain D is an antisymmetric and transitive binary relation on D. A *total order* on D satisfies the additional condition that, for all $d \neq d'$, either $d \prec d'$ or $d' \prec d$. A *chain* of partial order $\prec$ is a subset C of its domain D such that $\prec$ restricted to C is a total order. An *antichain* of $\prec$ is a subset A of D such that for no two elements a, a' of A we have $a \prec a'$. The $\prec$-minimal subset of $S \subset D$ is the set of elements $s \in S$ such that for no $s' \in S$ is $s' \prec s$, and is denoted $S \downarrow_\prec$. The following is known as Dilworth's theorem [9]:

Proposition 1. *The minimum number of chains of a partial order required to cover its domain is equal to the largest size of its antichains.*

The outer product of two partial orders $\prec_1$ and $\prec_2$ is the smallest partial order $\prec$ on the Cartesian product of their domains satisfying the following condition: $(d_1, d_2) \prec (d'_1, d'_2)$ if either $d_1 \prec_1 d'_1$ and not $d'_2 \prec_2 d_2$, or $d_2 \prec_2 d'_2$ and not $d'_1 \prec_1 d_1$. The definition is readily extended to outer products of several partial orders.

2.2 Algebras

In this paper we consider unsorted algebras. The use of sorted algebras, as in e.g., [1], is closer to the computer implementation of our method, but it entails cumbersome notation. There are no deep technical differences between sorted and unsorted algebras in this application. We consider thus a *signature* to be a set F of operations, where an operation $f \in F$ has *arity* n_f, which is a non-negative integer. An operation of arity 0 is called *nullary*. For a signature F, an F-*algebra* is a tuple $M = (D, (f_M)_{f \in F})$, where D is a set, the *domain* or *carrier* of M, usually denoted $|M|$, and for each operation f of F of arity n there is a function $f_M : D^n \to D$. A *finite algebra* is an algebra with a finite carrier. A *term* over F is an expression in the symbols of F respecting their arity, thus if $t_1, \ldots, t_i$ are terms and $f \in F$ is of arity i, then $f(t_1, \ldots, t_i)$ is a term. The smallest terms correspond to the nullary operations of F, thus if e is nullary in F, then $e()$ is a term. We denote by $T(F)$ the initial F-algebra (term algebra) of terms over F. Every F-algebra M generated by F is a homomorphic image of $T(F)$ and we denote the unique associated homomorphism $h_M : |T(F)| \to |M|$. We only consider algebras such that $h_M(|T(F)|) = |M|$, i.e., $|M|$ is generated by F. For a term t, $h_M(t)$ is thus the element of the carrier to which t evaluates in M. A *congruence* of M is an equivalence relation $\approx$ on $|M|$ such that whenever $d_i \approx d'_i$ for $i = 1, \ldots, n$ and $f \in F$ has arity n, then $f_M(d_1, \ldots, d_n) \approx f_M(d'_1, \ldots, d'_n)$. The equivalence classes of a congruence relation are also called *congruence classes*, and the congruence class containing element x is denoted $[x]_\approx$ or, when the congruence relation is understood, $[x]$. A congruence is of *finite index* if it has finitely many equivalence classes. A set $L \subset |M|$ is M-*recognizable* if it is the union of congruence classes of a congruence of finite index. Given an F-algebra M and a congruence $\approx$, the *quotient algebra* $M/\approx$ is the F-algebra whose elements are the congruence classes of $\approx$ and which satisfies $f_{M/\approx}([d_1], \ldots, [d_n]) = [f_M(d_1, \ldots, d_n)]$, for all $f \in F$ and all $d_i \in |M|$.

For a term t over $F \cup \{x\}$, where x is a nullary operation called a variable and occurs exactly once in t, $f[\ ] = \lambda x.t$ defines a function $D \to D$ that is called a *context*. (Actually, the context formally defines a mapping $|T(F)| \to |T(F)|$, but this is interpreted via the homomorphism h_M as a mapping from D to D). As a special case, the context $\lambda x.x$ defines the identity mapping. We will assume the existence of a *size* function $s : |M| \to \mathbf{N}^+$ such that (i) there are only finitely many elements of a given size, and (ii) if $s(x) < s(y)$, then, for every context $f[\ ]$ we have $s(f[x]) < s(f[y])$.

Let $L \subset |M|$ for an F-algebra M. We denote by $\sim_L$ the congruence on M defined by: $m \sim_L m'$ if and only if, for every context $f[\ ]$, $f[m] \in L$ if and only if $f[m'] \in L$.

Let $\prec_L$ be defined as follows: $m \prec_L n$ (read m *dominates* n *wrt.* L) if, for every context $g[\ ]$, $g[n] \in L \Rightarrow g[m] \in L$ and there is a context $g'[\ ]$ such that $g'[m] \in L$ but $g'[n] \notin L$. If neither $n \prec_L m$, $m \prec_L n$, nor $m \sim_L n$, then n and m are said to be independent *wrt.* L. The domination relation $\prec_L$ defines a partial order which is monotonic in the following sense.

Proposition 2. *For any context $f[\ ]$ in an algebra and any two objects $m \prec_L n$, if not $f[m] \sim_L f[n]$ then $f[m] \prec_L f[n]$.*

We can extend in the natural way the concept of domination to domination on congruence classes because of the following fact.

Proposition 3. *Given two objects, $x \prec_L y$, for every pair of objects $u \sim_L x$ and $v \sim_L y$ we have $u \prec_L v$.*

For two graphs such that $x \prec_L y$, we can thus say that $[x] \prec_L [y]$. We extend further the concept of context to sets of congruence classes by defining, for a set S of congruence classes, $f[S] = \cup_{c \in S} f[c]$. We will say that two sets S_1 and S_2 of congruence classes are *equivalent* if for any context $f[\ ]$, $f[S_1] \cap L = \emptyset$ if and only if $f[S_2] \cap L = \emptyset$.

Proposition 4. *If a set S of congruence classes contains two classes $[x]$ and $[y]$ such that $[x] \prec [y]$ then S is equivalent to the set $S - \{[y]\}$.*

Proof. Any context $f[\ ]$ that introduces an element of L as $f[u]$ for a $u \in [y]$, adds also an element of L as $f[v]$, $v \in [x]$, since v dominates u. Thus, excluding $[y]$ from S does not influence the membership in L of elements of $f[S]$.

2.3 Graph Algebras

We will consider graphs as values of algebraic expressions involving i-sourced graphs. An i-*sourced graph* g_i is a pair consisting of an unoriented graph called the *underlying graph* of g_i and a sequence of i of its distinct vertices called the *sources*. While we are concerned with recognizing classes of 0-sourced graphs, in the derivation of the appropriate reduction systems we will regard those as the result of removing the source designation from the corresponding vertices of i-sourced graphs (the operation r_i^* below).

In general, we will consider the following operations on i-sourced graphs:

P_i: the parallel composition of two i-sourced graphs, where the resulting i-sourced graph is obtained by identifying (fusing) the corresponding sources of the two argument graphs.

r_i: removes the last element from the sequence of sources of an i-sourced graph (but keeps the corresponding vertex).

l_i^j: the lifting of an $(i-1)$-sourced graph to an i-sourced graph by insertion of a new isolated vertex and inserting it at position j among the sources $(1 \leq j \leq i)$.

S_i: the series composition of i i-sourced graphs into a new i-sourced graph. For $i = 1$, $S_1(g_1)$ contains a new source vertex which is connected by an edge to the source of the argument which is not the source of the result. For $i = 2$, $S_2(g_1, g_2)$ is the common series composition of two two-terminal graphs, i.e. the second source of g_1 is fused with the first source of g_2, and the resulting graph has source 1 of g_1 as its first source and source two of g_2 as its second. The exact generalization to arbitrary i can be found in [1].

e: is a nullary operation that evaluates to an edge with its two end-vertices as sources.

v: is a nullary operation that evaluates to a one-source one-vertex graph.

r_i^*: is the operation that removes source designation from all vertices of an i-sourced graph (the result is thus an 0-sourced graph).

The union of the above operations for $i = 1, \ldots, k$, for some fixed k, is a signature we call F_k. When applied to i-sourced graphs, the operations defined above define an F_k-algebra that we will call M_k. However, our constructions will require multiple subscripts, and we prefer to use F and M instead of F_k and M_k henceforth. Since we deal with unsorted algebras and the number of sources of a graph acts as a sort in the definitions of the graph operations, we will apply the convention that if an operation is applied to a graph with the wrong number of sources, then the result is an *ill-formed* graph which we add to the carrier, one for each size and number of sources. It is shown in [1] that for any fixed k, the algebra M generates exactly the class of partial k-trees, the i-sourced graphs for $0 < i \leq k$ that have a tree-decomposition of width k with all sources in one of its bags X_n, and the set of ill-formed graphs.

We illustrate the above operations by their results for the 1- and 2-sourced graphs. In the following we will only consider connected graphs of treewidth ≤ 2, which can be obtained using a signature derived from F for $k = 2$. Operation P_1 results in fusing roots of its two arguments, S_2 and P_2 correspond to the series and parallel operations on *two-terminal series parallel graphs* (see, for instance [15]), and r_2 is related to the branch operation for such graphs (see [12]). Instead of using the l_j^i operations, we introduce a derived series composition operation S_2', which takes two two-sourced graphs g_1 and g_2, and one one-sourced graph g_3. This operation fuses the second source of g_1 with the first source of g_2 and the source of $g1$. The sources of the result are the first source of g_1 and the second source of g_2.

3 Basis for the method

Our methodology follows the result of [7] stating that for every class L of graphs with bounded treewidth that can be defined in Monadic Second Order Logic, MSOL, the congruence $\sim_L$ has finite index (finite number of congruence classes). The paradigm for constructing this set of congruence classes proposed in [1] consists of discovering new classes by applying the operations of the algebra to the representatives of the already constructed classes, starting with classes represented by the constant graphs (nullary operations). It follows from the finiteness of the set of congruence classes that representatives of all the classes will be generated as a result of a finite number of these operation applications.

In the above process, the computation of $x = f_M(m_1, \ldots, m_n)$ for some operation f and list of argument graphs $m_1, \ldots, m_n$ has two possible outcomes. Either the graph x is not congruent to any of the graphs already designated as the representative of a class, or there is a class represented by another graph $y \sim_L x$. In the former case, the set of classes is extended by $[x]$ with x as a representative, in the latter, there is a possibility of a new reduction rule, $x \to y$, if x has larger size than y.

The critical step in the construction process is the recognition of the congruence, the decision whether $x \sim_L y$ or not. In general, this decision has to be made based on the knowledge of the structure of the graph class (represented by, for instance, the logic formula defining the class). To aid this decision making we use the concept of attribution.

3.1 Derived graph algebras

From the signature F (where the treewidth k is understood to be a fixed integer), we derive several related graph algebras. Let $F^{\mathcal{A}}$ be the following derived signature to be applied to attributed graphs. Instead of the nullary edge operation e in F, $F^{\mathcal{A}}$ has one operation $\mathbf{e}^{abc}$ for every attribute a of an edge and pair (b, c) of attributes for vertices. Likewise, instead of the operation l_i^j of F, $F^{\mathcal{A}}$ has one operation $l_i^{j\,a}$ for every vertex attribute a. Similarly, we introduce the nullary operations $\mathbf{v}^a$ instead of $\mathbf{v}$, and unary operations S_1^a instead of S_1, for every vertex attribute a.

We define operations of the corresponding algebra $M^{\mathcal{A}}$ of attributed graphs as follows. Let, for some $f \in F$, $f_M(m_1, \ldots, m_n) = m_0$ and let $m_0', \ldots, m_n'$ be attributed graphs, such that m_i' is an attribution of the non-attributed graph m_i. In addition, we require that whenever f_M fuses an element (a vertex or an edge) x_i of m_i with an element x_j of m_j, then the attribution of x_i in m_i' is the same as that of x_j in m_j', and so is the attribution of the corresponding element in m_0'. Then $f_{M^{\mathcal{A}}}(m_1', \ldots, m_n') = m_0'$. These rules define a partial algebra, since the result is only defined when the operands "fit together". Just as with the case of sorted/unsorted algebra, we prefer total algebras for notational reasons although a computer implementation might be closer to the partial algebra. We assume that $M^{\mathcal{A}}$ is total by extending the carrier with ill-formed graphs, that are the results of operations with bad arguments. Again, this is just a technicality in this application.

Our construction of the powerset algebra follows that of [5]. The *powerset algebra* of $M^{\mathcal{A}}$ is $\mathcal{P}(M^{\mathcal{A}})$. The carrier of $\mathcal{P}(M^{\mathcal{A}})$ consists of sets of attributed graphs. For an operation $f \in F^{\mathcal{A}}$ we have $f_{\mathcal{P}(M^{\mathcal{A}})}(S_1, \ldots, S_n) = \{f_{M^{\mathcal{A}}}(s_1, \ldots, s_n) | s_1 \in S_1, \ldots, s_n \in S_n\}$. For the operations in $F - F^{\mathcal{A}}$ we introduce the meaning in $\mathcal{P}(M^{\mathcal{A}})$ by defining derived operations. If $f \in F - F^{\mathcal{A}}$ corresponds to the set $\{f^x\}_{x \in \mathcal{A}}$ in $F^{\mathcal{A}}$, then we define $f_{\mathcal{P}(M^{\mathcal{A}})}(G_1, \ldots, G_{n_f}) = \bigcup_{x \in \mathcal{A}} f^x_{M^{\mathcal{A}}}(G_1, \ldots, G_{n_f})$. (For instance, $l^j_{i\,\mathcal{P}(M^{\mathcal{A}})}(G) = \bigcup_{g \in G} l^{j\,x}_{i\,M^{\mathcal{A}}}(g)$ is the set of graphs obtainable from a graph $g \in G$ by adding a source vertex with every possible attribution.) In this way $\mathcal{P}(M^{\mathcal{A}})$ is extended to an $(F \cup F^{\mathcal{A}} \cup \{\oplus\})$-algebra, where the operation $\oplus$ will be interpreted as set union.

Proposition 5. *M is isomorphic to the subalgebra of $\mathcal{P}(M^{\mathcal{A}})$ generated by F.*

Proof. Each element of $|M|$ is an unattributed graph g. The corresponding element in $|\mathcal{P}(M^{\mathcal{A}})|$ is the set of all attributed graphs with the same underlying graph g.

Proposition 6. *[10] Let $\sim$ be a congruence on $M^{\mathcal{A}}$. Let us extend it to a congruence on $\mathcal{P}(M^{\mathcal{A}})$ by defining $S \sim S'$ if and only if for every $s \in S$ there is an element $s' \in S'$ such that $s \sim s'$, and similarly for every element $s' \in S'$. Then the quotient algebra $\mathcal{P}(M^{\mathcal{A}}/\sim)$ is isomorphic to $\mathcal{P}(M^{\mathcal{A}})/\sim$.*

If the attribution of graphs has been suitably choosen wrt. the set L of graphs, then it is possible to construct an algebra Q over signature $F \cup F^{\mathcal{A}}$ which can be used to decide membership in L for values of graph expressions generated by F.

To define the algebra, we specify a finite set C of *labels*, composite attributions that characterize possible proofs of graphs' membership in L or in graph classes related to L. These labels specify the roles that graph elements play in the unattributed graphs. Since some attributions may preclude other, these labels can be thought of as "resource usage".

We are interested in a larger set of labels than those corresponding to members of L because, in general, for some graphs not in L there might be contexts in which they can result in members of L. We define a subset A of C corresponding to $L^{\mathcal{A}}$, the set of attributed graphs admitting a proof of the underlying graphs' membership in L. As the possible attribution is not unique, we assume that for every graph in L, membership in $L^{\mathcal{A}}$ can be proved for at least one of the corresponding attributed graphs. Thus L is the set of underlying graphs of graphs in $L^{\mathcal{A}}$. The membership proof for a graph g proceeds inductively over $F^{\mathcal{A}}$-terms in an expression in $M^{\mathcal{A}}$ that evaluates to a graph g' obtained by an attribution of g.

Definition 7. Let M, $M^{\mathcal{A}}$, L and $L^{\mathcal{A}}$ be as above. An $(F \cup F^{\mathcal{A}} \cup \{\oplus\})$-algebra Q is a *committed-choice algebra* for M, $M^{\mathcal{A}}$, L and $L^{\mathcal{A}}$ if

(i) Q is isomorphic to a quotient algebra of $\mathcal{P}(M^{\mathcal{A}})$; and

(ii) There are sets C and A such that $|Q| \subset 2^C$ and $A \subset C$ and for every term t' over $F^{\mathcal{A}}$, $h_{M^{\mathcal{A}}}(t') \in L^{\mathcal{A}}$ if and only if $h_Q(t') \cap A \neq \emptyset$. The set C will be called the *label set* of Q.

In other words, if a graph g is the value of t in M, then we can prove membership of g in L by finding an attributed graph g' which is the value of t' in $M^{\mathcal{A}}$, and checking that t' evaluates in Q to a set non-disjoint with A. By the construction of Q, this is the case if and only if $h_Q(t) \cap A \neq \emptyset$, so that we do not have to actually find t'. The committed-choice algebra is typically designed as follows: Recall that $L^{\mathcal{A}}$ is a class of attributed graphs such that the attribution provides proof of membership in L for their underlying graphs. Let a subset A of C correspond to a proof of membership of a graph in $L^{\mathcal{A}}$. For every $f \in F^{\mathcal{A}}$ of arity n, define a function $\delta_f : C^n \to 2^C$. The algebra Q is now defined by its carrier 2^C and operations: $f_Q(C_1, \ldots, C_n) = \bigcup_{c_i \in C_i} \delta_f(c_1, \ldots, c_n)$.

This construction method, which is somewhat analogous to the treatment of the existential quantifier when deciding *MSOL* (see [2, 17]), is however not practical in general since the size of $|Q|$ is $2^{|C|}$. We can limit the magnitude of $|Q|$ by use of a domination relation. Recall that $g_1 \prec_A g_2$ if and only if there is a context $f[\ ]$ such that $f[g_1] \in A$ and $f[g_2] \notin A$, and for every context $f[\ ], f[g_2] \in A \Rightarrow f[g_1] \in A$.

Proposition 8. *Let Q be a committed-choice algebra for M, $M^{\mathcal{A}}$, L and $L^{\mathcal{A}}$, with C and A as in Definition 7. Define the equivalence relation $\sim_A$ on 2^C: If $D_1, D_2 \subset C$, then $D_1 \sim_A D_2$ if they have the same set of $\prec_A$-minimal elements. Then $Q/\sim_A$ is also a committed-choice algebra for M, $M^{\mathcal{A}}$, L and $L^{\mathcal{A}}$.*

Proof. Let $\widehat{M}$ be the algebra isomorphic to $Q/\sim_A$ but whose carrier is the set of $\prec_A$-minimal subsets of C. Consider the conditions in Definition 7: (i) is obviously satisfied: $\widehat{M}$ is a quotient algebra of Q which is a quotient algebra of $\mathcal{P}(M^{\mathcal{A}})$, thus $\widehat{M}$ is a quotient algebra of $\mathcal{P}(M^{\mathcal{A}})$; (ii) follows from the observations that $h_{\widehat{M}}(t')$ is the set of $\prec_A$-minimal subsets $D \downarrow_{\prec_A}$ of $D = h_Q(t')$ and that if $a \in A \cap D$, then a is also in the $\prec_A$-minimal subset of D unless there is some other element $a' \in D$ such that $a' \prec_A a$ and therefore $a' \in A$. In either case, $h_{\widehat{M}}(t') \cap A$ is non-empty if and only if $h_Q \cap A$ is non-empty.

This theorem bounds from above the size of a smallest committed-choice algebra. Moreover, we can generate $\widehat{M}$ from δ_f and $\prec_A$ without creating the large algebra Q as above. The complexity of computing it is a low-order polynomial in the size of the resulting algebra. But there is o general way to ascertain that this size is small - the size depends on luck but is small in our examples. In the particular cases where the set C is partially ordered by the relation $\prec_A$ that is a partial order and a product of total orders, the number of congruence classes becomes manageable, as follows from Proposition 1.

The committed-choice algebra can be used to decide membership of graph G in L as follows: Construct a term t over F such that $h_M(t) = G$. Then evaluate

$c_G = h_Q(t)$. G is in L if and only if $c_G \cap A$ is non-empty. By use of a graph reduction method, we can construct t and evaluate $h_Q(t)$ in parallel without actually retaining t, as we show in the next section.

3.2 Reduction of labeled graphs

For $k = 1, 2, 3$ there are reduction systems that recognize graphs of treewidth $\leq k$ and such that an i-sourced graph rewritten (a subgraph connected by i vertices to the rest of the graph) is always rewritten to a graph having a hyperedge containing the i sources. Such rewrite systems do not exist for larger k, but such systems recognizing other graph classes of large treewidth do exist. When the graph reduction system derived from the graph algebra M has this property, the removed part of the graph can always be regarded as a number of i-sourced graphs, each attached to the remaining part of the graph on a hyperedge of size i, for $i = 0, 1, \ldots, k$. Our reduction system for recognizing the graph set L using a committed-choice algebra Q for L will store the value $h_Q(t)$ as a label for the hyperedge into which $h_M(t)$ is reduced. We are here concerned with graphs of treewidth ≤ 2, so our system will introduce labels on vertices ($i = 1$) and edges ($i = 2$). For example, when a pendant edge between a leaf labelled a, an unlabelled edge and a source labelled b is reduced, the source will get the label $P_{1Q}(b, S_{1Q}(a))$ in the reduced graph. This rule is adequate for recognizing a class of acyclic graphs. When the class is series-parallel, we need a parallell edge reduction, which gives the edge created by parallell composition of edges labelled a and b the label $P_{2Q}(a, b)$, and a series edge reduction gives the label $S_{2Q}(a, b)$. For the class of graphs of treewidth ≤ 2 we need the union of these rules and also the modified series rule S_2. series reduction rule where the removed vertex (joining the operands) is labelled c. the remaining edge will then have label $S'_{2Q}(a, b, c)$. In the next section we will illustrate this procedure for a particular class L of graphs, namely partial 2-paths.

4 Acyclic partial 2-paths

In this section we illustrate the concepts introduced above on the example of the acyclic graphs with proper pathwidth 2 (partial 2-paths). For this class of graphs we define attributions and committed-choice algebras, including construction of quotient algebras and developement of a reduction system.

We leave the presentation of more exhaustive examples of several other subclasses of partial 2-trees to the full version of the paper.

The definition of a 2-path allows viewing it as a special case of a triangulation of a polygon. This is equivalent to a chordal planar graph consisting of two simplicial vertices (2-leaves), two paths between them (the *circumference*), and a number of *chords* (edges induced by minimal separators), each connecting vertices of the two paths (see Figure 1).

Consider an embedding of a graph G in the 2-path H, where G is connected and the embedding does not use the simplicial vertices of H. Each vertex of G

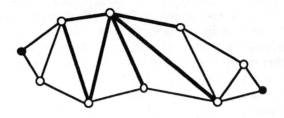

Fig. 1. A 2-path and its embedded partial 2-path

lies on the circumference of H, and each edge of G is either a chord or a path edge of H. The fact that each edge of a partial 2-path is either a chord or a path edge of an embedding 2-path gives us an attribution of graphs helpful in recognizing graphs of the class: each edge has an attribute **c** (for a chord edge) or **s** (for a path edge). There are no attributes for vertices.

In the complete example, we will consider the development of committed-choice algebras with the following signatures $F_\mathbf{t}$, $F_\mathbf{t}^\mathcal{A}$, $F_\mathbf{sp}$, $F_\mathbf{sp}^\mathcal{A}$, $F_\mathbf{c2}$, and $F_\mathbf{c2}^\mathcal{A}$:

$F_\mathbf{t}$	$\{\mathbf{v}, S_1, P_1\}$
$F_\mathbf{t}^\mathcal{A}$	$\{\mathbf{v}, S_1^\mathbf{s}, S_1^\mathbf{c}, P_1\}$
$F_\mathbf{sp}$	$\{\mathbf{e}, S_2, P_2\}$
$F_\mathbf{sp}^\mathcal{A}$	$\{\mathbf{e}^\mathbf{s}, \mathbf{e}^\mathbf{c}, S_2, P_2\}$
$F_\mathbf{c2}$	$\{\mathbf{v}, \mathbf{e}, S_1, P_1, S_2', P_2\}$
$F_\mathbf{c2}^\mathcal{A}$	$\{\mathbf{v}, \mathbf{e}^\mathbf{s}, \mathbf{e}^\mathbf{c}, S_1^\mathbf{s}, S_1^\mathbf{c}, P_1, S_2', P_2\}$

Let G be a partial 2-path embedded in a 2-path H. For 1-sourced graphs, all the interactions with the rest of graph take place through the source vertex. We will label this vertex to represent the graph's resource usage. An embedding of the 1-source graph that separates the source vertex from a simplicial vertex of the embedding 2-path H renders the source vertex "oblivious" to the actual embedding beyond that separator. Therefore, the set of possible non-equivalent embeddings is relatively small and so is the number of possible vertex labels, the label set C of a committed-choice algebra, or rather, that subset of C corresponsing to one-sourced graphs.

The algebra that we consider, $M_\mathbf{t}$, generates all trees. Its signature $F_\mathbf{t}$ consists of the constant (nullary operation) $\mathbf{v}$, a unary operation S_1, and a binary operation P_1. An acyclic partial 2-path G can be embedded in a 2-path H so that edges of G are either chord or path edges of H. These two values **c** and **s** are the attributes associated with edges of graphs that are in the carrier of the derived algebra $M_\mathbf{t}^\mathcal{A}$ with the signature $F_\mathbf{t}^\mathcal{A}$. In dealing with attributed graphs, we represent them by source labels reflecting committed embeddings of the underlying graphs. Thus, we can replace the operations on graphs with operations on labeled vertices. The constant of this algebra is the empty label ' ' (representing $\mathbf{v}$). For the series operation S_1 of $F_\mathbf{t}$, we define *two* 1-source series operations in $F_\mathbf{t}^\mathcal{A}$, $S_1^\mathbf{c}$ and $S_1^\mathbf{s}$ that correspond to different attributions of the added edge.

If G is a subgraph of 2-path H, it is also a partial graph of the 2-path H' obtained by fusing every vertex of $H - G$ with one of its neighbors on the circumference of H (fusing two vertices replaces them by a new vertex that inherits adjacencies of the two vertices). We will only consider embeddings that do not use the simplicial vertices of the embedding 2-path, since this reduces the complexity of subsequent case analyses. This is permissible since if G is a partial 2-path, it can be embedded in a 2-path without using its simplicial vertices. Moreover, we can treat the connected parts of G separately, because G is a partial 2-path if and only if each of its connected components is a partial 2-path.

The following simplifies the search over all contexts in the algebras of interest (used to compute relations like $\prec_L$ and $\sim_L$):

Proposition 9. *For each of the signatures F_k, $F_k^{\mathcal{A}}$, $F_{\mathbf{t}}$, $F_{\mathbf{t}}^{\mathcal{A}}$, F_{sp}, $F_{\mathrm{sp}}^{\mathcal{A}}$, F_{c2}, and $F_{\mathrm{c2}}^{\mathcal{A}}$, and their corresponding graph algebras, every context $f[\,]$ is equivalent to a context with the variable x occuring on the outermost level: $f[\,]$ can be written as $\lambda x.x$ or $\lambda x.f(t_1, \ldots, t_{i-1}, x, t_{i+1}, \ldots, t_n)$ for some operation f and terms t_j in the relevant signature.*

The above was proved for F_k in [1] (in which case only the operations P_i appear on the outermost level). For the other cases, the proofs are easy.

4.1 Committed-choice algebra for acyclic partial 2-paths

Starting with the algebra $M_{\mathbf{t}}$, we have defined the derived algebra $M_{\mathbf{t}}^{\mathcal{A}}$ based on edge attributes **c** (denoting a chord edge in a 2-path embedding) and **s** (denoting a circumference edge). We will now discuss the derivation of the quotient algebra $Q_{\mathbf{t}}$, starting with the representation of 1-source attributed graphs by source vertex labels.

The label set C of the committed-choice algebra describes "resource usage" for a possible embedding of a 1-sourced graph. Since a 2-path is outerplanar (all vertices are on the outer mesh of a plane embedding of the graph), an embedding of a given 1-sourced graph G in a 2-path could use both, one, or neither of the circumference edges incident with the source vertex (called simply source in this section). These cases are encoded as source labels ' ', 's' and 'ss', depending on the number of edges with attribute **s** that are incident with the source. (Figure 2 shows examples of the different codes.) An embedding in a 2-path can also use an edge incident with the source as the chord, which requires that the edge has the attribute **c**; this is encoded as the source label 't'. Several such 1-source edges can be put in parallel without using any more resources than one edge does. A qualitatively different resource usage is required by a series operation on a 1-source edge attributed **c**. It results in a two-edge path subgraph that starts at the source. Such a subgraph separates the source from one of the simplicial vertices in any embedding, and thus at most two such subgraphs can be incident with the source (one in the direction of each simplicial vertex). We encode graphs with one such subgraph by source label

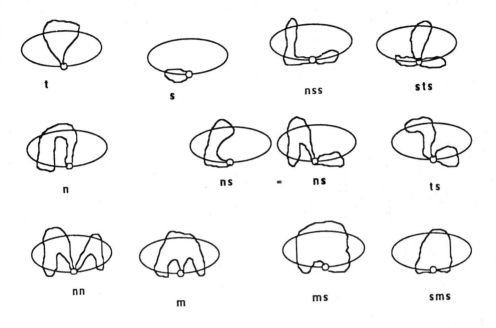

Fig. 2. Symbolic description of some resourse use codes for 1-source graphs

'n' and graphs with two such subgraphs by 'nn'. Another code, 'm', describes the situation where no new chord edge can be made incident with the source of G in any embedding.

Since the circumference edges and the chord edges can be used independently from each other, each combination of two codes from the two groups is possible as description of resource usage. This yields the following set of 15 possible codes: $\{$ ' ', 's', 'ss', 't', 'st', 'sst', 'n', 'sn', 'ssn', 'nn', 'snn', 'ssnn', 'm', 'sm', 'ssm'$\}$. We use an additional label $\perp$ denoting graphs with proper pathwidth greater than 2.

If we know the resource usage code of 1-sourced graphs G_1 and G_2, then we can decide the resource usage codes of $S_1^s(G_1), S_1^c(G_1)$ and $P_1(G_1, G_2)$. Table 4.1 shows the operation tables. It turns out that not all the resource codes have to be used in our committed-choice algebra. This is implied by the fact that some types of resource usage are "more parsimonious" than others and thus dominate them in the sense of section 2.3. For instance, if it is possible to embed a graph with an attribution encoded by 't', then it is unnecessary to try to embed it as encoded by 'n'. (An example of such a graph is the two-edge path with either one or both edges attributed **c**.) This domination is expressed by 't'$\prec$'n'.

	S_1^S	S_1^C	P_1	⟨⟩	t	st	s	ss	sst	n	ns
⟨⟩	s	t	⟨⟩	⟨⟩	t	st	s	ss	sst	n	ns
t	ns	n	t	t	t	st	st	sst	sst	n	ns
st	ns	n	st	st	st	sst	sst	⊥	⊥	ns	nss
s	s	t	s	s	st	sst	ss	⊥	⊥	ns	nss
ss	⊥	t	ss	ss	sst	⊥	⊥	⊥	⊥	nss	⊥
sst	⊥	n	sst	sst	sst	⊥	⊥	⊥	⊥	nss	⊥
n	ns	n	n	n	n	ns	ns	nss	nss	nn	nns
ns	ns	n	ns	ns	ns	nss	nss	⊥	⊥	nns	$nnss$
nss	⊥	m	nss	nss	nss	⊥	⊥	⊥	⊥	$nnss$	⊥
nn	ms	m	nn	nn	nn	nns	nns	$nnss$	$nnss$	⊥	⊥
nns	ms	m	nns	nns	nns	$nnss$	$nnss$	⊥	⊥	⊥	⊥
$nnss$	⊥	⊥	$nnss$	$nnss$	$nnss$	⊥	⊥	⊥	⊥	⊥	⊥
m	ms	⊥	m	m	⊥	⊥	ms	mss	⊥	⊥	⊥
ms	ms	⊥	ms	ms	⊥	⊥	mss	⊥	⊥	⊥	⊥
mss	⊥	⊥	mss	mss	⊥	⊥	⊥	⊥	⊥	⊥	⊥
⊥	⊥	⊥	⊥	⊥	⊥	⊥	⊥	⊥	⊥	⊥	⊥

Table 1. Operation tables for 1-sourced graphs: (i) series, and (ii) parallel.

The domination relation is an outer product of the following two orders '⟨⟩' $\prec_1$'s' $\prec_1$ 'ss' and '⟨⟩' $\prec_2$ 't' $\prec_2$ 'n' $\prec_2$ 'nn' $\prec_2$ 'm', as shown in Figure 3. The largest antichain has size 3 (which is also easy to see), and a committed-choice algebra has at most $\binom{5}{3}\binom{3}{3} + \binom{5}{2}\binom{3}{2} + \binom{5}{1}\binom{3}{1} = 55$ elements as opposed to 2^{15} if domination is not used. However, many of these 55 antichains are not reachable in the subsignature $F_t^{\mathcal{A}}$. The subsignature $F_t^{\mathcal{A}}$ generates only nine of them. As another example of the use of domination, note that $S_1^S(`n')$ can be both 'ms' and 'ns' because the new source of the graph can be embedded (on the circumference of the embedding 2-path H) on either side of the old source. However, 'ns' is preferable for the purpose of embedding the graph since it allows, e.g., for parallel composition with a chord, $P_1(`t',\cdot)$, which the former does not, while no composition permitted with 'ms' is prohibited by 'ns'. Thus the domination concept helps us to get a simpler committed-choice algebra $\widehat{M_t}$. For series-parallell andgeneral partial 2-paths,this method works even better. We have develoed committed-choice algebras for $\widehat{M_{sp}}$ and $\widehat{M_{c2}}$. These have label sets of size 43 and 84, respectively. For the series-parallel case,the committed choice algebra generated by F_{sp} has only 71 elements.

Reduction system to recognize acyclic partial 2-paths Labels that represent congruence classes of the corresponding committed-choice algebra allow us to construct a labeled reduction system. In this system, there is a single structural reduction rule corresponding to the elimination of a pendant vertex. All the information about the structure of the reduced graph is encoded in vertex labels. We can now reverse the process by which we have developed the

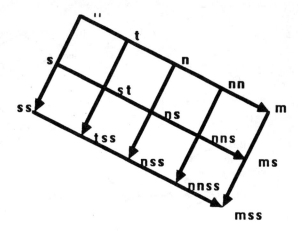

Fig. 3. Domination relation for congruence classes of 1-sourced partial 2-paths

committed-choice algebra and interpret the labeled reductions as reduction of unlabeled graphs.

S_1	$S_1^S \cup S_1^C$		P_1	$\{t,s\}$	$\{t,ss\}$	$\{t\}$	$\{n\}$	$\{nn\}$	$\{m\}$	$\{ms\}$	$\{mss\}$
$\{\,'\,\}$	$\{t,s\}$		$\{\,'\,\}$	$\{t,s\}$	$\{t,ss\}$	$\{t\}$	$\{n\}$	$\{nn\}$	$\{m\}$	$\{ms\}$	$\{mss\}$
$\{t,s\}$	$\{t,s\}$		$\{t,s\}$	$\{t,ss\}$	$\{t\}$	$\{t\}$	$\{n\}$	$\{nn\}$	$\{ms\}$	$\{mss\}$	$\{\bot\}$
$\{t,ss\}$	$\{t\}$		$\{t,ss\}$	$\{t\}$	$\{t\}$	$\{t\}$	$\{n\}$	$\{nn\}$	$\{mss\}$	$\{\bot\}$	$\{\bot\}$
$\{t\}$	$\{n\}$		$\{t\}$	$\{t\}$	$\{t\}$	$\{t\}$	$\{n\}$	$\{nn\}$	$\{\bot\}$	$\{\bot\}$	$\{\bot\}$
$\{n\}$	$\{n\}$		$\{n\}$	$\{n\}$	$\{n\}$	$\{n\}$	$\{nn\}$	$\{\bot\}$	$\{\bot\}$	$\{\bot\}$	$\{\bot\}$
$\{nn\}$	$\{m\}$		$\{nn\}$	$\{nn\}$	$\{nn\}$	$\{nn\}$	$\{\bot\}$	$\{\bot\}$	$\{\bot\}$	$\{\bot\}$	$\{\bot\}$
$\{m\}$	$\{ms\}$		$\{m\}$	$\{ms\}$	$\{mss\}$	$\{\bot\}$	$\{\bot\}$	$\{\bot\}$	$\{\bot\}$	$\{\bot\}$	$\{\bot\}$
$\{ms\}$	$\{ms\}$		$\{ms\}$	$\{mss\}$	$\{\bot\}$	$\{\bot\}$	$\{\bot\}$	$\{\bot\}$	$\{\bot\}$	$\{\bot\}$	$\{\bot\}$
$\{mss\}$	$\{\bot\}$		$\{mss\}$	$\{\bot\}$	$\{\bot\}$	$\{\bot\}$	$\{\bot\}$	$\{\bot\}$	$\{\bot\}$	$\{\bot\}$	$\{\bot\}$

Table 2. Operation tables in the committed-choice algebra for acyclic partial 2-paths.

The congruence classes of the comitted-choice algebra for acyclic partial 2-paths is shown in Table 2. The table of the S_1 operation determines the labeled graph reduction system rules in the following manner. For labels l_1 and l_2, if $S_1(l_1) = l_2$ then a pendant vertex labeled l_1 is reduced, and label l_2 is added to its neighbor. This label is subsequently combined with the neighbor's other label according to the operation P_1. By interpreting the operations as "adding a new source vertex adjacent to the old source of the argument graph" in S_1 and "fusing the sources of the argument graphs" in P_1, we will use this labeled graph reduction system to construct a graph reduction system for unlabeled graphs.

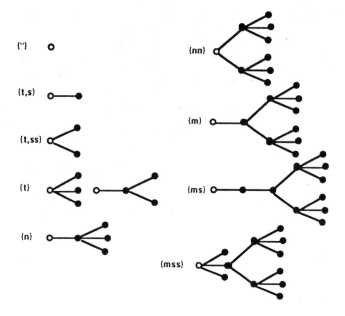

Fig. 4. Representatives for congruence classes for acyclic partial 2-paths

Recall that the labels (congruence classes) are results of applying the algebra's operations to generated congruence classes, starting with the constant of the algebra, $\mathbf{v}$, which can be evaluated as the trivial 1-source, one-vertex graph. Since $S_1(\mathbf{v}) = \{t, s\}$, this congruence class, easily shown not congruent to $\mathbf{v}$, can be evaluated as 1-source, one-edge graph. Similar evaluation pattern applies to the other congruence classes, see Figure 4. Notice that there are classes with several non-isomorphic, minimum-sized evaluations (graphs). In Figure 4, we show this for only one congruence class, $\{t\}$ that contains two non-isomorphic graphs with three vertices; this basic pair of equal-size, congruent graphs can be used to construct other such graphs for the remaining classes.

Application of an operation that results in an element of an existing congruence class gives rise to a reduction rule. The left-hand side of a rule is a graph which is the result of the operation, and the right-hand side of the rule is any minimum-size graph in the congruence class. For instance, cf. Table 2, the class $\{t\}$ is the result of operations $S_1(\{t, ss\}), P_1(\{t, ss\}, \{t, s\}), P_1(\{t, ss\}, \{t, ss\})$, $P_1(\{t\}, \{t, s\}), P_1(\{t\}, \{t, ss\})$ and $P_1(\{t\}, \{t\})$ Evaluation of the first two operations defines the two minimum-size graphs in this class. The reduction determined by the penultimate operation in the above sequence is superseded by the smaller reductions derived from the other operations.

The (almost) complete set of reduction rules obtained from Table 2 is given in Figure 5. Missing are six reduction rules derived from the same expressions as each of the last three reduction rules, with appriopriately modified left-hand sides involving the alternative representation of t.

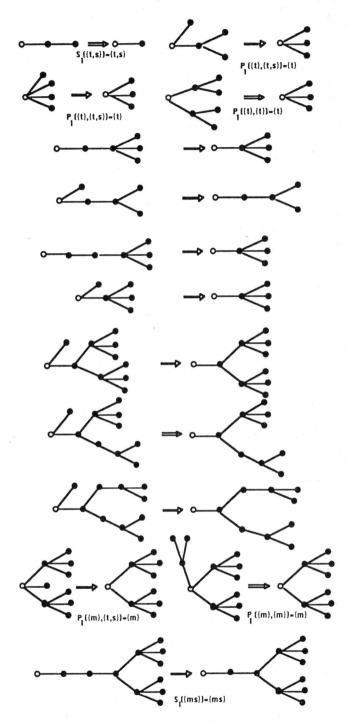

Fig. 5. Some reduction rules to recognize acyclic partial 2-paths

References

1. S. Arnborg, B. Courcelle, A. Proskurowski and D. Seese, An algebraic theory of graph reduction, *J. ACM* **40** (1993), 1134-1164;
2. S. Arnborg, J. Lagergren and D. Seese, Problems easy for tree-decomposable graphs, *J. of Algorithms* **12**, (1991), 308-340;
3. S. Arnborg and A. Proskurowski, Canonical representations of partial 2- and 3-trees, *BIT* **32** (1992), 197-214;
4. S. Arnborg, A. Proskurowski, and D. Seese, Monadic second order logic, tree automata and forbidden minors, in *Proceedings of the 4th Computer Science Logic Workshop*, E.Börger and H. Kleine Buning, Eds., Springer-Verlag *LNCS* **533**, 1-16 (1991);
5. M. Bauderon and B. Courcelle, Graph expressions and graph rewritings, *Mathematical Systems Theory* **20** (1987), 83-127;
6. H. Bodlaender, On reduction algorithms for graphs with small treewidth, *Proc. 19th Workshop on Graph-Theoretic Concepts in Computer Science*, Springer Verlag, *LNCS* **790**, 45-56 (1993);
7. B. Courcelle, The monadic second-order logic of graphs I: recognizable sets of finite graphs, *Information and Computation* **85** (1990), 12-75;
8. B. Courcelle, The Monadic Second-Order Logic of Graphs III: Tree-Decompositions, Minors and Complexity Issues, *Informatique Théorique et Applications* **26** (1992),257-286,
9. R.P. Dilworth, A decomposition theorem for partially ordered sets, *Ann. of Mathematics* **51** (1951), 161-166;
10. Graetzer, *Universal Algebra*, Springer Verlag, 1979;
11. N. Kinnersley and M. Langston, Obstruction set isolation for the gate matrix layout problem, *Discrete Applied Mathematics* **54**, 169-214 (1994);
12. N.M. Korneyenko, Combinatorial algorithms on a class of graphs, *Discrete Applied Mathematics* **54**, 215-218 (1994);
13. D. Kozen, On the Myhill-Nerode theorem for trees, *Bulletin of the EATCS* **47**, 170-173;
14. D. Rose, On simple characterization of k-trees, *Discrete Mathematics* **7**, 317-322 (1974);
15. J. Valdes, E. Lawler and R.E. Tarjan, The recognition of series-parallel digraphs, *SIAM J. Computing* **11**, 298-313 (1982);
16. A. Takahashi, S. Ueno, and Y. Kajitani, Minimal acyclic forbidden minors for the family of graphs with bounded path-width, in *Proceedings of SIGAL* (1991);
17. J.W. Thatcher and J.B. Write, Generalized finite automata theory with an application to a decision problem in second-order logic, *Mathematical Systems Theory* **2**, 57-81 (1968).
18. A. Wald and C.J. Colbourn,Steiner Trees, Partial 2-trees, and minimum IFI networks, *Networks* **13**, 159-167 (1983).
19. T.V. Wimer, Linear algorithms on k-terminal graphs, PhD Thesis *URI-030*, Clemson University, Clemson (1988).

The Definition in Monadic Second-Order Logic of Modular Decompositions of Ordered Graphs

Bruno Courcelle

Université Bordeaux-I, LaBRI[1]
351, Cours de la Libération
33405 TALENCE, France

Abstract: Every graph can be represented uniquely in a hierarchical way by means of its *modular decomposition*. We establish that the modular decomposition of a linearly ordered graph is definable in monadic second-order (MS) logic in the graph itself. The modular decomposition does not depend on the linear order of the given graph. A set of graphs is recognizable w.r.t. the operations associated with graph substitution iff it is definable by a formula of an extension of MS logic based on the use of auxilliary linear orderings. This paper is an extended abstract: complete proofs can be found in [7].

Introduction

Understanding the relationships between recognizability and MS-definablity is our main motivation. The notion of a recognizable set of graphs has been introduced in [3]. It is based on graph congruences with finitely many classes and is relative to operations on graphs that, typically, glue two graphs together or extend in some way a given graph. It is known from Büchi and Doner, (see Thomas [13]) that a set of words (or of binary trees) is recognizable iff it is MS-definable. For trees of unbounded degree, a result of this form is proved in [3] where definability is relative to an extension of MS logic called *Counting Monadic Second-order logic* (CMS in short). Its formulas are written with special "counting modulo q" existential first-order quantifiers: $\exists^{\mathbf{mod}\,q}x.\varphi(x)$ means that the number of

---

[1] Laboratoire associé au CNRS ; email : courcell@labri.u-bordeaux.fr
This work is supported by Working Group COMPUGRAPH II

elements x that satisfy φ is a multiple of q. We ask the following general question, already considered in [4,5,11,14] :

Question : *For which classes of finite graphs ℂ does there exist an extension ℒ of monadic second-order logic such that, for every $L \subseteq ℂ$, L is recognizable iff it is ℒ-definable.*

In [4] we proved that CMS is the appropriate extension of MS for the class of graphs of tree-width at most 2. We now sketch the proof method for this result and others of the same type established in the present paper.

Let ℂ be a class of graphs, let 𝔽 be the set of graph operations on ℂ involved in the intended notion of recognizability, let us also assume that every graph in ℂ is the value of an 𝔽-expression, i.e., of an algebraic expression over 𝔽. (The set 𝔽 is in some sense a parameter: different sets 𝔽 yield different notions of recognizability). Assume we have a language ℒ (say an extension of MS like CMS), for which we know that, if a subset of ℂ is ℒ-*definable* then it is recognizable. Assume finally that for every graph G in ℂ we can construct "in G" an 𝔽-expression that defines this graph. Then, if L is a recognizable subset of ℂ, there exists a finite tree-automaton recognizing the set of 𝔽-expressions the value of which is in L. Given a graph G we can express that $G \in L$ by means of a formula in ℒ that works at follows :

(1) it defines in G an 𝔽-expression, the value of which is G,

(2) it checks whether the automaton accepts this expression:

the graph G is in L iff the automaton accepts the expression.

So the language ℒ must not be too powerful (we want that every ℒ-definable set be recognizable) but it must be powerful enough to do two things

(1) to "parse" the graph (i.e., to define an 𝔽-expression for it)

(2) to simulate the behaviour of a finite automaton on the obtained 𝔽-expression.

A linear ordering of the given graph helps sometimes to "parse" it by MS-formulas. It follows that we are lead to introduce an extension of MS logic (denoted) by MS(≤), that uses *auxilliary* "invariant" *linear orders* on the vertices of the considered graphs. All properties expressible in CMS logic are of this form: a linear ordering helps to express that a set X has an even number of elements, because one can divide X into two parts, the elements of even rank and those of odd rank, where ranks are relative to this linear order; X has even cardinality iff the last element has even rank; the answer, namely the parity of the cardinality of X *does not depend on the chosen ordering* of the vertices although it is expressed logically in terms of this order: this is what means the term "invariant".

We prove that every MS($\leq$)-definable set of graphs is recognizable. For trees and related graphs, CMS-definability, MS($\leq$)-definability and recognizability are equivalent notions, all strictly larger than MS-definability.

We show that with the help of such an auxilliary "invariant" ordering, we can reconstruct the "internal structure" a tree from its leaves and a ternary relation on leaves that "projects" it on them. We can do the same reconstruction *without any linear order* for the class of trees having a degree bounded by a fixed constant.

We apply this to prove that the unique *modular decomposition of any graph* can be defined by an MS($\leq$) formula (i.e., MS formulas using an auxilliary invariant linear order). We obtain logical characterizations of recognizable sets of graphs having modular decompositions with prime graphs of bounded size.

Order invariant MS-definable graph properties.

An *ordered graph* is a pair consisting of a graph G and a linear order P on its set of vertices $\mathbf{V}_G$. If $\mathbb{G}$ is a class of graphs, we let $\mathbb{G}(\leq)$ denote the class of all ordered graphs of the form $<G, P>$ for $G \in \mathbb{G}$. A property $\mathcal{P}$ of ordered graphs is *order-invariant* if for every $G \in \mathbb{G}$, for every two linear orders P and P' on $\mathbf{V}_G$: $\mathcal{P}(<G, P>) \Leftrightarrow \mathcal{P}(<G, P'>)$.

Graphs are defined as {**edg**}-structures i.e., as relational structures of the form $<\mathbf{V}_G, \mathbf{edg}_G >$ where $\mathbf{edg}_G$ is a binary relation representing the edges in a natural way. We let $\leq$ be a new binary relation symbol. A property $\mathcal{P}$ of graphs of a class $\mathbb{G}$ is *MS($\leq$)-definable* iff it is an order-invariant MS-property of the graphs in $\mathbb{G}(\leq)$ iff there exists an MS-formula φ over {**edg**, $\leq$} such that
1) for every $G \in \mathbb{G}$, for any two linear orders P and P' on $\mathbf{V}_G$:
    $<G, P> \models \varphi$ iff $<G, P'> \models \varphi$ (where P and P' are values of $\leq$)

2) for every $G \in \mathbb{G}$:

$\mathcal{P}(G)$ holds iff $<G, P> \models \varphi$ for some linear order P on $\mathbf{V}_G$

(equivalently, iff $<G, P> \models \varphi$ for all linear orders P on $\mathbf{V}_G$).

A set of graphs is MS(≤)-*definable* iff it is the set of graphs satisfying an order-invariant MS-property. Every CMS-definable property is MS(≤)-definable. (We recall that CMS refers to *Counting Monadic Second-order* logic, see the introduction).

Theorem 1 [7, Cor.4.2]: *Every MS(≤)-definable set of graphs is recognizable.*

We have the following hierarchy of families of sets of graphs: MS-definable ⊆ CMS-definable ⊆ MS(≤)-definable ⊆ Recognizable.

However, it is not always strict:

Proposition 2 [3]: *For every set L of forests, the following properties are equivalent:*
(1) *L is CMS-definable,*
(2) *L is MS(≤)-definable,*
(3) *L is recognizable.*

The reconstruction of a tree from its leaves

The set of nodes of a (rooted) tree T is denoted by $\mathbf{N}_T$. A node that is not a leaf is *internal*. We shall denote by $\mathbf{L}_T$ the set of leaves of T. We shall denote by $\leq_T$ (or $\leq$ if the context makes T clear) the partial order defined by accessibility (leaves are maximal elements). Every two nodes x and y have a greatest lower bound for $\leq_T$ that we shall denote by $x \wedge y$ (or by $x \wedge_T y$ if necessary). Finally, for every triple of leaves of T, we let $\mathbf{R}_T(x, y, z)$ hold iff $x \wedge y \leq z$ i.e., iff z belongs to the smallest subtree containing x and y. We let $\lambda(T) = <\mathbf{L}_T, \mathbf{R}_T>$ (λ stands for "leaves"). A tree is *proper* if it has at least two leaves and no node has exactly one successor. The number of nodes of a proper tree is at most twice the number of its leaves.

Proposition 3 [7, Prop. 5.1]: *Given a finite set L of cardinality at least 2 and a ternary relation R on L, there is at most one proper tree T such that $\lambda(T) = <L, R>$.*

We refer the reader to [6] for MS-definable transductions. (Intuitively, an MS-definable transduction associates with a relational structure S one or more structures defined by fixed MS formulas "inside" the structure formed from the union of k disjoint copies of S, where k is also fixed).

Theorem 4 [7, Thm. 5.3]: *The transduction of relational structures*
$\{(<L, R, P>, <N, \leq>) \ / \ P$ *is a linear order on* L,
$<N, \leq>$ *is a proper tree* T *and* $<L, R> = \lambda(T)\}$
is definable. In other words, the transduction λ^{-1} *is definable provided the input structure* (namely $<L, R>$) *is ordered.*

Proof: Let T be proper, let $<L, R> = \lambda(T)$, let P be a linear order on L. We derive from P a strict linear order on the successors of the nodes of T: if y, z are two successors of x, we let $y <_P z$ iff
the P-smallest leaf below y is strictly P-smaller than
the P-smallest leaf below z.

Since P is proper, every internal node x has a first successor and a second successor (w.r.t. the strict order $<_P$), denoted respectively by **suc1**(x) and **suc2**(x). Every internal node x has a *representative leaf*, denoted by **rep**(x), that we define by: **rep**(x) = **suc2***(**suc1**(x)) where for every node y, **suc2**$^*(x)$ = **suc2**$^n(x)$ and n is the unique integer ($n \geq 0$) such that **suc2**$^n(x)$ is a leaf. (Intuitively, one finds the leaf representing an internal node x by going from x to its first successor and then to second successors until one reaches a leaf.) The mapping **rep** is thus a bijection of $N_T - L_T$ onto a subset of L_T.

Our next purpose it to find an MS formula $\theta \in MS(\{R, P\}, \{x, y, z\})$ such that, for every x, y, z in L:
$<L, R, P> \models \theta(x, y, z)$ iff $z = \textbf{rep}(x \wedge y)$.
One can construct first order formulas $\varphi_1, ..., \varphi_4$ such that, for all $u, v, u', v' \in L$:
$<L, R> \models \varphi_1(u, v, u', v')$ iff $u \wedge v \leq_T u' \wedge v'$,
$<L, R> \models \varphi_2(u, v, u', v')$ iff $u' \wedge v'$ is a successor of $u \wedge v$,
$<L, R, P> \models \varphi_3(u, v, u', v')$ iff $u' \wedge v'$ is the 1st successor of $u \wedge v$,
$<L, R, P> \models \varphi_4(u, v, u', v')$ iff $u' \wedge v'$ is the 2nd successor of $u \wedge v$.

We observe that for all $x, y, z \in L$, $z = \textbf{rep}(x \wedge y)$ iff there exists a nonempty sequence $x_1, x_2, ..., x_n$ of leaves such that :
$z \wedge x_1$ is the first successor of $x \wedge y$,

$z \wedge x_i$ is the second successor of $z \wedge x_{i-1}$ $(2 \leq i \leq n)$, $z = x_n$.

These conditions can also be written as follows :

$z = \mathbf{rep}(x \wedge y)$ iff there exists a set of leaves X and a leaf $x' \in X$ such that :

(i) $z \wedge x'$ is the first successor of $x \wedge y$

(ii) the graph $<X, \rightarrow>$ where $u \rightarrow v$ iff $z \wedge v$ is the second successor of $z \wedge u$ is a path with 1st element x' and last element z.

From this latter formulation, the construction of an MS-formula $\theta(x, y, z)$ defining the relation $z = \mathbf{rep}(x \wedge y)$ follows immediately. We now construct from $<L, R, P>$ the structure $<N, suc^*>$:

. $N = L \times \{1\} \cup L' \times \{2\}$ where $L' = \{\mathbf{rep}(x \wedge y) \ / \ x \neq y, \ x, y \in L\}$

. $((z, i), (z', j)) \in suc$ iff

either $j = 1$ and $i = 2$ and there exist u, v with $u \neq v$, such that z is a successor of $u \wedge v$ and $z' = \mathbf{rep}(u \wedge v)$ (this can be expressed by $\varphi_2(u, v, z, z) \wedge \theta(u, v, z')$)

or $i = j = 2$ and there exist u, v, u', v' with $u \neq v$, $u' \neq v'$ such that $z = \mathbf{rep}(u \wedge v)$, $z' = \mathbf{rep}(u' \wedge v')$ and $u' \wedge v'$ is a successor of $u \wedge v$ (this can be expressed by the formula $\varphi_2(u, v, u', v') \wedge \theta(u, v, z) \wedge \theta(u', v', z')$).

From the formulas $\varphi_1, ..., \varphi_4, \theta$ one obtains that the transformation $\tau = <L, R, P> \mapsto <N, suc^*>$: is definable.

If P is a linear order on L and $<L, R> = \lambda(T)$ then $<N, suc^*>$ is isomorphic to T. Note that $<N, suc^*> = \tau(<L, R, P>)$ may be well-defined, even if $<L, R>$ is not of the form $\lambda(T)$. An additional MS formula ψ can be written such that, for every $<L, R, P>$, we have $<L, R, P> \models \psi$ iff P is a linear order on L and the structure $\tau(<L, R, P>) = <N, suc^*>$ is a tree, the leaves of which are the elements of N of the form $(x, 1)$ and such that for every x, y, z in L we have:

(1) $(x, x, z) \in R$ iff $x = z$ and,

(2) if $x \neq y$, then $(x, y, z) \in R$ iff there exists $u \in L$ such that $\theta(x, y, u)$ holds, and $(u, 2)$ is an ancestor of $(z, 1)$ in the tree $<N, suc^*>$.

The restriction of τ to the structures that satisfy ψ is thus the desired transduction. (A similar idea is used in [12].) □

Theorem 5:[7, Prop. 5.5]: *Let $n \in \mathbb{N}, n \geq 2$. One can define by MS formulas a linear ordering on the structures $\lambda(T)$ such that T is a tree of outdegree at most n. The transduction*
$$\{(<L, R>, <N, suc^*>) \; / \; <L, R> = \lambda(T),$$
$$T = <N, suc> \text{ and } T \text{ is proper tree of outdegree at most } n\}$$
is definable.

Modular decompositions

We shall consider directed graphs. A *module* in a graph G is a subset X of $\mathbf{V}_G$ such that, every vertex $y \in \mathbf{V}_G - X$ "sees all vertices of X in the same way". Formally, X is a module iff for every $y \in \mathbf{V}_G - X$

    either $(x, y) \in \mathbf{edg}_G$ for all x in X

    or $(x, y) \in \mathbf{edg}_G$ for no x in X

and   either $(y, x) \in \mathbf{edg}_G$ holds for all $x \in X$

    or $(y, x) \in \mathbf{edg}_G$ for no $x \in X$

We say that X is a *prime module* if for every module Y :

    either $X \subseteq Y$ or $Y \subseteq X$ or $X \cap Y = \emptyset$

It follows that $\emptyset$, $\mathbf{V}_G$ and all singletons are prime modules. The *modular tree* of G is the tree $\mathbf{mt}(G) = <N, suc^*>$ defined as follows

. its set of nodes N is the set of nonempty prime modules,

. Y is a successor of X iff $Y \subseteq X$ and there is no prime module Z with $Y \subseteq Z \subseteq X$ and $Y \neq Z \neq X$.

The modular decomposition is based on the *substitution* of graphs for vertices of graphs. Let H be a graph with $\mathbf{V}_H = \{v_1,...,v_k\}$. For pairwise disjoint graphs $G_1,...,G_k$, we let $H \langle G_1,...,G_k \rangle$ be obtained in the following way. One takes the union of $G_1, ..., G_k$ and one adds an edge (x, y) whenever $x \in \mathbf{V}_{G_i}, y \in \mathbf{V}_{G_j}$, $i \neq j$, and (v_i, v_j) is an edge of H. Hence, we get a total k-ary operation on graphs (isomorphic graphs are considered as equal; see [7]). We shall use in particular the operations:

$G_1 \oplus G_2 = H\langle G_1, G_2 \rangle$ where $H = \{ \bullet v_1 \quad \bullet v_2\}$

$G_1 \otimes G_2 = H\langle G_1, G_2 \rangle$ where $H = \{ v_1 \bullet \rightleftarrows \bullet v_2\}$

$G_1 \overset{\rightarrow}{\otimes} G_2 = H\langle G_1, G_2 \rangle$ where $H = \{ v_1 \bullet \longrightarrow \bullet v_2\}$.

The first two are associative and commutative, the last one is only associative. A graph G is *prime* if it cannot be written $H \langle G_1, ...,G_k \rangle$ except in a trivial way with $H = G$ and G_i having a unique vertex for each i. This is equivalent to the condition that all modules are empty, singletons or equal to the set of vertices. (The above three operations are associated with the three graphs with two vertices, which are all prime.)

The modular decomposition of G is the labelled tree $\mathbf{mdec}(G)$ = $<\mathbf{mt}(G), lab>$ where we define lab as follows. (The nodes of $\mathbf{mt}(G)$ are the nonempty prime modules of G.) For every prime module X that is not a singleton we have exactly one of the following 4 cases, where $Y_1,\ldots,Y_k$ are the successors of X in $\mathbf{mt}(G)$ (they are prime modules and can be singletons) :

Case 1: $G[X] = G[Y_1] \oplus \ldots \oplus G[Y_k]$

Case 2: $G[X] = G[Y_1] \otimes \ldots \otimes G[Y_k]$

Case 3: $G[X] = G[Y_1] \overrightarrow{\otimes} \ldots \overrightarrow{\otimes} G[Y_k]$

(for some appropriate numbering of the successors of X, let us recall that the operation $\overrightarrow{\otimes}$ is not commutative)

Case 4: $G[X] = H\langle G[Y_1], \ldots, G[Y_k] \rangle$

for some prime graph H with at least 3 vertices; this graph H is obtained from $G[X]$ by the fusion of any two vertices in a same set Y_i, the deletion of the resulting loops and the fusion of the resulting multiple edges with the same direction.

The label of X is defined by: $lab(X) = \oplus$ in Case 1, $lab(X) = \otimes$ in Case 2, $lab(X) = \overrightarrow{\otimes}$ in Case 3, and $lab(X) = H$ in Case 4.

The modular decomposition of a graph G can be seen as the tree representing an algebraic expression denoting G and constructed with substitution operations. It has defined independently many times. See for instance [10], and [1] for a linear algorithm computing the modular decomposition of a directed or undirected graph. Here is an example showing that $\mathbf{mdec}(G)$ can be considered as an algebraic expression evaluating to G.

Example : See on next page a graph and its modular decomposition. The vertices are $a,b,\ldots,k$; the boxes show the prime modules. In the tree, the operation labelling node 4 is the substitution associated with the graph H : $v_1 \bullet \longleftarrow v_2 \bullet \longleftarrow v_3 \bullet \longrightarrow \bullet v_4$. Note the relevant ordering of the vertices of H. □

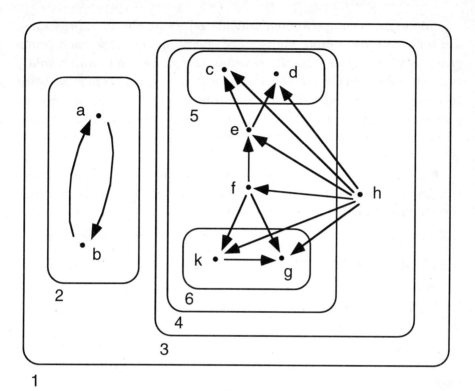

Modular decompositions, considered as algebraic expressions, are built with an infinite alphabet of operations because each prime graph is turned into a graph operation and there are infinitely many prime graphs. This is not convenient for our purposes since we want to consider all modular decompositions as relational structures using a same finite set of relations. We shall redefine them as graphs by means of the notion of *graph expansion* already used in Courcelle [8].

An ε-*graph* is a directed graph K some edges of which (called the ε-edges) are labeled by ε (the other being unlabelled), and such that the subgraph of K consisting of the ε-edges is acyclic. An ε-graph K will be represented by the relational structure $<\mathbf{V}_K, \mathbf{edg}_K, \mathbf{edg}_K^{\varepsilon}>$ where $\mathbf{edg}_K^{\varepsilon}$ represents the ε-edges and $\mathbf{edg}_K$ the other ones. A graph G called the *expansion* of K can be associated with an ε-graph K. We let $\mathbf{V}_K$ be the set of vertices of K that are not the source of any ε-edge. We let (x, y) be an edge of G iff $x \neq y$ and there exist an edge (x', y') in K such that there is an ε-path in K from x' to x and one from y' to y. (An ε-*path* is a directed path consisting of ε-edges.) We denote G by $\mathbf{exp}(K)$.

For every graph G, we let $\mathbf{gdec}(G)$ (read "the **g**raph representation of the modular **dec**omposition of G ") be the ε-graph K defined as follows from the modular tree $\mathbf{mt}(G)$:
. $\mathbf{V}_K$ is the set of nodes of $\mathbf{mt}(G)$,
. its ε-edges are the edges of $\mathbf{mt}(G)$,
. for every node X of $\mathbf{mt}(G)$, i.e., every prime module of G, we put edges between the successors $Y_1,...,Y_k$ of X in $\mathbf{mt}(G)$ according to the 4 cases considered in the definition of $\mathbf{mdec}(G)$:

in Case 1, we put no edge (X is a $\oplus$–*node*) ;
in Case 2, we put an edge from Y_i to Y_j for every $i,j \neq i$ (X is a $\otimes$–*node*) ;
in Case 3 we put an edge from Y_i to Y_j for every i,j with $1 \leq i < j \leq k$ (we assume that the successors $Y_1,...,Y_k$ are such that $G[X] = G[Y_1] \vec{\otimes} ... \vec{\otimes} G[Y_k]$; X is a- $\vec{\otimes}$ *node*)
in Case 4 we have $G[X] = H\langle G[Y_1],...,G[Y_k]\rangle$ where H is prime with at least 3 vertices, and we put an edge from Y_i to Y_j iff there is an edge in G from a vertex of Y_i to one of Y_j and $i \neq j$; (X is an H-node).

Proposition 6 [7, Prop. 6.5]: *For every graph G, **gdec**(G) is an ε-graph and G = **exp**(**gdec**(G)).*

Theorem 7 [7, Prop. 6.6]: *The transduction that associates with an ordered graph G the graph **gdec**(G) representing its modular decomposition is definable.*

Corollary 8 [7, Cor. 6.7] : *(1) Every graph property that is expressible as an MS-property of the modular decomposition of the considered graph is MS ($\leq$)-definable .*
(2) Every graph property that is expressible as a first-order property of the modular decomposition of the considered graph is MS.

We define the *modular width* of a graph G as the maximal number of vertices of a prime graph H appearing in an H-node in the modular decomposition of G. We shall denote it by **mwd**(G). If G is prime then **mwd**(G) = **card**(V_G). The modular width of a graph is either 0 or at least 3 (the graph: •←—•←—• is prime).

Proposition 9 [7, Prop. 6.8] : *For every $n \in \mathbb{N}$ it can be expressed in MS logic that the modular width of a graph is at most n.*

For each $n \in \mathbb{N}$, $n \geq 3$, we let PR_n be the set of prime graphs with at most n and at least 3 vertices. For each $H \in PR_n$, given with a fixed enumeration $v_1,...,v_k$ of its vertices ($k \leq n$), we define a function symbol $\mathbf{sub}_H$ intended to represent the operation that associates $H\langle G_1,...,G_k \rangle$ with graphs $G_1,...,G_k$.

We let $\mathcal{F}_n = \{\oplus, \otimes, \vec{\otimes}, \mathbf{1}\} \cup \{\mathbf{sub}_H / H \in PR_n\}$ and $\mathcal{F}_\infty = \cup\{\mathcal{F}_n / n \geq 0\}$. The expressions over $\mathcal{F}_n$ denote the graphs of modular width at most n. Expressions over $\mathcal{F}_n$, handled as trees, are convenient to represent graphs of bounded modular width.

A set of graphs L is $\mathcal{F}_\infty$-*recognizable* if it is recognizable with respect to the operations of $\mathcal{F}_\infty$.(See [3,9] for recognizability in general.) Since every graph is expressible as a finite combination of operations in $\mathcal{F}_\infty$ and since whenever $H = K\langle H_1,...,H_n \rangle$ we have $\mathbf{sub}_H = \mathbf{sub}_K \circ (\mathbf{sub}_{H_1},...,\mathbf{sub}_{H_n})$ it follows that L is $\mathcal{F}_\infty$-recognizable iff it is with respect to the the operations $\mathbf{sub}_H$ associated with all, not necessarily prime, graphs H. Equivalently, this means that L is saturated for an equivalence relation ~ on graphs with finitely many classes and such that, for all graphs H, $G_1,...,G_k$, $G'_1,...,G'_k$ (where H has k vertices)

$$G_1 \sim G'_1,...,G_k \sim G'_k \Rightarrow H\langle G_1,...,G_k\rangle \sim H\langle G'_1,...,G'_k\rangle$$

Proposition 10 [7, Prop. 6.10] : *Every MS($\leq$)-definable set of graphs is $\mathcal{F}_\infty$-recognizable.*

If L is a set of graphs, we denote by **Gdec**(L) the set of graphs **gdec**(G) for G in L.

Theorem 11 [7, Prop. 6.11] : *Let $n \geq 0$ and L be a set of graphs of modular width $\leq n$. The following are equivalent*
(1) L *is MS($\leq$) definable,*
(2) L *is $\mathcal{F}_\infty$-recognizable,*
(3) L *is $\mathcal{F}_n$-recognizable,*
(4) **Gdec**(L) *is CMS-definable.*

Corollary 12 [7, Cor. 6.12] : *Let L be a set of graphs with modular trees of degree at most some integer n. One can MS-define a linear order on these graphs. The statements of Theorem 11 hold with MS instead of MS($\leq$) and CMS.*

In the table on next page, we review the results we know concerning the comparison of the classes of MS-, CMS-, MS($\leq$)-definable and recognizable sets of graphs. By REC we mean the class of recognizable sets of graphs as defined in [3,4,9]. By $\mathcal{F}_\infty$-REC we mean the class of $\mathcal{F}_\infty$-recognizable sets.

Class of graphs	
all graphs	MS $\subset$ CMS $\subseteq$ MS($\leq$) $\subset$ REC =?
graphs of bounded tree-width	MS $\subset$ CMS $\subseteq$ MS($\leq$) $\subseteq$ REC =? =?
graphs of bounded modular width	MS $\subset$ CMS $\subseteq$ MS($\leq$) = $\mathcal{F}_\infty$-REC =?
discrete graphs, trees, graphs of tree-width at most 2	MS $\subset$ CMS = MS($\leq$) = REC by [4]
words, trees of bounded degree, traces,	MS = CMS = MS($\leq$) = REC (see [3,7,14])
graphs with modular trees of bounded degree	MS = CMS = MS($\leq$) = $\mathcal{F}_\infty$-REC

The stars and the discrete graphs establish the strictness of the inclusions MS $\subset$ CMS of this table. The equality CMS = REC is conjectured in [4] for graphs of bounded tree-width, and proved for those of tree-width at most 2.

Conjecture 13 : *It is not possible to reconstruct a proper tree T from $\lambda(T)$ by a definable transduction.*

Conjecture 14 : *CMS-definability is strictly weaker than MS ($\leq$)-definability for general graphs.*

We suggest an example for proving it. Let $E(G)$ be the graph property saying: "G has an even number of prime modules", equivalently, that the modular tree **mt**(G) has an even number of nodes. Then E is MS($\leq$)-definable, because the property $E(G)$ is a CMS-property of **mt**(G)).

Conjecture 15 : *The property that a graph has an even number of prime modules is not CMS-definable.*

We have the following implications : $15 \Rightarrow 14$, $14 \Rightarrow 13$, $15 \Rightarrow 13$.

References

[1] COURNIER A., HABIB M., A new linear algorithm for modular decomposition, Proceedings of CAAP'94, LNCS **787** (1994) 68-94.

[2] COURCELLE B., On recognizable sets and tree automata, in "Resolution of equations in Algebraic Structures , Vol.1" , M. Nivat and H. Ait-Kaci eds., Acadamic Press 1989, pp. 93-126.

[3] COURCELLE B., The monadic second-order logic of graphs I: Recognizable sets of finite graphs. Information and Computation **85** (1990) 12-75.

[4] COURCELLE B., The monadic second-order logic of graphs V: On closing the gap beween definability and recorgnizability, Theoret. Comput. Sci. **80** (1991) 153-202.

[5] COURCELLE B., The monadic second-order logic of graphs VIII: Orientations, Annals Pure Applied Logic, **72** (1995) 103-143.

[6] COURCELLE B., Monadic second-order definable graph transductions: a survey, Theoret. Comput. Sci. **126**(1994) 53-75.

[7] COURCELLE B., The monadic second-order logic of graphs X: Linear orders, 1994 , to appear in Theoret. Comput. Sci. in 1996, (see also http://www.labri.u-bordeaux.fr /~courcell).

[8] COURCELLE B., Structural properties of context-free sets of graphs generated by vertex-replacement, Information and Computation **116**(1995) 275-293.

[9] COURCELLE B., Recognizable sets of graphs: Equivalent definitions and closure properties, Math. Struct. in Comp. Sci. **4**(1994) 1-32.

[10] EHRENFEUCHT A., ROZENBERG G., Theory of 2-structures. Part 2. Representations through labeled tree families, Theoret. Comput. Sci. **70**(1990) 305-342.

[11] HOOGEBOOM H., ten PAS P., Recognizable text languages, Proceedings of MFCS 1994, LNCS **841** (1994) 413-422.

[12] POTTHOF A., THOMAS W., Regular tree languages without unary symbols are star-free, Proceedings FCT '93, LNCS **710** (1993) 396-405.

[13] THOMAS W., Automata on infinite objects, in "Handbook of Theoretical Computer Science, Volume B", J. Van Leeuwen ed., Elsevier, 1990, pp. 133-192.

[14] THOMAS W., On logical definability of trace languages, Proceedings of a workshop held in Kochel in October 1989 , V. Diekert ed., Report of "Technische Universität München " I-9002, 1990, pp. 172-182.

Group Based Graph Transformations and Hierarchical Representations of Graphs[†]

A. EHRENFEUCHT[1]   T. HARJU[2]   G. ROZENBERG[1,3]

[1] Department of Computer Science, University of Colorado at Boulder
Boulder, Co 80309, U.S.A.
[2] Department of Mathematics, University of Turku
FIN-20500 Turku, Finland
[3] Department of Computer Science, Leiden University
P.O.Box 9512, 2300 RA Leiden, The Netherlands

Abstract. A labeled 2-structure, ℓ2s for short, is a complete edge-labeled directed graph without loops or multiple edges. An important result of the theory of 2-structures is the existence of a hierarchical representation of each ℓ2s. A δ-reversible labeled 2-structure g will be identified with its labeling function that maps each edge (x, y), $x \neq y$, of the domain D into a group Δ so that $g(y, x) = \delta(g(x, y))$ for an involution δ of Δ. For each mapping (selector) $\sigma: D \to \Delta$ a δ-reversible 2-structure g^σ is obtained from g by $g^\sigma(x, y) = \sigma(x)g(x, y)\delta(\sigma(y))$. A dynamic δ-reversible 2-structure $G = [g]$ generated by g is the set $\{g^\sigma \mid \sigma \text{ a selector}\}$. We define the plane trees of G to capture the hierarchical representation of G as seen by individual elements of the domain. We show that all the plane trees are strongly related to each other. Indeed, they are all obtainable from one simple unrooted undirected tree - the form of G. Thus, quite surprisingly, all hierarchical representations of ℓ2s's belonging to one dynamic ℓ2s G can be combined into one hierarchical representation of G.

1 Introduction

A labeled 2-structure is a complete edge-labeled directed graph without loops or multiple edges. The decomposition theorem of [4] for 2-structures can be used to investigate various systems such as graphs, partial orders and communication networks, where the objects of a system (nodes of a graph, or processors of a network) are linked together by binary operations or relations. For these applications we refer to [1], [4] and [7] (in [7] these structures are called *binary relational structures*).

For a finite set D let

$$E_2(D) = \{(x, y) \mid x, y \in D, \ x \neq y\}$$

[†] The authors are grateful to BRA Working Groups ASMICS, COMPUGRAPH and Stiltjes Institute for their support.

be the complete set of (directed) edges between the elements of D.

A *labeled 2-structure* (ℓ2s, for short) $g = (D, \lambda, \Delta)$ is an edge-labeled directed graph, where D is the finite *domain* of g (consisting of *nodes*), $E_2(D)$ is the set of edges and $\lambda\colon E_2(D) \to \Delta$ is a function labeling the edges by elements of the set Δ. The set Δ of labels can be infinite, while the domain D is always assumed to be finite. In the later sections Δ will be a group. We shall denote the domain D of g by $\mathrm{dom}(g)$.

The labeling function of an ℓ2s g uniquely determines g, and therefore we shall identity g with its labeling function. Thus, in this paper, an ℓ2s will be a function $g\colon E_2(D) \to \Delta$.

We continue the study of *dynamic labeled 2-structures* from [5] by defining and investigating quotients of these structures. Our main interest here is in the decomposition of dynamic ℓ2s's, and this brings us to the plane trees of these systems in Section 5.

In the rest of this section we discuss the intuition behind the definition of a dynamic labeled 2-structure.

The dynamic ℓ2s was motivated in [5] (and in [3]) by evolution of networks and similar processes. Consider a network consisting of a set of processors D, where each $x \in D$ is connected by a channel to each $y \in D$ for $y \neq x$. A channel (x, y) may assume a certain value or state from the set Δ at a specific time, *e.g.*, if there are only two values $\Delta = \{0, 1\}$, then the channels may be interpreted as having a sleeping and an active state; on the other hand, if $\Delta = \mathbb{R}$, then the channels can be considered as weighted by reals.

The *concurrent* activities of the processors modify the states of the channels. The activity of each $x \in D$ will be described by two sets of actions, O_x and I_x, by which x changes the state of a channel from and to x, respectively. The actions of x are thus mappings of Δ into Δ.

Now, for each (x, y) there are two processors, x and y, that change the state of this channel simultaneously; x changes it by a mapping from O_x and y changes it by a mapping from I_y. In order to accomodate the concurrent behaviour of the processors, the mappings from O_x and I_y should commute, *i.e.*, for each $\varphi \in O_x$ and $\psi \in I_y$, $\varphi\psi = \psi\varphi$.

To avoid unnecessary sequencing of actions the composition of two actions from O_x (I_x, resp.) is again assumed to be an action, *i.e.*, the sets O_x and I_x form transformation semigroups on Δ under the operation of composition.

Further, to assure a minimal freedom of the actions for each x we assume that for each $a, b \in \Delta$ and $x \in D$ there exist a $\varphi \in O_x$ and a $\gamma \in I_x$ such that $\varphi(a) = b$ and $\gamma(a) = b$, *i.e.*, the semigroups O_x and I_x are transitive.

Now, after these natural assumptions, it was shown in [5] (see, also [2]) that if $|D| \geq 3$, then there are two isomorphic groups O and I of permutations on Δ such that for each $x \in D$, $O_x = O$ and $I_x = I$. Thus the actions come from two groups O and I, which are independent of the processors. Moreover, as shown in [5], we can define an operation on Δ such that Δ becomes a group isomorphic to O and I. In fact, O and I become defined by left and (involutive) right multiplication of the group Δ.

These observations lead to the definition of a dynamic ℓ2s, given formally in Section 3, in terms of *selectors*. A selector σ is a mapping, which captures the action of each processor $x \in D$ at a specific stage of the network. As the intuition above suggests a selector will be simply a function $\sigma: D \to \Delta$.

A global state of a network is represented by an ℓ2s, for which Δ is the set of labels. An evolution of the network is then presented as a set of ℓ2s's that represent possible global states of the network. The transitions from one ℓ2s g to another h (and therefore from one global state to another) are the transformations $g \mapsto h$ induced by the selectors.

2 Preliminaries on labeled 2-structures

We shall recall in this section some basic definitions and properties of labeled 2-structures from [4].

Let D be a nonempty finite set. For an edge $e = (x,y) \in E_2(D)$ we let $e^{-1} = (y,x)$ be the *reverse edge* of e.

We shall assume throughout the paper that the set Δ of labels forms a (possibly infinite) group. The group properties of Δ will be made use of by the dynamic labeled 2-structures defined in Section 3.

As shown in [4] reversibility provides a natural normal form for ℓ2s's. In order to define reversibility for group labeled ℓ2s's we need to recall the notion of involution.

A mapping $\delta: \Delta \to \Delta$ is an *involution* of the group Δ, if it is a bijection such that $\delta(ab) = \delta(b)\delta(a)$ for all $a, b \in \Delta$, and $\delta^2(a) = a$ for all $a \in \Delta$. An involution δ satisfies $\delta(a^{-1}) = \delta(a)^{-1}$ for all $a \in \Delta$.

Example 1. (1) For any group Δ the most obvious involution is the inverse function: $\delta(a) = a^{-1}$ for each $a \in \Delta$.

(2) If Δ is the group of nonsingular $n \times n$-matrices (over a field F), then transposition is an involution of Δ.

(3) If Δ is a group of even order, then it has an element a of order 2, and, as easily seen, the mapping $\delta(x) = ax^{-1}a$ ($x \in \Delta$) is an involution. □

Let δ be an involution of Δ. An ℓ2s $g: E_2(D) \to \Delta$ is δ-*reversible*, if $g(e^{-1}) = \delta(g(e))$ for all edges $e \in E_2(D)$. Hence the label of $e \in E_2(D)$ determines uniquely the label of its reverse edge e^{-1}.

The family of δ-reversible ℓ2s's on the domain D and with the group of labels Δ will be denoted by $\mathcal{R}^D(\Delta, \delta)$. We reserve the symbol D for a domain, the symbol Δ for a group and the symbol δ for an involution of Δ.

Let $X \subseteq D$ be nonempty. The *substructure* $\text{sub}_g(X)$ of g *induced by* X is defined to be the restriction of g onto $E_2(X)$, i.e., $\text{sub}_g(X): E_2(X) \to \Delta$ satisfies $\text{sub}_g(X)(e) = g(e)$ for all $e \in E_2(X)$. Clearly, for each nonempty $X \subseteq D$, $\text{sub}_g(X)$ is δ-reversible, i.e., $\text{sub}_g(X) \in \mathcal{R}^X(\Delta, \delta)$.

A subset $X \subseteq D$ is a *clan* of $g \in \mathcal{R}^D(\Delta, \delta)$, if for all $x, y \in X$ and all $z \notin X$, $g(z,x) = g(z,y)$ (and hence $g(x,z) = g(y,z)$ by the reversibility condition).

Thus X is a clan if and only if each node not in X sees the nodes of X by the same labels. Clearly, the sets $\emptyset$, $\text{dom}(g)$ and $\{x\}$ for all $x \in D$ are clans of g. We shall call these the *trivial clans* of g.

Let $C(g)$ denote the family of clans of g.

Example 2. Let $\Delta = \mathbb{Z}_3$ be the additive cyclic group $(\{0,1,2\},+)$ of three elements, and let δ be the inverse function of Δ. (Indeed, the inverse function is the only involution that $\mathbb{Z}_3$ has). Consider the reversible $g \in \mathcal{R}^D(\Delta, \delta)$ from Figure 1, where the reverse edges have not been drawn, since the labels of these are uniquely determined – they are the inverses of the labels of the drawn edges. Here the domain is $D = \{x_1, x_2, x_3, x_4, x_5\}$. The set $\{x_1, x_2\}$ is not a clan, because $g(x_3, x_1) = 2 \ (= -1 \ (\text{mod } 3))$ and $g(x_3, x_2) = 0$. The set $X_1 = \{x_1, x_4, x_5\}$ is a clan, since $g(x_2, x_1) = 1 = g(x_2, x_4) = g(x_2, x_5)$ and $g(x_3, x_1) = 2 = g(x_3, x_4) = g(x_3, x_5)$. Similarly, $X_2 = \{x_4, x_5\} \in C(g)$, and these two are the only nontrivial clans of g. □

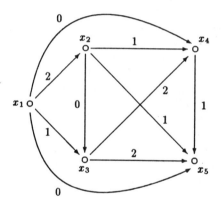

Fig. 1. A 2-structure g

The following lemma from [4] gives some basic results on clans. Here we say that the sets X and Y *overlap*, if $X \cap Y$, $X \setminus Y$ and $Y \setminus X$ are all nonempty.

Lemma 1. *Let $X, Y \in C(g)$ for a $g \in \mathcal{R}^D(\Delta, \delta)$. Then $X \cap Y \in C(g)$, and if X and Y overlap, then $X \cup Y \in C(g)$ and $X \setminus Y \in C(g)$.*

On the other hand, as seen from Example 2 the complement $\text{dom}(g) \setminus X$ of a clan X need not be a clan of g.

A label $a \in \Delta$ is called *symmetric*, if $\delta(a) = a$. Notice that for every $b \in \Delta$, the labels $b \cdot \delta(b)$ and $\delta(b) \cdot b$ are symmetric, and the identity 1_Δ of the group Δ is symmetric.

We say that g is *primitive* if it has only trivial clans; g is *complete*, if for all $e, e' \in E_2(D)$, $g(e) = g(e')$; g is *linear* if there is a label $a \in \Delta$ such that the relation $<_a$ defined by $x <_a y$ iff $g(x,y) = a$, is a linear order of the domain D.

An $\ell 2$s having a domain of two nodes is, by definition, always primitive and at the same time either complete or linear. For this reason we say that an $\ell 2$s g is *truly primitive*, if it is primitive and has at least three nodes.

If g is complete, then all its edges have the same symmetric label. If g is linear, then it has two labels a and $\delta(a)$ such that $a \neq \delta(a)$, and these labels give dual linear orderings of the domain D.

Example 3. Let g be the $\ell 2$s from Example 2 (see Figure 1). The substructure $\mathrm{sub}_g(\{x_2, x_4, x_5\})$ is linear, since it is ordered by label 1 and dually by 2, where $2 = \delta(1)$. The substructure $\mathrm{sub}_g(\{x_2, x_3\})$ is complete, since $0 = \delta(0)$. □

Let $g \in \mathcal{R}^D(\Delta, \delta)$ and $h \in \mathcal{R}^B(\Delta, \delta)$ be two reversible $\ell 2$s's on the domains D and B, respectively. We say that g and h are *(combinatorially) isomorphic*, if there are bijections $\alpha\colon D \to B$ and $\beta\colon \Delta \to \Delta$ such that for all $(x,y) \in E_2(B)$, $h(\alpha(x), \alpha(y)) = \beta(g(x,y))$. Also, g and h are *strictly isomorphic*, if they are isomorphic, $D = B$ and the bijection α above is the identity on D.

A partition $\mathcal{H} \subseteq \mathcal{C}(g)$ of the domain $D = \mathrm{dom}(g)$ into clans is called a *clan partition* of g. The *quotient* of g by $\mathcal{H}$ is defined to be the $\ell 2$s $g/\mathcal{H}$ on the domain $\mathcal{H}$ such that for all $X, Y \in \mathcal{H}$ (with $X \neq Y$), $(g/\mathcal{H})(X,Y) = g(x,y)$, whenever $x \in X$ and $y \in Y$. The following elementary result, see [4], ensures that the quotient $g/\mathcal{H}$ is well defined for all $g \in \mathcal{R}^D(\Delta, \delta)$.

Lemma 2. *Let $X, Y \in \mathcal{C}(g)$ be two disjoint clans of $g \in \mathcal{R}^D(\Delta, \delta)$. Then there is a label $a \in \Delta$ such that $g(x,y) = a$ and $g(y,x) = \delta(a)$ for all $x \in X$ and $y \in Y$. In particular, if $\mathcal{H}$ is a clan partition of g, then $g/\mathcal{H}$ is δ-reversible, i.e., $g/\mathcal{H} \in \mathcal{R}^\mathcal{H}(\Delta, \delta)$.*

Note that, by Lemma 2, every quotient $g/\mathcal{H}$ is isomorphic with a substructure of g, viz. with $\mathrm{sub}_g(X)$, where X is a set of representatives of $\mathcal{H}$ (i.e., $X \cap Y$ is a singleton for every $Y \in \mathcal{H}$); this is a rather unusual situation from the general algebraic point of view. In this isomorphism the bijection $\beta\colon \Delta \to \Delta$ is just the identity.

A nonempty clan $X \in \mathcal{C}(g)$ is *prime* if X does not overlap with any clan of g. Hence, X is a prime clan, if for each clan Y of g with $X \cap Y \neq \emptyset$, either $X \subseteq Y$ or $Y \subseteq X$. The set of all prime clans of g is denoted by $\mathcal{P}(g)$.

The following result, see [4], shows that the clans of an $\ell 2$s g are inherited by the substructures and by the quotients. For a family $\mathcal{A}$ of sets we write $\cup \mathcal{A} = \bigcup_{A \in \mathcal{A}} A$.

Lemma 3. *Let $g \in \mathcal{R}^D(\Delta, \delta)$.*
(1) *For each $X \in \mathcal{C}(g)$, $\mathcal{C}(\mathrm{sub}_g(X)) = \{Y \mid Y \in \mathcal{C}(g), Y \subseteq X\}$.*

(2) For each $X \in \mathcal{P}(g)$, $\mathcal{P}(\mathrm{sub}_g(X)) = \{Y \mid Y \in \mathcal{P}(g), Y \subseteq X\}$.

(3) If $\mathcal{H} \subseteq \mathcal{P}(g)$ is a clan partition of g into prime clans, then $\mathcal{A} \in \mathcal{C}(g/\mathcal{H})$ if and only if $\cup \mathcal{A} \in \mathcal{C}(g)$.

A prime clan X of $g \in \mathcal{R}^D(\Delta, \delta)$ is said to be *maximal*, if it is a proper clan (*i.e.*, $X \neq D$) and for any proper prime clan Y, $X \subseteq Y$ implies that $X = Y$. Let $\mathcal{P}_{\max}(g)$ denote the set of all maximal prime clans of g. For a g with a singleton domain $D = \{x\}$, we shall make the convention that $\mathcal{P}_{\max}(g) = \{\{x\}\}$.

Every singleton subset of D is a prime clan and hence each node belongs to a unique maximal prime. Therefore the maximal prime clans form a clan partition. In particular, the quotient

$$\Phi_{\max}(g) = g/\mathcal{P}_{\max}(g)$$

is well defined.

The following basic decomposition result was proved in [4].

Theorem 4. *For any g, $\Phi_{\max}(g)$ is linear, or complete, or truly primitive.*

The partially ordered set $(\mathcal{P}(g), \subseteq)$ of prime clans of g forms a rooted and directed tree $T(g)$, which we shall call the *prime tree* of g. The root of $T(g)$ is the domain D, the leaves are the singleton sets, and there is an edge (X, Y) in $T(g)$, if X is the minimal element of $\mathcal{P}(g)$ such that $Y \subseteq X$ and $X \neq Y$.

Example 4. Consider g from Example 2 (Figure 1), where the nontrivial clans of g are $X_1 = \{x_1, x_4, x_5\}$ and $X_2 = \{x_4, x_5\}$. In this example both of these clans are also prime clans of g, and X_1 is a maximal prime clan. We have $\mathcal{P}_{\max}(g) = \{X_1, \{x_2\}, \{x_3\}\}$, and hence the quotient $g/\mathcal{P}_{\max}(g)$ has three nodes, and it is primitive.

Figure 2 shows the prime tree $T(g)$ of g. There as usual we have identified a singleton $\{x\}$ with the element x.

The substructure $h_1 = \mathrm{sub}_g(X_1)$ has two maximal primes, namely X_2 and $\{x_1\}$, and the quotient $h_1/\mathcal{P}_{\max}(h_1)$ is complete. The substructure $h_2 = \mathrm{sub}_g(X_2)$ has maximal primes $\{x_4\}$ and $\{x_5\}$ and the quotient $h_2/\mathcal{P}_{\max}(h_2)$ is isomorphic to the linear $\ell 2$s h_2.

Notice that the partition $\mathcal{H} = \{\{x_1\}, \{x_2\}, \{x_3\}, X_2\}$ is also a clan partition of g into prime clans. The quotient $g/\mathcal{H}$ has four nodes, and since $X_1 \in \mathcal{C}(g)$, also $\{\{x_1\}, X_2\}$ is a clan of $g/\mathcal{H}$. □

3 Selectors and dynamic labeled 2-structures

The group Δ of labels of a $g \in \mathcal{R}^D(\Delta, \delta)$ becomes employed by the selectors, which label the nodes $x \in D$ by the elements of the group Δ.

Definition 1. Let $g \in \mathcal{R}^D(\Delta, \delta)$ for a group Δ and its involution δ.
(1) A function $\sigma: D \to \Delta$ is called a *selector*.

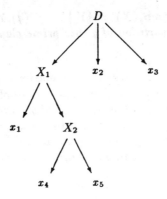

Fig. 2. The prime tree of g

(2) For a selector $\sigma: D \to \Delta$ define the *σ-transformation* of g, denoted g^σ, by
$$g^\sigma(x,y) = \sigma(x) \cdot g(x,y) \cdot \delta(\sigma(y))$$
for all $(x,y) \in E_2(D)$.

(3) The family $[g] = \{g^\sigma \mid \sigma \text{ a selector}\}$ is a *(single axiom) dynamic δ-reversible $\ell 2s$* (*generated by g*). The family of dynamic δ-reversible $\ell 2s$'s is denoted by $\mathcal{G}^D(\Delta, \delta)$.

Hence a selector $\sigma: D \to \Delta$ transforms each $g \in \mathcal{R}^D(\Delta, \delta)$ to g^σ by direct left and involutive right multiplication. The new value of an edge depends on the nodes and the label of the edge. Indeed, it can be that $g(e_1) = g(e_2)$, but $g^\sigma(e_1) \neq g^\sigma(e_2)$ for edges $e_1, e_2 \in E_2(D)$. In [5] the transformation $g \mapsto g^\sigma$ was denoted by $\text{tr}_\sigma(g)$.

Example 1. Let g be as in Example 2 (see, Figure 1). Define a selector σ by $\sigma(x_1) = 0$ and $\sigma(x_i) = g(x_1, x_i)$ for $i \neq 1$. Then g^σ is given in Figure 3.

We notice that g^σ is quite different from the original g. Now, g^σ has the nontrivial clans $\{x_2, x_3, x_4, x_5\}$, $\{x_2, x_3\}$ and $\{x_1, x_2, x_3\}$ in addition to the clans $\{x_4, x_5\}$ and $\{x_1, x_4, x_5\}$ of g. Of these clans, the first and last are not prime clans, since they overlap with each other. Also, $\{x_1, x_2, x_3\}$ is not a prime clan, since it overlaps with $\{x_2, x_3, x_4, x_5\}$. The maximal prime clans of g^σ are $\{x_1\}$, $X_0 = \{x_2, x_3\}$ and $X_2 = \{x_4, x_5\}$. In Figure 4 we have the prime tree for g^σ. □

The following lemma from [5] shows that the σ-transformation of g for of a $g \in \mathcal{R}^D(\Delta, \delta)$ is also δ-reversible.

Lemma 2. *For all $g \in \mathcal{R}^D(\Delta, \delta)$ and $\sigma: D \to \Delta$, also $g^\sigma \in \mathcal{R}^D(\Delta, \delta)$. In particular, $[g] \subseteq \mathcal{R}^D(\Delta, \delta)$.*

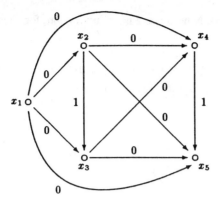

Fig. 3. The 2-structure g^σ

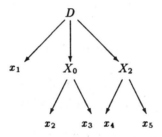

Fig. 4. The prime tree of g^σ

Proof. Let σ be a selector and $g \in \mathcal{R}^D(\Delta, \delta)$. For each edge $e = (x, y) \in E_2(D)$ we have $g(e^{-1}) = \delta(g(e))$, since g is δ-reversible. Thus

$$g^\sigma(e^{-1}) = \sigma(y) \cdot g(e^{-1}) \cdot \delta(\sigma(x)) = \sigma(y) \cdot \delta(g(e)) \cdot \delta(\sigma(x))$$
$$= \delta^2(\sigma(y)) \cdot \delta(g(e)) \cdot \delta(\sigma(x)) = \delta(\sigma(x) \cdot g(e) \cdot \delta(\sigma(y))) = \delta(g^\sigma(e)).$$

Therefore g^σ is also δ-reversible. □

The *product* of two selectors $\sigma_1, \sigma_2 \colon D \to \Delta$ is defined by $\sigma_2\sigma_1(x) = \sigma_2(x)\sigma_1(x)$ for all $x \in D$. In particular, $\sigma_2\sigma_1$ is a selector. The following lemma shows that the product is well defined.

Lemma 3. *Let $\sigma_1, \sigma_2 \colon D \to \Delta$. Then $g^{\sigma_2\sigma_1} = (g^{\sigma_1})^{\sigma_2}$ for all $g \in \mathcal{R}^D(\Delta, \delta)$.*

Proof. Let $g \in \mathcal{R}^D(\Delta, \delta)$. Now, since δ is an involution, for each edge $e = (x,y) \in E_2(D)$ we have that

$$g^{\sigma_2\sigma_1}(e) = \sigma_2(x)\sigma_1(x) \cdot g(e) \cdot \delta(\sigma_2(y)\sigma_1(y)) = \sigma_2(x)\sigma_1(x)g(e)\delta(\sigma_1(y))\delta(\sigma_2(y))$$
$$= \sigma_2(x) \cdot g^{\sigma_1}(e) \cdot \delta(\sigma_2(y)) = (g^{\sigma_1})^{\sigma_2}(e).$$

This proves the claim. □

It is rather immediate that the selectors form a group under this product. The inverse σ^{-1} of a selector σ becomes defined by $\sigma^{-1}(x) = \sigma(x)^{-1}$ for all $x \in D$. In group theoretic terms (see, e.g., [6]) the group of selectors *acts* on the δ-reversible ℓ2s's.

By the above result, if $g_1 \in [g]$ is such that $g^{\sigma} = g_1$, then $g = g_1^{\sigma^{-1}}$, and hence any ℓ2s from $[g]$ generates $[g]$. Indeed, if $g_2 = g^{\sigma_1} \in [g]$ for some selector σ, then $g_2 = g_1^{\sigma_1\sigma^{-1}} \in G(g_1)$. We have thus the following lemma.

Lemma 4. *Let $g \in \mathcal{R}^D(\Delta, \delta)$. Then $[g^{\sigma}] = [g]$ for all $\sigma: D \to \Delta$.*

4 Quotients of dynamic labeled 2-structures

Let $G = [g]$ be a dynamic δ-reversible ℓ2s. A clan of each individual $h \in G$ is called a *clan* of G. We shall write $\mathcal{C}(G) = \bigcup_{h \in G} \mathcal{C}(h)$ for the family of clans of G.

The closure properties of the clans of a dynamic δ-reversible ℓ2s G are quite different from those of ℓ2s's. It is rather easy to find an example, where the intersection of two clans of a G is no longer a clan of G. This should be contrasted to the basic result, Lemma 1, for ℓ2s's. On the other hand, as shown in [5] (see Lemma 1 below), the clans of G are closed under complement. As noticed above this is not generally true for ℓ2s's.

We let $\overline{X}$ denote the complement $D \setminus X$ of $X \subseteq D$.

A clan $X \in \mathcal{C}(g)$ is said to be *isolated*, if $g(x, y) = 1_\Delta$ for all $x \in X$ and $y \in \overline{X}$. In particular, if X is an isolated clan of g, then its complement is also a clan of g.

Lemma 1. *Let $g \in \mathcal{R}^D(\Delta, \delta)$ and $X \in \mathcal{C}(g)$. There exists a selector σ such that X is an isolated clan of g^{σ}.*

Proof. Consider a clan $X \in \mathcal{C}(g)$, and define σ as follows $\sigma(x) = 1_\Delta$ for all $x \in X$, and $\sigma(y) = \delta(g(x,y)^{-1})$ for all $y \notin X$. We have now for all edges $e = (x, y)$ with $x \in X$ and $y \notin X$, $g^{\sigma}(e) = 1_\Delta \cdot g(e) \cdot g(e)^{-1} = 1_\Delta$ as required. □

By Lemma 1, for each clan $X \in \mathcal{C}(G)$ there is a $g \in G$ such that X and $\overline{X}$ are clans of g.

Definition 2. Let $G \in \mathcal{G}^D(\Delta, \delta)$.

(1) A partition $\mathcal{H}$ of the domain D into clans of G is called a *clan partition* of G.

(2) The *quotient of G by* a clan partition $\mathcal{H}$ of G is the family

$$G/\mathcal{H} = \{g/\mathcal{H} \mid g \in G, \mathcal{H} \text{ a clan partition of } g\}.$$

By Lemma 2, the family $G/\mathcal{H}$ is well defined in the sense that each $g/\mathcal{H}$ is a δ-reversible ℓ2s. Below we show that for each clan partition $\mathcal{H} \subseteq \mathcal{C}(G)$, the quotient $G/\mathcal{H}$ is nonempty and, indeed, it is the dynamic δ-reversible ℓ2s generated by $g/\mathcal{H}$ for some $g \in G$.

We say that a selector $\sigma: D \to \Delta$ is *constant* on a subset $X \subseteq D$, if $\sigma(x_1) = \sigma(x_2)$ for all $x_1, x_2 \in X$.

The following lemmas show that the creation of a clan $X \in \mathcal{C}(G)$ in a dynamic G depends only on X. The first of these lemmas was stated in [5].

Lemma 3. *Let X be a proper clan of a $g \in \mathcal{R}^D(\Delta, \delta)$ and let σ be a selector. Then $X \in \mathcal{C}(g^\sigma)$ if and only if σ is constant on X.*

Proof. Let $y \notin X$, and $x_1, x_2 \in X$. Since $X \in \mathcal{C}(g)$, $g(y, x_1) = g(y, x_2)$. We have $g^\sigma(y, x_i) = \sigma(y)g(y, x_i)\delta(\sigma(x_i))$ for $i = 1, 2$. Hence $g^\sigma(y, x_1) = g^\sigma(y, x_2)$ holds if and only if $\sigma(x_1) = \sigma(x_2)$, which proves the claim. □

If $X = D$ (hence X is not a proper clan), and $\sigma(x) = a$ is a constant selector on X, then for all $e \in E_2(D)$, $g^\sigma(e) = a \cdot g(e) \cdot \delta(a)$. Clearly, g^σ is strictly isomorphic with g. Hence the following lemma holds.

Lemma 4. *Let $g \in \mathcal{R}^D(\Delta, \delta)$. If $h = g^\sigma$ for a selector σ constant on D, then h and g are strictly isomorphic. In this case, the substructures $\text{sub}_g(X)$ and $\text{sub}_h(X)$ and the quotients $g/\mathcal{H}$ and $h/\mathcal{H}$ are strictly isomorphic for all $X \subseteq D$ and clan partitions $\mathcal{H}$ of g.*

Our main result for quotients is the following.

Theorem 5. *Let $\mathcal{H}$ be a clan partition of a $G \in \mathcal{G}^D(\Delta, \delta)$. There exists a $g \in G$ such that $\mathcal{H}$ is a clan partition of g, and $G/\mathcal{H}$ is the dynamic δ-reversible ℓ2s generated by $g/\mathcal{H}$.*

Proof. Let for each $X \in \mathcal{H}$, $X \in \mathcal{C}(g^{\sigma_X})$ for a selector σ_X. Define $\sigma: D \to \Delta$ as follows: $\sigma(x) = \sigma_X(x)$ for all $x \in X$. Since $\mathcal{H}$ is a partition, the selector σ is well defined. By Lemma 3, we have that $X \in \mathcal{C}(g^\sigma)$ for each $X \in \mathcal{H}$, and thus $\mathcal{H}$ is a clan partition of $h = g^\sigma$.

Now, if $\mathcal{H}$ is a clan partition of some $g_1 \in [g]$, then from Lemma 4 we obtain that $g_1 = h^{\sigma_1}$ for a selector σ_1. Consequently, Lemma 3 yields that σ_1 is constant on each $X \in \mathcal{H}$, and, therefore $g_1/\mathcal{H} = (h/\mathcal{H})^{\sigma_2}$ for the selector $\sigma_2: \mathcal{H} \to \Delta$ defined by $\sigma_2(X) = \sigma_1(x)$, where $x \in X$, for all $X \in \mathcal{H}$.

On the other hand, each selector $\sigma_2: \mathcal{H} \to \Delta$ can be extended to a selector $\sigma_1: D \to \Delta$ by $\sigma_1(x) = \sigma_2(X)$ for each $X \in \mathcal{H}$ and $x \in X$. Evidently, now $h^{\sigma_1}/\mathcal{H} = (h/\mathcal{H})^{\sigma_2}$, and thus $G/\mathcal{H} = \{(h/\mathcal{H})^{\sigma_2} \mid \sigma_2 \text{ a selector}\}$ as required. □

5 Horizons and plane trees

In general, two given $g, h \in G$ can be quite different from each other. Indeed, the clan structures of g and h may be drastically different as shown in the following example.

Example 1. Let $D = \{x_1, x_2, x_3\}$, $\Delta = \mathbb{Z}_5$ and let δ be the inverse function of Δ. Let g be the complete $\ell 2s$ corresponding to the label 0, i.e., $g(e) = 0$ for all $e \in E_2(D)$. Hence all subsets of D are clans of g. Let then σ be a selector defined by $\sigma(x_1) = 0$, $\sigma(x_2) = 1$ and $\sigma(x_3) = 3$. Then $h = g^\sigma$ is specified by $h(x_1, x_2) = 4$, $h(x_2, x_3) = 3$ and $h(x_3, x_1) = 3$. Now, h is primitive, and has only the trivial clans. □

Horizons were introduced in [5] for the study of the clans of a dynamic $\ell 2s$ G. The horizons give rise to planes of G, which capture the essential structure of G.

Definition 1. Let $G \in \mathcal{G}^D(\Delta, \delta)$ and let $g \in G$.
(1) A node $x \in D$ is called a *horizon* of g, if $g(x, y) = 1_\Delta$ for all $y \neq x$.
(2) Let x be a horizon of g. The substructure $\pi_x(g) = \mathrm{sub}_g(D \setminus \{x\})$ is called the *x-plane* of g (or an *x-plane* of G).
We shall denote by G_π^x the family of x-planes of G with the horizon x.

Since for each node x of a $G \in \mathcal{G}^D(\Delta, \delta)$, $\{x\} \in \mathcal{C}(G)$, Lemma 1 implies that there exists a $g \in G$ such that x is a horizon of g. Hence G_π^x is nonempty for each $x \in D$.

Theorem 2. *Let $G \in \mathcal{G}^D(\Delta, \delta)$. The planes in G_π^x are strictly isomorphic.*

Proof. Let $g_1, g_2 \in G$ be such that x is a horizon in both of them, and let $h_i = \pi_x(g_i)$ for $i = 1, 2$. By Lemma 4, there exists a selector σ such that $g_2 = g_1^\sigma$. Clearly, $h_2 = h_1^{\sigma_1}$, where $\sigma_1 : D \setminus \{x\} \to \Delta$ is the restriction of σ. Since $D \setminus \{x\}$ is a common clan of g_1 and g_2, Lemma 3 implies that σ is constant on $D \setminus \{x\}$. By Lemma 4, h_1 and h_2 are strictly isomorphic, since $\mathrm{dom}(h_1) = D \setminus \{x\} = \mathrm{dom}(h_2)$. □

In particular, by Lemma 4, for two planes $h_1, h_2 \in G_\pi^x$ we have $\mathcal{C}(h_1) = \mathcal{C}(h_2)$ and $\mathcal{P}(h_1) = \mathcal{P}(h_2)$. It follows also that for all $X \in \mathcal{P}(h_1)$ the quotients $\Phi_{\max}(\mathrm{sub}_{h_1}(X))$ and $\Phi_{\max}(\mathrm{sub}_{h_2}(X))$ are of the same type, i.e. they are both primitive or both complete or both linear. This justifies the following definition, where an element $X \in \mathcal{P}(h_1)$ obtains a label p, c or ℓ according to the type of $\Phi_{\max}(\mathrm{sub}_{h_1}(X))$.

Definition 3. Let $x \in D$ for a $G \in \mathcal{G}^D(\Delta, \delta)$.
(1) Denote $\mathcal{C}(x) = \mathcal{C}(h_x)$, $\mathcal{P}(x) = \mathcal{P}(h_x)$ and $\mathcal{N}(x) = \mathcal{P}(h_x) \cup \{x\}$, where h_x is an arbitrary element of G_π^x.

(2) Define $\Phi_x(\{x\}) = c$, and for all $X \in \mathcal{P}(h_x)$ let $q_X = \Phi_{\max}(\text{sub}_{h_x}(X))$ and
$$\Phi_x(X) = \begin{cases} p, & \text{if } q_X \text{ is truly primitive,} \\ c, & \text{if } q_X \text{ is complete,} \\ \ell, & \text{if } q_X \text{ is linear and } |q_X| > 1. \end{cases}$$

(3) The *plane tree* $L(x)$ of the node x is the node-labeled directed rooted tree with the set of nodes $\mathcal{N}(x)$ labeled by Φ_x, which is obtained from the prime tree $T(h_x)$ by adding $\{x\}$ as a new root and $(\{x\}, D \setminus \{x\})$ as a new edge.

In the above definition we have the requirement $|q_X| > 1$ in the linear case in order to have Φ_x well defined, because, by definition, a singleton $\ell 2s$ is always both complete and linear.

Example 2. Consider g from Example 2 (see Figure 1) and the $g^\sigma \in [g]$ obtained in Example 1 (see Figure 3). In g^σ the node x_1 is a horizon. The x_1-plane h_1 of g^σ is given in Figure 5.

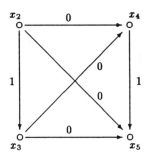

Fig. 5. Plane h_1 of $G(g)$

The nontrivial prime clans of h_1 are $\{x_2, x_3\}$ and $\{x_4, x_5\}$, and the plane tree of x_1 is given in Figure 6, where $X : d$ means that $\Phi_x(X) = d$. Note that in $X : d$, d is a label ($viz\ \Phi_x(X)$) of the node X.

Let us then make x_4 a horizon of g. This is done by defining a selector σ_1 as follows: $\sigma_1(x_4) = 0$ and $\sigma_1(x_i) = g(x_4, x_i)$ for $i \neq 4$. Now, the plane h_4 of g^{σ_1} with the horizon x_4 is given in Figure 7.

The nontrivial primes of h_4 are $\{x_1, x_2, x_3\}$ and $\{x_2, x_3\}$, and the plane tree of x_4 is given in Figure 8.

The plane trees $L(h_1)$ and $L(h_4)$ are nonisomorphic, but their underlying unrooted and undirected trees are node-isomorphic, and indeed, as we shall see, the quotients of the corresponding nodes are of the same size and type. □

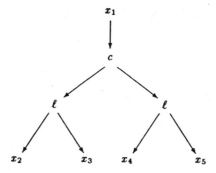

Fig. 6. Plane tree $L(x_1)$

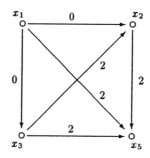

Fig. 7. Plane h_4 of $G(g)$

The next result is from [5].

Lemma 4. *Let $G \in \mathcal{G}^D(\Delta, \delta)$ and let $x \in D$ be a horizon of $g \in G$. Then $C(G) = \{X \mid X \in C(g) \text{ or } \overline{X} \in C(g)\}$.*

In particular, for the planes of G we have the following result.

Lemma 5. *Let $x \in D$ for a $G \in \mathcal{G}^D(\Delta, \delta)$. Then for each clan $X \in C(G)$ such that $X \notin \{\emptyset, D\}$, $X \in C(x)$, if $x \notin X$, and $\overline{X} \in C(x)$, if $x \in X$.*

Proof. Let x be a horizon of $g \in G$ and let $\pi_x(g) = h$. Assume $X \in C(G)$.

Assume $Y \in \{X, \overline{X}\}$ is such that $Y \in C(g)$, by Lemma 4 such a Y exists. If $x \notin Y$, then since $\text{dom}(h) = D \setminus \{x\}$ is a clan of g, Lemma 3 implies that $Y \in C(h)$.

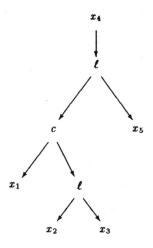

Fig. 8. The plane tree $L(x_4)$

On the other hand, if $x \in Y$, then $\overline{Y} \in \mathcal{C}(g)$. Indeed, for all $y \in Y$ ($y \neq x$) and $z \notin Y$ we have that $g(y,x) = 1_\Delta = g(z,x)$, since x is a horizon of g. Moreover, since $Y \in \mathcal{C}(g)$, $g(x,z) = g(y,z)$. Hence $g(y,z) = 1_\Delta$ for all $y \in Y$ and $z \notin Y$, which shows that $\overline{Y} \in \mathcal{C}(g)$. In this case $\overline{Y} \in \mathcal{C}(h)$ by Lemma 3. Now the claim follows from Lemma 4. □

Our purpose is to compare the planes of a dynamic G. For this we first study the relationships between the prime clans.

Lemma 6. *Let $G \in \mathcal{G}^D(\Delta, \delta)$ and $x, y \in D$. Assume that $X \in \mathcal{P}(x)$.*
  (1) *If $y \notin X$, then $X \in \mathcal{P}(y)$.*
  (2) *If $y \in X$, then $\overline{X} \in \mathcal{P}(y)$.*
In particular, $\mathcal{P}(x)$ and $\mathcal{P}(y)$ have the same number of prime clans.

Proof. The claim follows from Lemma 5, when we observe that a clan $Y \in \mathcal{C}(G)$ overlaps with a subset $Z \subseteq D$ if and only if Y overlaps with $\overline{Z}$. □

Example 3. Consider the plane trees $L(x_1)$ and $L(x_4)$ of Example 2 (see, Figures 6 and 8). Denote $A = \{x_2, x_3\}$ and $B = \{x_4, x_5\}$. We have redrawn these plane trees in Figure 9 in order to illustrate the 1-1 correspondence between them. Hence $L(x_4)$ is obtained from $L(x_1)$ by reversing the path from x_1 to x_4, and each inner node of this reversed path is the complement of its successor in the original path of $L(x_1)$; thus, e.g., the node B becomes $\overline{x_4}$. The singleton node x_4 that begins the reversed path does not change, because it does not have a successor in $L(x_1)$. Notice that the corresponding nodes are of the same type. □

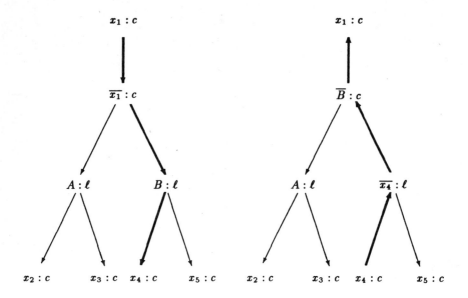

Fig. 9. Comparision of two plane trees

Following Lemma 6 (and the previous example) we define for all $x, y \in D$ a mapping $\eta_{xy} \colon \mathcal{N}(x) \to \mathcal{N}(y)$ as follows:

$$\eta_{xy}(X) = \begin{cases} X, & \text{if } y \notin X \text{ or } X = \{y\}, \\ \overline{Y}, & \text{if } y \in X, y \in Y \text{ and } (X, Y) \text{ is an edge in } L(x). \end{cases}$$

Lemma 7. *Let $G \in \mathcal{G}^D(\Delta, \delta)$ and let $x, y \in D$. Then η_{xy} is a bijective correspondence between $\mathcal{N}(x)$ and $\mathcal{N}(y)$.*

Proof. First of all η_{xy} is well defined. Indeed, for each $X \in \mathcal{N}(x)$ with $y \in X$ and $X \neq \{y\}$ there is a unique edge (X, Y) in $L(x)$ such that $y \in Y$ and $Y \subset X$. Now, by Lemma 6, $\overline{Y} \in \mathcal{N}(y)$ (with $y \notin \overline{Y}$), and, by the definition of η_{xy}, $\eta_{xy}(X) = \overline{Y}$.

The injectiveness claim follows from the above stated uniqueness of the edge (X, Y), since if $y \in X$ for some $X \in \mathcal{N}(x)$ with $X \neq \{y\}$, then for the unique $Y \in \mathcal{N}(x)$ with $\eta_{xy}(X) = \overline{Y}$ we have that $\overline{Y} \notin \mathcal{N}(x)$ (because $x \in \overline{Y}$). It follows that the mapping η_{xy} is surjective, because the sets $\mathcal{N}(x)$ and $\mathcal{N}(y)$ are finite and have the same number of elements. □

In the following result we give a simple construction how the plain tree $L(y)$ is obtained from $L(x)$. Indeed, $L(y)$ is obtained from $L(x)$ by first changing the nodes by the mapping η_{xy}, and then reversing the direction of the path from x to y.

Theorem 8. *Let $G \subseteq \mathcal{R}^D(\Delta, \delta)$ be a δ-reversible $\ell 2s$ and $x, y \in D$. If (X, Y) is an edge of $L(x)$, then*
 (1) $(\eta_{xy}(X), \eta_{xy}(Y))$ *is in* $L(y)$, *if* $y \notin Y$,
 (2) $(\eta_{xy}(Y), \eta_{xy}(X))$ *is in* $L(y)$, *if* $y \in Y$.

Proof. To simplify notation we let $\eta = \eta_{xy}$. Consider an edge (X, Y) of $L(x)$.

First of all, if $Y = \{y\}$, then $\eta(Y) = Y$, $y \in X$, and $\eta(X) = \overline{Y} = D \setminus \{y\}$. In this case $(\eta(Y), \eta(X))$ is an edge of $L(y)$ as required.

If $X = \{x\}$, we have $Y = D \setminus \{x\}$, and hence $\eta(X) = \{x\}$ and $\eta(Y) = \overline{Z}$, where Z is a maximal element in $\mathcal{P}(x)$ such that $Z \subset Y$ and $y \in Z$. Suppose $\{x\} \subset A \subset \overline{Z}$ for an $A \in \mathcal{P}(y)$. Since $x \notin \overline{A}$, Lemma 6 implies that $\overline{A} \in \mathcal{P}(x)$. Now, $Z \subset \overline{A} \subset D \setminus \{x\}$ contradicts maximality of Z. Consequently, $\{x\}$ is a maximal element in $\mathcal{P}(y)$ such that $\{x\} \subset Z$, and $(\eta(\{x\}), \eta(Y))$ is an edge in $L(y)$.

Suppose then that $X \neq \{x\}$ and $Y \neq \{y\}$.

Since (X, Y) is an edge of $L(x)$, Y is a maximal element in $\mathcal{P}(x)$ such that $Y \subset X$.

We shall consider separately three cases.

Assume first that $y \in Y$ and hence that $y \in X$. Directly from the definition of η we obtain that for all $A, B \in \mathcal{P}(x)$, we have $\eta(A), \eta(B) \in \mathcal{P}(y)$ and

(1) $A \subset B$ with $y \in A$ iff $\eta(B) \subset \eta(A)$ with $x \in \eta(B)$.

If $\eta(X) \subset F \subset \eta(Y)$ for an $F \in \mathcal{P}(y)$, then by Lemma 7 there exists a $H \in \mathcal{P}(x)$ such that $\eta(H) = F$. By (1), $\eta(X) \subset \eta(H)$ implies that $H \subset X$, and $\eta(H) \subset \eta(Y)$ implies that $Y \subset H$. Now, $Y \subset H \subset X$ contradicts the maximality of Y in X. We conclude that $(\eta(Y), \eta(X))$ is an edge of $L(y)$, whenever $y \in Y$.

Assume next that $y \notin Y$ and $y \in X$. By the definition of η we have the following equivalence for $A, B \in \mathcal{P}(x)$:

(2) $A \subset B$ with $y \notin A$ and $y \in B$ iff $\eta(A) \subset \eta(B)$ with $x \notin \eta(A)$ and $x \in \eta(B)$.

Using (2), we deduce similarly to the previous case that if $y \notin Y$ and $y \in X$, then $\eta(X)$ is a maximal element in $\mathcal{P}(y)$ such that $\eta(X) \subset \eta(Y)$. We conclude in this case that $(\eta(Y), \eta(X))$ is an edge of $L(y)$ as required.

As the last case assume that $y \notin Y$ and $y \notin X$. We have the following equivalence for $A, B \in \mathcal{P}(x)$:

(3) $A \subset B$ with $y \notin A$ and $y \notin B$ iff $\eta(B) \subset \eta(A)$ with $x \notin \eta(B)$ and $x \notin \eta(B)$.

Using (3), we conclude as above that if $y \notin Y$ and $y \notin X$, then $\eta(X)$ is a maximal element in $\mathcal{P}(y)$ such that $\eta(X) \subset \eta(Y)$. Hence, also in this case, $(\eta(Y), \eta(X))$ is an edge of $L(y)$. This concludes the proof of the theorem. □

We will now show that the correspondence η_{xy} preserves the types of the quotients.

Theorem 9. *Let $G \in \mathcal{G}^D(\Delta, \delta)$ and let $x, y \in D$. Then $\Phi_x(X) = \Phi_y(\eta_{xy}(X))$ for all $X \in \mathcal{N}(x)$.*

Proof. Let $h_x \in G_\pi^x$ and $h_y \in G_\pi^y$. Assume $h_x = \pi_x(g_x)$ and $h_y = \pi_y(g_y)$, where g_x and g_y have horizons x and y, respectively. Let σ be a selector such that $g_y = g_x^\sigma$.

Let $X \in \mathcal{N}(h_x)$. In order to simplify notation denote $\eta = \eta_{xy}$.

If X is a singleton, then also $\eta(X)$ is a singleton and in this case $\Phi_x(X) = \Phi_y(\eta(X)) = c$. Assume thus that X has at least two nodes, and thus that $X \in \mathcal{P}(h_x)$. By Lemma 3, $X \in \mathcal{C}(g_x)$.

Suppose first that $y \notin X$, and hence that $\eta(X) = X$. By Lemma 6, $X \in \mathcal{P}(h_y)$, and $X \in \mathcal{P}(g_y)$. From Lemma 3 it follows that σ is constant on the common clan X of g_x and g_y, and hence Lemma 4 implies that $\Phi_{\max}(\text{sub}_{h_x}(X))$ and $\Phi_{\max}(\text{sub}_{h_y}(X))$ are strictly isomorphic. In particular, these quotients are of the same type, i.e., $\Phi_x(X) = \Phi_y(\eta(X))$ as required.

Suppose then that $y \in X$. Let $\mathcal{P}_{\max}(\text{sub}_{h_x}(X)) = \{Y, Y_1, \ldots, Y_k\}$, where $y \in Y$. In this case $\eta(X) = \overline{Y}$, and by Theorem 8, $\mathcal{P}_{\max}(\text{sub}_{h_y}(\overline{Y})) = \{\overline{X}, Y_1, \ldots, Y_k\}$.

Assume $\Phi_{\max}(\text{sub}_{h_x}(X))$ is nonprimitive, and let $\mathcal{A}$ be a nontrivial clan of it. Let

$$\mathcal{B} = \{A \in \mathcal{P}(\text{sub}_{h_x}(X)) \mid A \notin \mathcal{A}\} \cup \{\overline{X}\}.$$

Hence $\cup \mathcal{B}$ is the complement of $\cup \mathcal{A}$ in D.

By Lemma 3, $\cup \mathcal{A} \in \mathcal{C}(\text{sub}_{h_x}(X))$ and $\cup \mathcal{A} \in \mathcal{C}(h_x)$. Now, by Lemma 5, either $\cup \mathcal{A}$ is a clan of h_y or its complement $\cup \mathcal{B}$ is a clan of h_y depending on whether or not y is in $\cup \mathcal{A}$.

If $y \notin \cup \mathcal{A}$ (and hence $\cup \mathcal{A} \in \mathcal{C}(h_y)$), then $\cup \mathcal{A} \subset \overline{Y}$, and thus, by Lemma 3 and the fact that $\overline{X} \notin \mathcal{A}$, $\mathcal{A}$ is a nontrivial clan of the quotient $\Phi_{\max}(\text{sub}_{h_y}(\overline{Y}))$.

If $y \in \cup \mathcal{A}$ (and hence $\cup \mathcal{B} \in \mathcal{C}(h_y)$), then $\cup \mathcal{B} \subset \overline{Y}$ and, by Lemma 3, $\mathcal{B}$ is a nontrivial clan of the quotient $\Phi_{\max}(\text{sub}_{h_y}(\overline{Y}))$.

A symmetric reasoning shows that if the quotient $\Phi_{\max}(\text{sub}_{h_y}(\overline{Y}))$ is nonprimitive, then also $\Phi_{\max}(\text{sub}_{h_x}(X))$ is nonprimitive. Consequently, if one of these quotients is primitive, then so is the other.

Suppose then that the quotient $q_X = \Phi_{\max}(\text{sub}_{h_x}(X))$ is complete. Let $a \in \Delta$ be a symmetric label, $a = \delta(a)$, such that $q_X(A, B) = a$ for all prime clans $A, B \in \mathcal{P}_{\max}(\text{sub}_{h_x}(X))$ with $A \neq B$.

Let $Y, A \in \mathcal{P}_{\max}(\text{sub}_{h_x}(X)$ be different prime clans such that $y \in Y$, $u \in A$. Since x is a horizon of g_x and y is a horizon of g_y, we have that $g_x(x, y) = 1_\Delta = g_x(x, u)$ and $g_y(x, y) = 1_\Delta = g_y(y, u)$. Therefore $g_y(x, y) = g_x^\sigma(x, y) = \sigma(x) \cdot g_x(x, y) \cdot \delta(\sigma(y))$, which implies that $\sigma(y) = \delta(b^{-1})$ for $b = \sigma(x)$. Furthermore, $g_y(y, u) = \delta(b^{-1}) \cdot a \cdot \delta(\sigma(u))$ implies that $\sigma(u) = b \cdot \delta(a^{-1})$, and thus $g_y(x, u) = b \cdot 1_\Delta \cdot a^{-1} \delta(b)$. Now, $g_y(u, x) = \delta(g_y(x, u)) = b \cdot \delta(a^{-1}) \cdot \delta(b) = b \cdot a^{-1} \cdot \delta(b) = g_y(x, u)$, since $\delta(a^{-1}) = a^{-1}$. We have thus deduced that the label $g_y(x, u)$ is symmetric. The quotient $\Phi_{\max}(\text{sub}_{h_y}(\overline{Y}))$ is nonprimitive (since $\Phi_{\max}(\text{sub}_{h_x}(X))$ is nonprimitive) and hence it is complete or linear. By above $\Phi_{\max}(\text{sub}_{h_y}(\overline{Y}))$ has a symmetric label on one of its edges, and this is possible only if it is complete. A symmetric reasoning shows that if $\Phi_{\max}(\text{sub}_{h_y}(\overline{Y}))$ is complete then so is $\Phi_{\max}(\text{sub}_{h_x}(X))$.

From the above cases it also follows that $\Phi_{\max}(\text{sub}_{h_x}(X))$ is linear if and only if $\Phi_{\max}(\text{sub}_{h_y}(\overline{Y}))$ is linear, because, by Theorem 4, these quotients are primitive, complete or linear. □

Let $x \in D$ for a dynamic $G \subseteq \mathcal{R}^D(\Delta, \delta)$. We define $F(x) = (\mathcal{N}(x), \Phi_x)$ as the *form* of x, that is, $F(x)$ is an undirected and unrooted tree such that $\{X, Y\}$ is an edge in $F(x)$ if and only if either (X, Y) or (Y, X) is an edge of $L(x)$. The labeling $\Phi_x : \mathcal{N}(x) \to \{p, c, \ell\}$ of the nodes remains the same as in $L(x)$.

Theorem 10. *Let $x, y \in D$ be two nodes in a $G \in \mathcal{G}^D(\Delta, \delta)$. The forms $F(x)$ and $F(y)$ are isomorphic. Indeed, $\eta_{xy} : \mathcal{N}(x) \to \mathcal{N}(y)$ is a label preserving isomorphism from $F(x)$ onto $F(y)$.*

Proof. By Theorem 8, if $\{X, Y\}$ is an edge of $F(x)$, then $\{\eta_{xy}(X), \eta_{xy}(Y)\}$ is an edge of $F(y)$. Moreover, η_{xy} is a bijection from $\mathcal{N}(x)$ onto $\mathcal{N}(y)$. This shows that η_{xy} is an isomorphism between $F(x)$ and $F(y)$.

By Theorem 9, $\Phi_x(X) = \Phi_y(\eta_{xy}(X))$ for all $X \in \mathcal{N}(x)$, and hence η_{xy} preserves the labels of the nodes. This proves the claim. □

Theorem 10 allows us to construct a tree $F(G)$, the *form* of G, for each $G \in \mathcal{G}^D(\Delta, \delta)$ as follows. Consider the tree $F(x)$ for an $x \in D$ and leave the nodes unidentified, but labeled by Φ_x. Hence a tree $F(G)$ becomes an abstract tree in the sense that the names of the nodes are omitted.

Example 4. For the dynamic $\ell 2s$ G from Example 2 (see, Figure 1), we obtain the form $F(G)$ in Figure 10. By Theorem 10 the tree $F(G)$ is independent of the choice of the node $x \in D$. □

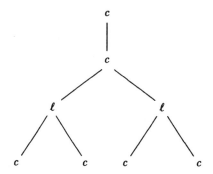

Fig. 10. The form $F(G)$

References

1. Buer, H. and R.H. Möhring, A fast algorithm for the decomposition of graphs and posets, *Math. Oper. Res.* 8 (1983) 170 – 184.
2. Ehrenfeucht, A., T. Harju and G. Rozenberg, Permuting transformation monoids, *Semigroup Forum* 47 (1993) 123 – 125.
3. Ehrenfeucht, A., T. Harju and G. Rozenberg, Invariants of 2-structures on groups of labels, Manuscript (1994).
4. Ehrenfeucht, A. and G. Rozenberg, Theory of 2-structures, Parts I and II, *Theoret. Comput. Sci.* 70 (1990) 277 – 303 and 305 – 342.
5. Ehrenfeucht, A. and G. Rozenberg, Dynamic labeled 2-structures, *Mathematical Structures in Computer Science*, to appear.
6. Rotman J.J., *The Theory of Groups: An Introduction*, Allyn and Bacon, Boston, 1973.
7. Schmerl, J. H. and W. T. Trotter, Critically indecomposable partially ordered sets, graphs, tournaments and other binary relational structures, *Discrete Math.* 113 (1993) 191 – 205.

Integrating Lineage and Interaction for the Visualization of Cellular Structures

F. David Fracchia

Graphics and Multimedia Lab, School of Computing Science
Simon Fraser University, Burnaby, B.C. V5A 1S6 CANADA

Abstract. This paper introduces *context-sensitive cell systems* as an extension of context-free cell systems for the simulation of pattern formation in two-dimensional cellular structures. The integration of interaction and lineage within the production rules allows for the modelling of division patterns that cannot solely be explained using lineage rules, such as the differentiation of epidermal leaf cells into stomata. The concept is illustrated using inhibition and reaction-diffusion models to simulate the development of a spot pattern.

1 Introduction

An important issue in developmental biology is the exploration of *morphogenesis*, or mechanisms responsible for the development of complex forms found in living organisms. These mechanisms are categorized by the way in which information is exchanged between cells, and form two fundamental classes – *lineage* and *interaction*. Cell lineage refers to the flow of information from an ancestor cell to its descendants. In contrast, interaction deals with the transfer of information between cells.

Although it has been shown that cell lineage is sufficient to model specific developmental stages of some fern gametophytes and animal embryos [4], it cannot solely explain the irregular division patterns observed in other structures, such as the differentiation of epidermal leaf cells into a periodic pattern of uniformly distributed pores surrounded by specialized cells, called *stomata* [1, 6, 14]. In such cases the flow of information between cells is an important factor.

Cell interaction plays a major role in modelling biological pattern formation by *cellular automata* and *reaction-diffusion* models. In these methods, structures are represented as regular arrays of interconnected cells, and information is passed between cells based upon some specified criterion. For example, Meinhardt [11] uses a variety of activator-inhibitor models to simulate patterns such as animal stripes and leaf venation. However, cell connections are static, cells cannot divide, and any expansion is limited to the border.

Map L-systems, proposed by Lindenmayer and Rozenberg [8], allow for cell divisions within a structure using rules which operate on cell walls. However, observations which capture the behavior of cells must be translated to wall productions using a laborious, nonintuitive procedure. Also, representing cell interaction at the level of walls is difficult and highly nonintuitive, and has met with limited success in *map IL-systems* [12].

Double-wall map L-systems, which allow labels on either side of the wall (thus, each cell has its own labels for shared walls), have been extensively used by Lück and Lück [9, 10] to model various organisms. However, as in the single-walled case, they are not easily extended to include context.

Context-sensitive L-systems have been successfully employed to visualize many plants and trees by Prusinkiewicz and Lindenmayer [13], but are not readily applicable to cyclic structures such as cells.

Context-free cell systems [3], are based solely on lineage. They were shown to include all bounded map 0L-systems, while utilizing a fraction of the number of productions. Also, they produce a structure which exhibits variability.

This paper introduces *context-sensitive cell systems* as a framework for the simulation of pattern and shape formation in two-dimensional cellular structures. They are an extension of context-free cell systems (lineage) and cellular automata (interaction). The geometry of the generated structures is determined by the mechanical interaction of cells due to cell pressure and wall tension [4].

It is important to note that this research focuses on organisms which preserve cell neighbourhoods (mainly plants). In this case, the mechanisms of lineage and interaction influence the spatial and temporal organization of cell divisions; the underlying *cell division patterns* which determine cellular development. The timing and orientation of divisions are believed to be the main factors determining the form of a tissue [2].

This paper is organized as follows. Section 2 provides the formal definition of context-sensitive cell systems, introduces a notation for rule specification, and presents a simple example based on the *game of life* [5]. Several more elaborate examples, intended to simulate a spot pattern using inhibition and reaction-diffusion models, are presented in Sect. 3. Section 4 discusses future research.

2   Context-Sensitive Cell Systems

Context-sensitive cell systems operate on cellular structures developing in a vector field. Since they are an extension of context-free cell systems, several of the initial definitions are borrowed from [3].

Definition 2.1. A *cellular structure* is a finite set of cells, each characterized by its *shape* and *state*. The cells are bounded, simple, non-overlapping polygons that tile a surface, thus forming a *polygon mesh*.

Definition 2.2. A vector field $\mathcal{Z}$ over a surface X is an assignment of a vector $\mathbf{Z}$ to each point of X.

A vector field, regarded as a *polarity field*, provides directional information to the cells. Meinhardt [11, pages 6-7] describes a polarity field as a "distribution of a particular property over a field of cells". He characterizes such fields as having at least three origins: defined by cell composition changes, subdivided into distinguishable substructures, and specified by the polarity and alignment

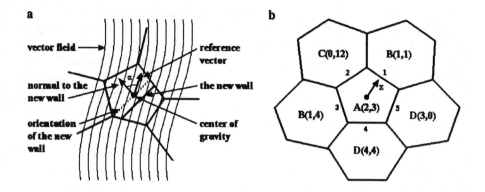

Fig. 1. Context-sensitive cell systems. (a) Specification of the division wall with respect to the vector field. (b) Indexing of neighbour cells for context.

of cells. Vector fields may be specified independently of the structure or be controlled locally by the structure itself. Examples include light direction, gravity, and concentration gradient of a morphogen.

Definition 2.3. A *context-sensitive cell system* $\mathcal{H}$ over a vector field $\mathcal{Z}$ consists of:

- a finite *alphabet* of cell labels Σ,
- a finite set of *formal parameters* Γ,
- a finite set of *indices* $\Pi = I \cup \lambda$, where I is the set of positive integers,
- a *starting cellular structure* ω with labels from Σ and parameters from $\Re$ (real numbers), and
- a finite *set of cell productions*

$$P \subset (\Sigma \times \Gamma^*) \times \mathcal{C}(\Gamma \times \Pi) \times (\Sigma \times \mathcal{E}(\Gamma \times \Pi)) \times (\lambda \cup (\mathcal{E}(\Gamma \times \Pi) \times \mathcal{E}(\Gamma \times \Pi) \times (\Sigma \times \mathcal{E}(\Gamma \times \Pi))))$$

where $\mathcal{C}$ and $\mathcal{E}$ denote the sets of all correctly constructed *logical* and *arithmetic expressions* which consist of: numeric constants; arithmetic operators $+,-,\star,\wedge$; relational operators $<,>,=$; logical operators $!,\&,|$; and parentheses. The operators : and $\rightarrow$ separate the components of a production: the *predecessor, condition,* and *successor*. The orientation of a division wall is specified in the successor by the symbol $\uparrow$ and two parameters: the division wall angle and the relative size of the first daughter cell to its mother.

The new division wall is positioned such that the angle between its normal and the reference vector $\mathbf{Z}$, measured at the center of gravity of the cell, is equal to the division angle specified in the successor of the matching production rule. The wall is then adjusted to account for differences in the relative sizes of the daughter cells (Fig. 1a).

In order to extend context-free cell systems to include interaction it is assumed that each cell has associated with it numerical parameters which represent various characteristics, such as the age of a cell, size of a cell, concentration of a morphogen, number of neighbouring cells, and so on (model dependent). The new label and parameters of a cell, or those of its daughters, are determined by taking into account the labels and parameter values of neighbouring cells (Fig. 1b). The reference vector **Z** points to neighbour 1 with others numbered counterclockwise. Neighbour cell parameters are referred to by neighbour number. These parameters can also influence the orientation of division walls and the relative sizes of daughter cells.

Furthermore, it is assumed that all cells have the same set of formal parameters. This simplifies the task of referring to parameters of neighbour cells. If a parameter is not applicable to a specific cell type, its value is considered undefined. For example, if parameters a and d of cell $A(a, b, c, d)$ are unnecessary (considered to be NULL), the cell is written simply as $A(b, c)$.

For example, the production:

$$A(a,b) : b_2 > 10 \rightarrow B(a_1 + a, b_2 - b) \uparrow (90 + a/b_1, 0.7) \, C(0, b)$$

divides cell A into two cells B and C if the actual value of parameter b of neighbour 2 is greater than 10. Values of parameters in cells B and C, as well as the angle used to position the division wall with respect to **Z**, are calculated by substituting the actual values into the formal parameters. Given the cellular structure shown in Fig. 1b, cell $A(2, 3)$ splits into cells $B(3, 9)$ and $C(0, 3)$. Cell B is 7/10 the size of A, and the division wall angle is 92°.

Definition 2.4. A cell system $\mathcal{H}$ operates in a sequence of *derivation steps*, noted $\mu_i \Longrightarrow \mu_{i+1}$, during which the *predecessor structure* μ_i yields the *successor structure* μ_{i+1}. Each derivation step consists of three phases. First, a structure μ'_{i+1} is obtained from μ_i by applying productions from P *simultaneously* to all cells using vector field $\mathcal{Z}_i$. Second, a structure μ_{i+1} is obtained from μ'_{i+1} by applying the dynamic model to determine geometry. Finally, the vector field is recomputed ($\mathcal{Z}_{i+1}$).

The sequence of structures starting with $\mu_0 = \omega$ is called the *developmental sequence* generated by $\mathcal{H}$.

Example 2.1. Figure 2 shows the development of a simple pattern on a 12×12 array of cells, using context-sensitive cell systems, based on the *game of life* as invented by Conway and reported in Gardner [5]. In this case, the fate of a cell relies entirely on the states of neighbouring cells.

Each cell has eight neighbours, corresponding to the cells which share its vertices, and may be in one of two states: 0 (dead) or 1 (alive). The rules for determining a cell's state are as follows:

- If a cell is dead and three of its neighbours are alive, then the cell will be alive in the new structure.

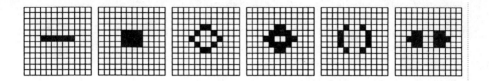

Fig. 2. The *game of life*. White cells are dead while dark cells are alive.

- If a cell is alive and less than two or more than three of its neighbours are alive, then the cell will be dead in the new structure.

In all other cases, the state of a cell remains unchanged.
The rules are applied to all cells in parallel such that the new state of a cell is based on the previous states of its neighbours. The rules are expressed by the following cell system:

ω : 12 × 12 array of alive A(1) and dead A(0) cells
p_1 : A(s) : s = 0 & S = 3 $\rightarrow$ A(1)
p_2 : A(s) : s = 1 & (S < 2|S > 3) $\rightarrow$ A(0)

where $S = \sum_{j=1}^{n} s_j$ is the sum of the n neighbours' states (in this case, $n = 8$ for all cells).

In this case, the cellular structure remains static. However, unlike cellular automata, cell systems have the ability to locally modify cell-cell connections in any part of the structure.

Example 2.2. Figure 3 presents a variant of the game of life that allows for cell division. The rules are the same as in the previous example with the following exception: cells that stay alive over one derivation step give rise to two (alive) daughter cells. As a consequence, the number of neighbours n varies between cells. Initially, all reference vectors are pointing upwards. Then the vectors are assumed to be normal to the division wall for each of the subsequent daughter cells. The rules are expressed by the following cell system:

ω : 12 × 12 array of alive A(1) and dead A(0) cells
p_1 : A(s) : s = 0 & S = 3 $\rightarrow$ A(1)
p_2 : A(s) : s = 1 & (S < 2|S > 3) $\rightarrow$ A(0)
p_3 : A(s) : s = 1 & (S = 2|S = 3) $\rightarrow$ A(1) $\uparrow$ (90.0) A(1)

In these two examples the parameter values of the cells do not influence the orientation of the division walls. However, in the next section, it is shown that such interaction is necessary to simulate the development of a specific pattern.

3 Spot Pattern Formation

The interaction between cells has been postulated to play a major role in the formation of cell differentiation patterns in developing tissues. One possible method

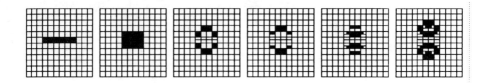

Fig. 3. The *game of life* with cell divisions.

of interaction is by the diffusion of morphogens whose concentrations influence the differentiation of cells into distinct types. Differentiation patterns then result from the spatial distribution of the concentrations. A variety of models for pattern formation have appeared in the literature. Many such models are presented, and others surveyed, in [11].

A frequently occurring pattern exhibited by many organisms is the periodic distribution of differentiated cells, or simply spots. Examples of such distributions are patterns of stomata on the surface of a leaf [1, 6, 14], celia on an embryo [7], and bristles on the cuticle of an insect [18].

This section explores two models which generate a relatively uniform spacing of differentiated cells in a structure in which undifferentiated cells divide. The first uses a single morphogen to inhibit the differentiation of cells, while the second employs two morphogens to regulate the size of the "spots".

Example 3.1. This example explores the effects of a model based on the diffusion of an inhibitory morphogen on cell differentiation patterns in a growing structure. The developmental sequence is composed of synchronous cell divisions separated by periods of morphogen diffusion and cell differentiation.

The model operates as follows:

- there are two types of cells: undifferentiated cells (U) and differentiated cells (D),
- the concentration c of inhibitor in a cell (x, y) over time t is determined by the production p, decay b, and diffusion D_f of the inhibitor according to Fick's law [17]:

$$\frac{dc}{dt} = (\frac{d^2c}{dx^2} + \frac{d^2c}{dy^2})D_f + p - bc$$

and is discretized as follows. Let c_j, $j = 1, \ldots, n$ denote the concentrations in neighbouring cells. In a time step Δt the change in concentration Δc of a cell is:

$$\Delta c = D_f \frac{\Delta t}{(\Delta l)^2} \sum_{j=1}^{n} (c_j - c) + (p - bc)\Delta t \qquad (1)$$

where Δl is the diameter of a circle inscribed by the cell. Δl and Δt are assumed to be constant $= 1$ (to simplify the calculations, cells are treated as single points in space of equal distance apart). Note however that this disregards the geometry of the structure, which may play a crucial role. The

simplification made is based on the assumption that the cells will maintain a fairly regular geometric pattern.
- inhibitor production occurs only in D cells at a constant rate p while the concentration c is below a maximum M ($p = 0$ for U cells). This feedback approach assumes that high concentrations stop the production of inhibitor, thus eliminating an unnaturally large build-up,
- the inhibitor in U cells decays at a constant rate b ($b = 0$ for D cells), and a U cell differentiates into a D cell when its inhibitor concentration is below a fixed threshold T,
- D cells do not divide, while U cells divide synchronously according to the *Korn-Spalding* pattern (as proposed in [8]). The division angle has been computed to be $-79.11°$ [3],
- the initial inhibitor concentration in daughter cells equals that of the mother cell, and
- the orientation of the division wall is determined by cell lineage if the concentration c is low. For increasing concentrations the orientation biases towards being perpendicular to the local concentration gradient. This guarantees that the space between D cells increases. That is:

$$division\ angle = \begin{cases} -79.11\left(1 - \frac{c}{K}\right) + \gamma\frac{c}{K} & c < K \\ \gamma & c \geq K \end{cases} \qquad (2)$$

where γ is the angle between the reference vector and the morphogen gradient vector, and K is the concentration at which the division wall is perpendicular to the gradient.

The orientation of division walls of undifferentiated cells is specified relative to the vector field which is perpendicular to the previous division wall. The gradient vector $\mathbf{G}$ at a particular cell reflects the direction and magnitude of the average difference in concentration between the cell and its neighbours. It is calculated as follows:

$$\mathbf{G} = \frac{1}{n}\sum_{j=1}^{n}\mathbf{G_j}$$

where $\mathbf{G_j}$ is directed towards the cell of higher concentration, for each cell-neighbour pair, with length $|c - c_j|$.

The initial cellular structure ω consists of six hexagonal undifferentiated cells $U(4, 0, 0.05, 0.1, 0)$ surrounding a differentiated hexagonal cell $D(10, 2)$ (short for $D(10, 2, NULL, NULL, NULL)$). The parameters are concentration c, production rate p, decay rate b, rate of diffusion D_f, and number of iterations N (in that order). Note that although the values of p, b, and D_f are constant, they differ for each cell type and thus are carried as cell parameters rather than treated as constants such as T (although the latter is possible of course).

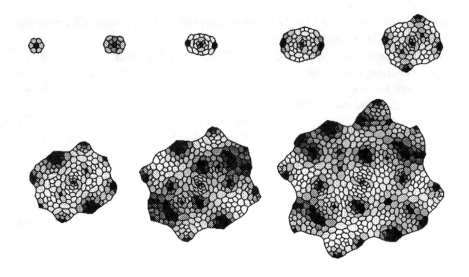

Fig. 4. Simulated development of a spacing pattern using an inhibition model. Colours indicate the concentration of inhibitor in each cell; the darker, the higher the concentration (differentiated cells are black).

The following productions capture the assumptions of the inhibition model.

```
#define T  1.75   /* concentration threshold */
#define CI 5      /* iterations between differentiation */
#define DI 50     /* iterations between divisions */
```
$p_1 : U(c,p,b,D_f,N) : N = CI \;\&\; c \leq T \rightarrow D(c,2)$
$p_2 : U(c,p,b,D_f,N) : N < DI \rightarrow U(c',p,b,D_f,N+1)$
$p_3 : U(c,p,b,D_f,N) : N = DI \rightarrow U(c,p,b,D_f,0) \uparrow (\alpha, 0.5) \; U(c,p,b,D_f,0)$
$p_4 : D(c,p) : \star \rightarrow D(c',p)$

where c' denotes the new concentration $c' = c + \Delta c$ (Eqn. 1), and α is the division angle (Eqn. 2). The $\star$ in production p_4 denotes an empty condition. The maximum concentration M is set to 10 and K is 10. The values of parameters were derived by intuition and trial-and-error.

Productions p_2 and p_4 are responsible for inhibitor production, decay, and diffusion in U and D cells, respectively. If after CI derivation steps, a U cell's concentration drops below or is equal to the concentration threshold, then the cell differentiates into a D cell by production p_1. The delay of five steps allows for the diffusion of morphogen from new D cells so that neighbouring cells have less chance of differentiating. This reduces the frequency of two neighbouring D cells. Production p_3 divides U cells into two undifferentiated daughter cells every DI derivation steps. The developmental sequence is composed of synchronous cell divisions separated by periods of morphogen diffusion and cell differentiation.

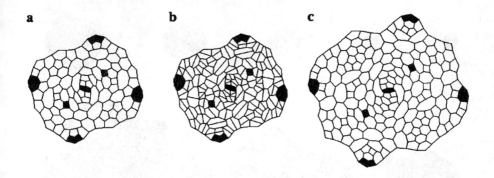

Fig. 5. Spacing of differentiated cells due to cell division and expansion: (a) structure with a distribution of differentiated cells (black), (b) cell divisions oriented by lineage and gradient, and (c) the resulting expansion of the structure.

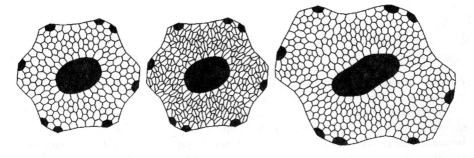

Fig. 6. Division angle kept constant. Notice the increase in size of differentiated cells due to subsequent incident division walls.

The simulation showing successive division steps, each corresponding to fifty derivation steps, is illustrated in Fig. 4. The differentiated cells do exhibit a relatively uniform distribution throughout the structure. Also, the closer an undifferentiated cell is to a differentiated cell, the more biased the orientation of the new division wall is towards being perpendicular to the path between the cells, since the difference in concentration of the two types of cells is relatively high. The result is an expansion of the space (occupied by undifferentiated cells) between differentiated cells, as illustrated in Fig. 5. This also keeps the size of differentiated cells low by minimizing the number of intersections with the division walls of neighbouring undifferentiated cells. This is quite apparent in the case when the division angle is kept constant and not affected by the concentration gradient (Fig. 6).

However, the differentiated cells form clusters when groups of cells with similar concentrations all fall below the specified threshold and differentiate. By adjusting the inhibition model parameters and concentration threshold, it is

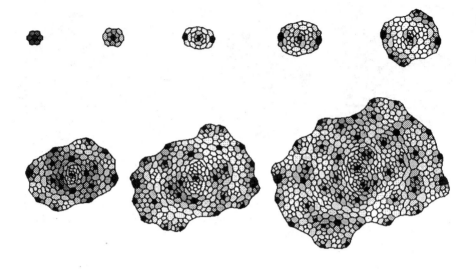

Fig. 7. Inhibition model constrained to produce single-celled spots.

possible to reduce the size of these groups. However, after numerous attempts, it was not possible to limit the spot size to a single cell.

Specifying additional constraints could limit the spot size. Korn [6] implemented a model of stomata formation in which each cell within a group of neighbouring cells ready to differentiate is assigned an equal probability of becoming a stoma. Those destined to differentiate into stomatal cells are selected randomly such that a specified ratio of stomata to total number of cells is maintained. A simpler alternative is to modify the production rule controlling differentiation such that an undifferentiated cell with an inhibitor concentration below the specified threshold can only differentiate if its concentration is lower than that of its neighbouring cells. The modified production rule is:

$$p_1 : \mathtt{U}(c,p,b,D_f,N) : \mathtt{N} = \mathtt{CI} \ \& \ c \leq \mathtt{T} \ \& \ c < c_j \ (j = 1 \ldots n) \rightarrow \mathtt{D}(c,2)$$

The resulting pattern is composed of spots no larger than a single cell (Fig. 7).

Example 3.2. Although the constrained inhibition model produced a relatively uniform distribution of single-celled spots, it does not prevent the emergence of pairs of differentiated cells, separated by a single undifferentiated cell. Thus it masks the problem of closely located differentiated cells. A potential solution is the use of two morphogens rather than one. These morphogens diffuse throughout the structure and react with each other to either increase or decrease their concentrations until a stable pattern is formed. This process is known as *reaction-diffusion* and was first described by Turing [15].

The model in the previous example is modified as follows:

- the concentration c, production p, decay b, and diffusion D_f of the inhibitor are replaced by the concentrations a and i, diffusion rates D_a and D_i, scale s, and concentration irregularity β. Concentrations a and i denote the amount of *activator* and *inhibitor*. The discrete form of the change in concentration according to Turing [15] and described in Turk [16] is as follows. Let $a_j, i_j, j = 1, \ldots, n$ denote the concentrations in neighbouring cells. In a single step the change in concentrations Δa and Δi of a cell is:

$$\Delta a = s(ai - a - \beta) + D_a \sum_{j=1}^{n}(a_j - a) \qquad (3)$$

$$\Delta i = s(16 - ai) + D_i \sum_{j=1}^{n}(i_j - i) \qquad (4)$$

Once again, to simplify the calculations, cells are treated as single points in space of equal distance apart.

Parameter s is constant for all cells and can be used to control the size of spots since a small value slows down the reaction part of the system relative to the diffusion (large spots) and a large value speeds it up (small spots). Parameter β varies from cell to cell and introduces slight irregularities in the activator concentration in order to initially destabilize the structure.

The initial cellular structure ω consists of six hexagonal undifferentiated cells $U(4, 4, 0.03125, 0.125, 12 \pm 0.05, 0)$ surrounding a differentiated hexagonal cell $D(7, 0)$. The parameters are concentrations a and i, diffusion rates D_a and D_i, random variation β, and number of iterations N (in that order).
The following productions capture the assumptions of the reaction-diffusion model.

```
#define I    30.0    /* inhibitor threshold */
#define CI   10      /* iterations between differentiation */
#define DI   400     /* iterations between divisions */
p1 : U(a, i, Da, Di, β, N) : N = CI & i > I  → D(7, 0)
p2 : U(a, i, Da, Di, β, N) : N < DI          → U(a', i', Da, Di, β, N + 1)
p3 : U(a, i, Da, Di, β, N) : N = DI          → U(a, i, Da, Di, β, 0) ↑ (α, 0.5)
                                                U(a, i, Da, Di, β, 0)
p4 : D(a, i)                 : *             → D(a, i)
```

where $a' = a + \Delta a$ and $i' = i + \Delta i$ (Eqns. 3 and 4), and α is the division angle (Eqn. 2 – K is 15). Parameter s was chosen to be 0.05.

Parameter values were derived by first experimenting with the reaction-diffusion equations on a regular hexagonal grid. By increasing the value of s it was possible to generate small spots a short distance apart. However, once introduced into a growing cellular structure, the system behaved quite erratically - often concentrations quickly reached values approaching infinity. Finally, after numerous attempts, a compromise was achieved whereby cells at new peaks (in

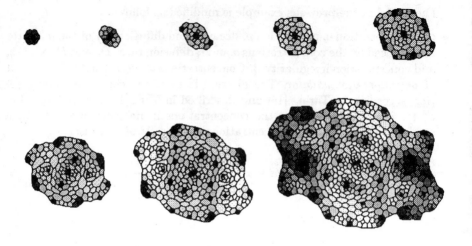

Fig. 8. Simulated development of a spacing pattern using a reaction-diffusion model. Colours indicate the concentration of inhibitor in each cell; the darker, the lower the concentration (differentiated cells are black).

this case, inhibitor) were differentiated into cells with minimum inhibitor concentration. With further analyses of the behaviour of reaction-diffusion models on developing structures, this model can be improved significantly.

The simulation showing successive division steps, each corresponding to four hundred derivation steps (it took much longer to reach a stable state than in the previous example), is illustrated in Fig. 8.

Although the differentiated cells do exhibit a relatively uniform distribution throughout the structure, a few (13 out of 50 - only 4 with more than two cells) multi-celled clusters still appeared. However, irregularly shaped spots can also occur in the static regular hexagonal grid case (Fig. 9). By imposing a similar constraint as in the previous example, the spots could be all be reduced to a single cell (Fig. 10).

4 Conclusions

This paper presents a new extension to cell systems for integrating lineage and interaction. Context is supplied via a mechanism for identifying and retrieving information from neighbouring cells.

Context-sensitive cell systems are important for several reasons. First, they can be used to study the effects of lineage and interaction on development. Second, they can be applied to visualize the development of many multicellular

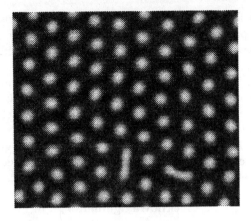

Fig. 9. Reaction-diffusion on a regular hexagonal grid also produces irregular spots.

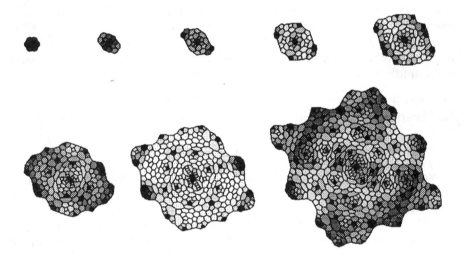

Fig. 10. Reaction-diffusion model constrained to produce single-celled spots.

structures using a relatively small number of rules. Third, the model can be validated easily with observations of the real organism. Finally, the methods provide an environment for conducting further research.

The choice of appropriate reaction-diffusion equations is also important. For example, the problem of concentration values approaching infinity could probably be avoided with the introduction of nonlinear terms controlling saturation (as in [11]). Also, the examples shown do not solve the problem of the emergence of closely located differentiated cells.

Furthermore to simplify calculations cells are treated as single points in space of equal distance apart, based on the assumption that the cells will maintain a fairly regular geometric pattern. However, diffusion is dependent on the geometry of the space.

Future research includes the current development and implementation of a more general and powerful extension of the cell system described here. The focus is on integrating the derivation steps with the dynamic model of cell shape and expanding the functionality of the rule specification.

Our ongoing study includes investigating the effect of growth on pattern formation, and the effect of such patterns on growth. A better understanding of how morphogens behave in developing structures is critical to such research.

Acknowledgements

Many thanks to Dr. Przemyslaw Prusinkiewicz for his insights into and contributions to this research, and to both he and Dr. Martin de Boer for their contributions to the initial context-free cell systems. Thanks to Sheelagh Carpendale as well, for reading previous drafts and providing helpful suggestions. This research was supported by research and equipment grants from the Natural Sciences and Engineering Research Council of Canada, and the SFU President's Research grant. Thanks also to the Graphics and Multimedia Lab and the School of Computing Science, Simon Fraser University.

References

1. E. Bünning. Die entstehung von mustern in der entwicklung von pflanzen. *Hand. Pflanzenphysiol*, XV(1):383–408, 1965.
2. M. J. M. de Boer. *Analysis and computer generation of division patterns in cell layers using developmental algorithms*. PhD thesis, University of Utrecht, the Netherlands, 1989.
3. M. J. M. de Boer, F. D. Fracchia, and P. Prusinkiewicz. A model for cellular development in morphogenetic fields. In G. Rozenberg and A. Salomaa, editors, *Lindenmayer Systems. Impacts on Theoretical Computer Science, Computer Graphics, and Developmental Biology*, pages 351–370. Springer-Verlag, Berlin, 1992.
4. F. D. Fracchia, P. Prusinkiewicz, and M. J. M. de Boer. Visualization of the development of multicellular structures. In *Proceedings of Graphics Interface '90*, pages 267–277, 1990.
5. M. Gardner. The fantastic combinations of John Conway's new solitaire game 'life'. *Scientific American*, 223(4):120–123, 1970.
6. R. W. Korn. A neighboring-inhibition model for stomate patterning. *Dev. Biol.*, 88:115–120, 1981.
7. U. Landström. On the differentiation of perspective ectoderm to a ciliated cell pattern in embryos of *Ambystoma mexicanum*. *J. Embryol. Exp. Morph.*, 41:23–32, 1977.
8. A. Lindenmayer and G. Rozenberg. Parallel generation of maps: Developmental systems for cell layers. In V. Claus, H. Ehrig, and G. Rozenberg, editors, *Graph grammars and their application to computer science; First International Workshop*,

Lecture Notes in Computer Science 73, pages 301–316. Springer-Verlag, Berlin, 1979.
9. H. B. Lück and J. Lück. Vers une metrie des graphes evolutifs, representatifs d'ensembles cellulaires. In H. Le Guyader and T. Moulin, editors, *Actes du premier seminaire de l'Ecole de Biologie Théorique du CNRS*, pages 373–398. Ecole Nat. Sup. de Techn. Avanc., Paris, 1981.
10. J. Lück and H. B. Lück. Sur la structure de l'organisation tissulaire et son incidence sur la morphogenèse. In Hervé Le Guardier, editor, *Actes du deuxième séminaire de l'Ecole de Biologie Théorique du CNRS*, pages 385–397. Publications de l'Université de Rouen, Abbaye de Solignac, 1982.
11. H. Meinhardt. *Models of biological pattern formation*. Academic Press, London & New York, 1982.
12. A. Nakamura, A. Lindenmayer, and K. Aizawa. Some systems for map generation. In G. Rozenberg and A. Salomaa, editors, *The Book of L*, pages 323–332. Springer-Verlag, Berlin, 1986.
13. P. Prusinkiewicz and A. Lindenmayer. *The Algorithmic Beauty of Plants*. Springer-Verlag, New York, 1990. With J. Hanan, F. D. Fracchia, D. R. Fowler, M. J. M. de Boer and L. Mercer.
14. T. Sachs. *Pattern formation in plant tissues*. Cambridge University Press, Cambridge, 1991.
15. A. Turing. The chemical basis of morphogenesis. *Phil. Trans. Roy. Soc. B*, 237(32):5–72, 1952.
16. G. Turk. Generating synthetic textures on arbitrary surfaces using reaction-diffusion, 1991. Proceedings of SIGGRAPH '91 (Las Vegas, Nevada, July 28 - August 2, 1991), in *Computer Graphics* 25,4 (July 1991), pages 289–298, ACM SIGGRAPH, New York, 1991.
17. A. H. Veen and A. Lindenmayer. Diffusion mechanism for phyllotaxis: Theoretical physico-chemical and computer study. *Plant Physiology*, 60:127–139, 1977.
18. V. B. Wigglesworth. Local and general factors in the development of a pattern in *Rhodnius prolixus*. *J. Exp. Biol.*, 17:180–200, 1940.

Cellworks with Cell Rewriting and Cell Packing for Plant Morphogenesis

Jacqueline Lück & Hermann B. Lück

Laboratoire de Botanique analytique et Structuralisme végétal, Faculté des Sciences et Techniques de St-Jérôme, CNRS, URA 1152, 13397 Marseille cedex 13, France

Abstract. The development of plant meristems is simulated by 3D-cellworks. They are topological systems based on a sub-alphabet of labels designating cells and a corresponding set of convex polyhedra with face and node labels, a set of cell productions, a table of cell contacts, and an initial cellwork of one or several cells. A cellwork derivation is realized, conformable to cell productions and the table, by a new packing of cells with node and wall fusion. An algorithm defines, for a given polyhedral cellular conformation, all possible ways of auto-reproductive cell divisions. The generation age of walls is used for the construction of such cellworks.

1. Introduction

Cellwork systems are extensions of L-systems [1] which were found to be useful for the generation of 3D-patterns that describe the development of plant meristems. The term was introduced by [2] for cycle edge label control. They are also investigated as marker systems [3]. In our approach, these systems characterized by double walls and face labels were previously based on wall matching rules [4, 5]. We now propose a new and more intuitive version of these systems with pattern generation based on parallel derivation of cells but, for the geometrical representation, a sequential packing of cells with glueing of cell walls.

2. The Biological Background

The biological background is emphasized by Fig.1a which represents the apical region of a simple fern shoot. At the top of the shoot figures an apical tetrahedral cell from which derive all other cells of the shoot. The apical cell, say a, is composed of a triangular surface wall and three other walls which are embedded in the shoot tissue and touch neighbour cells. Each division of the apical cell furnishes, with one of its daughter cells, a new apical a cell in central position and sideways a sister cell, say a b cell. The division wall separating these two daughter cells is parallel to one of the embedded walls.

Seen from above, the division walls (dw) of successive partitions of the a cell are positioned in a helicoidal way (Fig.1b). Such a pattern is called a 3-sided helicoidal segmentation. Each *b* cell produces by further divisions a huge number of cells which constitute a segment, also called a *merophyte*. Leaves may appear in direct relation to the helicoidal arrangement of merophytes. But this is not always the case: a shoot constructed on the base of a 3-sided helicoidal segmentation may bear leaves arranged in a 2-5 phyllotactic pattern.

We admit that the regularity of such a division pattern of an apical cell results from the repetition of a specific positioning of the division wall with respect to the position of the former division wall of a cell. This procedure can be expressed by some rules describing the transformation of the cell walls which perform such a pattern.

3. The Cellwork System

The cellwork system proposed presently, $G = \{\Sigma, P, T, c_0\}$, is based on a finite alphabet Σ of n labelled convex polyhedra, representing cells. The alphabet Σ is composed of a sub-alphabet C of labels designating cell types, and a sub-alphabet F of polyhedral representations, $\Sigma = C \cup F$, $C = \{c_1,...,c_n\}$ and $F = \{f_1,...,f_n\}$. To each label c_i corresponds a polyhedron f_i. The index of c and f indicates the polyhedral degree of the cell. A set P of cell productions such as $c \rightarrow \omega$, $c \in C$, $\omega \in C^*$ and $C \subset \Sigma$ specifies how cell types are derived one from another, ω being composed of one or two cells. The corresponding derivations of polyhedra specify how the polyhedral wall boundary of a cell changes during a developmental step. The starting cellwork c_0 consists of one or several cells.

For the cellwork derivation we use first a parallel cell derivation using cell productions of P and generating strings of cell labels like in an ordinary PD0L system. Secondly, for the geometrical representation, cells are packed sequentially, conformably to the sequence given by the derived strings of cell labels. The way this is achieved finds further specifications in the *table of cell contacts T* indicating how the corresponding polyhedra are associated in a 3D-space. Using its glueing device,

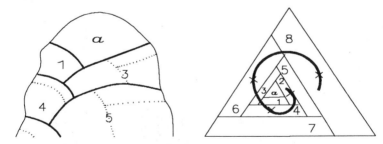

Fig. 1: Representation of the apical region of the shoot of the fern *Ceratopteris* (redrawn from [6]. (at left): external view, apical cell (*a*) and merophytes 1 to 5. (at right): schematic representation of the 3-sided segmentation of the apical cell with derivation of successive segments or *merophytes;* x: position of leaves, related by a helicoidal line.

the assemblage itself is, at each derivation step, realized by a new packing of cells with node fusion leading to wall fusion. One cell after the other is replaced and added to the cellwork in construction. At first topological, the cellular assemblage becomes at the end geometrical.

Each polyhedron of the sub-alphabet F has face and node labels. Face labels are either simple colors or integers which represent the generation age of the respective cell walls. That is, at each cell division the new partition wall takes the age 1 and the generation age of the other walls of the cell boundary are incremented by one unit (more details in section 6). Obviously, with consecutive cell divisions all walls of a cell have different ages. Node labels are introduced by a similar procedure (details in section 6).

In the table T of cell contacts, the line and column entries are given by the elements of the sub-alphabet C, i.e. the labels of the cell types. This table indicates how the cells of the sub-alphabet F are glued together by bringing into correspondence a given wall of each of two cells. For two given cells, c_i and c_j, which touch one another, the corresponding element in the table states that a wall of age x of c_i has to be associated and glued to a wall of age y of c_j ; x and $y \in N$. The table T indicates also a sequence of node labels (small integers) that delimits the boundary of these walls. Walls are associated conformably to the given node labels which are either identical on both walls to be associated, or have the same order-relation on both walls.

4. Examples

4.1 Example 1 - Cellwork with an Apical Tetrahedral Autoreproductive Cell

A first example simulates the construction of the fern shoot illustrated in Fig.1. There is an **alphabet** Σ with 4 cells, the **sub-alphabet** $C = \{a_4, b_5, b_6, b_7\}$, a_4 the apical cell and its derivatives b_5, b_6, and b_7. They are represented by the following **sub-alphabet F of polyhedra** (with wall labels, wall colours, and node labels).

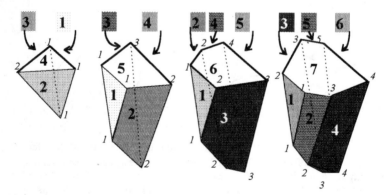

The **set P of cell productions**, $P = \{ a_4 \rightarrow a_4 b_5, b_5 \rightarrow b_6, b_6 \rightarrow b_7, b_7 \rightarrow b_7 \}$ states that the apical cell a_4 divides into a new apical cell and a second daughter cell b_5. We admit for the present case that all b cells do not divide further on. As the index represents the polygonal degree of cells, in a production like $b_6 \to b_7$, the cell b obtains one wall more.

Table T of Cell Contacts:

Line and column entries: elements of the sub-alphabet C; e: environment; bold integers: generation age of a wall; sequence of small integers: node labels of a wall; each element of the table indicates a contact (*cont*) between two walls, written, e.g., *cont*(1-4) between the wall 1 of a cell b_7 and the wall 4 of a cell b_5. Instead of generation ages, walls belonging to two cells and having the same colour (*col*) are associated, written, e.g. *cont*(*col*4); such two walls are fused by bringing into correspondence the cycles of small integers (ex.: for 112 and 223, node 1 corresponds to 2 and 2 to 3). Symmetric cases of cell contacts are not indicated. X: non existing contacts. For example, the 6 walls of b_6 are found in the table in the contacts with double boundaries (walls 1 and 2 in line b_6 and walls 3, 4 and 5 in column b_6).

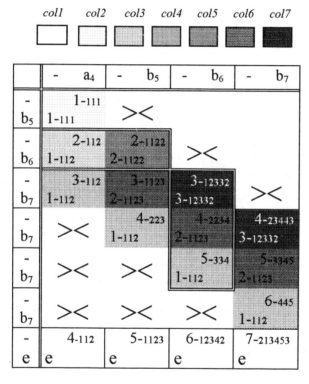

A cellwork derivation: We suppose that at step i-1 there is a cellwork composed of three cells, b_6, b_7, and b_7. After one derivation step, and conformably to the string derivation of cell labels, cell b_6 is changed into cell b_7.

$$b_6 \, b_7 \, b_7 \quad \to \quad b_7 \, b_7 \, b_7$$

The construction of the corresponding cellworks is done with the help of both the table T of contacts and the sequentiality of symbols in the string (here from right to left).

The assemblage of cells for the cellwork of step i-1 uses contacts from this table differentiated in the following figure by hatchings:

- The first b_7 is glued to the second b_7 cell : *cont*(3-4), i.e. 4th line and 4th column entry). Node 1 of wall 3 of the first b_7 cell corresponds to node 2 of wall 4 in the second b_7 cell.
- The second b_7 is glued to the cell b_6 by *cont*(3-3), age difference 1, corresponding in the table to the 3rd line entry and the 3rd column. Between both walls nodes with the same label correspond.
- The first b_7 is glued to the cell b_6 by *cont*(2-4), a difference of 2 generations.

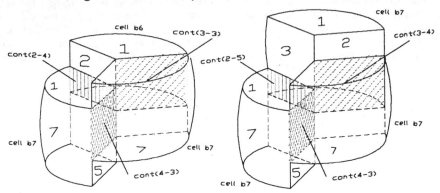

Deriving now the cellwork for timestep i, we restart the packing for the string $b_7 b_7 b_7$. In fact the first two glueings are done and it is sufficient to replace the cell b_6 by a cell b_7. For this we replace contacts :
- for the cell with age difference 1: *cont*(3-3) between $b_6 b_7$ is replaced by *cont*(3-4) between two b_7 cells,
- for the cell with age difference 2: *cont*(2-4) between $b_6 b_7$ is replaced by *cont*(2-5) between two b_7 cells.

As polyhedra of the alphabet are topological objects, the geometry of the cellwork is established by geometrical reorganisation after cell glueing. In fact, as soon as the geometry of the b_7 cell is well known, all other cells in the cellwork can be placed accurately.

In fact, as labels are only necessary for the system construction (cf. section 6) they can be replaced by colors. The table shows then which walls of a given color of two cells are associated by each glueing. Practically, in a 3D representation, each polyhedral object of the alphabet is moved so that a wall of a given color is brought into contact with the wall of the neighbour cell having a corresponding color.

There are seven colors used in the example 1. All walls are different in a cell except for the basic construction items given by b_7 cells. In these cells, two walls of the same color are nevertheless distinguished by their sequence of node labels.

The cellwork, in the given example, represents a plant shoot composed of 3 files of b_7, i.e. 7-sided cells associated in a helicoidal fashion. At the top, between the apical cell and such b_7 cells, are placed two intermediary states of b cells which bring the top of the files to the same level. These three top walls of the files represent the surface in which the apical cell is embedded. The correspondingly constructed plant body has been illustrated by Fig. 5 in [4]. Similar illustrations can be found presently in example 2.

4.2 Example 2 - Cellwork with an Apical 5-sided Autoreproductive Cell and its 4-sided Segmentation.

A second example is the construction of a plant body which consists of four compact files of 9-sided cells becoming, later on, merophytes and which are in a helicoidal arrangement. It results from the activity of a 5-sided apical cell with a 4-sided helicoidal segmentation (cf. Table section 5, Class IV, Group 2).

The system uses an **alphabet** Σ with 5 cells, the sub-alphabet $C = \{a_5, b_6, b_7, b_8, b_9\}$, a_5 the apical cell and its derivatives. They are represented by the sub-alphabet F of polyhedra (cf. on next page). Cell walls have generation age labels or colours. The set of **cell productions is** $P = \{ a_5 \to a_5b_6, , b_6 \to b_7, b_7 \to b_8, b_8 \to b_9, b_9 \to b_9\}$. We admit that b cells do not divide further on.

Table T of Cell Contacts: (explanations like in Table T of example 1). 11 colours are used to differentiate walls to be associated in cellular contacts. Table elements with double boundaries show how all walls from a cell (f. ex. b_7) are found in the table.

	col1	col2	col3	col4	col5	col6	col7	col8	col9	col10	col11

	- a_5	- b_6	- b_7	- b_8	- b_9
- b_6	1-11111 1-11111	><			
- b_7	2-112 1-112	2-11222 2-11222	><		
- b_8	3-1123 1-1123	3-112 2-112	3212333 3-212333	><	
- b_9	4-113 1-113	4-11224 2-11224	4-223 3-223	4-3213444 4-3213444	><
- b_9	><	5-224 1-113	5-22335 2-11224	5-334 3-223	5-4234555 4-3213444
- b_9	><	><	6-335 1-113	6-33445 2-11224	6-445 3-223
- b_9	><	><	><	7-446 1-113	7-55447 2-11224
- b_9	><	><	><	><	8-557 1-113
- e	5-1132 e	6-1124 e	7-12352 e	8-23453 e	9-145543 e

Alphabet Σ of example 2:

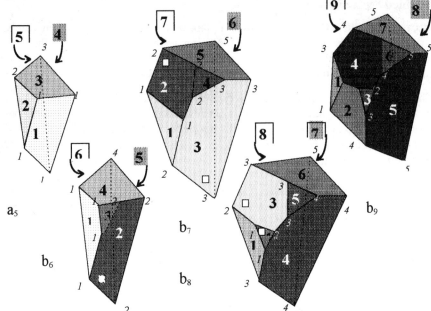

a_5

b_6

b_7

b_8

b_9

On the right, parts of cellworks give an insight into the construction of the generated shoot , i.e. into the assemblage of the apical cell and two tiers of cells underneath. The uppermost tier emphasizes how the intermediary b cells, one b_9 , one b_8, one b_7 and   one b_6 cell, respectively, bring the four cell files to the same level on which lies the apical cell. Underneath, the helicoidal disposition of b_9 cells in a normal tier shows the existence not only of contacts between adjacent files but also of contacts between opposite files (*cont(3-6)*). 3D constraints confer to the b_9 cells an obligatory skewing of walls labelled 2, 4, 5 and 7.

5. Classification of Plant Meristems

The cellworks can be classified with respect to the form of the initial cell, its number of faces that do not touch a neighbour cell, and the set of cell productions.

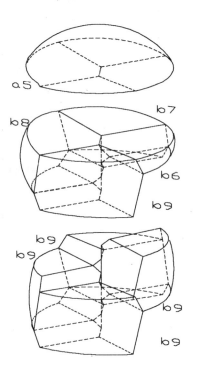

Table of the classification:
I-VII: Classes of meristems with respect to the conformation of the initial cell. Groups: classification with respect to the number of walls of the initial cell touching the environment. Apical cell: n, e and v: number of walls, edges and nodes, respectively. Dw: division wall of a mother cell. Blank cases not yet investigated.

I.	II.	III.	IV.	V.	VI.	VII
\multicolumn{7}{c}{Conformation of the initial cell}						
$n=1$	$n=2$	$n=3$	$n=4$	$n=5$	$n=6$	$n=7$
$e=0$	$e=1$	$e=3$	$e=6$	$e=9$	$e=12$	$e=15$
$v=0$	$v=0$	$v=2$	$v=4$	$v=6$	$v=8$	$v=10$
		Polygonal degree of the dw				
0	1	2 (1+dw)	3 (2+dw)	4 (3+dw)	5 (4+dw)	6 (5+dw)
0 environmental faces - **Group of Internal Meristems** - Productions of cell types						
$a_1 \to a_1 b_2$	$a_2 \to a_2 b_3$	$a_3 \to a_3 b_4$	$a_4 \to a'_4 b_5$			
$b_2 \to b_2$	$b_3 \to b_4$	$b_4 \to b_5$	$a'_4 \to a_4 b_5$			
	$b_4 \to b_4$	$b_5 \to b_6$	$b_5 \to b_6$			
		$b_6 \to b_6$	$b_6 \to b_7$			
			$b_7 \to b_8$			
			$b_8 \to b_8$			
		Structure of the generated cellular body				
Vesicles (invagination)		3-lateral segmentation	4-lateral segmentation			
1 environmental face - **Group of Apical Meristems** - Productions of cell types						
$a_1 \to a_1 a_1$	$a_2 \to a_2 b_3$	$a_3 \to a_3 b_4$	$a_4 \to a_4 b_5$	$a_5 \to a_5 b_6$	$a_6 \to a_6 b_7$	$a_7 \to a_7 b_8$
	$b_3 \to b_3$	$b_4 \to b_5$	$b_5 \to b_6$	$b_6 \to b_7$	$b_7 \to b_8$	$b_8 \to b_9$
		$b_5 \to b_5$	$b_6 \to b_7$	$b_7 \to b_8$	$b_8 \to b_9$	$b_9 \to b_{10}$
			$b_7 \to b_7$	$b_8 \to b_9$	$b_9 \to b_{10}$	$b_{10} \to b_{11}$
				$b_9 \to b_9$	$b_{10} \to b_{11}$	$b_{11} \to b_{12}$
					$b_{11} \to b_{11}$	$b_{12} \to b_{13}$
						$b_{13} \to b_{13}$
		Structure of generated cellular body				
Bud (extrusion)	Filament	2-lateral segmentation	3-lateral helicoidal segmentation	4-lateral helicoidal segmentation	5-lateral helicoidal segmentation	2/5-helicoidal segmentation
2 environmental faces - **Group of Plate Meristems** - Productions of cell types						
				$a_5 \to a_5 b_6$	$a_6 \to a_6 b_7$	
				$b_6 \to b_7$	$b_7 \to b_8$	
				$b_7 \to b_8$	$b_8 \to b_9$	
				$b_8 \to b_8$	$b_9 \to b_{10}$	
					$b_{10} \to b_{10}$	
		Structure of generated cellular body				
				Tunica	Tunica	

6. Construction of a Cellwork

To construct a specific system, we have first to choose a specific *dw* positioning with respect to the last formed *dw*. This is supposed to be repeated in the same way at each forthcoming division of the apical cell. Secondly, the *a* cell must be autoreproductive, i.e. at least one daughter cell has to be equal to its mother cell.

6.1 Face Labelling

In order to specify *wall contacts* between cells, walls are distinguished by labels that represent the generation age at which a given wall has been added to the cell as a new division wall. The following schematic representation of a tetrahedral apical cell a_4 has four walls labelled from 1 to 4. The wall 1 is the youngest one, i.e. the last *dw* added to the cell boundary during the last cell division. The wall labelled 2 is the *dw*, entirely or a part from it, added two generation earlier, and so on.

Wall label transformation by division $a_4 \rightarrow a_4\ b_5$:

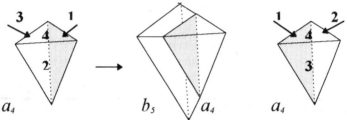

When the cell a_4 divides such that $a_4 \rightarrow a_4\ b_5$, the scheme shows that the two daughter cells are separated by a new *dw* which, conformably to our definition, takes the label 1. All wall labels still present are incremented by 1 unit except walls which disappear from the cell boundary by integration into a neighbour cell, or free walls with regard to the environment. So, the new daughter cell a_4 loses wall 3 now integrated in the boundary of the cell b_5. The free wall 4 does not change its label as it remains in any case the eldest one.

It results that in the filiation of apical cells, the successive a_4 cells are identical, only their orientation with respect to the cellwork in which they are embedded changes. At each third cell division, the orientation is the same. This leads to the 3-sided helicoidal segmentation of the tetrahedral apical cell.

The edges by which a new partition wall is attached to older walls divide these into two parts which take age labels conformably to the given definition. Further, by contact, each partition wall insertion edge induces also a splitting of the wall in the neighbour cell which it touches. The resulting two faces have the same age (nevertheless, for clarity, it may be necessary to distinguish them).

6.2 Node Labelling

Node labels are necessary to orient accurately the two walls to be glued together. The nodes introduced at the sites where the cycle of division wall edges cuts older wall edges take the label 1 as shown in the following schematic representation of a_4.

After a cell division, all node labels, still present in the daughter cells, are incremented by one. Once a cell does not change its configuration (neither by proper division nor by divisions in neighbour cells), i.e. $c_i \rightarrow c_i$, node labels remain unchanged.

Node labels transformation by division $a_4 \rightarrow a_4\, b_5$:

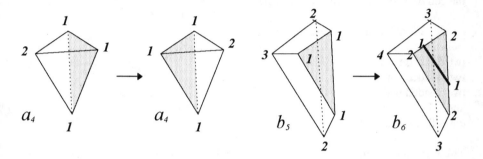

6.3 Cell Alphabet and Cell Production Rules

The changes in wall and node labels are pursued until cells during further divisions remain unchanged. This procedure permits us to define the cell alphabet needed for a system based on a specific positioning of the division wall. Once the alphabet and hence the cell productions are established, it is not necessary for a cellwork derivation to take into account explicitly the mentioned changes in wall ages and node labels, as they are implicitly given by the polyhedra.

The geometry of the cells of the alphabet is established by means of an inverse procedure: at first the shape of the basic cells (like b_7 in example 1) is roughly determined from their assemblage. The shape of cells with fewer walls than the basic ones is succesively determined by removing one by one the last added edges. So, the cell b_6 can be derived from cell b_7 by removing one edge and its nodes, i.e. by fusion of the two edges which were separated by a removed node.

6.4 Cell Contacts

During the establishment of the cycle of possible orientations of an apical cell, the successive wall contacts are noted and grouped into a table. For simplicity, wall labels can be replaced by colors, a same color designates walls to be glued together (cf. the table T in example 1).

7. An Algorithm to Search for Possible Systems

An algorithm is given to define all divisions of polyhedra which are self-reproducing, i.e. in which at least one of the two daughter cells has exactly the same configuration as the mother cell. By help of a *wall neighbourhood graph* the cycles of walls which will support the partition wall can be defined. For autoreproduction, the cycle length equals the polygonal degree of the division wall. Eliminating inverse cycles, the few remaining possible partitions are self-reproducing. To each structure of the wall derivation graph is associated a type of plant morphology.

Conformation of the initial cell: Consider a cell c_n with n faces or walls, its polyhedral conformation and the division wall (dw) (i.e. the last added face during the cell division giving rise to this cell).

ex.: c_6 : 3D polyhedral representation

two 5-sided walls
two 4-sided walls
two 3-sided walls
and.
dw with 5 sides

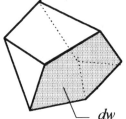

The wall-neighbourhood graph of a cell c_n: The cell wall boundary is described by the neighbourhood relationships of the walls. For this, the polyhedral representation of c_n is opened by removing partially the face corresponding to the dw, letting it be attached by one edge. It results in a 2D-map representation of the wall of the cell, the cut being the border of the map.
   The neighbourhood relationship of the walls is obtained by
        - an arbitrary labelling of the map regions (different from age labels; the label A is here reserved for the dw),
        - and by the establishment of the matrix of this neighbourhood.
Example of cell c_6 (continued):
labelled map representation   matrix of neighbourhood

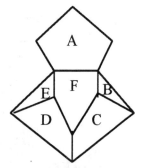

	A	B	C	D	E	F
A	0	1	1	1	1	1
B	1	0	1	0	0	1
C	1	1	0	1	0	1
D	1	0	1	0	1	1
E	1	0	0	1	0	1
F	1	1	1	1	1	0

(for ex.: the 4th line means that the wall labelled D touches walls A, C, E, and F.)

Partition of a wall-map: During its division, a cell is partitioned by the introduction of a new dw. On the cell envelope, this dw is visualized by a closed sequence of edges. On the map, such a line is a path, going once through some map regions, and so that in 3D it would correspond to a cycle. The length of such a *partition cycle* equals the number of map regions crossed. All possible map bipartitions are searched for. These bipartitions of a map represent the possible ways of dw insertions in a dividing cell.
   In the given example, the sequence A.C.F.B.(A) is a cycle; wall C is crossed coming from wall A and going to wall F, and so on.

Conserving partitions for cycle lengths equal to the polygonal degree of the dw. By autoreproductive divisions, the new *dw* has to be identical to the previous one, i.e. the length of the partition cycle is equal to the polygonal degree of the previous *dw*. Consequently, in the example, only cycles of length 5 are retained.

Elimination of redundancies: equal cycles (identical sequences but different starting points) and inverse cycles are equal and consequently ignored.

Retaining one of symmetric partitions: they count nevertheless as distinct partitions as they lead to segmentations turning in opposite directions.

Conserving only a daughter wallmap identical to that of the mother cell considered without its *dw*. Map parts separated by the partition cycle are analyzed with respect to the polygonal degree of their faces. Only parts equal to the wallmap of the mother cell considered without its *dw* are convenient situations.

Conformable partition parts are relabelled: The part of the map in a partition conformable to all the previous constraints (in grey, in the following figure) are now relabelled arbitrarily. The label A is reserved for the *dw* which is not represented as a region but with its insertion line.

Writing wall productions: Between the labelled mother wallmap and the partitioned map, wall productions can be written. The following example means that

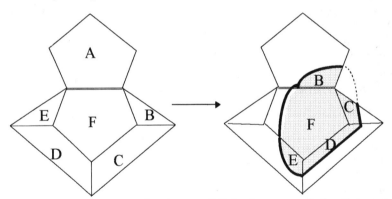

the mother wall A produces a daughter wall B in the new apical cell (arrow to the right) and a second unnamed wall in the other daughter cell (arrow downwards).

$$A \to B \to C \to D \to E \qquad F \to F$$
$$\downarrow \quad \downarrow \quad \downarrow \quad \downarrow \quad \downarrow \qquad \downarrow$$

Now age labels can be attributed to the derivation graph:

$$1 \to 2 \to 3 \to 4 \to 5 \qquad 6 \to 6$$
$$\downarrow \quad \downarrow \quad \downarrow \quad \downarrow \quad \downarrow \qquad \downarrow$$

For the given example, there are 17 different wallmap partitions of length 5 (Fig. X) from which only three correspond to autoreproductive cell divisions. One of them leads to a 2-sided segmentation giving rise to two rows of merophytes. Such a morphology can still been simulated with simpler apical cells such as Fig. 7 in [5] and which figures in the classifications of meristems (cf. section 5) in Class III Group 2. The two others correspond one, to the generation of a shoot composed of 5 compact files of cells or files of imbricated merophytes (Class VI Group 2). Whereas the third one generates a tunica-like shoot (like a finger of a glove) with cell files round about a free space (Class VI Group 3). This example is particularly interesting on the morphological level.

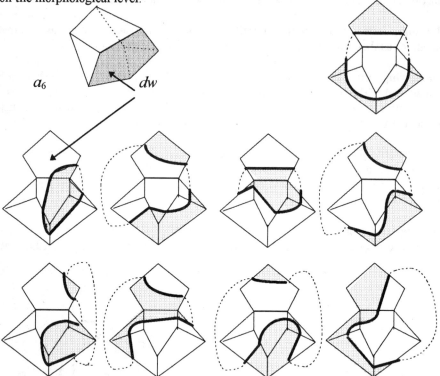

Fig. 2 : Map partitions enabling autoreproduction of initial cells: a_6, the initial cell. 9 map partitions represented from 17 possibilities, cases 2-8 have symmetric partitions. The 5-sided division wall (dw) figures on the top of each map. Heavy line: partition line; stippled lines show that the partition lines are cycles; in grey: part of the partition which correspond to the autoreproductive daughter cell.

6. Conclusion

The table of cell contacts represents, for the biologist, a convenient tool to construct for himself the cellwork derivations. An element of this table states that a cell c_i touches a cell c_j, and this by a wall of color x in the first cell which is fused to the wall of the same color of the latter cell. Two such walls are always polygons with the same degree and with the same or a similar sequence of node labels, i.e. with an unambiguous correspondence between the two faces to be glued together. The proposed insight into the way a plant body could be constructed is especially useful in the case of 3-dimensional cell bodies. The alphabet of polyhedra and the table of cell contacts are suitable for a future interactive animation, as in plant development animation becomes more and more of interest for biologists [9].

This table also represents an interface by which a given system can be defined as a cellwork system based either on simple wall labels, edge labels, or node labels. For such a purpose, age labels of walls are more instructive than simple colours.

It is also of interest that the 3D developmental systems based on different division wall insertions fit into a table of morphological realizations repealling well known plant developments of increasing complexity. The present table extends the preliminary versions [7,8] of this classification to three groups of meristems with regard to the number of faces by which the initial cell faces the environment. Further illustrations of theoretical plant bodies in [4, 5, 7, 8].

References

1. G. Rozenberg, A. Salomaa: The mathematical theory of L systems. New-York: Academic Press 1980
2. A. Lindenmayer: Models for plant tissue development with cell division orientation regulated by preprophase bands of microtubules. Differentiation 26: 1-10 (1984)
3. F.D. Fracchia, P. Prusinkiewicz: Physically-based graphical interpretation of marker cellwork L-systems. In: H. Ehrig, H.J. Kreowski, G. Rozenberg (eds): Graph grammars and their application to computer science. Lecture Notes in Computer Science 532. Berlin: Springer 1991, pp. 363-377
4. J. Lück, H.B. Lück: Double-wall cellwork systems for plant meristems. In: H. Ehrig, H.J. Kreowski, G. Rozenberg (eds): Graph grammars and their application to computer science. Lecture Notes in Computer Science 532. Berlin: Springer 1991, pp. 564-581
5. J. Lück, H.B. Lück: Cellworks: an application to plant morphogenesis. In: G. Rozenberg, A. Salomaa (eds): Lindenmayer systems. Impacts on theoretical computer science, computer graphics, and developmental biology. Berlin: Springer 1992, pp. 385-404
6. R. Hébant-Mauri: Segmentation apicale et initiation foliaire chez Ceratopteris thalictroides (Fougère leptosporangiée). Can. J. Bot. 55: 1820-1828 (1977)
7. J. Lück, H.B. Lück: Parallel rewriting dw-cellwork systems for plant development. In: J. Demongeot, V. Capasso (eds): Mathematics applied to biology and medicine. Winnipeg, Canada: Wuerz Publishing Ltd 1993, pp. 461-466
8. H.B. Lück, J. Lück: De la théorie des automates vers la phyllotaxie. In: B. Millet (ed): Mouvements, rythmes et irritabilité chez les végétaux, hommage à Lucien Baillaud. Besançon (Fr): Univ. de Franche-Comté 1993, pp. 85-94
9. P. Prusinkiewicz, M.S. Hammel, E. Mjolsness: Animation of plant development. Computer Graphics Proceedings, Annual Conference Series 1993, pp. 351-360
10. B. Mayoh: Patterns, graphs and DNA. Aarhus University 1994

Subapical Bracketed L-Systems

Przemyslaw Prusinkiewicz[1] and Lila Kari[2]

[1] Department of Computer Science, University of Calgary
2500 University Drive N.W., Calgary, Alberta, Canada T2N 1N4
e-mail: pwp@cpsc.ucalgary.ca
[2] Department of Mathematics, University of Western Ontario
London, Ontario, Canada N6A 5B7
e-mail: lkari@julian.uwo.ca

Abstract. This paper characterizes the development of modular branching structures that satisfy three assumptions: (a) subapical branching, meaning that new branches can be created only near the apices of the existing branches, (b) finite number of module types and states, and (c) absence of interactions between coexisting components of the growing structure. These assumptions are captured in the notion of subapical bracketed deterministic L-systems without interactions (sBDOL-systems). We present the biological rationale for sBDOL-systems and prove that it is decidable whether a given BDOL-system is subapical or not. In addition, using the assumption that modules, once created, continue to exist, we show that (propagating) sBDOL-systems are too weak to generate acrotonic and mesotonic branching structures, which are often observed in nature. Their development must therefore be controlled by more involved mechanisms, overriding at least one of the assumptions (a-c) above.

1 Introduction

Bracketed L-systems, introduced by Lindenmayer [8, 9] to model the development of branching structures, have been investigated to a lesser degree from the theoretical point of view than the L-systems without brackets (*c.f.* [10, page 138]). In contrast, most practical applications of L-systems fall in the areas of modeling, simulation, and visualization of higher plants with branches (for example, see [15]). Consequently, theoretical results pertinent to this class of structures are needed.

We analyze the class of branching structures and developmental sequences generated by *subapical* deterministic bracketed L-systems without interactions (subapical BDOL-systems, or sBDOL-systems in short), formalized and first studied by Kelemenová [7]. For the class of non-branching structures, a related notion of filamentous systems with apical growth was introduced by Nirmal and Krithivasan [13] (see also [1, 14, 20]). Intuitively, a BDOL-system is subapical if new branches are created only near the apices (tips) of the existing branches. This notion captures the fundamental biological observation that new structural components of a growing plant, such as branches, leaves, or flowers, can only be initiated by *apical meristems*, that is the zones of actively dividing cells situated near the apices of branch axes [19, page 1].

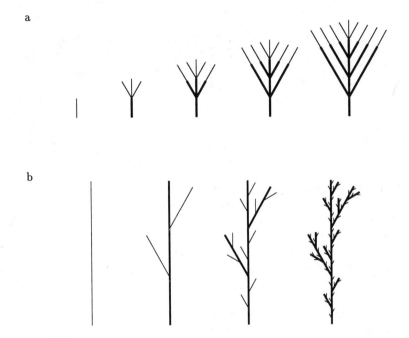

Fig. 1. A comparison of the development of two branching structures. Structure (a) develops with subapical branching; structure (b) does not (branches everywhere). The thin lines indicate branches created in the current derivation step.

A sample developmental sequence generated by a subapical BDOL-system is shown in Figure 1a. For a comparison, Figure 1b shows the development of an "everywhere branching" structure that violates the assumption of subapical branching. In Section 2 we restate Kelemenová's definition of subapical BDOL-systems, provide a corrected proof of her assertion that it is decidable whether a given BDOL-system is subapical or not, and illustrate the discussed notions using several biologically motivated examples.

Frijters and Lindenmayer [4] observed that in structures generated using subapical BDOL-systems, branches closer to the apex are less developed than those positioned closer to the base. Structures satisfying this property are called *basitonic* [2, page 248] (Figure 2a). In nature one also finds *mesotonic* and *acrotonic* structures, with the most developed branches situated near the middle or the top of the mother branch (Figures 2b and 2c). The problem of generating mesotonic and acrotonic structures using L-systems has been first discussed by Frijters and Lindenmayer in reference to the inflorescences of *Aster novae-angliae* [3], and extended to other inflorescences by Janssen and Lindenmayer [5] (see also [15, Section 3.3.3]). Lück, Lück, and Bakkali [12] (see also [11]) presented a detailed study of acrotonic, mesotonic and basitonic branching patterns using a formalism related to L-systems. In general, the proposed mechanisms for modeling mesotonic and acrotonic structures can be divided into two categories: those postulat-

Fig. 2. Basitonic (a), mesotonic (b), and acrotonic (c) branching patterns

ing control of development using *signals* [3, 5], and those introducing numerical parameters to characterize *growth potential* or *vigor* of individual apices [12]. Both mechanisms require a departure from the class of DOL-systems, either by introducing context-sensitive rules to represent signals, or by assuming an infinite alphabet to represent the set of vigor values. In Section 3 we prove that at least in the case of propagating development (where modules, once created, remain in the structure) these departures are indeed necessary, as neither acrotonic nor mesotonic developmental sequences can be generated by propagating sBDOL-systems. This result is related to Theorem 3.2 of Kelemenová [7], which, however, was formulated without an explicit reference to the notion of acrotony.

2 Subapical bracketed DOL-systems

Let Σ denote a finite nonempty *alphabet*, the brackets [and] be two symbols outside of Σ called *branch delimiters*, and # be another symbol outside of Σ called the *branch marker*. We will denote the respective extensions of Σ by $\Sigma_E = \Sigma \cup \{[,]\}$ and $\Sigma_\# = \Sigma \cup \{\#\}$.

Definition 1. A word over Σ_E is *well nested* iff it can be specified by finitely many applications of the following rules:

– every word $u \in \Sigma^*$ is well nested;
– if $u, v \in \Sigma_E^*$ are well nested then $[u]$ and uv are also well nested.

A word $[w] \in \Sigma_E^*$ such that w is well nested is called a *branch*.

Definition 2. The *standard decomposition* of a branch $[w] \in \Sigma_E^*$ is a word of the form:
$$[w] = [x_1[\alpha_1]x_2[\alpha_2]\ldots x_n[\alpha_n]x_{n+1}],$$
where the subwords $x_1, x_2, \ldots, x_{n+1} \in \Sigma^*$ do not contain brackets, and the subwords $\alpha_1, \alpha_2, \ldots, \alpha_n \in \Sigma_E^*$ are well nested. The words $x_1 x_2 \ldots x_n x_{n+1}$ and $x_1 \# x_2 \# \ldots x_n \# x_{n+1}$ are called the (main) *axis* and the *marked axis* of $[w]$, respectively. Within these axes, the subwords $x_1, x_2, \ldots, x_n$ are called the *internodes*, and the subword x_{n+1} is called the *apex*. The words $[\alpha_1], [\alpha_2], \ldots, [\alpha_n]$ are called the (first-order) *lateral branches* of $[w]$.

It is known that the standard decomposition of a branch is unique, thus the above definition is unambiguous [6]. The terminology corresponds to the standard interpretation of well nested bracketed words as string representations of branching structures [8, 9]. As the "empty branch" [] appears to have no biological interpretation, we assume in practice that the word w in any branch $[w]$ is not empty. This assumption, however, is not essential to the mathematical reasoning presented in this paper.

Example 1. Figure 3 shows a branching structure represented by the word

$$[w] = [\ \overbrace{ab}^{x_1}\ \underbrace{[cd]}_{[\alpha_1]}\ \overbrace{ef}^{x_2}\ \underbrace{[g[h]i]}_{[\alpha_2]}\ \overbrace{j}^{x_3}\ \underbrace{[k]}_{[\alpha_3]}\ \overbrace{lm}^{x_4}\ ].$$

The word $abefjlm$ is the axis of $[w]$, $ab\#ef\#j\#lm$ is the marked axis, $x_1 = ab$, $x_2 = ef$ and $x_3 = j$ are the internodes, and $x_4 = lm$ is the apex. The subwords $[\alpha_1] = [cd]$, $[\alpha_2] = [g[h]i]$ and $[\alpha_3] = [k]$ denote the lateral branches, where $[\alpha_1]$ and $[\alpha_3]$ have only apices, whereas $[\alpha_2]$ has an internode g, an apex i, and a (second-order) lateral branch $[h]$.

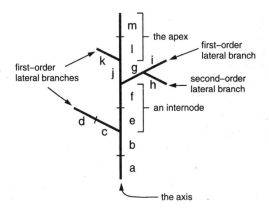

Fig. 3. Example of a branching structure

Diagrams such as that shown in Figure 3 can be thought of as graphs representing the branching *topology* of modeled organisms, for example algae, herbaceous plants, or trees. Depending on the complexity of the organism and the abstraction level of the model, symbols may represent individual cells or larger *modules* [16] of the structure. The bracketed words may be also assigned a *geometric* interpretation, needed to automatically visualize the models using computer graphics [15]. Within this paper we focus on the topological interpretation.

In order to describe the development of a structure over time we use the formalism of L-systems. We assume that the reader is familiar with the fundamental notions of L-system theory, such as a D0L-system, developmental sequence, and language generated by an L-system (see [15, 17, 18] for a reference), and only recall the definition of a bracketed DOL-system, which is essential to this paper.

Definition 3. A *bracketed DOL-system* (BDOL-system) is a DOL-system $S = \langle \Sigma_E, [w_0], P \rangle$, where the *axiom* $[w_0]$ is a well nested word over the alphabet Σ_E, and each *production* in the *production set* $P \subset \Sigma_E \times \Sigma_E^*$ has one of the following forms:

- $a \longrightarrow \alpha$, where $a \in \Sigma$, $\alpha \in \Sigma_E^*$, and α is well nested,
- $[\longrightarrow [$, or
- $] \longrightarrow]$.

Example 2. The DOL-system $S = \langle \{1, 2, \ldots, 9, [,]\}, [1], P \rangle$ with productions:

$$1 \longrightarrow 23, \; 2 \longrightarrow 2, \; 3 \longrightarrow 24, \; 4 \longrightarrow 25, \; 5 \longrightarrow 65,$$
$$6 \longrightarrow 7, \; 7 \longrightarrow 8, \; 8 \longrightarrow 9[3], \; 9 \longrightarrow 9, \; [\longrightarrow [, \;] \longrightarrow]$$

is a BDOL system. The developmental sequence generated by S begins with the following words:

$$[w_0] = [1]$$
$$[w_1] = [23]$$
$$[w_2] = [224]$$
$$[w_3] = [2225]$$
$$[w_4] = [22265]$$
$$[w_5] = [222765]$$
$$[w_6] = [2228765]$$
$$[w_7] = [2229[3]8765]$$
$$[w_8] = [2229[24]9[3]8765]$$
$$[w_9] = [2229[225]9[24]9[3]8765]$$
$$[w_{10}] = [2229[2265]9[225]9[24]9[3]8765]$$
$$[w_{11}] = [2229[22765]9[2265]9[225]9[24]9[3]8765]$$
$$[w_{12}] = [2229[228765]9[22765]9[2265]9[225]9[24]9[3]8765]$$
$$[w_{13}] = [2229[229[3]8765]9[228765]9[22765]9[2265]9[225]9[24]9[3]8765]$$

This L-system was proposed by Lindenmayer [8] as a mathematical model of the development of a red alga *Callithamnion roseum*. Symbols of the alphabet Σ denote individual cells, and the matching pairs of brackets delimit branches. Selected developmental stages obtained by this model with the addition of *turtle geometry* symbols [15] to indicate the direction of branching are shown in Figure 4.

It is easy to notice that BDOL-systems always generate well nested words. The converse, however, is not true, as languages of well nested words can also be generated by DOL-systems that do not satisfy the requirements of Definition 3.

Example 3. The DOL-system $S = \langle \{A, B, C, [,]\}, [AB], P \rangle$ with productions

$$A \longrightarrow A[C, \; B \longrightarrow C]B, \; C \longrightarrow C, \; [\longrightarrow [, \;] \longrightarrow]$$

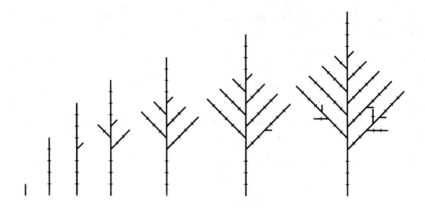

Fig. 4. Developmental stages $[w_0], [w_4], [w_7], [w_9], [w_{11}], [w_{13}]$, and $[w_{15}]$ of the model of *Callithamnion roseum*

is not a BDOL-system, because the successors of two productions are not well nested words. Nevertheless, it generates well nested words, as suggested by the following initial elements of the developmental sequence:

$$[w_0] = [AB]$$
$$[w_1] = [A[CC]B]$$
$$[w_2] = [A[C[CC]C]B]$$
$$\ldots$$

In this paper we adhere to the biologically well justified original notion of bracketed L-systems [8, 9], where branches can be initiated only by individual parent modules. In this context, the notion of subapical development is formalized as follows.

Definition 4. Given a BDOL-systems $S = \langle \Sigma_E, [w_0], P \rangle$, a letter $a \in \Sigma$ is called *branching* iff it produces a word including a lateral branch $[\beta]$:

$$a \longrightarrow \alpha[\beta]\gamma, \quad \text{where} \quad \alpha, \beta, \gamma \in \Sigma_E^*.$$

The subset of Σ consisting of all branching letters in S is denoted Σ_B.

Definition 5. A BDOL-system $S = \langle \Sigma_E, [w_0], P \rangle$ is *subapical with respect to the main axis* (of the generated branches) iff for any $[w] \in L(S)$ with the standard decomposition

$$[w] = [x_1[\alpha_1]x_2[\alpha_2]\ldots x_n[\alpha_n]x_{n+1}],$$

the internodes $x_1, x_2, \ldots, x_n$ do not contain branching letters:

$$x_1, x_2, \ldots, x_n \in (\Sigma \backslash \Sigma_B)^*.$$

The class of BDOL-systems subapical with respect to the main axis is obviously included in the class of unrestricted BDOL-systems.

Example 4. The BDOL-system $S = \langle \Sigma_E, [F], P \rangle$ with alphabet $\Sigma = \{F\}$ and productions
$$F \longrightarrow F[F]F, \ [\longrightarrow [, \ ] \longrightarrow]$$
generates the sequence of everywhere branching structures shown in Figure 1b. This L-system is not subapical with respect to the main axis, because the branching letter F appears in the internode of the branch $[F[F]F]$ generated by S.

Example 5. The BDOL-system $S = \langle \{A, B, C\}, [ABC], P \rangle$ with productions
$$A \longrightarrow C, \ B \longrightarrow B, \ C \longrightarrow [B], \ [\longrightarrow [, \ ] \longrightarrow]$$
is not subapical with respect to the main axis, because in the developmental sequence
$$[ABC] \Longrightarrow [CB[B]] \Longrightarrow [[B]B[B]],$$
the branching letter C appears in the internode. Note that the technique used in [7, page 188] to decide whether a BDOL-systems is subapical or not would misclassify this L-system as subapical.

Theorem 1. It is decidable whether a given BDOL-system is subapical with respect to the main axis.

Proof. Consider a mapping $f : \Sigma_E^* \to \Sigma_\#^*$ that substitutes branches $[w] \in \Sigma_E^*$ by their marked axes. Thus, assuming the standard decomposition of $[w]$, we have:
$$f([w]) = f([x_1[\alpha_1]x_2[\alpha_2]\ldots x_n[\alpha_n]x_{n+1}]) = x_1 \# x_2 \# \ldots x_n \# x_{n+1}.$$
Given a BDOL-system $S = \langle \Sigma_E, [w_0], P \rangle$, construct a DOL-system
$$S_\# = \langle \Sigma_\#, f([w_0]), P_\# \rangle,$$
where $P_\# = \{\# \longrightarrow \#\} \cup P'$, and productions in P' are obtained by replacing the successors of productions in $P \setminus \{[\longrightarrow [,] \longrightarrow]\}$ with their marked axes:

if $a \longrightarrow \alpha$ belongs to P then $a \longrightarrow f(\alpha)$ belongs to P'.

It is clear that the L-system $S_\#$ generates the set of marked axes of the words in $L(S)$:
$$L(S_\#) = \{f([w]) : [w] \in L(S)\} = f(L(S)).$$
According to Definition 5, a BDOL-system S is subapical with respect to the main axis iff no branching letters appear in the internodes of words $[w] \in L(S)$. Since the internodes of words $[w]$ and $f([w])$ are the same, the criterion for subapicality can be expressed as
$$L(S_\#) \cap \Sigma_\#^* \Sigma_B \Sigma_\#^* \# \Sigma^* = \emptyset.$$
Thus, a BDOL-system S is subapical with respect to the main axis if and only if the intersection of the language $L(S_\#)$ generated by a related DOL-system $S_\#$ and the regular language $\Sigma_\#^* \Sigma_B \Sigma_\#^* \# \Sigma^*$ is empty. This problem is decidable, because:

- the class of languages generated by DOL-systems is included in the class of languages generated by extended 0L-systems (EOL-systems) [17, page 54],
- the class of EOL languages is closed with respect to intersection with regular languages [17, Theorem 1.8],
- the emptiness problem is decidable for EOL languages [17, Theorem 5.6]. ♣

The reference to the general properties of EOL-systems and regular languages makes the proof of Theorem 1 concise, but does not lead to a straightforward algorithm for testing subapicality of given BDOL-systems. Since subapicality is an essential property of biologically motivated models of branching structures, we also present a more direct test.

We say that the letter a *occurs to the left of* b in a word $w \in \Sigma^*$, and note $(a, b) \in \Gamma(w)$, iff $w = xaybz$ for some $x, y, x \in \Sigma^*$. We extend this definition to languages using the equation:

$$\Gamma(L) = \bigcup_{w \in L} \Gamma(w).$$

In order to construct the relation $\Gamma(L)$ for the language L generated by an arbitrary non-bracketed DOL-system $T = \langle \Sigma, w_0, P \rangle$, we consider the initial elements of the developmental sequence generated by T,

$$w_0 \Longrightarrow w_1 \Longrightarrow \ldots w_i \Longrightarrow \ldots,$$

and construct the family of relations $\Gamma_i \in \Sigma \times \Sigma$:

$$\Gamma_0 = \emptyset,$$
$$\Gamma_{i+1} = \Gamma_i \cup \Gamma(w_i) \quad \text{for} \quad i = 0, 1, 2, \ldots.$$

Since the alphabet Σ is finite, there exists a natural number $k \leq (\text{card}(\Sigma))^2$ such that the k-th iteration of the above formula will not add new elements to Γ_k. Due to the context-free character of derivations in DOL-systems, the subsequent iterations also will not add new elements: $\Gamma_k = \Gamma_{k+j}$ for all $j \geq 0$. Thus, the set Γ_k contains all pairs of letters $(a, b) \in \Sigma$ such that a occurs to the left of b in some word of $L(T)$, and $\Gamma_k = \Gamma(L(T))$.

In order to apply this result to test the subapicality of a given BDOL-system $S = \langle \Sigma_E, [w_0], P \rangle$, we construct the relation $\Gamma(L(S_\#))$ for the language $L(S_\#)$ generated by the DOL-system $S_\#$ associated with S. The L-system S is subapical with respect to the main axis iff there is no branching letter $a \in \Sigma_B$ such that $(a, \#) \in \Gamma(L(S_\#))$.

Example 6. We will show that the BDOL-system S from Example 2 is subapical with respect to the main axis. To this end, we first create a DOL-system $S_\# = \langle \{1, 2, \ldots 9, \#\}, 1, P_\# \rangle$ associated with S. The set $P_\#$ consists of the production $\# \longrightarrow \#$ and productions of P except for $[\longrightarrow [$ and $] \longrightarrow]$; production $8 \longrightarrow 9[3]$ is replaced by $8 \longrightarrow 9\#$. The relation *occurs to the left* for the language $L(S_\#)$ is given by Figure 5. An arrow from node a to node b indicates that letter a occurs to the left of b in a word of $L(S_\#)$. The arrows that can

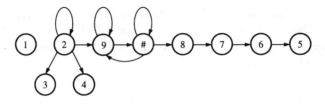

Fig. 5. Simplified graph of the relation $\Gamma(L(S_\#))$ for the developmental model of *Callithamnion roseum*

be reconstructed as a transitive closure of the graph shown in Figure 5 have been omitted for clarity (the relation $\Gamma(L(S_\#))$ is transitive in this example). We observe that the branching symbol 8 does not occur to the left of the branch marker #, thus the L-system S is subapical with respect to the main axis.

We will now extend the notion of subapicality from the main axis to the entire branching structure.

Definition 6. Given a BDOL-systems S, a *branch of order N* is characterized recursively as follows:

- a word $[w] \in L(S)$ is a branch of order $N = 0$,
- if $[w] = [x_1[\alpha_1]x_2[\alpha_2]\ldots x_n[\alpha_n]x_{n+1}]$ is a branch of order $N \geq 0$ then the subwords $[\alpha_1], [\alpha_2], \ldots, [\alpha_n]$ are branches of order $N + 1$.

The set of all branches (of any order $N \geq 0$) generated by S is denoted $L_B(S)$.

Definition 7. A BDOL-system S is called *subapical with respect to all branches*, or in short *subapical*, iff in the standard decomposition of any branch

$$[\alpha] = [y_1[\beta_1]y_2[\beta_2]\ldots y_m[\beta_m]y_{m+1}] \in L_B(S)$$

the internodes $y_1, y_2, \ldots, y_m$ do not contain branching letters.

The following extension of Theorem 1 provides an effective method for deciding whether a BDOL-system is subapical or not.

Theorem 2. Let $S = \langle \Sigma_E, [w_0], P\rangle$ be a BDOL-system, and $[w_1], [w_2], \ldots, [w_p]$ be the branches that occur in successors of productions in P accessible from the axiom $[w_0]$. Denote by S_i the BDOL-system $\langle \Sigma_E, [w_i], P\rangle$, where $i = 1, 2, \ldots, p$. The L-system S is subapical with respect to all branches iff each of the L-systems $S, S_1, S_2, \ldots, S_p$ is subapical with respect to the main axis.

Proof. Each branch $[\alpha]$ in the set $L_B(S)$ belongs to a developmental sequence that starts with one of the words $[w_0], [w_1], [w_2], \ldots, [w_p]$. Consequently, the set of the axes of all branches generated by S is the same as the set of the main axes of the branches generated by L-systems $S, S_1, S_2, \ldots, S_p$, and the requirement that no branching letters occur in the internodes of branches $[\alpha] \in L_B(S)$ is equivalent to the requirement that no branching letters occur in the main axes of words $[w]$ generated by L-systems $S, S_1, S_2, \ldots, S_p$. ♣

Example 7. We will show that the BDOL-system from Example 2 is subapical with respect to all branches. To this end we consider both the original L-system S and its modification $S_1 = \langle \{1, 2, \ldots, 9, [,]\}, [3], P \rangle$, in which the original axiom [1] has been replaced by branch [3] created by production $8 \longrightarrow 9[3]$. The graph of the relation $\Gamma(L(S_\#))$ for the L-system $S_\#$ associated with S was constructed in Example 6 and shown in Figure 5. The graph for the L-system $S_{\#1}$ associated with S_1 is similar, except that node 1 is absent and there is no arrow between nodes 2 and 3. The branching letter 8 does not occur to the left of the branch marker # in either graph, thus both L-systems S and S_1 are subapical with respect to the main axis, and the L-system S is subapical with respect to all branches.

3 Acrotonic languages

According to Section 1, mesotonic or acrotonic structures share the property that the most developed lateral branches are not situated near the bottom of the mother branch. In this section, we introduce a definition of acrotonic languages intended to formalize this intuitive characterization. We then prove that the acrotonic languages cannot be generated by subapical BDOL-systems.

Definition 8. A language $L \in \Sigma_E^*$ is called *acrotonic* iff for every natural k there exists a branch $[w] \in L$ with the standard decomposition

$$[x_1[\alpha_1]x_2[\alpha_2]\ldots x_n[\alpha_n]x_{n+1}]$$

and a sequence of indices

$$1 \leq i_1 < i_2 < \ldots < i_k \leq n$$

such that

$$\lg(\alpha_{i_1}) < \lg(\alpha_{i_2}) < \ldots < \lg(\alpha_{i_k}).$$

The notation $\lg(\alpha_{i_j})$ denotes the *size* of the branch $[\alpha_{i_j}]$, measured as the number of letters other than [and] in the word α_{i_j}.

According to this definition, a language L is acrotonic if it contains branches with arbitrarily long (sub)sequences of lateral branches, the size of which increases while traversing the main axis from the bottom up.

Theorem 3. An acrotonic language L cannot be generated by a propagating subapical BDOL-system.

Proof. Suppose, by contradiction, that there exists an acrotonic language L generated by a propagating subapical BDOL-system $S = \langle \Sigma_E, [w_0], P \rangle$. Introduce the term *branch initial* to denote any branch appearing in a production successor, and let N stand for the total number of branch initials that appear in the successors of productions in P. Consider a word $[w] \in L$ satisfying conditions set forth in Definition 8 for some $k > N$. Since $k > N$, there are at least two branches in the sequence $\{[\alpha_{i_1}], [\alpha_{i_2}], \ldots, [\alpha_{i_k}]\}$, say $[\alpha_p]$ and $[\alpha_q]$, produced from different occurrences of the same branch initial $[v]$:

$$a \longrightarrow \mu_1[v]\mu_2 \quad \text{and} \quad [v] \Longrightarrow^{m_p} [\alpha_p],$$
$$b \longrightarrow \nu_1[v]\nu_2 \quad \text{and} \quad [v] \Longrightarrow^{m_q} [\alpha_q].$$

Assuming that $p < q$, we have $\lg(\alpha_p) < \lg(\alpha_q)$. According to the principle of subapical branching (Definition 5), branch $[\alpha_p]$ must have been initiated not later than $[\alpha_q]$, thus the derivation length m_p is not less than m_q (Figure 6).

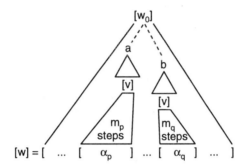

Fig. 6. Derivation tree illustrating the proof of Theorem 3

Since the BDOL-system S is deterministic, the following derivation exists:

$$[v] \Longrightarrow^{m_q} [\alpha_q] \Longrightarrow^{m_p - m_q} [\alpha_p].$$

The L-system S is propagating, thus $\lg(\alpha_q) \leq \lg(\alpha_p)$. This contradicts the inequality $\lg(\alpha_p) < \lg(\alpha_q)$ obtained earlier and leads to the conclusion that an acrotonic language L cannot be generated by a propagating subapical BDOL-system. ♣

4 Conclusions

Subapical branching is an essential characteristic of plant development. In Section 2, we recalled the definition of subapical branching formulated by Kelemenová [7] within the conceptual framework of L-systems, and showed that subapicality is a decidable property of BDOL-systems. We also presented a practical method for testing whether a given BDOL-system is subapical or not, and illustrated it using a model of a red alga *Callithamnion roseum* as an example. It is interesting to note that some models of plant-like structures devised for computer graphics purposes, such as the everywhere-branching structure considered in Example 4 and structures shown in [15, Figures 1.24 b,c, 1.25 and 1.26]), do not develop according to the principle of subapical branching and therefore cannot be considered biologically correct.

Branching structures can be generally categorized as basitonic, mesotonic, or acrotonic, depending on whether the most vigorous lateral branches are situated near the base of the plant, in its middle zone, or near the top. In Section 3, we applied notions of L-systems to propose a formal definition of acrotony (including mesotony as a special case) and proved that acrotonic structures cannot be generated by propagating subapical BDOL-systems. Consequently, control mechanisms beyond those expressible using propagating sBDOL-system are necessary to model the development of acrotonic structures. Two types of mechanisms, based on information flow between coexisting plant modules [5, 15] or unlimited number of states that may characterize individual modules [11, 12] have been considered in the literature.

The results reported in this paper leave many questions open for further research. The first group of questions focuses on the formal properties of bracketed L-systems.

- We conjecture that Theorem 3 can be extended to non-propagating subapical BDOL-systems. Is this conjecture true?
- Consider two L-systems with the alphabet $\{A, B, I, L\}$ and axiom $[BA]$. The only non-identity production of L-system S_1 is $A \longrightarrow I[L][L]A$, and of L-systems S_2 is $B \longrightarrow BI[L][L]$. Both L-systems generate the same developmental sequence $\{B(I[L][L])^n A : n = 0, 1, 2, \ldots\}$, but L-system S_1 is subapical while S_2 is not. This raises the following questions:
  - Given an arbitrary BDOL-system S, is it decidable whether there exists a subapical BDOL-system S' generating the same developmental sequence (or language)? If the answer is positive, is there an effective procedure for finding S', given S?
  - Is it decidable whether a given developmental sequence or language can be generated by a subapical BDOL-system?
- The class of L-systems considered in this paper has been limited to BDOL-systems, but similar questions are pertinent to other classes of L-systems as well. For example, is it decidable whether a given bracketed context-sensitive L-system (BIL-system) is subapical or not? Is it decidable whether a given BIL-system generates an acrotonic structure?

A mathematical definition that aims at capturing the essence of a natural phenomenon is always somewhat arbitrary; we cannot prove that it is "right" or "wrong", although we can argue whether it faithfully reflects our intuition of the phenomenon and is operative as an element of a theory. In the context of this paper, such questions are related to the notions of subapical branching and acrotony.

- Definitions of subapical branching (5 and 7) present only one possible formalization of the underlying biological concept. What are the implications of alternative definitions, for example based on the conditions that:

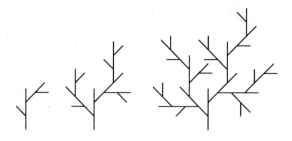

Fig. 7. Development of a sympodial branching structure generated by a BDOL-system S with axiom $[A]$ and productions $A \longrightarrow E[C]B$, $B \longrightarrow E[A]E$, $C \longrightarrow D$, $D \longrightarrow A$, $E \longrightarrow E$, $[\longrightarrow [,\ ] \longrightarrow]$. Biologists would qualify this structure as acrotonic, although it does not satisfy Definition 8. Developmental stages w_3, w_6, and w_9 are shown.

- branch creation is restricted to a predefined number of distal (topmost) modules in branches of all orders, or
- no new branch can be created below a predefined number of existing branches (counting from the apex down)?

– The proposed definition of acrotony (8) refers to the number of modules in a branch as a measure of the branch size. Other measures are also possible, for example based on the number of modules along the axes of lateral branches, the maximum branching order within the lateral branches, or the lengths of the axes in case of models operating on the geometric level. What are the implications of these definitions?

– Definition 8 requires that an arbitrarily long sequence of branches of increasing size be found while traversing the axis of a developing plant from the base up (acropetally). In nature, however, we always deal with finite structures, thus assumptions describing what an underlying developmental mechanism would produce if it was allowed to operate indefinitely cannot be experimentally verified. Is there an alternative formal definition of acrotony that would capture the increased complexity of acrotonic structures compared to basitonic structures, yet would apply to finite structures?

– The proposed definition of acrotony is more appropriate for *monopodial* than *sympodial* branching structures. A monopodial structure is characterized by a pronounced main axis that may carry an arbitrarily large number of lateral branches. In contrast, a sympodial structure has a repetitive branching pattern in which the number of lateral branches carried by each axis may be small, even if the whole structure is potentially unlimited (Figure 7). Biologists would qualify a structure that has increasingly long (or better developed) branches along each axis as an acrotonic structure irrespective of the number of branches carried by each axis. How should the definition of acrotony be improved to encompass both monopodial and sympodial structures?

In spite of the spectacular progress of the L-system theory since its inception almost thirty years ago, many fundamental problems pertinent to plant modeling remain open. They point to a fertile area for a continuing research bridging biology, computer graphics, and theoretical computer science.

Acknowledgements

We would like to thank Prof. Drs. Joost Engelfriet, Jacqueline and Hermann Lück, Gheorghe Păun, Arto Salomaa, Gabriel Thierrin, and an anonymous referee for useful discussions and suggestions. We are also indebted to Mark Hammel for his editorial help. This work has been supported in part by research and equipment grants from the Natural Sciences and Engineering Research Council of Canada.

References

1. T. Bălănescu, M. Gheorghe, and Gh. Păun. Three variants of apical growth filamentous systems. *Int. J. of Computer Mathematics*, 22:227–238, 1987.
2. A. Bell. *Plant form: An illustrated guide to flowering plants.* Oxford University Press, Oxford, 1991.
3. D. Frijters and A. Lindenmayer. A model for the growth and flowering of *Aster novae-angliae* on the basis of table (1,0)L-systems. In G. Rozenberg and A. Salomaa, editors, *L Systems*, Lecture Notes in Computer Science 15, pages 24–52. Springer-Verlag, Berlin, 1974.
4. D. Frijters and A. Lindenmayer. Developmental descriptions of branching patterns with paracladial relationships. In A. Lindenmayer and G. Rozenberg, editors, *Automata, languages, development*, pages 57–73. North-Holland, Amsterdam, 1976.
5. J. M. Janssen and A. Lindenmayer. Models for the control of branch positions and flowering sequences of capitula in *Mycelis muralis* (L.) Dumont (Compositae). *New Phytologist*, 105:191–220, 1987.
6. H. Jürgensen, H. Shyr, and G. Thierrin. Monoids with disjunctive identity and their codes. *Acta Mathematica Hungarica*, 47(3–4):299–312, 1986.
7. A. Kelemenová. Complexity of L-systems. In G. Rozenberg and A. Salomaa, editors, *The book of L*, pages 179–191. Springer-Verlag, Berlin, 1986.
8. A. Lindenmayer. Mathematical models for cellular interaction in development, Parts I and II. *Journal of Theoretical Biology*, 18:280–315, 1968.
9. A. Lindenmayer. Developmental systems without cellular interaction, their languages and grammars. *Journal of Theoretical Biology*, 30:455–484, 1971.
10. A. Lindenmayer. Models for multicellular development: Characterization, inference and complexity of L-systems. In A. Kelemenová and J. Kelemen, editors, *Trends, techniques and problems in theoretical computer science*, Lecture Notes in Computer Science 281, pages 138–168. Springer-Verlag, Berlin, 1987.
11. H. B. Lück and J. Lück. Approche algorithmique des structures ramifiées acrotone et basitone des végétaux. In H. Vérine, editor, *La biologie théorique à Solignac*, pages 111–148. Polytechnica, Paris, 1994.
12. J. Lück, H. B. Lück, and M. Bakkali. A comprehensive model for acrotonic, mesotonic, and basitonic branching in plants. *Acta Biotheoretica*, 38:257–288, 1990.
13. N. Nirmal and K. Krithivasan. Filamentous systems with apical growth. *Int. J. of Computer Mathematics*, 12:203–215, 1983.

14. Gh. Păun. Some further remarks on regular controlled apical growth filamentous systems. *Bull. Math. de la Soc. Sci. Math. de la R. S. de Roumanie*, 32(4):341–344, 1988.
15. P. Prusinkiewicz and A. Lindenmayer. *The algorithmic beauty of plants*. Springer-Verlag, New York, 1990. With J. S. Hanan, F. D. Fracchia, D. R. Fowler, M. J. M. de Boer, and L. Mercer.
16. P. M. Room, L. Maillette, and J. Hanan. Module and metamer dynamics and virtual plants. *Advances in Ecological Research*, 25:105–157, 1994.
17. G. Rozenberg and A. Salomaa. *The mathematical theory of L systems*. Academic Press, New York, 1980.
18. A. Salomaa. *Formal languages*. Academic Press, New York, 1973.
19. T. A. Steeves and I. M. Sussex. *Patterns in plant development*. Cambridge University Press, Cambridge, 1989.
20. K. G. Subramanian. A note on regular-controlled apical-growth filamentous systems. *Intern. J. Compu. Inf. Sci.*, 14:235–242, 1985.

Author Index

Alberich, R. 1
Arnborg, S. 469
Aßmann, U. 321

Banach, R. 16
Barthelmann, K. 225
Bauderon, M. 27
Blostein, D. 38
Brandenburg, F.J. 336
Burmeister, P. 1

Corradini, A. 56, 240, 257
Courcelle, B. 351, 487

Dassow, J. 412
DeBrunner, L.S. 185
Derk, M.D. 185
Drewes, F. 196

Ehrenfeucht, A. 502
Ehrig, H. 56, 137, 240
Engelfriet, J. 368
Engels, G. 137

Fahmy, H. 38
Fracchia, F.D. 521

Grbavec, A. 38

Habel, A. 75, 412
Harju, T. 502

Janssens, D. 271
Jeong, T.E. 383

Kari, L. 550
Kim, C. 383
Klempien-Hinrichs, R. 397
Korff, M. 288
Kreowski, H.-J. 89
Kuske, S. 89

Löwe, M. 56, 240
Lück, H.B. 536
Lück, J. 536

Maggiolo-Schettini, A. 107
McCreary, C.L. 443
Montanari, U. 56, 240

Nagl, M. 155

Padberg, J. 56
Parisi-Presicce, F. 428
Peron, A. 107
Plump, D. 75
Proskurowski, A. 469
Prusinkiewicz, P. 550

Ribeiro, L. 288
Rosselló, F. 1
Rossi, F. 240, 257
Rozenberg, G. 502

Schürr, A. 122, 155
Sénizergues, G. 351
Skodinis, K. 211, 336

Taentzer, G. 304
Taubenberger, S. 412

Valiente, G. 1
Vereijken, J.J. 368

Wanke, E. 211
Willis, L.M. 170
Wojdylo, B. 1

Zinßmeister, G. 443
Zündorf, A. 454

Springer-Verlag and the Environment

We at Springer-Verlag firmly believe that an international science publisher has a special obligation to the environment, and our corporate policies consistently reflect this conviction.

We also expect our business partners – paper mills, printers, packaging manufacturers, etc. – to commit themselves to using environmentally friendly materials and production processes.

The paper in this book is made from low- or no-chlorine pulp and is acid free, in conformance with international standards for paper permanency.

Lecture Notes in Computer Science

For information about Vols. 1–1006

please contact your bookseller or Springer-Verlag

Vol. 1007: A. Bosselaers, B. Preneel (Eds.), Integrity Primitives for Secure Information Systems. VII, 239 pages. 1995.

Vol. 1008: B. Preneel (Ed.), Fast Software Encryption. Proceedings, 1994. VIII, 367 pages. 1995.

Vol. 1009: M. Broy, S. Jähnichen (Eds.), KORSO: Methods, Languages, and Tools for the Construction of Correct Software. X, 449 pages. 1995. Vol.

Vol. 1010: M. Veloso, A. Aamodt (Eds.), Case-Based Reasoning Research and Development. Proceedings, 1995. X, 576 pages. 1995. (Subseries LNAI).

Vol. 1011: T. Furuhashi (Ed.), Advances in Fuzzy Logic, Neural Networks and Genetic Algorithms. Proceedings, 1994. (Subseries LNAI).

Vol. 1012: M. Bartošek, J. Staudek, J. Wiedermann (Eds.), SOFSEM '95: Theory and Practice of Informatics. Proceedings, 1995. XI, 499 pages. 1995.

Vol. 1013: T.W. Ling, A.O. Mendelzon, L. Vieille (Eds.), Deductive and Object-Oriented Databases. Proceedings, 1995. XIV, 557 pages. 1995.

Vol. 1014: A.P. del Pobil, M.A. Serna, Spatial Representation and Motion Planning. XII, 242 pages. 1995.

Vol. 1015: B. Blumenthal, J. Gornostaev, C. Unger (Eds.), Human-Computer Interaction. Proceedings, 1995. VIII, 203 pages. 1995.

VOL. 1016: R. Cipolla, Active Visual Inference of Surface Shape. XII, 194 pages. 1995.

Vol. 1017: M. Nagl (Ed.), Graph-Theoretic Concepts in Computer Science. Proceedings, 1995. XI, 406 pages. 1995.

Vol. 1018: T.D.C. Little, R. Gusella (Eds.), Network and Operating Systems Support for Digital Audio and Video. Proceedings, 1995. XI, 357 pages. 1995.

Vol. 1019: E. Brinksma, W.R. Cleaveland, K.G. Larsen, T. Margaria, B. Steffen (Eds.), Tools and Algorithms for the Construction and Analysis of Systems. Selected Papers, 1995. VII, 291 pages. 1995.

Vol. 1020: I.D. Watson (Ed.), Progress in Case-Based Reasoning. Proceedings, 1995. VIII, 209 pages. 1995. (Subseries LNAI).

Vol. 1021: M.P. Papazoglou (Ed.), OOER '95: Object-Oriented and Entity-Relationship Modeling. Proceedings, 1995. XVII, 451 pages. 1995.

Vol. 1022: P.H. Hartel, R. Plasmeijer (Eds.), Functional Programming Languages in Education. Proceedings, 1995. X, 309 pages. 1995.

Vol. 1023: K. Kanchanasut, J.-J. Lévy (Eds.), Algorithms, Concurrency and Knowlwdge. Proceedings, 1995. X, 410 pages. 1995.

Vol. 1024: R.T. Chin, H.H.S. Ip, A.C. Naiman, T.-C. Pong (Eds.), Image Analysis Applications and Computer Graphics. Proceedings, 1995. XVI, 533 pages. 1995.

Vol. 1025: C. Boyd (Ed.), Cryptography and Coding. Proceedings, 1995. IX, 291 pages. 1995.

Vol. 1026: P.S. Thiagarajan (Ed.), Foundations of Software Technology and Theoretical Computer Science. Proceedings, 1995. XII, 515 pages. 1995.

Vol. 1027: F.J. Brandenburg (Ed.), Graph Drawing. Proceedings, 1995. XII, 526 pages. 1996.

Vol. 1028: N.R. Adam, Y. Yesha (Eds.), Electronic Commerce. X, 155 pages. 1996.

Vol. 1029: E. Dawson, J. Golić (Eds.), Cryptography: Policy and Algorithms. Proceedings, 1995. XI, 327 pages. 1996.

Vol. 1030: F. Pichler, R. Moreno-Díaz, R. Albrecht (Eds.), Computer Aided Systems Theory - EUROCAST '95. Proceedings, 1995. XII, 539 pages. 1996.

Vol.1031: M. Toussaint (Ed.), Ada in Europe. Proceedings, 1995. XI, 455 pages. 1996.

Vol. 1032: P. Godefroid, Partial-Order Methods for the Verification of Concurrent Systems. IV, 143 pages. 1996.

Vol. 1033: C.-H. Huang, P. Sadayappan, U. Banerjee, D. Gelernter, A. Nicolau, D. Padua (Eds.), Languages and Compilers for Parallel Computing. Proceedings, 1995. XIII, 597 pages. 1996.

Vol. 1034: G. Kuper, M. Wallace (Eds.), Constraint Databases and Applications. Proceedings, 1995. VII, 185 pages. 1996.

Vol. 1035: S.Z. Li, D.P. Mital, E.K. Teoh, H. Wang (Eds.), Recent Developments in Computer Vision. Proceedings, 1995. XI, 604 pages. 1996.

Vol. 1036: G. Adorni, M. Zock (Eds.), Trends in Natural Language Generation - An Artificial Intelligence Perspective. Proceedings, 1993. IX, 382 pages. 1996. (Subseries LNAI).

Vol. 1037: M. Wooldridge, J.P. Müller, M. Tambe (Eds.), Intelligent Agents II. Proceedings, 1995. XVI, 437 pages. 1996. (Subseries LNAI).

Vol. 1038: W: Van de Velde, J.W. Perram (Eds.), Agents Breaking Away. Proceedings, 1996. XIV, 232 pages. 1996. (Subseries LNAI).

Vol. 1039: D. Gollmann (Ed.), Fast Software Encryption. Proceedings, 1996. X, 219 pages. 1996.

Vol. 1040: S. Wermter, E. Riloff, G. Scheler (Eds.), Connectionist, Statistical, and Symbolic Approaches to Learning for Natural Language Processing. IX, 468 pages. 1996. (Subseries LNAI).

Vol. 1041: J. Dongarra, K. Madsen, J. Waśniewski (Eds.), Applied Parallel Computing. Proceedings, 1995. XII, 562 pages. 1996.

Vol. 1042: G. Weiß, S. Sen (Eds.), Adaption and Learning in Multi-Agent Systems. Proceedings, 1995. X, 238 pages. 1996. (Subseries LNAI).

Vol. 1043: F. Moller, G. Birtwistle (Eds.), Logics for Concurrency. XI, 266 pages. 1996.

Vol. 1044: B. Plattner (Ed.), Broadband Communications. Proceedings, 1996. XIV, 359 pages. 1996.

Vol. 1045: B. Butscher, E. Moeller, H. Pusch (Eds.), Interactive Distributed Multimedia Systems and Services. Proceedings, 1996. XI, 333 pages. 1996.

Vol. 1046: C. Puech, R. Reischuk (Eds.), STACS 96. Proceedings, 1996. XII, 690 pages. 1996.

Vol. 1047: E. Hajnicz, Time Structures. IX, 244 pages. 1996. (Subseries LNAI).

Vol. 1048: M. Proietti (Ed.), Logic Program Syynthesis and Transformation. Proceedings, 1995. X, 267 pages. 1996.

Vol. 1049: K. Futatsugi, S. Matsuoka (Eds.), Object Technologies for Advanced Software. Proceedings, 1996. X, 309 pages. 1996.

Vol. 1050: R. Dyckhoff, H. Herre, P. Schroeder-Heister (Eds.), Extensions of Logic Programming. Proceedings, 1996. VII, 318 pages. 1996. (Subseries LNAI).

Vol. 1051: M.-C. Gaudel, J. Woodcock (Eds.), FME'96: Industrial Benefit and Advances in Formal Methods. Proceedings, 1996. XII, 704 pages. 1996.

Vol. 1052: D. Hutchison, H. Christiansen, G. Coulson, A. Danthine (Eds.), Teleservices and Multimedia Communications. Proceedings, 1995. XII, 277 pages. 1996.

Vol. 1053: P. Graf, Term Indexing. XVI, 284 pages. 1996. (Subseries LNAI).

Vol. 1054: A. Ferreira, P. Pardalos (Eds.), Solving Combinatorial Optimization Problems in Parallel. VII, 274 pages. 1996.

Vol. 1055: T. Margaria, B. Steffen (Eds.), Tools and Algorithms for the Construction and Analysis of Systems. Proceedings, 1996. XI, 435 pages. 1996.

Vol. 1056: A. Haddadi, Communication and Cooperation in Agent Systems. XIII, 148 pages. 1996. (Subseries LNAI).

Vol. 1057: P. Apers, M. Bouzeghoub, G. Gardarin (Eds.), Advances in Database Technology — EDBT '96. Proceedings, 1996. XII, 636 pages. 1996.

Vol. 1058: H. R. Nielson (Ed.), Programming Languages and Systems – ESOP '96. Proceedings, 1996. X, 405 pages. 1996.

Vol. 1059: H. Kirchner (Ed.), Trees in Algebra and Programming – CAAP '96. Proceedings, 1996. VIII, 331 pages. 1996.

Vol. 1060: T. Gyimóthy (Ed.), Compiler Construction. Proceedings, 1996. X, 355 pages. 1996.

Vol. 1061: P. Ciancarini, C. Hankin (Eds.), Coordination Languages and Models. Proceedings, 1996. XI, 443 pages. 1996.

Vol. 1062: E. Sanchez, M. Tomassini (Eds.), Towards Evolvable Hardware. IX, 265 pages. 1996.

Vol. 1063: J.-M. Alliot, E. Lutton, E. Ronald, M. Schoenauer, D. Snyers (Eds.), Artificial Evolution. Proceedings, 1995. XIII, 396 pages. 1996.

Vol. 1064: B. Buxton, R. Cipolla (Eds.), Computer Vision – ECCV '96. Volume I. Proceedings, 1996. XXI, 725 pages. 1996.

Vol. 1065: B. Buxton, R. Cipolla (Eds.), Computer Vision – ECCV '96. Volume II. Proceedings, 1996. XXI, 723 pages. 1996.

Vol. 1066: R. Alur, T.A. Henzinger, E.D. Sontag (Eds.), Hybrid Systems III. IX, 618 pages. 1996.

Vol. 1067: H. Liddell, A. Colbrook, B. Hertzberger, P. Sloot (Eds.), High-Performance Computing and Networking. Proceedings, 1996. XXV, 1040 pages. 1996.

Vol. 1068: T. Ito, R.H. Halstead, Jr., C. Queinnec (Eds.), Parallel Symbolic Languages and Systems. Proceedings, 1995. X, 363 pages. 1996.

Vol. 1069: J.W. Perram, J.-P. Müller (Eds.), Distributed Software Agents and Applications. Proceedings, 1994. VIII, 219 pages. 1996. (Subseries LNAI).

Vol. 1070: U. Maurer (Ed.), Advances in Cryptology – EUROCRYPT '96. Proceedings, 1996. XII, 417 pages. 1996.

Vol. 1071: P. Miglioli, U. Moscato, D. Mundici, M. Ornaghi (Eds.), Theorem Proving with Analytic Tableaux and Related Methods. Proceedings, 1996. X, 330 pages. 1996. (Subseries LNAI).

Vol. 1072: R. Kasturi, K. Tombre (Eds.), Graphics Recognition. Proceedings, 1995. X, 308 pages. 1996.

Vol. 1073: J. Cuny, H. Ehrig, G. Engels, G. Rozenberg (Eds.), Graph Grammars and Their Application to Computer Science. Proceedings, 1994. X, 565 pages. 1996.

Vol. 1074: G. Dowek, J. Heering, K. Meinke, B. Möller (Eds.), Higher-Order Algebra, Logic, and Term Rewriting. Proceedings, 1995. VII, 287 pages. 1996.

Vol. 1075: D. Hirschberg, G. Myers (Eds.), Combinatorial Pattern Matching. Proceedings, 1996. VIII, 392 pages. 1996.

Vol. 1076: N. Shadbolt, K. O'Hara, G. Schreiber (Eds.), Advances in Knowledge Acquisition. Proceedings, 1996. XII, 371 pages. 1996. (Subseries LNAI).

Vol. 1077: P. Brusilovsky, P. Kommers, N. Streitz (Eds.), Mulimedia, Hypermedia, and Virtual Reality. Proceedings, 1994. IX, 311 pages. 1996.

Vol. 1078: D.A. Lamb (Ed.), Studies of Software Design. Proceedings, 1993. VI, 188 pages. 1996.

Vol. 1079: Z.W. Raś, M. Michalewicz (Eds.), Foundations of Intelligent Systems. Proceedings, 1996. XI, 664 pages. 1996. (Subseries LNAI).

Vol. 1080: P. Constantopoulos, J. Mylopoulos, Y. Vassiliou (Eds.), Advanced Information Systems Engineering. Proceedings, 1996. XI, 582 pages. 1996.